Modeling Damage, Fatigue and Failure of Composite Materials

About the Series

The Woodhead Publishing Series in Composites Science and Engineering is a suite of professional reference books providing comprehensive coverage of recent developments in composite materials research for application now and in the future.

Woodhead Publishing Series in Composites Science and Engineering

Modeling Damage, Fatigue and Failure of Composite Materials

Second Edition

Edited by

Ramesh Talreja

Texas A&M University, College Station, TX, United States

Janis Varna

Luleå University of Technology, Luleå, Sweden

ELSEVIER

WP WOODHEAD PUBLISHING

An imprint of Elsevier

Woodhead Publishing is an imprint of Elsevier
50 Hampshire Street, 5th Floor, Cambridge, MA 02139, United States
The Boulevard, Langford Lane, Kidlington, OX5 1GB, United Kingdom

ISBN: 978-0-443-18489-5

For information on all Woodhead Publishing publications visit our website at https://www.elsevier.com/books-and-journals

Publisher: Matthew Deans
Acquisitions Editor: Gwen Jones
Editorial Project Manager: Tom Mearns
Production Project Manager: Manju Paramasivam
Cover Designer: Vicky Pearson Esser

Typeset by TNQ Technologies

Contents

List of contributors

A. Barroso School of Engineering, University of Seville, Seville, Spain

A. Blázquez University of Seville, Seville, Spain

P.A. Carraro University of Padova, Vicenza, Italy

E. Correa School of Engineering, University of Seville, Seville, Spain

Larissa Gorbatikh Department of Materials Engineering, KU Leuven, Leuven, Belgium

E. Graciani School of Engineering, University of Seville, Seville, Spain

M. Hajikazemi Department of Materials, Textiles and Chemical Engineering, Faculty of Engineering and Architecture, Ghent University, Ghent, Belgium

Endel V. Iarve University of Texas at Arlington, Mechanical and Aerospace Engineering, Arlington, TX, United States

Stepan V. Lomov Department of Materials Engineering, KU Leuven, Leuven, Belgium

Wei-Tsen Lu University of Texas at Arlington Research Institute, Institute for Predictive Performance Methodologies, Fort Wort, TX, United States

V. Mantič School of Engineering, University of Seville, Seville, Spain

L.N. McCartney Department of Electromagnetic and Electrochemical Technologies, National Physical Laboratory, Teddington, United Kingdom; Department of Engineering, Materials and Electrical Science, National Physical Laboratory, Teddington, Middlesex, United Kingdom

Minh Hoang Nguyen Department of Aerospace Engineering, University of Michigan, Ann Arbor, MI, United States

F. París School of Engineering, University of Seville, Seville, Spain

Andrejs Pupurs Riga Technical University, Riga, Latvia; Lulea University of Technology, Lulea, Sweden

M. Quaresimin University of Padova, Vicenza, Italy

J. Reinoso University of Seville, Seville, Spain

Chandra Veer Singh Department of Materials Science and Engineering, University of Toronto, Toronto, ON, Canada

C. Soutis Aerospace Research Institute, The University of Manchester, Manchester, United Kingdom

Bent F. Sørensen Composites Analysis and Mechanics Section, Department of Wind and Energy Systems, Technical University of Denmark, Roskilde, Denmark

Ramesh Talreja Department of Aerospace Engineering, Department of Materials Science and Engineering, Texas A&M University, College Station, TX, United States

Janis Varna Luleå University of Technology, Luleå, Sweden

M.L. Velasco School of Engineering, University of Seville, Seville, Spain

Anthony M. Waas Department of Aerospace Engineering, University of Michigan, Ann Arbor, MI, United States

Linqi Zhuang ANSYS, Houston, TX, United States

Woodhead Publishing Series in Composites Science and Engineering

1 **Thermoplastic aromatic polymer composites**
F. N. Cogswell
2 **Design and manufacture of composite structures**
G. C. Eckold
3 **Handbook of polymer composites for engineers**
Edited by L. C. Hollaway
4 **Optimisation of composite structures design**
A. Miravete
5 **Short-fibre polymer composites**
Edited by S. K. De and J. R. White
6 **Flow-induced alignment in composite materials**
Edited by T. D. Papthanasiou and D. C. Guell
7 **Thermoset resins for composites**
Compiled by Technolex
8 **Microstructural characterisation of fibre-reinforced composites**
Edited by J. Summerscales
9 **Composite materials**
F. L. Matthews and R. D. Rawlings
10 **3-D textile reinforcements in composite materials**
Edited by A. Miravete
11 **Pultrusion for engineers**
Edited by T. Starr
12 **Impact behaviour of fibre-reinforced composite materials and structures**
Edited by S. R. Reid and G. Zhou
13 **Finite element modelling of composite materials and structures**
F. L. Matthews, G. A. O. Davies, D. Hitchings, and C. Soutis
14 **Mechanical testing of advanced fibre composites**
Edited by G. M. Hodgkinson
15 **Integrated design and manufacture using fibre-reinforced polymeric composites**
Edited by M. J. Owen and I. A. Jones
16 **Fatigue in composites**
Edited by B. Harris
17 **Green composites**
Edited by C. Baillie
18 **Multi-scale modelling of composite material systems**
Edited by C. Soutis and P. W. R. Beaumont
19 **Lightweight ballistic composites**
Edited by A. Bhatnagar

20 **Polymer nanocomposites**
Y-W. Mai and Z-Z. Yu
21 **Properties and performance of natural-fibre composite**
Edited by K. Pickering
22 **Ageing of composites**
Edited by R. Martin
23 **Tribology of natural fiber polymer composites**
N. Chand and M. Fahim
24 **Wood-polymer composites**
Edited by K. O. Niska and M. Sain
25 **Delamination behaviour of composites**
Edited by S. Sridharan
26 **Science and engineering of short fibre reinforced polymer composites**
S-Y. Fu, B. Lauke and Y-M. Mai
27 **Failure analysis and fractography of polymer composites**
E. S. Greenhalgh
28 **Management, recycling and reuse of waste composites**
Edited by V. Goodship
29 **Materials, design and manufacturing for lightweight vehicles**
Edited by P. K. Mallick
30 **Fatigue life prediction of composites and composite structures**
Edited by A. P. Vassilopoulos
31 **Physical properties and applications of polymer nanocomposites**
Edited by S. C. Tjong and Y-W. Mai
32 **Creep and fatigue in polymer matrix composites**
Edited by R. M. Guedes
33 **Interface engineering of natural fibre composites for maximum performance**
Edited by N. E. Zafeiropoulos
34 **Polymer-carbon nanotube composites**
Edited by T. McNally and P. Pötschke
35 **Non-crimp fabric composites: Manufacturing, properties and applications**
Edited by S. V. Lomov
36 **Composite reinforcements for optimum performance**
Edited by P. Boisse
37 **Polymer matrix composites and technology**
R. Wang, S. Zeng and Y. Zeng
38 **Composite joints and connections**
Edited by P. Camanho and L. Tong
39 **Machining technology for composite materials**
Edited by H. Hocheng
40 **Failure mechanisms in polymer matrix composites**
Edited by P. Robinson, E. S. Greenhalgh, and S. Pinho
41 **Advances in polymer nanocomposites: Types and applications**
Edited by F. Gao
42 **Manufacturing techniques for polymer matrix composites (PMCs)**
Edited by S. Advani and K-T. Hsiao
43 **Non-destructive evaluation (NDE) of polymer matrix composites: Techniques and applications**
Edited by V. M. Karbhari
44 **Environmentally friendly polymer nanocomposites: Types, processing and properties**
S. S. Ray

45 **Advances in ceramic matrix composites**
Edited by I. M. Low
46 **Ceramic nanocomposites**
Edited by R. Banerjee and I. Manna
47 **Natural fibre composites: Materials, processes and properties**
Edited by A. Hodzic and R. Shanks
48 **Residual stresses in composite materials**
Edited by M. Shokrieh
49 **Health and environmental safety of nanomaterials: Polymer nanocomposites and other materials containing nanoparticles**
Edited by J. Njuguna, K. Pielichowski, and H. Zhu
50 **Polymer composites in the aerospace industry**
Edited by P. E. Irving and C. Soutis
51 **Biofiber reinforcement in composite materials**
Edited by O. Faruk and M. Sain
52 **Fatigue and fracture of adhesively-bonded composite joints: Behaviour, simulation and modelling**
Edited by A. P. Vassilopoulos
53 **Fatigue of textile composites**
Edited by V. Carvelli and S. V. Lomov
54 **Wood composites**
Edited by M. P. Ansell
55 **Toughening mechanisms in composite materials**
Edited by Q. Qin and J. Ye
56 **Advances in composites manufacturing and process design**
Edited by P. Boisse
57 **Structural integrity and durability of advanced composites: Innovative modelling methods and intelligent design**
Edited by P. W. R. Beaumont, C. Soutis and A. Hodzic
58 **Recent advances in smart self-healing polymers and composites**
Edited by G. Li and H. Meng
59 **Manufacturing of nanocomposites with engineering plastics**
Edited by V. Mittal
60 **Fillers and reinforcements for advanced nanocomposites**
Edited by Y. Dong, R. Umer and A. Kin-Tak Lau
61 **Biocomposites: Design and mechanical performance**
Edited by M. Misra, J. K. Pandey and A. K. Mohanty
62 **Numerical modelling of failure in advanced composite materials**
Edited by P. P. Camanho and S. R. Hallett
63 **Marine applications of advanced fibre-reinforced composites**
Edited by J. Graham-Jones and J. Summerscales
64 **Smart composite coatings and membranes: Transport, structural, environmental and energy applications**
Edited by M. F. Montemor
65 **Modeling damage, fatigue and failure of composite materials**
Edited by R. Talreja and J. Varna
66 **Dynamic deformation, damage and fracture in composite materials and structures**
Edited by V. V. Silberschmidt

Preface to the second edition

The original book volume, now referred to as the First Edition, composed of 19 chapters assembled in four parts. The current Second Edition has 22 chapters, also assembled in the same four parts, with significant updating and additions. The First Edition published in 2016 was well received and prompted publishing this edition to reflect the high level of activity evidenced by contributions in the literature.

In Part One, the damage development is addressed by four chapters devoted to observations and understanding of the underlying mechanisms in straight-fiber and textile composites under different loading modes, with two chapters focusing on uniaxial and multiaxial cyclic loads.

Part Two is concerned with modeling of failure mechanisms. Four of the seven chapters in this part treat modeling approaches aimed at initiation and progression of fiber-matrix debonding in unidirectional composites under axial and transverse loads. Failure mechanisms in composite laminates under tension are addressed in two chapters, one on multiple transverse cracking and another on delamination fracture. A separate chapter is devoted to compression failure of composite laminates.

Modeling of the materials response induced by damage is taken up in Part Three. Here the treatments are necessarily of multiscale nature, connecting the microlevel damage to the macrolevel response. Five chapters in this part address different aspects of multiscale modeling. One chapter analyzes scale effects and examines how reducing the ply thickness in laminates alters the damage behavior. Several analytical and semi-analytical approaches using micromechanics and continuum damage mechanics are presented in other chapters.

Part Four of the book is focused on failure analysis of composite structures and joints. The chapters in this part address static and fatigue failure of structural parts and joints. Computational simulation techniques dealing with failure due to stress gradients and at joints are treated here. Incorporating manufacturing defects in failure analysis is also included.

We thank all contributors to this book and the staff at Elsevier for facilitating the publication process. It is our hope that this edition of the book will also be useful to researchers in academia and industry.

Ramesh Talreja
Janis Varna

Part One

Damage development in composite materials

Failure mechanisms: observations

1

Ramesh Talreja
Department of Aerospace Engineering, Department of Materials Science and Engineering, Texas A&M University, College Station, TX, United States

1.1 Introduction

In monolithic materials such as metals, ceramics, and polymers, failure initiates at a point when the stress state there reaches certain criticality. For instance, in metals, the failure process begins with yielding and progresses to the final event of failure by material separation. The heterogeneous nature of composite materials leads to a multitude of failure events triggered by the failure of the constituents and their interfaces. The first failure event, the subsequent failure events, and the final failure event occur differently depending on the composition of the composite and the imposed loading. This chapter is concerned with static loading, while the next three chapters will review the failure processes under repeated loads.

The composition of composite materials can be in many forms. Our focus here is on those forms where continuous fibers act as reinforcement in a polymeric matrix. Most studies motivated by applications in the aerospace industry have been on straight fibers placed within layers that are stacked in different orientations. This class of composites called laminates displays failure processes that are relatively well understood and form the fundamentals on which studies of other composite compositions are commonly built. This chapter will review these fundamentals systematically by considering the basic unit of laminates, that is, a unidirectional (UD) composite, subjected to elementary loading modes of tension, compression and shear, and their combinations, followed by the failure in laminates. In failure of laminates, the basic features of the failure process are captured by constrained failure in laminas and the failure at ply interfaces. These failures are displayed in essence in laminate configurations of cross-ply and angle-ply types. This review will consider these configurations under different loading modes.

In the following sections, the failure mechanisms will be described by reviewing the observations reported at different resolutions of the techniques used. Inferences will be made concerning the initiation and progression of the observed mechanisms without analyzing the driving forces for initiation and interactions of the initiated failure events, and any criteria for attainment of the final failure. Later chapters in the book will address these aspects by modeling efforts. The current chapter is meant to provide the basic knowledge on which to build those models. A full understanding of the failure process comes from synergism of the observations and models to incorporate those.

Modeling Damage, Fatigue and Failure of Composite Materials. https://doi.org/10.1016/B978-0-443-18489-5.00013-8

1.2 Failure in UD composites

Logically, in any two-phase material such as a UD composite, the three possible components of failure are the failure of the two phases and the interfaces where they are bonded. However, under a given loading, in which phase the failure initiates first, how it impacts other failure components, how on further loading the three failure components interact and progress, and finally, what brings about final failure are the questions that have different answers in different situations. Here, we shall describe which failure mechanisms operate in UD composites under different simple loading modes (axial tension, axial compression, transverse tension, transverse compression, and in-plane shear) and their pairwise combinations. In the next section, discussions of the failure process will be continued for laminates of straight fibers.

1.2.1 Failure under longitudinal tension

Among the earliest studies of failure under axial tension of UD composites was the study reported by Fuwa et al. (1975). In this study, the authors tested dry bundles of carbon fibers as well as UD composites made of carbon/epoxy prepregs of 0.35 fiber volume fraction. The dry bundles were obtained by dissolving the resin from the prepregs, while the composites were prepared in fully cured or semicured forms. The fully cured specimens were manufactured by hot pressing in a normal way, while the semicured specimens were prepared by reducing the cure temperature in the gauge section. The tensile strength was found to show larger scatter for fully cured composites than for semicured composites and for fiber bundles. Based on acoustic emissions and scanning electron microscope (SEM) observations of failed composites, the authors found that the successive failure of fibers was localized within the composite in what may be described as subbundles and that several such failed subbundles connected through sheared planes between them to cause final failure. Another study by Purslow (1981) independently confirmed most of these observations. In Purslow (1981), a carbon/epoxy composite of 0.65 fiber volume fraction was made by curing the prepregs in an autoclave and then testing it in axial tension. SEM was performed on the failed surfaces of the specimens, and the failed fibers in those surfaces were painstakingly traced. The authors inferred from those observations that small zones of failed fibers were formed independently.

Observing the tensile failure process in UD composites by optical microscopy and SEM has severe limitations. Therefore, not much progress was made in understanding the process until high-resolution X-ray computed tomography (X-CT) was employed. Using this technique, the first study that focused on the sequential fiber failure process in the tensile failure of UD composites was by Aroush et al. (2006). This study used 125 quartz fibers in an epoxy matrix to make a model composite that was tested in a tensile tester, while X-CT scans were taken at increasing stress. Fig. 1.1 shows a series of X-ray radiographs reported by this study where single fiber breaks (singlets), two neighboring fiber breaks (doublets), and increasing number of neighboring fiber breaks (k-plets) are marked. A critical cluster of fiber breaks before imminent failure is also marked.

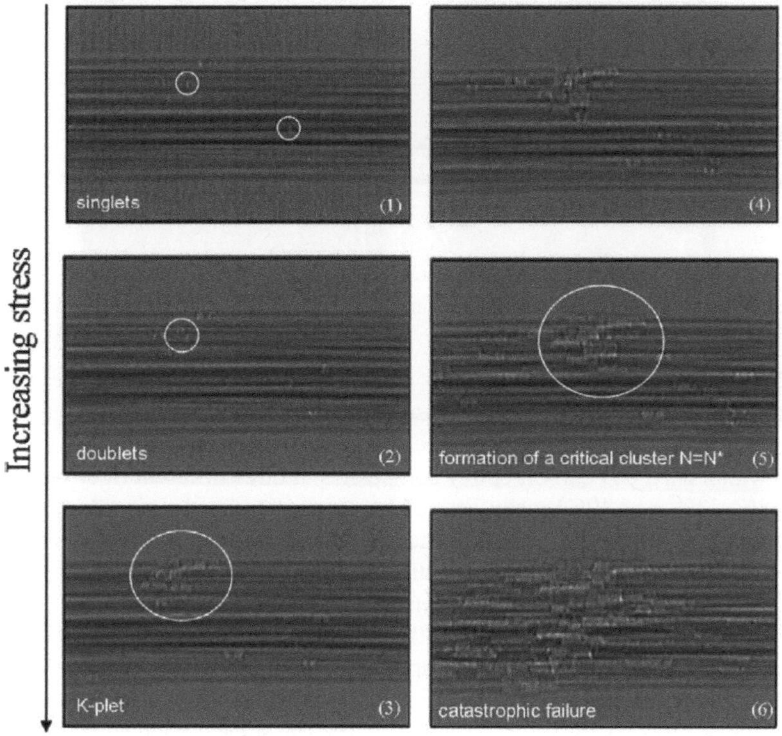

Figure 1.1 A series of X-CT radiographs showing increasing number of neighboring fibers breaking to form a critical cluster that produced failure (Aroush et al., 2006).

More studies of the fiber failure process in axial tension were conducted later using X-CT on carbon/epoxy [90/0]s laminates by Scott et al. (2011, 2012). In Scott et al. (2011), the laminate was given a double notch and the notch region was CT-scanned from the edge. The observed area showed damage in the form of transverse ply cracks, 0/90 delamination, and axial splits. Focusing on the fiber breaks, it was found that the sequence of fiber failures showed similar behavior to that described in Aroush et al. (2006). The number of fiber breaks was plotted against the nominal stress in the 0 degree plies and showed an exponential increase. On separating the fiber breaks as singlets, doublets, etc., it was found that the singlets started earliest and accumulated to the largest number, while doublets, triplets, etc., started increasingly later and reached decreasing final number. The largest k-plet recorded was for $k = 14$. Fig. 1.2 (Scott et al., 2011) displays CT scans of transverse slices normal to the fibers indicating clear details of the broken fiber clusters. A similar study by Garcea et al. (2017) found by using fast X-CT that a substantial number of fiber breaks occurred very close to the final failure. That study also found specimen-to-specimen variation in fiber break accumulation. Another study by Swolfs et al. (2015) on the same material as in Scott et al. (2011, 2012) focused on cluster formation. It was found that clusters of fiber breaks formed in one load step but did not increase in size (number of

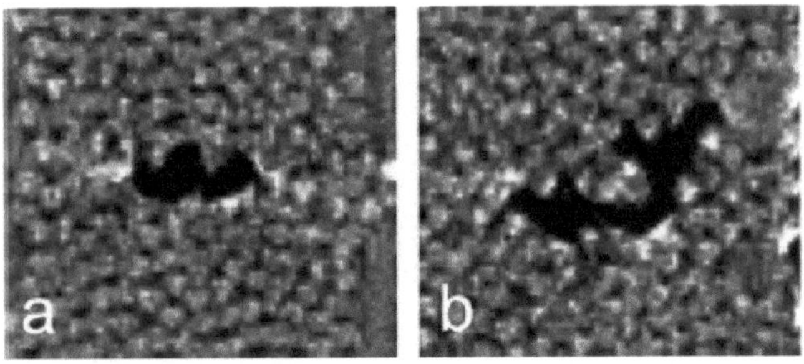

Figure 1.2 Planar sections taken with X-CT near clusters of (a) 5 and (b) 14 broken fibers (Scott et al., 2011).

broken fibers) as the load was increased. Instead, more clusters were formed of different sizes. Yet another study of fiber breakage clusters was reported in Na et al. (2018), which used UD carbon/epoxy composites in three different fiber volume fractions (38%, 43%, and 49%) manufactured by vacuum-assisted resin transfer molding. X-CT was performed in this study on the specimens, while in a tensile tester and stereology was used on the images to quantify the fiber breakage cluster size. The cluster size was found to increase with the fiber volume fraction.

The roles, if any, played by failure within the matrix and at the fiber/matrix interface have not been reported in the X-CT observations. An attempt was made to measure the axial strain along one or more broken fibers using digital volume correlation (DVC) in conjunction with X-CT by Schöberl et al. (2020). The same carbon epoxy [90/0]s was tested in axial tension, and X-CT technique was employed as in Scott et al. (2011, 2012) except this time the DVC was used to measure the volumetric displacement field around broken fibers. While the DVC technique cannot directly detect fiber/matrix debonding, the axial strain in the volume around broken fibers provided the length from the broken fiber end beyond which the strain was recovered to the prefailure level. The data gathered suggested that as a singlet evolved into doublet, the recovery length increased substantially.

1.2.2 Failure under longitudinal compression

This case of the failure process is where experimental observations and modeling efforts have been closely tied together. Before any evidence of failure of UD composites under longitudinal compression could be gathered, modeling studies by Rosen (1965) suggested microbuckling of fibers as a likely mechanism and pointed to the role of matrix shear stiffness as a critical factor. However, experimental studies later showed that kink bands formed under longitudinal compression and motivated the modeling efforts to analyze these bands. Argon's analysis of kink band formation (Argon, 1972) explained the role of fiber misalignment in this process, and since then more experimental studies have focused on this manufacturing induced defect.

While the role of preexisting fiber misalignment in giving rise to a nucleus for the formation of a kink band was confirmed, an experimental study by Narayanan and Schadler

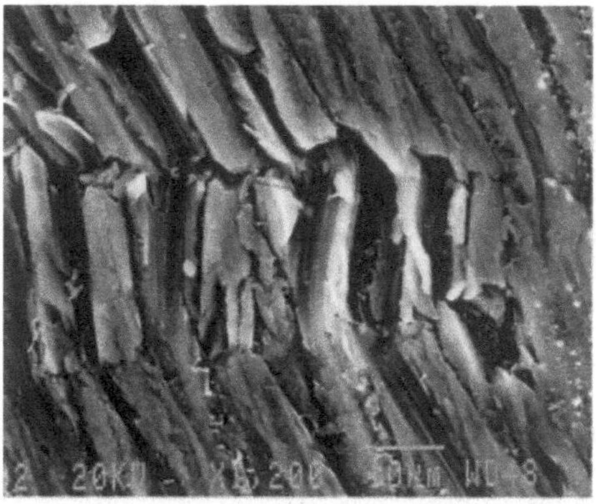

Figure 1.3 An SEM image of a kink band formed in a single tow carbon/epoxy composite (Narayanan and Schadler, 1999).

(1999) found evidence of initially broken or crushed fibers during processing to also be a source of local fiber microbuckling resulting in kink band initiation. Once initiated, a kink band appears as depicted by a SEM image in Fig. 1.3. In fact, the kink bands observed vary significantly depending on the constituent and interface properties in addition to the defects triggering their initiation. More recently, high-resolution X-ray tomography has provided many details not previously seen with clarity. The tomogram in Fig. 1.4 (Wang et al., 2017) displays the presence of multiple kink bands of different orientations in a damage zone of a partially failed specimen. It also clearly shows buckled and broken fibers as well as splitting (matrix cracks) along fibers within a kink band.

1.2.3 Failure under transverse tension

When a UD composite of glass/epoxy or carbon/epoxy is loaded in tension transversely, that is, normal to the fiber axis, it deforms in linear elastic manner and fails

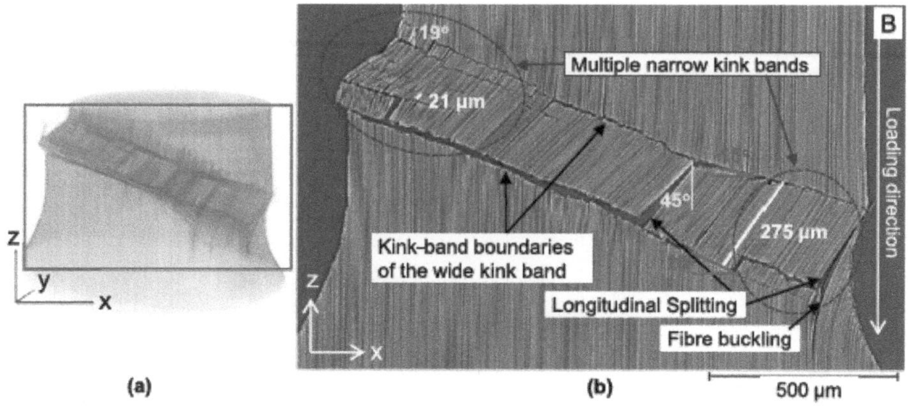

Figure 1.4 The zone of a partially failed specimen scanned by x-CT (a), and (b) tomogram showing multiple kink bands, buckled and broken fibers as well as splitting along fibers in a kink band (Wang et al., 2017).

abruptly. This makes it difficult to observe the stages of its failure process. Therefore, it is common to study this process by embedding the UD composite as a 90 degree layer within a cross-ply laminate where the failure is stabilized by the constraining longitudinal plies. If relatively thick 90 degree layer is used, then the failure can be studied in the mid-thickness region of the layer where the transverse stress is approximately uniform. A good example of this was reported in Gamstedt and Sjögren (1999) from which Fig. 1.5 is taken. The authors of this study manufactured a [0/90]s laminate with glass fibers and vinyl-ester resin by an RTM process with the outer 0 degree layer thickness of 0.5 mm and the inner 90 degree layers of 1.0 mm. They applied a tensile load in the 0 degree direction and observed the edge of the specimen under an optical microscope. They found that the fibers debonded from the matrix initially, and as the load was increased, some of the debond cracks coalesced to form a transverse crack, as depicted in Fig. 1.5a. On further loading, a coalesced crack showed crack surface separation, indicating that a continuous crack had formed, as seen in Fig. 1.5b. Similar observations were reported earlier by Harrison and Bader (1983) with less clarity

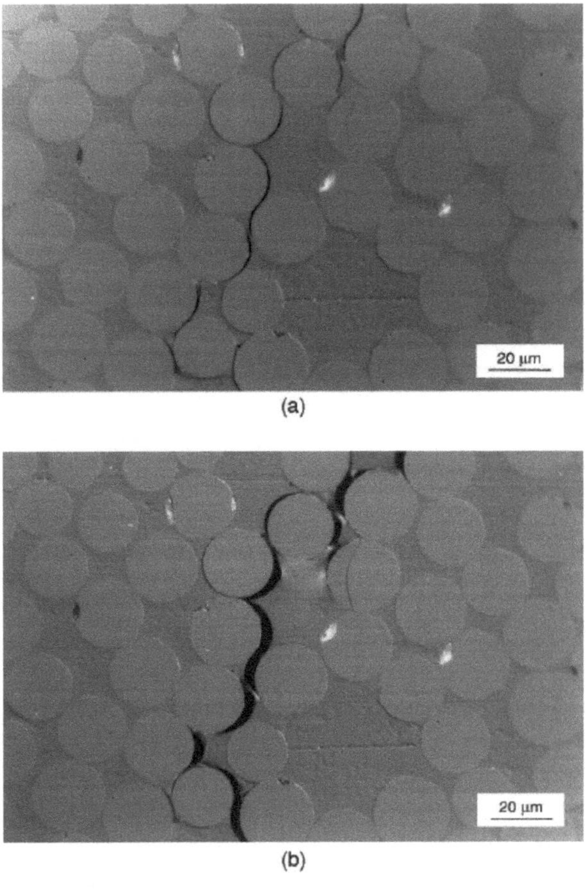

Figure 1.5 (a) Fiber/matrix debonding and (b) coalescence of debonds into a transverse crack (Gamstedt and Sjögren, 1999).

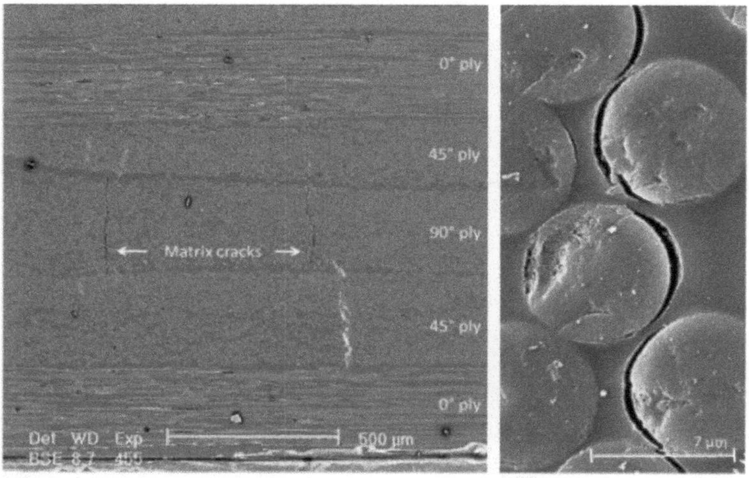

(a) (b)

Figure 1.6 SEM images of (a) matrix cracks in a 90 degree ply of a 0/45/90 carbon-epoxy laminate and (b) a zoomed-in region of a matrix crack showing fiber/matrix debonding (Ramanov et al., 2015).

for a carbon-epoxy laminate. Later, the fiber/matrix debonding was reported at a higher resolution by SEM in Ramanov et al. (2015). Fig. 1.6, taken from that paper, shows a SEM image of a transverse crack spanning the 90-degree layer thickness, Fig. 1.6a and a zoomed-in image of the crack showing fiber/matrix debonding Fig. 1.6b.

1.2.4 Failure under transverse compression

Testing of UD composites in transverse compression faces difficulties due to possible buckling and limited opportunities of observing the early stages of the failure process. One study by Argonès (2007) whose results were also reported in Gonzales and Llorca (2007) showed that final failure under transverse compression occurred along an angle of 56 degree to the loading direction, as indicated in Fig. 1.7. It also found fiber/matrix debonding and shear failure in the matrix as evidenced by SEM images of the fracture surfaces shown in Fig. 1.7 where debonded fibers and hackles in the matrix can be seen.

Postfailure observations such as those in Fig. 1.7 do not provide direct evidence of the first failure event and the early progression of failure. A study by Zumaquero et al. (2018) attempted to achieve this by testing a carbon-epoxy UD composite under compressive loads of 25%, 50%, 75%, and 80% of the average failure load and observing the polished specimen edge under optical microscope. Based on those observations, the authors described the failure process as occurring in stages of fiber/matrix interface debonding and debond crack kinking culminating in a macrocrack that leads to final failure. Their observations are summarized in Fig. 1.8. It is noted that these images are selected from different specimens to illustrate the failure advanced to increasing levels prior to final failure.

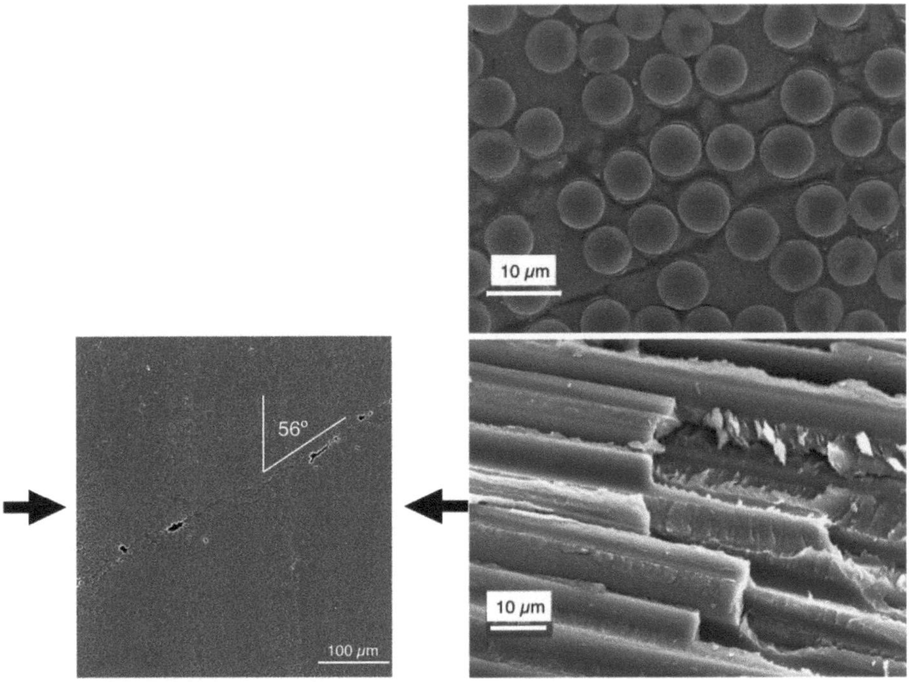

Figure 1.7 Left: SEM image of the lateral surface of a carbon-epoxy UD composite taken before the failure load showing the formation of a failure plane inclined to the loading direction indicated by arrows. Top: a close-up of the lateral surface near the failure plane showing interface debonding. Right: fracture surface showing the presence of matrix shear (hackles) (Gonzalez and LLorca, 2007).

A recent study by Flores et al. (2022) took pains to machine out a small sample from UD composites of carbon/epoxy, where direct SEM observations could be made at the fiber scale. This study found evidence of fiber/matrix debonding followed by matrix cracks in the loading direction that connected the debonds. The final failure occurred on planes inclined to the loading direction (Fig. 1.9).

1.2.5 Failure under in-plane shear

One of the challenges of testing to determine the behavior of a UD composite layer under in-plane shear is to induce this stress uniformly into a specimen. For flat specimens, the complex stress state and its nonuniformity caused by the gripping fixtures extend far into the specimen making it difficult to determine the deformation and failure response caused by pure shear. Many testing methods have emerged over the years, and a good review of these can be found in Carlsson et al. (2014). Among these methods, described as Iosipescu shear testing, two- and three-rail shear testing, [+ 45/− 45]s axial tension testing, and short beam shear testing, none has been found suitable to examine the shear failure process in its entirety, while they can be used with

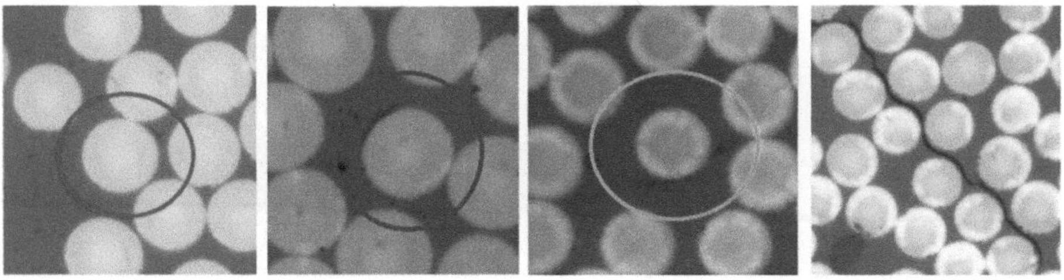

(a) (b) (c) (d)

Figure 1.8 Optical micrographs of a UD carbon-epoxy composite under transverse compression in vertical direction taken to illustrate evidence of different stages of failure. (a) Initiation of fiber/matrix debond, (b) Growth of debond, (c) Kink-out of a debond crack, and (d) Macrocrack before final failure.
Source: Zumaquero et al. (2018).

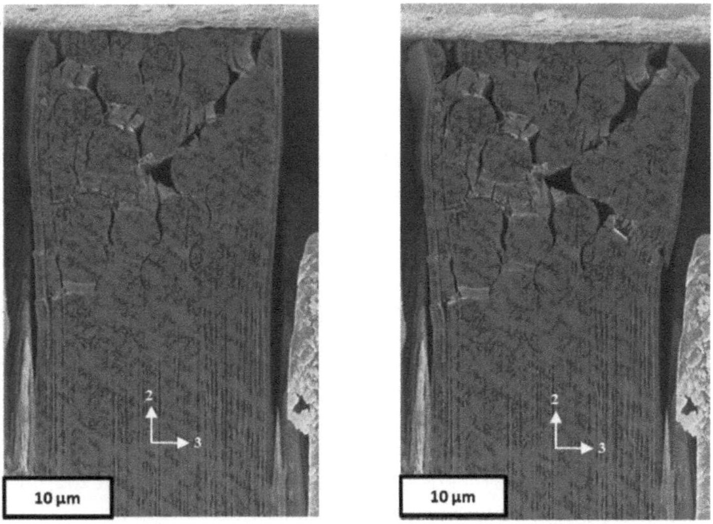

a) b)

Figure 1.9 SEM images of a microscopic specimen loaded in compression in the transverse (vertical) direction. (a) Prior to ultimate failure and (b) After ultimate failure (Flores et al., 2022).

more or less accuracy to find the deformation response. Testing of thin-walled tubes in torsion offers the best solution to pure shear testing, and it has been used for UD composites by Quaresimin and Carraro (2014) for fatigue.

While the prefailure deformation response of a UD composite is the same for $\tau_{xy} = \tau_{yx}$, where x-axis is in the fiber direction and y-axis is in the transverse direction in the plane of the UD composite layer, the failure response differs for "axial" shear τ_{yx} from that for "transverse" shear τ_{xy} due to the fiber/matrix interface that fails under the axial shear. It is thus expected that the axial shear failure will begin with fiber/matrix debonding. Since the fibers are discrete, the question remains as to how these discrete

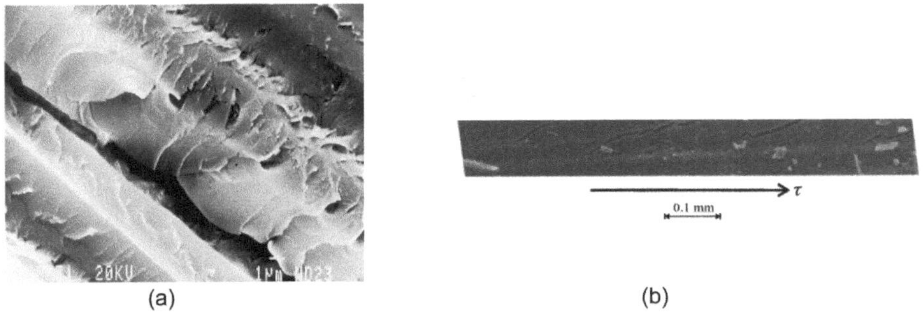

Figure 1.10 (a) Image of fracture surface showing hackles caused by failure in shear (Plumtree and Shi, 2002), and (b) Sigmoidal shape cracks in the matrix produced by axial shear between fibers (Redon, 2000).

interface failures are connected to create a failure plane. In other words, how does the matrix polymer fail to bridge the fiber/matrix debonds. Evidence has been found by Plumtree and Shi (2002) that in the presence of axial shear, the matrix between the fibers fails by plastic deformation leaving hackles on the fracture surface, as shown in Fig. 1.10a. The cracks that form in the matrix appear as axial cracks forming the final fracture plane parallel to fibers. However, these axial cracks seem to result from coalescence of much smaller sigmoidal shape cracks that connect neighboring fiber/matrix debonds, as seen in Fig. 1.10b. This image was obtained by Redon (2000) where a local axial shear shown as τ in Fig. 1.10b was induced by cutting fibers that were extended by tension and on cutting retracted to produce the shear stress.

1.2.6 Failure under combined loading

It is commonly assumed that when two elementary loading modes are applied together, for example, longitudinal tension and in-plane shear, the failure mechanisms induced by the two modes individually undergo interaction. The nature of the interaction is, however, not always clear. In each loading mode, the failure process consists of a sequence of failure events, as described above, and the question to ask is: How are the failure events and their progression corresponding to one loading mode affected by those corresponding to the other loading mode? In the following, we shall address this question, taking two loading modes at a time. It is noted that most studies of composite failure interaction have been focused on the final event of failure (strength) motivated by the need to describe a failure surface in the stress space in the context of a failure theory. The multiaxial testing for this purpose is fraught with practical issues of specimen shapes and gripping fixtures. A survey of the test methods can be found in Olsson (2011).

1.2.6.1 Longitudinal tension and in-plane shear

As described above, the failure events under longitudinal tension consist of fiber breakage starting as single fiber breaks and culmination in clusters of fiber breaks.

These fiber events are statistical and are presumably dictated by fiber defects and the matrix deformation characteristics. The latter can be affected by the presence of the superimposed in-plane shear. However, systematic experimental studies to clarify the combined loading effect on the fiber failure process do not appear to be done likely because of the difficulties in introducing combined loads and in observing the failure process at the fiber scale.

Under in-plane shear, the observations do not show a dominant effect of fibers on the axial shear failure process, as described above. However, the presence of the superimposed longitudinal tension is likely to affect the growth of axial cracks by altering the driving force for these cracks. In any case, experimental observations to quantify such effects do not appear to be reported.

1.2.6.2 Longitudinal tension and transverse tension

As discussed above for the case of longitudinal tension and axial shear, the combination of longitudinal tension and transverse tension has the same situation of two distinctly different failure mechanisms where interaction opportunity is small. The transverse stress values at which failure in UD polymer matrix composites occur is usually much smaller than the values at which longitudinal tensile failure occurs. This limits the range of combined loading in which interaction between failures in the two loading modes can be observed.

1.2.6.3 Longitudinal tension and transverse compression

The strength values of UD composites in longitudinal tension and transverse compression are sufficiently large for providing a good range of combined loading in which the interaction between failures in the two loading modes can be studied. However, test fixtures for applying the combined loading limit observations, and therefore not much information concerning failure mode interaction is available. It is conceivable that under longitudinal tension, the local load sharing between broken fibers and their intact neighbors is affected by transverse compression via plastic deformation in the matrix at sufficiently high compressive stresses. The interactive effect then would be like the case of applying in-plane shear in addition to the longitudinal tension, although to a lesser degree.

1.2.6.4 Longitudinal compression and in-plane shear

Failure under pure longitudinal compression has been found to induce kink bands, as discussed above. This failure is typically catastrophic and therefore it is difficult to track the growth of kink bands. In a study of combined longitudinal compression and in-plane shear (Vogler and Kyriakidis, 2001), using a fixture developed for the purpose, the authors found that the kink bands could be stabilized to follow their growth. Fig. 1.11 depicts the geometry of the kink band as it grows across the fibers. Contrary to the case of pure compression, the kink band in combined loading was found to have few broken fibers. As shown in Fig. 1.11, the fiber curvature and rotation at the front edge of the kink band reduce to zero, making the applied compression

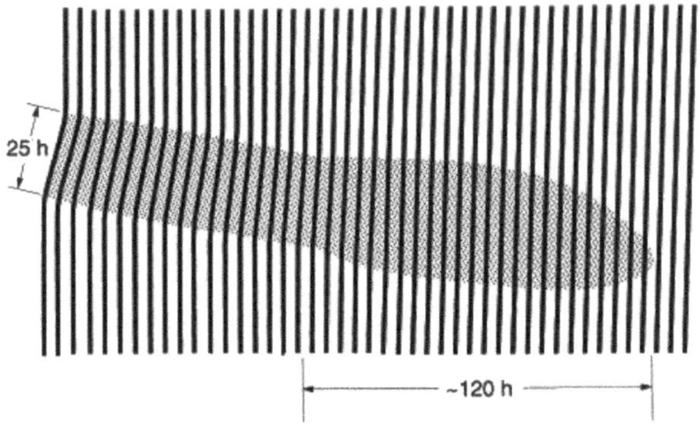

Figure 1.11 A sketch of a kink band growing in combined longitudinal compression and in-plane shear (Vogler and Kyriakidis, 2001). The dimensions shown are in terms of the fiber diameter h.

insufficient to buckle the fibers. The kink band then grows due to the transverse displacement induced by shear. A study (Wilhelmsson et al., 2019) performed X-CT to examine the effect of combining shear with longitudinal compression for carbon-epoxy UD composites made of noncrimp fabric by testing specimens cut at off-axis angles and tested under longitudinal compression. It was found that kink bands formed out-of-plane when no shear was present but became increasingly oriented in the plane as the shear (off-axis angle) was increased.

1.2.6.5 Longitudinal compression and transverse tension

Due to the low strength in transverse tension, only small values of this stress can be superimposed on longitudinal compression. Also, the fixture needed for applying this biaxial loading mode is bound to be complicated if observations of failure initiation and progression are to be made. For these reasons, this combined loading mode is studied by computational simulation.

1.2.6.6 Longitudinal compression and transverse compression

The effect of multiaxial compression was investigated in a study by Oguni et al. (2000) where an E-glass/vinylester UD composite was made as a rod and was confined concentrically. Axial compression then produced a proportional axial and radial compression. The specimens were loaded to failure and examined after failure. The observations showed that a failure mode transition occurred from axial splitting to kink band formation going from uniaxial to multiaxial compression. Under longitudinal compression, axial splitting-induced kink bands were found, while under proportional multiaxial compression, only kink bands were observed.

The type of postmortem observations reported in Oguni et al. (2000) is useful in the absence of in situ tracking of failure events.

Testing in biaxial compression on UD composites is a challenge if the objective is to avoid buckling while also allowing in situ observation of the failure process. The design considerations for cruciform specimens and loading fixtures have been addressed in Horta Muñoz (2022).

1.2.6.7 Transverse tension and in-plane shear

The specimen geometry most suited for this combined loading mode is a thin-walled cylinder with circumferential fibers loaded in axial tension and torsion. This specimen geometry was used in Queresimin and Carraro (2014) to study failure initiation and progression under cyclic loading. By choosing the level of loading in the two modes and their relative values, the failure process could be studied in detail. To stabilize the failure process so it could be observed in its entirety, it was found useful to constrain the process by adding longitudinal fabrics to the tubes. Observations of the failure process were made by an infrared camera and visual inspection aided by lighting in the interior of the tubular specimen. The first failure event was assessed to be crack initiation along fibers, and it was found to be affected by the presence of defects. Once a crack initiated, its growth was in mixed mode as depicted in Fig. 1.12 (Quaresimin and Carraro, 2014).

1.2.6.8 Transverse compression and in-plane shear

As described above, the failure in the two loading modes applied individually has been found to be dominated by shear-driven inelasticity. When the two loading modes are combined, the local stress states will depend on the load combination and consequently the inelastic deformation in the interfiber regions will govern failure. An experimental study (Vogler and Kyriakidis, 1999) tested carbon-PEEK composites in a biaxial shear/compression device and found differences in the mechanical response in different loading paths. This can be attributed to the nonlinear inelastic response. The failure mechanisms as observed on fracture surfaces in pure compression, pure shear, and combined compression/shear were found to be similar. The hackles found on the fracture surfaces were more for combined loading than for pure loading cases.

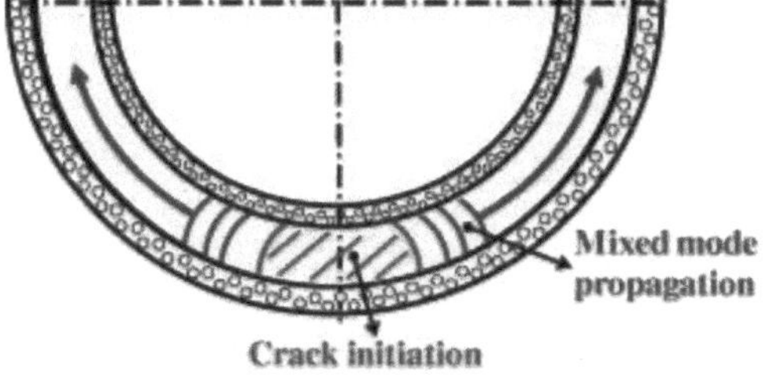

Figure 1.12 Crack initiation and its mixed mode propagation along circumferential fibers in a tubular specimen subjected to longitudinal tension and torsion (Quaresimin and Carraro, 2014).

1.3 Failure modes in laminates

The combinations in multidirectional laminates are many when lamina orientation and sequence (layup) and lamina thickness are taken as parameters. However, practical laminates in most cases are limited to balanced angle-ply laminates and quasiisotropic laminates. Cross-ply laminates are used in experimental studies as a simple case of the effects of constraint by a neighboring lamina on a lamina undergoing cracking. In the following, the failure modes in cross-ply laminates will be reviewed followed by the description of additional features observed in other (general) laminates.

1.3.1 Cross-ply laminates

Fig. 1.13 (Garrett and Bailey, 1977) depicts the multiple transverse cracking process which results from the ply constraint (deformation of transverse plies restricted by the presence of axial plies) and is a characteristic feature of cracking in plies of laminates. The images shown illustrate how the transverse crack spacing increases as the thickness of the outer axial plies is decreased. In the limit, as the thickness of the outer plies tends to zero, the crack spacing tends to infinity, that is, the laminate approaches a UD composite that develops a single transverse crack.

The multiple cracking shown in Fig. 1.13 is seen in quasistatic loading, that is, in monotonically increasing load. However, if a load below the failure load is applied repeatedly, other local failure mechanisms become visible. Fig. 1.14 (Reifsnider and Jamison, 1982) depicts the additional mechanisms of axial splits and interior delamination that are observed under cyclic loading.

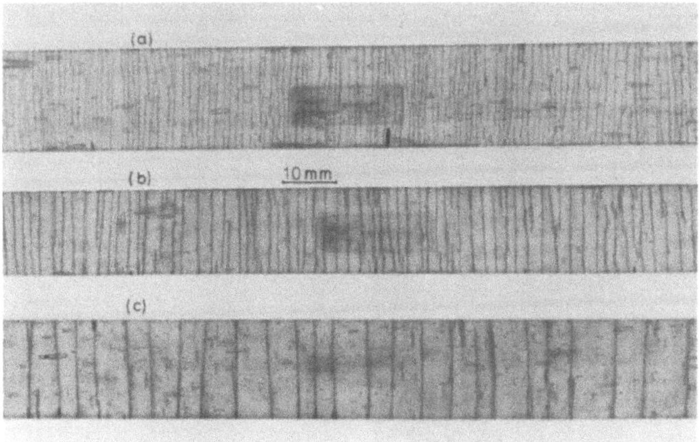

Figure 1.13 Multiple cracking in the transverse plies of cross-ply laminates of glass fiber reinforced polymer composites under axial tension (Garrett and Bailey, 1977).

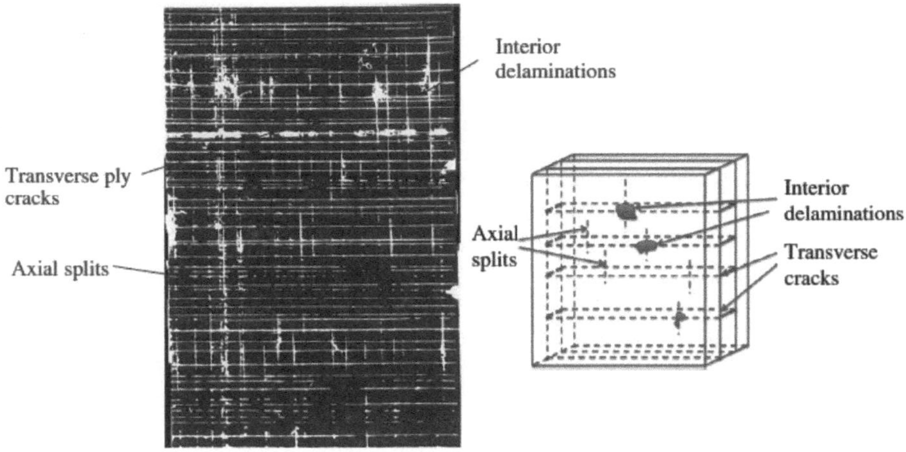

Figure 1.14 X-ray radiograph of a carbon/epoxy cross-ply laminate subjected to tension–tension cyclic loading (Reifsnider and Jamison, 1982). The sketch to the right shows a three-dimensional depiction for clarity.

1.3.2 General laminates

A systematic study (Varna et al., 1999) reported the constraint effect on ply cracking by varying the off-axis angle while keeping the axial plies as constraining plies. In that study, laminates of $[0/+\theta/-\theta/0_{1/2}]s$ layup tested in axial tension showed that as the angle θ was decreased from 90 degree, the initiation of ply cracks was delayed, and the crack spacing reached at the same applied load was also higher. In fact, for θ approaching 0 degree, ply cracks were not seen before failure of the laminate occurred. Axial splits and interior delamination cracks also appear to diminish as the off-axis angle reduces from 90 degree.

1.4 Conclusion

This chapter has reviewed observations reported over the years on the initiation, progression, and criticality of failure in UD composites and laminates made by using them as building blocks. The attempt has been to outline the main features of the failure processes revealed by the observations for the purpose of supporting the modeling efforts to be described in later chapters. The complexity of the failure processes necessitates that the modeling efforts and the observations go hand in hand to produce a full understanding of the processes.

The observations described in this chapter have been limited to straight fiber composites. Woven fiber and other fiber architectures are treated in later chapters.

References

Argon, A.S., 1972. Fracture of composites. In: Treatise of Materials Science and Technology, vol. 1. Academic Press, New York, pp. 79–114.

Argones, D., 2007. Fracture Micromechanisms in C/epoxy Composites Under Transverse Compression. Masters thesis. Polytechnical University of Madrid.

Aroush, D.R.-B., Maire, E., Gautheir, C., Youseff, S., Cletens, P., Wagner, H.D., 2006. A study of fracture of unidirectional composites using in situ high-resolution synchrotron X-ray microtomography. Compos. Sci. Technol. 66, 1348–1353.

Carlsson, L.A., Adams, D.F., Pipes, R.B., 2014. Experimental Characterization of Advanced Composite Materials. CRC press.

Flores, M., Sharits, A., Wheeler, R., Sesar, N., Mollenhauer, D., 2022. Experimental analysis of polymer matrix composite microstructures under transverse compression loading. Compos. Appl. Sci. Manuf. 156, 106859.

Fuwa, M., Bunsell, A.R., Harris, B., 1975. Tensile failure mechanisms in carbon fibre reinforced plastics. J. Mater. Sci. 10, 2062–2070.

Gamstedt, E.K., Sjögren, B.A., 1999. Micromechanisms in tension-compression fatigue of composite laminates containing transverse plies. Compos. Sci. Technol. 59, 167–178.

Garcea, S.C., Sinclair, I., Spearing, S.M., Withers, P.J., 2017. Mapping fibre failure in situ in carbon fibre reinforced polymers by fast synchrotron X-ray computed tomography. Compos. Sci. Technol. 149, 81–89.

Garrett, K.W., Bailey, J.E., 1977. Multiple transverse fracture in 90° cross-ply laminates of a glass fibre-reinforced polyester. J. Mater. Sci. 12, 157–168.

Gonzalez, C., LLorca, J., 2007. Mechanical behavior of unidirectional fiber-reinforced polymers under transverse compression: microscopic mechanisms and modeling. Compos. Sci. Technol. 67, 2795–2806.

Harrison, R.P., Bader, M.G., 1983. Damage development in CFRP laminates under monotonic and cyclic stressing. Fibre Sci. Technol. 18, 163–180.

Horta Muñoz, S., Serna Moreno, M.D.C., 2022. Advances in cruciform biaxial testing of fibre-reinforced polymers. Polymers 14 (4), 686.

Na, W., Kwon, D., Yu, W.-R., 2018. X-ray computed tomography observation of multiple fiber fracture in unidirectional CFRP under tensile loading. Compos. Struct. 188, 39–47.

Narayanan, S., Schadler, L.S., 1999. Mechanism of kink-band formation in graphite/epoxy composites: a micromechanical experimental study. Compos. Sci. Technol. 59, 2201–2213.

Oguni, K., Tan, C.Y., Ravichandran, G., 2000. Failure mode transition in unidirectional e-glass/vinylester composites under multiaxial compression. J. Compos. Mater. 34, 2081–2097.

Olsson, R., 2011. A survey of test methods for multiaxial and out-of-plane strength of composite laminates. Compos. Sci. Technol. 71, 773–783.

Plumtree, A., Shi, L., 2002. Fatigue damage evolution in off-axis unidirectional CFRP. Int. J. Fatig. 24, 155–159.

Purslow, D., 1981. Some fundamental aspects of composites fractography. Composites 12, 241–247.

Quaresimin, M., Carraro, P.A., 2014. Damage initiation and evolution in glass/epoxy tubes subjected to combined tension–torsion fatigue loading. Int. J. Fatig. 63, 25–35.

Redon, O., 2000. Fatigue Damage Development and Failure in Unidirectional and Angle-Ply Glass Fibre/Carbon Fibre Hybrid Laminates. Risø National Laboratory, Roskilde, Denmark. Technical Report Risø-R-1168.

Reifsnider, K.L., Jamison, R., 1982. Fracture of fatigue-loaded composite laminates. Int. J. Fract. 4, 187–197.

Romanov, V.S., Lomov, S.V., Verpoest, I., Gorbatikh, L., 2015. Modelling evidence of stress concentration mitigation at the micro-scale in polymer composites by the addition of carbon nanotubes. Carbon 82, 184–194.

Rosen, V.W., 1965. Mechanics of composite strengthening. In: Fiber Composite Materials. American Society of Metals, Metals Park, Ohio, pp. 37–75.

Schöberl, E., Breite, C., Melnikov, A., Swolfs, Y., Mavrogordato, M., Sinclair, I., Spearing, S.M., 2020. Fibre-direction strain measurement in a composite ply under quasi-static tensile loading using digital volume correlation and in situ synchrotron radiation computed tomography. Compos. Part A 137, 105935.

Scott, A.E., Mavrogordato, M., Wright, P., Sinclair, I., Spearing, S.M., 2011. In situ fibre fracture measurement in carbon-epoxy laminates using high resolution computed tomography. Compos. Sci. Technol. 71, 1471–1477.

Scott, A.E., Sinclair, I., Spearing, S.M., Thionnet, A., Bunsell, A.R., 2012. Damage accumulation in carbon/epoxy composite: comparison between a multiscale model and computed tomography experimental results. Compos. Part A 43, 1514–1522.

Swolfs, Y., Morton, H., Scott, A.E., Gorbatikh, L., Reed, P.A.S., Sinclair, I., Spearing, S.M., Verpoest, I., 2015. Synchrotron radiation computed tomography for experimental validation of a tensile strength model for unidirectional fibre-reinforced composite. Compos.Part A 77, 106–113.

Varna, J., Joffe, R., Akshantala, N.V., Talreja, R., 1999. Damage in composite laminates with off-axis plies. Compos. Sci. Technol. 59, 2139–2147.

Vogler, T.J., Kyriakidis, S., 2001. On the initiation and growth of kink bands in fiber composites: Part 1. Experiments. Int. J. Solid Struct. 38, 2639–2651.

Vogler, T.J., Kyriakidis, S., 1999. Inelastic behavior of an AS4/PEEK composite under combined transverse compression and shear. Part I: Experiments. Int. J. Plast. 15, 783–806.

Wang, Y., Burnett, T.L., Chai, Y., Soutis, C., Hogg, P.J., Withers, P.J., 2017. X-ray computed tomography study of kink bands in unidirectional composites. Compos. Struct. 160, 917–924.

Wilhelmsson, D., Mikkelsen, L.P., Fœster, S., Asp, L.E., 2019. Influence of in-plane shear on kink-plane orientation in a unidirectional fibre composite. Compos. Part A 119, 283–290.

Zumaquero, P.L., Correa, E., Justo, J., París, F., 2018. Microscopical observations of interface cracks from inter-fibre failure under compression in composite laminates. Compos. Appl. Sci. Manuf. 110, 76–83.

Fatigue damage mechanisms

2

Ramesh Talreja
Department of Aerospace Engineering, Department of Materials Science and Engineering, Texas A&M University, College Station, TX, United States

2.1 Introduction

In understanding fatigue of materials, one must confront the question: What is the source of irreversibility in the material's behavior? Said differently, what in the material state is different when the load applied the first time is reduced to zero compared to when the load was zero the first time? This difference holds the key to the so-called damage accumulation rate per load cycle.

In metal fatigue, the source of irreversibility in the material state has been identified as plasticity. The crystalline slip underlying plasticity is what initiates a crack, and the plastic deformation localized at the crack tip governs the crack growth rate per cycle. The dependency of this rate on the range of stress intensity factor in the form of a power law, known as the Paris law, is widely used in fatigue life prediction. The basic reason for the ability of this relationship to describe experimental data lies in the fact that the stress intensity factor is a measure of the stress field at a crack tip where the plasticity is localized.

Fatigue of continuous fiber-reinforced polymer matrix composites, abbreviated as PMCs, has little in common with metal fatigue. Although irreversibility due to inelastic deformation of the polymer matrix exists, its role in fatigue damage accumulation is not the same as that of plasticity in metal fatigue. Still, methodologies for life prediction developed for metals have been carried over to fatigue of composite materials, resulting in heavy reliance on empirical data and lack of reliability.

In this chapter, the author will attempt to address the fundamentals of fatigue damage in PMCs, starting with the unidirectional (UD) composites that form the building block of composite laminates. A framework for interpretation of the fatigue mechanisms, called fatigue life diagrams (FLDs), will be used to clarify the roles of the fibers, the matrix, and the fiber–matrix interface in the progression of the mechanisms and the consequent cycles to failure. Fatigue of multidirectional laminates will be discussed briefly, leaving a more detailed discussion of the subject to other publications.

2.2 Axial tension fatigue of UD composites

2.2.1 Experimental observations

Gamstedt and Talreja (1999) reported microscopy observations made on the surface of a carbon-epoxy (AS4/8552) UD composite subjected to axial tension–tension

Modeling Damage, Fatigue and Failure of Composite Materials. https://doi.org/10.1016/B978-0-443-18489-5.00007-2

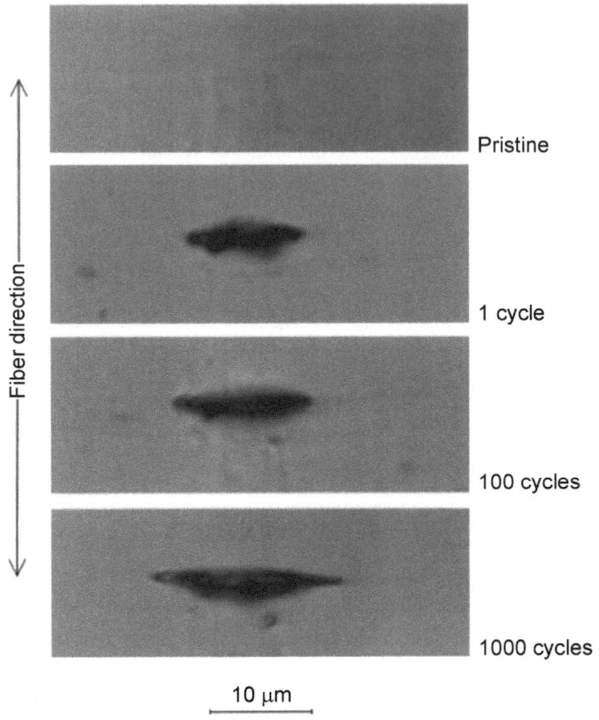

Figure 2.1 Micrographs of surface replicas showing a matrix crack at different applied load cycles (Gamstedt and Talreja, 1999).

cycling at an R-ratio of 0.1 and a frequency of 10 Hz. Surface replicas were collected periodically at different load levels using cellulose acetate films. After replication, the films were hardened and coated with a layer of carbon by sputtering. They were then observed under a microscope. Fig. 2.1 shows a series of micrographs showing the progression of a crack on the specimen surface, and Fig. 2.2 shows more clearly how a crack is bridged by fibers. The crack opening profile evident in Fig. 2.1 has the crack-tip region squeezed, which can be attributed to the closing pressure on crack surfaces induced by the bridging fibers. While Figs. 2.1 and 2.2 are for a case where the specimen failed in fatigue, Fig. 2.3 shows the case of a crack formed at a lower load level that did not fail the specimen until 10.5 million load cycles. The axial cracking at the crack tip initiating at a large number of cycles appears to be the cause of arresting the further growth of the lateral crack. Although not possible to confirm, the axial cracking is likely the fiber−matrix debonding. This cracking slows down and eventually stops. This could be the underlying cause of a possible fatigue limit.

It is noted that the surface observations illustrated in Figs. 2.1−2.3 do not form the complete picture of the fatigue mechanisms. It is not possible in a nontransparent material such as carbon-epoxy to view the progression of failure events that take place in the interior of a specimen. The surface observations show cracks often initiating

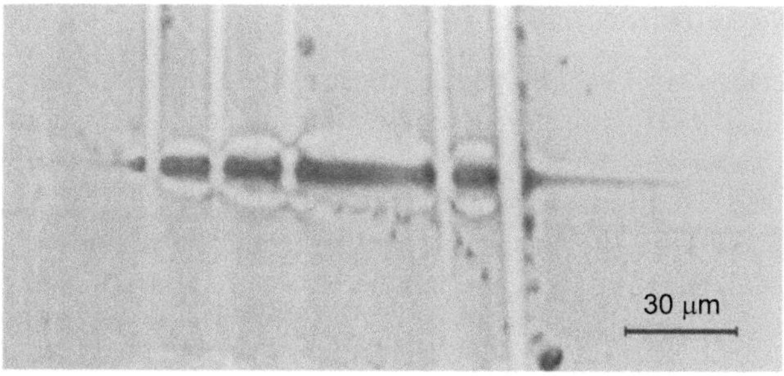

Figure 2.2 Micrograph showing a matrix crack bridged by fibers (Gamstedt and Talreja, 1999).

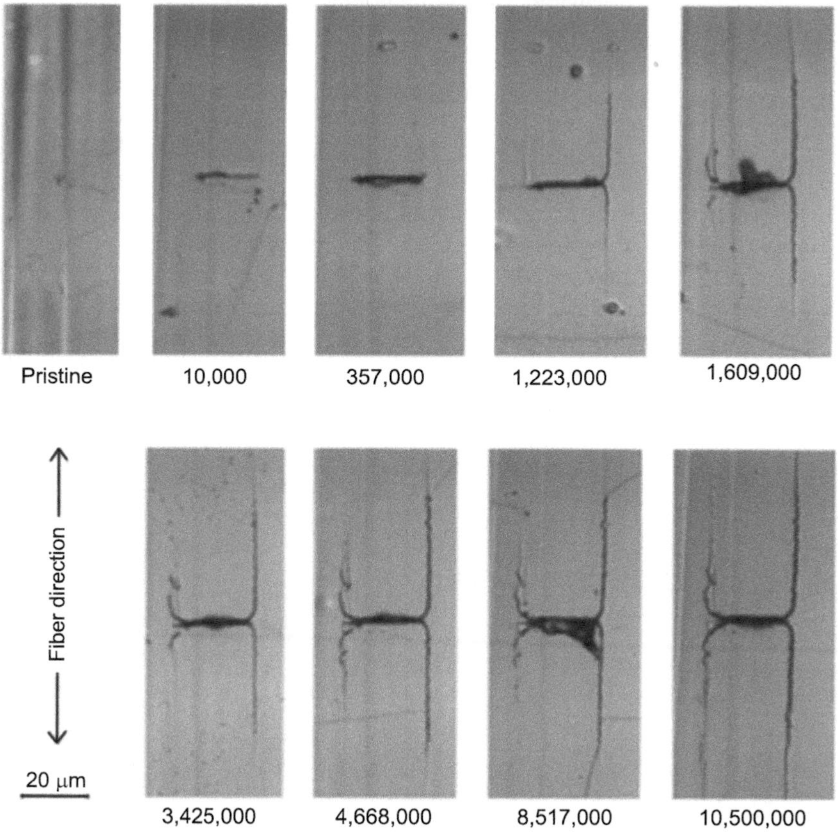

Figure 2.3 A series of micrographs of surface replicas showing axial cracking emanating at the fronts of a transverse matrix crack (Gamstedt and Talreja, 1999).

from broken fiber ends. The fiber breakage could have happened during processing or could have been induced by the first load applied. Broken fibers or other defects induced by manufacturing could also exist within the material volume and could be initiation sites for cracks.

2.2.2 Fatigue life diagrams

Long before detailed observations, such as those by Gamstedt and Talreja (1999), were reported, Talreja (1981) conjectured progression of fatigue damage and proposed a conceptual framework for interpretation of the roles of composite constituents in governing fatigue life. The framework was named FLDs, to note that the plots were not based on empirical data (S—N curves) but depictions of regions of dominance of different mechanisms. The concept behind these diagrams has its roots in the deformation-mechanism maps (Ashby, 1972). The FLDs are maps of the regions of dominance of progressive and nonprogressive mechanisms on a two-dimensional plot with a reference-loading parameter as the ordinate and the logarithm of the number of cycles to failure as the abscissa. The baseline FLD is that for a UD composite subjected to cyclic axial tension. This will be described next.

2.2.2.1 Nonprogressive fiber failure (Region I)

In order to understand the nature of the fiber-dominated damage under axial cyclic tension of a UD composite, let us first consider a dry bundle of fibers loaded by a tensile force *P*, as illustrated in the sketch to the left in Fig. 2.4. Owing to the inevitable defects in fibers, the weakest fiber in the bundle will fail first. If the applied load that caused this first fiber failure is reduced to zero and reapplied to its previous maximum value, no further fiber failures will occur. Thus, unless the fibers have a progressive cycle-dependent failure mechanism (which we assume not to exist), none of the fibers that survived the first load will fail under any number of load repetitions.

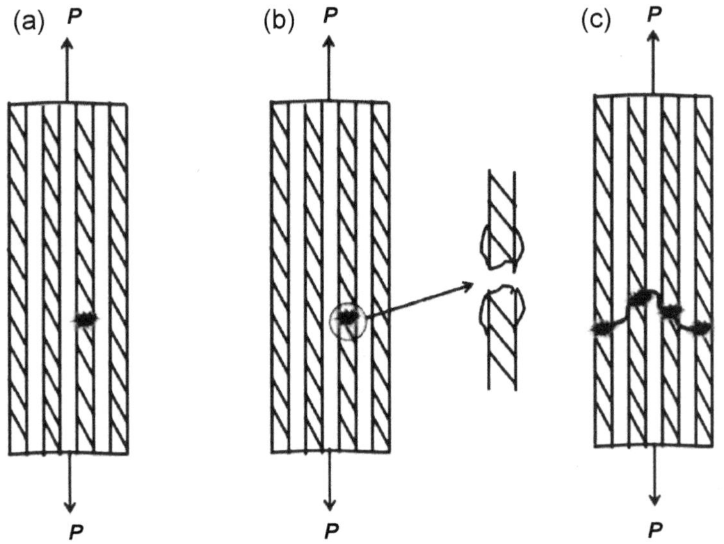

Figure 2.4 Fiber failure scenarios under a tensile load P. (a) A dry fiber bundle with the weakest fiber failure. (b) A fiber composite with the weakest fiber failure and debonding over a short length. (c) Composite failure from linkage of failed fiber regions.

Consider now a polymer matrix UD composite reinforced by the same fibers as in the dry bundle with perfect fiber—matrix bonding. The axial tensile load P applied to this composite will at some value of P cause the weakest fiber to fail. However, the consequence of the fiber failure in the composite will be quite different from that in the dry bundle. First, only the fibers in the immediate neighborhood of a broken fiber will carry the additional stress released by the broken fiber. Second, the broken fiber will debond locally from the matrix, as illustrated in the zoomed-in sketch next to the composite. Third, the matrix surrounding the broken fiber ends will be stressed at a significantly higher level than prior to the fiber failure. Most importantly, at this new stress level, the matrix polymer will likely deform in elastically, producing the irreversibility that is the key to the fatigue process. Thus, on unloading the load at which the fiber failed and reloading to this value will now change the stress states in the matrix as well as in the surrounding fibers in the immediate neighborhood of the broken fiber. Repeated application of this load can result in an accumulative process that can cause fatigue failure of the composite if a critical failure condition for composite failure is reached. A scenario for such criticality is sketched to the right in Fig. 2.4, indicating failure from linkage of broken fibers in a local neighborhood of the weakest fiber.

In a UD composite of a given volume, the fiber defects will be distributed in the volume depending on the manufacturing process. Consequently, under axial tensile load, the average failure stress (or strain) will not be deterministic but will have a probability distribution. Since the average stress in the fibers differs from that in the matrix, we will refer to the average strain in the composite at failure, as this is also the nominal fiber failure strain irrespective of the fiber volume fraction. Thus, let the mean value of the composite failure strain be denoted by ε_c and let the scatter in the failure strain be described by certain convenient extreme values such as those at 5% and 95% probabilities of failure.

Now let the composite be subjected to an axial tensile load such that the resulting maximum strain, $\varepsilon_{\max}$, is within the scatter band of the composite failure strain, as indicated in Fig. 2.5, where the maximum applied strain is plotted along the vertical axis. If the composite survives this load, then it can be concluded that none of the fiber failure regions reached the failure condition discussed above and illustrated in the sketch to the right in Fig. 2.4. For this condition to occur, an adequate number of fibers must fail in a local region such that the matrix crack formed grows unstably at the maximum applied load. Since the load applied is high enough for the maximum strain to be within the failure scatter band, it is reasonable to assume that many fibers fail in different regions at this load, although the failure condition is not reached in any of those regions. Consider now unloading and reloading to the same maximum load value. In the reapplication of load, each of the fiber failure regions will undergo stress redistribution due to the inelastic deformation of the matrix. However, the resulting stress fields in the regions will be different, and the consequence in terms of more failing fibers will also be different because of the random distribution of defects (weak points) in the fibers. If the reapplication of the maximum load is repeated, then with each repetition, a new scenario of the fiber failure regions will form, with the region most likely to fail the composite in the next application of the maximum

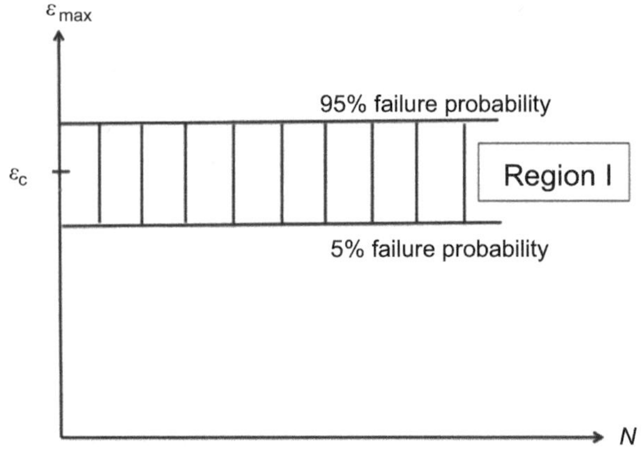

Figure 2.5 Region I of the fatigue life diagram.

load shifting from region to region. Thus, which of the failure regions will reach the composite failure condition first cannot be predicted, as a progressive mechanism that depends on the maximum applied load does not exist. Consequently, the composite failure is not a function of the number of load cycles, N, and the probability of this failure is also independent of N. In other words, the scatter band of initial failure strain remains unchanged with the number of load cycles N. This is depicted in Fig. 2.5, where the scatter band is labeled as Region I of the composite fatigue process.

2.2.2.2 Progressive failure process (Region II)

Let the UD composite be loaded in axial tension such that the maximum strain in the first application of load is below the scatter band of ε_c, as shown in Fig. 2.5. Under this load, few, if any, fibers will be expected to fail. However, broken fibers often exist as a consequence of the manufacturing process. The broken fiber ends become sites of stress concentration in the matrix, and consequently the matrix deforms into the inelastic regime under a sufficiently high load. Thus, under repeated load application, these sites will be a source of irreversibility needed to initiate and advance fatigue cracks. The role of fibers now will be to slow down, and possibly arrest, the growth of the fatigue cracks, as indicated by the observations reported in Gamstedt and Talreja (1999) and described above. The crack growth rate, and consequently, the number of load repetitions needed to reach the failure condition, will depend on the applied load. The inherent randomness of the crack growth rate results in the fatigue life scatter band, as depicted in Fig. 2.6, where the fiber-bridged crack is sketched to signify the underlying fatigue mechanism. The fatigue process represented by this scatter band will be called Region II.

2.2.2.3 Fatigue limit and Region III

Fatigue limit is of practical importance for designing durable structures. The concept of a limiting condition for fatigue, similar to a yield stress for metal plasticity, came

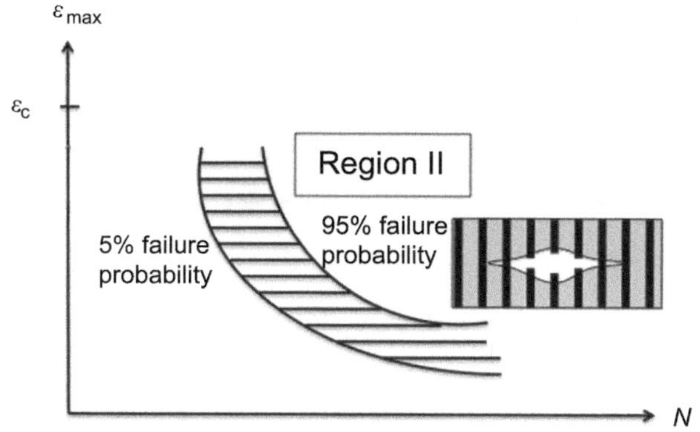

Figure 2.6 Region II of fatigue life diagram.

from observing that the fatigue life of some metals, particularly steels, was essentially infinite (or much longer than needed for safe performance in most structures) if the maximum stress was kept below a threshold value. If the metallic structure develops a crack, whose unstable growth defines failure, then the fatigue threshold is expressed in terms of the minimum range of the stress intensity factor below which no crack growth occurs. Studies of metal fatigue have indicated that the fatigue limit is sensitive to the microstructure, for example, the grain size, and the mechanisms responsible for limiting fatigue have to do with creating microstructure barriers to the formation of cracks from cyclic slip within grains and, if a crack initiates, to block its advance by such barriers.

In a UD composite undergoing Region II fatigue, the microstructure barriers are the fibers bridging the matrix cracks. These fibers provide resistance to crack growth by reducing the energy release rate (via reducing the crack surface displacement). The fibers ahead of the crack front provide obstacles to the crack growth. In spite of these mechanisms of crack growth retardation, it is likely that the failure condition is not avoided but is reached at very large number of load cycles. While in metals 10^6 cycles is viewed as a large number of cycles and is often used to define the fatigue limit, polymer-based composites are used in applications, such as wind turbine blades, where 10^7 or more cycles are expected in the design life. Since testing to such large number of cycles is time-consuming and costly, it is desirable that estimates for the fatigue limit can be made from considerations of the mechanisms.

Fig. 2.7 schematically depicts a scenario for arresting the fatigue crack growth in a composite. These cracks are assumed to form by the fatigue process in the matrix polymer under the applied cyclic load. Assuming the applied load level to be low enough that essentially no fibers fail, the matrix polymer surrounding the fibers undergoes cyclic stressing dictated by the cyclic deformation of the fibers. This cyclic deformation produces the same strain in the matrix as in the fibers (and the composite). In order for the matrix to form fatigue cracks, this cyclic strain must be equal to or greater than the fatigue limit of the matrix (measured in terms of strain). Thus, one estimate of the fatigue limit of the composite would be the fatigue limit of the matrix. Obviously, this

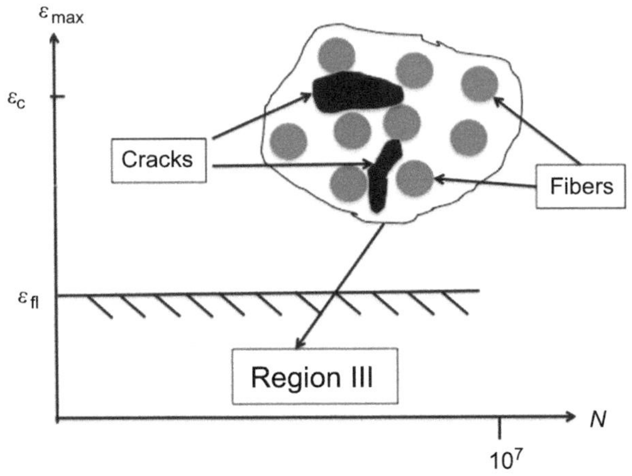

Figure 2.7 Region III of the fatigue life diagram indicating the mechanism of matrix crack arrest by fibers.

estimate of the composite fatigue limit, ε_{fl}, is an approximation since the local strain in the matrix is assumed to be the same as that in the composite, neglecting any strain enhancement caused by fibers. Thus, the matrix fatigue limit, ε_m, could be viewed as the lower bound to the composite fatigue limit, that is, $\varepsilon_{fl} < \varepsilon_m$. On the other hand, because of the crack growth arrest by the fibers, depicted in Fig. 2.7, the actual composite fatigue limit may be higher than the matrix fatigue limit, that is, $\varepsilon_{fl} > \varepsilon_m$. Fundamental studies are needed to determine how the fiber architecture (straight fibers vs. woven fabric, for example) as well as the fiber volume fraction affects the composite fatigue limit. For glass fiber-reinforced epoxy, data produced by Dharan (1975) suggested that the fatigue limit of epoxy, found to be at 0.6%, was a good approximation to the UD composite fatigue limit at three different fiber volume fractions (0.16, 0.33, and 0.50). However, for composites with stiffer carbon fibers, the fatigue limit is usually higher than 0.6%, as will be discussed later.

The region of the FLD lying below the progressive fatigue damage region, Region II, signifies the mechanisms that have no potential for reaching the critical failure conditions within a large number of load cycles, for example, 10^7. This region is labeled Region III, as indicated in Fig. 2.7. The matrix crack growth arresting mechanism depicted in Fig. 2.7 and discussed above is most likely, but other mechanisms are also possible. Gamstedt and Talreja (1999) described the role of fiber—matrix debonding ahead of an advancing fiber-bridged matrix crack in slowing down and eventually stopping the matrix crack. This mechanism was found to be significant in carbon—PEEK composites where the fiber—matrix interfacial region is brittle.

2.2.2.4 Roles of constituents in the fatigue process

The three regions of fatigue have been defined individually based on the dominant mechanisms operating in those regimes. Depending on the fiber, matrix, and interface

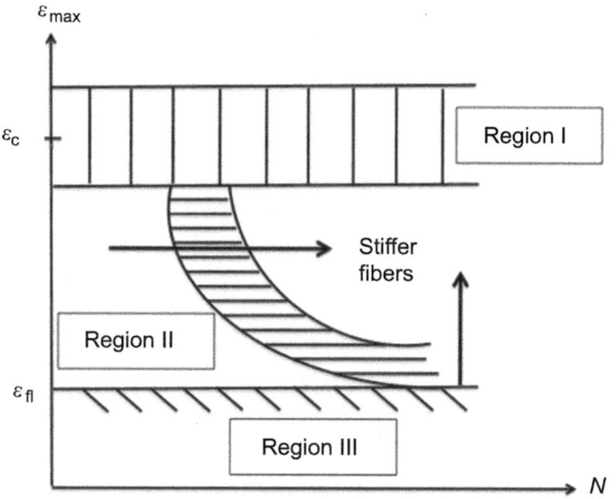

Figure 2.8 The total fatigue life diagram indicating the three regions.

properties, the regions will place themselves differently on the FLD. As a baseline construction, the three regions are placed as depicted in Fig. 2.8. It is noted that Region I in the FLD is the quasistatic failure scatter band, independent of load cycling, and therefore it does not depend only upon the matrix. The other two regions, Regions II and III, are primarily dependent upon the matrix fatigue properties, but they are affected by fiber properties such as the fiber stiffness.

Effect of fiber stiffness

Considering the mechanisms operating in Region II of the FLD, it can be surmised that with stiffer fibers, the cyclic growth of a fiber-bridged crack will be retarded, leading to longer fatigue life. Furthermore, the arresting of matrix fatigue cracks by stiffer fibers is expected to be more effective, resulting in enhancement of the fatigue limit. The arrows in Fig. 2.8 mark these trends. Fig. 2.9 shows fatigue life data and the

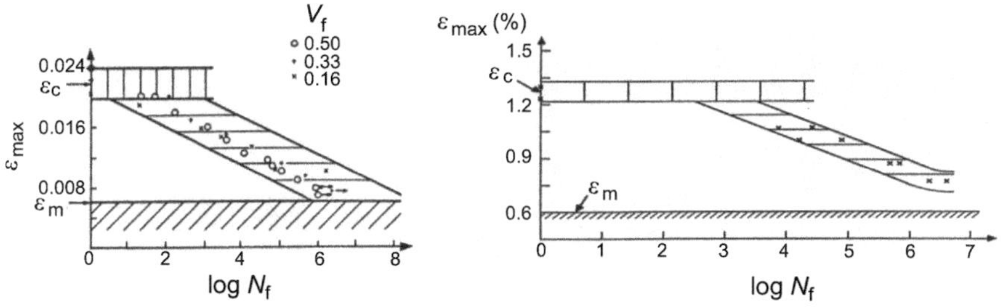

Figure 2.9 Fatigue life data and superimposed FLD for a glass–epoxy composite (left) at three fiber volume fractions and for a carbon–epoxy composite (right).
Data for glass–epoxy from Dharan (1975); and for carbon–epoxy courtesy of P.T. Curtis (private communication).

superimposed FLDs to support these trends. As seen in Fig. 2.9, the test data for a glass–epoxy composite of three different fiber volume fractions come together in the same FLD when plotted with the strain as the ordinate. The fatigue limit of the epoxy matrix is marked at 0.6% strain, as reported by Dharan (1975). The test data for a carbon–epoxy composite and its FLD are also shown in Fig. 2.9. By comparison of the FLDs of glass–epoxy and carbon–epoxy, the shift of Region II to higher fatigue life from glass fibers to carbon fibers is evident. Also, the enhancement in the fatigue limit of carbon–epoxy in comparison to glass–epoxy can be noted.

Effect of fiber–matrix interface

The nature of the fiber–matrix interface bond will play a role in fatigue crack progression, for example, by affecting the growth of the fiber-bridged matrix crack. In a comparative study of fatigue of UD carbon–epoxy and carbon–PEEK, Gamstedt and Talreja (1999) found that the nature of the interface bond in the two composites had a significant effect on Region II and III behavior. Their observations are summarized in the schematic in Fig. 2.10, which illustrates the roles of strong bond (left) and weak bond (right) in fiber-bridged cracking. The strong bond, typical of carbon–epoxy, results in relatively low debond growth, and as a consequence, allows the growing crack to break the fibers ahead of the crack tip and in the wake of the crack.

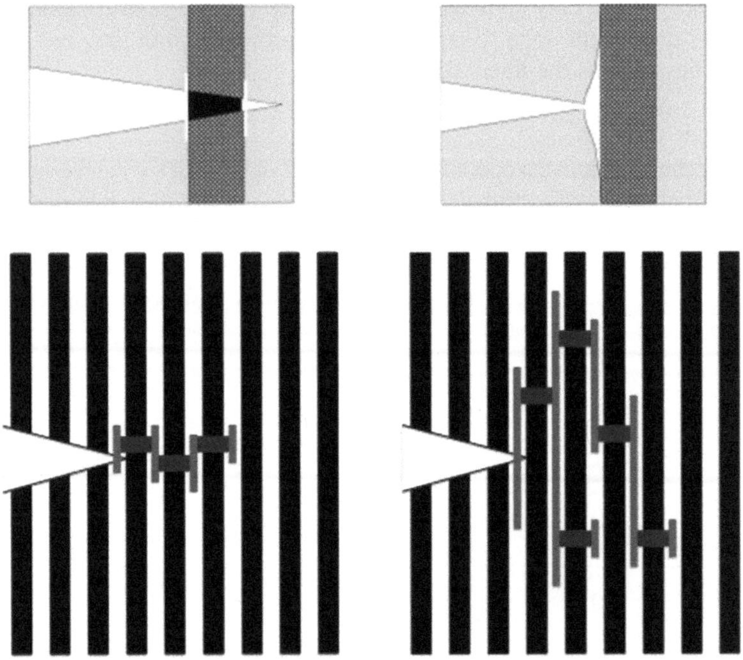

Figure 2.10 Schematic illustrating the effect of the fiber–matrix interface on the growth of a matrix crack. A strong interface (left) allows crack growth by breakage of fibers, while a weak interface (right) diverts the crack front and reduces crack advance.

A weak bond, on the other hand, leads to more extensive debond growth, which releases the strain energy at the crack tip in this failure mode, resulting in retardation of the growth of fiber-bridged matrix crack. However, the propensity to debond has another consequence: the individual cracks formed are easily connected by debonded interfaces, which promote failure by providing sufficient surface area needed for unstable growth.

For the extreme case of a very weak fiber—matrix bond, or defective bonding, the mode of fatigue failure progression could change from a fiber-bridged crack growth process to an accumulation of the fiber breakage process, promoted by a spread of stress enhancement from a broken fiber end to a larger neighborhood. Forming a well-defined fiber-bridged crack would then be less likely; instead, widespread fiber breakage would occur.

Effect of matrix inelasticity

Matrix inelasticity is in fact the main source of progressive fatigue failure in composites. If an elastic material, for example, a ceramic, is used as the matrix, then Region II in the FLD would not exist, unless another source of irreversibility is present. As discussed in Talreja (1990), in fatigue of ceramic matrix composites, the frictional sliding at the fiber—matrix interface provides the necessary irreversibility for progressive fatigue failure. Returning to matrix polymers, which are the subject of focus here, the stiff, elastic fibers of glass or carbon restrict the inelastic deformation of the matrix when the tensile load direction is aligned with the fibers. However, at the broken fiber ends, localized inelasticity occurs, as discussed above. If by material selection a given matrix is replaced by a softer (more ductile) matrix, then the local inelastic strain in the matrix at the broken fiber ends will be more intense and more extensive (i.e., spread over a greater volume). As a consequence, the fatigue-failure progression will be more effective, resulting in fewer load cycles to failure. Thus, Region II of the FLD will shift to the left. This trend is opposite to the trend discussed earlier in the "Effect of fiber stiffness" section.

2.3 Fatigue of UD composites in other loading modes

The FLD discussed above was constructed for the baseline case of UD composites under tension load cycling along the fibers. Variations in the diagram from this case will be induced by other loading modes. The cases of compression loading and off-axis loading are discussed next.

2.3.1 Compression loading

Under cyclic loading in the fiber direction, load excursions in tension versus compression have fundamentally different effects on the fatigue process. The tension case was considered in constructing the baseline FLD. The variation to this diagram due to compression loading will be discussed next.

For the case of compression along the fibers in a UD composite, Rosen (1964) conceived and analyzed fiber microbuckling as a possible failure mechanism. Experimental observations later indicated that the fibers tend to buckle under the influence of local shear at stress concentration sites such as defects in the matrix and misaligned or wavy fibers. If totally elastic (reversible) conditions persist surrounding fiber microbuckling, then no further changes can be expected under repeated load excursions in compression. However, if the polymer matrix deforms inelastically at the microbuckled fiber, then this irreversible deformation process will intensify the fiber buckling amplitude, thereby inducing microbuckling of the neighboring fibers, leading eventually to the formation of kink-band. This progressive failure mechanism is illustrated in Fig. 2.11, highlighting essential features without secondary aspects such as broadening of kink-bands during their progression. Final failure (often described as shear crippling) in fatigue is expected when sufficient fibers break at the kinks in the kink-bands, leading to the loss of load-carrying capacity. For details of microbuckling and kink-bands, and their analysis by mechanics models, the reader is referred to studies in Jelf and Fleck (1992), Budiansky and Fleck (1994), Moran et al. (1995), Vogler et al. (2001).

As described above, the failure process in UD composites under cyclic compression has phases of initiation, progression, and criticality. If fiber microbuckling occurs under the first application of a compressive load, then under subsequent load applications, progression of microbuckling, kink-band formation, and fiber

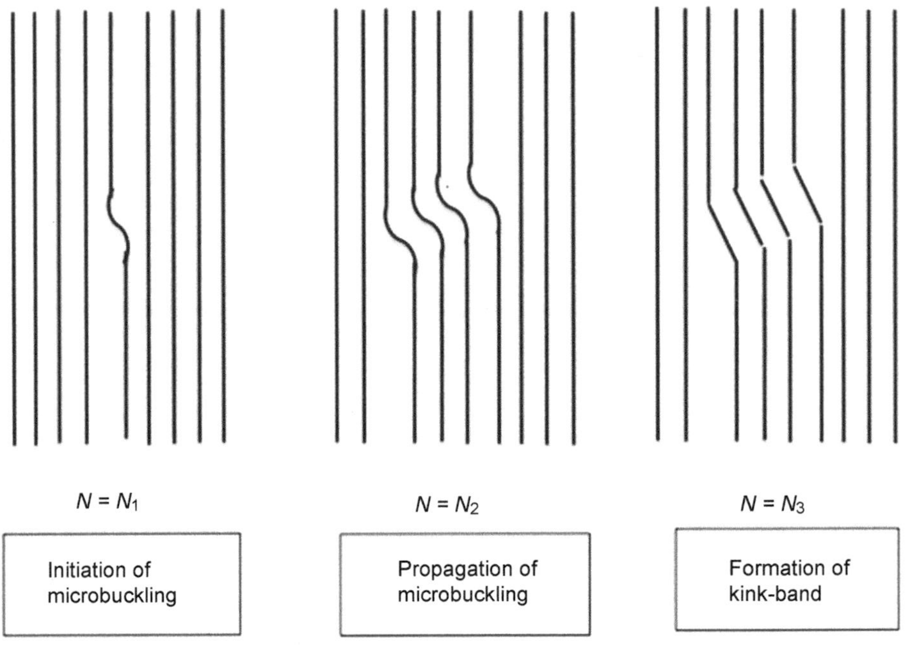

Figure 2.11 Schematic illustration of compressive failure mechanism in a UD composite.

breakage are possible. The rates at which the postmicrobuckling mechanisms occur will depend on the maximum level and range of the repeating stress. If the maximum stress level is within the scatter band of the compressive strength, and the composite survives, then failure will be expected in a few subsequent load applications. Thus, conditions for nonprogressive fiber failure that characterizes Region I in the baseline FLD will not be present in compression fatigue. In other words, Region II of the FLD (corresponding to progressive failure in compression) will start from the first application (static) strength and will slope downwards. A continuous slope in Region II will be expected if the nature of progressive failure mechanisms does not change with the load level. The fatigue limit will then be given by the threshold stress below which no fiber microbuckling occurs in the first application of load. This stress level is expected not only to depend on the fiber properties (e.g., stiffness) but also on the presence of manufacturing-induced defects such as fiber misalignment and waviness.

2.3.2 Off-axis loading

If the applied load is not aligned with fibers in a UD composite, then the failure mechanisms underlying the construction of the baseline FLD do not hold. Under loading at off-axis angles of a few degrees from the fiber direction, the role of fibers reduces drastically, while the matrix and interface govern the failure. Region I in the baseline FLD does not exist in this case, as fiber failures are not the controlling mechanism, and Region II consists of progressive failure of the matrix and/or the interface. The mechanism of progressive failure in this case is illustrated in Fig. 2.12. The first application of an off-axis load produces microcracks at different sites in the matrix and/or at the fiber—matrix interfaces, depending on the defects present. The inelasticity of the matrix provides the necessary irreversibility at the crack fronts to incrementally advance the cracks in subsequent load applications. The cracks advance at different rates depending on the local conditions, for example, the distribution of fibers ahead of the crack fronts, and final failure results from the first unstable growth of any of the cracks. The 2D schematic in Fig. 2.12 is only for illustration of the cracking process; the actual cracks formed have generally irregular planes lying in regions between fibers. The irregularly shaped curves drawn in the figure are intended to represent intersection of those planes with the projected 2D plane. The major direction of advance of the cracks is expected to follow the inclined fibers.

The set of cracks that are formed in the first application of the applied load grow at different rates under load reapplication, as stated above. The manufacturing-induced microstructure, including defects (irregular fiber distribution, fiber misalignment, matrix voids, disbonded interfaces, etc.), produces differing conditions for advancement of the cracks, and attainment of the critical condition for failure, resulting in the scatter in the fatigue life. The fatigue life scatter band (Region II) starts at the scatter band of static failure (i.e., failure under first application of load), slopes downwards, and ends in the fatigue limit. This is illustrated in Fig. 2.13, where the baseline FLD is also shown for reference.

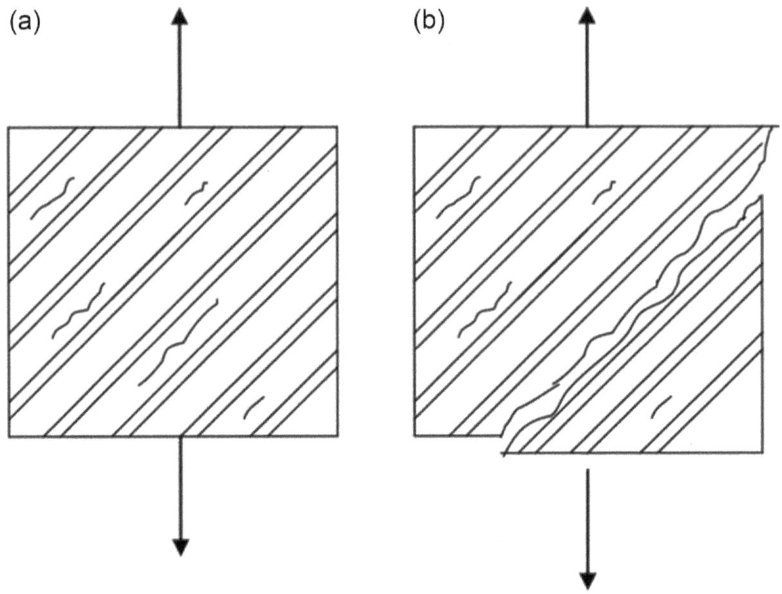

Figure 2.12 Schematic illustration (2D) of cracking along the fibers under off-axis loading. (a) Cracks initiated at the first application of the maximum load. (b) Failure from a crack that grew unstably.

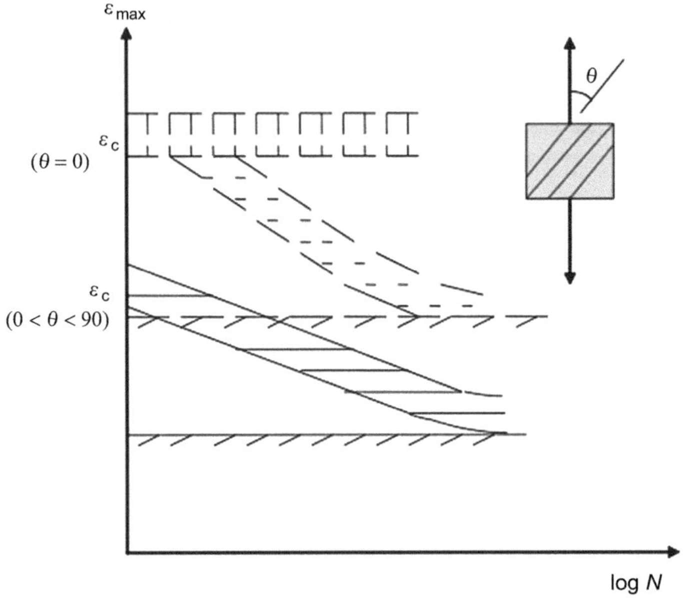

Figure 2.13 The fatigue life diagram for off-axis loading along with the baseline FLD (dashed lines) for reference.

2.4 Conclusions

This chapter has provided a review of fatigue damage mechanisms in UD composites. Based on a wealth of observations and conjectures, a framework for describing the dominant mechanisms, called FLDs, has been presented. This framework facilitates interpretation of the roles of fibers, matrix, and fiber–matrix interfaces in affecting the damage development and fatigue life. From the baseline diagram for axial tension fatigue of UD composites, diagrams for other cases, such as compression and off-axis loading, can be derived.

Fatigue in composite laminates involves mechanisms such as multiple intralaminar cracking and interlaminar cracking (delamination). These mechanisms can be viewed as subcritical mechanisms that occur before the fatigue failure of the last load-bearing ply. Thus, fatigue of UD composites in the context of laminates can be viewed as the final (critical) stage in the failure process. The baseline FLD then allows interpretation of the roles of ply properties and laminate layup in the laminate fatigue process. For further discussion of composite laminate fatigue, the reader is referred to Talreja (1981, 2000, 2003), Talreja and Singh (2012).

References

Ashby, M.F., 1972. A first report on deformation-mechanism maps. Acta Metall. 20 (7), 887–897.

Budiansky, B., Fleck, N.A., 1994. Compressive kinking of fiber composites: a topical review. Appl. Mech. Rev. 47, S246–S250.

Dharan, C.K.H., 1975. Fatigue failure in graphite fibre and glass fibre-polymer composites. J. Mater. Sci. 10, 1665–1670.

Gamstedt, E.K., Talreja, R., 1999. Fatigue damage mechanisms in unidirectional carbon-fibre-reinforced plastics. J. Mater. Sci. 34, 2535–2546.

Jelf, P.M., Fleck, N.A., 1992. Compression failure mechanisms in unidirectional composites. J. Compos. Mater. 26, 2706–2726.

Moran, P.M., Liu, X.H., Shih, C.F., 1995. Kink band formation and band broadening in fiber composites under compression loading. Acta Metall. Mater. 43, 2943–2958.

Rosen, B.W., 1964. Mechanics of composite strengthening: fiber composites. Am. Soc. Metals 37–45.

Talreja, R., 1981. Fatigue of composite materials: damage mechanisms and fatigue life diagrams. Proc. Roy. Soc. Lond. A378, 461–475.

Talreja, R., 1990. Fatigue of fibre-reinforced ceramics. In: Bentzen, J.J., et al. (Eds.), Structural Ceramics-Processing, Microstructure and Properties. Riso/National Laboratory, Roskilde, Denmark.

Talreja, R., 2000. Fatigue of polymer matrix composites. In: Talreja, R., Månson, J.-A.E., Kelly, A., Zweben, C. (Eds.), Comprehensive Composite Materials, vol. 2. Elsevier, Oxford, p. 529.

Talreja, R., 2003. Fatigue of composite materials. In: Altenbach, H., Becker, W. (Eds.), Modern Trends in Composite Laminates Mechanics. Springer.

Talreja, R., Singh, C.V., 2012. Damage and Failure of Composite Materials. Cambridge University Press.

Vogler, T.J., Hsu, S.-Y., Kyriakides, S., 2001. On the initiation and growth of kink bands in fiber composites. Part II: analysis. Int. J. Solid Struct. 38, 2653–2682.

The process of cracking in textile composites subjected to quasistatic tensile and tension–tension fatigue loadings

3

Larissa Gorbatikh and Stepan V. Lomov
Department of Materials Engineering, KU Leuven, Leuven, Belgium

3.1 Introduction

Polymer composites based on textile reinforcements are an important class of high-performance materials. They offer improved damage tolerance in comparison with unidirectional (UD) cross-ply laminates. In production, textile reinforcements are easier to handle and apply in closed-mold processes. Because of the yarn crimp, inherent to textile reinforcements, the stiffness and strength of these composites are, however, lower than those achieved in UD-based composites. For a general introduction to textile reinforcements and their composites, the reader is referred to classical works (Chou and Ko, 1989; Naik, 1994; Bogdanovich and Pastore, 1996), more recent books (Long 2005, 2007; Lomov, 2011; Boisse, 2021), an on-line encyclopedia (Lomov and Verpoest, 2010), and proceedings of the TexComp conference series (1995–2022). Modeling of the dry behavior during forming is also covered in these works.

The mechanical performance of fiber-reinforced composites and particularly their final failure strongly depend on the material structural organization. The latter, together with properties of the matrix and the fiber/matrix interface, controls the damage initiation and development in the material. The present chapter introduces different stages of the damage process in textile composites loaded in quasistatic and fatigue tensile tests. It highlights similarities and differences in the way cracks develop in textile composites and UD cross-ply laminates. Similarities in patterns are mainly due to comparable directionality of yarns in textiles and fibrous plies in laminates. On the other hand, textile composites also possess unique features in the way cracks are formed that are not present in the UD cross-ply laminates. These are attributed to special geometrical features of textile reinforcement such as crimp of the yarns; the presence of resin-rich pockets; variable fiber volume fraction inside the yarns, etc. The chapter starts off with generalities of this behavior and continues with examples for 2D and 3D woven, braided, and noncrimp fabric (NCF) composites, reinforced with carbon, glass, and flax fibers.

When subjected to cyclic loading (fatigue), the types of cracks and sequence of their appearance are quite close to the ones observed under quasistatic loading. The analogy that exists in UD cross-ply laminates between damage development in quasistatic

Modeling Damage, Fatigue and Failure of Composite Materials. https://doi.org/10.1016/B978-0-443-18489-5.00021-7

loading with load increase and in fatigue loading with progression of fatigue cycles (Talreja, 2012) is also present in textile composites (Carvelli and Lomov, 2015, 2017). The reader can find experimental observations of this in Adden et al. (2005), Carvelli et al. (2010), Karahan et al. (2011), Vallons et al. (2009).

The reported data are obtained following a methodology described in Lomov et al. (2008, 2018), Carvelli and Lomov (2015), Oz et al. (2018). This methodology involves carrying out quasistatic tests conducted on a displacement-controlled testing machine and fatigue tests under the load control. Composites are loaded in the machine (warp), cross (weft) and bias directions. Strains are measured on the specimen surface with the use of an optical full field strain measurement in the central part of the specimen. The damage is also observed with optical microscope on the polished edge of the specimen. The tests are carried out till composite failure and till different strain levels to allow for subsequent characterization of damage. The damage is examined by means of (I) continuous monitoring of acoustic emission (AE) signals during the test, (II) X-ray or micro-computed tomography (CT) characterization after the test, and (III) optical or scanning electron microscopy (SEM) examinations of sample cross-sections after the test. This methodology has been applied to composite materials with different textile architectures (woven, braided, multiaxial multiply warp-knitted, structurally stitched) as reported in Truong Chi et al. (2005), Ivanov et al. (2009), Ivanov et al. (2009, 2017), Koissin et al. (2009), Lomov et al. (2009), Daggumati et al. (2010), De Greef et al. (2011), De Greef et al. (2011), Bogdanovich et al. (2013), Kersani et al. (2014), Carvelli et al. (2017), Ivanov et al. (2017), Gigliotti et al. (2018). The data and observations collected in these studies are the source of the information presented in this chapter.

Once a textile composite is subjected to loads that go beyond its elastic deformation, the supply of energy is absorbed through the formation of free surfaces (cracks) and plastic deformation. The more complex the network of cracks and the higher their density, the more energy can be absorbed by the composite. Table 3.1 reviews different types of cracks encountered in textile composites after they are subjected to tensile loading in the fiber or bias directions. This review is based on an example of a 3D woven carbon/epoxy composite (Bogdanovich et al., 2013), but most of the crack types, like transverse cracks inside the yarns, splitting, and delaminations, are also observed in composites with other textile architectures. The cracks are classified according to their orientation and position vis-à-vis the reinforcement geometry. Cracks of type Z are only relevant for 3D or stitched reinforcements, when through-the-thickness directed yarns (Z-yarns) are present.

3.2 Overview of the cracking process

A typical sequence of damage events in a composite loaded in the direction of one of the yarn systems (called below "longitudinal"), with an increased tensile load, is as follows (Fig. 3.1):

- The first damage in the composite typically appears in the form of transverse cracks. These cracks usually initiate inside yarns where the local fiber volume fraction is the highest, but

Table 3.1 Different types of cracks observed in textile composites loaded in tension.

Crack type	Notation	Orientation	Schematic representation in the composite cross-section
Transverse matrix cracks inside yarns	T	Normal to the loading direction and sample surface	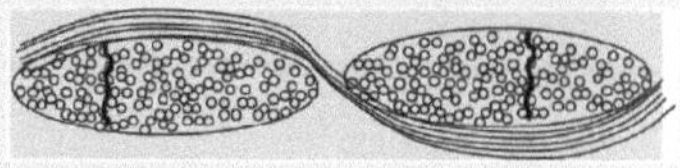
Transverse cracks at yarn boundaries	B	Normal to the loading direction and sample surface	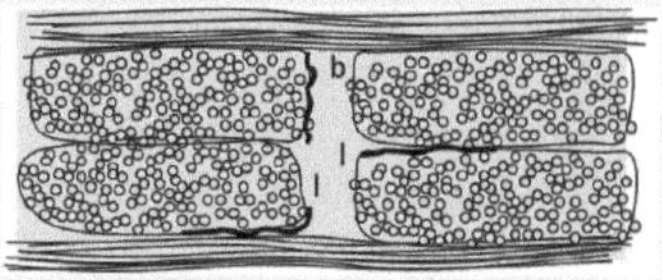
Debonding of yarns and local delaminations inside yarns	L	Parallel to the loading direction and sample surface	
Transverse cracks in the matrix pockets	Mt	Normal to the sample surface and to the loading direction	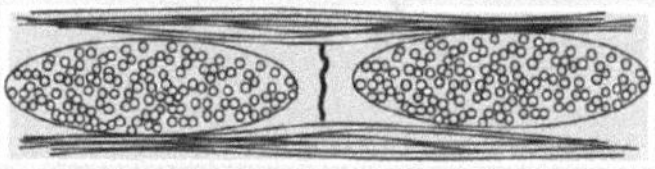
Shear matrix cracks	S	Inclined toward the loading direction and normal or inclined to the sample surface	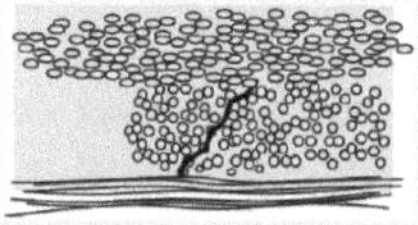
Splitting of the yarns	Sp	Normal to the sample surface and parallel to the loading direction	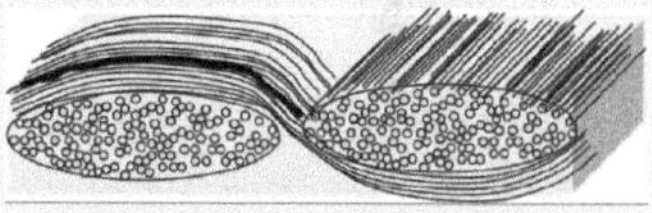
Cracks on the boundary of or inside Z-yarns	Z	Follow the direction of Z-yarns, normal to the loading direction	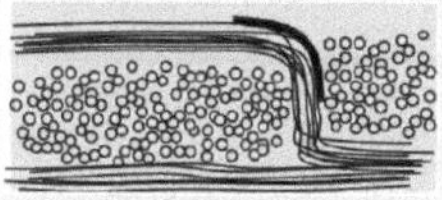
Longitudinal cracks in the matrix pockets	ml	Parallel to the sample surface and to the loading direction	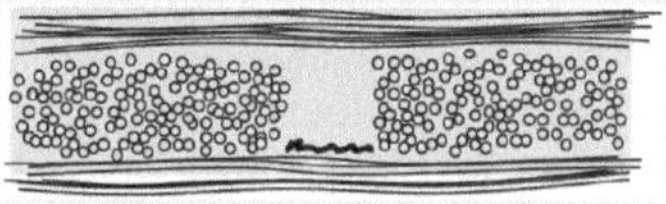
Delaminations	d	Parallel to the loading direction and sample surface	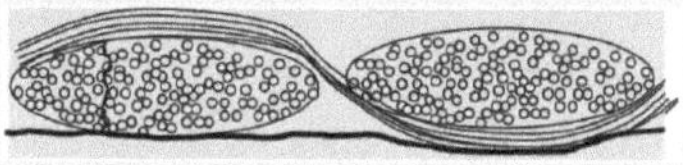

Note: All schematics are shown such that the loading and horizontal directions coincide.

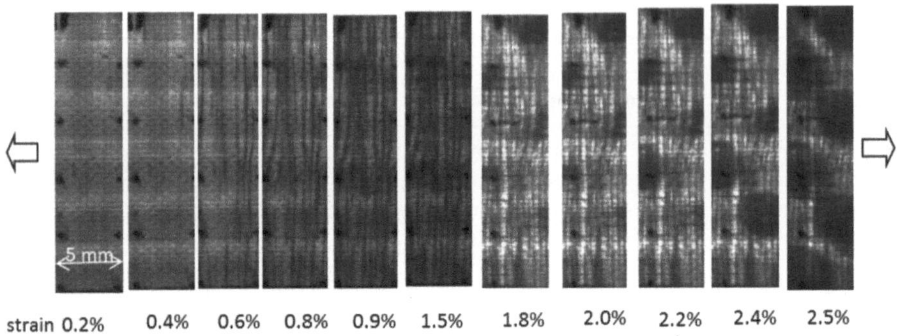

Figure 3.1 Backlight images showing development of cracks in a plain weave glass/epoxy composite as a function of strain. Loading is in the warp direction.
Adapted from Li et al. (2015).

for some textile architectures, they can also form at yarn boundaries. The crack position is likely to be controlled by both the meso-structure of the textile and the fiber distribution inside the yarn.

- Upon further load increase, the transverse cracks propagate along the yarn length. They also increase in numbers until a certain critical number is reached. This level is sometimes referred to as a saturation level. There may be several transverse cracks per yarn or only one, depending on the material. Transverse cracks in resin-rich pockets are usually cracks that originated inside yarns but then propagated into these areas.
- When transverse cracks are well developed, they further propagate as local delaminations. These delaminations are triggered by shear stresses resulting from interaction of the transverse cracks with longitudinal layers. The local delaminations progress, leading to larger inter-ply delaminations. The onset and propagation of delaminations between plies are dependent on the interlaminar fracture toughness of the composite.
- At a certain strain level, longitudinal yarns will experience tensile transverse stresses caused by the Poisson's effect due to the constraint by the transverse yarns. When the transverse strength of these longitudinal yarns is exceeded by the local stress, they can start splitting.
- In the final stage, breakage of fibers in longitudinal yarns begins. The strain at which this happens in textile composites is typically below the ultimate strain of fibers. This reduction is caused by the fiber crimp and developed delaminations, which prohibit efficient stress transfer inside fiber bundles.

For loading in the bias direction, the damage development is affected by prevailing shear deformations inside fiber bundles and in matrix pockets. The sequence of the damage process described above still holds, with an exception that transverse cracks become shear cracks, and yarn splitting does not occur. Because of rotation of the yarns, additional local delaminations can appear at the yarn crossings.

Formation of free surfaces can be captured with the AE registration and analyzed by means of certain AE thresholds. For this, two AE sensors are placed at the boundaries of the gauge length region. Signals that occur outside the sensors are filtered out by the AE system. The energy of AE events is registered, and the dependency of cumulative energy of AE events vs. tensile strain is plotted. The change in the rate of AE event accumulation indicates a switch to another damage mechanism. This is used to identify thresholds of the damage development as illustrated in Fig. 3.2a.

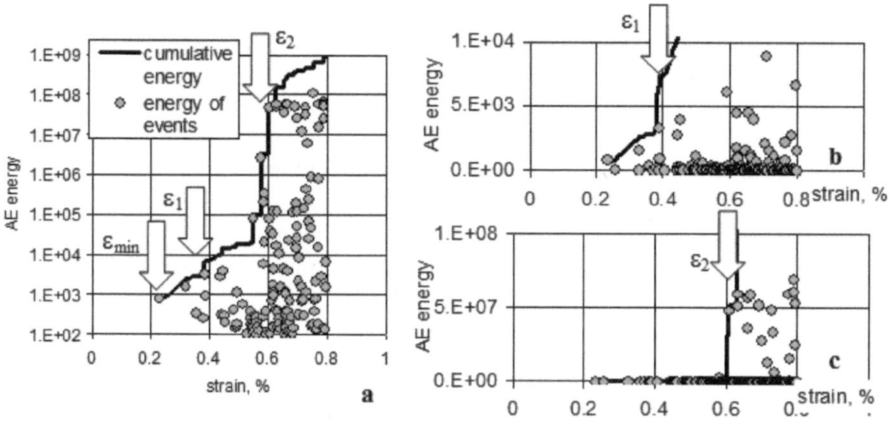

Figure 3.2 Damage thresholds identified with AE registration on a typical acoustic emission diagram (the data are for a 3D orthogonal woven glass/epoxy composite, tested in warp direction): (a) Full AE diagram, logarithmic scale; (b) Low energy events, the jump corresponds to ε_1; (c) High energy events, the jump corresponds to ε_2 (Lomov et al., 2009).

In the elastic regime at small strains, no acoustic events are registered, as no cracks are yet formed. Starting from a certain strain level, characterized as ε_{min} (*AE threshold strain*), low energy events start to occur with a low frequency. Then the frequency of the events increases sharply, and the energy content reaches higher levels. This is reflected by the increase of the slope of the cumulative AE event energy curve. The corresponding strain is designated as ε_1 (*first damage threshold strain*). This strain level is also considered as *damage initiation threshold*. At a certain higher level of the applied strain, a second "knee" on the AE cumulative energy curve may appear. This is considered to be the *second damage threshold strain*, designated ε_2. Above this strain level, but well below strain level corresponding to the final failure of the sample, the AE sensors were removed to avoid their damage. Fig. 3.2b and c illustrates how the strain thresholds ε_{min}, ε_1, and ε_2 are identified using separate plots for low and high AE ranges.

This AE energy-based analysis provides indirect information about damage development in terms of the thresholds and allows comparative analysis between materials. It is, however, much more complicated to interpret the AE data in terms of crack types. The AE analysis, therefore, needs to be accompanied with direct characterization of damage. The first threshold strain, ε_1, is typically associated with a general onset of damage, and the second threshold strain, ε_2, with the start of significant damage, like transverse cracks spanning full yarn thickness and considerable length (type *t*), or local debonds (type *l*), or yarn splitting (type *s*). More detailed analysis, involving frequency features of the AE-signal, may allow better interpretation of the AE signals and identification of specific damage features associated with these signals (de Groot et al., 1995; Gutkin et al., 2011; Sause et al., 2012; Kersani et al., 2014; Li et al. 2014, 2015).

Studies of AE in multidirectional laminates (Baker et al., 2015; Maillet et al., 2015; Oz et al., 2017, 2018) have demonstrated that frequency characteristics of the AE signal may depend on the laminate structure and internal damage, hypothetically causing dispersion of the AE signal. This may lead to the appearance of high-frequency signals caused by matrix cracks as well as by fiber breaks. The identification of high-frequency AE events as related to the fiber breakage, often stated in the literature, should be taken with caution. AE registration is not sufficient for damage mode identification in carbon fiber–reinforced composites. Even if a clustering algorithm can recognize well-separated and dense clusters of AE events in the frequency/amplitude parameters space, those clusters may not be unambiguously used to identify the type of the damage which caused the event. Multi-instrument in situ observations are a must for reliable damage mode identification.

The presence of cracks in the composite may affect its performance, particularly stiffness. In some composites (for example, glass fiber composites), the detrimental effect of accumulated cracks on the material stiffness becomes evident in the interval between thresholds ε_1 and ε_2. The damage recorded prior to ε_1 is normally not sufficient to reduce the composite stiffness. The effect of transverse cracks on the composite stiffness is hardly noticeable in carbon fiber composites due to a much lower transverse stiffness of carbon fibers in comparison with glass fibers.

3.3 Initiation of matrix cracks

A typical mechanism for the initiation of matrix cracks in textile composites on the microlevel is similar to the one in cross-ply laminates. The onset usually starts as microdebonding at the interface between fibers and matrix. It is driven by tensile and shear stress concentrations at the interface. These stress concentrations appear due to the mismatch in the transverse stiffness of the fiber and matrix. The ratio of the fiber stiffness to polymer stiffness is 5–6 for carbon fibers, 20–25 for glass fibers, and 60–70 for steel fibers. An irregular shape of the fiber cross-section additionally contributes to the stress concentrations (Fig. 3.3a). The high stiffness mismatch would lead to high stress concentrations near the fiber–matrix interface and can cause premature debonding (Fig. 3.3c). The debonds may then propagate into the matrix, coalesce, and form transverse matrix cracks (Fig. 3.3b); this process can be modeled with finite elements (FEs) models involving cohesive elements (Sharma et al., 2020). The latter then run through the yarns of the textile reinforcement and through the matrix pockets. Not all debonds become transverse cracks.

This mechanism is essentially the same as for the crack formation in transverse plies of UD laminates. A new feature appearing in textile composites is the presence of matrix cracks on yarn boundaries. If the reinforcement geometry includes yarn boundaries, which are (approximately) normal to the loading direction, then they become a preferred site for the initial fiber debonding. Interestingly, this does not necessarily imply that these initial boundary cracks will develop into "full" transverse cracks. This is an indication that the location of transverse cracks is also controlled by the stress distribution on the meso-level, namely by the architecture of the textile reinforcement. Fiber debonds, on the other hand, observed prior to the onset of cracking, are formed as a result of local inhomogeneities in the distribution of fibers and yarn geometry.

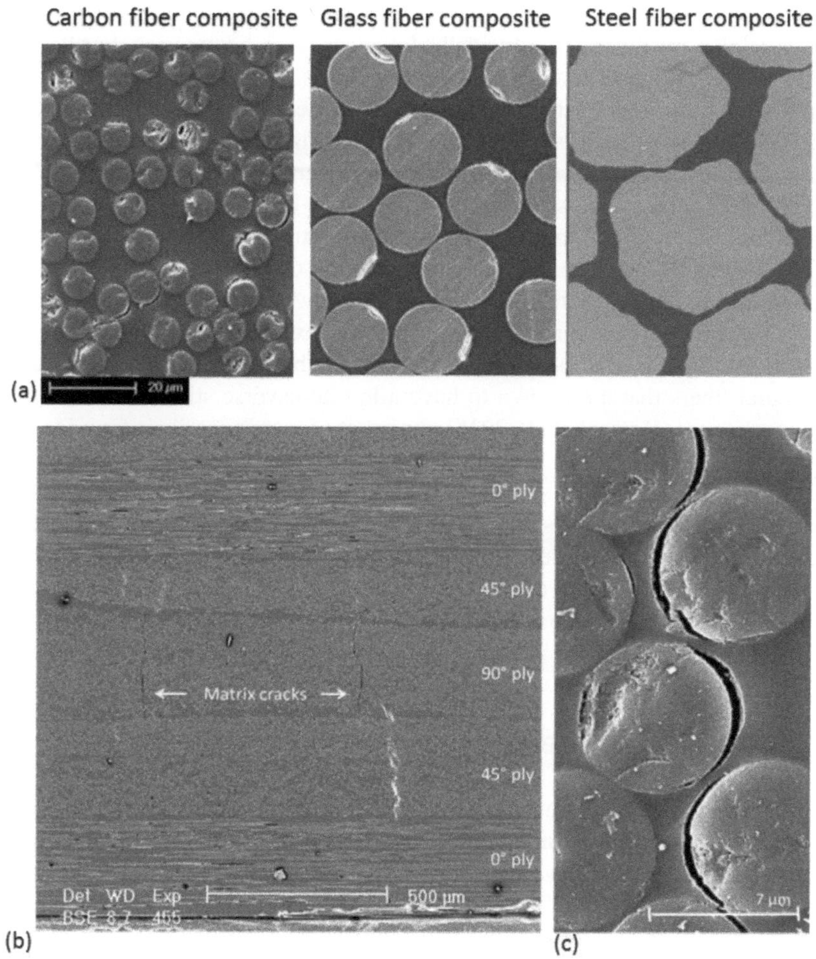

Figure 3.3 SEM images showing (a) Microstructure of composites with carbon, glass, and steel fibers (Sabuncuoglu et al. 2014 in print); (b) Matrix cracks in transverse plies; (c) Zoom in on a matrix crack revealing debonding at the fiber/matrix interface in a carbon fiber/epoxy composite. The composite was loaded under tension along the 0 degree fibers (Romanov et al., 2015).

This was observed in a carbon/epoxy twill composite (De Greef et al., 2011), with yarn cross-sections having a quasielliptical shape. Fiber interface debonds appeared first on yarn boundaries, while debonds inside yarns were formed later. The debonds inside yarns grew later into transverse cracks. Those at the yarn boundaries typically remained as individual debonds without progressing into a large crack. This was also the case in a carbon/epoxy 3D woven orthogonal noncrimp composite with quasirectangular yarn cross-sections and large quasiflat yarn surfaces normal to the loading direction (Bogdanovich et al., 2013).

Under bias loading, damage in the form of microdebonding at the fiber–matrix interface is also observed. It is typically found on yarn boundaries. Further loading leads to the formation of matrix cracks inside yarns. These cracks have meandering surfaces indicating ductile fracture that was preceded by high shear deformation in the matrix. This is in contrast to the appearance of transverse cracks that have clean fracture surfaces—typical for brittle fracture. In certain places, massive debonding and pull-out of fibers can also be observed.

As stated earlier, initiation of matrix cracks is controlled by the stress concentration at the fiber–matrix interface, as well as by strength of the interface (or the matrix, if cracks are formed in the matrix). In the case of a low contrast between the transverse stiffness of fibers and the matrix, the onset of matrix cracks can be delayed till later stages of loading. An example of such behavior is observed in composites reinforced with natural fibers that are known to have a low transverse stiffness. In woven flax/epoxy composites (Kersani et al., 2015), the onset of matrix cracks is found to be quite peculiar: it starts just before the entire composite fails. Under bias loading, matrix cracks never appear at all, in spite of occasional debonding of flax fibers from epoxy caused by the residual stresses after cooling the composite in production and the weak flax/epoxy adhesion. The onset of debonds can be delayed if the strength of the interface is increased. This can be done by depositing nanoreinforcements on the fiber surface or in the matrix, among other approaches (Gorbatikh et al., 2016; Karger-Kocsis et al., 2020).

3.4 Influence of the yarn crimp

The yarn crimp in textile reinforcements has an important influence on the initiation and development of damage in composites based on these reinforcements. This influence is due to a different stress state that yarns experience locally in comparison with straight, noncrimped plies of the same orientation in UD composites. Even for the yarns oriented in 0 or 90 degree directions, this stress state becomes complex with a nonnegligible shear component. The first detectable difference due to crimp is an earlier threshold for the initiation of transverse cracks in textile composites in comparison with UD cross-ply laminates. Fig. 3.4a shows comparisons of the damage initiation thresholds observed in tensile loading for different types of carbon/epoxy composites, accompanied by images of the internal structure of some of these composites (Fig. 3.4b–d). The data are collected from studies of cross-ply–laminated (CPL) composites made of UD layers (Yokozeki et al., 2007; Truong Chi et al., 2008) and textile composites, reinforced with noncrimp stitch-bonded fabric (NCF) (Truong Chi et al., 2005, 2008; Vallons et al., 2009); 2D triaxial braided (2DTB) fabric (Ivanov et al., 2009); 2D woven (2DW) fabric (De Greef et al., 2011; Karahan, 2011); steel fiber woven fabric (Callens et al., 2014, 2015) 3D angle interlock weave (3DAIW) fabric (El Hage, 2006; Lomov et al., 2011); and 3D noncrimp orthogonal woven (3DNCOW) fabric (Bogdanovich et al., 2013). Although these comparisons should be taken with some reservation, because fibers, resin materials, composite fabrication methods, fiber volume fractions, and relative fiber volume contents in different

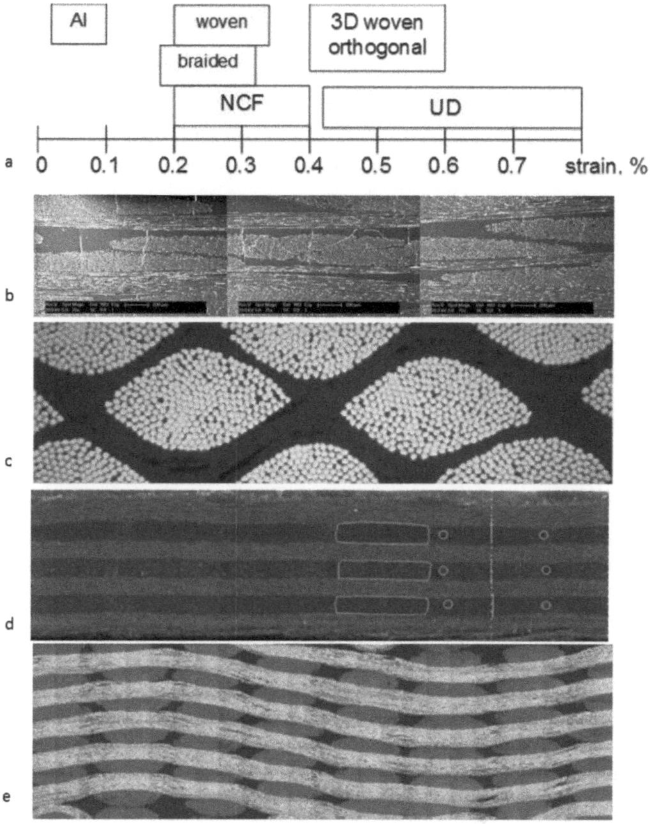

Figure 3.4 (a) A typical range of the damage initiation threshold in carbon fiber composites during tensile on-axis loading and (b–e) mages of cross-sections revealing typical yarn crimp. Reinforcement types: (b) Woven twill fabric (De Greef et al., 2011); (c) Woven quasiunidirectional steel fiber fabric (Callens et al., 2015); (d) 3D noncrimp orthogonally woven fabric (Bogdanovich et al., 2013); (e) 3D angle interlock woven fabric (Lomov et al., 2011).

reinforcement directions varied broadly in these studies, certain persistent trends could be noted.

The damage initiation threshold for UD cross-ply laminates is in the range of 0.4–0.8% strain. Noncrimp 3D fabrics (Karahan et al., 2010), with extremely small waviness of warp and weft tows (below 0.1%), exhibit a damage initiation threshold on the lower range of this interval (0.4–0.6% strain). Once crimp is introduced in the reinforcement structure, the damage initiation strain levels become significantly lower. The typical range for 2D woven, braided, and NCF composites is 0.2–0.4% strain. Contrary to the name "non-crimp fabric," these stitched textiles have considerable in-plane deviations of fiber directions from the nominal directions in supposedly UD plies. These deviations are caused by the stitching. The difference in angles can reach 5–8 degrees (Lomov et al., 2002; Lomov, 2011; Schneider, 2011). For 3D angle

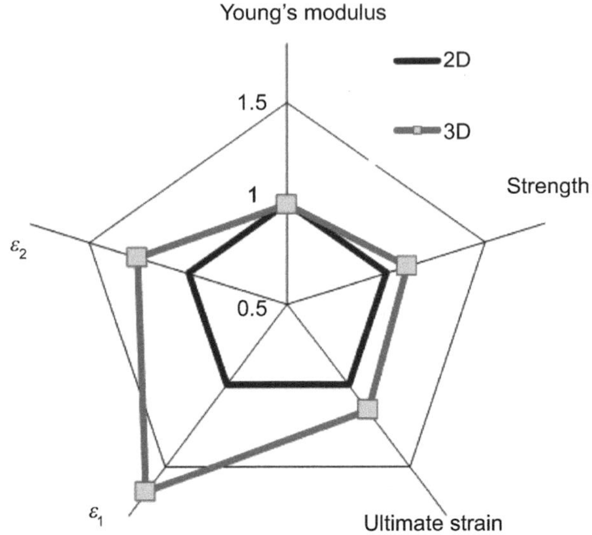

Figure 3.5 Comparison of relative mechanical properties of 2D and 3D glass/epoxy woven laminates. Properties of 2D laminate = 1.
Adapted from Lomov et al. (2009).

interlock composites, with extreme crimp of their yarns, the damage threshold is the lowest, below 0.1% strain.

The composite tensile strength is also affected by crimp. One textbook (Tsai, 1992) suggests a decrease of the woven laminate strength by 15–20% in comparison with cross-ply laminates with the same fiber volume fraction. Fig. 3.5 compares mechanical properties of 2D and 3D glass/epoxy laminates studied in Lomov et al. (2009). The composite samples have the same thickness and the same fiber volume fraction. The reinforcements were made of E-glass fibers of the same type. An important difference between the 2D and 3D materials is the yarn crimp (see insets in Fig. 3.5): in 2D laminate yarns have considerable crimp; in 3D reinforcement their crimp is negligible. This difference in crimp, apart from affecting the composite stiffness, strongly influences the composite strength and damage thresholds: the strength and ultimate strain of the 3D noncrimp composite is 10–15% higher, and the damage initiation strain is 65% higher than the corresponding properties of the 2D laminate.

Apart from affecting the strength and the damage initiation threshold, the yarn crimp also limits geometrical characteristics of transverse matrix cracks like their length and spacing. At initiation, the crack length is limited by the crimp intervals of the yarns. Under further loading, the cracks propagate along the yarns till they span full width of the specimen. In contrast to cross-ply laminates, the cracks do not always start at the specimen edges.

Fig. 3.6a and Fig. 3.1 illustrate a gradual growth of matrix cracks in a braided carbon/epoxy (Ivanov et al., 2009) and a woven glass fiber/epoxy composite (Li et al., 2015), respectively. The braided composite has high crimp and intensive interlacing

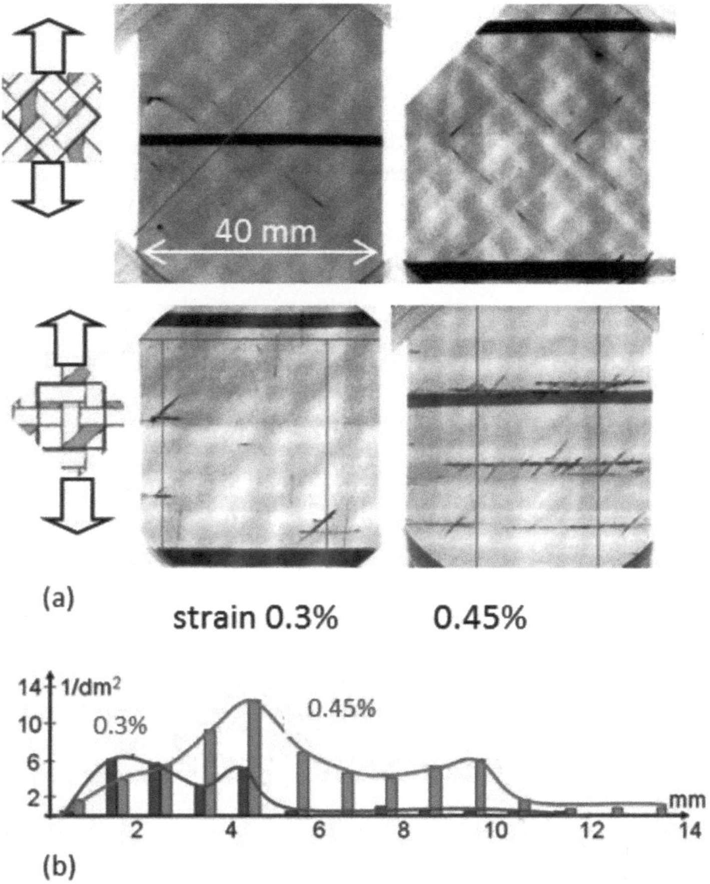

Figure 3.6 Development of cracks in (a) A three-axial 0°/45°/-45° braided carbon/epoxy composite under loading in axial and bias directions, X-ray images; (b) Distribution of the crack length at two strain levels under axial loading (Ivanov et al., 2009).

of the yarns. Fig. 3.6b shows changes in the crack length distribution with increase of the applied axial strain from 0.3% to 0.45%. The total number of cracks and their length increase, but they still stay relatively short, not extending across the specimen width (40 mm). The cracks are confined inside unit cells of the textile. For the glass fiber/epoxy with the plain weave (Fig. 3.1), cracks start both in the middle of the specimen and at the edges and are initially limited in length. They propagate through the specimen width quite fast, creating finally a saturated crack system that is quite similar to the one observed in cross-ply laminates.

Additionally, stress concentrations and a complex 3D stress state due to the yarn crimp can affect localization of cracks inside the yarns. The literature gives contradictory evidence on the "rules" of such a localization. For example, for woven laminates studied in Daggumati et al. (2010), transverse cracks are observed to start at the edges of the weft yarns when loaded in the warp direction. At the same time, other studies

(Zako et al., 2003) report transverse cracks initiated in the middle of the yarn. Optical measurements of the strain on the surface of a textile composite specimen (Lee et al., 2004; Lomov et al., 2008; Daggumati et al., 2011) also show different possibilities for localization of transverse strains over a yarn. Apparently, the location of the crack initiation depends on details of the stress distribution over the yarn cross-section and on geometrical details of the yarn cross-section and yarn interlacing. The importance of these details makes FE predictions for crack localization difficult. Typically, the meso-FE models use a simplified geometrical representation of the reinforcement (Lomov et al., 2007; Grail et al., 2013).

3.5 Influence of through-the-thickness reinforcement

Some textile composites may have through-the-thickness reinforcements like 3D textiles or structurally stitched reinforcement. The NCF composites, which have intralayer nonstructural stitching, can also be considered belonging to this category. In such textile composites, the localization of damage and its pattern are significantly influenced by the presence of this additional reinforcing component. The damage tends to start in the stitching sites and in or near Z-yarns. Fig. 3.7b illustrates this phenomenon.

In NCF composites, which consist of UD plies stitched with a thin polyester yarn (6–10 tex), the stitching creates resin-rich pockets with fibers around them deviating from the nominal directions of the plies. These local imperfections in the laminate structure dictate the location of first damage. The sites of the damage initiation, as seen in Fig. 3.7a, clearly correspond to the periodicity of the stitching, shown in the insert on the same figure. More information of the damage processes in NCF composites can be found in Truong Chi et al. (2005, 2008), Mikhaluk et al. (2008), Lomov (2011). Fig. 3.7a (Koissin et al., 2009) illustrates the same phenomenon. It shows a crack pattern in a nonstructurally stitched (with polyester yarn) and structurally stitched (with 1K carbon tows) NCF laminates. In the nonstructurally stitched laminates, the damage sites correspond to the positions of the stitching on the NCF. When structural stitching is present, it creates, in addition to the nonstructural stitching in NCF layers, another set of damage sites, much more intensive. These two sets are well visible in Fig. 3.7a and perfectly correspond to the placement of the tufts.

In 3D orthogonal woven noncrimp composites (Bogdanovich et al., 2013), the damage is associated with positions of Z-yarns (Fig. 3.7b). The first cracks are Z-type, as shown in Table 3.1, which lead to cracks of b- and t-type and to shear cracks in matrix pockets, as seen in X-ray images of Fig.3.7b. The same applies to the case of thermal loading of 3D composites (Gigliotti et al., 2018).

3.6 Crack saturation and development of delaminations

In cross-ply laminates, new matrix cracks continue to appear until a saturation crack density is reached as discussed previously. After saturation, no more transverse cracks can be created and the energy has to be dissipated in some other way. The crack density

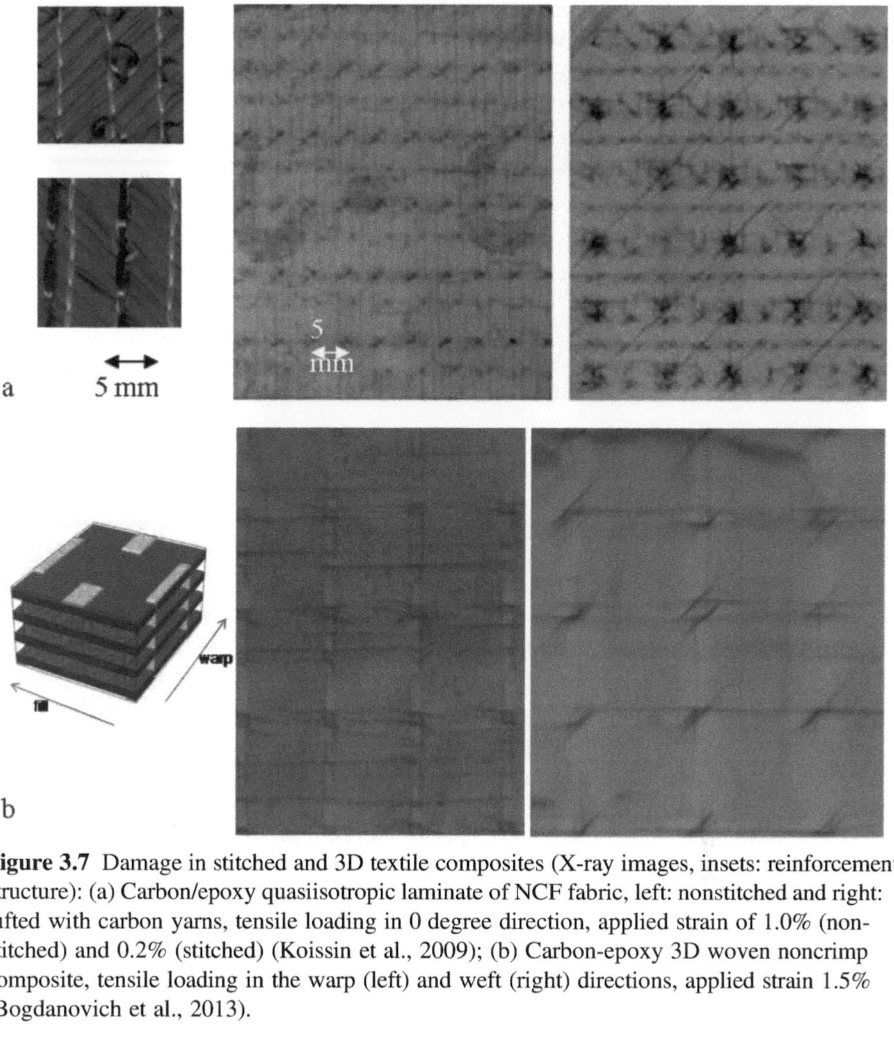

Figure 3.7 Damage in stitched and 3D textile composites (X-ray images, insets: reinforcement structure): (a) Carbon/epoxy quasiisotropic laminate of NCF fabric, left: nonstitched and right: tufted with carbon yarns, tensile loading in 0 degree direction, applied strain of 1.0% (non-stitched) and 0.2% (stitched) (Koissin et al., 2009); (b) Carbon-epoxy 3D woven noncrimp composite, tensile loading in the warp (left) and weft (right) directions, applied strain 1.5% (Bogdanovich et al., 2013).

at the saturation is controlled by stress redistribution in cracked plies (Hashin, 1985; Varna and Berglund, 1991).

In textile composites, the crack saturation phenomenon is similar. It can be qualitatively seen in Fig. 3.8a. A more detailed investigation, however, reveals differences in the development of the crack system in composite with different textile architecture, as illustrated in Fig. 3.8a and b. The crack development in textile composites strongly depends on the properties of the matrix. For the same textile architecture but different polymers, the saturation either can happen early on or very close to the final failure of the composite.

The crack saturation is observed in glass/epoxy plain weave composites (Fig. 3.8a). At the applied strain of about 1%, new transverse cracks stop to appear. Some of the

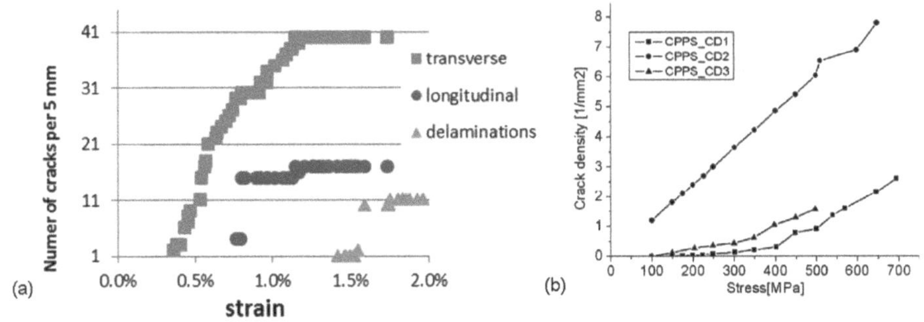

Figure 3.8 Development of crack systems in (a) A plain weave glass/epoxy composite as a function of strain. Loading is in the warp direction (Li et al., 2015); (b) A 5H satin carbon/PPS composite, tensile loading in warp direction, three different samples (Daggumati et al., 2010).

existing transverse cracks continue growing till they span the whole width of the specimen. The same saturation phenomenon is also seen for longitudinal cracks. The latter start appearing at an applied strain of about 0.8%, when tensile strains in the transverse direction are sufficiently high to exceed the transverse strength of the yarns. It is interesting to note that these longitudinal cracks are confined to the crimp intervals and their length stays limited to a float length of the yarn in the weave.

On the other hand, there are studies where no saturation of transverse cracks is observed. Daggumati et al. (2010) report an absence of the transverse crack saturation in 5-harness satin carbon/PPS composites (Fig. 3.8b) up to the specimen failure (stress 700 MPa corresponds to the ultimate stress). This controversy can be resolved if instead of a crack density, a distance between cracks is considered. Let us assume that a minimum distance between cracks exists, and this distance defines the crack density at the saturation strain (Varna and Berglund, 1991). Transverse cracks are confined inside yarns. Because of the limited width of the yarns, the minimum crack distance limits the number of transverse cracks per yarn. This hypothesis is supported by microscopy images of the cross-sections, demonstrating approximately equal distance between cracks and approximately the same (three) number of transverse cracks per yarn. Then the presence or absence of the saturation level for crack density depends on whether all yarns are fractured to the maximum number of cracks before the final failure. If yes, the saturation is present. If not, the number of cracks will increase till the sample failure.

Development of transverse cracks leads to delaminations. In contrast with the delaminations in cross-ply laminates, in textile composites, these delaminations are localized. Development of such local delaminations, appearing at the yarn intersection cites, is clearly seen in Fig. 3.1 for a plain weave glass/epoxy laminate. The delaminations propagate with an increasing load, but they stay confined to the crimp interval, revealing the weave pattern of the reinforcement.

The onset and propagation of delaminations play a decisive role in transition from damage accumulation in the matrix to fiber failure in 0 degree layers and the final failure of the specimen. The dramatic effect, which delaminations have on the specimen strength, is demonstrated in Mattsson et al. (2007, 2008) and also discussed in Lomov et al. (2007). In two carbon/epoxy NCF composites with a different stacking sequence of 0 and 90 degree plies, subjected to the 0° tensile load, one develops delaminations at about 1% strain, apparently caused by a more pronounced fiber waviness in this material, while the other shows no delaminations till very late stages of loading. As a result, the strength of the first material is about 25% lower than that of the second one.

3.7 Analogy between the damage development under quasistatic tension and tension–tension fatigue loadings

The development of damage during tension–tension fatigue loading of textile composites with brittle matrix is analogous to the sequence of damage events outlined in Section 3.2 for the quasistatic loading. This is illustrated in Fig. 3.9 for a 3D woven

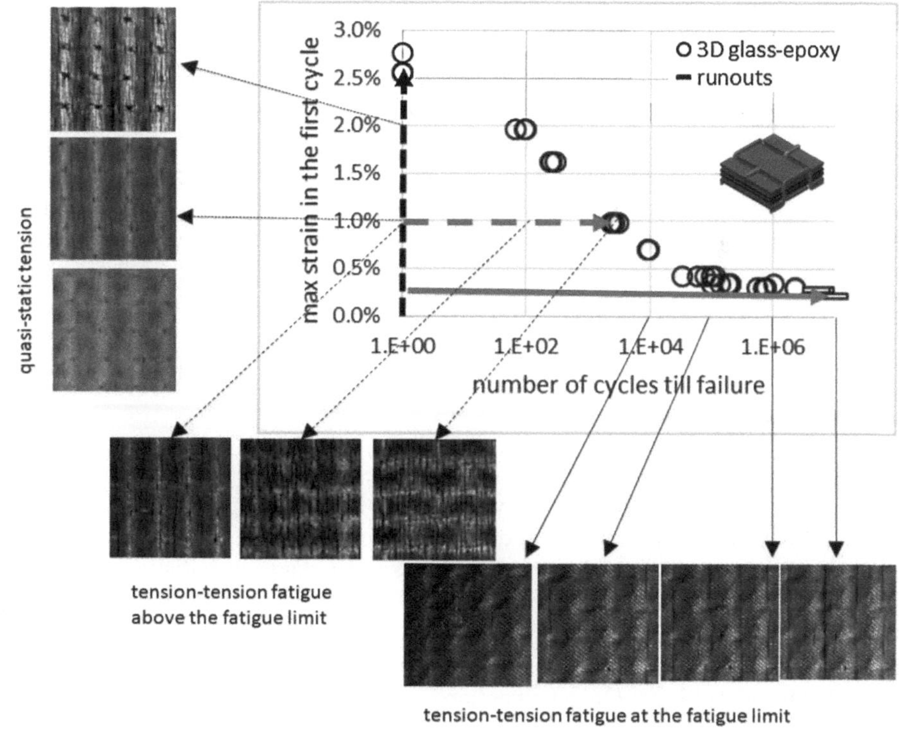

Figure 3.9 Damage development in quasistatic tension and in tension–tension fatigue. 3D woven glass-epoxy composites, loading in warp direction (horizontal direction of the images) (Ivanov et al., 2009; Carvelli et al., 2010).

glass/epoxy composite under three loading cases: quasistatic tension in warp direction, tension–tension fatigue with $R = 0.1$ at the maximum load corresponding to 1.0% (fatigue failure at 2000 cycles), and to 0.25% strain in the first cycle (no fatigue failure up to 5,000,000 cycles), studied in Ivanov et al. (2009), Carvelli et al. (2010).

The sequence of the crack development under fatigue loading is very much the same as under quasistatic tension: initiation, growth, and multiplication of transverse cracks. In case of the high fatigue load (1% maximum strain), the development of the transverse cracks system leads to local delaminations, associated with the positions of the "crowns" of the binding Z-yarns in the 3E weave. In the case of the lower load of 0.25% maximum strain, these delaminations do not appear.

The driving force of the damage development under quasistatic loading is the increasing load and corresponding influx of the deformation energy, which can be used for and released during the new cracks formation and propagation of the existing cracks. The "decision," if a new crack will appear and if the existing crack will propagate, is defined by the level of the local stress. Under fatigue loading, the deformation energy is supplied again and again under loading cycles. The appearance of new cracks and propagation of existing cracks under the local stress state, defined by the maximum of the loading cycle, are controlled by the local fatigue strength of the matrix.

The damage thresholds under quasistatic loading are known to be inked to the fatigue life limits for composite laminates (Talreja and Singh, 2012). In textile composites, this link is maintained, but the behavior is observed (Carvelli et al., 2017) to be more complex than a simple hypothesis "a good estimate [of the fatigue limit] will be given by the strain at which transverse cracking initiates."

This statement was observed to correspond to reality for glass-reinforced textile composites: the fatigue life limit (corresponding to 1 million cycles) is close to the quasistatic damage threshold ε_1. For carbon-reinforced textile composites, the ε_1 damage threshold is a serious underestimation of the fatigue life limit. The ε_2 threshold serves better as a conservative estimation of it.

3.8 Conclusions

In comparison with UD-based composites, textile composites possess a richer variety of internal structures. They affect the ways cracks initiate and propagate in the composite. As this review shows, damage accumulation in textile composites subjected to quasistatic and cyclic loadings goes through similar stages of initiation and development as in cross-ply laminates with UD plies. This similarity is attributed to the fact that textile composites are composed of structural elements—impregnated yarns, tows, or fiber bundles—which locally are morphologically equivalent to the fibrous structure of an UD ply in a laminate. On the other hand, textile composites exhibit some distinct damage features that are not found in cross-ply laminates. These new damage patterns originate from the presence of yarn crimp and three-dimensional nature of the reinforcement (especially in the presence of through-the-thickness oriented yarns in 3D or stitched textiles). These unique features in textile composites include:

- Richer "taxonomy" of damage types, which apart from transverse and longitudinal cracks in fiber bundles, and delaminations between plies, includes cracks at yarn boundaries, localized delaminations, and matrix cracks in resin-rich pockets.
- Earlier damage initiation and lower strength of the textile composites.
- Limited length of transverse and longitudinal cracks in the fiber bindles.
- Saturation of the transverse cracks system governed by the size of the fiber bundles.
- Localized delaminations as a decisive phenomenon for the onset of fiber failure in the surrounding bundles.

The earlier damage initiation in static tensile tests can be an indication of a lower fatigue limit of textile composites in comparison with their UD cross-ply counterparts. Decreasing yarn crimp in a textile reinforcement structure is thus viewed to be beneficial for the load-carrying ability of the material.

The process of cracking under tension–tension fatigue loading in textile composites is similar to the one during quasistatic loading with increasing applied strain. The quasistatic damage threshold limits can be used as guidance for determination of the fatigue limit.

References

Adden, S., Horst, P., 2005. Damage propagation in non-crimp fabrics under bi-axial static and fatigue loading. Compos. Sci. Technol. 66, 626–633.

Baker, C., Morscher, G.N., Pujar, V.V., Lemanski, J.R., 2015. Transverse cracking in carbon fiber reinforced polymer composites: modal acoustic emission and peak frequency analysis. Compos. Sci. Technol. 116, 26–32.

Bogdanovich, A.E., Karahan, M., Lomov, S.V., Verpoest, I., 2013. Quasi-static tensile behavior and progressive damage in carbon/epoxy composite reinforced with 3D non-crimp orthogonal woven fabric. Mech. Mater. 62, 14–31.

Bogdanovich, A.E., Pastore, C.M., 1996. Mechanics of Textile and Laminated Composites. Chapman and Hall, London.

Boisse, P. (Ed.), 2021. Composite Reinforcements for Optimum Performance, 2nd edition. Woodhead Publishing, Oxford.

Callens, M.G., Gorbatikh, L., Verpoest, I., 2014. Ductile steel fibre composites with brittle and ductile matrices. Compos. Part A 61, 235–244.

Callens, M.I.G., Gorbatikh, L., Bertels, E., Goderis, B., Smet, M., Verpoest, I., 2015. Tensile behaviour of stainless steel fibre/epoxy composites with modified adhesion. Compos. Part A 69, 208–218.

Carvelli, V., Gramellini, G., et al., 2010. Fatigue behaviour of non-crimp 3D orthogonal weave and multi-layer plain weave e-glass reinforced composites. Compos. Sci. Technol. 70 (14), 2068–2076.

Carvelli, V., Tomaselli, V.N., et al., 2010. Fatigue and post-fatigue tensile behaviour of non-crimp stitched and unstitched carbon/epoxy composites. Compos. Sci. Technol. 70, 2216–2224.

Carvelli, V., Lomov, S.V. (Eds.), 2015. Fatigue of Textile Composites. Elsevier – Woodhead, Cambridge.
Carvelli, V., Jain, A., Lomov, S.V. (Eds.), 2017a. Fatigue of Textile and Short Fiber Reinforced Composites. Wiley - ISTE, London.
Carvelli, V., D'Ettore, A., Lomov, S.V., 2017b. Acoustic emission and damage mode correlation in textile reinforced PPS composites. Compos. Struct. 163, 399–409.
Chou, T.-W., Ko, F. (Eds.), 1989. Textile Structural Composites. Elsevier, Amsterdam.
Daggumati, S., De Baere, I., Van Paepegem, W., Degrieck, J., Xu, J., Lomov, S.V., Verpoest, I., 2010. Local damage in a 5-harness satin weave composite under static tension: Part I - experimental analysis. Compos. Sci. Technol. 70 (13), 1926–1933.
Daggumati, S., Voet, E., Van Paepegem, W., Degrieck, J., Xu, J., Lomov, S.V., Verpoest, I., 2011. Local strain in a 5-harness satin weave composite under static tension: Part I - experimental analysis. Compos. Sci. Technol. 71 (8), 1171–1179.
De Greef, N., Gorbatikh, L., Godara, A., Mezzo, L., Lomov, S.V., Verpoest, I., 2011a. The effect of carbon nanotubes on the damage development in carbon fiber/epoxy composites. Carbon 49, 4650–4664.
De Greef, N., Gorbatikh, L., Lomov, S.V., Verpoest, I., 2011b. Damage development in woven carbon fiber/epoxy composites modified with carbon nanotubes under tension in the bias direction. Compos. Part A 42, 1635–1644.
de Groot, P.J., Wijnen, P.A.M., Janssen, R.B.F., 1995. Real-time frequency determination of acoustic emission for different fracture mechansms in carbon/epoxy composites. Compos. Sci. Technol. 55, 405–412.
El Hage, C., 2006. Modelisation Du Comportement Elastique Endommageable De Materiaux Composites a Renfort Tridimensionel. PhD thesis, Laboratoire Roberval. Compiegne, L'Universite de Technologie de Compiegne.
Gigliotti, M., Pannier, Y., Antoranz Gonzalez, R., Lafarie-Frenot, M.C., Lomov, S.V., 2018. X-ray micro-computed-tomography characterization of cracks induced by thermal cycling in non-crimp 3D orthogonal woven composite materials with porosity. Compos. Part A 112, 100–110.
Gorbatikh, L., Wardle, B.L., Lomov, S.V., 2016. Hierarchical lightweight composite materials for structural applications. MRS Bull. 41, 672–677.
Grail, G., Hirsekorn, M., Wendling, A., Hivet, G., Hambli, R., 2013. Consistent finite element mesh generation for meso-scale modeling of textile composites with preformed and compacted reinforcements. Compos. Appl. Sci. Manuf. 55, 143–151.
Gutkin, R., Green, C.J., Vangrattanachai, A., Pinho, S.T., Robinson, P., Curtis, P.T., 2011. On acoustic emission for failure investigation in CFRP: pattern recognition and peak frequency analyses. Mech. Syst. Signal Process. 25, 1393–1407.
Hashin, Z., 1985. Analysis of cracked laminates: a variational approach. Mech. Mater. 4, 121–136.
Ivanov, D.S., Lomov, S.V., Baudry, F., Xie, H., Van Den Broucke, B., Verpoest, I., 2009a. Failure analysis of triaxial braided composite. Compos. Sci. Technol. 69, 1372–1380.
Ivanov, D.S., Lomov, S.V., Bogdanovich, A.E., Karahan, M., Verpoest, I., 2009b. A comparative study of tensile properties of non-crimp 3D orthogonal weave and multi-layer plain weave e-glass composites. Part 2: comprehensive experimental results. Compos. Part A 40, 1144–1157.
Ivanov, S.G., Beyens, D., Gorbatikh, L., Lomov, S.V., 2017. Damage development in woven carbon fibre thermoplastic laminates with PPS and PEEK matrices: a comparative study. J. Compos. Mater. 51, 637–647.

Karahan, M., 2011. Investigation of damage initiation and propagation in 2x2 twill woven carbon-epoxy multi-layer composites. Textil. Res. J. 81, 412–428.

Karahan, M., Lomov, S.V., Bogdanovich, A.E., Mungalov, D., Verpoest, I., 2010. Internal geometry evaluation of non-crimp 3D orthogonal woven carbon fabric composite. Compos. Part A 41, 1301–1311.

Karahan, M., Lomov, S.V., et al., 2011. Fatigue tensile behavior of carbon/epoxy composite reinforced with non-crimp 3D orthogonal woven fabric. Compos. Sci. Technol. 71, 1961–1972.

Karger-Kocsis, J., Mahmood, H., Pegoretti, A., 2020. All-carbon multi-scale and hierarchical fibers and related structural composites: a review. Compos. Sci. Technol. 186.

Kersani, M., Lomov, S.V., Van Vuure, A., Bouabdallah, A., Verpoest, I., 2014. Damage Analysis Based on the Correlation between Acoustic Emission and E Modulus Degradation in Flax/epoxy Quasi Unidirectional Woven Laminates Proceedings of the 16th European Conference on Composite Materials (ECCM-16), Seville, Electronic edition (s.p).

Kersani, M., Lomov, S.V., Van Vuure, A.W., Bouabdallah, A., Verpoest, I., 2015. Damage in flax/epoxy quasi unidirectional woven laminates under quasi-static tension. J. Compos. Mater. 49, 403–413.

Koissin, V., Kustermans, J., Lomov, S.V., Verpoest, I., Van Den Broucke, B., Witzel, V., 2009. Structurally stitched NCF preforms: quasi-static response. Compos. Sci. Technol. 69 (15–16), 2701–2710.

Lee, J.R., Molimard, J., Vautrin, A., Surrel, Y., 2004. Digital phase-shifting grating shearography for experimental analysis of fabric composites under tension. Compos. Part A 35, 849–859.

Li, L., Lomov, S.V., Yan, X., 2015. Correlation of acoustic emission with optically observed damage in a glass/epoxy woven laminate under tensile loading. Compos. Struct. 123, 45–53.

Li, L., Lomov, S.V., Yan, X., Carvelli, V., 2014. Cluster analysis of acoustic emission signals for 2D and 3D woven glass/epoxy composites. Compos. Struct. 116, 286–299.

Lomov, S.V., Belov, E.B., Bischoff, T., Ghosh, S.B., Truong Chi, T., Verpoest, I., 2002. Carbon composites based on multiaxial multiply stitched preforms. Part 1: geometry of the preform. Compos. Part A 33 (9), 1171–1183.

Lomov, S.V., Bogdanovich, A.E., Ivanov, D.S., Mungalov, D., Karahan, M., Verpoest, I., 2009. A comparative study of tensile properties of non-crimp 3D orthogonal weave and multilayer plain weave E-glass composites. Part 1: materials, methods and principal results. Compos. Part A 40, 1134–1143.

Lomov, S.V., 2011. In: Non-crimp Fabric Composites: Manufacturing, Properties and Applications. Woodhead Publisher Ltd, Cambridge.

Lomov, S.V., Ivanov, D.S., Truong Chi, T., Verpoest, I., Baudry, F., Vanden Bosche, K., Xie, H., 2008a. Experimental methodology of study of damage initiation and development in textile composites in uniaxial tensile test. Compos. Sci. Technol. 68, 2340–2349.

Lomov, S.V., Ivanov, D.S., Vallons, K., Verpoest, I., Klimshin, D.V., Truong Chi, T., 2007a. Peculiarities of damage behaviour of NCF carbon/epoxy laminates under tension. In: 16th International Conference On Composite Materials (ICCM-16). CD edition, Kyoto.

Lomov, S.V., Ivanov, D.S., Verpoest, I., Zako, M., Kurashiki, T., Nakai, H., Hirosawa, S., 2007b. Meso-FE modelling of textile composites: road map, data flow and algorithms. Compos. Sci. Technol. 67, 1870–1891.

Lomov, S.V., Ivanov, D.S., Verpoest, I., Zako, M., Kurashiki, T., Nakai, H., Molimard, J., Vautrin, A., 2008b. Full field strain measurements for validation of meso-FE analysis of textile composites. Compos. Part A 39, 1218–1231.

Lomov, S.V., Perie, G., Ivanov, D.S., Verpoest, I., Marsal, D., 2011. Modelling 3D fabrics and 3D reinforced composites: challenges and solutions. Textil. Res. J. 81 (1), 26–41.

Lomov, S.V., Verpoest, I., 2010. Textile composite materials: polymer matrix composites. In: Blockley, R., Shyy, W. (Eds.), Encyclopedia of Aerospace Engineering. John Wiley and Sons, Ltd., Chichester, pp. 2159–2176.

Lomov, S.V., Breite, C., Carrella-Payan, D., Carvelli, V., Ersoy, N., Gigliotti, M., Antoranz-Gonzalez, R., Ivanov, S., Kersani, M., Lafarie-Frenot, M.-C., Mavrogordato, M., Mehdikhani, M., Melnikov, A., Mesquita, F., Oz, F., Pannier, Y., Pyl, L., Schöberl, E., Sinclair, I., Spearing, S.M., Swolfs, Y., Van Hemelrijk, D., Zhu, M., Gorbatikh, L., 2018. Multi-instrument multi-scale experimental damage mechanics for fibre reinforced composites. In: IOP Conference. Series: Materials Science and Engineering (13th International Conference on Textile Composites (TEXCOMP-13)) 406: 012057.

Long, A. (Ed.), 2005. Design and Manufacture of Textile Composites. Woodhead Publishing Ltd, Cambridge.

Long, A. (Ed.), 2007. Composite Forming Technologies. Woodhead publishing, Cambridge.

Maillet, E., Baker, C., Morscher, G.N., Pujar, V.V., Lemanski, J.R., 2015. Feasibility and limitations of damage identification in composite materials using acoustic emission. Compos. Appl. Sci. Manuf. 75, 77–83.

Mattsson, D., Joffe, R., Varna, J., 2007. Methodology for characterization of internal structure parameters governing performance in NCF composites. Compos. B Eng. 38 (1), 44–57.

Mattsson, D., Joffe, R., Varna, J., 2008. Damage in NCF composites under tension: effect of layer stacking sequence. Eng. Fract. Mech. 75 (9), 2666–2682.

Mikhaluk, D.S., Truong, T.C., Borovkov, A.I., Lomov, S.V., Verpoest, I., 2008. Experimental observations and finite element modelling of damage and fracture in carbon/epoxy non-crimp fabric composites. Eng. Fract. Mech. 75 (9), 2751–2766.

Naik, N.K., 1994. Woven Fabric Composites. Lancaster-Basel, Technomic.

Oz, F., Ersoy, N., Lomov, S.V., 2017. Do high frequency acoustic emission events always represent fibre failure in CFRP laminates? Compos. Part A 103, 230–235.

Oz, F., Ersoy, N., Mehdikhani, M., Lomov, S.V., 2018. Multi-instrument in-situ damage monitoring in quasi-isotropic CFRP laminates under tension. Compos. Struct. 196, 163–180.

Romanov, V., Lomov, S.V., Verpoest, I., Gorbatikh, L., 2015. Modelling evidence of stress concentration mitigation at the micro-scale in polymer composites by the addition of carbon nanotubes. Carbon 82, 184–194.

Sabuncuoglu, B., Orlova, S., Gorbatikh, L., Lomov, S.V., Verpoest, I., 2015. Micro-scale finite element analysis of stress concentrations in steel fiber composites under transverse loading. J. Compos. Mater. 49 (9), 1057–1069. Published online.

Sause, M.G.R., Gribov, A., Unwin, A.R., Horn, S., 2012. Pattern recognition approach to identify natural clusters of acoustic emission signals. Pattern Recogn. Lett. 33, 17–23.

Schneider, M., 2011. Automated System for characterisation of defects. In: Lomov, S.V. (Ed.), NCF. Non-Crimp Fabric Composites: Manufacturing, Properties and Applications. Woodhead Publishing, Oxford, pp. 103–114.

Sharma, A., Daggumati, S., Gupta, A., Van Paepegem, W., 2020. On the prediction of the bi-axial failure envelope of a UD CFRP composite lamina using computational micro-mechanics: effect of microscale parameters on macroscale stress–strain behavior. Compos. Struct. 251, 112605.

Talreja, R., Singh, C.V., 2012. Damage and Failure of Composite Materials. Cambridge University Press, Cambridge.

Truong Chi, T., Ivanov, D.S., Klimshin, D.V., Lomov, S.V., Verpoest, I., 2008. Carbon composites based on multiaxial multiply stitched preforms. Part 7: mechanical properties and damage observations in composite with sheared reinforcement. Compos. Part A 39, 1380–1393.

Truong Chi, T., Vettori, M., Lomov, S.V., Verpoest, I., 2005. Carbon composites based on multiaxial multiply stitched preforms. Part 4: mechanical properties of composites and damage observation. Compos. Part A 36, 1207–1221.

Tsai, S.W., 1992. Theory of Composites Design. Dayton, Think Composites.

Vallons, K., Lomov, S.V., Verpoest, I., 2009. Fatigue and post-fatigue behaviour of carbon/epoxy non-crimp fabric composites. Compos. Part A 40, 251–259.

Varna, J., Berglund, L., 1991. Multiple transverse cracking and stiffness reduction in cross-ply laminates. J. Compos. Technol. Res. 13 (2), 97–106.

Yokozeki, T., Iwahori, Y., Ishiwata, S., 2007. Matrix cracking behaviors in carbon fiber/epoxy laminates filled with cup-stacked carbon nanotubes (CSCNTs). Compos. Part A 38, 917–924.

Zako, M., Uetsuji, Y., Kurashiki, T., 2003. Finite element analysis of damaged woven fabric composite materials. Compos. Sci. Technol. 63, 507–516.

Damage accumulation under multiaxial fatigue loading

4

M. Quaresimin and P.A. Carraro
University of Padova, Vicenza, Italy

4.1 Introduction: parameters influencing the fatigue behavior of composites

In many structural components, such as turbine blades and airplane wings, composite laminates are unavoidably subjected to cyclic multiaxial loads. In spite of the increasing trend in the application of composite materials for structural components, reliable models and design tools against fatigue failure or degradation are still not available or recognized by the scientific and industrial communities. Indeed, the problem is not trivial at all, since, as highlighted in the previous chapters, the fatigue behavior of composite laminates is characterized by several competing and/or sequential damage mechanisms involving different length scales. The presence of multiaxial loads complicates this scenario and adds a certain number of parameters, which may strongly influence the initiation and evolution of damage.

In the work by Quaresimin et al. (2010), more than 1000 data from the literature were collected with the aim to obtain information on the influence of some design parameters on the fatigue behavior of composite laminates. The most relevant external parameters are defined in the following.

1. The presence of shear stress (i.e., the multiaxial stress state): Under external multiaxial loads, all the three in-plane stress components are in general different from zero in the plies of a laminate. Quaresimin et al. (2010) proposed to describe the degree of multiaxiality of the stress state by means of the following biaxiality ratios between stress amplitudes:

$$\lambda_1 = \frac{\sigma_{2,a}}{\sigma_{1,a}}, \lambda_2 = \frac{\sigma_{6,a}}{\sigma_{1,a}} \tag{4.1}$$

where σ_1 is the fiber-direction (longitudinal) stress, σ_2 is the transverse stress, σ_6 is the in-plane shear stress in a ply, and the subscript a denotes the amplitude. These two parameters are enough to describe general loading conditions in the presence of a nonzero longitudinal stress. However, a third parameter, λ_{12}, has been later defined by Quaresimin and Carraro (2013), which is mostly related to the matrix-dominated behavior:

$$\lambda_{12} = \frac{\sigma_{6,a}}{\sigma_{2,a}} \tag{4.2}$$

As highlighted by Quaresimin et al. (2010), it is important to express the bi-axiality ratios in terms of stresses in the material coordinates system, as defined by Eqs. (4.1) and (4.2). It will

Modeling Damage, Fatigue and Failure of Composite Materials. https://doi.org/10.1016/B978-0-443-18489-5.00019-9

be shown later in the chapter that λ_2 and λ_{12} have a strong detrimental influence on the fatigue behavior of composite laminates.

2. The load ratio: When external loads in multiple directions are applied, the three stress components can be written as functions of time as:

$$
\begin{aligned}
\sigma_1(t) &= \sigma_{1,m} + \sigma_{1,a} \cdot sin(\omega \cdot t) \\
\sigma_2(t) &= \sigma_{2,m} + \sigma_{2,a} \cdot sin(\omega \cdot t - \delta_{2,1}) \\
\sigma_6(t) &= \sigma_{6,m} + \sigma_{6,a} \cdot sin(\omega \cdot t - \delta_{6,1})
\end{aligned}
\tag{4.3}
$$

and load ratios can be defined for each stress component as:

$$
R_1 = \frac{\sigma_{1,min}}{\sigma_{1,max}}, R_2 = \frac{\sigma_{2,min}}{\sigma_{2,max}}, R_6 = \frac{\sigma_{6,min}}{\sigma_{6,max}} \tag{4.4}
$$

Again, it will be shown later that the load ratios have a very important influence on damage initiation and evolution in laminates, as well as on their fatigue life.

3. The phase shift between stress components: In Eq. (4.3), $\delta_{2,1}$ and $\delta_{6,1}$ represent the phase shift of the transverse and shear stresses, respectively, with respect to the longitudinal stress. Though these parameters should be accounted for when designing composite components against fatigue, it will be shown that their effect seems to be quite limited, at least for the configurations analyzed in the literature, mainly with respect to the effects of the biaxiality and load ratios.

Finally, it is worth pointing to a peculiarity of composite materials, which is caused by their anisotropic elastic properties. In fact, in the material coordinate system, composite plies are in general subjected to a multiaxial stress state even if the external load is uniaxial. This happens in the presence of off-axis plies and therefore in most of practical applications in which pure unidirectional (UD) laminates are not suitable. This kind of multiaxiality, originated by the material anisotropy, has been defined by Quaresimin and Carraro (2013) as internal multiaxiality, different from the external multiaxiality that is achieved by means of external loads in multiple directions. In principle, if the fatigue behavior is controlled by the local stresses, there should be no difference between these two conditions, even if some contradictory results were reported by Quaresimin et al. (2010). Some new and more detailed results in this sense will be shown later.

4.2 Biaxial testing of composite laminates

Different test procedures and specimen geometries can be adopted for multiaxial fatigue testing of composites. Comprehensive reviews on this subject can be found in Found (1985), Chen and Matthews (1993), Olsson (2011). The most common sample geometries, as shown in Fig. 4.1 are:

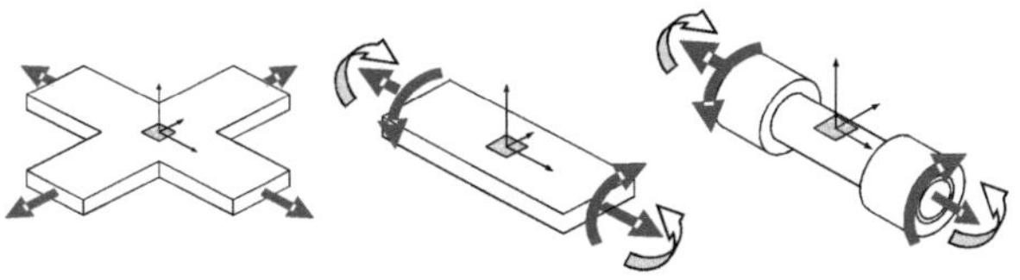

Figure 4.1 Schematic of the most common specimen configurations for multiaxial testing of composites.

1. Cruciform flat specimens under biaxial normal loads.
2. Bars/rods under tension/torsion or bending/torsion.
3. Thin-walled tubes subjected to combined tension/torsion (or bending/torsion) even in the presence of internal pressure.

By testing flat cruciform specimens, a multiaxial stress state can be obtained by applying external forces to the orthogonal arms of the specimen. When using UD laminates, if the material coordinate system is aligned with the global one, an external multiaxiality is achieved, with the presence of normal stresses σ_1 and σ_2. Conversely, if plies are laid up with an off-axis angle with respect to the global coordinate system, the stress state results in a combination of external and internal multiaxiality, with the presence of all three in-plane stress components. The design of the specimen geometry requires a lot of care to avoid arms failure and high stress concentrations at the corners, as well as to obtain a uniform stress distribution in the central part of the specimen, which is indeed the region of interest. Makris et al. (2010a) proposed an optimization procedure to identify the best geometrical parameters for two configurations of cruciform specimens, matching the requirements mentioned above. As a further problem, the manufacturing of cruciform specimens is quite complicated since the optimized geometry requires rounded corners and a different number of layers in the central region with respect to the arms. Finally, a four-actuator testing machine, suitable to apply loads in two normal directions, is needed. In spite of these practical difficulties, some results are available in the literature, in particular for static loadings (Jones et al., 1985; Mailly and Wang, 2008; Makris et al., 2010b; Antoniou et al., 2010). Only few papers report fatigue test results obtained from cruciform specimens in the case of multidirectional laminates made of UD (Jones et al., 1985; Radon and Wachnicky, 1985) and fabric (Smith and Pascoe, 1989) reinforcements. Some interesting guidelines for biaxial testing with cruciform specimens have been presented by Ramault (2012), where it appears that fatigue testing with these specimens is indeed not an easy task.

The possibility of using bar or rod specimens tested under combined bending and torsion (Atcholi et al., 1992; Ferry et al., 1999; El-Assal and Khashaba, 2007) or combined tension and torsion (Ogasawara et al., 2007) loading conditions was investigated in the past. The interest in this geometry was limited, however, and this may be due to the non-uniform stress distributions in the specimen cross-section.

Considering the amount of published data available, the most common sample geometry for multiaxial fatigue loading of composites seems to be a thin-walled tube, subjected to a biaxial/triaxial external loading obtained by applying combined tension (or bending), internal pressure, and torsion loading. Even in this case, an external multi-axiality is achieved, and, when using UD laminates, two main combinations of local stresses can be obtained. If the tubes are manufactured with the fibers oriented only at 0 degree (parallel to the tube's axis), a tension/torsion external loading condition leads to a combination of longitudinal and in-plane shear stresses (σ_1 and σ_6). On the other hand, if the fibers are oriented at 90 degree (normal to the tube axis), the presence of the stress components σ_2 and σ_6 is achieved. The required values of the biaxiality ratios λ_1 and λ_{12} can be easily obtained by properly combining the external tension and torsion loading. In any case, the *external* multiaxial condition can be combined also with an *internal* multiaxiality if the winding angle of the fibers is different from 0 or 90 degree with respect to the tube axis, or if multidirectional laminates are adopted, as in most cases in the literature.

Fatigue test results on tubes subjected to internal pressure and tensile loading have been presented in Owen and Griffiths (1978), Foral and Humpherey (1984), Perreux and Joseph (1997). The introduction of internal pressure is not easy indeed, because it requires dedicated equipment to pump the oil inside the specimen to produce the hoop stress, therefore complicating the testing.

Several papers in the literature report fatigue results obtained by applying a combined tension/torsion loading on tubes (see Francis et al., 1977; Wang et al., 1982a,b; Krempl et al., 1988; Amijima et al., 1991; Fujii et al., 1994; Fujii and Lin, 1995; Kawakami et al., 1996; Aboul Wafa et al., 1997; Schmidt et al., 2012a,b among the others). Testing tubular samples is rather attractive because of the relative ease of specimen preparation and test execution, a biaxial (tension/torsion) testing machine being sufficient. Another advantage is that free edge effects, which can influence the results on flat coupons, are avoided. In addition, with this test procedure, it is simple to obtain whatever stress state of interest, by simply combining the ratio of the external loads and the specimen lay-up.

4.3 Experimental results for the main test methods

In this section, some representative experimental results from the literature are reported for the three test methods presented above. Results are discussed to highlight the effect of the main design parameters (biaxiality and load ratios and phase shift) on fatigue life.

4.3.1 Results of tests on cruciform specimens

One of the pioneering works on biaxial fatigue testing of fiber-reinforced plastics by means of cruciform specimens was by Smith and Pascoe (1989). They tested woven glass fibers–polyester laminates made of two orthogonal layers, oriented at several angles with respect to the loading arms. External loads were proportionally applied with a load ratio $R = -1$, which is therefore the same for all the stress components. In Fig. 4.2, S–N curves for fatigue failure are reported in terms of $\sigma_{1,a}$ for two stacking sequences under biaxial loading. The multiaxial condition is expressed by the biaxiality ratios defined in Eq. (4.1). It can be seen that, while λ_1 has a weak effect on fatigue

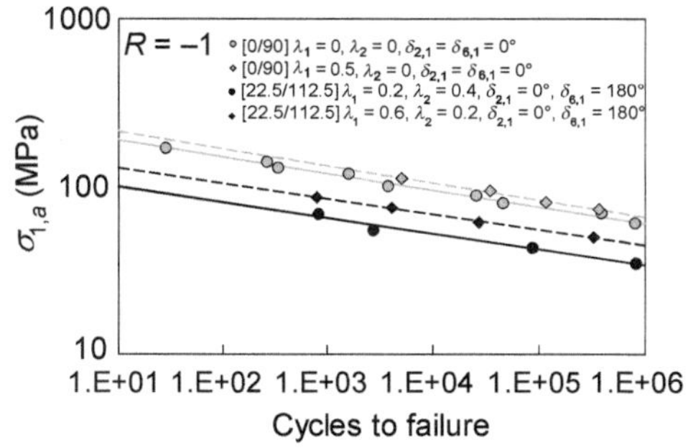

Figure 4.2 Influence of biaxiality ratios on the fatigue strength of woven glass—polyester cruciform specimens under combined tension/tension (Smith and Pascoe, 1989).
Reproduced from Quaresimin et al. (2010).

strength, increasing values of λ_2 lead to a significant degradation of the performance, confirming the already mentioned detrimental effect of the shear stress. This is even clearer in Fig. 4.3, where results for [$\pm$ 45] laminates are also plotted.

4.3.2 Results of tests on bars/rods

Atcholi et al. (1992) tested bars made of 0 degree oriented glass fibers and epoxy matrix under in-phase combined bending and torsion, obtaining stress states characterized by the presence of σ_1 and σ_6 only, both with a load ratio $R = -1$. By changing the ratio between the external loads, λ_2 was changed from 0 (pure longitudinal stress) to 0.15, while λ_1 was kept equal to 0. This represented a sound way of investigating the interactions between two stress components only (σ_1 and σ_6), and therefore giving a deeper insight into the influence of the shear stress on the fiber-dominated fatigue behavior. Results are plotted in Fig. 4.4, where the fatigue strength is clearly seen to decrease when λ_2 is increased. Also the slope of the S—N curves changes, so that the shear stress effect is more pronounced at high stress levels (low fatigue lives).

El-Assal and Khashaba (2007) reported results on glass/polyester rods with the fibers aligned with the rod axis, tested under fully reversed cyclic in-phase bending and torsion, producing the stress components σ_1 and σ_6, respectively. They tested specimens with fiber volume fraction between 15.8% and 44.7%. The fatigue strength, both under pure bending and combined loads, was found to increase with the volume

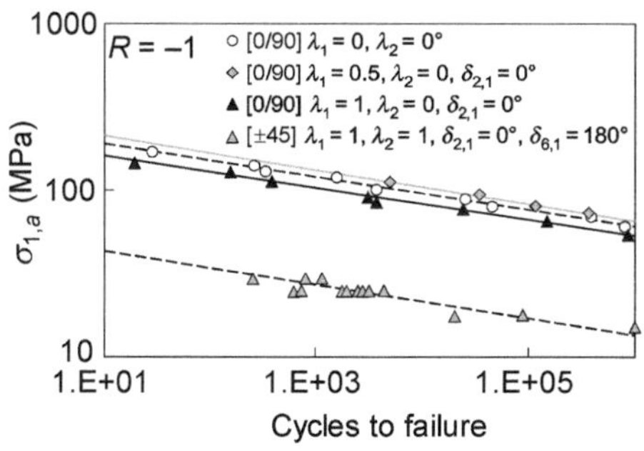

Figure 4.3 Influence of biaxiality ratio λ_2 on the fatigue strength of woven glass—polyester cruciform specimens under combined tension/tension (Smith and Pascoe, 1989).
Reproduced from Quaresimin et al. (2010).

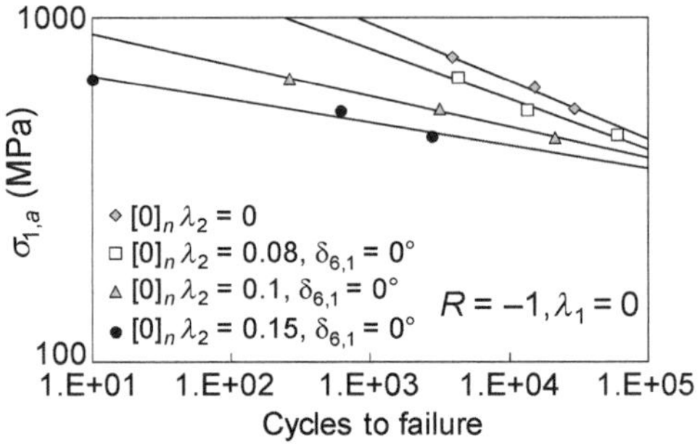

Figure 4.4 Influence of biaxiality ratio λ_2 on the fatigue strength of glass–epoxy [0]n bars under bending and torsion (Atcholi et al., 1992).
Reproduced from Quaresimin et al. (2010).

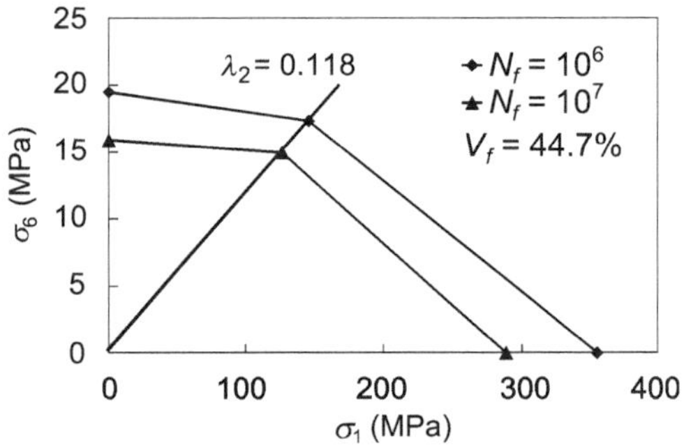

Figure 4.5 Constant-life curves for 0 degree UD rods under bending/torsion (El-Assal and Khashaba, 2007).

fraction. In Fig. 4.5, results are reported for specimens with a volume fraction of 44.7% tested under pure bending, pure torsion, and combined loads with $\lambda_2 = 0.118$. Results are presented in the form of constant-life curves in the $\sigma_{1,a}$–$\sigma_{6,a}$ plane. It is clear that the fatigue strength, in terms of $\sigma_{1,a}$ is strongly decreased as the shear stress contribution is increased. The fatigue strength under pure torsion is much lower with respect to that in pure bending, and this is due to the fact that if only the shear stress is applied, the material behavior is controlled by the matrix.

4.3.3 Results of tests on tubular specimens

As already mentioned, thin-walled tubes are the most common solution for biaxial testing of composites.

In the 1990s, an extensive experimental investigation on multiaxial fatigue of composite laminates by means of tubular specimens was carried out by the Japanese composite group coordinated by Toru Fujii (Amijima et al., 1991; Fujii et al., 1994; Fujii and Lin, 1995; Kawakami et al., 1996). Some results are reported in Fig. 4.6a for woven glass–polyester resin tubes with the lay-up $[0/90]_n$ tested under combined proportional tension/torsion loading with load ratio $R = 0$ (Amijima et al., 1991). This loading condition gives rise to the stress components σ_1 and σ_6 only. From the S–N curves (related to the final failure) plotted in Fig. 4.6a, it is evident, again, that the shear stress has a strong detrimental effect on fatigue strength. This is confirmed

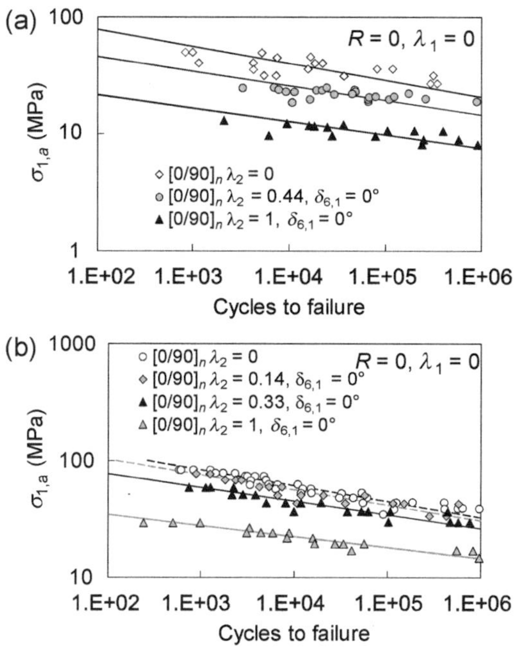

Figure 4.6 Influence of biaxiality ratio λ_2 on the fatigue strength of (a) glass–polyester $[0/90]_n$ tubes loaded in combined tension/torsion (Amijima et al., 1991) and (b) of glass–polyester $[0/90]_n$ tubular samples subjected to tension/torsion (Fujii et al., 1994). Reproduced from Quaresimin et al. (2010).

also for notched tubes tested by Fujii et al. (1994), whose results are shown in Fig. 4.6b.

Aboul Wafa et al. (1997) tested woven glass–epoxy tubes under combined bending/torsion loading, resulting, again, in the stress components σ_1 and σ_6, both with a load ratio $R = -1$. The detrimental influence of the shear stress is again confirmed by the S–N curves for final failure reported in Fig. 4.7 for [0/90] tubes, whereas no influence on the fatigue strength can be observed if the phase shift $\delta_{6,1}$ is changed from 0 to 90 degree (Aboul Wafa et al., 1997).

Qi and Cheng (2007) tested filament-wound glass fiber-reinforced epoxy composite tubes under tension/torsion loading. External loads were proportionally applied, and two values of the load ratio were adopted ($R = 0, \ -1$) in order to investigate its

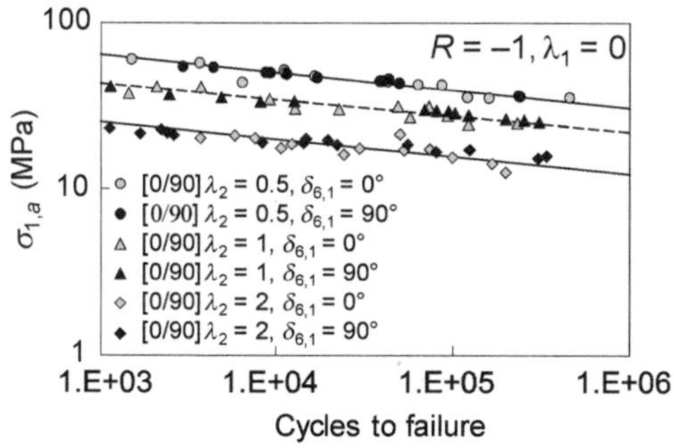

Figure 4.7 Influence of nonzero out-of-phase angles on the fatigue strength of glass–polyester tubes subjected to bending and torsion (Aboul Wafa et al., 1997). Reproduced from Quaresimin et al. (2010).

Figure 4.8 Maximum stress S−N curves for glass/epoxy [± θ] tubes under tension/torsion and R = 0, − 1 (Qi and Cheng, 2007).

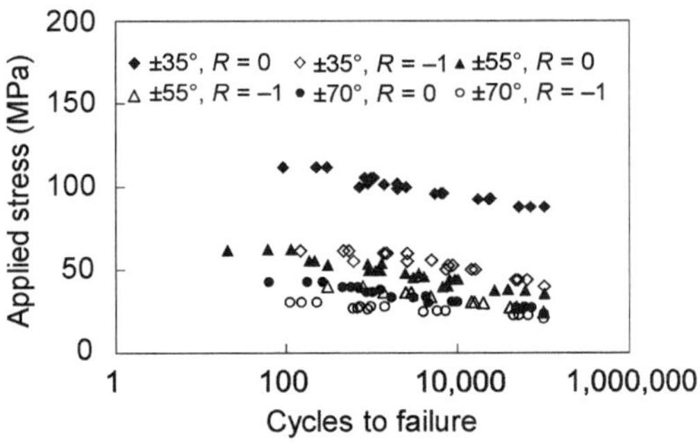

influence on the fatigue strength of [± θ] tubes ($\theta = 35$, 55, and 70 degrees). The presence of off-axis plies produced a combination of *external* and *internal* multiaxiality, leading to the presence of all three in-plane stress components. Some results are shown in Fig. 4.8 for biaxially loaded tubes in terms of the maximum cyclic value of the applied global stress in the tube axis direction ($\sigma_{x,\max}$). A significant drop in the fatigue strength can be observed if the load ratio is changed from 0 to −1 , which is particularly pronounced for low off-axis angles (35 degree).

All the experimental data shown so far give some useful indications on the influence of the biaxiality and load ratios, as well as of the phase shift, on the fatigue strength of composite laminates. However, attention has been focused only on the final failure of the specimens, connected to the incapability to carry the prescribed load.

Adden and Horst (2006, 2010) instead focused on the influence of multiaxial loads on damage evolution by testing tubes made of glass noncrimp fabric reinforced epoxy, subjected to combined tension and torsion loading with $R = 0$. In Fig. 4.9, results for [0/45/90/−45/45/90/−45/0] tubes presented by Adden and Horst (2006) are shown in terms of the crack density evolution in the plies. Fig. 4.9a is related to a pure torsion loading condition. Similar to what has been frequently observed in uniaxially loaded multidirectional laminates, the crack density increases quite rapidly in the first part of the fatigue life and then reaches a sort of plateau, indicating that saturation has occurred. In Fig. 4.9b, the crack density in the − 45 degree ply is plotted for pure torsion and combined loading, revealing a significant influence of the multiaxial condition on damage evolution.

4.3.4 Discussion

As a conclusion of the analysis of experimental data from the literature, it is clear that the presence of a multiaxial stress state strongly influences the fatigue behavior of composites. In particular, it was shown that the presence of shear stress in the load-bearing plies has a detrimental effect on the cycles to the final failure of laminates.

Also the load ratio, which can be different for different stress components in case of *external* multiaxiality, has been shown to affect the fatigue strength, a particularly detrimental condition being represented by the presence of shear and compressive normal stresses.

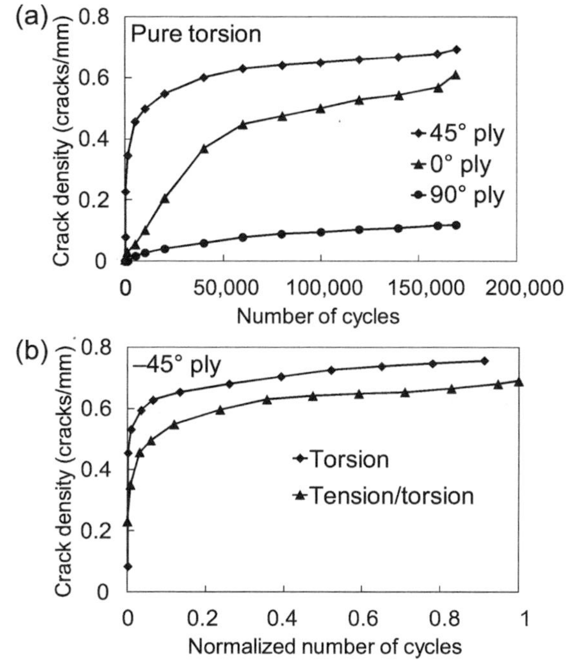

Figure 4.9 Crack density evolution in the plies of [0/45/90/−45/45/90/−45/0] tubes under (a) torsion and (b) comparison with combined tension/torsion loading for the − 45 degree ply (Adden and Horst, 2006).
Reproduced from Adden and Horst (2010).

It has also emerged that, as it happens for uniaxially loaded coupons, the fatigue behavior of multidirectional laminates under multiaxial loading is characterized by a progressive damage evolution from the first cycles up to the final failure. As shown by Adden and Horst (2006), the first stages of damage evolution are represented by the initiation and propagation of multiple off-axis cracks. The accumulation of such cracks leads to the degradation of the global elastic properties, as shown by Adden and Horst (2010), as well as to the onset of delaminations, also promoting the failure of the load-bearing fibers and therefore the final separation of the laminate.

From the analysis of the literature, it is clear that the majority of experimental investigations under multiaxial loads is focused on the final failure of laminates, whereas quantitative results on damage evolution can be found in a limited number of contributions (Wang et al., 1982a; Adden and Horst, 2006; Schmidt et al., 2012a,b). However, also in those cases, results are reported for particular stacking sequences and therefore it is hard to draw reliable conclusions of general validity, which are needed for the development of suitable criteria to predict fatigue damage initiation and evolution and failure of composite components.

Keeping in mind that the first stages of damage are characterized by the initiation of off-axis cracks, it is clear that it is of fundamental importance to characterize and understand the matrix-dominated fatigue behavior under multiaxial loading. As a consequence, investigating the influence of the biaxiality ratio λ_{12} on fatigue crack initiation and propagation in UD plies is essential. Clear data and general observations cannot be found in the literature on this particular topic, which therefore has been the object of extensive investigations in the recent years by the present authors at the University of Padova.

4.4 Recent results from the University of Padova

In order to fill some gaps encountered in the literature and to provide a deeper insight into damage initiation and evolution in laminates under multiaxial loading, an extensive experimental investigation has been carried out in the recent years by the present authors. As already mentioned, the attention has been focused on the matrix-dominated behavior as it controls the early stages of damage evolution in laminates. Tubular specimens were chosen as the best solution for multiaxial testing, thanks to their several advantages extensively discussed in Quaresimin and Carraro (2013, 2014).

Initially $[90_n]$ UD tubes were tested under combined tension/torsion loading producing the stress components σ_2 and σ_6, combined according to different values of the biaxiality ratio λ_{12}. This specimen configuration allowed us to analyze the influence of λ_{12} on the first crack initiation process, as the UD tubes suddenly failed immediately after this event (Quaresimin and Carraro, 2013).

Later, tubular specimens with lay-up $[0_F/90_{U,3}/0_F]$ were tested, where three 90 degree UD layers were constrained between external and internal thin fabric plies, in order to stabilize crack propagation in the transverse plies. The attention was focused on crack initiation and propagation in the 90 degree layers. The main results of the experimental work originally presented in Quaresimin and Carraro (2014), Quaresimin et al. (2015) are summarized below.

Results are reported for $[0_F/90_{U,3}/0_F]$ glass/epoxy tubes, produced by mandrel wrapping and cured in an autoclave.

The following prepregs were used:

- UE400-REM produced by Saati S.p.A. (Italy), glass/epoxy UD tape, thickness = 0.38 mm, for the 90 degree UD plies.
- EE106-ET443 produced by Saati S.p.A. (Italy), glass/epoxy fabric, thickness = 0.13 mm, for the 0 degree fabric plies.

The tubes were fatigue tested on an axial-torsional MTS 809 testing system. Proportional tension/torsion cycles were applied in load control, with load ratio $R = \sigma_{2,min}/\sigma_{2,max} = \sigma_{6,min}/\sigma_{6,max}$ of values 0.05, 0.5, and -1 and a frequency of 10 Hz. Three values of the biaxiality ratio λ_{12} (computed in the 90 degree plies) were adopted for all the load ratios ($\lambda_{12} = 0$, 1, 2), and also $\lambda_{12} = 0.5$ has been considered for $R = 0.05$ only.

4.4.1 Description of damage evolution

Damage initiation and evolution were monitored during the tests by an FLIR SC7600 MW infrared camera with temperature accuracy of 20 mK and lock-in analysis, as well as by eye observations with the aid of an internal lighting system, as described in Quaresimin and Carraro (2013, 2014).

The typical damage pattern in biaxially loaded tubes is shown schematically in Fig. 4.10.

The first macroscopic (visible) damage event is the initiation of a crack in the 90 degree ply. When a crack is detected, it covers an angle in the circumferential direction which is typically between 10 and 30 degree.

The initiation of the first crack is followed by its tunneling propagation in the circumferential direction, as well as by further crack initiation in other regions of the specimen.

Not all the specimens were brought to the final separation, and when this happened, the phenomenon was controlled by the 0 degree fabric plies, whose behavior was not of interest in the works presented in Quaresimin and Carraro (2014), Quaresimin et al. (2015). However, it is worth noting that the plane of separation was usually the one containing the longer crack, which was not always the first crack initiated, since other regions could have had a lower apparent fracture toughness, thus giving rise to a faster propagation. Finally, it is remarked that no cracks were observed to reach a propagation angle of 360 degree (full circle), the final failure occurring before.

A picture of a specimen showing the presence of multiple transverse cracks is presented in Fig. 4.11.

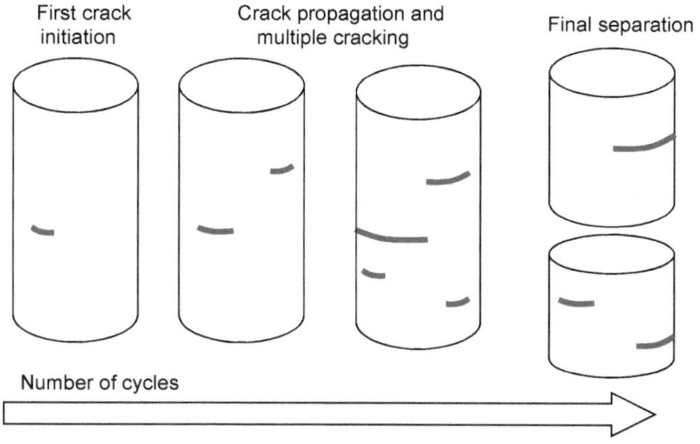

Figure 4.10 Schematic of damage evolution in $[0_F/90_{U,3}/0_F]$ tubes under tension/torsion loading.

Figure 4.11 $[0_F/90_{U,3}/0_F]$ tube with multiple cracks.

In the following sections, the crack initiation and propagation results will be shown and discussed in more detail.

4.4.2 Fatigue crack initiation results

The results of fatigue tests are presented as S–N curves, relating the maximum cyclic transverse stress in the 90 degree plies, $\sigma_{2,max}$, to the cycles at the initiation of the first visible transverse crack. Results were originally presented in Quaresimin and Carraro (2014) and in Quaresimin et al. (2015). Data are shown in Fig. 4.12 in double logarithmic scales and fitted by straight lines. This figure highlights the influence of the multiaxial condition on the cycles to crack initiation at a fixed load ratio of 0.05. A detrimental effect of the increasing shear stress component can be seen when the results are plotted in terms of the maximum cyclic transverse stress in the 90 degree plies. The same trend is found for other load ratios, and the worst condition is represented by the combination of a high shear stress and fully reversed cycles.

The influence of the load ratio on the life to crack initiation is highlighted in Fig. 4.13 for $\lambda_{12} = 1$. Qualitatively, it can be seen that a higher value of R leads to a higher number of cycles to first crack initiation for the same maximum transverse stress, while a tension–compression loading condition is strongly detrimental for the fatigue strength in terms of the maximum cyclic stress. It can also be seen that

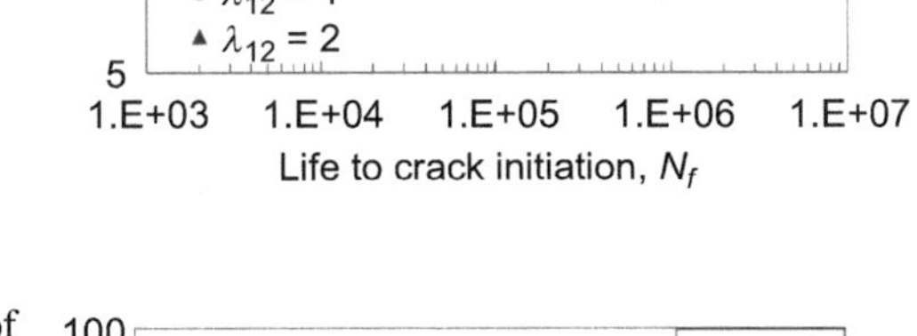

Figure 4.12 S–N curves for nucleation of the first crack on [0F/90U,3/0F] tubes, for $\lambda_{12} = \sigma_6/\sigma_2 = 0, 0.5, 1$, and 2, and R = 0.05 (Quaresimin and Carraro, 2014).

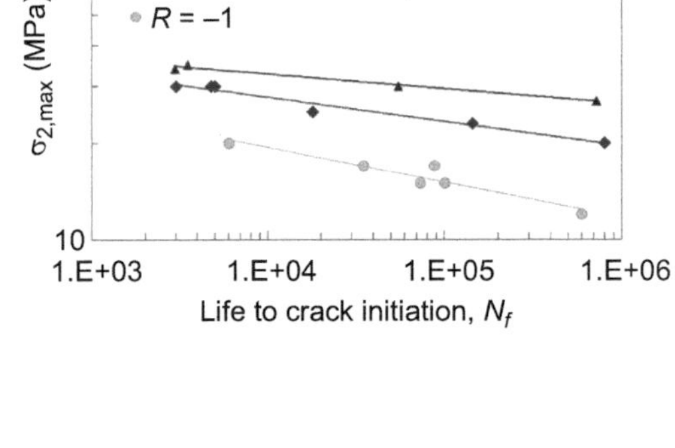

Figure 4.13 S–N curves for nucleation of the first crack on $[0_F/90_{U,3}/0_F]$ tubes, for R = 0.5, 0.05, and – 1, and $\lambda_{12} = 1$ (Quaresimin et al., 2015).

the S–N curves are steeper as R decreases, and therefore its effect is higher for lower load levels (or longer fatigue lives to crack initiation). This trend is confirmed for the other values of λ_{12} (Quaresimin et al., 2015).

4.4.3 Fatigue crack propagation results

The initiation of a crack in the 90 degree plies was followed by its circumferential propagation, which was made stable by the presence of the fabric layers. As highlighted in Fig. 4.14, tension and torsion loadings give rise to a mixed mode I + II (opening + sliding) propagation.

During fatigue tests, crack propagation was analyzed by measurements taken directly on the specimens. As already mentioned, when a crack was observed for the first time, it had already extended for an initial angle $2\alpha_i$ between 10 and 30 degree, and then propagated as the number of cycles was increased. An example of circumferential crack propagation is schematically shown in a top view in Fig. 4.15 for a $[0_F/90_{U,3}/0_F]$ tube tested with $\lambda_{12} = 2$, $R = 0.05$, and a transverse stress on the 90 degree plies of 14 MPa. N_p represents the number of propagation cycles, that is the total number of cycles minus those spent for the initiation of the considered crack. In this case, an initial angle of 23 degree was observed, and then the crack propagated, both at the right and left tips, up to an angle of 67 degree after about 66,000 cycles.

A plot of the crack angle against the number of propagation cycles is also shown in Fig. 4.15, showing an approximately linear trend.

Some other examples of crack propagation curves are shown in Fig. 4.16a and b, where the relative crack angle $2\alpha - 2\alpha_i$ is plotted against N_p.

Fig. 4.16a shows the crack propagation curves for the same transverse stress level ($\sigma_{2,\max} = 28$ MPa) and increasing biaxiality ratios. From a qualitative point of view, the curves are steeper as λ_{12} is increased, suggesting a detrimental effect of the shear stress also on the crack propagation phase, as will be discussed more later. Fig. 4.16b shows instead the propagation curves for $R = -1$ and several cracks nucleated in the

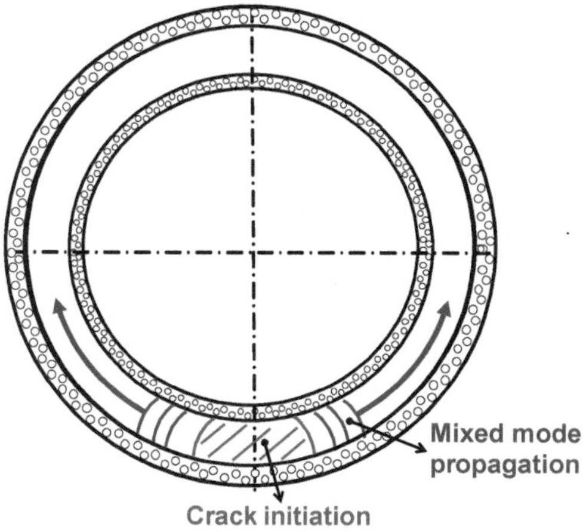

Figure 4.14 Crack initiation and propagation for a $[0_F/90_{U,3}/0_F]$ tube. Reproduced from Quaresimin and Carraro (2014).

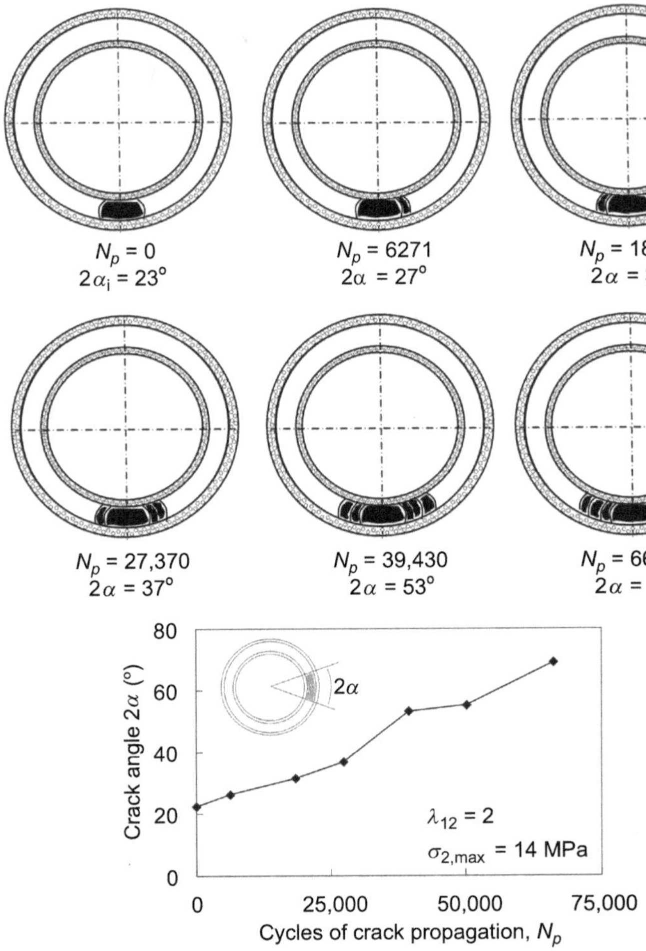

Figure 4.15 Example of crack propagation for a $[0_F/90_{U,3}/0_F]$ tube with $\lambda_{12} = 2$ and $\sigma_{2,max} = 14$ MPa (Quaresimin and Carraro, 2014).

same specimen tested with $\lambda_{12} = 1$. In both Fig. 4.16a and b, it can be observed, again, that a large part of the propagation curves can be reasonably described by a straight line. This is physically justified by the fact that the mode I and II components (G_I and G_{II}) of the energy release rate (ERR) do not depend on the crack angle 2α when it is higher than about 30 degree, as shown by finite element analyses presented in Quaresimin and Carraro (2014). The constancy of the ERR components leads to a steady-state crack propagation (with a uniform growth rate), with limited local variations due to defects or changes in the microstructure in the crack path. Therefore, the crack growth rate (CGR) for each crack can be calculated as the slope of the straight line fitting each propagation curve. Referring again to Fig. 4.16a, it can be said now that a higher shear stress component results in a higher CGR.

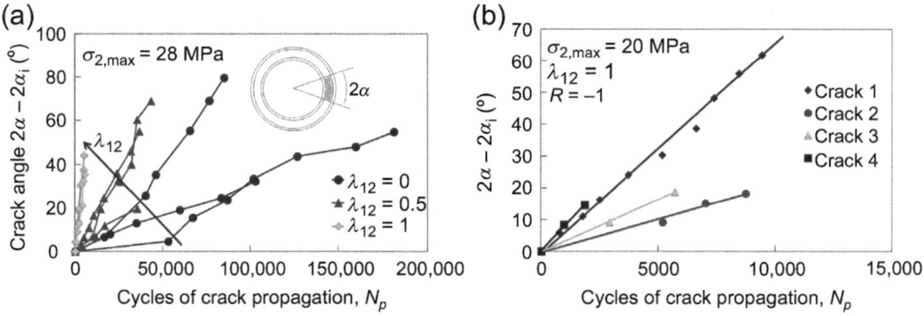

Figure 4.16 Crack propagation curves for $[0_F/90_{U,3}/0_F]$ tubes with (a) $R = 0.05$ and (b) $R = -1$.

The crack propagation phenomenon was extensively analyzed by measuring the cracks initiated in all the tested specimens at different number of cycles. The CGR was then calculated for each of them fitting the propagation curves by means of a straight line, and its value was associated with the ERR, as explained above. With this procedure, the crack propagation data could be presented in Paris-like charts relating the CGR to the mode I ERR.

Most of the specimens developed multiple cracking; therefore, one or more propagating cracks could be measured against the number of cycles for each specimen, as far as they were not interacting, as highlighted also by Quaresimin and Carraro (2014).

As the analysis of crack propagation is presented here in terms of the ERR, a better parameter for quantifying the multiaxial condition is the mode mixity, $MM = G_{II}/G_{tot}$, where $G_{tot} = G_I + G_{II}$.

To highlight the influence of the MM on the CGR, Paris-like data are presented in Fig. 4.17 in terms of the maximum cyclic value of G_I, $G_{I,max}$, for $R = 0.05$ (Quaresimin and Carraro, 2014). Power laws are used to fit the data, presented in double logarithmic scales.

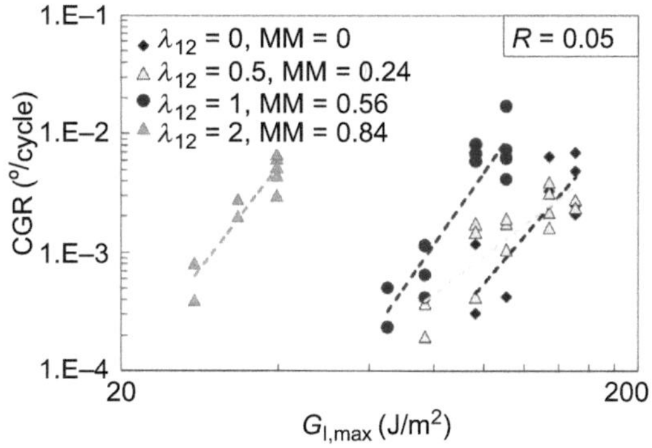

Figure 4.17 Paris-like curves for $[0_F/90_{U,3}/0_F]$ tubes in terms of the mode I ERR for $R = 0.05$ (Quaresimin and Carraro, 2014).

It is evident that the Paris-like curves are shifted from higher to lower values of G_I as the mode II contribution increases. This means that a lower mode I ERR is necessary to reach a given CGR. This effect, confirmed for the other values of R, is particularly evident for $\lambda_{12} \geq 1$ (MM ≥ 0.56). Conversely, data points for $\lambda_{12} = 0$ and 0.5 (MM = 0 and 0.24) seem to be reasonably described by the same scatter band in terms of G_I, which can be assumed to control the crack propagation phenomenon in this range of mode mixity.

To highlight the influence of the load ratio, the crack growth data are plotted in Fig. 4.18 for MM = 0.56 ($\lambda_{12} = 1$) (Quaresimin et al., 2015).

A larger value of R brings to lower CGR for the same values of $G_{I,max}$, whereas the presence of compressive loading leads to higher CGR. Also, the slope of the curves is influenced by the load ratio, as it is lower for $R = -1$ and higher for $R = 0.5$, indicating that the effect of R is more pronounced at lower ERR levels.

4.4.4 Damage mechanisms at the microscopic scale

As shown above, a strong influence of λ_{12} and R was found on the S−N curves for crack initiation and on Paris-like curves for crack propagation, highlighting the need of suitable criteria to account for these parameters in the design process. With the aim to provide a solid basis for the development of physically based models, the damage mechanisms at the microscopic scale were analyzed by means of SEM analyses of the fracture surfaces of the tubes. Furthermore, the damage mechanisms were found to depend on the multiaxial stress state. In particular, the fracture surfaces of tubes tested with positive load ratios and a high enough shear stress contribution ($\lambda_{12} \geq 1$) were characterized by the wide presence of shear cusps (see Fig. 4.19b). They are probably the result of the onset of microcracks in the matrix, the accumulation of which is presumably the cause of the initiation and propagation of a macrocrack. Conversely, when the shear stress was very low or zero, quite smooth fracture surfaces, typical of tensile or mode I failure, were observed (see Fig. 4.19a). It is therefore reasonable to assume a change of the leading damage mechanism shifting from a macroscopically tension-dominated to a shear-dominated loading condition.

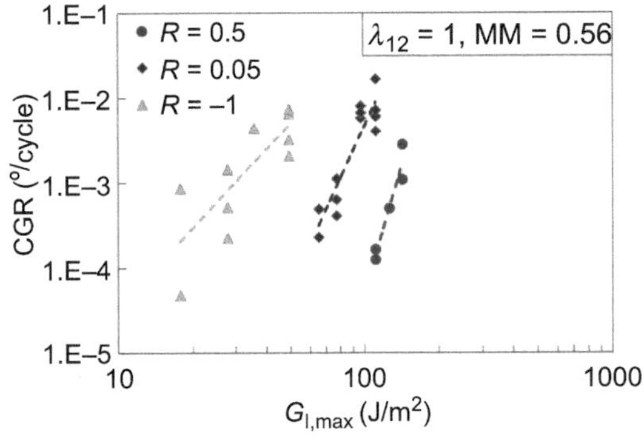

Figure 4.18 Paris-like curves for $[0_F/90_{U,3}/0_F]$ tubes in terms of the mode I ERR for MM = 0.56 (Quaresimin et al., 2015).

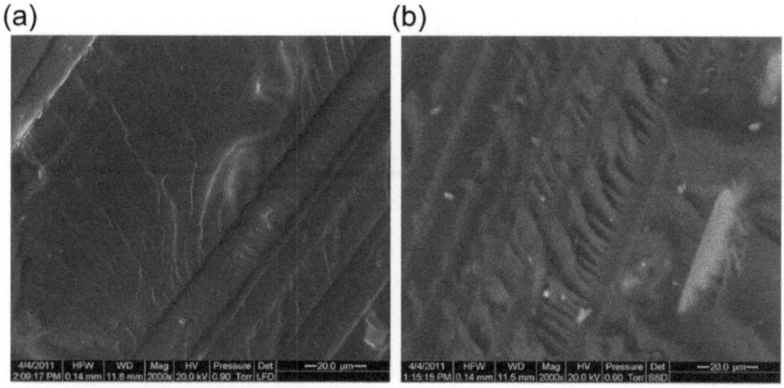

Figure 4.19 SEM images of the fracture surfaces for $[0_F/90_{U,3}/0_F]$ tubes with (a) $\lambda_{12} = 0$ and (b) $\lambda_{12} = 2$ (Quaresimin and Carraro, 2014).

4.5 Comparison with results on flat laminates

As shown so far, tubular specimens subjected to tension/torsion loading represent a very suitable testing condition for characterizing the initiation and evolution of damage under multiaxial stress states. However, their use is contingent on the availability of a biaxial testing machine, which is not so common in the industrial environment. For this reason, a particularly attractive possibility is represented by testing composite materials with uniaxial loads exploiting the anisotropy-induced *internal* multiaxiality.

Remembering the contradictory results highlighted by Quaresimin et al. (2010), with the aim of verifying the equivalence between fatigue damage evolution under *external* and *internal* multiaxial stress states, a dedicated experimental study was recently carried out on multidirectional flat laminates.

The same UD glass–epoxy prepreg material adopted for the tubes was used for manufacturing $[0/\theta_2/0/-\theta_2]_s$ flat specimens with off-axis angles θ of 50 and 60 degree, in order to achieve biaxiality ratios $\lambda_{12} = 0.57$ and 1.15 in the off-axis plies. The laminates were subjected to a uniaxial cyclic load with $R = 0.1$. The results of the fatigue tests, originally presented by Quaresimin et al. (2014), are compared to those of the tubes for $\lambda_{12} = 0.5$ and 1, both in terms of crack initiation and propagation.

In Fig. 4.20a and b, the S–N data for the initiation of the first crack in the 90 degree layers of the tubes are compared to those for the first cracks initiation in the $[0/\theta_2/0/-\theta_2]_s$ laminates. All data are presented in terms of the maximum cyclic transverse stress in the off-axis plies.

It can be seen that the results for similar values of the biaxiality ratio (i.e., similar local multiaxial stress states) are in reasonable agreement. In particular, the S–N curves are only slightly shifted, and this can be due to small differences in the cured ply properties (fiber volume fraction, for instance); the constraining effect provided by the surrounding layers, which are different for the two kinds of samples, and the presence of a certain amount of longitudinal stress, σ_1, in the flat coupons. However, it is very important to note that the slope of the curves is consistent, and this is a very important issue since it represents the rate of the damage evolution during fatigue life, which is found to depend on the local multiaxial stress state, no matter if it is originated by external multiaxial loads (tubes) or by the material anisotropy (flat specimens).

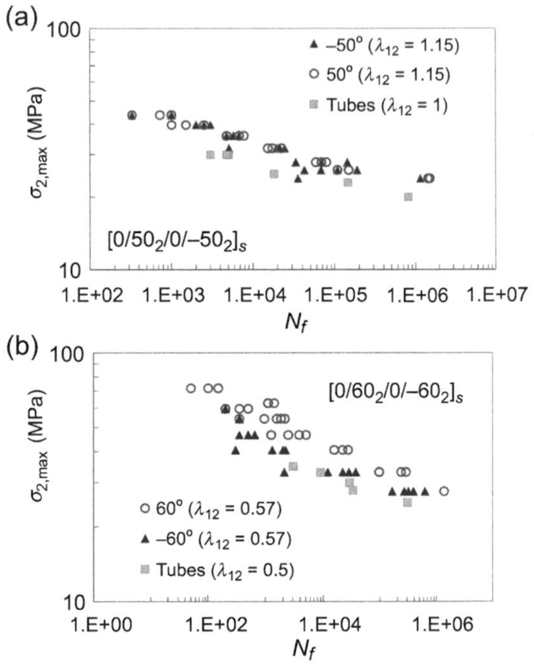

Figure 4.20 Comparison between the S–N data for the first cracks initiation for tubes and flat laminates results for (a) $[0/50_2/0/-50_2]_s$ and (b) $[0/60_2/0/-60_2]_s$ specimens (Quaresimin et al., 2014).

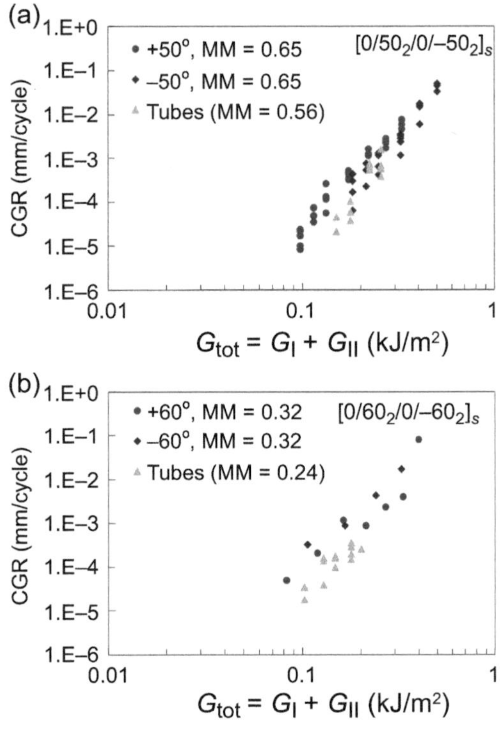

Figure 4.21 Comparison between the crack growth data for tubes and results on flat laminates for (a) $[0/50_2/0/-50_2]_s$ and (b) $[0/60_2/0/-60_2]_s$ specimens (Quaresimin et al., 2014).

The steady-state crack growth data for the flat specimens in terms of the total ERR are also compared, in Fig. 4.21, to those obtained for the tubes with a similar multiaxial condition, that is, $\lambda_{12} = 1$ and 0.5 (MM = 0.56 and 0.24). Also in this case, it can be seen that the results for similar values of the MM are reasonably consistent.

It can be concluded, at least for positive load ratios, that the fatigue behavior of a UD ply under *external* and *internal* multiaxial conditions is consistent, in terms of crack initiation and propagation, provided that the local multiaxial stress state is the same, independently of the way it is originated. As a consequence, experimental results and observations can be extended from *external* to *internal* multiaxial stress states and vice versa, allowing one to adopt easier test configurations for the characterization of damage onset and evolution under multiaxial stress states.

4.6 Conclusions

As a conclusion for this chapter, it can be said that the fatigue behavior of composite laminates is strongly dependent on the multiaxial stress state and on the load ratio. This is confirmed by several works in the literature concerning the total fatigue life of multidirectional laminates tested under biaxial loads. It was also shown that damage evolution, in the form of initiation and propagation of off-axis cracks, strongly depends on the local multiaxial stress state. This is a very important issue mainly with reference to a stiffness-based design against fatigue, since the accumulation of this kind of damage leads to the progressive degradation of the global elastic properties of a composite laminate. With the aim to provide a deeper insight into the off-axis crack initiation and propagation phenomena under multiaxial loading, an extensive experimental activity was carried out by the present authors testing glass–epoxy tubes under tension/torsion loading. A strong influence of crack initiation (in terms of S–N curves) and propagation (in terms of Paris-like curves) was found and quantified on the amount of shear stress with respect to the transverse stress.

Damage mechanisms at the microscopic scale were also observed to be dependent on the biaxial stress state, and this represents a fundamental starting point for the development of damage-based criteria for crack initiation and propagation under multiaxial loads.

Finally, comparing the results on tubes with those obtained for the off-axis plies of multidirectional flat laminates, it was possible to say that crack initiation and propagation phenomena are consistent under *internal* and *external* multiaxial stress states, provided that the local stress state is the same. This opens the possibility of testing the fatigue behavior under multiaxial stress states also by means of uniaxial loads, at least for positive load ratios.

References

Aboul Wafa, M.N., Hamdy, A.H., El-Midany, A.A., 1997. Combined bending torsional fatigue of woven rowing GRP. J. Eng. Mater. Technol. 119, 181–185. https://doi.org/10.1115/1.2805991.

Adden, S., Horst, P., 2006. Damage propagation in non-crimp fabrics under bi-axial static and fatigue loading. Compos. Sci. Technol. 66, 626–633. https://doi.org/10.1016/j.compscitech.2005.07.034.

Adden, S., Horst, P., 2010. Stiffness degradation under fatigue in multiaxially loaded non-crimped-fabrics. Int. J. Fatig. 32, 108–122. https://doi.org/10.1016/j.ijfatigue.2009.02.002.

Amijima, S., Fujii, T., Hamaguchi, M., 1991. Static and fatigue tests of woven glass fabric composite under biaxial tension-torsion loading. Composites 22 (4), 281–289. https://doi.org/10.1016/0010-4361(91)90003-Y.

Antoniou, A.E., Van Hemelrijck, D., Philippidis, T.P., 2010. Failure prediction for a glass/epoxy cruciform specimen under static biaxial loading. Compos. Sci. Technol. 70, 1232–1241. https://doi.org/10.1016/j.compscitech.2010.03.011.

Atcholi, K.E., Oytana, C., Varchon, D., Perreux, D., 1992. Superposed torsion-flexure of composite materials: experimental method and example of application. Composites 23, 327–333. https://doi.org/10.1016/0010-4361(92)90332-O.

Chen, A.S., Matthews, F.L., 1993. A review of multiaxial/biaxial loading tests for composite materials. Composites 24 (5), 395–40610, 1016/0010-4361(93) 90247-6.

El-Assal, A.M., Khashaba, U.A., 2007. Fatigue analysis of unidirectional GFRP composites under combined bending and torsional loads. Compos. Struct. 79, 599–605. https://doi.org/10.1016/j.compstruct.2006.02.026.

Ferry, L., Perreux, D., Varchon, D., Sicot, N., 1999. Fatigue behaviour of composite bars subjected to bending and torsion. Compos. Sci. Technol. 59, 575–582. https://doi.org/10.1016/S0266-3538(98)00103-1.

Foral, R.F., Humpherey, W.D., 1984. Biaxial stress behaviour of graphite and Kevlar 49 fibre/epoxy composites and hybrids. AIAA J. 22 (1), 111–11610, 2514/3.8347.

Found, M.S., 1985. Review of the Multiaxial Fatigue Testing of Fibre Reinforced Plastics. ASTM Special Technical Publication, Philadelphia, pp. 381–395.

Francis, P.H., Walrath, D.E., Sims, D.F., Weed, D.N., 1977. Biaxial fatigue loading of notched composites. J. Compos. Mater. 11, 488–501. https://doi.org/10.1177/002199837701100410.

Fujii, T., Lin, F., 1995. Fatigue behavior of a plain-woven glass fabric laminate under tension/torsion biaxial loading. J. Compos. Mater. 29 (5), 573–59010.1177/002199839502900502.

Fujii, T., Shina, T., Okubo, K., 1994. Fatigue notch sensitivity of glass woven fabric composite having a circular hole under tension/torsion biaxial loading. J. Compos. Mater. 28 (3), 234–25110.1177/002199839402800303.

Jones, D.L., Poulose, P.K., Liebovitz, H., 1985. Effect of Biaxial Loads on the Static and Fatigue Properties of Composite Materials, vol 853. ASTM Special Technical Publication, Philadelphia, pp. 413–427.

Kawakami, H., Fujii, T., Morita, Y., 1996. Fatigue degradation and life prediction of glass fabric polymer composite under tension/torsion biaxial loadings. J. Reinforc. Plast. Compos. 15 (2), 15 183–19510.1177/073168449601500204.

Krempl, E., Elzey, D.M., Hong, B.Z., Ayar, T., Loewy, R.G., 1988. Uniaxial and biaxial fatigue properties of thin-walled composite tubes. J. Am. Helicopter Soc. 33 (3), 3–10. https://doi.org/10.4050/JAHS.33.3.

Mailly, L., Wang, S.S., 2008. Recent development of planar cruciform experiment on biaxial tensile deformation and failure of unidirectional glass/epoxy composite. J. Compos. Mater. 42, 1359–1379. https://doi.org/10.1177/0021998308092203.

Makris, A., Vandenbergh, T., Ramault, C., Van Hemelrijck, D., Lamkanfi, E., Van Paepegem, W., 2010a. Shape optimisation of a biaxially loaded cruciform specimen. Polym. Test. 29, 216–223. https://doi.org/10.1016/j.polymertesting.2009.11.004.

Makris, A., Ramault, C., Van Hemelrijck, D., Zarouchas, D., Lamkanfi, E., Van Paepegem, W., 2010b. An investigation of the mechanical behaviour of carbon epoxy cross ply cruciform specimens under biaxial loading. Polym. Compos. 31, 1554–1561. https://doi.org/10.1002/pc.20943.

Ogasawara, T., Onta, K., Yokozeki, T., Ogihara, S., 2007. Tension/torsion fatigue behavior of unidirectional GFRP and CFRP. In: Proceedings of the 16th International Conference on Composite Materials – ICCM16. Kyoto.

Olsson, R., 2011. A survey of test methods for multiaxial and out-of-plane strength of composite laminates. Compos. Sci. Technol. 71, 773–783. https://doi.org/10.1016/j.compscitech.2011.01.022.

Owen, M.J., Griffiths, J.R., 1978. Evaluation of biaxial stress failure surfaces for a glass fabric reinforced polyester resin under static and fatigue loading. J. Mater. Sci. 13, 1521–1537.

Perreux, D., Joseph, E., 1997. The effect of frequency on the fatigue performance of filament wound pipes under biaxial loading: experimental results and damage mechanics. Compos. Sci. Technol. 57, 353–364. https://doi.org/10.1016/S0266-3538(96)00155-8.

Qi, D., Cheng, G., 2007. Fatigue behaviour of filament-wound glass fibre reinforced epoxy composite tubes under tension/torsion biaxial loading. Polym. Compos. 28 (1), 116–123. https://doi.org/10.1002/pc.20275.

Quaresimin, M., Carraro, P.A., 2013. On the investigation of the biaxial fatigue behaviour of unidirectional composites. Compos. B Eng. 54, 200–208. https://doi.org/10.1016/j.compositesb.2013.05.014.

Quaresimin, M., Carraro, P.A., 2014. Damage initiation and evolution in glass/epoxy tubes subjected to combined tension-torsion fatigue loading. Int. J. Fatig. 63, 25–35. https://doi.org/10.1016/j.ijfatigue.2014.01.002.

Quaresimin, M., Susmel, L., Talreja, R., 2010. Fatigue behaviour and life assessment of composite laminates under multiaxial loadings. Int. J. Fatig. 32, 2–16. https://doi.org/10.1016/j.ijfatigue.2009.02.012.

Quaresimin, M., Carraro, P.A., Pilgaard Mikkelsen, L., Lucato, N., Vivian, L., Brøndsted, P., Sørensen, B.F., Varna, J., Talreja, R., 2014. Damage evolution under internal and external multiaxial cyclic stress state: a comparative analysis. Compos. B Eng. 61, 282–290. https://doi.org/10.1016/j.compositesb.2014.01.056.

Quaresimin, M., carraro, P.A., Maragoni, L., 2015. Influence of load ratio on the biaxial fatigue behaviour and damage evolution in glass/epoxy tubes under tension-torsion loading. Compos. Appl. Sci. Manuf. https://doi.org/10.1016/j.compositesa.2015.08.009.

Radon, J.C., Wachnicky, C.R., 1985. Biaxial fatigue of glass fibre reinforced polyester resin. In: Miller, K.J., Brown, M. (Eds.), Multiaxial Fatigue, ASTM STP 853. American Society for Testing and Materials, Philadelphia, pp. 396–412.

Ramault, C., 2012. Guidelines for Biaxial Testing of Fibre Reinforced Composites Using a Cruciform Specimen (Ph.D. Thesis). Vrije Universiteit Brussel.

Schmidt, F., Rheinfurth, M., Horst, P., Busse, G., 2012a. Multiaxial fatigue behaviour of GFRP with evenly distributed or accumulated voids monitored by various NDT methodologies. Int. J. Fatig. 43, 207–216. https://doi.org/10.1016/j.ijfatigue.2012.03.013.

Schmidt, F., Rheinfurth, M., Horst, P., Busse, G., 2012b. Effects of local fibre waviness on damage mechanisms and fatigue behaviour of biaxially loaded tube specimens. Compos. Sci. Technol. 72, 1075–1082. https://doi.org/10.1016/j.compscitech.2011.09.012.

Smith, E.W., Pascoe, K.J., 1989. Biaxial fatigue of a glass-fibre reinforced composite. Part 1: fatigue and fracture behaviour. In: Brown, M., Miller, K.J. (Eds.), Biaxial and Multiaxial Fatigue. EGF 3. Mechanical Engineering Publications, London, pp. 367–396.

Wang, S.S., Chim, E.S.-M., Socie, D.F., 1982. Biaxial fatigue of fiber-reinforced composites at cryogenic temperature. Part I fatigue fracture life and damage mechanisms. J. Eng. Mater. Technol. 104, 128–136. https://doi.org/10.1115/1.3225047.

Wang, S.S., Chim, E.S.-M., Socie, D.F., 1982. Stiffness degradation of fiber-reinforced composites under uniaxial tensile, pure torsional, and biaxial fatigue at cryogenic temperature. In: Composite Materials: Testing and Design. ASTM STP 787 EB, pp. 287–301.

Part Two

Modeling of failure mechanisms in composite materials

Matrix and fiber—matrix interface cracking in composite materials

5

Ramesh Talreja
Department of Aerospace Engineering, Department of Materials Science and Engineering, Texas A&M University, College Station, TX, United States

5.1 Introduction

Failure in a lamina lying within a laminate initiates when critical conditions generated at one or more local sites in the lamina are reached under loading of the laminate. Depending on the laminate loading (in-plane, bending, torsion, or combinations of these), failure within a given lamina may initiate in fibers or in the regions between fibers. The mechanisms involving fiber failure are distinctly different from those that initiate cracking in the matrix and/or at the fiber—matrix interfaces. For a lamina, considered as a thin layer of homogeneous solid, the stress state may with good approximation be described as in-plane, given by the stresses σ_x, σ_y, and σ_{xy}, where the normal stresses σ_x and σ_y act along fibers and across fibers in the lamina plane, respectively, and σ_{xy} is the lamina in-plane shear stress (Fig. 5.1). These stresses can be viewed as remote stresses with respect to the local sites within the lamina (a unidirectional [UD] composite) where failure initiates. Fig. 5.1 illustrates the triaxial stress state developing locally at a point in the matrix of the lamina subjected to the transverse normal stress σ_y and the in-plane shear stress σ_{xy}. The axial normal stress σ_x is not shown, to indicate that its influence on the failure initiation in the matrix and/or at the fiber—matrix interfaces will be neglected here. This chapter will limit itself to modeling of the conditions for the initiation of failure in the regions excluding fibers in a UD composite.

In Chapter 1 of this book, observations of the failure mechanisms in UD composites have been described. Those mechanisms will be referred to as they pertain to failure initiation in the matrix and at the fiber—matrix interface. The modeling of the mechanisms will be addressed here, separating these in three elementary cases: brittle and ductile cracking of the matrix, and debonding of the fiber—matrix interface.

5.2 Modeling of failure initiation

In modeling initiation of failure, it is important to recognize that the initiation is a "point failure process," that is, failure initiates at a material point when critical (favorable) conditions are reached at that point. Because of failure initiation, different scenarios can develop. If the material point where failure initiates lies in a brittle

Modeling Damage, Fatigue and Failure of Composite Materials. https://doi.org/10.1016/B978-0-443-18489-5.00018-7

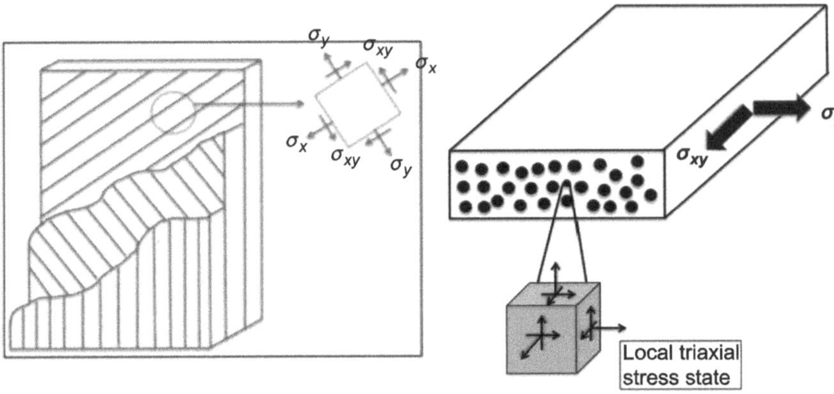

Figure 5.1 A multi-directional composite laminate loaded in an arbitrary manner develops an in-plane stress state in a lamina, as illustrated in the figure to the left. The sketch to the right illustrates the local tri-axial stress state developed at a point within a lamina loaded by transverse normal stress σ_y and in-plane shear stress σ_{xy}.

region, which has little capacity to deform inelastically (i.e., irreversibly), then the consequence of failure will be a brittle crack. On the other hand, if the region deforms in elastically, then crazes or shear bands, or both, can form (Donald and Kramer, 1982). A proper way to examine the propensity of brittle or ductile failure conditions to prevail is to assess the density of stored strain energy at the considered point during deformation prior to failure. If the strain energy density is totally or largely dilatational, the incentive to induce inelastic (shear-driven) processes at the point will be nonexistent or low. If, on the other hand, the distortional energy density dominates at a point, then ductile failure mechanisms involving shear banding and crazing will be likely. These fundamental concepts will guide the modeling considerations for the failure mechanisms in the discussion to follow.

5.2.1 Brittle matrix failure

In a polymer matrix UD composite of high fiber volume fraction (typically 0.5 or greater), transverse tension loading produces triaxial stress state locally in the matrix. Asp et al. (1996a,b) showed that at certain points in the matrix close to the fiber surfaces, the strain energy density under this loading becomes nearly or totally dilatational. They assumed that what appeared to be fiber–matrix debonding was indeed a consequence of the unstable growth of cavitation close to the fiber–matrix interface. They proposed that the expansion of a cavity, presumably formed from the free volume or from a preexisting defect in the polymer matrix, was governed by the energy density of dilatation. Later, Neogi et al. (2018) clarified the role of stress triaxiality in the cavitation formation mechanism by molecular dynamics simulation. While direct evidence of the breakage of bonds at the fiber–matrix interface because of unstable cavitation growth does not exist, it seems highly likely that the energy released by the cavity "burst" in the matrix close to the fiber–matrix interface will result in the local failure of the interface. Thus, as assumed in Asp et al. (1996a,b), debonding caused by unstable growth of cavitation occurs when the dilatational energy density given by

$$U_v = \frac{1-2v}{6E}(\sigma_1 + \sigma_2 + \sigma_3)^2 \tag{5.1}$$

reaches a critical material-specific value. In Eq. (5.1), σ_1, σ_2, and σ_3 are principal stresses and E and v are the Young's modulus and Poisson ratio, respectively, of the polymer matrix.

The sites at which fiber–matrix debonding is observed experimentally also correspond to the sites of the largest dilatation energy density values (Asp et al., 1996a,b).

The fiber–matrix debonding mechanism is also possible without the cavitation failure as a precursor. This will be the case if favorable conditions for cavitation do not exist or if the fiber–matrix interface is sufficiently weak or is sufficiently weakened by defects. In that case, the radial tensile stress on the fiber surface will break the fiber–matrix interface bonds, initiating debonding. For modeling purposes, it is difficult to know what interface failure properties to use, as these depend on the actual quality of the bond formed during the manufacturing process. The interface bond strength or toughness (critical energy release rate) cannot be determined accurately by theoretical means. Several experimental methods have therefore been devised (for a review, see Zhandarov and Mader, 2005). However, applying the interface toughness criterion for evaluating initiation of fiber–matrix debonding faces difficulties due to uncertainty of knowing the flaw size and its variability.

5.2.2 Ductile matrix failure

5.2.2.1 Initiation of yielding

The polymer matrix behaves in a ductile manner at a given point in the composite if the distortional part of the strain energy density at that point is sufficiently large. For isotropic metals, the initiation of yielding is satisfactorily given by the critical value of the distortional energy density obtained experimentally for the metal at hand. Equivalently, the yield criterion for metals can be expressed in terms of the second invariant of the deviatoric stress tensor, as done in the von Mises criterion. The initiation of yielding (more correctly, inelastic deformation) in polymers is governed by molecular phenomena that differ significantly from the dislocation motion underlying yielding of crystalline metals. Still, it is common to describe the onset of inelastic deformation in polymers by the approaches used for metal yielding. In contrast to metals, the inelastic response of glassy polymers displays pressure sensitivity, as discussed by Rottler and Robbins (2001). This has prompted modifying the metal yield criteria by including the hydrostatic stress, for example, by adding to the threshold of the octahedral shear stress τ_0 a constant α times the mean pressure P, as:

$$\tau_{oct}^{y} = \tau_0 + \alpha p \tag{5.2}$$

where P is the average of the three principal stresses, $P = -(\sigma_1 + \sigma_2 + \sigma_3)/3$.

Eq. (5.2) is the modified von Mises yield criterion, which in energy terms states that the dilatational energy density contributes to the shear-driven onset of inelastic deformation in polymers. Additionally, temperature and strain rate are also found to affect the inelastic deformation (Arruda et al., 1995).

5.2.2.2 Ductile crack formation

In a polymer matrix within a UD composite, the stress triaxiality is generally high except in resin-rich regions. The inelastic deformation will thus tend to occur away from the fiber−matrix interfaces in these regions. Once initiated, the inelastic deformation can lead to shear banding before crack formation. Estevez et al. (2000) have studied the crack formation process in glassy polymers by considering the competition between shear banding and crazing. Based on their study, it can be stated that the role of the distortional part of the strain energy density at a point is to localize inelastic deformation in shear bands, while the dilatational component is responsible for cavitation leading to craze formation, craze widening, and breakdown of craze fibrils. The mix of the two energy components determines the ease or difficulty of ductile crack formation in the matrix within the composite.

Models for ductile crack formation in polymers are complex and require many material constants for implementation. Huang and Talreja (2006) have shown that the Rice−Tracey ductile fracture model (Rice and Tracey, 1969) can predict ductile cracking in polymers with high accuracy. The Rice−Tracey model assumes that preexisting microvoids (from defects or inclusions) grow within plastically deformed material to a critical size, at which point neighboring microvoids coalesce and cause fracture. The microvoid growth is controlled by the stress tri-axiality and equivalent plastic strain, given by Eq. (5.3):

$$D = ln\left(\frac{R}{R_0}\right) = \int_{\varepsilon_{p0}}^{\varepsilon_p} \alpha \, exp\left(\frac{3}{2}\frac{\sigma_m}{\sigma_p}\right) d\varepsilon p \tag{5.3}$$

where σ_m and σ_p are mean stress and von Mises equivalent stress, respectively; ε_p is equivalent plastic strain; and R and R_0 are the current and initial void radii, respectively. The lower limit of integration is often taken as zero for simplicity. In implementation, the damage variable $D = \ln(R/R_0)$ is used instead of the void size. When D reaches a critical value D_c at the material point, a crack is considered to occur or to pass through that point by void coalescence. Rice and Tracey (1969) calculated the constant α to be 0.283 under assumptions of spherical voids and an infinite medium. The critical void growth ratio D_c is believed to be a material constant that does not change with geometry or loading conditions. However, it cannot be directly measured. Instead, a calibration procedure is carried out to obtain its value by matching the simulation-predicted results with experimental measurements for either a smooth or a precracked specimen under selected loading conditions. The value of D_c thus obtained is then used to predict the ductile fracture of the same material under other geometries and loading conditions.

5.3 Failure plane formation

As described in Chapter 1, the observed failure of a UD composite with a polymer matrix under compression transverse to fibers indicates that final failure occurs on a plane inclined to the loading direction. The inclination of this plane is not along the plane of maximum shear stress, that is, at 45 degree. This has prompted the assumption that the yielding on that inclined plane is governed by cohesion and friction acting on the plane. The criterion for this type of yielding is the classical Mohr–Coulomb criterion, expressed by Eq. (5.4):

$$\tau = c - \tan\phi \tag{5.4}$$

where τ is the shear stress acting on the critical plane, and c and ϕ are material constants representing cohesion and friction angle, respectively. This criterion is like the modified von Mises yield criterion, Eq. (5.2), if c is considered as the yield stress in pure shear and ϕ is viewed as the effect of hydrostatic stress. The angle made by the inclined failure plane with respect to the uniaxial compression is given by Eq. (5.5):

$$\beta = \pi/4 + \phi/2. \tag{5.5}$$

It should be noted that the Mohr–Coulomb criterion (5.4) is not a "point-failure" criterion like the von Mises criterion for initiation of yielding at a point, or for that matter, initiation of brittle cavitation at a point. Instead, it applies to the final stage of failure which has culminated from a sequence of failure events ranging from initiation of the first event (at a point) to progression of events into formation of the plane of final failure.

5.4 Conclusions

This chapter has focused on the initiation of failure within UD composites in the matrix and at the fiber–matrix interfaces. On continued loading beyond the point of failure initiation, a range of failure progression mechanisms occur. These are affected by the presence of constraint on a given UD composite layer (lamina) imposed by other layers in a laminate. Describing the failure process from the initiation until attainment of a critical condition that signifies loss of functionality of a laminated composite structure involves treating failure progression at different length scales. Subsequent chapters will treat multiscale modeling for this purpose.

The mechanisms of failure in the matrix have been separated as brittle and ductile failure initiation. The brittle failure initiation has been identified as cavitation when it occurs close to a fiber surface leading to fiber–matrix interface failure, while ductile failure initiation is attributed to yielding confined to the matrix. In both cases, the modeling considerations have been made in terms of the strain energy density at the considered point. Depending on the tri-axiality of the stress state, the mechanism triggered is assumed to be cavitation, if the dilatation component of the strain energy density is high, and ductile failure initiation is assumed when the distortional component dominates.

The formation of a crack in the polymer matrix by ductile failure mechanisms has been discussed, and one approach for treating this has been described as illustration. Finally, criteria related to failure plane formation have been discussed with emphasis placed on the fact that such criteria treat the final event of failure and are not to be viewed as describing failure initiation.

References

Arruda, E.M., Boyce, M.C., Jayachandran, R., 1995. Effects of strain rate, temperature and thermomechanical coupling on the finite strain deformation of glassy polymers. Mech. Mater. 19, 193−212.

Asp, L., Berglund, L.A., Talreja, R., 1996a. Prediction of matrix initiated transverse failure in polymer composites. Compos. Sci. Technol. 56, 1089−1097.

Asp, L., Berglund, L.A., Talreja, R., 1996b. A criterion for crack initiation in glassy polymers subjected to a composite-like stress state. Compos. Sci. Technol. 56, 1291−1301.

Donald, A.M., Kramer, E.J., 1982. The competition between shear deformation and crazing in glassy polymers. J. Mater. Sci. 17, 1871−1879.

Estevez, R., Tijssens, M.G.A., Van der Giessen, E., 2000. Modeling of the competition between shear yielding and crazing in glassy polymers. J. Mech. Phys. Solid. 48, 2585−2617.

Huang, H., Talreja, R., 2006. Numerical simulation of matrix micro-cracking in short fiber reinforced polymer composites: initiation and propagation. Compos. Sci. Technol. 66, 2743−2757.

Neogi, A., Mitra, N., Talreja, R., 2018. Cavitation in epoxies under composite-like stress states. Compos. Appl. Sci. Manuf. 106, 52−58.

Rice, J.R., Tracey, D.M., 1969. On the ductile enlargement of voids in triaxial stress fields. J. Mech. Phys. Solid. 17, 201−217.

Rottler, J., Robbins, M.O., 2001. Yield conditions for deformation of amorphous polymer glasses. Phys. Rev. 64, 051801.

Zhandarov, S., Mader, E., 2005. Characterization of fiber/matrix interface strength: applicability of different tests. Approaches and parameters. Compos. Sci. Technol. 65, 149−160.

Fiber–matrix debonding in composite materials: transverse loading

6

E. Correa, F. París and V. Mantič
School of Engineering, University of Seville, Seville, Spain

6.1 Introduction

Fiber-reinforced composite materials are habitually designed to work in the longitudinal direction (the direction of the fibers), and this makes them especially given to failure in the transverse direction. In particular, transverse failure is seen in the case of multidirectional laminates containing 90-degree plies and in the case of laminates subjected to impact loads.

Focusing on the first situation, the inclusion of 90-degree plies in multidirectional laminates is very common due to the extra stiffness they confer to the whole structure and their capacity for preventing it from splitting. However, as a consequence, they are the first plies within the laminates to show cracks due to the high stresses generated by the transverse loads acting upon them. The development of this type of failure grows to be very significant in the global damage mechanism of the whole laminate.

Alluding to the second situation, impact loads cause stresses that are transmitted to a given laminate in different directions, which in turn may provoke the generation of transverse failure at particular plies oriented in a specific direction within the laminate.

The mechanism of failure associated with transverse loads is commonly known as interfiber failure or matrix failure. The failure planes characterizing this mechanism are oriented parallel to the direction of the fibers, as schematically represented in Fig. 6.1. At a micromechanical level, broken specimens can be observed connecting the

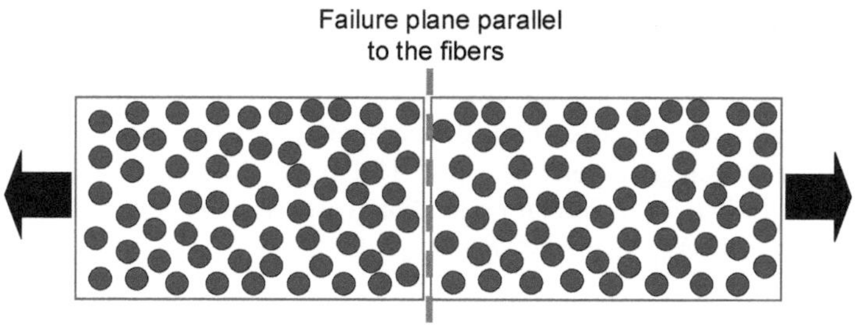

Figure 6.1 Diagram of the failure plane associated with matrix–interfiber failure in a unidirectional laminate.

Modeling Damage, Fatigue and Failure of Composite Materials. https://doi.org/10.1016/B978-0-443-18489-5.00016-3

generation of damage to the presence of debonds between the fibers and the matrix surrounding them (Gamstedt, 1997).

A profound knowledge of the generation of this mechanism of damage at the micromechanical level is essential for a more accurate and efficient prediction of failure occurrence (París, 2001). This aspect is becoming even more essential in view of the growth in usage of composite materials in primary structures, which in turn increases the weight of responsibility that these materials bear in such constructions.

In the present study, the investigation of interfiber failure under tension transverse to the fibers is undertaken, revisiting the different stages of the mechanism of damage identified by the authors (París et al., 2007) and analyzing the influence on it of a secondary transverse load (París et al., 2003; Correa et al., 2013) and of thermal residual curing stresses (Correa et al., 2011a). The studies were performed using the boundary element method (BEM) (París and Cañas, 1997) and employing interfacial fracture mechanics concepts (Mantič et al., 2006). Different aspects of the same mechanisms of damage have already been analyzed by means of finite fracture mechanics (Mantič, 2009). The mechanism of damage under compression has also been the authors' object of investigation in previous studies (Correa et al., 2008a,b, 2011b, 2012, 2014).

In this chapter, the main features of the numerical models employed, including the material properties and parameters under analysis, are described in Section 6.2. Sections 6.3−6.5 are devoted to the study of the three initial stages of the interfiber failure under (uniaxial) tension. Section 6.6 contains a summary of the main conclusions obtained in these sections. A numerical analysis aiming to study the development of interfiber failure under biaxial loads is included in Section 6.7. Finally, Section 6.8 focuses on the study of the effect of thermal residual curing stresses on the development of the mechanism of damage.

6.2 Micromechanical view: numerical model

The numerical study was carried out using a tool based on BEM that makes it possible to perform the numerical analysis of plane elastic problems taking contact and interface cracks into consideration. Two BEM models were used in this analysis. The basic model employed is shown in Fig. 6.2a and, using the plane strain hypothesis, represents the case of a crack that grows along the interface.

The second model employed (Fig. 6.2b) is used to study a different stage of growth in which the interface crack has kinked and extends through the matrix.

The energy release rate (ERR), G, is the parameter used to characterize the problem from the fracture mechanics point of view. The expression employed to calculate it, based on Irwin (1957), for a circular crack that grows from a certain debonding angle, θ_d (Fig. 6.2a) to $\theta_d + \Delta\theta_d$ ($\Delta\theta_d << \theta_d$), is:

$$G(\theta_d, \Delta\theta_d) = \frac{1}{2\Delta\theta_d} \int_{\theta}^{\theta_d+\Delta\theta_d} [\sigma_{rr}(\theta_d + \alpha)\Delta u_r(\theta_d - \Delta\theta_d + \alpha) + \sigma_{r\theta}(\theta_d + \alpha)\Delta u_\theta(\theta_d - \Delta\theta_d + \alpha)]d\alpha \tag{6.1}$$

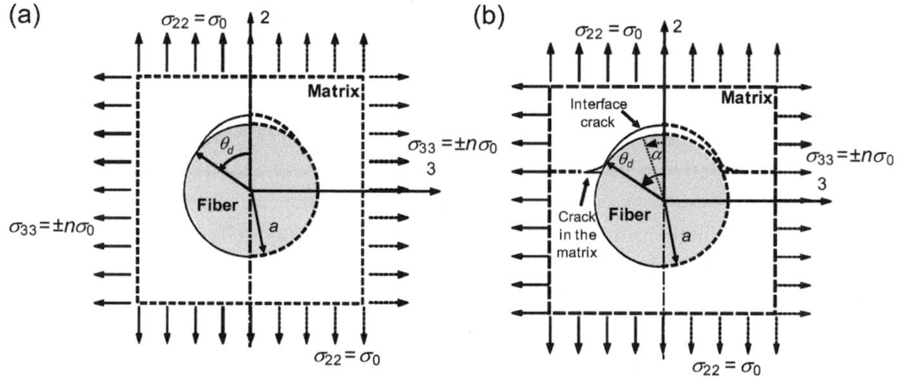

Figure 6.2 Single-fiber BEM model (a) with interface crack and (b) with kinked interface crack. Source: Correa et al. (2013).

where α is the circumferential coordinate with reference to the 2-axis. σ_{rr} and $\sigma_{r\theta}$ represent, respectively, radial and shear stresses along the interface. Δu_r and Δu_θ represent the relative displacements of the crack faces. Both modes of fracture, I (associated with σ_{rr}) and II (associated with $\sigma_{r\theta}$), are considered in Eq. (6.1). For this study, $\Delta\theta_d$ has been taken equal to 0.5 degrees.

The materials chosen for the analysis correspond to a typical configuration among fiber-reinforced materials: a glass fiber–epoxy matrix system, whose elastic properties are included in Table 6.1. The fiber radius considered was $a = 7.5 \times 10^{-6}$ m.

Dimensionless results for G will be presented. These values are obtained, following Toya (1974) and Murakami (1988), by dividing the values of G by $G_0 = (1 + \kappa^m / 8\mu^m)\sigma_0^2 a\pi$, where $\kappa^m = 3 - 4\nu^m$, μ^m is the shear modulus of the matrix, and σ_0 is the external applied tension.

6.3 Failure initiation

The first step in the micromechanical study of the generation of damage for the case of interfiber failure under tension is the detection of the origin of failure. In order to localize the origin of the damage, the stress state at the fiber–matrix interface under

Table 6.1 Elastic properties of the bimaterial system.

Material	Poisson coefficient, ν	Young's modulus, E
Matrix (epoxy)	$\nu^m = 0.33$	$E^m = 2.79 \times 10^9$ Pa
Fiber (glass)	$\nu^f = 0.22$	$E^f = 7.08 \times 10^{10}$ Pa

Source: Correa et al. (2013).

transverse tension (Goodier, 1933) is obtained analytically just before any damage is detected—that is, an undamaged interface situation.

The results obtained are presented in Fig. 6.3, where radial stresses, σ_{rr}, and shear stresses, $\sigma_{r\theta}$, at the interface are plotted versus the angular position, α (see Fig. 6.2b); for reasons of symmetry, only results for one-half of the fiber are presented.

It can be observed in Fig. 6.3 that the radial stress shows a tensile character affecting almost the entire boundary of the fiber, apart from an area of reduced size centered at $\alpha = 90$ degrees, where compressions of low value are acting. The maximum values of σ_{rr} take place at $\alpha = 0$ degrees and 180 degrees. It can also be observed that the radial stress remains quite constant in a zone corresponding to a value of α, of about 10 degrees from the aforementioned peaks. With reference to the $\sigma_{r\theta}$ distribution, it reaches its maximum value at $\alpha = 45$ degrees and 135 degrees; the value of these peaks is lower than the radial stress peaks (approximately 50% lower).

It is often difficult to find good tabulations on shear strength in the literature. With reference to epoxy resin, the shear strength values are usually a little bit lower than the tensile ones (Soden et al., 1998). Based on this, if both strengths are considered to be of the same order, the role of shear stresses is not underestimated. Taking also into account the results presented in Fig. 6.3, it can be concluded that the maximum radial stress–maximum shear stress ratio is, in any case, higher than the tensile strength–shear strength ratio.

Therefore, it is quite reasonable to assume that, when the radial stress reaches the rupture value of the interface in tension, it will produce a debond from which interface fracture mechanics is able to control the growth of such a defect. The conclusions obtained from the subsequent analysis are obviously independent of the specific value of the initial debonding angle taken within the range of initial values considered.

Based on the former reasoning and choosing arbitrarily from the two possible areas of damage initiation ($\alpha = 0$ degrees, 180 degrees), a 20 degrees length debond

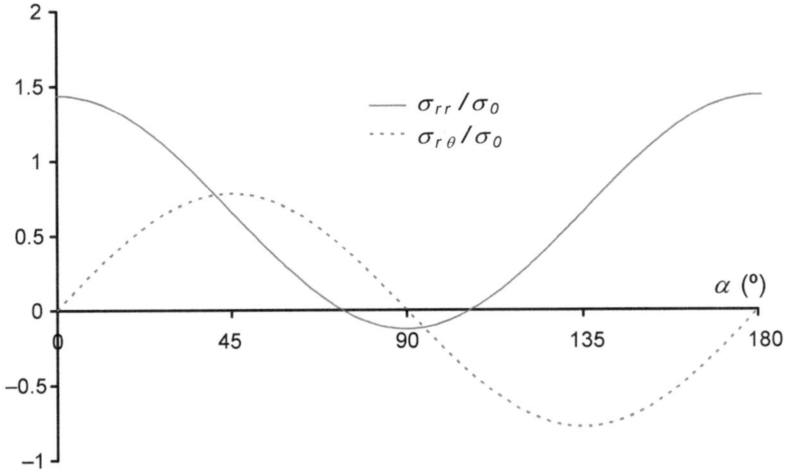

Figure 6.3 Radial and shear stress (σ_{rr} and $\sigma_{r\theta}$) distribution at the undamaged interface.

centered at 0 degrees (10 degrees debonding angle, θ_d, see Fig. 6.2a) was considered as the smallest initial debond, and its growth is studied in the next section by means of the BEM model presented in Fig. 6.2a.

6.4 The interface crack

6.4.1 Energy release rate

This section deals with the energetic study of the growth along the interface of the assumed initial debond under (uniaxial) tension (T-0 case in what follows). In this way, the calculation of G versus θ_d is followed. To this end, the single-fiber BEM model presented in Fig. 6.2a, which considers the sole application of σ_{22}, is employed. The results obtained are shown in Fig. 6.4.

For the bimaterial system considered in this study, the maximum value of G occurs at $\theta_d = 60$ degrees, although it is roughly constant along the interval of θ_d between 50 and 70 degrees. When this peak is reached, it coincides with the detection of a physically relevant near-tip contact zone.

With reference to the distribution of modes, G_I is only significant versus G_{II} for small values of θ_d; in these cases, the crack tips are oriented almost perpendicularly to the external load direction. As the debond grows, G_I starts to disappear and G_{II} dominates. This happens for $\theta_d \geq 30$ degrees, coinciding with the appearance of an extremely small near-tip contact zone, which is detected by the present numerical model.

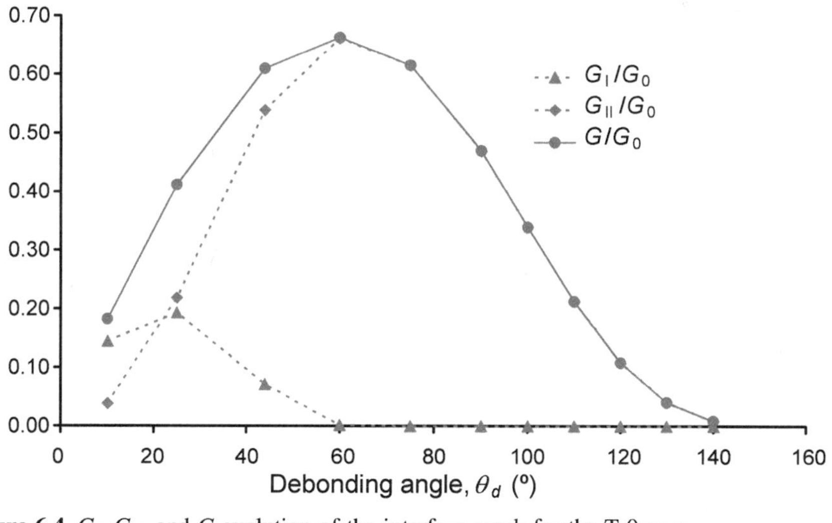

Figure 6.4 G_I, G_{II}, and G evolution of the interface crack for the T-0 case.

Looking again at the G evolution, it seems clear that a potential growth of the crack in pure Mode II (from $\theta_d = 60$ degrees onwards) would have stable character. This deduction is based on the decreasing shape of G versus θ_d for $\theta_d > 60-75$ degrees.

6.4.2 Prediction of growth

In order to perform energetic predictions about the interface crack growth, it is, first of all, necessary to have an estimation of the critical value of G, G_c. The law considered in this study for G_c is based on the simplified empirical proposal by Hutchinson and Suo (1992):

$$G_c(\psi_K) = G_{1c}\left(1 + \tan^2(1 - \lambda)\psi_K\right) \tag{6.2}$$

where G_{1c} is the critical value of G_c for Mode I, λ is the fracture mode sensitivity parameter, and ψ_K is the local phase angle. This angle has been calculated following Mantič and París (2004):

$$\psi_K = 0.5 \text{ arccos}\left[F(\varepsilon)^{-1}\frac{G_\text{I} - G_\text{II}}{G_\text{I} + G_\text{II}}\right] \tag{6.3}$$

where $F(\varepsilon) = 1 + (\pi^2/3 - 2)\varepsilon^2 + O(\varepsilon^4)$ and ε is the oscillatory index defined as:

$$\varepsilon = \frac{1}{2\pi}\ln\frac{1 - \beta}{1 + \beta} \tag{6.4}$$

β being Dundurs parameter (Dundurs, 1968). For the bimaterial system employed, $\varepsilon = -0.074$.

Therefore, G_c depends on the evolution of the fracture mode mixity, which in the case of the interface crack means that it is a function of θ_d.

$\psi_K(\theta_d)$ is schematically represented in Fig. 6.5 for θ_d values within the range 5–75 degrees. As observed in this figure, $\psi_K = 90$ degrees, associated with pure Mode II, is reached for a value of θ_d slightly greater than 60 degrees. This result was expected in view of the evolutions of G_I and G_II represented in Fig. 6.4.

In order to evaluate G_c, three different values for parameter λ were selected: 0.2, 0.25, and 0.3. The G_{1c} value, based on the dimensionless character of the problem, was fixed to reach (for each value of λ considered) the equality $G = G_c(\psi_K)$ at the initial stage of growth, that is, $\theta_d = 5$ degrees.

Once the evolution of both G and G_c is known, and based on the relationship between parameters from fracture mechanics stress and energetic approaches, it is possible to calculate the value of the critical load, σ_c, which produces the growth of the interface crack. The expression employed is the following:

$$\frac{\sigma_c}{\sigma_0} = \frac{\sqrt{G_c}}{\sqrt{G}} \tag{6.5}$$

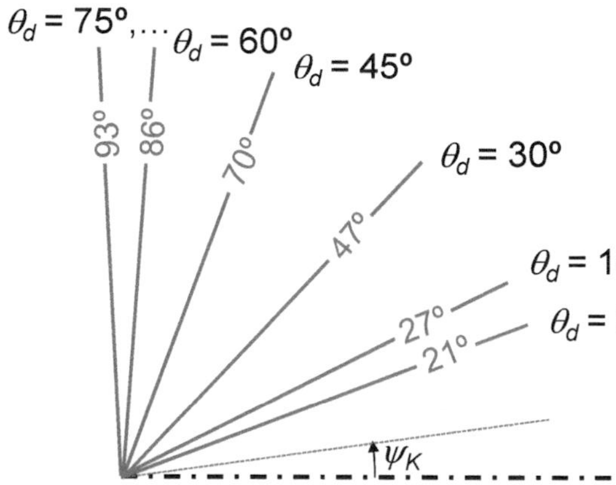

Figure 6.5 ψ_K evolution versus θ_d for the T-0 case.

The results obtained are shown in Fig. 6.6, where it can be checked that σ_c is lower than σ_0 within the θ_d ranges detailed in Table 6.2 for the three values of λ considered. These ranges determine the corresponding intervals of unstable growth of the crack along the interface.

For debonding angles greater than those specified in Table 6.2, the growth character becomes stable, which means that an external load increment would be necessary in order to promote additional crack extension.

Focusing on the intermediate value of λ chosen, that is, 0.25, this change in growth character from unstable to stable in the neighborhood of $\theta_d \cong 60-70$ degrees (see Table 6.2) favors the change to a different stage of the mechanism of damage. This new

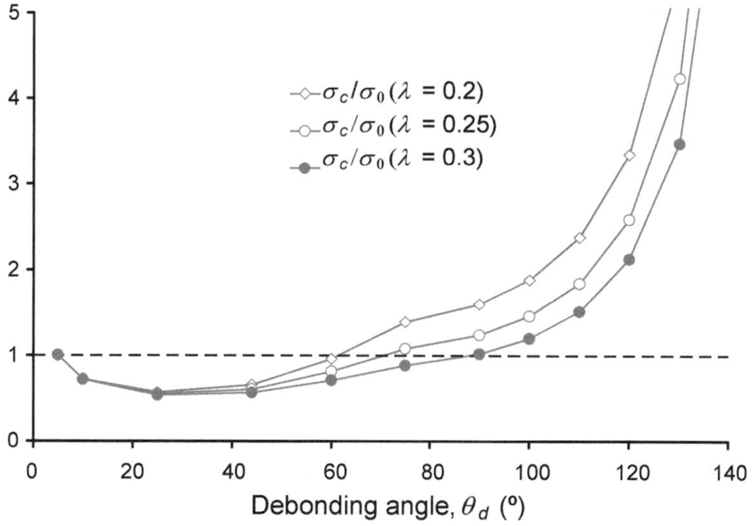

Figure 6.6 Critical load, σ_c, versus the debonding angle, θ_d, for the T-0 case.

Table 6.2 Results of the predicted intervals of unstable growth of the interface crack.

λ	**Interval of unstable growth**
0.2	$\theta_d = 5-60$ degrees
0.25	$\theta_d = 5-70$ degrees
0.3	$\theta_d = 5-82$ degrees

stage consists of the change of crack path, that is, kinking into the matrix and its further growth through it, which is studied in the next section.

6.5 Growth through the matrix

The prediction of the interface crack kinking toward the matrix, once the period of unstable growth at the interface has finished, consists of two steps: (1) the search for the preferential direction of the incipient crack in the matrix and (2) the evaluation of the energetic possibility of this change.

6.5.1 *Kinking orientation*

With reference to the first step, the application of a kinking criterion, such as the maximum circumferential stress criterion (Erdogan and Sih, 1963), gives

$$\sigma_{\theta\theta}^{\max} = \sigma_{\theta\theta}(r, \theta_{\text{kink}}) = max_{\theta}\sigma_{\theta\theta}(r, \theta) \tag{6.6}$$

in the neighborhood of the interface crack tip within the θ_d range of termination of unstable growth, allowing the prediction of the most favorable direction of the incipient crack to penetrate into the matrix, θ_{kink}.

$\sigma_{\theta\theta}$ distribution versus θ (Fig. 6.7a) needs to be evaluated from the results of the BEM model (Fig. 6.2a) at a certain distance, r, from the crack tip. This distance needs to be small enough for the stresses to be able to control the possible change of direction of the crack, but at the same time large enough to maintain the physical meaning of the matrix as a continuum medium. Two values for parameter r satisfying these conditions were considered, specifically: $r = 0.001a$ and $r = 0.01a$, a being the fiber radius.

The results associated with both values of r for each debonding angle define a range of maximum circumferential stress, which has been drawn up in graphical form in Fig. 6.7b. The results show a weak dependence on r, and the direction of maximum circumferential stress, θ_{kink}, reaches the expected macromechanical orientation of failure, 90 degrees (Fig. 6.1), for $\theta_d \cong 60-70$ degrees; that is, it is within the end-range of unstable growth of the interface crack.

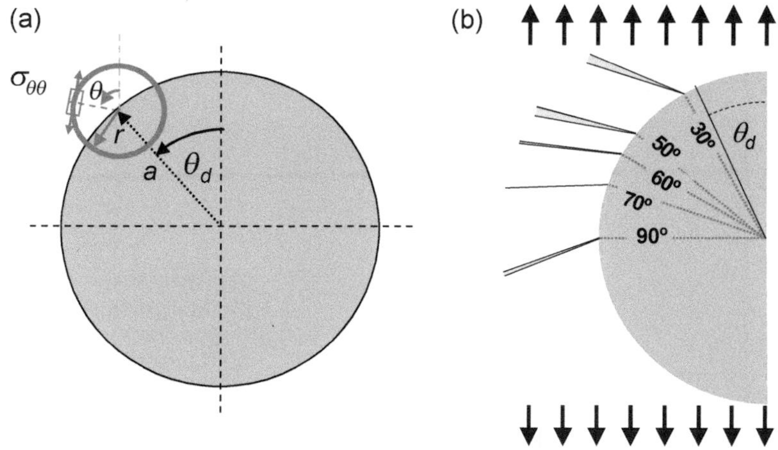

Figure 6.7 (a) Circumferential stress definition in the neighborhood of the interface crack tip. (b) Evolution of the direction of maximum circumferential stress versus the debonding angle.

Additionally from the results obtained, it can also be seen that, for both values of the inspection radius, the maximum value of the circumferential stress for all values of the debonding angle considered takes place at $\theta_d = 60$ degrees, though this value is also approximately achieved by the maximum associated with $\theta_d = 50$ degrees and $\theta_d = 70$ degrees.

6.5.2 Prediction of growth

Once the potential kinking direction has been estimated, it is necessary to study the possibility of kinking from an energetic point of view.

To this end, the G value was calculated for an incipient crack in the matrix, oriented parallel to the direction perpendicular to the external tension applied, G^{kink}. This study was performed using the BEM model shown in Fig. 6.2b.

The results of this study are presented in Fig. 6.8, where G_{I}, G_{II}, and G of the incipient-kinked crack in the matrix (i.e., $G_{\text{I}}^{\text{kink}}$, $G_{\text{II}}^{\text{kink}}$, and G^{kink}) are plotted against θ_d. The evolution of G of the interface crack, G^{int}, is also included in the figure. The results obtained show that, for the whole range of positions along the interface analyzed, G^{kink} is greater than the value corresponding to the interface crack, G^{int}. In addition, the maximum value of G^{kink} occurs at $\theta_d = 60$ degrees (a position at the interface within the end-range of unstable interfacial growth, and therefore a position of expected kinking occurrence).

Moreover, for $\theta_d = 60$ degrees, the evaluation in energetic terms of a further growth through the matrix after kinking, following the same transverse direction, shows that matrix crack extension would be unstable and have Mode I character, as shown in Fig. 6.9.

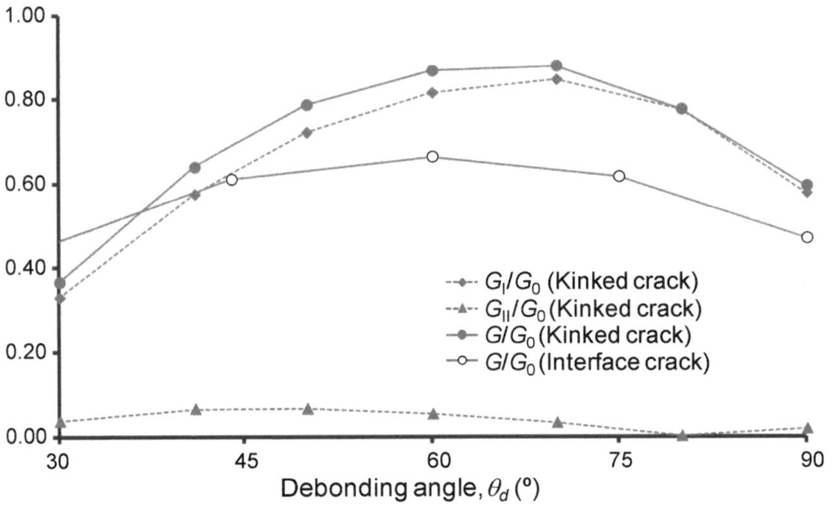

Figure 6.8 Comparison between the energy release rate of the kinked crack and the interface crack for the T-0 case.

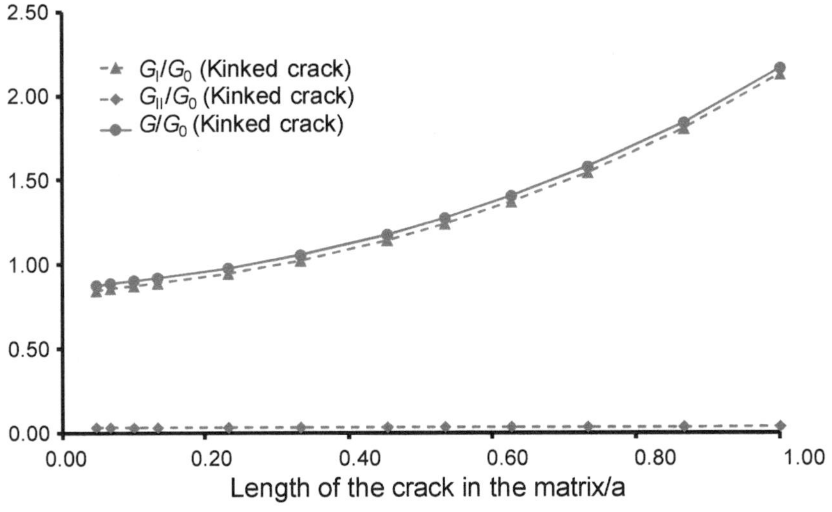

Figure 6.9 G_{I}, G_{II}, and G evolution of the kinked crack in the matrix for the T-0 case. Source: Mantič et al. (2006).

Once the evolution of G^{kink} is known, He and Hutchinson criterion (1989) can be employed to make a prediction about crack growth in the matrix. Based on this criterion, growth out of the interface (and thus across the matrix) would occur if the following expression is fulfilled:

$$\frac{G^{\mathrm{int}}}{G_c^{\mathrm{int}}} < \frac{G^{\mathrm{kink}}}{G_c^{\mathrm{kink}}} \tag{6.7}$$

where G_c^{int} is the critical value of the ERR of the interface crack, and G_c^{kink} is the critical value of the ERR of the kinked crack in the matrix.

Focusing on the $\theta_d = 60$ degrees case and taking into account the results already obtained, it can be assumed that $G^{\text{int}}(\theta_d = 60 \text{ degrees}) \cong G_{\text{II}}^{\text{int}}(\theta_d = 60 \text{ degress})$ and $G^{\text{kink}}(\theta_d = 60 \text{ degrees}) \cong G_{\text{I}}^{\text{kink}}(\theta_d = 60 \text{ degrees})$. Thus, Eq. (6.7) particularized for the $\theta_d = 60$ degrees, becomes:

$$\frac{G_{\text{II}}^{\text{int}}}{G_{\text{II}c}^{\text{int}}} < \frac{G_{\text{I}}^{\text{kink}}}{G_{\text{I}c}^{\text{kink}}} \tag{6.8}$$

Besides, as expressed in Fig. 6.8, $G_{\text{II}}^{\text{int}}(\theta_d = 60 \text{ degrees}) \langle G_{\text{I}}^{\text{kink}}(\theta_d = 60 \text{ degrees})$.

With reference to the critical values of the ERR included in the former expression, G_c^{int} and G_c^{kink}, and for the bimaterial system under study, it can be assumed that $G_{\text{II}c}^{\text{int}} \geq G_{\text{I}c}^{m}$. This assumption is based on the fact that $G_{\text{II}c}^{\text{int}}$ values found in the literature (Wagner et al., 1995; Varna et al., 1997; Kim and Nairn, 2002) cover a wide range, 2—220 J/m^2 (though most authors tend toward the greater values of the interval), whereas the $G_{\text{I}c}^{m}$ values of an epoxy resin are included in the range 30—80 J/m^2.

Finally, based on the relative values of G and G_c for the interface and the matrix and the different character of crack growth for both possibilities (unstable growth for the kinked crack in the matrix and stable growth along the interface in the range of interest), the kinking of the interface crack toward the matrix is expected, once this crack has reached a stable position at the interface, in the direction perpendicular to the external tension applied, σ_{22} (Fig. 6.2).

6.6 Micromechanical stages of the mechanism of damage under tension

The results derived from the analysis performed in the previous sections allows three different stages of the mechanism of damage under tension to be identified. They are graphically depicted in Fig. 6.10 and explained in what follows.

Stage 1: Failure initiation

Assuming that the interface between fiber and matrix is initially free of defects, crack nucleation is supposed to be governed by the radial stress generated between fiber and matrix. The maximum values of this stress are detected at the angles 0 and 180 degrees, which refer to the direction of the external tension applied. Therefore, the first debonds at the interfaces are predicted to appear at any of these positions.

Stage 2: Interface crack growth

The initial debonds grow symmetrically and unstably along the interfaces and then become interface cracks. This stage finishes when these cracks reach a certain length at the interface (corresponding to a debonding angle of 60—70 degrees, that is, a total length of 120—140 degrees), coinciding with the appearance of a physically relevant contact zone at the interface crack tips.

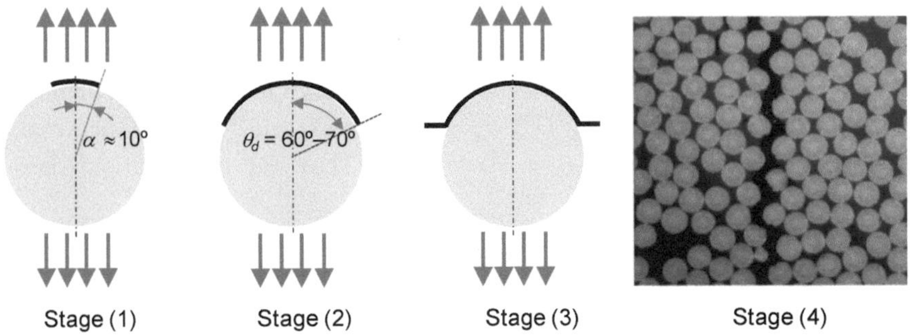

Figure 6.10 Micromechanical stages identified for the interfiber failure under tension.

Stage 3: Kinking toward the matrix

Once a stable position at the interface is reached, the crack abruptly changes its direction of growth, kinking into the matrix and following the direction perpendicular to the load applied.

Stage 4: Macrofailure

The coalescence between different cracks in the matrix coming from different interfaces may lead to final macrofailure (i.e., transverse failure), which is supported by experimental observations of specimens tested under uniaxial tension seen through the microscope.

6.7 Effect of a secondary transverse load

Many of the existing proposals for the prediction of interfiber failure at lamina level are based on the hypothesis that the failure taking place at a plane is governed by the components of the stress vector associated with that plane (Hashin, 1980). In the present section, this assumption is revised for the interfiber failure under tension. An analysis of the influence of an out-of-failure plane stress component (tension or compression) on the generation of the damage dominated by a transverse tension is carried out.

The key idea is to check if the presence of an out-of-(assumed) plane of failure external traction σ_{33} (Fig. 6.2) acting simultaneously with the tension nominally responsible for the failure, σ_{22} alters in any way the initial stages of the mechanism of damage identified for the uniaxial case studied in the previous sections (Fig. 6.10). To this end, the methodology implemented in the study of the interfiber failure under uniaxial tension (Sections 6.3–6.5) was followed in order to study the development of the mechanism of damage in the case of biaxial load.

The notation employed to distinguish between the different biaxial cases is as follows: T-nC and T-nT, where the first T represents the tension nominally responsible for the failure applied parallel to axis 2 (i.e., σ_{22}), n (see Fig. 6.2) is the coefficient that specifies the value of the secondary traction (i.e., σ_{33}) with reference to σ_{22} ($0 \leq n \leq 1$), and the second T or C represents the sign of the secondary traction (T = tension, C = compression).

Next, some of the most noteworthy conclusions of the study performed are briefly reported. In addition, they are illustrated (only from a qualitative point of view) in Fig. 6.11 for two representative cases of the T-nT and T-nC situations (in particular, T-0.5T and T-C), and the reference case, T-0.

Stage 1: Failure initiation

The appearance of the first debonds at the interface seems to be controlled by the radial stress for both the T-nT and the T-nC cases, as occurred for the T-0 case. Assuming an initial debond of length α for the T-0 case (Fig. 6.11), a slightly greater initial debond is predicted for the T-nT cases versus a slightly smaller one for the T-nC ones. With reference to the location of the initial debond, it coincides with the T-0 one for both biaxial cases, with the exception of the limit case T–T (any point of the interface could be the origin of damage in this case). Finally on analyzing the load level required for the generation of the first debond, a tensile secondary load (T-nT cases) seems to minimally reduce the risk of failure, whereas a compressive secondary load (T-nC cases) slightly increases that risk.

Stage 2: Interface crack growth

In this step of the analysis, the most interesting results are obtained. In particular, the interface crack growth is predicted to be larger for the T-nT cases and slightly smaller for the T-nC cases (Fig. 6.11). In addition, crack growth occurs at a lower level of the external load in the T-nC cases and at an almost undetectable higher load level in the T-nT ones. This conclusion is inferred from Figs. 6.12 and 6.13, where the evolutions of G versus θ_d for the T-nT and T-nC cases are included. Crack morphology also suffers some relevant alterations, mainly associated with the delay in the appearance of the finite contact zone at the crack tip in the T-nT cases, and with its advance in the T-nC ones.

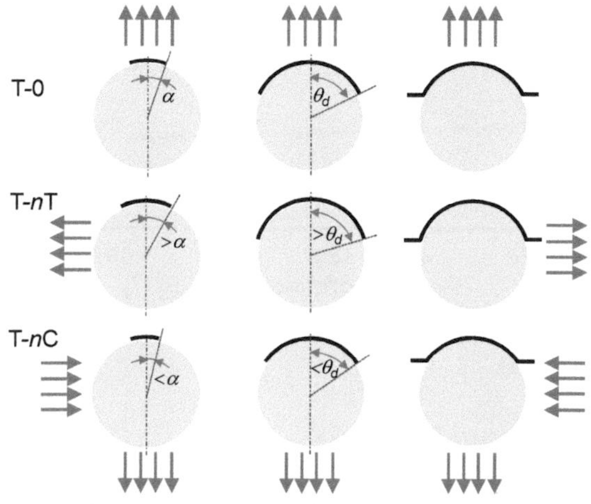

Figure 6.11 Stages comparison between the biaxial cases (T-nT and T-nC) and the uniaxial case (T-0).
Source: Correa et al. (2013).

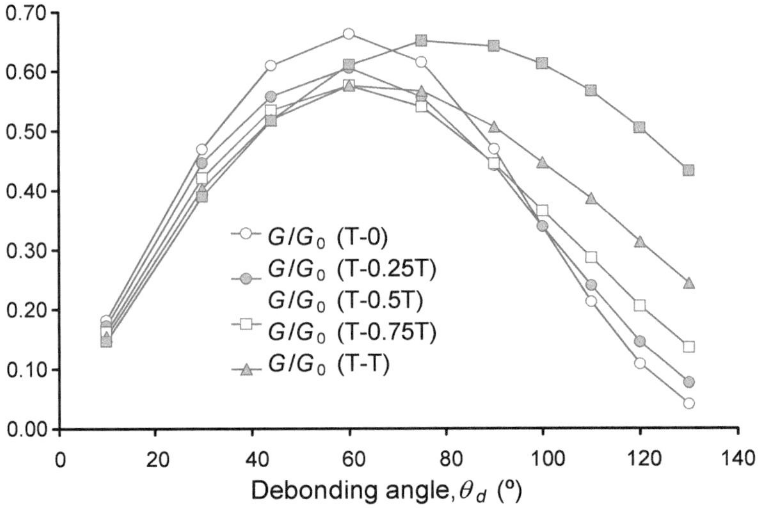

Figure 6.12 G evolution versus θ_d of the T-nT cases ($n = 0, 0.25, 0.5, 0.75, 1$).
Source: Correa et al. (2013).

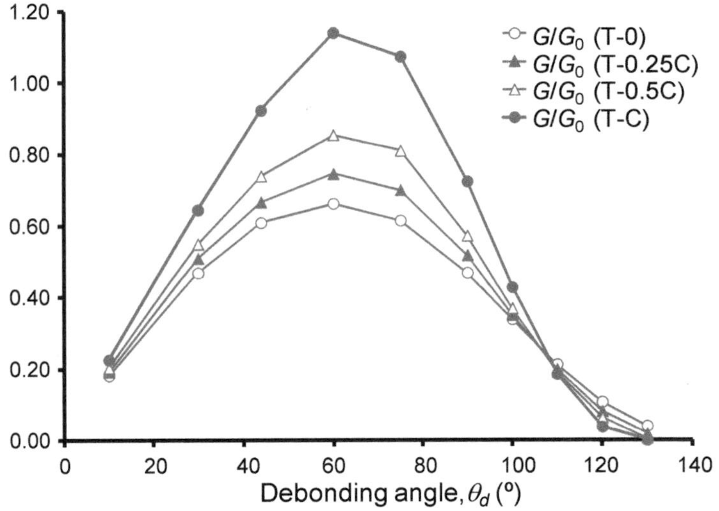

Figure 6.13 G evolution versus θ_d of the T-nC cases ($n = 0, 0.25, 0.5, 1$).
Source: Correa et al. (2013).

Stage 3: Kinking toward the matrix

This phase shows no relevant differences between both biaxial cases (T-nT and T-nC) and the uniaxial one; thus, kinking toward the matrix is predicted in the direction perpendicular to the external tension nominally responsible for the failure, σ_{22}, and within the corresponding end-range of unstable growth at the interface for each case (Fig. 6.11).

As a global conclusion to the study performed in this section, it can be stated that the results obtained predict the influence of the stress parallel to the plane of failure on the appearance of such failure. Specifically, a slightly inhibitive effect of the external secondary tensile load against failure is foreseen, versus an accelerative effect of the external secondary compressive load.

The aforementioned conclusions relative to the T-*n*C and T-nT cases are in agreement with the previous experimental evidence obtained by the authors (París et al., 2003; Zumaquero et al., 2022).

6.8 Effect of thermal residual stresses

The final stage of the manufacturing process of fibrous composite materials is the curing of the material. The differences in free contraction between the fibers and the matrix during curing cause the appearance of thermal residual stresses. The presence of these stresses may affect the strength of the laminate and also influences the development of failure mechanisms in the material.

The studies presented in Sections 6.3—6.6 have made it possible to understand the initial stages of the interfiber failure under uniaxial tension but without considering the presence of thermal residual stresses. In order to complete the study of this mechanism of damage, this section analyzes the influence of these stresses (produced by the cooling down associated with the curing process) on the conclusions obtained so far concerning the different micromechanical stages identified in the above sections.

The BEM tool employed in the investigations presented in this study (see Section 6.2) is able to include the presence of a thermal decrease or increase affecting the solids included in the model. In this sense, the curing process used for the matrix considered in this work (epoxy resin) commonly consists of a first step at room temperature followed by one or more steps at a higher temperature. Independently of the temperature at which the curing process takes place, the contraction of the matrix (always higher than that of the fiber) dominates the generation of thermal residual stresses. This fact allows the inclusion of thermal residual stresses in the analysis to be modeled by means of an adequate temperature decrease that captures the real curing contraction of the material. In the present study, a longitudinal transverse contraction of 0.4% has been examined, which corresponds to an 80°C decrease, París et al. (1998).

The methodology already implemented in the study of the interfiber failure under (uniaxial) tension has been followed again in this section to analyze the effect of the presence of thermal residual stresses on each stage of damage. The main conclusions are summarized below.

Stage 1: Failure initiation

Thermal residual compressive radial stresses are detected at the undamaged interface when analyzing its stress state generated by a thermal decrease. This result is obtained from the solution of the analytical problem presented in Correa et al. (2013), that is, a plane problem of a fiber embedded in an infinite matrix. The presence of these compressive stresses would inhibit the appearance of the interfiber failure, since they involve an apparent additional strength having to be surmounted by the stress state to produce the first debonds. The location of these debonds would not be altered, as it coincides with that represented in Fig. 6.10.

Stage 2: Interface crack growth

The results obtained, in terms of G versus the debonding angle θ_d, are presented in Fig. 6.14 for the case under study (named as $\Delta T = -80°C$ in the figure and which considers both thermal residual stresses and those derived from the load σ_0) and also for the case of the single action of σ_0 ($\Delta T = 0°C$ in the figure, i.e., the uniaxial case studied in Sections 6.3–6.5). The evolutions observed for both cases are similar, not only for the global value G but also for the modes distribution, though lower values in all evolutions are obtained for the $\Delta T = -80°C$ case. This result makes it possible to predict a protective effect of the thermal residual stresses on the interface against the evolution of the interface crack; this coincides with the defensive action against the initiation of failure already deduced in Stage 1. The results obtained also predict unstable growth for the interface cracks up to a debonding angle contained in the range 50–70 degrees.

Stage 3: Kinking toward the matrix

The study of the kinking toward the matrix for the $\Delta T = -80°C$ case provides results that coincide with those obtained for the reference case, that is, $\Delta T = 0°C$. Thus, the most favorable orientation of kinking, for a debonding angle within the end-range of unstable growth at the interface, coincides with the direction perpendicular to the external load. Moreover, the comparison between the energetic state of the interface crack and the kinked crack favors the kinking toward the matrix.

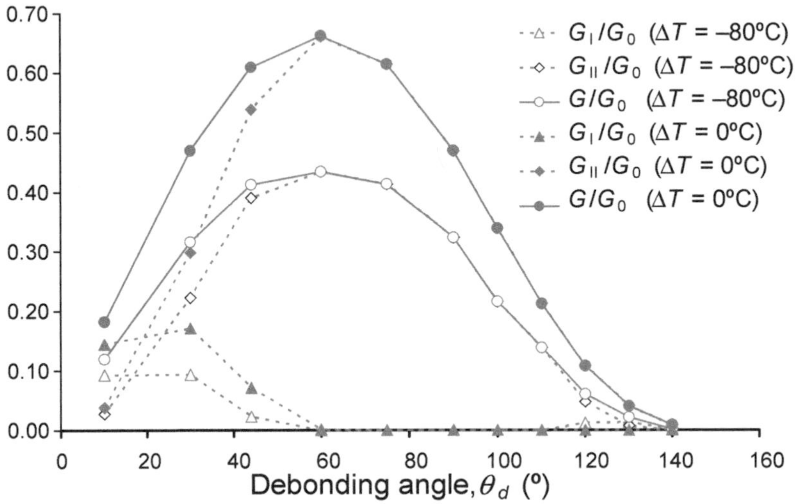

Figure 6.14 G_I, G_{II}, and G evolution of the interface crack for the T-0 case ($\Delta T = 0°C$ and $\Delta T = -80°C$ cases).
Source: Correa et al. (2011).

As a global conclusion of the results presented in this section, the presence of thermal residual curing stresses reduces the risk of interfiber failure development under tension at micromechanical level.

Macromechanical tests were performed in order to experimentally check this global conclusion. To this end, two groups of 90 degrees unidirectional specimens were manufactured, each of them cured following a different curing cycle with different curing temperatures (177 and 121°C) and later tested under uniaxial tension.

The curing at different temperatures promotes a different level of thermal residual stresses, which, based on the conclusions derived from the numerical model (already presented), could make the strength differ between the two sets of specimens (Correa et al., 2011a). The selection of the duration of each step of the curing cycles was carefully performed in order to ensure a similar degree of curing between both sets of specimens after the process. This fact was checked by measuring the glass transition temperature using a dynamic mechanical analyzer. The results of this test were 149 and 156°C for the laminates cured at 121 and 177°C, respectively. Later, during the mechanical tests, this point was also confirmed through the measurement of the elasticity modulus of both sets of specimens (presented in Table 6.3). This table includes the mean values, μ, and also the standard deviation, σ.

The strength results measured in the tests are compiled in Table 6.4. It can be seen that the mean value associated with the specimens cured at 177°C is 27% higher than that of the specimens cured at 121°C. As a consequence, and in agreement with the conclusion inferred from the micromechanical numerical results, the macromechanical experimental evidence shows that higher thermal residual stresses are associated with a higher value of transverse strength in unidirectional laminates, and this in turn has a protective effect against failure occurring.

Table 6.3 Values of the elasticity modulus, E, measured for specimens cured following two different cycles.

	E (MPa)	
Curing temperature (°C)	μ	σ
177	8210	828
121	8507	913

Table 6.4 Values of the tensile strength, Y_T, measured for specimens cured following two different cycles.

	Y_T (MPa)	
Curing temperature (°C)	μ	σ
177	48.6	5
121	38.4	4.9

6.9 Conclusions

Tension-dominated interfiber failure has been studied at micromechanical level by means of single-fiber BEM models and in the light of interfacial fracture mechanics. Three stages of damage have been identified (failure initiation, interface crack growth, and kinking toward the matrix), and their main features are established.

A complementary numerical study investigating a secondary external transverse load acting simultaneously with the tension nominally responsible for failure was performed in order to clarify if out-of-failure plane stresses play any role in the development of the interfiber failure. The results obtained demonstrate that the presence of this secondary load alters several aspects of the mechanism of damage under study, leading to this global conclusion: The generation of failure seems to be slightly inhibited in the presence of an external secondary tensile load and accentuated in the presence of an external secondary compressive one. This conclusion agrees with the experimental macromechanical evidence obtained so far, which questions whether failure at a plane is only controlled by the stress vector associated with that plane.

A final numerical study aimed at identifying the role of thermal residual stresses (generated during the cooling down associated with the curing process of the composite material) on the generation of the interfiber failure under tension has been presented. The effect of the presence of these thermal stresses on the three stages of the mechanism of damage has been studied. The outcomes derived from the numerical analysis predict a protective effect of the thermal residual curing stresses against failure. Results from macromechanical experimental tests supporting this conclusion have also been presented.

Finally, the results presented here are the basis for further, more complex investigations, involving several regularly or randomly distributed fibers (Sandino et al., 2016; Távara et al. 2016).

Acknowledgments

This study was supported by the Spanish Ministry of Education and Science/Economy and Competitiveness, Junta de Andalucía, and European Social Fund (Projects MAT2013-45069-P, MAT2012-37387, DPI 2012–37187, TEP-4051, and TEP-7093).

References

Barroso, A., Correa, E., Freire, J., Pérez, M.D., París, F., 2012. Biaxial testing of composites in uniaxial machines: manufacturing of a device, analysis of the specimen geometry and preliminary experimental results. In: ECCM15-European Conference on Composite Materials, Venice (Italy), 24–28 June.

Correa, E., Mantič, V., París, F., 2008a. Numerical characterisation of the fibre–matrix interface crack growth in composites under transverse compression. Eng. Fract. Mech. 75 (14), 4085–4103.

Correa, E., Mantič, V., París, F., 2008b. A micromechanical view of inter-fibre failure of composite materials under compression transverse to the fibres. Compos. Sci. Technol. 68 (9), 2010–2021.

Correa, E., Mantič, V., París, F., 2011a. Effect of thermal residual stresses on matrix failure under transverse tension at micromechanical level: a numerical and experimental analysis. Compos. Sci. Technol. 71 (5), 622–629.

Correa, E., París, F., Mantič, V., 2011b. BEM analysis of inter-fibre failure under compression in composites: comparison between carbon and glass fibre systems. Plast., Rubber Compos. 40 (6–7), 333–341.

Correa, E., París, F., Mantič, V., 2012. Effect of thermal residual stresses on the matrix failure under transverse compression at micromechanical level – a numerical and experimental study. Compos. Part A Appl. Sci. 43 (1), 87–94.

Correa, E., Mantič, V., París, F., 2013. Effect of the presence of a secondary transverse load on the inter-fibre failure under tension. Eng. Fract. Mech. 103, 174–189.

Correa, E., París, F., Mantič, V., 2014. Effect of a secondary transverse load on the inter-fibre failure under compression. Compos. B Eng. 65, 57–68.

Dundurs, J., 1968. Discussion of a paper by D.B. Bogy. J. Appl. Mech. 36, 650–652.

Erdogan, F., Sih, G.C., 1963. On the crack extension in plates under plane loading and transverse shear. J. Basic Eng. 85, 519–527.

Gamstedt, E.K., 1997. Fatigue Damage Mechanisms in Polymer Matrix Composites (Ph.D. Thesis). Lulea University of Technology.

Goodier, J.N., 1933. Concentration of stress around spherical and cylindrical inclusions and flaws. Appl. Mech. 55, 39–44.

Hashin, Z., 1980. Failure criteria for unidirectional fiber composites. J. Appl. Mech. 47, 329–334.

He, M.Y., Hutchinson, J.W., 1989. Crack deflection at an interface between dissimilar elastic materials. Int. J. Solid Struct. 25 (9), 1053–1067.

Hutchinson, J.W., Suo, Z., 1992. Mixed mode cracking in layered materials. Adv. Appl. Mech. 29, 63–191.

Irwin, G.R., 1957. Analysis of stresses and strain near the end of a crack transversing a plate. J. Appl. Mech. 24, 361–364.

Kim, G.W., Nairn, J.A., 2002. Experimental verification of the effects of friction and residual stress on the analysis of interfacial debonding and toughness in single fiber composites. J. Mater. Sci. 37 (18), 3965–3972.

Mantič, V., París, F., 2004. Relation between SIF and ERR based measures of fracture mode mixity in interface cracks. Int. J. Fract. 130, 557–569.

Mantič, V., Blázquez, A., Correa, E., París, F., 2006. Analysis of interface cracks with contact in composites by 2D BEM. In: Guagliano, M., Aliabadi, M.H. (Eds.), Fracture and Damage of Composites. WIT Press, Southampton, pp. 189–248.

Murakami, Y., 1988. Stress Intensity Factor Handbook. Pergamon Press, Oxford.

Mantič, V., 2009. Interface crack onset at a circular cylindrical inclusion under a remote transverse tension. Application of a coupled stress and energy criterion. Int. J. Solid Struct. 46, 1287–1304.

París, F., Cañas, J., 1997. Boundary Element Method. Fundamentals and Applications. OUP, Oxford.

París, F., del Caño, J.C., Varna, J., 1998. BEM analysis of the contact problem in fibres debonded of a matrix. Effect of curing stresses. In: Kassub, A., Brebbia, C.A., Chopra, M. (Eds.), Boundary Elements XX. CMP, pp. 145–156.

París, F., 2001. A Study of Failure Criteria of Fibrous Composite Materials. NASA/CR-2001-210661.

París, F., Correa, E., Cañas, J., 2003. Micromechanical view of failure of the matrix in fibrous composite materials. Compos. Sci. Technol. 63, 1041–1052.

París, F., Correa, E., Mantič, V., 2007. Kinking of transversal interface cracks between fiber and matrix. J. Appl. Mech. 74, 703–716.

Sandino, C., Correa, E., París, F. Numerical analysis of the influence of a nearby fibre on the interface crack growth under transverse tensile load. Eng. Fracture Mech. 2016; 168B: 58–75.

Soden, P.D., Hinton, M.J., Kaddour, A.S., 1998. Lamina properties, lay-up configurations and loading conditions for a range of fibre-reinforced-composite laminates. Compos. Sci. Technol. 58, 1011–1022.

Távara, L., Mantic, V., Graciani, E., París, F., 2016. Modelling of interfacial debondings in unidirectional fibre–reinforced composites under biaxial transverse loads. Compos. Struct. 136, 305–312.

Toya, M., 1974. A crack along the interface of a circular inclusion embedded in an infinite solid. J. Mech. Phys. Solid. 22, 325–348.

Varna, J., Berglund, L.A., Ericson, M.L., 1997. Transverse single fibre test for interfacial debonding in composites 2: modelling. Composites Part A 28, 317–326.

Wagner, H.D., Nairn, J.A., Detassis, M., 1995. Toughness of interfaces from initial fiber-matrix debonding in a single fiber composite fragmentation test. Appl. Compos. Mater. 2, 107–117.

Zumaquero, P.L, Correa, E., Justo, J., París, F., 2022. Transverse biaxial tests on long fibre reinforced composites. Compos. Struct. 297, 115868.

Fiber—matrix debonding in composite materials: axial loading

7

E. Graciani [1], V. Mantič [1], F. París [1] and Janis Varna [2]
[1]School of Engineering, University of Seville, Seville, Spain; [2]Luleå University of Technology, Luleå, Sweden

7.1 Introduction

Today, composite laminates are widely used in industrial applications, especially in fields such as the aeronautical and space industries, where lightweight load-carrying structures are required. The complex internal architecture of composite laminates makes them fail in many different ways, which may include the failure of the fibers, the failure of the matrix, the debonding of the fiber—matrix interface, or a combination of several of these mechanisms.

In order to study the propagation of debond cracks along a fiber—matrix interface when a composite laminate is loaded in tension, the failure properties of the interface, as well as suitable modeling techniques capable of simulating crack onset and propagation, are required.

In this chapter, first, a novel technique is presented to determine the fiber—matrix Mode II interfacial fracture toughness. Next, a quantitative value of this failure property is calculated for a particular material combination, employing test results taken from the literature. And then, the measured property is employed, using an alternative modeling technique, to simulate debond crack onset and propagation.

Many experimental techniques have been proposed in the past to determine the fiber—matrix interfacial failure properties. Among the most popular, the push-out test, pull-out test, microbond test, microindentation test, and single-fiber fragmentation test (SFFT) can be cited (see, for example, Herrera-Franco and Drzal (1992) for a comparison of these methods). All these tests are carried out using single-fiber samples where the load is applied in the fiber direction. In the first tests, the load is applied directly to the fiber, and therefore low loads can be employed, which in turn makes the determination of reliable failure properties difficult. In contrast, in the last test (the SFFT), the load is applied to the sample, which results in a more repeatable test procedure.

In this chapter, the SFFT will be employed to calculate the fiber—matrix Mode II interfacial fracture toughness. Therefore, in Section 7.2, the SFFT is described and a set of experimental measurements obtained from the literature are summarized. These measurements are employed as a benchmark to illustrate quantitatively the data reduction technique developed for determining the fiber—matrix Mode II interfacial fracture toughness.

Modeling Damage, Fatigue and Failure of Composite Materials. https://doi.org/10.1016/B978-0-443-18489-5.00004-7

The data reduction technique is detailed in Section 7.3. This technique employs a numerical solution for the stress state within the SFFT sample in order to compute the energy release rate (ERR) associated with debond propagation. The major novelty associated with the evaluation of the fiber–matrix Mode II interfacial fracture toughness, using the procedure presented here, is that the effect of friction on the debonded fiber–matrix interface is considered in the analysis.

In Section 7.4, the fiber–matrix Mode II interfacial fracture toughness measured in Section 7.3 is employed to define a different numerical model capable of simulating the debond crack onset and propagation along the interface. There are two reasons for developing this model: (1) to illustrate how the properties measured can be employed for the analysis of debond propagation along the fiber–matrix interface and (2) to serve as a further validation of the data reduction technique presented in Section 7.3.

Finally, in Section 7.5, the main conclusions that can be extracted from this chapter are summarized.

7.2 Single-fiber fragmentation test

The SFFT was introduced by Kelly and Tyson (1965) and employs samples consisting of a sufficiently long fiber, surrounded by polymeric matrix, loaded in tension, as depicted in Fig. 7.1.

Initially, using the shear lag stress distribution model proposed by Cox (1952), the SFFT was employed to determine fiber–matrix interfacial strength. Using different assumptions, equilibrium was imposed on the final average fragment to determine the strength of the interface from the strength properties of the fiber. This approach was followed, for example, in the initial studies by Ohsawa et al. (1978), Whitney and Drzal (1987), Netravali et al. (1989), Herstemburg and Phoenix (1989), Lacroix et al. (1992, 1995), and Tripathi and Jones (1997), and in the reviews carried out by Verpoest et al. (1990), Herrera-Franco and Drzal (1992), and Tripathi and Jones (1998). All these studies are based on simplified solutions for the stress state along

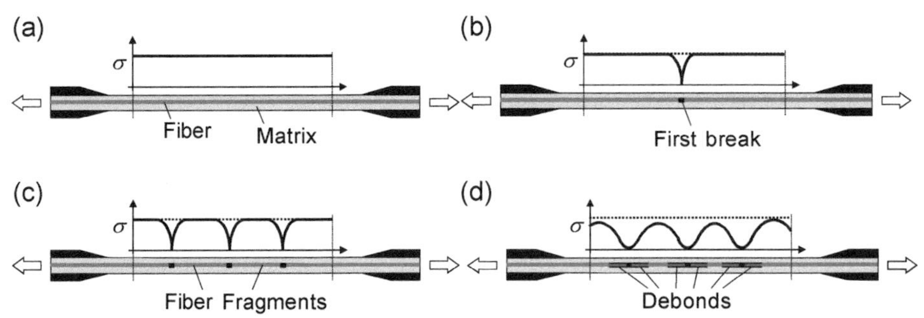

Figure 7.1 Stages of the SFFT. (a) Prior to the first fiber break. (b) After the first fiber break. (c) During fragmentation. (d) During fragmentation with simultaneous interfacial debond growth.

the fiber–matrix interface and, as a consequence, the fiber–matrix interfacial strengths measured do not necessarily lead to accurate predictions when they are employed to model debond propagation under different loading conditions.

An alternative approach is to characterize the fiber–matrix interface failure using either an energy balance or fracture mechanics. Using these methods, the fiber–matrix Mode II interfacial fracture toughness is obtained from the SFFT data, which results in an easier application of the failure properties measured to distinct loading conditions; see, for example, the studies by Nairn (1992), Varna et al. (1996), Nairn and Liu (1997), and Wu et al. (1997, 1998a,b, 2000).

To determine the fiber–matrix Mode II interfacial fracture toughness, the ERR associated with crack propagation needs to be determined. However, as will be explained later on, in order to obtain a proper definition of the fiber–matrix interfacial fracture toughness, the effect of interfacial friction along the debonded zone must be taken into account.

In Section 7.2.1, the stages occurring during the SFFT are described, focusing on the failure in the fiber and the interface during the experiment and on the load carried by the fiber and transferred through the fiber–matrix interface. Subsequently, in Section 7.2.2, a set of experimental measurements taken from the literature (Kim and Nairn, 2002a,b) are summarized. These measurements are taken from glass fiber–epoxy samples and will be used as a benchmark for describing the procedure developed to determine the fiber–matrix Mode II interfacial fracture toughness.

7.2.1 Experimental procedure

As mentioned in the previous section, the SFFT sample consists of a single fiber surrounded by a matrix loaded in tension. The SFFT sample is depicted in Fig. 7.1a, where the axial stress borne by the fiber can also be seen. Initially, prior to the first failure event, axial stresses along the fiber and the matrix are roughly constant and can be easily estimated analytically by taking the thermal residual stresses resulting from the manufacturing of the sample and the externally applied load into account.

Since elongation of the fiber and matrix are identical at this stage, no shear load transfer takes place through the fiber–matrix interface. In the direction perpendicular to the interface, a radial compression is created when glass fiber–epoxy samples are used; this happens (1) because the radial shrinkage of the epoxy matrix is greater than the radial shrinkage of the glass fiber during the cooling of the sample after solidification and (2) because the Poisson effect is larger in the epoxy matrix during the loading of the sample.

Since elongation to failure is much lower in the glass fiber than in the epoxy matrix, if the external load is steadily increased, at a certain value the fiber fails at its weakest point. The situation of the SFFT sample after this failure is depicted in Fig. 7.1b where, once again, the axial stress borne by the fiber can also be seen.

Once the fiber is broken, axial stress drops to zero at the point of failure. However, as the fiber is attached to the matrix, the axial load is transferred from the matrix to the fiber (inducing shear stresses along the fiber–matrix interface). As a result of this load transfer, the axial stress in the fiber steadily increases as the distance to the failure point increases. At a certain distance from the failure point, the fiber has practically recovered all the axial stress. As the fiber is very long when compared to its radius, most of the fiber still carries almost the same axial stress as it was carrying prior to the first failure.

Fiber strength is defined by a two-parameter Weibull distribution (see Wisnom (1999)). Therefore, after the first fiber break takes place, a fragmentation process begins during which the fiber splits into an increasing number of fragments as the external load increases. This situation is depicted in Fig. 7.1c.

As can be observed in the axial stress distribution along the fiber shown in Fig. 7.1c, the axial stress drops to zero at each failure point. Therefore, a certain zone exists in the vicinity of each failure point that will not be able to recover the stresses that were carried by the fiber prior to the breakage. As a consequence, if no failure takes place at the interface, fragmentation will continue until the fragments are too short to allow new fiber failures to appear. Obviously, SFFT cannot be employed in this scenario to determine failure properties of the interface, since no interfacial failure takes place during the experiment.

However, in many cases during the stage of fiber fragmentation, a second failure mechanism can be detected in the sample (see, for instance, Sjögren et al. (1999)). Since an increasing external load is needed to produce new fragments, shear stresses induced along the fiber–matrix interface at the vicinity of each fiber break will increase steadily during the test. After a certain load is applied, a series of debond cracks can be detected in the fiber–matrix interface. These cracks start at the fiber failure points (i.e., at fragment ends), and grow steadily toward the central part of the fragment. This situation is depicted in Fig. 7.1d.

As will be shown later, the growth of the debond cracks makes shear transfer through the interface less efficient. Therefore, the extent of the zone near each fiber failure that is not prone to fiber failure increases as the debond cracks increase, which results in a larger final fragment length when compared to that which would have been obtained if no interface failure had taken place.

As was mentioned previously, many of the studies found in the literature are devoted to determining the fiber–matrix interfacial strength from SFFT measurements that try to correlate, under certain simplifying assumptions, the strength of the final average fragment obtained from the SFFT with the interfacial failure properties. Therefore, the vast majority of experimental SFFT measurements found in the literature correspond to the evolution of the average fragment length with the load applied at sample ends.

However, most of the studies in the literature neglect the effect of interfacial friction along the debonded zone. As was mentioned above, both residual stress and the Poisson effect induce radial compressive stresses along the fiber–matrix interface. Therefore, the effect of friction should be taken into account in the analysis of debond crack propagation.

As will be shown in Section 7.3, debond crack propagation along the interface is strongly influenced by friction while, on the contrary, fragmentation of the fiber is only slightly influenced by friction.

Therefore, it is questionable whether it is possible to determine interfacial failure properties from the measurements of the average fragment length during SFFT. In the current investigation, a different approach will be used, in which the fiber–matrix Mode II interfacial fracture toughness and the interfacial friction coefficient will be simultaneously determined from the measurements of the average debond length during SFFT.

7.2.2 Experimental results

As already stated, failure properties of the interface will be obtained from the study of the stable propagation of the debond cracks during the SFFT. Unfortunately, little information can be found in the literature about the evolution of the debond crack length as a function of the applied load in the SFFT.

In this section, some of the results presented by Kim and Nairn (2002a,b) will be described. These results refer to the SFFT of glass fiber–epoxy samples and will be employed, in the following sections, in order to illustrate the methods developed in this chapter.

The material properties employed are listed in Table 7.1. As can be seen, both the Poisson ratio and the thermal expansion coefficient are higher in the matrix than in the fiber. Therefore, as has already been mentioned, the shrinkage of the matrix during cooling after the solidification of the sample and the Poisson effect when the external load is applied result in a compressive radial stress state along the fiber–matrix interface. According to Kim and Nairn (2002a), the solidification of SFFT samples occurs at 100 K above room temperature, which needs to be taken into account when evaluating the stress state within the sample since (as will be detailed later) it will affect the frictional stresses along crack faces.

Experimental measurements of the fiber crack density (number of cracks per unit fiber length) and average debond crack length are shown in Fig. 7.2a and b, respectively, as a function of the externally applied strain, ε. It can be seen that the first fragments appear around $\varepsilon = 1.8\%$ and saturation is reached (according to the authors) around $\varepsilon = 3\%$. Notice that only the range $1.5\% < \varepsilon < 3.5\%$ is shown in the graphs, in order to have a clearer view of the measurements.

It must be pointed out that although the results provided in both graphs shown in Fig. 7.2 correspond to the same samples, there is no indication in the above-mentioned studies of the correlation between the measurements shown in the two graphs. For example, for $\varepsilon = 1.8\%$, there are two measurements for the fiber crack density and two measurements for the average debond crack length, but it is impossible to

Table 7.1 Material properties and geometrical parameters (Kim and Nairn, 2002a,b).

Property	E-Glass	Epoxy
Outer radius	7 μm	1000 μm
Young's modulus	72.5 GPa	2.6 GPa
Poisson ratio	0.2	0.34
Thermal expansion coefficient	5.4×10^{-6} K^{-1}	40×10^{-6} K^{-1}

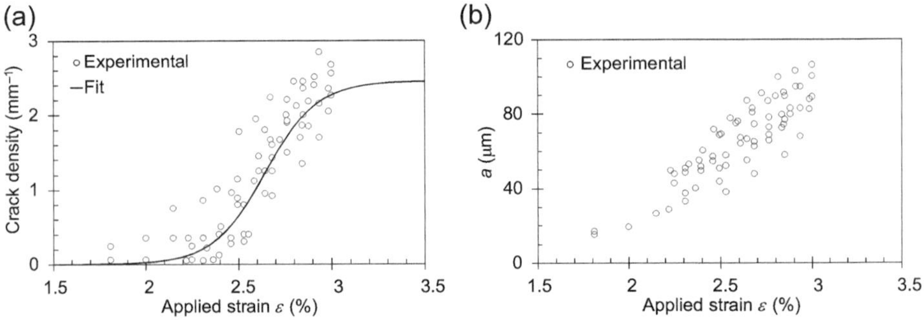

Figure 7.2 Evolution of (a) fiber crack density and (b) average debond crack length with the externally applied strain in the SFFT.
Data taken from Kim and Nairn (2002a,b).

identify which value of the fiber crack density corresponds to which value of the average debond crack length.

For the sake of simplicity, in the following analyses, values for the fiber crack density have been averaged, while with the intention of keeping a certain measurement of the dispersion of the results, individual measurements of average debond crack length have been employed.

Therefore, the average fiber crack density f has been defined as

$$f(\varepsilon) = A\left(1 - \frac{1}{1 + \exp[B(\varepsilon - C)]}\right), \tag{7.1}$$

with $A = 2.452\ \text{mm}^{-1}$, $B = 7.122$, and $C = 2.645\%$. Average values for f are shown in Fig. 7.2a using a continuous line, while actual individual measurements are shown with circles.

Average fragment length can be obtained by inverting the measurements of the fiber crack density. Notice that each fragment has two ends and a debond crack grows symmetrically from each fragment end toward the fragment center. In the following sections, only one side of the fragment will be studied and therefore the reference length employed to determine the size of the fragment is only one-half of the actual fragment length. This reference length is denoted as L_f in the following and can be easily obtained from the fiber crack density as

$$L_f(\varepsilon) = [2f(\varepsilon)]^{-1}. \tag{7.2}$$

Individual measurements of L_f, derived from the actual measurements of the fiber crack density, are shown in Fig. 7.3 using dots, while average values are shown with a continuous line obtained by introducing Eq. (7.1) into Eq. (7.2).

Notice that the circles presented in Figs. 7.2b and 7.3 contain some degree of homogenization. The evaluation of the average fragment length from the crack density assumes that all fragments within the sample have the same length, whereas

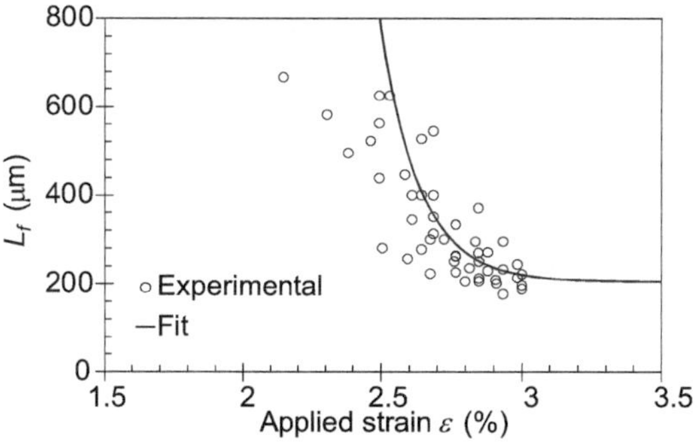

Figure 7.3 Evolution of the averaged semilength of the fragments, L_f, with the externally applied strain in the SFFT.

measurements of the debond crack length are in fact an average of the lengths of all cracks appearing in a certain reference length. Typically, the dispersion of individual debond lengths within a single sample is lower than the dispersion of the average debond lengths corresponding to distinct samples, whereas the dispersion of individual fragment lengths within one sample is similar to the dispersion of averaged fragment lengths corresponding to different samples.

Finally, it is important to mention that the results of carbon fiber–epoxy samples are also presented by Kim and Nairn (2002a,b). These results have not been included in this chapter for the sake of conciseness, although the differences arising from the fact that carbon fibers are transversely isotropic, in contrast with the isotropic character of the E-glass fibers, are commented on in Graciani et al. (2011b).

In what follows, two approaches are employed to define interface failure properties. In the first approach (Section 7.3), linear-elastic fracture mechanics (LEFM) is used to determine the fiber–matrix Mode II interfacial fracture toughness and the interfacial friction coefficient. In order to obtain quantitative values of these properties, experimental measurements already presented in this section, along with numerical models, are used. In the second approach (Section 7.4), cohesive elements are employed to show how the aforementioned properties can be used to model debond crack propagation.

The values for the fiber–matrix Mode II interfacial fracture toughness and the interfacial friction coefficient corresponding to the measurements shown in Fig. 7.2 and discussed in Section 7.3 are obtained using the first approach. These values are then employed to define the properties of the fiber–matrix interface in the numerical models used in Section 7.4 for simulating the SFFT, following the second approach. For the sake of validation, the results of these simulations are compared with the experimental measurements presented in Fig. 7.2b.

7.3 Numerical simulation of debond crack propagation using LEFM

To determine the fiber–matrix Mode II interfacial fracture toughness, G_{IIc}, from the measurements of the debond crack propagation in the SFFT, LEFM is employed.

Since debond crack growth is stable, at least at a certain stage during the test, it is assumed that, at this stage, the ERR associated with debond crack growth, G_{II}, equals the fiber–matrix Mode II interfacial fracture toughness.

The ERR has been calculated utilizing a generalization of Irwin's (1957) virtual crack closure technique (VCCT), which takes into account the effect of interfacial friction (Graciani et al., 2011a). This technique is described in Section 7.3.1. The ERR is calculated using the stresses and relative displacements in the vicinity of the debond crack tip, in the same way as the classical application of VCCT. In order to determine these, a set of numerical models have been developed where geometry and loads are taken from the experimental measurements described in the previous section.

The main assumptions considered in this chapter are the following:

- The linear elastic isotropic behavior of the constituents.
- Axial symmetry (with respect to the fiber axis).
- The periodicity of the solution in the vicinity of all fragment ends (which implies local symmetry with respect to the plane containing the fiber crack).
- The frictional contact between debond crack lips (using Coulomb's friction model with a friction coefficient μ).

Bearing these assumptions in mind, this chapter is limited to characterizing the propagation of a debond crack originating at one of the fragment ends and growing toward the fragment center. Therefore, the stress calculations can be made by analyzing the radial section of one-half of a typical fragment, with the geometry and boundary conditions depicted in Fig. 7.4. Notice that the representation is not proportional to the actual dimensions of the sample. The fiber radius, r_f, is much smaller than the sample radius, r_m (as shown in Table 7.1), and fragment semilength, L_f.

The stress analysis is carried out numerically, using the Boundary Element Method (BEM) (see París and Cañas (1997) and Graciani et al. (2005)). Details of the mesh employed can be found in Section 7.3.2, while the results obtained from the BEM models are described in Section 7.3.3.

Notice that in order to linearize the problem, the effect of load history has been neglected. That is, a set of models have been developed in which two preexisting cracks are assumed: one fiber crack on the right-hand side of the model ($0 < r < r_f$, $z = 0$), according to the coordinate system shown in Fig. 7.4, and one interfacial debond crack arising from it ($r = r_f$, $0 < z < a$). The solution in all cases has been obtained linearly using a uniform temperature decrement $\Delta T_0 = -100$ K and an average

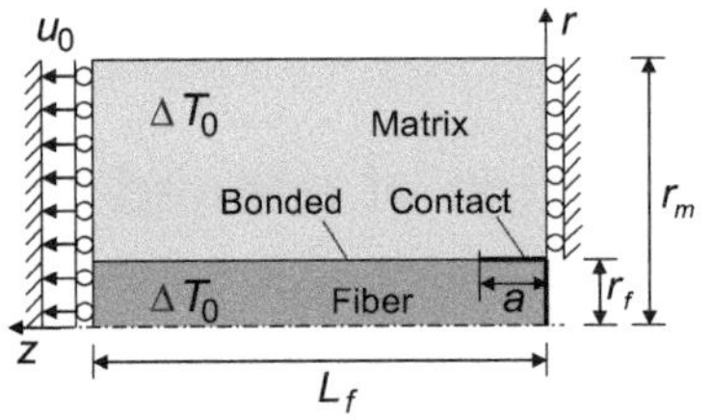

Figure 7.4 Geometry of the problem under study.

strain, applied as a uniform axial displacement u_0, on the left-hand side of the fragment. This uniform displacement, u_0, takes into account the average shrinkage of the sample during the cooling stage after solidification $\varepsilon\Delta_T \approx \alpha_m \Delta T_0 = -0.4\%$ and the externally applied strain ε. Therefore, $u_0 = (\varepsilon\Delta_T + \varepsilon)L_f$.

As is well known, the solution of a frictional contact problem depends on load history, because frictional stresses oppose instantaneous relative displacements and these depend on the load history. Nonetheless, in a problem like the one considered here, where loads and relative displacements increase monotonically, the dependence of the stress solution on load history is very weak and can be neglected. In any case, a different modeling technique, which more accurately takes into account the effect of load history, will be employed in Section 7.4 to validate the results obtained in the current section.

Since the friction coefficient along the debond crack faces is unknown, evaluation of interfacial fracture toughness requires its best fit estimate from the experimental measurements. This procedure can be summarized as follows. First, a set of models are created in which dimensions L_f and a, and load ε, are taken from the experimental measurements shown in Section 7.2.2. Next, the ERR is evaluated, varying the friction coefficient parametrically. Finally, the fiber–matrix Mode II interfacial fracture toughness and the friction coefficient are simultaneously determined by best fit with the propagation criteria. The correct friction coefficient is the one that makes the ERR become constant during crack propagation.

7.3.1 Theoretical background

As mentioned above, the ERR during interfacial debond crack propagation has been evaluated using a generalization of Irwin's VCCT. Since the crack is closed along its whole length (Graciani et al., 2009a), crack propagation takes place in pure Mode II. The elastic solution for a debond crack of length a, which extends by a certain length Δa, is sketched in Fig. 7.5 both before and after crack propagation. σ_{rz} refers to interfacial shear stresses, Δu_z is the relative sliding between crack faces, and $\rho = z - a$ is a local coordinate. Superscript + denotes a value corresponding to the final state

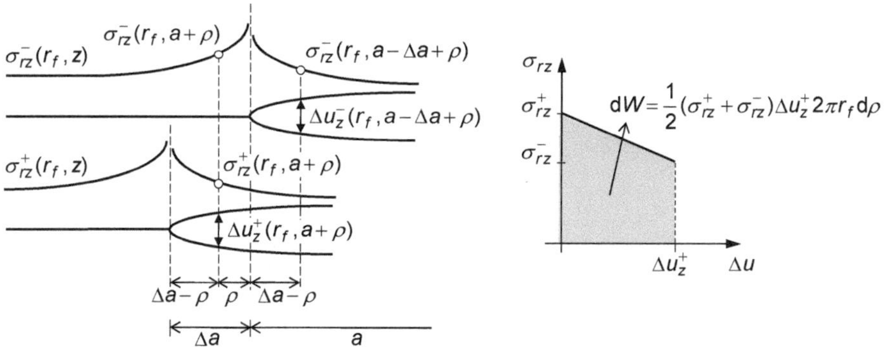

Figure 7.5 Illustration of crack propagation in the presence of friction.

after propagation (i.e., when the crack length is $a + \Delta a$), whereas superscript $-$ denotes a value corresponding to the initial state (i.e., when the crack length is a). For the sake of clarity, relative sliding has been represented as opening.

Initially, the points situated along the crack extension zone, $a < z < a + \Delta a$, are under a shear stress state $\sigma_{rz}^{-}(r_f, a+\rho)$ which becomes singular at the crack tip, as in the case of classical cracks growing in Mode II. After crack propagation, when friction is present, these points, instead of being free of stresses, are under a different shear stress state $\sigma_{rz}^{+}(r_f, a+\rho)$ which becomes singular at the new crack tip (Comninou, 1977). Consequently, during a virtual crack increment, the work done by the shear stresses corresponds to a linear change from $\sigma_{rz}^{-}(r_f, a+\rho)$ to $\sigma_{rz}^{+}(r_f, a+\rho)$, while the relative displacements change from 0 to $\Delta u_z^{+}(r_f, a+\rho)$.

Therefore, the work done by the shear stresses during crack growth along an infinitesimal length $\mathrm{d}\rho$ located inside the crack extension zone (Fig. 7.5) is $\mathrm{d}W = \frac{1}{2}\left(\sigma_{rz}^{+} + \sigma_{rz}^{-}\right)\Delta u_z^{+} 2\pi r_f \mathrm{d}\rho$. By integrating along the crack extension and dividing by $2\pi r_f \Delta a$, the ERR associated with crack propagation, G_{II}, is obtained from the near-tip elastic solution as follows:

$$G_{\mathrm{II}}(a, \Delta a) = \frac{1}{2\Delta a}\int_0^{\Delta a}\left[\sigma_{rz}^{+}(r_f, a+\rho) + \sigma_{rz}^{-}(r_f, a+\rho)\right]\Delta u_z^{+}(r_f, a+\rho)\mathrm{d}\rho \tag{7.3}$$

Notice that Eq. (7.3) can be split into two integrals, one corresponding to the frictional stresses σ_{rz}^{+} and the other corresponding to the singular stresses in front of the crack tip σ_{rz}^{-}. The second integral seems to be identical to that which appears in the evaluation of the ERR in a frictionless crack. However, this is not the case, since stresses in front of the crack tip σ_{rz}^{-} and relative displacements between crack faces Δu_z^{+} are strongly influenced by the presence of friction between crack faces. Actually, it can be verified that the change in these magnitudes, and not the appearance of the frictional stresses in Eq. (7.3), is the main cause for the change in the ERR associated with the presence of friction between crack faces.

When the crack extension is small in comparison with the crack length, $\Delta a \ll a$, self-similarity can be assumed, since the solution in the vicinity of the crack tip of length $a + \Delta a$ is very similar to the solution in the vicinity of the crack tip of length a:

$$\sigma_{rz}^{-}(r_f, a+\rho) = \sigma_{rz}(r_f, a+\rho), \tag{7.4}$$

$$\sigma_{rz}^{+}(r_f, a+\rho) \approx \sigma_{rz}^{-}(r_f, a-\Delta a+\rho) = \sigma_{rz}(r_f, a-\Delta a+\rho), \tag{7.5}$$

$$\Delta u_z^{+}(r_f, a+\rho) \approx \Delta u_z^{-}(r_f, a-\Delta a+\rho) = \Delta u_z(r_f, a-\Delta a+\rho). \tag{7.6}$$

Therefore, ERR G_{II} is expressed as

$$G_{\mathrm{II}}(a,\Delta a)=\frac{1}{2\Delta a}\int_{0}^{\Delta a}\left[\sigma_{rz}\left(r_f,a-\Delta a+\rho\right)+\sigma_{rz}\left(r_f,a+\rho\right)\right] \Delta u_z\left(r_f,a-\Delta a+\rho\right)\mathrm{d}\rho. \tag{7.7}$$

As shown by Comninou (1977), the singularity exponent of the stresses in the vicinity of the crack tip differs from the classical square root singularity and, as a consequence, the ERR $G_{\mathrm{II}}(a, \Delta a)$ vanishes for vanishing values of Δa. Therefore, a small but finite value of Δa must be employed in the calculations. The asymptotic character of near-tip stresses and relative displacements is detailed in Section 7.3.3, where the accuracy of the numerical solution employed is also stated. In Section 7.3.4, the results obtained for the ERR, using Eq. (7.7) and the aforementioned numerical solution of stresses and relative displacements, are presented along with further details of the calculation procedure.

7.3.2 Boundary element method model of the SFFT sample

The elastic solution in the SFFT sample was obtained using a self-developed BEM code for the analysis of axisymmetric frictional contact problems, Graciani et al. (2005). The code is extremely accurate in the evaluation of stresses and displacements in problems involving contact and singularities, which typically appear in the vicinity of interfacial crack tips; see Graciani et al. (2009b, 2010b, 2014).

Details of the BEM implementation are beyond the scope of this chapter. However, it should be emphasized that the numerical model used does not include any assumptions beyond those described at the beginning of Section 7.3. That is, material properties required are those listed in Table 7.1, and dimensions and loads are taken from the experimental measurements presented in Section 7.2.2.

These models are only employed to determine the stress state in the sample, since the sample's reduced size and the complexity of the problem do not allow physical measurements to be accurately carried out. Since the interfacial friction coefficient is not known, it is parametrically varied ($\mu = 0$, 0.2, 0.4, ...), and the correct value is determined at a later stage by best fit of the experimental measurements.

One of the BEM meshes employed for the numerical analysis is shown in Fig. 7.6. Since the domain to be modeled consists of the radial section of one-half of a fragment, the boundary of the problem is defined by a set of lines that can be meshed with linear elements. Due to the very different dimensions of the sample, the fiber, and the crack, the element size changes dramatically within the different zones of the model. As can be seen in Fig. 7.6, when the whole mesh is represented, the fiber is almost indistinguishable. Therefore, a mesh refinement is carried out in the matrix boundary from the external side of the sample to the zone near the fiber. Since only the boundary mesh is required, mesh refinement can be easily carried out. Notice also that an extremely fine mesh is carried out in the vicinity of the crack tip. In this zone, the

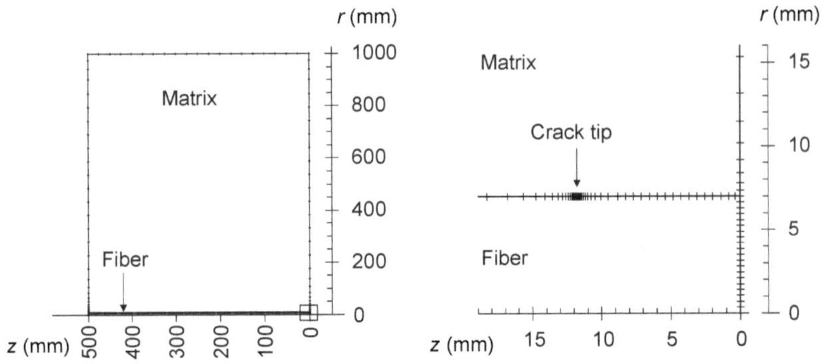

Figure 7.6 BEM mesh employed in the stress analysis of an SFFT fragment with semilength $L_f = 500$ μm and an interfacial debond crack with length $a = 11.8$ μm.

element size steadily decreases toward the crack tip, until it reaches the smallest element size of 10^{-7} μm, while keeping a constant 1.3 ratio between the lengths of consecutive elements.

The primary results obtained from the BEM model are the components of the displacements and tractions in the nodes of the mesh. Moreover, with the contact formulation employed (Graciani et al., 2005), relative displacements between the contacting boundaries are also obtained as a direct result of the BEM analysis. In addition, in the postprocessing stage, the components of the displacements and the stress tensor can be determined at any point along the boundary or inside the domain of the problem. In the next section, the solutions obtained, first, along the fiber axis and, second, at the interface in the vicinity of the crack tip, are presented.

7.3.3 Stress state within the sample

Employing the numerical models described in the previous section, the numerical solution of the stress state within the SFFT sample is now presented. Special care is taken in describing, firstly, the influence of fiber–matrix debonding on the axial stresses carried out by the fiber and, secondly, the stresses causing fiber–matrix debond crack growth. In both cases, the impact of interfacial friction on the solution presented is also stated.

7.3.3.1 Axial stresses within the fiber

In this section, the evolution of longitudinal stresses, with crack propagation, along the fiber axis $\sigma_{zz}(0, z)$ is presented. For the sake of clarity, although one model was created for each of the experimental measurements presented in Section 7.2.2, only a set of eight models have been selected in this section. The dimensions and loads corresponding to these models can be found in Table 7.2. The selected models are roughly evenly distributed where crack propagation takes place, in the interval $2\% < \varepsilon < 3\%$. The semifragment length employed in each model is given by the averaged semifragment length $L_f(\varepsilon)$ defined in Eq. (7.2). Finally, the value used for the debond crack length a has been selected from among those that correspond to the distinct measurements

Table 7.2 Dimensions and loads utilized in the models selected to describe the evolution of axial stresses in the fiber.

Model	1	2	3	4	5	6	7	8
ε (%)	2.00	2.15	2.31	2.50	2.69	2.77	2.85	2.99
L_f (μm)	>500	>500	>500	>500	363.2	291.2	250.3	221.6
a (μm)	19.3	26.6	37.5	50.8	65.1	72.8	77.0	82.6

presented for each value of ε and at the same time trying to keep the value close to the average value of these measurements.

Notice that when fragments are very long, the longitudinal stresses along the fiber axis become insensitive to fragment length, since at a certain distance from the fiber crack the stresses are fully restored and become constant. Therefore, models having a semifragment length $L_f > 500$ μm have been truncated at $z = 500$ μm to reduce the number of nodes in the model.

The evolution of $\sigma_{zz}(0, z)$ is represented in Fig. 7.7 for the eight cases listed in Table 7.2, in the range $0 < z < 500$ μm, where $z = 0$ corresponds to the plane containing the fiber crack, neglecting the effect of friction along crack faces (i.e., considering $\mu = 0$).

As can be seen, the results are in agreement with the qualitative description presented in Section 7.2.1. Axial stress is null in the plane of failure and increases as the fragment center is approached. For long fragments (models 1–4), it can be clearly appreciated that, at a certain distance from the plane of failure, axial stress reaches a plateau and remains roughly constant and equal to the stress corresponding to the intact fiber. Therefore, an increase in the load applied to the sample results in an increase in the stress sustained by the fiber, thus making new fragments appear. Once the length of the fragment decreases (models 5–8), this does not allow the axial stress to recover the stress value prior to failure. Therefore, although the external load is increased, axial stresses within the whole fragment are lower than those which caused the previous failure, and therefore no new fiber breaks are produced inside these short fragments.

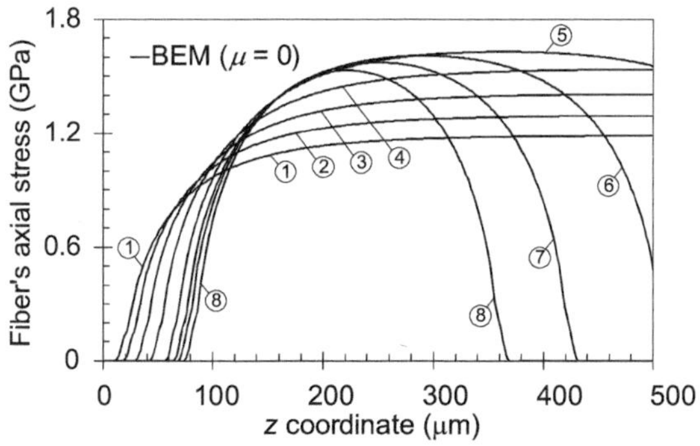

Figure 7.7 Evolution of longitudinal stresses along the fiber axis $\sigma_{zz}(0, z)$ for models listed in Table 7.2 in the absence of friction.

It can also be appreciated in Fig. 7.7 that, as the debond grows, the extent of the zone in which axial stresses decay increases, thus making it more difficult for new fragments to appear. Notice that this fact implies that the weaker the interface is, the longer the final average fragment will be. The majority of the studies found in the literature analyzing the SFFT in order to determine interfacial failure properties try to determine these properties from the final average fragment length. As will be shown in the following discussion, this approach is highly complicated, since debond growth is not only controlled by the failure properties of the interface but also by the friction between crack faces. Separating both effects, only taking into consideration the evolution of the average fragment length with the externally applied load, requires the use of strong simplifying assumptions (typically neglecting the effect of friction), which makes it difficult to apply the failure properties obtained to other problems with different loading conditions.

In Fig. 7.8a, the evolution of $\sigma_{zz}(0, z)$ in the absence of friction is compared to that obtained with $\mu = 1$ (for model 5), which has an intermediate fragment length. As can be seen, the effect of friction can be clearly appreciated in the zone close to the fiber crack, $z < 200$ μm, where the stress increase is gradual in the frictional case and abrupt in the frictionless case. On the other hand, stresses in the middle part of the fragment, 200 μm $< z <$ 500 μm, are very similar in the two cases. Therefore, an increase in the external load produces a similar increase in the axial stress within the fiber in the zone where the fragment is able to break. As a consequence, the presence of friction does not have a direct relation to the final average fragment length.

As shown in previous studies by the authors (Graciani et al., 2009a, 2011a) and summarized in Section 7.3.4, friction clearly influences debond crack propagation. Therefore, the final average fragment length in SFFT depends mostly on how fast the debond crack growth is, and crack growth depends both on the interfacial failure properties and on the friction coefficient, which makes it very difficult to determine interfacial failure properties from the final average fragment length.

For this reason, in this chapter, the interfacial failure properties are determined from the measurements of the debond crack growth evolution with the applied load, employing a technique that allows the interfacial friction coefficient to be determined simultaneously.

For the sake of comparison, Fig. 7.8b shows the same results represented in Fig. 7.7, the only difference being that a friction coefficient, $\mu = 1$, is employed in the calculations. It can be clearly appreciated that, similar to the frictionless case, the maximum stress in the fragment increases proportionally to the externally applied strain when fragments are long (models 1−4), but for shorter fragments (models 5−8), the increase to the externally applied strain does not produce an increase in the maximum axial stress within the fragment.

The results presented in Fig. 7.8b can be employed for further experimental validation of the technique presented here, using experimental techniques that already exist, such as Raman spectroscopy (see Galiotis (1991, 1993) or Huang and Young (1994)), which are capable of determining with a reasonable level of accuracy the axial stresses along a glass fiber in the SFFT. Accordingly, the stresses calculated numerically, employing the friction coefficient taken from the analysis of crack propagation, should

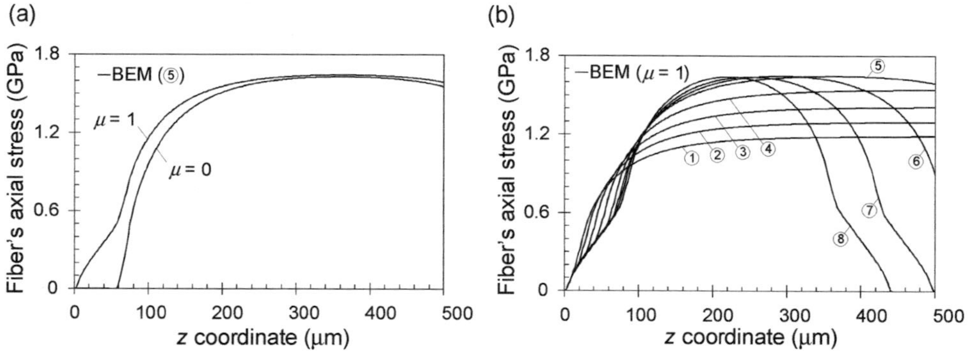

Figure 7.8 Effect of friction on the longitudinal stresses along the fiber axis. (a) Comparison of the frictional ($\mu = 1$) and frictionless solution for model 5. (b) Frictional solution ($\mu = 1$) for models listed in Table 7.2.

be in agreement with those measured experimentally. The comparison can be easily and precisely carried out, since the slope of the stress increase in the zone near to the fiber crack is strongly influenced by the friction coefficient.

7.3.3.2 Interfacial stresses

In this section, the solution is shown for the interfacial stresses along the fiber–matrix interface, $\sigma_{rr}(r_f, z)$ and $\sigma_{rz}(r_f, z)$, obtained using the BEM models presented in Section 7.4.2. Focus is placed on describing the influence of interfacial friction on the solution of near-tip shear stresses, since these stresses are needed to evaluate the ERR associated with crack propagation.

Notice that frictional contact conditions were imposed on the debonded side of the interface. According to Coulomb's friction law which was employed here, normal contact stresses must be compressive at all the points which are in contact and these points can be either in adhesion (with null relative sliding and $|\sigma_{rz}| < \mu|\sigma_{rr}|$) or sliding (with shear stresses $|\sigma_{rz}| = \mu|\sigma_{rr}|$ opposed to the relative sliding). In all cases solved, the solutions show that the crack is closed and sliding along the whole length.

The solutions for $\sigma_{rr}(r_f, z)$ and $\sigma_{rz}(r_f, z)$, corresponding to model 5 of those listed in Table 7.2, are represented, at both sides of the crack tip, in Fig. 7.9a. Notice that the debonded zone corresponds to the lower values of z, while the bonded zone corresponds to the higher values of z (the crack tip being situated at $z = 65.1$ μm). Notice also that although a sufficiently large extent of the interface is shown, only the vicinity of the crack tip is represented, the solution being truncated at both sides.

A friction coefficient, $\mu = 1$, was used in the calculation and, as a consequence, normal and shear stresses are coincident along the debonded side of the interface. It is also very interesting to note that all stresses represented are singular at the crack tip with the exception of the normal stresses on the bonded side, which remain bounded.

This result is in agreement with the asymptotic behavior of stresses in the vicinity of the tip of an interfacial crack in the presence of friction (Comninou, 1977). It must be highlighted that, in contrast with open cracks or frictionless closed cracks, the singularity exponent of stresses, λ, is not equal to 0.5, but is dependent on the friction coefficient and on the elastic properties of the materials at both sides of the interface, according to the following law:

$$\tan(\lambda\pi) = (\mu\beta)^{-1}, \tag{7.8}$$

where β is a Dundurs bimaterial parameter, which Dundurs (1969) defined as

$$\beta = \frac{\mu_f(\kappa_m - 1) - \mu_m(\kappa_f - 1)}{\mu_f(\kappa_m + 1) + \mu_m(\kappa_f + 1)} \tag{7.9}$$

with subindex f and m, respectively, referring to fiber and matrix. κ_f and κ_m are Kolosov's constants, and μ_f and μ_m are the shear stiffnesses of the constituents that can be drawn from their respective elastic moduli E_f and E_m and Poisson ratios ν_f and ν_m, shown in Table 7.1, using

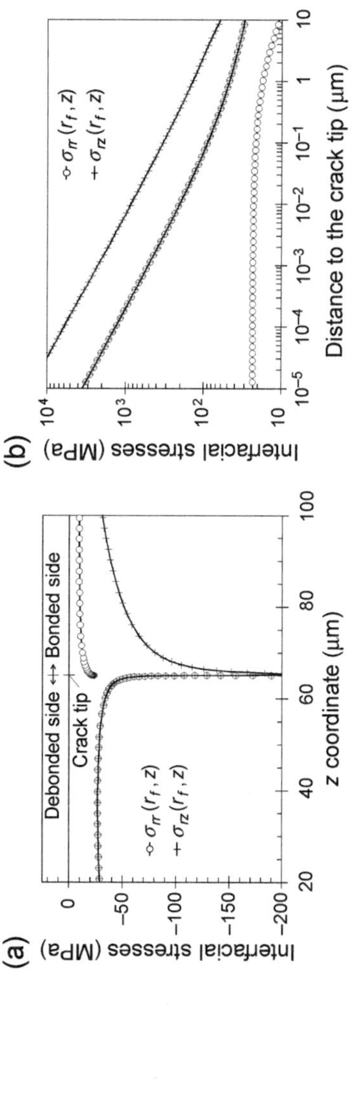

Figure 7.9 Interfacial stresses in the vicinity of the crack tip (where $\varepsilon = 2.69\%$, $a = 65.1$ μm, $L_f = 363.2$ μm, and $\mu = 1$). (a) Linear scale. (b) Log–log scale.

$$\kappa_f = 3 - 4\nu_f, \kappa_m = 3 - 4\nu_m, \mu_f = \frac{E_f}{2(1+\nu_f)} \text{ and } \mu_m = \frac{E_m}{2(1+\nu_m)}. \quad (7.10)$$

The asymptotic behavior of interfacial stresses can be easily appreciated in Fig. 7.9b, where log–log coordinates have been employed for the representation. As can be seen, at points sufficiently close to the crack tip, the representation of the singular stresses becomes a straight line whose slope should correspond to the singularity exponent. In the present case, the theoretical singularity exponent is $\lambda = 0.431$, whereas the exponent obtained from best fit of the numerical solution of shear stresses σ_{rz} on the bonded side, in the range 10^{-5} μm $< \rho < 10^{-4}$ μm (with ρ being the distance to the crack tip) is $\lambda = 0.429$, which results in a relative error lower that 0.5%.

The effect of friction on the near-tip singular shear stresses is shown in Fig. 7.10, which also includes the same results plotted for the $\mu = 1$ case in Fig. 7.9 for several frictional cases with $\mu = 0$, 0.2, 0.4, 0.6, 0.8, 1.

As can be seen in Fig. 7.10a, at points relatively far from the crack tip, shear stresses along the debonded interface increase with the friction coefficient, while on the contrary, shear stresses along the bonded interface decrease when the friction coefficient increases.

In Fig. 7.10b, the behavior of the interfacial stresses in the vicinity of the crack tip can be clearly appreciated. Along the bonded interface, not only the shear stresses decrease, but also the singularity exponent (the slope in the log–log representation of the stresses) decreases. Along the debonded side, the singularity exponent also decreases, but the behavior of shear stresses changes with the distance to the crack tip. Far from the crack tip, shear stresses increase with the friction coefficient, but close to the crack tip shear stresses decrease when the friction coefficient increases (as a consequence of the change in the singularity exponent). Notice that there is an intermediate zone in which shear stresses do not significantly change with the friction coefficient.

As can be seen in Fig. 7.10b, at the same distance from the crack tip, shear stresses along the bonded zone are at least one order of magnitude higher than shear stresses along the debonded zone. Therefore, in view of Eq. (7.7), the effect of friction on the shear stresses will cause a decrease in the ERR associated with debond crack propagation. Although not presented here for the sake of brevity, the relative displacements between crack faces obviously diminish when the friction coefficient increases, thus supporting the decrease of the ERR associated with the increase of the interfacial friction coefficient.

7.3.4 Evaluation of the ERR and determination of the fiber–matrix mode II interfacial fracture toughness

The ERR is evaluated using the near-tip solution of interfacial shear stresses and relative displacements, employing Eq. (7.7), where $\Delta a = 1$ μm. Notice that $G_{II}(a, \Delta a)$ tends to zero when Δa tends to zero, and therefore a small but finite value of Δa has to be used. Although $G_{II}(a, \Delta a)$ is slightly dependent on Δa, the dependence is

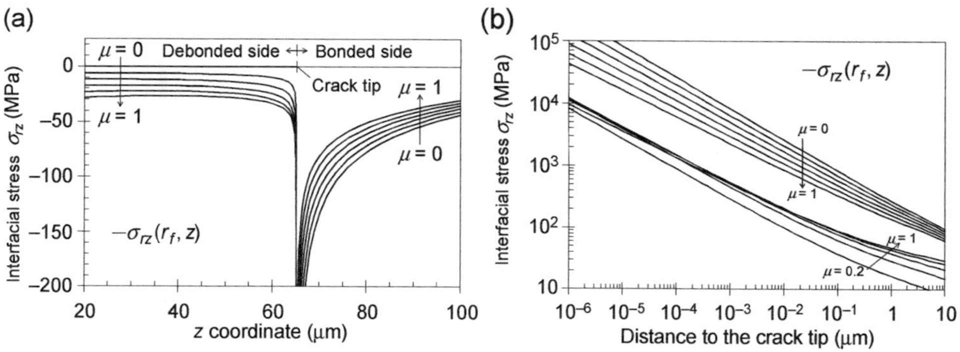

Figure 7.10 Effect of friction on the interfacial stresses in the vicinity of the crack tip (where $\varepsilon = 2.69\%$, $a = 65.1$ μm, $L_f = 363.2$ μm, and $0 \leq \mu \leq 1$). (a) Linear scale. (b) Log–log scale.

very weak, $G_{\text{II}}(a, \Delta a) \sim \Delta a^{2\lambda-1}$, and results do not vary significantly if a different value of Δa is utilized.

Results obtained for a set of values of the friction coefficient ($\mu = 0$, 0.4, 0.8, 1.2) are shown in Fig. 7.11. As can be clearly appreciated, the evolution of the ERR with the externally applied strain is roughly linear, the slope of this linear evolution decreasing while the friction coefficient increases.

To achieve the results shown in Fig. 7.11, a BEM model was created for each of the points shown in Fig. 7.2, as described in Section 7.3.2. In each model, the crack length, a, and the externally applied strain, ε, were taken from the experimental measurements plotted in Fig. 7.2, and the fragment length was defined by introducing ε into Eq. (7.2). The models are repeatedly solved for each of the desired values of the friction coefficient and the ERR associated with debond crack propagation is determined from the near-tip solution.

Since the debond crack growth is stable, the ERR associated with crack propagation should be equal to the fiber–matrix Mode II interfacial fracture toughness during crack propagation. Therefore, the correct value of the friction coefficient is the one that yields a horizontal line on the linear fit of the computed values of the ERR. That is, the one yielding a result most similar to the condition

$$G_{\text{II}}(a, \Delta a) = G_{\text{II}c}. \tag{7.11}$$

For a better view of the results, the evolution of ERR, $G_{\text{II}}(a, \Delta a)$, with the externally applied strain is plotted in Fig. 7.12 for $\mu = 0.8$, 1.0, 1.2, which are the values of the friction coefficient that best fit the expected behavior.

As can be seen, the best fit with a horizontal line is obtained for $\mu = 1.0$. For this value of the friction coefficient, the average value of the ERR gives an estimation of the fiber–matrix Mode II interfacial fracture toughness $G_{\text{II}c} = 12.12$ J/m^2.

As has been shown, the data reduction technique developed allows the fiber–matrix Mode II interfacial fracture toughness to be extracted from the experimental measurements of the evolution of the average fragment length and the debond crack length with the externally applied strain during the SFFT. Moreover, the technique developed takes into account the effect of friction during crack propagation and permits the simultaneous calculation of the interfacial friction coefficient.

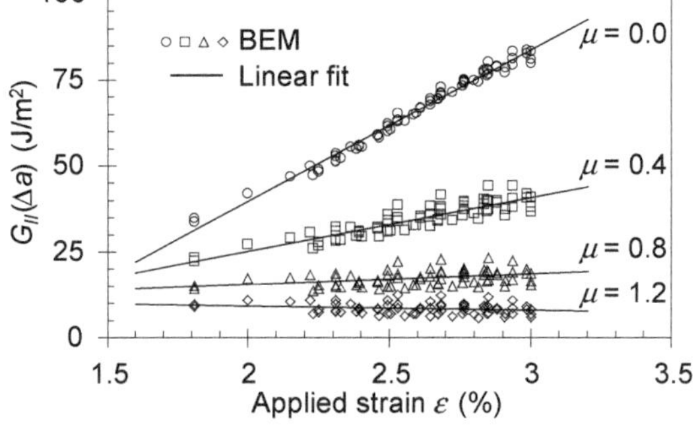

Figure 7.11 Evolution of ERR, $G_{\text{II}}(a, \Delta a)$, with the externally applied strain, $\mu = 0$, 0.4, 0.8, 1.2.

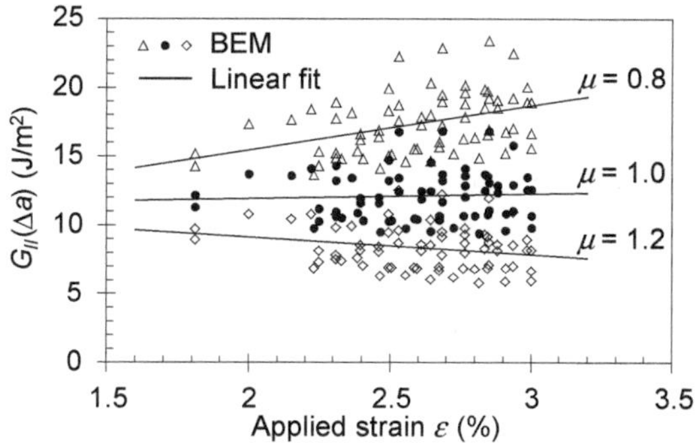

Figure 7.12 Evolution of the ERR, $G_{II}(a, \Delta a)$, with the externally applied strain, $\mu = 0.8, 1.0, 1.2$.

The calculation of the fiber–matrix Mode II interfacial fracture toughness is based on the evaluation of the ERR associated with debond crack propagation, which has been carried out numerically using a BEM code. However, a different numerical or analytical technique may be employed for this calculation, proving that the effect of friction is correctly taken into account during crack propagation.

7.4 Numerical simulation of debond propagation using cohesive elements

The most questionable hypotheses employed in the data reduction technique developed for obtaining the fiber–matrix Mode II interfacial fracture toughness from the experimental measurements of the SFFT are (1) the use of a set of linear models with preexisting debond cracks (thus neglecting the effect of load history on the elastic solution) and (2) the selection of a constant virtual crack increment, Δa, to determine the ERR associated with crack propagation.

In this section, the SFFT is simulated using a completely different numerical technique, based on the use of the finite element method (FEM) with cohesive elements (Graciani et al., 2010a), which does not require the above-mentioned assumptions. There is a double objective in this analysis. On the one hand, the intent is to show how the interfacial properties obtained can be employed in the analysis of other problems involving fiber–matrix interfacial debond growth; and on the other hand, the intent is to serve as a further validation of the hypothesis considered in the previous section.

In the present analysis, a set of models are developed, in which fiber and matrix are connected with cohesive elements to model the onset and propagation of the debond cracks. As detailed in Section 7.4.1, cohesive properties of the interface are in agreement with the fiber–matrix Mode II interfacial fracture toughness and the friction coefficient measured in the previous section.

In this case, the models do not have a preexisting crack and, therefore, debond crack propagation takes place following the actual load history of the SFFT sample. In order to simplify the analysis, a single fragment is considered (in fact, as in the previous section, one-half of a single fragment). In order to take into account the decreasing evolution of the average fragment length with the externally applied strain, a set of models have been created, as described in Section 7.4.2, with decreasing values for the fragment length.

Using a single nonlinear analysis in each model, the average evolution of crack length with the externally applied strain is achieved (see Section 7.4.3), which is in perfect agreement with the experimental measurements shown in Fig. 7.2b.

7.4.1 Theoretical background

To simulate debond crack onset and propagation in the SFFT, bilinear cohesive elements are employed with the cohesive law shown in Fig. 7.13. According to this law, the fiber–matrix interface behaves elastically (with a linear relationship between the shear stresses and the relative sliding) until a limit stress, τ_1, is reached. From then on, the fiber–matrix interface stiffness decreases as damage increases until the interface is fully broken at a point where the relative displacements reach a limit value, δ_{s2}.

However, the behavior described above will permit relative displacements between crack faces without frictional opposing loads. Therefore, in addition to the cohesive elements, frictional contact conditions are defined between fiber and matrix nodes, which are responsible for creating the frictional forces opposed to relative displacements once the cohesive elements are sufficiently degraded (as shown in Fig. 7.13).

According to the cohesive formulation employed, the interface is considered perfectly bonded at nodes that have $\Delta u_z < \delta_{s1}$, while the debond crack is constituted by the nodes that have $\Delta u_z > \delta_{s2}$. Between these two zones, a *damage process zone* exists, which includes all interface nodes that have relative displacements in the range $\delta_{s1} < \Delta u_z < \delta_{s2}$. At a certain value of the externally applied strain, the crack tip is located at the last node where the limit value $\Delta u_z = \delta_{s2}$ was reached.

The following values were used to define the bilinear cohesive law: $\tau_1 = 110$ MPa, $G_{IIc} = 12.12$ J/m^2, and $\mu = 1.0$. The last two were defined in view of the results obtained in the previous section. The value chosen for τ_1 does not significantly affect the debond crack propagation, as it is mainly related to the size of the damage process zone.

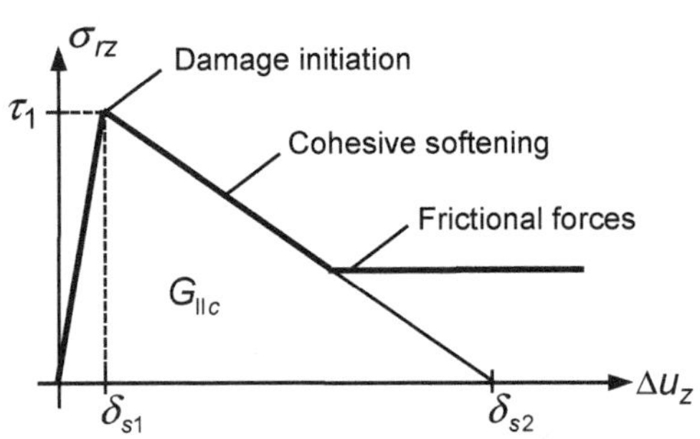

Figure 7.13 Cohesive law employed for the simulation of debond crack propagation in the SFFT.

7.4.2 FEM modeling of the SFFT sample

The FEM model of the SFFT sample is shown in Fig. 7.14. For a clearer view, some of the elements have been removed from the full-length view of the model shown in Fig. 7.14a. As can be seen, element size is larger in the area of the sample located far away from the fiber and is reduced until it reaches an element size equal to the fiber radius in the vicinity of the fiber.

As will be described later, fiber crack takes place at the left end of the fiber fragment. Therefore, a further refinement of the mesh was employed in this zone, in which an element size equal to one-sixteenth of the fiber radius was used.

Boundary conditions prescribe a null value for the horizontal displacements along the left side of the model and a constant value (detailed next) of the displacements along the right side. Loads are applied in two steps. During the first step, a decrease in temperature $\Delta T_1 = -100$ K and an end shortening $u_1 = \varepsilon\Delta_T L_f$, with the $\varepsilon\Delta_T$ value defined in Section 7.3, are imposed to simulate cooling of the SFFT sample after solidification. Subsequently, during the second step, the temperature remains constant and an end displacement $u_2 = \varepsilon\, L_f$ is enforced, with ε being the increasing externally applied strain.

To simulate the fiber break that triggers the debond crack, a second set of cohesive elements are placed on the left hand side of the fiber, using a cohesive law that produces a fiber failure at $\varepsilon \cong 1.5$. Once the fiber is broken, shear stresses are induced along the fiber–matrix interface, causing the interfacial debond to appear.

7.4.3 Simulation of crack propagation

Results obtained with the FEM models described in Section 7.4.2 are plotted in Fig. 7.15 along with the experimental measurements of the debond crack length. As can be seen, the predictions obtained with FEM are in excellent agreement with the experimental measurements, thus validating the applicability of the properties determined in Section 7.3.4.

Notice that results shown in Fig. 7.15 are, in fact, obtained from the solution of a set of FEM models with different fragment lengths. Therefore, for each value of the

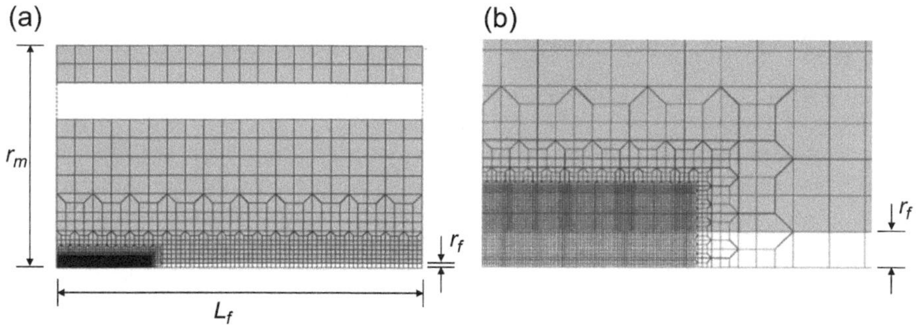

Figure 7.14 FEM model of the SFFT sample. (a) Full length. (b) Detail of the fine mesh.

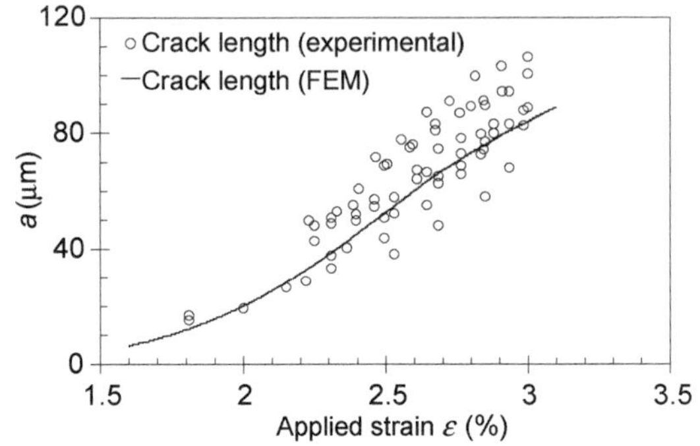

Figure 7.15 Simulation of debond crack propagation in an SFFT sample.

applied strain, the average crack length is obtained from the model that has the correct average fragment length (see Eq. 7.2).

In principle, the FEM models presented can also be used to determine the failure properties of the interface, parametrically varying the fiber—matrix Mode II interfacial fracture toughness defined in the cohesive law and the interfacial friction coefficient employed in the fiber—matrix interface. The change of each of these parameters produces a distinct effect, which simplifies the fitting procedure (Graciani et al., 2010a). On the one hand, the change in the fiber—matrix Mode II interfacial fracture toughness horizontally moves the prediction of $a(\varepsilon)$, that is, shifts the curve to higher or lower values of the applied strain; and on the other hand, the change in the interfacial friction coefficient changes the slope of the $a(\varepsilon)$ curve.

Notwithstanding, this procedure is much more time and resource consuming than the technique described in Section 7.3, since the nonlinear analysis required to simulate debond crack propagation with cohesive elements requires an extremely fine time discretization. As an indication, all linear models employed in Section 7.3 were solved over a period of several hours on a typical portable computer, while a single model of those required in Section 7.4 took about a week to solve using a 16-processor computer cluster.

7.5 Discussion and concluding remarks

In the present investigation, the fiber—matrix Mode II interfacial fracture toughness was calculated by analyzing the propagation of fiber—matrix debond cracks in the SFFT.

A novel data reduction technique has been described, in which the fiber—matrix Mode II interfacial fracture toughness is obtained from the evolution of the average debond crack length during the experiment.

To determine the ERR associated with fiber—matrix interfacial debond crack growth, a set of boundary element models have been employed in which the loads, material properties, and dimensions are taken from the experimental measurements. The near-tip elastic solution is determined with the numerical models and employed to evaluate the ERR, using the interfacial friction coefficient as a parameter. The value of the interfacial friction coefficient providing a constant value of the ERR during crack propagation is the correct one and, thus, the average ERR corresponds to the fiber–matrix Mode II interfacial fracture toughness.

The aforementioned technique was applied to a set of experimental measurements taken from the literature, and yielded a friction coefficient $\mu = 1.0$ and a fiber–matrix mode II interfacial fracture toughness $G_{IIc} = 12.12$ J/m^2 for glass fiber–epoxy matrix samples.

Finally, a set of finite element models with cohesive elements were created in which the above-mentioned interfacial properties were employed to model the interfacial interactions (i.e., the cohesive law and frictional forces after interfacial failure). Results from the numerical simulations are in excellent agreement with the experimental measurements.

Acknowledgments

This research was funded by the Junta de Andalucía and European Social Fund (Proyecto de Excelencia P08-TEP-4051) and the Spanish Ministry of Economy and Competitiveness and European Regional Development Fund (MAT2012-37387).

References

Comninou, M., 1977. The interface crack with friction in the contact zone. J. Appl. Mech. 44, 780–781.

Cox, H.L., 1952. The elasticity and strength of paper and other fibrous materials. Br. J. Appl. Phys. 3, 72–79.

Dundurs, J., 1969. Edge-bonded dissimilar orthogonal elastic wedges under normal and shear loading. J. Appl. Mech. 36, 650–652.

Galiotis, C., 1991. Interfacial studies on model composites by laser Raman-spectroscopy. Compos. Sci. Technol. 42, 125–150.

Galiotis, C., 1993. A study of mechanisms of stress transfer in continuous-fiber and discontinuous-fiber model composites by laser Raman-spectroscopy. Compos. Sci. Technol. 48, 15–28.

Graciani, E., Mantič, V., París, F., Blázquez, A., 2005. Weak formulation of axi-symmetric frictionless contact problems with boundary elements: application to interface cracks. Comput. Struct. 83, 836–855.

Graciani, E., Mantič, V., París, F., Varna, J., 2009a. Numerical analysis of debond propagation in the single fibre fragmentation test. Compos. Sci. Technol. 69, 2514–2520.

Graciani, E., Mantič, V., París, F., 2009b. Critical study of existing solutions for a penny-shaped interface crack, comparing with a new boundary element solution allowing for frictionless contact. Eng. Fract. Mech. 76, 533–547.

Graciani, E., Blázquez, A., París, F., Varna, J., 2010a. Numerical analysis of the single fibre fragmentation test using cohesive elements. In: Proceedings of the 14th European Conference of Composite Materials, 7–10 June 2010, Budapest, Hungary.

Graciani, E., Mantič, V., París, F., 2010b. A BEM analysis of a penny-shaped interface crack using the open and the frictionless contact models: range of validity of various asymptotic solutions. Eng. Anal. Bound. Elem. 34, 66–78.

Graciani, E., Varna, J., Mantič, V., Blázquez, A., París, F., 2011a. Evaluation of interfacial fracture toughness and friction coefficient in the single fiber fragmentation test. Procedia Eng. 10, 2478–2483.

Graciani, E., Varna, J., Mantič, V., Blázquez, A., París, F., 2011b. Influence of transverse properties in the modeling of debond propagation in single carbon fiber composites. In: Proceedings of the 32nd Risoe International Symposium on Material Science. Roskilde, Denmark.

Graciani, E., Mantič, V., París, F., 2014. Effect of friction on the size of the near-tip contact zone in a penny-shaped interface crack. Key Eng. Mater. 618, 179–201.

Henstenburg, R.B., Phoenix, S.L., 1989. Interfacial shear-strength studies using the single-filament-composite test. 2. A probability model and Monte-Carlo simulation. Polym. Compos. 10, 389–408.

Herrera-Franco, P.J., Drzal, L.T., 1992. Comparison of methods for the measurement of fibre/matrix adhesion in composites. Composites 23, 2–27.

Huang, Y.L., Young, R.J., 1994. Analysis of the fragmentation test for carbon-fiber epoxy model composites by means of Raman spectroscopy. Compos. Sci. Technol. 52, 505–517.

Irwin, G.R., 1957. Analysis of stresses and strains near the end of a crack traversing a plate. J. Appl. Mech. 24, 361–364.

Kelly, A., Tyson, W.R., 1965. Tensile properties of fibre-reinforced metals: copper/tungsten and copper/molybdenum. J. Mech. Phys. Solid. 13, 329–350.

Kim, B.E., Nairn, J.A., 2002a. Observations of fiber fracture and interfacial debonding phenomena using the fragmentation test in single fiber composites. J. Compos. Mater. 36, 1825–1858.

Kim, B.E., Nairn, J.A., 2002b. Experimental verification of the effects of friction and residual stress on the analysis of interfacial debonding and toughness in single fiber composites. J. Mater. Sci. 37, 3965–3972.

Lacroix, T., Tilmans, B., Keunings, R., Desaeger, M., Verpoest, I., 1992. Modeling of critical fiber length and interfacial debonding in the fragmentation testing of polymer composites. Compos. Sci. Technol. 43, 379–387.

Lacroix, T., Tilmans, B., Keunings, R., Desaeger, M., Verpoest, I., 1995. A new data reduction scheme for the fragmentation testing of polymer composites. J. Mater. Sci. 30, 683–692.

Nairn, J.A., 1992. A variational mechanics analysis of the stresses around breaks in embedded fibers. Mech. Mater. 13, 131–154.

Nairn, J.A., Liu, Y.C., 1997. Stress transfer into a fragmented, anisotropic fiber through an imperfect interface. Int. J. Solid Struct. 34, 1255–1281.

Netravali, A.N., Henstenburg, R.B., Phoenix, S.L., Schwartz, P., 1989. Interfacial shear-strength studies using the single-filament-composite test. 1. Experiments on graphite fibers in epoxy. Polym. Compos. 10, 226–241.

Ohsawa, T., Nakayama, A., Miwa, M., Hasegawa, A., 1978. Temperature-dependence of critical fiber length for glass fiber-reinforced thermosetting resins. J. Appl. Polym. Sci. 22, 3203–3212.

París, F., Cañas, J., 1997. Boundary Element Method, Fundamentals and Applications. Oxford University Press, Oxford.

Sjögren, A., Joffe, R., Berglund, L., Mäder, E., 1999. Effects of fiber coating (size) on properties of glass fiber vinyl ester composites. Compos. Part A 30, 1009–1015.

Tripathi, D., Jones, F.R., 1997. Measurement of the load-bearing capability of the fibre/matrix interface by single-fibre fragmentation. Compos. Sci. Technol. 57, 925–935.

Tripathi, D., Jones, F.R., 1998. Single fibre fragmentation test for assessing adhesion in fibre reinforced composites. J. Mater. Sci. 33, 1–16.

Varna, J., Joffe, R., Berglund, L.A., 1996. Interfacial toughness evaluation from the single-fiber fragmentation test. Compos. Sci. Technol. 56, 1105–1109.

Verpoest, I., Desaeger, M., Keunings, R., 1990. Critical review of direct micromechanical test methods for interfacial strength measurements in composites. In: Ishida, H. (Ed.), Controlled Interphases in Composite Materials. Elsevier Science Publishing, pp. 653–666.

Whitney, J.M., Drzal, L.T., 1987. Axisymmetric Stress Distribution Around an Isolated Fiber Fragment. ASTM Special Technical Publication.

Wisnom, M.R., 1999. Size effects in the testing of fibre-composite materials. Compos. Sci. Technol. 59, 1937–1957.

Wu, W., Desaeger, M., Verpoest, I., Varna, J., 1997. An improved analysis of the stresses in a single-fibre fragmentation test. 1. Two-phase model. Compos. Sci. Technol. 57, 809–819.

Wu, W., Verpoest, I., Varna, J., 1998a. An improved analysis of the stresses in a single-fibre fragmentation test. 2. Three-phase model. Compos. Sci. Technol. 58, 41–50.

Wu, W., Verpoest, I., Varna, J., 1998b. A novel axisymmetric variational analysis of stress transfer into fibres through a partially debonded interface. Compos. Sci. Technol. 58, 1863–1877.

Wu, W., Verpoest, I., Varna, J., 2000. Prediction of energy release rate due to the growth of an interface crack by variational analysis. Compos. Sci. Technol. 60, 351–360.

Fiber failure and debonding in axially loaded UD composite materials

8

Andrejs Pupurs [1,2] *and Linqi Zhuang* [3]
[1]Riga Technical University, Riga, Latvia; [2]Lulea University of Technology, Lulea, Sweden; [3]ANSYS, Houston, TX, United States

8.1 Introduction

One of the reasons behind the suitability of polymeric composites for applications in high-performance structures is their apparent insensitivity to fatigue when loaded in fiber direction even at stress levels close to static fracture strength. However, the practical use of the high design stress value is limited because prediction of the final fatigue failure of composites is very complex compared to fatigue of metallic materials (Talreja, 1981). The reasons for the complexity of composite fatigue behavior are several fundamentally distinct and load-dependent damage mechanisms that govern the final failure. The occurrence and growth of a specific damage mechanism in composites also depend on the loading mode (tensile, compressive, etc.) and on whether the loading is parallel or inclined to the fiber direction.

It has been shown that the fatigue life of the off-axis plies is much shorter than for the 0 degree plies aligned in the main loading direction. In fact, the fatigue life of a composite is often indistinguishable from fatigue life of the 0 degree plies which are the last ones to fail in a composite (Andersen et al., 1996; Liu and Lessard, 1994; Talreja and Chou, 1993).

It has been estimated in Dauskardt et al. (1993) that more than 80% of all service failures of structures are due to fatigue. Therefore, in order to design safe and fatigue-resistant composite structures, it is essential to properly understand the involved micro-damage mechanisms and to develop models that could accurately predict the damage propagation under arbitrary loading conditions.

This implies that analyzing the microdamage mechanisms in 0 degree plies of composite is important for improving the performance of composites and for accurate prediction of ultimate failure of composite in static and cyclic loading.

Different damage mechanisms may exist in unidirectional (UD) composites as it will be described ahead; however, the focus in the present chapter is on modeling fiber failure and fiber/matrix interface debonding. This mechanism, compared to other possible damage mechanisms, was shown to have higher driving forces for propagation (Pupurs and Varna, 2010). A methodology for debond growth analysis in UD composites subjected to quasistatic and cyclic tensile loading is presented in the sections to follow. In addition, the effects of neighboring fibers and proximity of surface on debond growth are presented and analyzed.

Modeling Damage, Fatigue and Failure of Composite Materials. https://doi.org/10.1016/B978-0-443-18489-5.00020-5

8.2 Damage mechanisms in UD composites under quasistatic loading

In polymeric UD composites, the fiber strain to failure is much smaller than the matrix strain to failure and when they are loaded in quasistatic tension in fiber direction, the first fiber breaks occur in somewhat random positions. This is because of statistical defect size distribution in fibers leading to randomness in the strength which can be described by the well-known Weibull distribution, see, for example, Chi et al. (1984), Curtin and Takeda (1998), Harlow and Phoenix (1981), and Pitt and Phoenix (1983). Fiber breaks reduce the axial stiffness of the UD composite (Laws and Dvorak, 1987). Stress transfer through the fiber/matrix interface ensures the possibility of multiple fracture (breaks) of the same fiber.

Depending on the properties of fibers, matrix, and the fiber/matrix interface, several scenarios of events may follow the formation of the fiber breaks, see Fig. 8.1a). (a) The crack may propagate from fiber into matrix until it is arrested by the neighboring fiber (Beyerlein and Phoenix, 1997; Laws and Dvorak, 1987) where it may cause a fiber break or may be deflected as a debond crack growing along the interface of the fiber; (b) shear yielding of matrix may occur at the fiber break tip blunting the crack (Behzadi et al., 2009); and (c) debond cracks can grow from the fiber break along fiber/matrix interface (Gamstedt, 2000).

The fiber/matrix debond growth is often observed and quantified using single-fiber fragmentation test (SFFT), see, for example, Goutianos and Peijs (2001), Graciani et al. (2011), and Pupurs et al. (2012). In a composite, the fiber/matrix debond crack may grow until it meets another debond crack growing along the same or neighboring fiber. Finally, the microcracks merge forming a macroscale crack leading to ultimate (final) failure of the UD composite as schematically shown in Fig. 8.2.

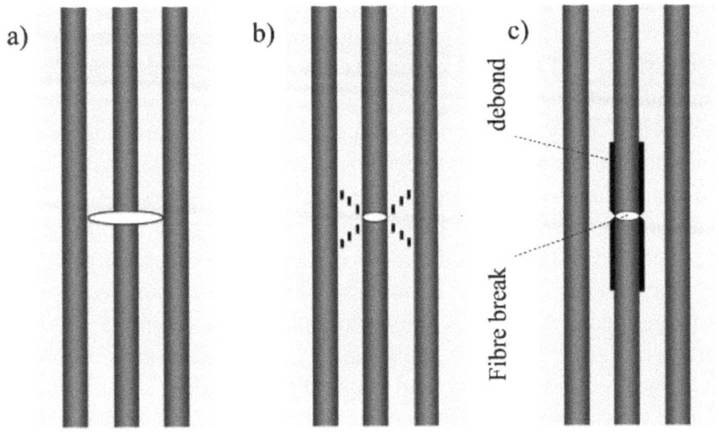

Figure 8.1 Damage development scenarios after fiber break formation. (a) Crack propagation in matrix, (b) matrix yielding, and (c) debonding of the fiber/matrix interface.

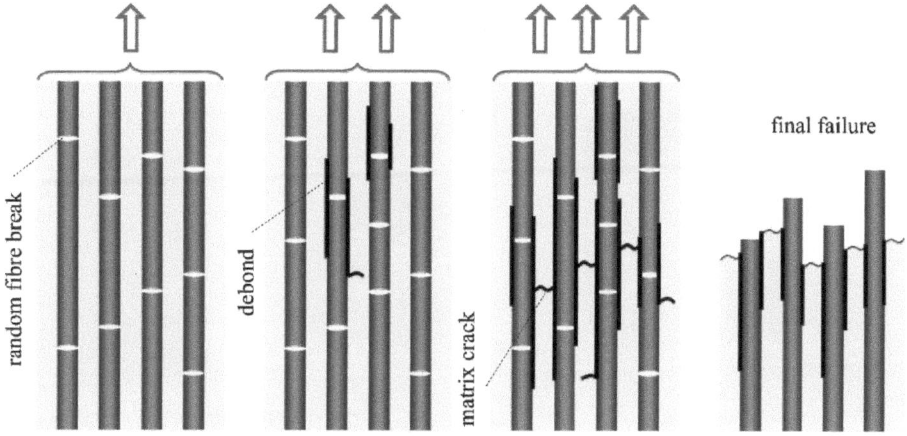

Figure 8.2 Schematic sequence of damage events leading to the ultimate (final) failure of a UD composite subjected to increasing tensile load.

8.3 Fiber/matrix interface debond growth

In the present chapter, we consider only fiber/matrix debond growth initiating from fiber breaks as shown in Fig. 8.1c. The fiber/matrix debond crack propagation is typically analyzed by applying linear elastic fracture mechanics (LEFM). The crack growth analysis in terms of LEFM requires calculation of the energy release rate (ERR). In quasistatic loading, the crack propagation criterion is formulated as when the debond growth ERR reaches the critical ERR value. Calculation of ERR has been widely applied for debond growth characterization in SFFT. The used calculation methods cover a wide spectrum from approximate analytical (Wu et al., 2000) to numerical such as finite element method (FEM) (Pupurs and Varna, 2013a; Pupurs et al., 2013) or boundary element method (BEM) (Graciani et al., 2009, 2011). Using BEM, a very accurate numerical local stress state distributions at the debond crack tip (stress intensity factors and degree of singularity) can be obtained (Graciani et al., 2009). However, this method is limited to isotropic constituents and, hence, not applicable for carbon fibers (transversally isotropic) or other anisotropic constituents in the model.

The ERR for debond crack propagation in a UD composite has been previously analyzed, for example, in McCartney (2013), Pupurs and Varna (2010, 2013a, 2013b), and Pupurs et al. (2013). In McCartney (2013), frictional sliding of debond faces was analyzed using a fiber/matrix unit with free outer surface without including in the analysis the surrounding composite. It was later shown in Pupurs and Varna (2017) that neglecting the presence of the surrounding composite and using only a fiber/matrix unit leads to large overestimation of the ERR.

The UD composite may be represented by a model with axial symmetry shown in Fig. 8.3: a broken and partially debonded fiber surrounded by matrix embedded in the effective composite.

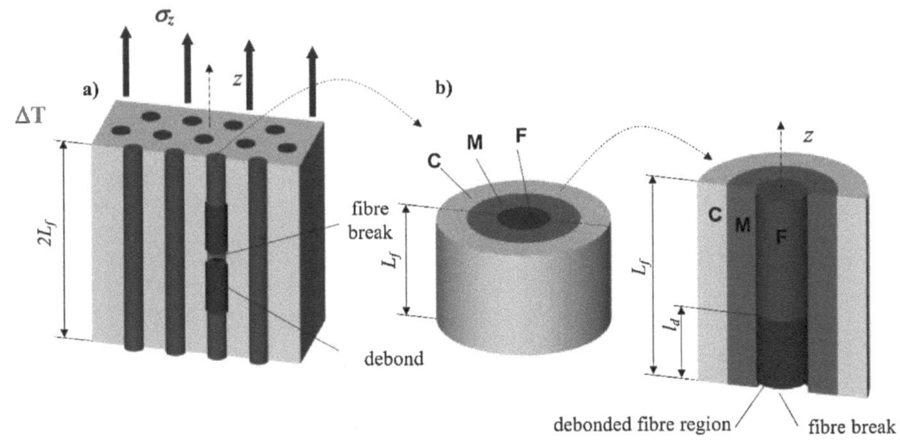

Figure 8.3 (a) UD composite with random fiber break and partial interface debonding. (b) Representation of UD composite by concentric cylinder assembly model. *C*, effective composite; *F*, fiber; *M*, matrix.
Redrawn from Pupurs and Varna (2010).

In general, thermal stresses that form during the cool-down from manufacturing to room temperature should also be taken into account additionally to mechanical loading. Due to larger thermal expansion coefficient of the matrix than that of the fibers in polymeric composites, the interfacial radial thermal stress is compressive in an axisymmetric model. Additional compressive radial stresses arise from the Poisson's effect in tensile loading.

Regarding the local stresses at the debond tip, reference Graciani et al. (2009) showed that in single-fiber composite, only the shear stress is singular. Consequently, there is always a contact between the debonded fiber and the matrix and the debond growth is in pure Mode II.

It has been shown that in quasistatic loading (Graciani et al., 2009; Kim and Nairn, 2002a), the friction reduces the ERR and hence the debond growth rate is lower. However, to find a steady-state solution for debond growth, the friction at the interface has to be neglected.

The applicable analysis methods for fiber/matrix debond growth depend on the debond length. For long debonds, where the debond crack tip is far from the fiber break, both analytical and numerical models can be used. Such debond cracks propagate in steady-state manner as described in Section 8.3.1.1. For short debonds, the stress state at the debond crack tip interacts with the stress state in the vicinity of the fiber break. Therefore, numerical models are typically used for the ERR calculation. In the present chapter, the analysis of steady-state (long) and short-debond growth in UD composites is described in separate subsections.

8.3.1 Steady-state debond growth

8.3.1.1 Analytical models

When the debond crack tip is far from the fiber break, the local stress profile at the crack front moves along the fiber axis without changing its shape and value, that is, the debond growth will be steady state.

Exact analytical solution has been derived for the ERR for steady-state debond growth, for example, in Pupurs and Varna (2017). The analytical model is briefly described here. Axial tensile load (average stress $\overline{\sigma}$) is applied to the composite with a broken and partially debonded fiber as shown in Fig. 8.4. In addition, negative temperature change, $\Delta T < 0$ can be applied to account for the manufacturing-induced residual thermal stresses. The model shown in Fig. 8.4 has an axisymmetric geometry. In Fig. 8.4, z (or 3) is the axial direction, r, θ (or 1, 2) are the radial and hoop directions, respectively. Indices f, m, and C are used to denote for fiber, matrix, and the homogenized composite, respectively. For accurate representation of the surrounding material, the thermoelastic properties of the effective composite phase can be calculated using Hashin's concentric cylinder assembly (CCA) model (Hashin, 1983) and Christensen's self-consistent scheme (Christensen and Lo, 1979).

Mathematically the condition for steady-state growth in terms of Fig. 8.4 can be written as $l_d \gg r_f$ and $L_f \gg l_d$. Here r_f is the fiber radius and L_f is half of the fiber fragment length. The outer radii of the matrix cylinder and the homogenized composite are R and R_E, respectively, and $R_E \to \infty$ represents an infinite composite. When the debond crack grows by dl_d:

(a) the debond crack tip (and the corresponding singular stress state in its vicinity) shifts in the z—direction by dl_d;
(b) the complex stress state in the region close to the fiber break does not change;
(c) the bonded region volume in the model reduces by $\pi R_E^2 dl_d$;
(d) the debonded region volume increases by the same amount.

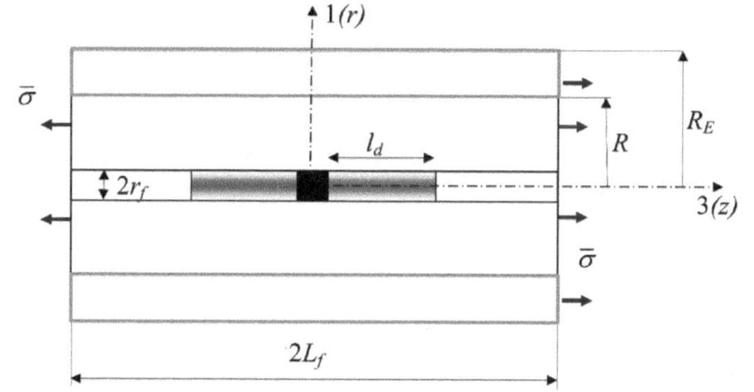

Figure 8.4 Schematic image of a 3-phase model of length $2L_f$ and outer radius R_E containing a fiber break at $z = 0$ and debonds of length l_d propagating symmetrically in axial direction. The *black* (*black* in print version) area represents the fiber crack opening.

The ERR is found to be:

$$G_{II} = \frac{\Delta U}{2\pi r_f dl_d} = \frac{\Delta U_f + \Delta U_m + \Delta U_c}{2\pi r_f dl_d} \tag{8.1}$$

In Eq. (8.1), ΔU is the potential energy change due to debond growth by dl_d. The representative bonded and debonded regions are shown in Fig. 8.5a and b, respectively.

Fig. 8.5a represents a fiber, matrix, and homogenized composite CCA of length dl_d with perfect bonding between phases subjected to the average stress $\overline{\sigma}$ and ΔT. In Fig. 8.5b, the fiber is debonded from the matrix and (in the absence of friction) the axial stress in the fiber is zero. In Fig. 8.5, upper indices b and d on stress and displacement stand for the bonded and the debonded regions, respectively.

Since local stresses and strains are proportional to the macroscopically applied axial strain and the applied temperature, the ERR is a quadratic function of the applied temperature change ΔT and the mechanical strain, denoted as ε_{mech}. It was shown in Pupurs and Varna (2017) that the ERR can be written as a square of the linear combination of these variables in the following form:

$$G_{II} = \frac{E_z^f r_f}{4}\left(k_m^\infty \varepsilon_{mech} + k_{th}^\infty \left(\alpha_z^c - \alpha_z^f\right)\Delta T\right)^2 \tag{8.2}$$

where E_z^f is the longitudinal elastic modulus of the fiber. Expressions for parameters k_m^∞ and k_{th}^∞, related to mechanical and thermal response, respectively, have been derived in Pupurs and Varna (2017).

The ERR solution presented in Pupurs and Varna (2017) can also be used for coated or uncoated single-fiber composites and for any three-phase composite with finite external radius. Since the solution is exact, it may be useful for the evaluation of numerical techniques for ERR determination by, for example, virtual crack closure technique (VCCT) and the J-integral.

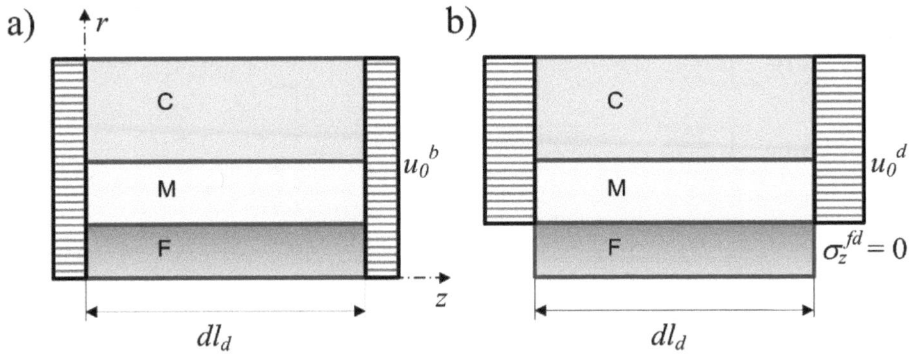

Figure 8.5 Geometrical representation of the bonded region (a), which due to debond crack growth by dl_d changes to debonded region (b) with length dl_d.

For a single-fiber composite case, the analytical solution for parameters k_m^∞ and k_{th}^∞ presented in Pupurs and Varna (2017) is identical to those presented in Nairn and Liu (1996).

8.3.1.2 FEM models

The ERR for the steady-state debond growth can also be calculated using numerical methods, for example, FEM and Eq. (8.1). The change of potential energy U in the system is equal to the difference between the change in strain energy U_S and the additional work W performed during the crack length increase, that is, $\Delta U = \Delta W - \Delta U_S$. In the steady-state growth case, the potential energy change is that at fixed applied force F on the bonded region of length dl_d in Fig. 8.5a becoming debonded as shown in Fig. 8.5b. Here ΔU_S is found as the strain energy difference in these two states, and the work $\Delta W = F\Delta u$, where Δu is the length change of the considered region in Fig. 8.5.

FEM calculations can be performed using commercial software, for example, ANSYS (ANSYS Release 13, 2011) to validate the analytical solution (Eq.8.2) from Pupurs and Varna (2017). A 2-D FEM model with axisymmetric element behavior can be generated according to Fig. 8.5 specifying the elastic properties of fiber, matrix, and the effective composite. For the case of a single-fiber composite, a similar model to the one shown in Fig. 8.5 can be used by considering fiber surrounded by a large matrix phase.

In FEM, the bonded model (Fig. 8.5a) should be generated so that the neighboring areas share the common interface. In the debonded model (Fig. 8.5b), exactly the same geometry as in the bonded model is used. However, the fiber/matrix interface is then represented by two coinciding but separate lines one belonging to the fiber and the other to the matrix area. Since the debonded interface is always in contact, as described above, radial displacement coupling can be used for the nodes belonging to the fiber/matrix interface, maintaining the capability for different axial displacements while eliminating the use of computationally more expensive contact elements (Pupurs and Varna, 2017).

Despite the accuracy of the analytical models, many of the previous studies (Pupurs and Varna, 2010, 2017; Pupurs et al., 2013; Wu et al., 2000) have analyzed an idealized geometry without taking into account the possible nonuniformity of the local fiber distribution, which is present in all practical applications. The local microstructure can affect the stress state around the broken fiber and hence affects the debond growth rate.

A simple approach to study the effect of the neighboring fibers on the ERR related to debond growth is, for example, to use a 5-phase FEM model (Zhuang and Pupurs, 2014) shown in Fig. 8.6. The 2-D axisymmetric FEM model shown in Fig. 8.6 can be used to calculate steady-state debond growth ERR. It is similar to a 4-phase model proposed in Case et al. (1995) simplifying a hexagonal fiber arrangement by a CCA. The 5-phase axisymmetric model shown in Fig. 8.6 consists of a fiber as a central phase (denoted as F), surrounded by a matrix phase (M), neighboring fiber phase (F), another matrix phase (M), and effective composite phase (C); z is the symmetry axis showing the axial direction and r is the radial direction; r_f denotes the radius of the central fiber, ID is the arbitrary interfiber distance between the central and neighboring fiber

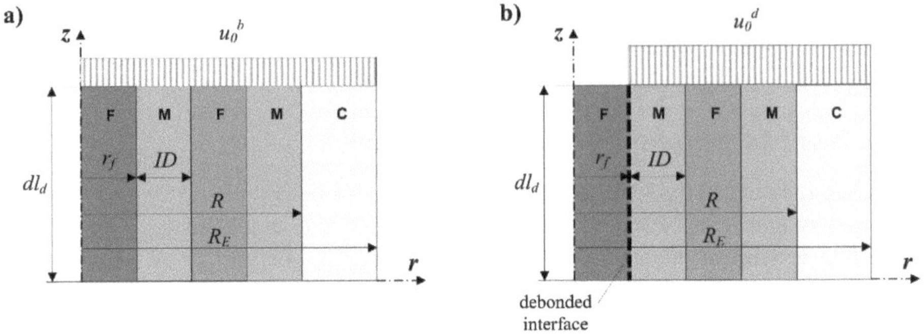

Figure 8.6 Schematic representation of a 5-phase FEM model: *C*, effective composite; *F*, fiber; *M*, matrix. (a) BBonded region; (b) debonded region.

cylinder, R is the outer radius of fiber/matrix unit assembly, R_E is the external radius of the concentric cylinder model including the effective composite phase, and dl_d is the length of the model. Fig. 8.6a shows the bonded model, and Fig. 8.6b shows the debonded model with boundary conditions defined similarly as for 3-phase models described above (Fig. 8.5).

A more detailed analysis of the effect of neighboring fiber on fiber/matrix debond growth can be performed using, for example, a 3-D FEM model with neighboring fibers arranged in hexagonal packing (Zhuang et al., 2016a). In such model, the broken and partially debonded fiber is located at the center of the model and the neighboring fibers are situated in a hexagonal pattern surrounding the broken central fiber. The cross-section view of such model and geometrical parameters are shown in detail in Fig. 8.7. The steady-state ERR can be calculated without knowing the exact shape

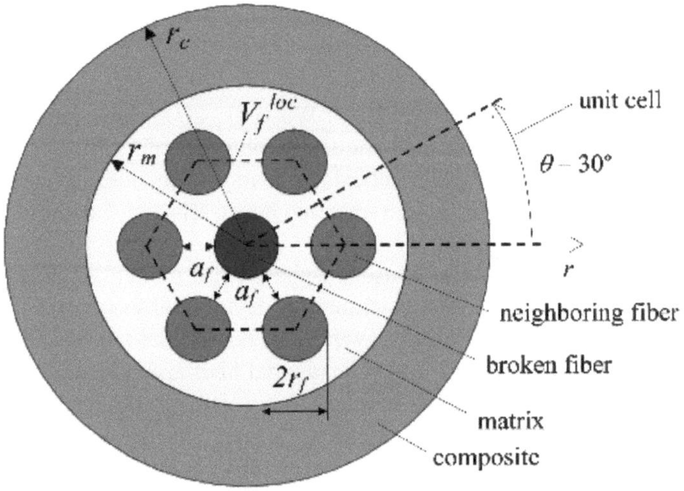

Figure 8.7 Cross-section and geometrical parameters of a hexagonal 3-D FEM model for analysis of the effect of neighboring fibers on debond growth.

of the debond front by considering representative volumes from bonded and debonded regions far from the crack front. The ERR can be calculated from potential energy change due to debond growth by dl_d, see Eq. (8.1). 3-D FEM models of representative bonded and debonded regions are schematically shown in Fig. 8.8a and Fig. 8.8b, respectively. The representative regions are circular sectors corresponding to $\theta = 0 - 30$ degree following the division of a unit cell in Fig. 8.7. In Fig. 8.8, F, M, and C denote the central fiber, matrix, and effective composite phases, respectively. As shown in Fig. 8.7, the central and neighboring fibers are embedded in a matrix phase with the outer radius r_m. The fiber volume fraction V_f in this region is equal to the fiber content in the effective composite surrounding the unit with seven fibers. The effective (homogenized) composite phase with volume fraction V_f has the outer radius r_c. As shown by a parametric study, the size of the outer composite phase radius r_c should be at least $r_c = 5r_m$ to satisfactorily represent an infinite effective composite without magnification of ERR values. The 3-D bonded and debonded models in Fig. 8.8a and b, respectively, are defined similarly as in the axisymmetric model (Fig. 8.5). Debonded interface in the debonded model (Fig. 8.8b) is defined using contact elements. To calculate the potential energy change ΔU due to debond growth by dl_d and to calculate ERR using Eq. (8.1), a distributed force P is applied on bonded and debonded models as shown in Fig. 8.8.

The effect of the clustering of the neighboring fibers on steady-state debond growth ERR can be evaluated by a parametric analysis studying different interfiber distance values, a_f, without changing the radius of the fiber/matrix unit r_m. For each interfiber distance a_f, a local fiber volume fraction V_f^{loc} can be defined, which represents the volume fraction of fibers inside the hexagon joining the centers of the surrounding fibers, as illustrated in Fig. 8.7. Thus, local volume fraction V_f^{loc} can be used as a measure of the local fiber distribution, meaning that a lower local volume fraction represents larger interfiber distance a_f and vice versa:

$$V_f^{loc} = \frac{2\pi}{\left(2 + \frac{a_f}{r_f}\right)^2 \sqrt{3}} \tag{8.3}$$

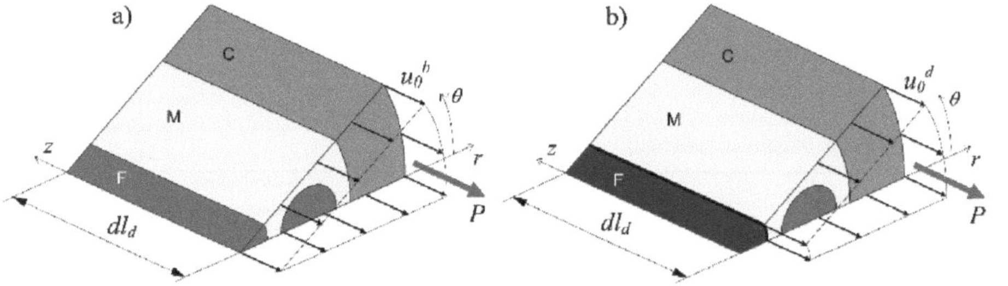

Figure 8.8 Representative volume element of hexagonal model of bonded region (a) with length dl_d, which due to debond growth changes to debonded region (b) with length dl_d. C, effective composite; F, fiber; M, matrix.

The 3-D FEM model shown in Fig. 8.8 and described above allows for a more accurate analysis of the effect of neighboring fibers. However, it cannot predict how the stress state at the debond crack tip and the debond crack propagation modes are affected by the neighboring fibers. This can be evaluated by a full-length hexagonal 3-D FEM model with a broken, partially debonded fiber and debond crack front far away from the fiber break, where the fiber/matrix debond crack initiated, as discussed later. To estimate the steady-state debond growth ERR, the debond crack front must be located sufficiently far from the fiber break and the load application surface should exclude stress perturbations. Different modes of the steady-state ERR can be calculated using, for example, VCCT (Irwin, 1958; Zhuang et al., 2016a). The VCCT can be applied for short debonds (Zhuang et al., 2016b), as described later in Section 8.3.2.1.

8.3.1.3 Results

The exact analytical solution for steady-state debond growth (Pupurs and Varna, 2017) can be compared with FEM results on different material systems of single-fiber composites and UD composites with a certain fiber volume content V_f. In FEM models (see Fig. 8.5), the size of the outer effective composite phase is always finite. Therefore, the influence of the chosen size of the composite radius on the ERR has to be evaluated.

The constituent elastic properties used in the calculation examples taken in the present chapter are given in Table 8.1 unless stated otherwise. In Table 8.1, CF and GF denote carbon and glass fibers, respectively. Epoxy matrix is denoted as EP. The elastic properties of the isotropic epoxy matrix, unless stated otherwise, are $E_m = 3.5$ GPa, $\nu_m = 0.40$, and $\alpha_m = 60 \cdot 10^{-6}$ 1/°C.

Properties for carbon fibre/epoxy (CF/EP) and glass fiber/epoxy (GF/EP) composites with $V_f = 0.6$ calculated using Christensen and Lo (1979) and Hashin (1983) are presented in Table 8.2.

The same constituent and composite notations and properties as listed above are used throughout all calculation examples presented in this chapter. Coordinate axes for material properties given in Tables 8.1 and 8.2 are given in coordinate system shown in Fig. 8.4.

In Table 8.3, analytical and FEM results are presented for single-fiber composites. Results are presented in the form of coefficients in Eq. (8.2), from which ERR for any arbitrary loading case can be calculated. The large number of significant digits is used to validate the accuracy of the FEM model.

Table 8.1 Fiber elastic properties.

	E_3	E_1	G_{31}	ν_{31}	ν_{12}	α_3	α_1
Material	**[GPa]**	**[GPa]**	**[GPa]**	**[-]**	**[-]**	**[1/°C]**	**[1/°C]**
CF	500	30	20.0	0.20	0.45	$-1 \cdot 10^{-6}$	$7.8 \cdot 10^{-6}$
GF	70	70	29.2	0.20	0.20	$4.7 \cdot 10^{-6}$	$4.7 \cdot 10^{-6}$

Table 8.2 Composite elastic properties.

	V_f	E_3	E_1	G_{31}	G_{12}	ν_{31}	ν_{12}	α_3	α_1
Material	[-]	[GPa]	[GPa]	[GPa]	[GPa]	[-]	[-]	[1/°C]	[1/°C]
CF/EP	0.6	301.4422	11.0389	4.0625	3.5767	0.2734	0.5432	$-0.6631 \cdot 10^{-6}$	$35.8513 \cdot 10^{-6}$
GF/EP	0.6	43.4425	13.7145	4.3140	4.6808	0.2726	0.4650	$6.8605 \cdot 10^{-6}$	$32.2140 \cdot 10^{-6}$

Table 8.3 Effect of finite outer radius of the single-fiber composite model on k_m^∞ and k_{th}^∞.

		CF/EP		GF/EP	
R_E/r_f	Method	k_m^∞	k_{th}^∞	k_m^∞	k_{th}^∞
1.5	FEM	10.27675462	11.57522280	3.93984010	4.52243286
	Analytical	10.27675464	11.57522282	3.93984011	4.52243287
5.0	FEM	2.62432555	2.66562097	1.34351554	1.38878643
	Analytical	2.62432555	2.66562098	1.34351554	1.38878643
20.0	FEM	1.16442272	1.16820928	1.02027832	1.04114888
	Analytical	1.16442272	1.16820928	1.02027832	1.04114889
Infinite	Analytical	0.99942620	1.00182807	0.99582850	1.01530001

Results for UD composites modeled as a 3-phase model (Fig. 8.5) with fiber content $V_f = 0.6$ are presented in Table 8.4.

The quadratic dependence of ERR (see Eq.8.2) on temperature change ΔT and the subsequent mechanical strain ε_{mech} were validated in Pupurs and Varna (2017) by performing an arbitrary combined thermomechanical loading case showing excellent agreement.

From Tables 8.3 and 8.4, it is seen that a small outer radius of the model can lead to significant overestimation of the ERR both for the case of single-fiber composites and UD composites. It was shown in Pupurs and Varna (2017) that the values of parameters k_m^∞ and k_{th}^∞ for both cases are insensitive to material properties. If the size of the outer composite phase is sufficiently large, the values of these parameters are practically equal to one.

The results in Table 8.3 for single-fiber composites show that the accuracy of FEM calculations is extremally high even for the largest matrix radius, $R_E/r_f = 20$, and the ERR parameters are much more sensitive to the matrix radius for carbon fiber composites. For the results to be representative, a matrix radius R_E/r_f >20 has to be used in FEM models, especially for carbon fibers.

Table 8.4 Effect of finite outer radius of the effective composite on k_m^∞ and k_{th}^∞.

		CF/EP		GF/EP	
R_E/R	Method	k_m^∞	k_{th}^∞	k_m^∞	k_{th}^∞
1.5	FEM	1.33818107	1.55207622	1.31718960	1.55605484
	Analytical	1.33818110	1.55207620	1.31718960	1.55605480
5.0	FEM	1.01999386	1.18302994	1.01556215	1.19972889
	Analytical	1.01999390	1.18302990	1.01556210	1.19972890
20.0	FEM	1.00073940	1.16069764	0.99699047	1.17778937
	Analytical	1.00073930	1.16069770	0.99699050	1.17778940
Infinity	Analytical	0.99949410	1.15925340	0.99578830	1.17636920

In numerical calculations, the matrix region is always finite, which introduces an unknown error in the ERR. Using the presented analytical solution, this error can be evaluated.

In Pupurs and Varna (2017), parameters k_m^∞ and k_{th}^∞ for UD composites were shown to be practically insensitive to the fiber volume fraction V_f. In contrast to the single-fiber composite case (Table 8.3), the k-parameters for UD composite (Table 8.4) are much less sensitive to R_E and in FEM modeling, value $R_E/R = 5$ would give sufficient accuracy.

More details of the exact solution and results for ERR for steady-state debond growth are given in Pupurs and Varna (2017).

Numerical results for steady-state debond growth ERR using the 5-phase model (see Fig. 8.6 for reference) showing the effect of the neighboring fibers are presented in Fig. 8.9 for CF/EP composites with different volume fractions. The results show ERR as a function of normalized interfiber distance, which is the interfiber distance normalized with respect to the fiber radius, $ID_n = ID/r_f$, see Fig. 8.6 for reference. The vertical lines in Fig. 8.9 indicate the interfiber distances that would correspond to cases of uniform hexagonal packing. The continuous red (gray in print version) line represents the case with global volume fraction $V_f = 0.6$ and the dashed black (black in print version) line represents the case with $V_f = 0.4$.

For the results shown in Fig. 8.9, the ERR was calculated using ANSYS (ANSYS Release 13, 2011) and Eq. (8.1). In all calculations, fiber radius was fixed to $r_f = 4$ μm, the model length was $dl_d = 2$ μm, and ID was arbitrarily chosen. The radius of the neighboring fiber phase was determined from the condition that it represents the area of six fibers surrounding the central fiber in a hexagonal fiber arrangement. The radius R of the fiber/matrix unit was determined from the given volume fraction V_f and the previously defined geometry. The outer radius of the concentric cylinder model was $R_E = 10R$ following suggestion of a similar parametric analysis performed in Pupurs and Varna (2017).

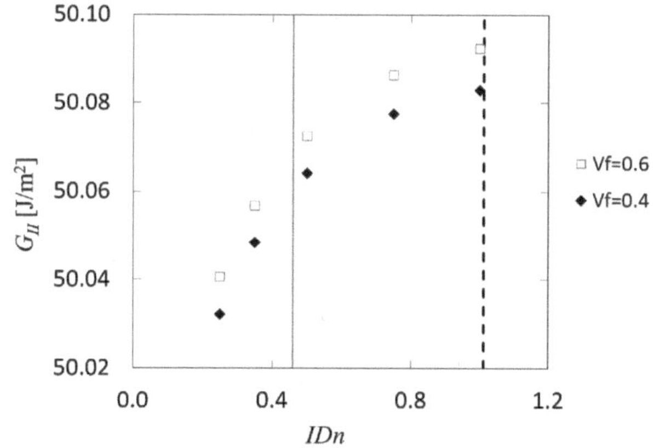

Figure 8.9 The effect of normalized interfiber distance ID_n (dimensionless) on steady-state debond growth ERR in CF/EP composite according to 5-phase model.

In general, results in Fig. 8.9 show that for both studied volume fractions, the ERR slightly increases with the increase of the normalized interfiber distance ID_n. The corresponding analytical result for a 3-phase composite obtained using Eq. (8.1) for $V_f = 0.6$ and $V_f = 0.4$ is equal to 50.45 J/m^2 which is higher than the results obtained with a 5-phase composite model. More detailed results regarding the effect of neighboring fibers on ERR using 5-phase model are provided in Zhuang and Pupurs (2014).

The steady-state debond growth ERR results obtained using hexagonal 3-D FEM models shown in Fig. 8.8a and b are summarized in Table 8.5 as data in the first row, denoted as G_{en}. The ERR is calculated from potential energy change due to debond growth by dl_d, Eq. (8.1). In the calculation examples shown here, the fiber radius was equal to $r_f = 4$ μm, the length of the FEM models was $dl_d = 2$ μm, and the outer radius of the effective composite phase (see Fig. 8.7) was $r_c = 5r_m$. For brevity, results only for CF/EP composite with $V_f = 0.6$ subjected to mechanical loading $\varepsilon_z = 1\%$ are shown with elastic properties of constituents and effective composite given in Tables 8.1, 8.2. The local volume fraction V_f^{loc} of fibers was varied by changing the interfiber distance a_f. Composites with three different local volume fractions were studied $V_f^{loc} = 0.66$, 0.72, and 0.78 corresponding to $a_f = 0.35r_f$, $0.25r_f$, and $0.15r_f$ (see Fig. 8.7).

G_{en} results in Table 8.5 show that increasing the local fiber content leads to the decrease of the ERR, but the effect is negligible. The ERR change is of the same magnitude and even smaller than the one calculated in Zhuang and Pupurs (2014) using the 5-phase model, thus supporting the conclusions of the 5-phase model in the mechanical loading case. The ERR values for uniform fiber distribution obtained using 3-phase assembly model (Pupurs and Varna, 2017) shown in Fig. 8.5, also presented in Table 8.5 in the fourth column, are slightly higher which is consistent with the above-discussed ERR decrease due to fiber clustering. It can be noted that in calculations using the 3-phase model, the size of the outer effective composite cylinder was 10 times the radius of the fiber matrix unit. According to Zhuang et al. (2016a) for CF/EP composite, the ERR values at $V_f = 0.6$ are in average 0.01% higher than for $V_f = 0.4$ which is a similar trend as found in Zhuang and Pupurs (2014). However, the main conclusion from data summarized in Table 8.5 is that the total ERR due to steady-state debond growth is very insensitive to the local variation (increase) of the fiber content.

Table 8.5 Energy release rate values for CF/EP $V_f = 0.6$ in purely mechanical loading $\varepsilon_z = 1\%$.

	V_f^{loc}			
	0.66	**0.72**	**0.78**	**0.60 (3-phase)**
G_{en} [J/m^2]	50.22728	50.22197	50.21351	50.45131
J [J/m^2]	49.38511	49.35217	49.27961	50.17184
G_I [J/m^2]	−0.00238	−0.00241	−0.00210	0.01534
G_{II} [J/m^2]	50.16155	50.21543	50.28309	49.08337
G_{III} [J/m^2]	0.00460	0.00585	0.00500	0.000000

For a more detailed analysis of the steady-state debond propagation modes, a 3-D FEM model with broken, partially debonded fiber and debond front far away from the fiber break was generated according to Fig. 8.13c. The model is a circular sector corresponding to $\theta = 0 - 30$ degree following the division of a unit cell in Fig. 8.7. Although it was discussed in Zhuang et al. (2016a) that due to the considered hexagonal arrangement of neighboring fibers, the shape of the debond front in the steady-state region is most probably a function of the angular coordinate θ, for simplicity, the shape was assumed circular in the presented calculation examples.

To conform with conditions of steady-state debond growth (see description in Section 8.3.1.1), the length of the model $L_f = 80r_f$ and the debond length $l_d = 25r_f$ were used in calculations. These model values were found from a parametric study corresponding to their independence of the ERR values. The geometric parameters of the cross-section of the model with debond front were identical to those of the steady-state models described before in Section 8.3.1.2.

The debonded fiber/matrix interface was modeled by generating coinciding areas with contact elements of the same type as described in Section 8.3.1.2.

Symmetry conditions were applied on nodes corresponding to angular coordinates $\theta = 0$ degree and $\theta = 30$ degree. These symmetry conditions were also applied on nodes of the matrix, neighboring fiber, and the effective composite at $z = 0$. The area of the central fiber at $z = 0$ was free of constraints representing the fiber break.

For brevity, the case with only mechanical loading is presented here with a uniform displacement corresponding to strain $\varepsilon_z = 1\%$ applied in z axis direction on all nodes at $z = L_f$ (Fig. 8.13c).

To obtain more details regarding the possible crack propagation modes in steady-state debond growth, calculations at the front of a long debond were performed. The model details are described in Section 8.3.2. It is noteworthy that the applied 1% strain to the model correspond to slightly lower load than in the steady-state modeling (Fig. 8.8) where the load was selected to give 1% strain in the bonded zone and not in average as it is in the model with debond front. Therefore, the ERR values are slightly lower using the debond front model. Assuming that the debond front is circular ($l_d \neq f(\theta)$), the J-integral was calculated and the VCCT was used to determine G_I (Mode I), G_{II} (Mode II), and G_{III} (Mode III). As shown in Fig. 8.10, the calculated values depend strongly on the angular coordinate θ for CF/EP with $V_f = 0.6$. The variation is large, and the maximum amplitude increases with increasing local fiber content V_f^{loc}. If one would use the critical values of the J-integral as a crack propagation criterion the debond propagation at $\theta = 0$ would start before the propagation starts at other angles. This leads to the conclusion that the debond front in steady-state growth will not be circular. Hence the presented trends should be considered as indicative only.

The average of the J-integral over θ was calculated, and the average values are presented in Table 8.5, along with the averaged G_I, G_{II}, and G_{III} calculated by VCCT. The corresponding results were calculated also for the 3-phase model (see Fig. 8.11, described in Section 8.3.2), where it is independent of the angle θ. It should be noted that the negative G_I values in Table 8.5 are numerical artifacts as a result of small crack

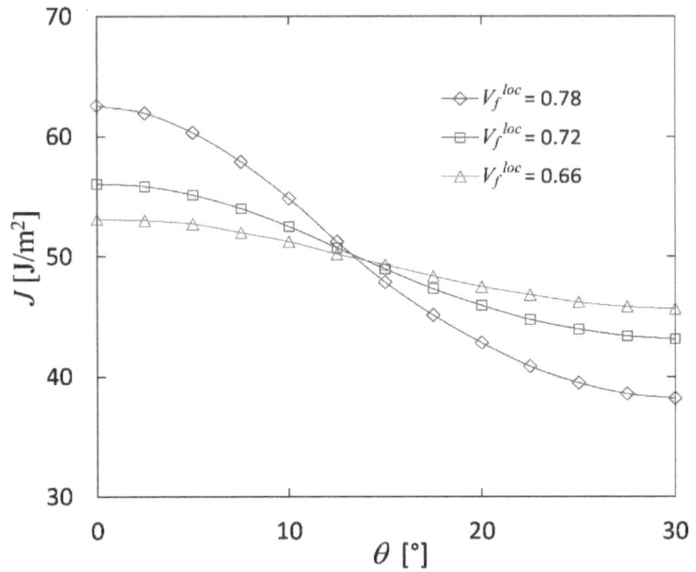

Figure 8.10 Steady-state ERR values for CF/EP $V_f = 0.6$ in purely mechanical loading $\varepsilon_z = 1\%$ obtained using the debond front model in Fig. 8.13c and J-integral calculation routine.

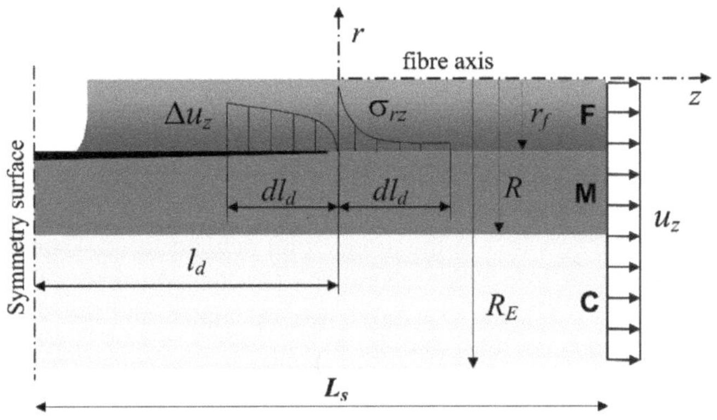

Figure 8.11 Schematic description of a 3-phase FEM model for short debond growth ERR calculation.

tip interpenetrations in FE model. The J-integral values in Table 8.5 are slightly lower than the G_{en}, but the difference is small. Since the value of the J-integral depends on the selected profile of the crack front and since the used circular shape is not the right one for the steady-state case, it can be concluded that the shape of the crack front has a relatively small effect on the total ERR. The values of the J-integral in Table 8.5 have exactly the same trends with respect to the local fiber content as the G_{en}, but the magnitude of the change is slightly larger.

More details on steady-state debond growth ERR for different fiber volume fractions, material systems, and also for thermal loading cases are available in Zhuang et al. (2016a).

8.3.2 Short debond growth

8.3.2.1 FEM models

For relatively short debond cracks, the steady-state crack-growth conditions described above do not apply due to the interaction between stress states at the debond crack tip and the fiber break, where the debond initiated. Due to perturbation of stress, an exact analytical solution for ERR is not attainable. Therefore, for short debond growth analysis, numerical modeling is a more appropriate approach. For axisymmetric problems with singular stresses (at the tip of debond crack), the application of BEM is advantageous for isotropic constituents (Graciani et al., 2009). In most cases, however, the application of FEM is the best choice as it can be applied for any material system subjected to arbitrary loading. When using FEM, the obtained results may be sensitive to mesh size at the tip of debond crack, hence a proper mesh refinement convergence analysis must be performed.

A 3-phase axisymmetric FEM model with broken and partially debonded fiber, schematically shown in Fig. 8.11, can be conveniently applied for calculation of the ERR. For mechanical loading, the boundary conditions are the following: axial symmetry on the central axis (z), symmetry boundary conditions on the fiber break plane (applied to matrix and effective composite phase), zero tractions on the fiber break and on the outer free surface of the effective composite phase (at $r = R_E$), and uniform axial displacement at $z = L_f$, as shown in Fig. 8.11.

There are several numerical techniques for calculating the ERR with FEM. One of the most widely used methods is the VCCT (Irwin, 1958) already mentioned above.

The principle of the VCCT is that the energy released due to debond crack growth by dA is equal to the work which is required to close the newly created surface from size $A + dA$ back to size A. The expression for calculation of the ERR is:

$$G_{II}(l_d) = \lim_{dl_d \to 0} \frac{1}{2dl_d} \int_{l_d}^{l_d+dl_d} \Delta u^{l_d}(z - dl_d)\sigma_{rz}^{l_d}(z)dz \tag{8.4}$$

where the distribution of the relative tangential displacement Δu_z behind the debond crack tip and the shear stress σ_{rz} values ahead of the crack tip can be obtained directly from the FEM model, as shown schematically in Fig. 8.11.

The VCCT technique, despite its simplicity requires performing accurate operations with FEM calculation results including the convergence analysis for the chosen integration length dl_d. An increasing number of automated techniques and specialized elements for fracture mechanics parameter calculations are implemented in the latest releases of the commercial FEM softwares. In the current versions of ANSYS software

(ANSYS Release 13, 2011 and later releases), for example, automated methods for calculating ERR are the J-integral method and the VCCT method, which are performed based on the specified crack tip nodes and crack extension plane direction.

The FEM model shown in Fig. 8.11 consists of three phases: fiber (denoted in Fig. 8.11 as F), matrix (M), and the effective composite (C). It represents a UD composite with certain fiber volume fraction V_f. The importance of the chosen size of the outer effective composite was demonstrated for steady-state debond growth results, see Table 8.4. In the calculation examples presented further, the outer composite radius with respect to the fiber/matrix unit cell was equal to $R_E/R = 10$, which according to the data in Table 8.4 is a representative size for ERR calculations. Contact elements must be generated on the debonded part of the fiber/matrix interface, and friction can be neglected to allow comparison with analytical results presented earlier.

To perform a parametric analysis of short debond growth ERR by varying the debond length l_d, proper length of the model, L_f must be chosen. In Pupurs et al. (2013), a minimum length of $90r_f$ was suggested.

Similarly as described above, short debond growth ERR can be calculated using the 5-phase model (Fig. 8.12) and the hexagonal packing model (Fig. 8.13) which allow estimation of the effect of neighboring fibers on debond growth. The details of both models are described in the following.

The 5-phase axisymmetric FEM model for the calculation of short debond growth ERR is schematically shown in Fig. 8.12. In principle, the model is very similar to the 5-phase model used for steady-state debond ERR calculation (see Fig. 8.6) with the difference that the fiber break is included in the model, the fiber is partly debonded (with debond length denoted as l_d in Fig. 8.12), and the length of the model L_f is

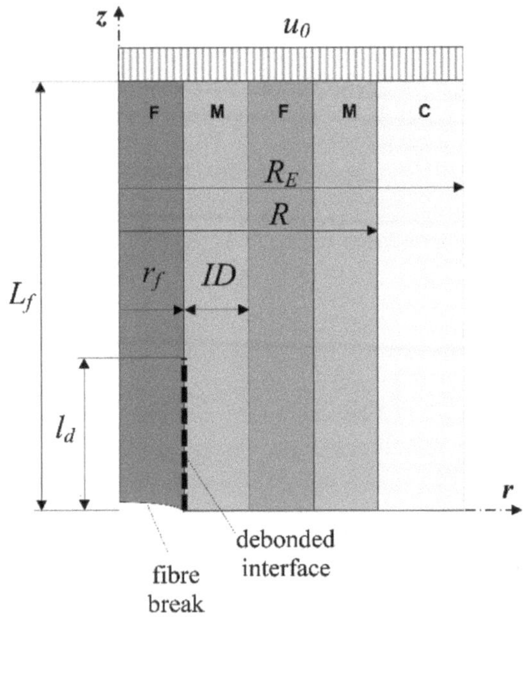

Figure 8.12 Schematic representation of a 5-phase concentric cylinder assembly FEM model for short debond growth ERR calculation: C, effective composite; F, fiber; M, matrix.

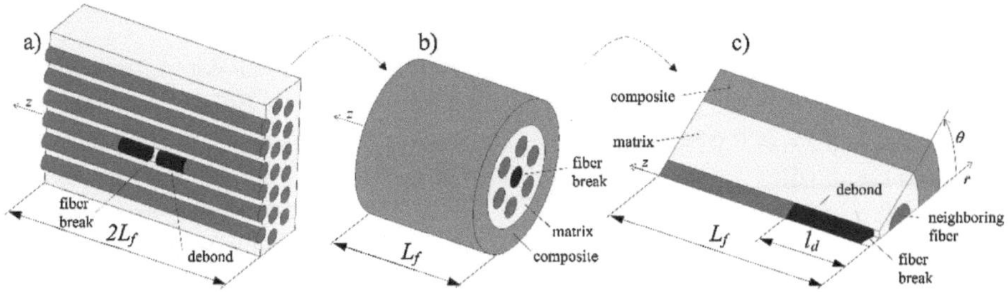

Figure 8.13 (a) UD composite with a broken and partially debonded fiber; (b) hexagonal model in full size; (c) unit cell corresponding to $\theta = 0 - 30$ degree.

significantly larger than that in Fig. 8.6. Regarding boundary conditions in case of mechanical loading, a uniform axial displacement u_0 is applied on the top lines and symmetry boundary conditions are applied on the fiber break plane ($z = 0$) on all phases except the central fiber phase, for which no tractions are applied as shown in Fig. 8.12. The ERR can be calculated, for example, using VCCT. The geometrical variables r_f, ID, R, and R_E were the same as in the case of steady-state debonds shown in Fig. 8.6 and described in Section 8.3.1.2. The length of the 5-phase FEM model must be sufficiently large to allow parametric analysis of the debond length. In Zhuang and Pupurs (2014), for example, model length of $L_f = 200r_f$ was used.

A hexagonal 3-D FEM model for short debond growth ERR calculations is schematically shown in Fig. 8.13. The model is similar to the steady-state debond models shown in Fig. 8.8a and b with cross-section details provided in Fig. 8.7, but with much larger length L_f. Following the symmetry of the cross-section in Fig. 8.7, the FEM model in Fig. 8.13c is a representative unit cell region (circular sector) corresponding to $\theta = 0 - 30$ degree.

Using FEM software, such as, ANSYS (ANSYS Release 13, 2011), the ERR can be calculated at the debond crack front using the built-in VCCT or J-integral routines. 3-D solid elements must be used, and an appropriate mesh refinement in the vicinity of the debond front must be applied to ensure the accuracy of J-integral and VCCT calculations.

For the boundary conditions, the FE model (Fig. 8.13c) is constrained as follows. Nodes at $r = 0$ and $\theta = 0$ degree are fixed. Nodes on $\theta = 0$ degree and $\theta = 30$ degree planes are fixed in the θ-direction. At $z = 0$ plane, symmetric boundary condition is applied except for fiber break surface which is traction free. At the far end $z = L_f$ plane, uniform displacement which equals to 1% average tensile strain ε_z along z direction is applied to all nodes belonging to that plane. The outer cylindrical surface of the model ($r = r_c$) is traction free. The debonded fiber/matrix interface is represented by 3-D contact elements. Previous studies have shown sensitivity of results on the chosen contact algorithm. In Zhuang et al. (2016a, 2016b), for example, pure Lagrange multipliers on normal and tangential contact were generated on the debonded fiber/matrix interface in order to minimize the interpenetration due to radial compression on the contact surface. More details of FEM models including mesh refinement at the debond crack front can be found in Zhuang et al. (2016b).

8.3.2.2 Numerical results for short debond growth

The 3-phase model results for short debond growth ERR in CF/EP composites are presented in Fig. 8.14. In the demonstrated calculation examples, CF/EP composites with volume fraction $V_f = 0.6$ were analyzed. The elastic properties of the studied constituents and composites are the same as those used for steady-state debond growth case and specified in Tables 8.1 and 8.2.

Purely mechanical loading case corresponding to uniform displacement causing 1% average tensile strain of the model is presented in calculation examples.

The outer size of the model was selected according to the ratio $R_E/R = 10$ suggested in Pupurs et al. (2013), see Fig. 8.11 for reference. The length of the model was in all cases equal to $L_f = 100r_f$ also following suggestions in Pupurs et al. (2013).

In Fig. 8.14, the ERR is presented as a function of the normalized debond length, $l_{dn} = l_d/r_f$. Results obtained with different numerical techniques are compared with analytical solutions from Pupurs and Varna (2017), that is, (Eq.8.2). To distinguish between different methodologies, the following notation is used in Fig. 8.14 for presenting the calculation example results: "J-int" represents automated J-integral calculations in ANSYS (ANSYS Release 13, 2011), "GII ANSYS" is a notation for automated VCCT calculations in ANSYS (ANSYS Release 13, 2011), and "GII VCCT" denotes calculations performed using manual VCCT, where a user manually postprocesses the stress and displacement components and selects the integration length (Eq.8.4). Finally, "GII en" denotes results obtained from FEM calculations, which use the change of the strain energy, when the debond crack propagates from length l_d to $l_d + dl_d$ (Eq.8.1). Obviously, with the latter method, two FEM calculations and Eq. (8.1) are necessary for the calculation of the ERR. The strain energy change needed for ΔU in Eq. (8.1) can be calculated in the postprocessing stage of FEM simulation. The results in Fig. 8.14 clearly show the magnification of the ERR when the debond length is short. It is also notable that at normalized debond length of approximately $l_{dn} = 10$, the ERR becomes constant indicating the onset of steady-state growth. Regarding the accuracy of

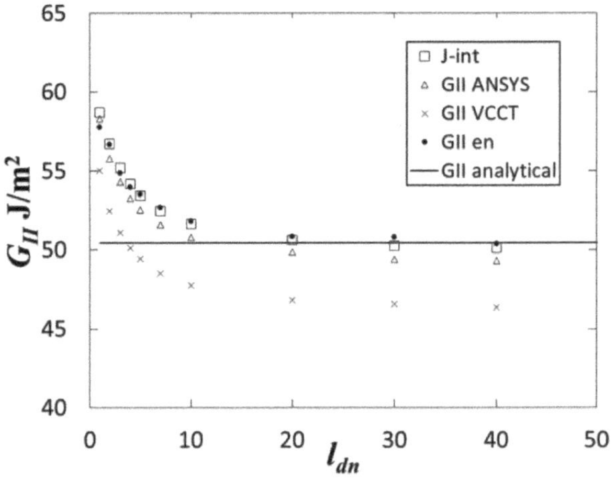

Figure 8.14 Short debond growth ERR with respect to normalized debond length for CF/EP UD composite obtained using 3-phase FEM model.

different numerical techniques, it can be clearly seen from Fig. 8.14 that the manual calculations with VCCT underestimate the ERR significantly, while the other techniques give good agreement with the analytical solution (Pupurs and Varna, 2017) when $l_{dn} \geq 10$. In particular, J-integral calculations ("J-int" in Fig. 8.14) and the energy difference method ("GII en" in Fig. 8.14) give the best agreement with analytical results.

As a fracture mechanics criterion for debond propagation, the calculated values of available ERR presented in Fig. 8.14 are compared to experimental values of interface fracture toughness, typically the Mode II fracture toughness G_{IIc}. The experimental values of G_{IIc} typically obtained from SFFT are available in the literature (Graciani et al., 2011; Kim and Nairn, 2002a, 2002b; Varna et al., 1996). A noteworthy observation from results in Fig. 8.14 is that applying the crack propagation criterion $G_{II} = G_{IIc}$, the debond crack growth is found to be stable in the initial stage (short debonds) as a result of decreasing available ERR, and the propagation becomes unstable at the onset of steady-state debond growth stage, where G_{II} becomes constant with respect to debond length. However, it should be noted that friction on the fiber/matrix interface was neglected in the presented calculation examples. In an experimental test, the potentially unstable steady-state debond crack growth would be limited by the frictional forces at the interface that increase with the debond length. Parametric analysis of the effect of friction on fiber/matrix debond growth ERR for single GF/EP composite has been performed in Graciani et al. (2011).

The 5-phase model results for short debond growth ERR obtained using model schematically shown in Fig. 8.12 are summarized in Fig. 8.15. The ERR was calculated for various debond lengths l_d and interfiber distances ID. The model length in the presented calculation examples was $L_f = 200r_f$, and the outer model size was $R_E/R = 10$ following suggestions described in Pupurs and Varna (2017) and Zhuang and Pupurs (2014). The same size of the 5-phase model outer radius was also used in the steady-state debond growth ERR calculation examples described in Section 8.3.1.

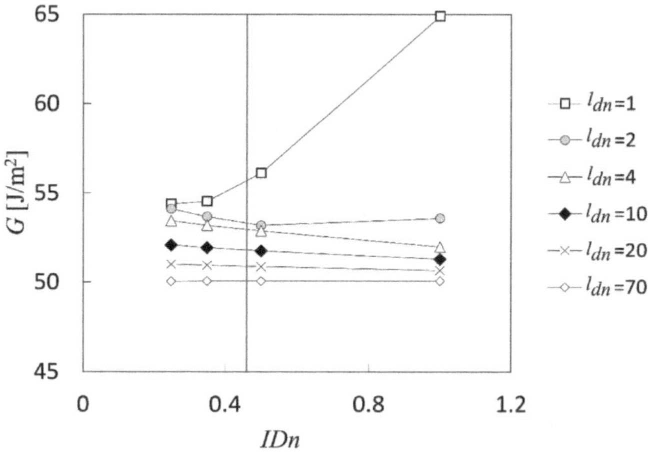

Figure 8.15 The effect of normalized interfiber distance ID_n on the total short debond growth ERR according to the 5-phase model.

The results in Fig. 8.15 show the effect of the interfiber distance on the total ERR (denoted as G) containing both Mode I and Mode II components. ERR is plotted against the normalized interfiber distance $ID_n = ID/r_f$. Curves corresponding to different normalized debond lengths $l_{dn} = l_d/r_f$ are presented. In most cases, G is dominated by Mode II ERR. However, it was found that for very short debond lengths, Mode I ERR is noticeable.

The results only for CF/EP composite with volume fraction $V_f = 0.6$ subjected to mechanical loading $\varepsilon_z = 1\%$ are presented in the following for sake of brevity. The solid vertical line in Fig. 8.15 indicates the interfiber distance that corresponds to uniform hexagonal packing for $V_f = 0.6$. The results in Fig. 8.15 show that when the debond length is very small, for example, $l_{dn} = 1$, the ERR significantly increases with the interfiber distance ID_n. It was also found (not shown here) that for very small debond lengths, the contribution of Mode I is larger than that for longer debond lengths.

More detailed results for short debond growth ERR obtained using 5-phase model are given in Zhuang and Pupurs (2014).

Calculation results for short debond growth ERR using 3-D hexagonal FEM model (shown in Fig. 8.13) are briefly summarized in the following.

For the demonstrated calculation examples, the radius of composite phase was $r_c = 5r_m$ and the length of the composite model was $L_f = 80r_f$, see Fig. 8.7 and Fig. 8.13 for reference to model geometry. These values were chosen based on a previous parametric study reported in Zhuang et al. (2016a). According to Zhuang et al. (2016a), the selected values of the geometrical parameters ensure that the calculated ERR values are not affected by the model boundaries.

It was shown in Table 8.5 that in mechanical loading, the Mode II ERR is much larger than the Mode I and Mode II components for the steady-state debond growth. The same observation has been reported for short debond growth ERR in Zhuang et al. (2016b) where it was shown that the Mode II was the dominant component. Hence in the calculation examples presented in the following, J-integral calculation routine was used to represent the ERR. It was demonstrated in Fig. 8.14 and also discussed in Zhuang et al. (2016b) that J-integral calculation routine gives very good agreement with asymptotic value of the analytical solution for steady-state debond growth ERR. Furthermore, the J-integral calculation routine is typically less sensitive to mesh refinement compared to the VCCT (Zhuang et al., 2016b).

The main results from 3-D hexagonal FEM model for short debond growth ERR are summarized in Fig. 8.16.

Results are shown for mechanical loading $\varepsilon_z = 1\%$, $V_f^{loc} = 0.78$, and they indicate variation of J-integral with respect to the angle θ, see Fig. 8.7 and Fig. 8.13 for reference to the model geometry. Similarly as for steady-state ERR shown in Fig. 8.10, Fig. 8.16 clearly shows that ERR for short debonds is not constant along the debond front, which indicates that the debond front might not be circular during propagation as it was assumed in the FEM model (Fig. 8.13). Results in Fig. 8.16 indicate that the debond crack would grow faster in the directions with the highest ERR. For each debond length, maximum ERR occurs at $\theta = 0$ degree where the distance between the debond front and the neighboring fiber surface is the smallest. The ERR decreases with increasing angle

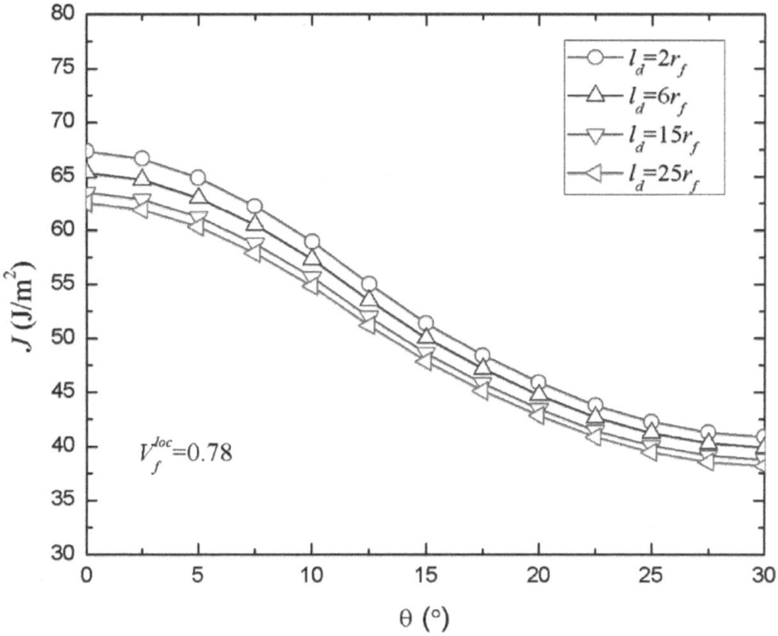

Figure 8.16 Angular dependence of ERR values for CF/EP composite in mechanical loading $\varepsilon_z = 1\%$. The results obtained using hexagonal 3-D FEM model with $V_f = 0.6$, $V_f^{loc} = 0.78$.

until $\theta = 30$ degree. In Zhuang et al. (2016b), it was shown that with decreasing local fiber volume fraction V_f^{loc}, the variation of the ERR along the debond front (θ direction) becomes smaller, as the perturbation from the neighboring fiber becomes smaller with neighboring fibers moving away from the broken fiber. It would be expected that the ERR would be constant along the debond front when neighboring fibers are sufficiently far away where the effect of the neighboring fibers cannot be felt.

Regarding the effect of the debond length, l_d, it can be seen from Fig. 8.16 that the ERR decreases with increasing debond length at each angular location along the debond front similarly as obtained with 3-phase and 5-phase models (Figs. 8.14 and 8.15). It was further discussed in Zhuang et al. (2016b) that for any debond length, the maximum ERR is higher in larger local volume fraction case suggesting debonds tend to grow first in the fiber clustering regions. Reference Zhuang et al., 2016b presents more detailed results of ERR for short debond growth including the effect of material properties.

8.4 FEM models for surface/edge effects

8.4.1 Model formulation

For validation of analytical and numerical models described above, a proper experimental data of debond growth for a given material system are necessary. However, experimental observation of debond growth in UD composites with multiple fibers is rather complex. In the bulk of a composite, for example, microtomography technique could be used. However, since the diameter of carbon and glass fibers is typically in the range between 4 and 20 μm, a high scanning resolution is required which at present significantly limits the scanning speed. An interesting alternative is to perform debond growth measurements on the surface of a UD composite specimen as follows. By using optical microscopy, a certain fiber break on the specimen surface can be identified and the debond growth on that fiber with increasing strain or with increasing number of loading cycles can be recorded. At present, such experimental data are very rare in the literature, see, for example, in Gamstedt (2000). Nevertheless, given the overall simpler procedure of preparing UD specimen compared to SFFT specimen, this could be a feasible experimental method for extraction of the fiber/matrix interface fracture toughness and for validation of numerical models. However, in such an experiment, the fibers are located close to the specimen surface and the stress state of the fiber therefore does not have axial symmetry as would have if embedded in the bulk of composite (see Figs. 8.3 and 8.13). Hence, to understand the stress state and to accurately represent the experimental test, a 3-D FEM model must be used. Such model is schematically shown in Fig. 8.17 and is briefly described in the following.

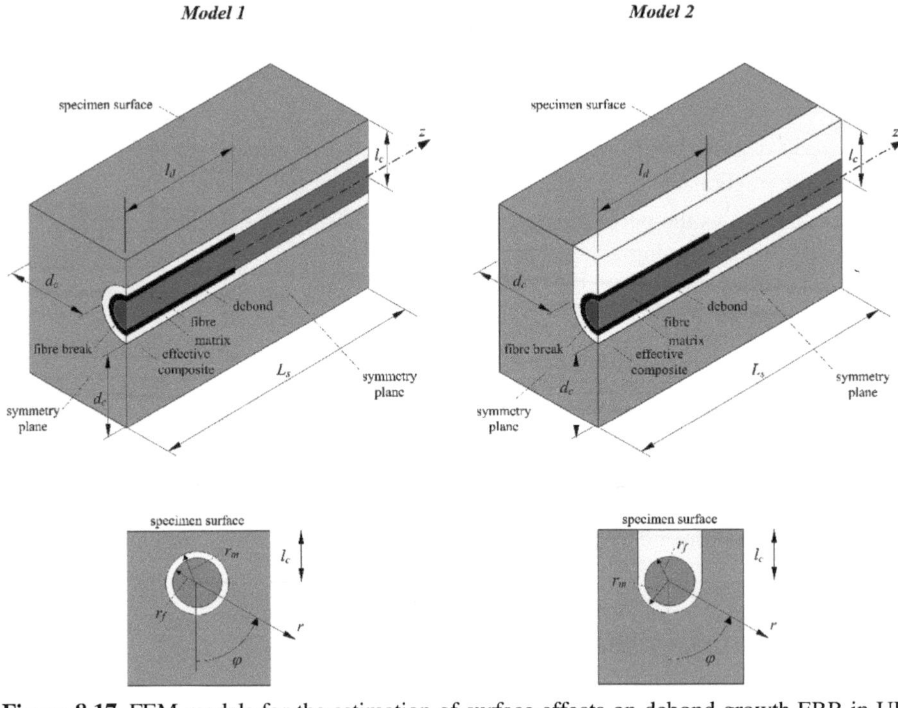

Figure 8.17 FEM models for the estimation of surface effects on debond growth ERR in UD composite: Model 1—the distance from the fiber center to the surface, $l_c > 2.5r_f$; model 2—the distance $l_c \leq 2.5r_f$.
Redrawn from Pupurs and Varna (2013b).

The FEM model consists of three phases: a fiber phase surrounded by matrix phase and the effective composite phase. The elastic constants of the effective composite can be calculated using Hashin's CCA model (Hashin, 1983) and Christensen's generalized self-consistent scheme (Christensen and Lo, 1979). Since a UD composite consists of many fibers, the thermal compressive stresses are much smaller in a UD composite compared to a single-fiber composite. Hence, it is acceptable that they are not included in the numerical analysis.

For the case when a fiber is near the composite specimen surface, axisymmetric approximation is not applicable due to the edge (surface) effects. Two separate cases could be analyzed. Model 1 (Fig. 8.17), where it is assumed that the fiber/matrix unit is surrounded from all sides (including the specimen's surface direction) by the homogenized effective composite phase. When the distance from the composite specimen surface l_c is very small, the use of the effective composite phase on top of the analyzed fiber/matrix unit assumed in Model 1 in Fig. 8.17 is not justified. It is more realistic then that the fiber is surrounded by a neat matrix region as shown in Model 2 in Fig. 8.17.

As shown in Fig. 8.17, a half model with symmetry boundary conditions applied on the symmetry plane can be used to represent the problem. When the debond is close to the specimen surface, the stress state is also not axisymmetric.

For an angle dependent 3-D stress state, the VCCT equation can be formulated as:

$$G_{II}(l_d) = \lim_{dl_d \to 0} \frac{1}{4\pi r_f dl_d} \int_0^{2\pi} \int_{l_d}^{l_d+dl_d} \Delta u^{l_d}(z - dl_d, \phi)\sigma_{rz}^{l_d}(z, \phi) r_f d\phi dz \tag{8.5}$$

The G_{II} defined according to Eq. (8.5) represents the ERR value averaged over all angles and over the distance dl_d. The use of this average value is consistent with the employed assumption that the crack front is circular during propagation.

The debonded fiber/matrix interface with debond length l_d can be modeled by defining coinciding surfaces with contact elements. More details regarding the model geometry, modeling, and meshing are given in Pupurs and Varna (2013b).

8.4.2 Results

In the calculation examples given here, 3-D FEM models schematically shown in Fig. 8.17 were generated in FEM software ANSYS (ANSYS Release 13, 2011). Standard contact elements were generated on the debonded interface, and the friction at the interface was neglected. Eq. (8.5) was implemented in the postprocessing stage of the FEM calculations to calculate the ERR. In the present chapter, only a brief summary of results is given for the sake of brevity. Instead of presenting calculated ERR as a function of the debond length, results are given as debond growth simulations using the Paris law, similarly as in Pupurs et al. (2012). Simulation results summarized in Fig. 8.18 reveal characteristics of ERR for different studied cases—higher growth rate of debonds means higher corresponding values of ERR. The debond growth

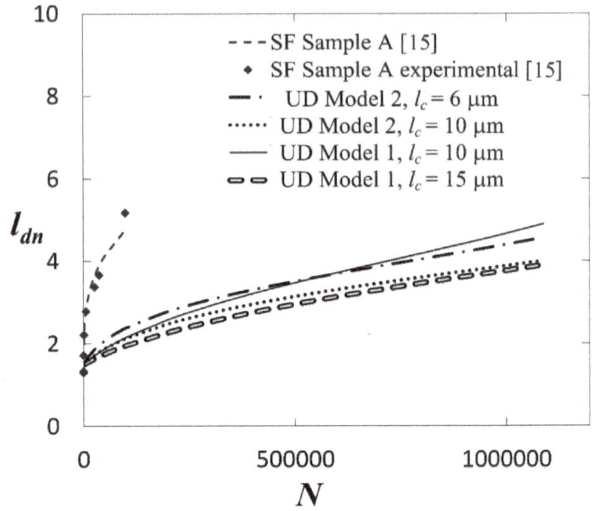

Figure 8.18 Debond growth simulation results for GF/EP composites. l_{dn} is the normalized debond length, N is the number of loading cycles, and l_c denotes the distance from the composite surface according to Fig. 8.17. Redrawn from Pupurs and Varna (2013b).

simulations shown in Fig. 8.18 were performed for GF/EP composite with $r_f = 8.5\mu m$, and the maximum tensile strain level was $\varepsilon_z = 1.76\%$ with cyclic load ratio $R = 0.1$. The geometrical, loading and material properties for GF/EP were taken from Pupurs et al. (2012) in order to compare debond growth rate with experimental data for single-fiber composites also given in Pupurs et al. (2012). Description of the application of Paris law for simulation of debond growth with increase of loading cycles is given in Section 8.5.2.

Thus, fiber radius, loading conditions and constituent properties for all simulation cases shown in Fig. 8.18 are the same as those in Pupurs et al. (2012).

Regarding the effect of the distance from the specimen surface, the results in Fig. 8.18 show that the growth of the debond is the slowest for the fiber located most distant from the composite specimen surface ($l_c = 15\mu m$, calculated using Model 1). For the case of $l_c = 10\mu m$, the debond growth simulations were performed using both Model 1 and Model 2. The results obtained using Model 1 show faster debond growth than when using Model 2. According to Model 2, the debond around the fiber closer to the specimen surface ($l_c = 6\mu m$) grows faster than the same debond at $l_c = 10\mu m$.

Notably, the debond growth rate is much faster in the SFFT specimen subjected to the same strain as in the UD composite. It is also notable that slightly higher ERR for debonded fibers close to the specimen surface leads to significantly higher debond growth rate. Since debond growth is faster in the vicinity of the surface, a method of studying debond growth on specimen surface can be suggested for accelerated testing and characterization of interfaces compared with testing in the bulk of composite.

Detailed parametric analysis results showing the ERR values, stress distributions, and how they depend on the distance from specimen surface are presented in Pupurs and Varna (2013b).

8.5 Debond growth in cyclic loading

8.5.1 Damage mechanisms

Most of the composite structures in their lifetime are subjected to cyclic mechanical and thermal loading. Unlike in quasistatic loading described in the previous sections, where fiber/matrix debond growth can be attributed to reaching the critical ERR (fracture toughness) of the interface, in cyclic loading, fiber/matrix debonds can grow at ERR values below the static fracture toughness of the fiber/matrix interface. The present section describes the damage mechanisms, modeling approaches, and experimental determination of parameters that characterize debond growth in UD composites under cyclic loading.

When a UD composite is subjected to uniaxial tension–tension cyclic loading along the fiber direction, the damage initiation and its further evolution strongly depend on the level of the applied stress. Possible damage initiation and evolution mechanisms related to composites have been widely described in Harris (2003) and Talreja (1987). The four conceptual damage development scenarios in tension–tension fatigue of UD composites are illustrated in Fig. 8.19.

The most extreme development shown in Fig. 8.19a is composite failure at very high applied stress levels, which cause the UD composite to fail within a few load cycles. The stress level for such case exceeds the average composite tensile strength ($\sigma \geq \sigma_c$), and the failure takes place in a manner almost identical to quasistatic failure.

It is known that carbon and glass fibers are brittle with a probabilistic distribution of strength. The weakest fibers may fail much earlier than when the UD composite fails. If a stress level larger than the strength of the weakest fiber but smaller than the composite strength ($\sigma_f \leq \sigma < \sigma_c$) is applied in tension–tension fatigue, fibers may break in random positions already during the first applied load cycle as shown in Fig. 8.19b. Carbon and glass fibers are usually not prone to degradation in mechanical fatigue. An isolated fiber break causes shear stress concentration at the interface close to the tip of the broken fiber. Therefore, when the maximum load is repeated, further damage evolution may be in form of growing interface debond cracks (Fig. 8.19b) starting from the tips of the fiber breaks.

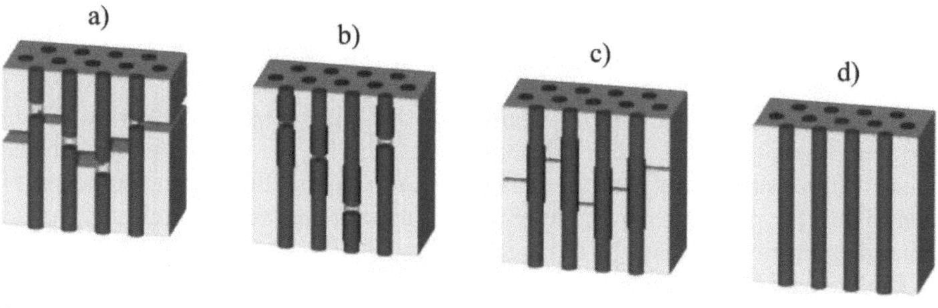

Figure 8.19 Damage evolution scenarios in UD composites under tension–tension cyclic loading. (a) Overload stress $\sigma \geq \sigma_c$, (b) high stress loading ($\sigma_f \leq \sigma < \sigma_c$), (c) medium stress ($\sigma_{FL} \leq \sigma < \sigma_f$), and (d) low stress ($\sigma \leq \sigma_{FL}$).

The above described is not the only possible damage evolution scenario in high stress tension–tension fatigue. Since stress concentration occurs near the debond crack tip, a possibility of cracks propagating from debond crack tips into matrix also exists, as shown in Fig. 8.2.

In medium stress tension–tension fatigue loading, where the maximum stress is smaller than the stress necessary to cause breakage of the weakest fibers during the first cycle ($\sigma_{FL} \leq \sigma < \sigma_f$), the mechanism illustrated in Fig. 8.19c can occur. While the fibers are assumed as nondegrading in mechanical fatigue, the polymer matrix is known to have fatigue beyond a certain fatigue limit (FL), above which matrix cracking can occur. In cyclic loading matrix cracks may develop from manufacturing defects (e.g., voids, fiber clusters, etc.). These cracks after reaching the fiber/matrix interface may deviate and propagate as debonds. The debond growth for such case was analyzed in Pupurs and Varna (2010) showing that the ERR and hence the crack growth driving force is several orders of magnitude lower than in case of debonds initiating from fiber breaks.

The final possible scenario, shown in Fig. 8.19d, is related to the infinite fatigue life and is the opposite of the extreme scenario of nearly instant failure shown in Fig. 8.19a. The applied strain level for the case shown in Fig. 8.19d is below the FL of the polymer matrix ($\sigma \leq \sigma_{FL}$). Initial damage or defects may be present also in this case, but here either the defect size is too small or the arresting mechanisms prevent any propagation.

The damage evolution scenarios described above are somewhat idealized. In practice, depending on the material properties and the level of applied stress, loading history, etc., these mechanisms may as well occur simultaneously.

Because different damage mechanisms occur depending on the level of applied stress, analyzing the fatigue life of UD composites can be complex. A comprehensive way to demonstrate the fatigue life is through fatigue-life diagrams. The fatigue-life diagrams were first introduced by Talreja (Talreja, 1981). These diagrams are useful as a framework to interpret fatigue properties of UD composite materials. In Fig. 8.20, a typical fatigue-life diagram for UD composites loaded in the fiber direction is shown. On the vertical axis of a fatigue-life diagram, the maximum applied fatigue strain is plotted, while on the horizontal axis logarithm of the number of cycles to failure is plotted. Strain instead of stress is typically used in fatigue-life diagrams since both fibers and matrix are subjected to the same strain, while stresses in the two phases differ depending on the volume fraction and the elastic properties of the two phases (Talreja, 1981).

The regions of the fatigue-life diagram defined in Fig. 8.20 each correspond to damage mechanisms previously described and depicted in Fig. 8.19. The presented mechanisms differ substantially from each other and require different approaches for analysis. In the present chapter, the focus is only on fiber breakage-related interface debonding mechanism (Fig. 8.19b).

Fatigue damage mechanisms under loading inclined to fiber direction (off-axis fatigue) including different lay-ups of composites are described in detail in Talreja (1981) and Talreja and Singh (2012).

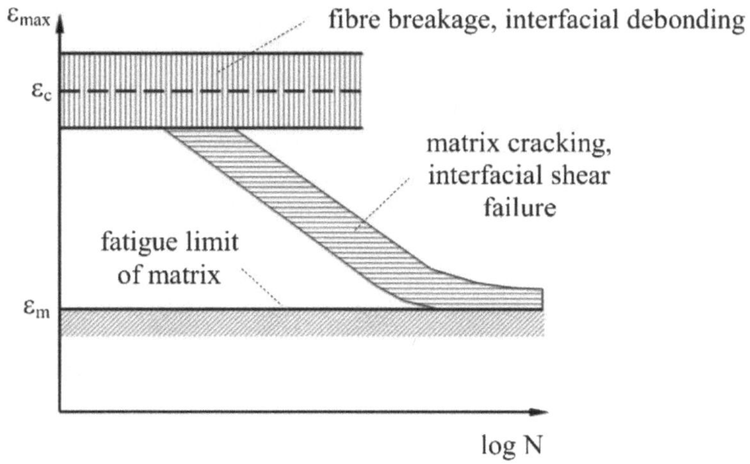

Figure 8.20 Fatigue-life diagram for UD composites under loading parallel to fibers. Redrawn from Talreja (1981).

8.5.2 Modeling

In cyclic loading with constant amplitude, we usually assume that fibers do not experience fatigue and all fiber breaks occur during the first cycle. With an increasing number of cycles, the growth of debonds along the fiber/matrix interface is expected and thus the same LEFM concepts of ERR may be used.

A hypothesis was validated in Pupurs et al. (2012) that the growth of each individual debond can be characterized by Paris' type of law, where the debond growth rate is a power function of the ERR change in one cycle. To validate this hypothesis, it was shown in Pupurs et al. (2012) that experimental data on debond growth in a single-fiber composite under cyclic loading can be described by a model based on a Paris (power) law. Material parameters in this law were determined from a fitting procedure in Pupurs et al. (2012).

The identified power law parameters are material properties and could also be used to model the debond growth in fatigue in UD composites of the same fiber/matrix system.

Experimental debond growth observation in UD composites is however difficult, and with optical microscopy, it is possible only on specimen surface (Gamstedt, 2000) as described in Section 8.4.

An alternative experimental method is to measure the debond length and its increase in a single-fiber composite, which is a system consisting of a single-fiber embedded in a large block of resin.

Subjecting a single-fiber composite with fragmented fiber to axial cyclic loading, we may observe debond length increase with the number of cycles (Pupurs et al., 2012). Similar fatigue test on model composites with only two to five fibers has been reported in Goutianos and Peijs (2001).

In the procedure to extract the power law parameters, friction at the interface may be neglected because at present it is not clear how to include it in cyclic loading, where the sliding cyclically changes direction and the friction effect is very dependent on the loading history.

The most straightforward approach is to consider tension−tension cyclic loading with load ratio $R = \varepsilon_{mech}^{min} / \varepsilon_{mech}^{max}$. Since the size of the resin block is very large compared to the fiber diamater, the residual thermal strain needs to be taken into account by applying a fixed temperature change ΔT. A law that governs debond crack propagation in cyclic loading below the critical ERR values needs to be adopted to characterize debond growth. One of the most common approaches is to use a power law function, where it is assumed that the fiber/matrix interface debond length l_{dn} (normalized with respect to fiber radius) increases with the number of applied cycles N according to following:

$$\frac{dl_{dn}}{dN} = B^*(\Delta G_{II})^m \tag{8.6}$$

where: B^* and m are unknown material parameters, which have to be determined experimentally; ΔG_{II} is the ERR difference between the values corresponding to ε_{mech}^{max} and ε_{mech}^{min}.

The quadratic form of the ERR with respect to the applied strain ε_{mech} and the temperature change ΔT (see Eq.8.2) is also valid for short debonds. For a single-fiber composite, it can be written as:

$$G_{II} = \frac{E_z^f r_f}{4}\left(k_m \varepsilon_{mech} + k_{th}\left(\alpha_m - \alpha_z^f\right)\Delta T\right)^2 \tag{8.7}$$

where

$$k_m = k_m^*(l_{dn})k_m^\infty, \; k_{th} = k_{th}^*(l_{dn})k_{th}^\infty \tag{8.8}$$

As demonstrated before, for example, in Figs. 8.14−8.16, the ERR is magnified for shorter debond lengths and approaches a steady-state value, when the debond length increases. The coefficients $k_m^*(l_{dn})$ and $k_{th}^*(l_{dn})$ in Eq. (8.8) are defined as magnification coefficients, and they are equal to one, when the debond growth is steady state.

Coefficients k_m^∞ and k_{th}^∞ can be found analytically in Pupurs and Varna (2017), whereas magnification coefficients $k_m^*(l_{dn})$ and $k_{th}^*(l_{dn})$ are determined from FEM calculations. The accuracy of the magnification coefficient values obtained by FEM was investigated in Pupurs and Varna (2011) comparing the BEM results (Graciani et al., 2009) for the case of a purely mechanical loading obtaining excellent agreement.

For the convenience of simulations, it is useful to apply simple mathematical expressions to describe the magnification coefficients $k_m^*(l_{dn})$ and $k_{th}^*(l_{dn})$ as functions of the debond length, $k_0(l_{dn})$. In Pupurs and Varna (2011), it was shown that a simple hyperbolic function is rather accurate.

Hence, using Eq. (8.5) for the case of mechanical tension–tension cyclic loading at fixed temperature change, the ERR range ΔG_{II} can be calculated as:

$$\Delta G_{II}(l_{dn}) = \frac{r_f E_z^f k_0^2(l_{dn})}{4}\left(\left(\varepsilon_{mech}^{max}\right)^2 - \left(\varepsilon_{mech}^{min}\right)^2 + 2\left(\alpha_m - \alpha_z^f\right)\Delta T\left(\varepsilon_{mech}^{max} - \varepsilon_{mech}^{min}\right)\right) \tag{8.9}$$

ΔG_{II} expression in Eq. (8.9) can be implemented in Eq. (8.6) and debond growth simulations can be performed. A detailed parametric analysis showing the effect of material properties, geometry, and initial temperature change on debond growth rate is reported in Pupurs et al. (2012).

8.5.3 Results and parameter identification

In a calculation example demonstrated here, results for GF/EP single-fiber composite are shown, similarly as in Pupurs et al. (2012). A two-phase (fiber/matrix) axisymmetric FEM model similar to Fig. 8.11 was used in order to calculate G_{II}. It has been shown in Pupurs et al. (2012) that it is important to account for residual thermal and curing stresses in single-fiber composites in order to accurately determine the power law parameters.

The experimental measurements of the debond length in cyclic mechanical loading were performed in Pupurs et al. (2012) for three samples. For samples A and B, the applied maximum mechanical tensile strain ε_{mech}^{max} was 1.76%, while for sample C, ε_{mech}^{max} was 1.32%. Other loading parameters for all samples were: $R = 0.1$, $\Delta T = -7.9°C, f = 2Hz$ (loading frequency).

Typical optical microscopy images from the study Pupurs et al. (2012) showing the fiber/matrix debonds after a certain number of cycles are given in Fig. 8.21.

The detailed procedure of the determination of power law parameters is presented in Pupurs et al. (2012).

The power law parameters were found in Pupurs et al. (2012) for samples A and B. Since power law parameters are fiber/matrix interface-related properties, the parameters determined for samples A and B were validated for sample C, for which the level of applied mechanical strain was lower.

Experimental and modeling results for all three samples are presented in Fig. 8.22. The overall agreement between experimental data and simulations for samples A, B, and C is acceptable justifying the assumption of a power law Eq. (8.6). Modeling data for samples A and B are self-predictive, while modeling data for sample C were obtained using the constants from sample A.

8.6 Summary and conclusions

Fiber/matrix interface debonding is one of the key damage mechanisms in UD composites and in 0 degree plies of multilayered laminates. Understanding the debond

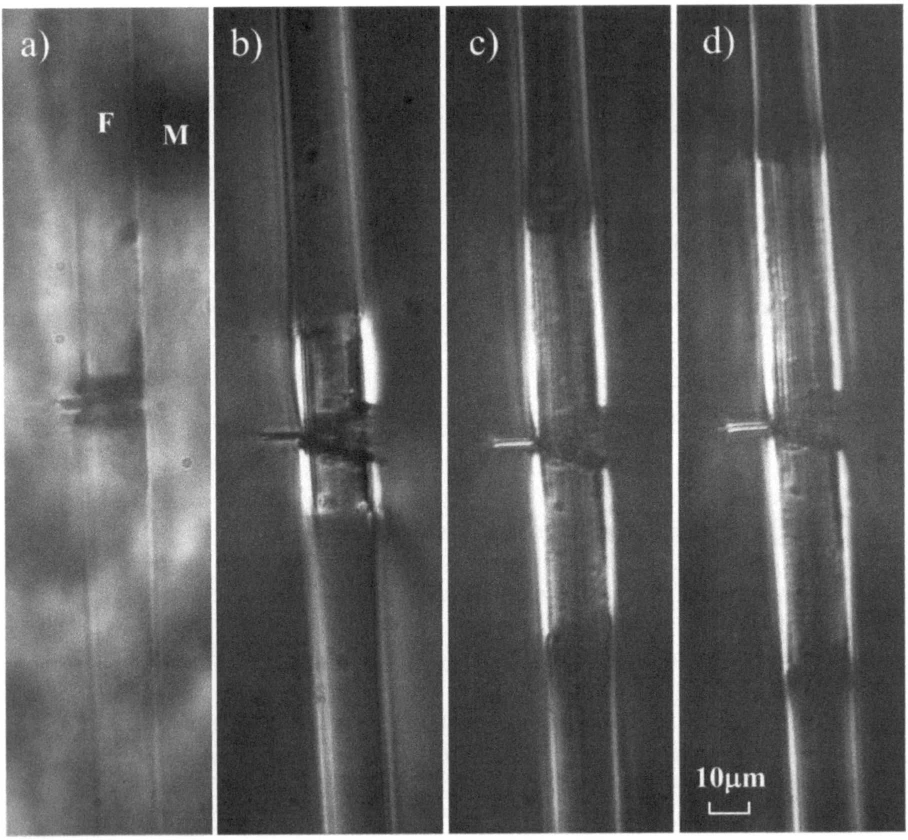

Figure 8.21 Optical microscopy images showing the debond length increase with the number of load cycles in tension–tension cyclic loading. (a) $N = 1$, (b) $N = 1,000$, (c) $N = 5,000$, (d) $N = 10,000$. *F*, fiber; *M*, matrix.
Redrawn from Pupurs et al. (2012).

growth in composites is essential for improved modeling and prediction of the ultimate failure and fatigue life of composites. In general, two distinct stages of debond growth have been identified, namely steady-state (long) and short debond growth, each requiring a different approach for analysis. For steady-state debond growth, exact analytical models have been developed for highly accurate calculation of the debond growth ERR. For more complex geometries taking into account local fiber spacing, FEM models with 5-phase 2-D axisymmetric assembly and 3-D hexagonal fiber arrangement can be used to calculate the steady-state debond growth ERR. For the estimation of short debond growth ERR, which is governed by stress perturbations between fiber break and debond crack tip regions, FEM models with partially debonded fiber/matrix interface can be used with excellent accuracy. Compared to steady-state debond growth, the short debond growth ERR increases significantly for shorter debonds.

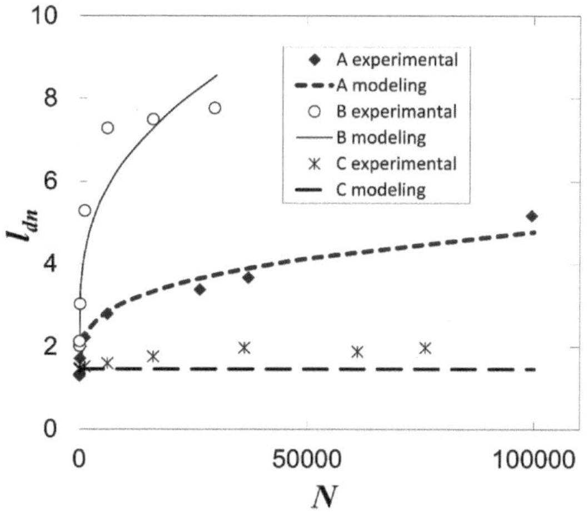

Figure 8.22 Average normalized debond length l_{dn} as a function of applied load cycles N for test samples A, B, and C. Comparison between experimental and modeling results.
Redrawn from Pupurs et al. (2012).

It was found that local fiber clustering can affect the shape of the debond crack front, and it generally increases the debond growth ERR, but the increase compared to uniformly spaced fiber arrangement is relatively small. It was also found that debond growth ERR increases slightly in the vicinity of the composite surface as compared with the bulk of composite.

Finally, fatigue damage mechanisms in UD composites have been analyzed: they may be different than the ones observed in quasistatic loading. However, fiber/matrix debonding is the key damage mechanism with the highest driving force. It was shown that debond growth in composites subjected to cyclic tension–tension loading follows a power law with respect to ERR change for which the parameters can be obtained experimentally using single-fiber composite tests.

References

Andersen, S.I., Lilholt, H., Lystrup, A.A., 1996. Properties of composites with long fibers. In: Meyer, R.M. (Ed.), Design of Composite Structures against Fatigue, Bury St.Edmunds. Mechanical Engineering Publications, pp. 15–31.

ANSYS Release 13.0, 2011. ANSYS Academic Research, ANSYS Inc., Canonsburg, Pennsylvania.

Behzadi, S., Curtis, P.T., Jones, F.R., 2009. Improving the prediction of tensile failure in unidirectional fibre composites by introducing matrix shear yielding. Compos. Sci. Technol. 69, 2421–2427.

Beyerlein, I.J., Phoenix, S.L., 1997. Stress profiles and energy release rates around fiber breaks in a lamina with propagating zones of matrix yielding and debonding. Compos. Sci. Technol. 57, 869–885.

Case, S.W., Carman, G.P., Lesko, J.J., Fajardo, A.B., Reifsnider, K.L., 1995. Fiber fracture in unidirectional composites. J. Compos. Mater. 29 (2), 208–228.

Chi, Z., Chou, T.W., Shen, G., 1984. Determination of single fiber strength distribution from fiber bundle testings. J. Mater. Sci. 19, 3319–3324.

Christensen, R.M., Lo, K.H., 1979. Solutions for effective shear properties in three phase sphere and cylinder models. J. Mech. Phys. Solid. 27, 315–330.

Curtin, W.A., Takeda, N., 1998. Tensile strength of fiber-reinforced composites: I. Model and effects of local fiber geometry. J. Compos. Mater. 32, 2042–2059.

Dauskardt, R.H., Ritchie, R.O., Cox, B.N., 1993. Fatigue of advanced materials: Part 1. Adv. Mater. Process. 7, 26–31.

Gamstedt, E.K., 2000. Effects of debonding and fiber strength distribution on fatigue damage propagation in carbon fiber reinforced epoxy. J. Appl. Polym. Sci. 76, 457–474.

Goutianos, S., Peijs, T., 2001. Experimental and numerical investigation into fatigue damage mechanisms in multifiber microcomposites. Plast. Rubber Compos. 30, 222–232.

Graciani, E., Mantic, V., Paris, F., Varna, J., 2009. Numerical analysis of debond propagation in the single fiber fragmentation test. Compos. Sci. Technol. 69, 2514–2520.

Graciani, E., Varna, J., Mantič, V., Blázquez, A., París, F., 2011. Evaluation of interfacial fracture toughness and friction coefficient in the single fiber fragmentation test. Procedia Eng. 10, 2478–2483.

Harlow, D.G., Phoenix, S.L., 1981. Probability distributions for the strength of composite materials II: a convergent sequence of tight bounds. Int. J. Fract. 17, 601–630.

Harris, B. (Ed.), 2003. Fatigue in Composite Materials. Woodhead Publishing Ltd., Cambridge.

Hashin, Z., 1983. Analysis of composite materials—a survey. J. Appl. Mech. 50, 481–505.

Irwin, G.R., 1958. Fracture, Handbuch der Physik, vol. 5. Springer Verlag, Berlin.

Kim, B.W., Nairn, J., 2002. Experimental verification of the effects of friction and residual stress on the analysis of interfacial debonding and toughness in single fiber composites. J. Mater. Sci. 37, 3965–3972.

Kim, B.W., Nairn, J., 2002. Observations of fiber fracture and interfacial debonding phenomena using the fragmentation test in single fiber composites. J. Compos. Mater. 36 (15), 1825–1858.

Laws, N., Dvorak, G.J., 1987. The effect of fiber breaks and aligned penny-shaped cracks on the stiffness and energy release rates in unidirectional composites. Int. J. Solid Struct. 23, 1269–1283.

Liu, B., Lessard, L.B., 1994. Fatigue damage-tolerance analysis of composite laminates: stiffness loss, damage-modelling, and life prediction. Compos. Sci. Technol. 51, 43–51.

McCartney, L.N., 2013. Analytical model for debonded interfaces associated with fiber fractures or matrix cracks. Comput. Mater. Contin. 35 (3), 183–227.

Nairn, J.A., Liu, Y.C., 1996. On the use of energy methods for interpretation of results of single-fiber fragmentation experiments. Compos. Interf. 4, 241–267.

Pitt, R.E., Phoenix, S.L., 1983. Probability distributions for the strength of composite materials IV: localized load-sharing with tapering. Int. J. Fract. 22, 243–276.

Pupurs, A., Varna, J., 2010. Unidirectional composite in mechanical fatigue: modelling debond growth from fiber breaks. Plast. Rubber Compos. 39, 128–136.

Pupurs, A., Varna, J., 2011. Fracture mechanics analysis of debond growth in single fiber composite under cyclic loading. Mech. Compos. Mater. 47 (1), 109–124.

Pupurs, A., Varna, J., 2013. Energy release rate based fiber/matrix debond growth in fatigue. Part I: self-similar crack growth. Mech. Adv. Mater. Struct. 20, 276–287.

Pupurs, A., Varna, J., 2013. FEM modeling of fiber/matrix debond growth in tension-tension cyclic loading of unidirectional composites. Int. J. Damage Mech. 22, 1144–1160.

Pupurs, A., Varna, J., 2017. Steady−state energy release rate for fiber/matrix interface debond growth in unidirectional composites. Int. J. Damage Mech. 26, 560−587.

Pupurs, A., Goutianos, S., Brøndsted, P., Varna, J., 2012. Interface debond crack growth in tension-tension cyclic loading of single fiber polymer composites. Compos. Appl. Sci. Manuf. 44, 86−94.

Pupurs, A., Krasnikovs, A., Varna, J., 2013. Energy release rate based fiber/matrix debond growth in fatigue. Part II: debond growth analysis using Paris law. Mech. Adv. Mater. Struct. 20, 288−296.

Talreja, R., 1981. Fatigue of composite materials: damage mechanisms and fatigue-life diagrams. Proc. Roy. Soc. Lond. 378, 461−475.

Talreja, R., 1987. Fatigue of Composite Materials. Technomic Publishing Inc., Lancaster, Pennsylvania.

Talreja, R., 1993. Fatigue of composites. In: Chou, T.W. (Ed.), Structure and Properties of Composites. VCH, Weinheim, pp. 583−606.

Talreja, R., Singh, C.V., 2012. Damage and Failure of Composite Materials. Cambridge University Press.

Varna, J., Joffe, R., Berglund, L., 1996. Interfacial toughness evaluation from the single-fiber fragmentation test. Compos. Sci. Technol. 56 (9), 1105−1109.

Wu, W., Verpoest, I., Varna, J., 2000. Prediction of energy release rate due to the growth of interface crack by variational analysis. Compos. Sci. Technol. 60, 351−360.

Zhuang, L., Pupurs, A., 2014. Effect of neighboring fibers on energy release rate during fiber/matrix debond growth (conference paper). In: 16th European Conference on Composite Materials. ECCM, Seville; Spain. June 22-26, 2014; Code 109290.

Zhuang, L., Pupurs, A., Varna, J., Ayadi, Z., 2016. Effect of fiber clustering on debond growth energy release rate in UD composites with hexagonal packing. Eng. Fract. Mech. 161, 76−88.

Zhuang, L., Pupurs, A., Varna, J., Ayadi, Z., 2016. Fiber/matrix debond growth from fiber break in unidirectional composite with local hexagonal fiber clustering. Compos. B Eng. 101, 124−131.

Evolution of multiple matrix cracking

9

Chandra Veer Singh
Department of Materials Science and Engineering, University of Toronto, Toronto, ON, Canada

9.1 Introduction

Composite laminates are tailored to provide designed stiffness properties in longitudinal and off-axis directions through a mix of on-axis and off-axis plies. When they are subjected to quasistatic loading, the matrix typically fails first leading to microcrack formation. Being unstable in nature, these microcracks grow quickly, spanning the thickness and width of plies transverse to the loading direction. On further increase in applied loading, more such cracks appear in the transverse plies forming an array of nearly parallel (through lamina thickness and width), same size, and equidistant crack surfaces. Therefore, they can be considered as *fully grown.* For damage mechanics, the growth of an individual ply crack is not very important, rather the increase in their density, as further loads are applied on the composite laminate, is of particular concern as these cracks decrease the overall stiffness of the composite structure. Because of their equal size, nearly uniform spacing, and same orientation, ply cracks can be regarded as *self-similar,* allowing formulation of boundary value problems that become possible to solve analytically in some cases, or simplified significantly for numerical solutions. This ply cracking process for a typical composite laminate is illustrated in Fig. 9.1. The ply cracks can be experimentally detected and analyzed through a variety of techniques such as edge replication, acoustic emission, X-ray radiography, ultrasonic C-scan, and Raman spectroscopy (see Talreja and Singh, 2012 for more details).

This chapter summarizes the main models for predicting initiation and progression of damage in the form of off-axis ply cracks in composite laminates. In the next section, we will outline analytical models to predict the evolution of matrix cracking. Section 9.3 will treat transverse cracking in cross-ply laminates. In Section 9.4, an energy-based approach applicable to the case of cracking in multidirectional laminates will be described. Statistical aspects of crack initiation and evolution will be described in Section 9.5. Damage evolution under fatigue loading will be covered next. Finally, we will discuss some open issues in predicting damage evolution and outline expected future trends. Some sources for further information and advice on this topic are provided at the end of the chapter.

Modeling Damage, Fatigue and Failure of Composite Materials. https://doi.org/10.1016/B978-0-443-18489-5.00022-9

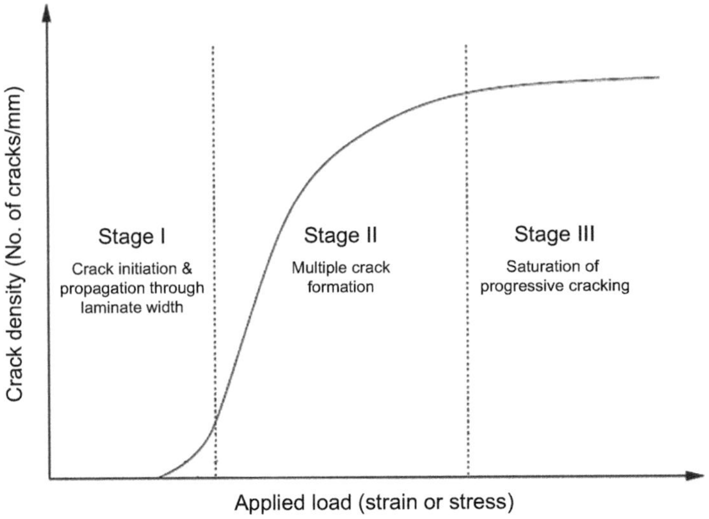

Figure 9.1 A typical damage evolution curve for ply cracking in composite laminates. Initially, microcracks initiate from matrix failure and quickly grow through the transverse ply thickness and width (Stage I). On further loading, more and more cracks appear in those plies forming a periodic array (Stage II). When sufficient crack density has been achieved, stress fields of adjacent cracks interact and reduce the rate of increase in crack density, which eventually approaches a saturation state (Stage III).

9.2 Analytical models for evolution of multiple matrix cracking

The techniques to predict damage progression due to multiple matrix cracking can be categorized into two groups: (a) strength-based and (b) energy-based. Strength-based approaches utilize a point failure criterion for crack initiation and multiplication, whereas the energy-based methods consider the balance of energy during crack formation and are conceptually similar to the energy release rate concept in linear elastic fracture mechanics. Most of the models developed so far have been for the case of 90° ply cracking in cross-ply laminates. More recently, there has been some progress in dealing with off-axis ply cracking in multidirectional laminates.

9.2.1 *Transverse cracking in cross-ply laminates*

Assuming self-similarity of fully developed ply cracks, we can define the boundary value problem as shown in Fig. 9.2, where State 1 represents the damage state with a crack spacing, $s = 2l$, while State 2 represents twice the crack density, that is, $s = l$. The damage evolution process usually involves determining the stress field in a composite laminate with a periodic array of preexisting equal cracks in the transverse plies. From this, required increase in applied far-field stress (σ_0) to cause further cracks to form in between existing cracks is determined. In an energy-based approach, this crack multiplication is expressed in terms of energy required to form the additional cracks. Assuming self-similarity of stress fields around existing cracks and constant

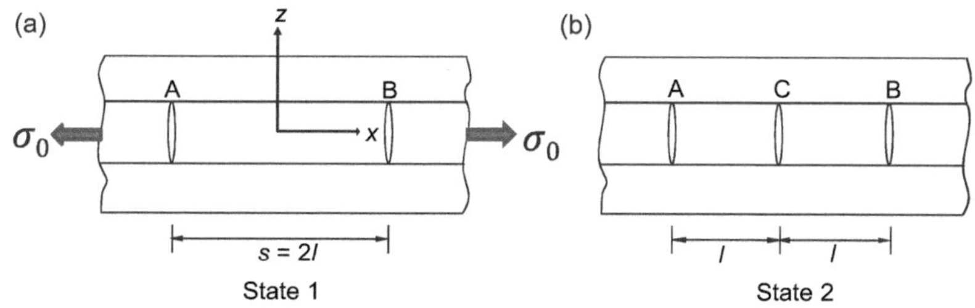

Figure 9.2 Schematics showing progressive multiplication of transverse ply cracks in a cross-ply laminate loaded in uniaxial tension. (A) State 1 represents damage state with crack spacing of 2l, while (B) State 2 has an additional crack that developed midway between points A and B.

resistance to fracture in the material microstructure, the new crack shall form in the middle (Point C) between locations for existing cracks (Points A and B). However, in reality, local fracture toughness varies spatially due to material heterogeneities and presence of manufacturing induced defects. Thus, the crack multiplication process tends to be statistical in nature and the new crack may not form in the middle. These statistical aspects are described in detail in Section 9.5. For now, we assume a deterministic process for crack density evolution.

The boundary value problem for stress analysis of a cracked cross-ply composite laminate for a given damage state is defined as: to determine σ_{ij} which satisfy following conditions:

1. Force balance:

$$N_{xx} = \sigma_c A \tag{9.1}$$

2. Equilibrium conditions:

$$\sigma_{ij,j} = 0 \tag{9.2}$$

3. Boundary and continuity conditions:

$$\begin{aligned}
&\text{Laminate mid}-\text{plane symmetry}: \sigma_{xz}^{90}(x,0) = 0 \\
&\text{Traction continuity across interface}: \sigma_{xz}^{90}(x,t_{90}) = \sigma_{xz}^{0}(x,t_{90}) \\
&\sigma_{zz}^{90}(x,t_{90}) = \sigma_{zz}^{0}(x,t_{90}) \\
&\text{Traction}-\text{free boundary}: \sigma_{xz}^{0}(x,h) = 0 \\
&\sigma_{zz}^{0}(x,h) = 0 \\
&\text{Traction}-\text{free crack surfaces}: \sigma_{xz}^{90}(\pm l,z) = 0, -t_{90} \le z \le t_{90} \\
&\sigma_{xx}^{90}(\pm l,z) = 0, -t_{90} \le z \le t_{90}.
\end{aligned} \tag{9.3}$$

where 2 t_{90} is the total thickness of the central 90°-plies, and $2h$ is the total laminate thickness.

9.2.1.1 Shear lag model

The shear lag theory provides a simple one-dimensional solution to the above BVP. According to this approach, it is assumed that the transverse ply does not carry any axial load in the crack-plane, while away from the crack a part of this load is transferred back to the transverse ply through axial shear. Following the analysis conducted by Garret and Baily (1977), the transferred load over a distance y is given by:

$$F = 2t_0 w \Delta\sigma_0 \left[1 - e^{-\beta y}\right] \tag{9.4}$$

where t_0 denotes thickness of the supporting 0° ply, w is lamina width, $\Delta\sigma_0$ is the maximum additional stress placed on the longitudinal ply due to cracking in the transverse ply, and β, known as the shear lag parameter, is defined as:

$$\beta^2 = G_{xz0}^{90}\left[\frac{1}{E_{x0}^{90}} + \frac{1}{\lambda E_{x0}^{0}}\right] \tag{9.5}$$

where G_{xz0}^{90} is the initial (undamaged) in-plane shear modulus of the 90° ply, $\lambda = \frac{t_0}{t_{90}}$ is the ply thickness ratio, and E_{x0}^{0} and E_{x0}^{90} are the longitudinal moduli for 0 and 90° plies, respectively. Using maximum normal stress point failure criterion, the transverse ply will fail in tension when:

$$F = 2t_{90} w \sigma_{tu} \tag{9.6}$$

where σ_{tu} is the transverse ply strength. The first crack is assumed to form in the middle of the specimen length at $\Delta\sigma_0 = \frac{t_{90}}{t_0}\sigma_{tu}$, that is, at an applied stress of $\sigma_0 = E_c\varepsilon_{tu}$ where E_c is longitudinal modulus for the composite and ε_{tu} is the transverse ply cracking strain. After first crack formation, an increase in applied loading causes further cracking through simultaneous formation of two additional cracks, one left and another right of the first crack. The additional far-field stress required to form these new cracks,$\Delta\sigma_0$, can be obtained by equating F in Eq. (9.6) to that in Eq. (9.4) with $y = l$, where $2l$ is the initial crack spacing, as:

$$\Delta\sigma_0 = \frac{1}{\lambda}\frac{\sigma_{tu}}{1 - e^{-\beta l}}. \tag{9.7}$$

Further increase in loading will cause more cracks to appear, when:

$$\Delta\sigma_0 = \frac{1}{\lambda}\frac{\sigma_{tu}}{1 + e^{-\beta l} - 2e^{-\beta l/2}}. \tag{9.8}$$

Following this procedure, the $(N+2)^{th}$ crack will form when:

$$\Delta\sigma_0 = \frac{1}{\lambda} \frac{\sigma_{tu}}{1 + e^{-\frac{\beta l}{N}} - 2e^{-\frac{\beta l}{2N}}}. \tag{9.9}$$

Eqs. (9.7)–(9.9) thus provide an iterative way to determine the evolution of crack density as a function of applied load, as plotted in Fig. 9.3.

Energy-based shear lag analysis

The above shear lag model was strength-based. For predicting damage evolution, energy-based models are preferred. Most of such models utilize the energy release rate concept from linear elastic fracture mechanics. Laws and Dvorak (1988) carried out the first major energy analysis for progressive multiplication of ply cracks in cross-ply laminates. They again considered two stages of cracking as shown in Fig. 9.1. If the laminate has a width of w and total thickness equal to h, the energy released during formation of a new crack at point C, between preexisting cracks at points A and B, can be derived as:

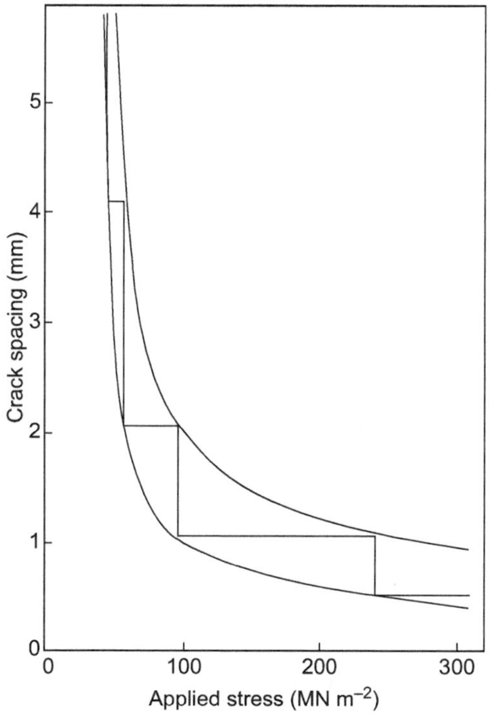

Figure 9.3 Progressive evolution of crack spacing for increasing applied load according to the one-dimensional shear lag theory.
From Garret and Bailey (1977). Reprinted, with the kind permission, from Spring Science + Business Media.

$$\Delta\Gamma = \frac{2t_{90}^2 whE_c}{\beta t_0 E_{x0}^0 E_{x0}^{90}}\left(\sigma_{xxR}^{90} + \frac{E_{x0}^{90}}{E_{x0}^0}\sigma_c\right)^2\left[\tanh\frac{\beta l_1}{2t_{90}} + \tanh\frac{\beta l_2}{2t_{90}} - \tanh\frac{\beta l}{t_{90}}\right] \tag{9.10}$$

where E_c is the longitudinal modulus for the composite laminate, t_0 denotes the ply thickness, and σ_c is the external stress applied on the composite. The shear lag parameter is defined in this analysis as $\beta^2 = Kt_{90}\left|\frac{1}{E_{x0}^{90}} + \frac{1}{\lambda E_{x0}^0}\right|$, σ_{xxR}^{90} is the thermal residual stress in transverse ply, and l_1 and l_2 represent distance of point C from points A and B, respectively. According to this analysis, the condition for new crack formation is related to the critical energy release rate, γ, as:

$$\frac{t_{90}hE_c}{\beta t_0 E_{x0}^0 E_{x0}^{90}}\left(\sigma_{xxR}^{90} + \frac{E_{x0}^{90}}{E_{x0}^0}\sigma_c\right)^2\left[\tanh\frac{\beta l_1}{2t_{90}} + \tanh\frac{\beta l_2}{2t_{90}} - \tanh\frac{\beta l}{t_{90}}\right] \geq \gamma \tag{9.11}$$

The applied stress to initiate first crack, that is, first ply failure stress, is given by the limit $l \to \infty$. Therefore, the transverse crack initiation stress is determined by:

$$\sigma_c^{fpf} = \left(\frac{\beta t_0 E_{x0}^0 E_c \gamma}{t_{90} h E_{x0}^{90}}\right)^{1/2} - \frac{E_c}{E_{x0}^{90}}\sigma_{xxR}^{90} \tag{9.12}$$

In order to characterize the location of new crack formation, point C, Laws and Dvorak introduced statistical concepts concerning spatial randomness of resistance to crack formation. If the probability density function for the next crack to occur at a given location is given by p in a laminate which already contains cracks, the expected value of the applied stress to cause additional cracking is then:

$$E[\sigma_c(\rho_c)] = \int_0^{2l} p(x)\sigma_c(x)\mathrm{d}x \tag{9.13}$$

where $\rho_c = \frac{t_{90}}{l}$ represents the normalized crack density. The authors considered three possible choices for $p(x)$:

Case 1: Next crack forms midway, giving:

$$p(x) = \delta(x - l) \tag{9.14}$$

where $\delta(x)$ is the Dirac delta function.

Case 2: All locations are equally likely. In this case, $p(x)$ follows a uniform probability distribution:

$$P(x) = \frac{1}{2l} \tag{9.15}$$

Case 3: The location with higher stress has higher probability of being the site for the formation of a new crack, that is, the probability of *p(x)* is proportional to the stress at the location. Thus:

$$P(x) = \left(\sigma_{xxR}^{90} + \frac{E_{x0}^{90}}{E_{x0}^{0}} \sigma_c \right) \left[1 - \frac{\cosh \frac{\beta x}{t_{90}}}{\cosh \frac{\beta l}{t_{90}}} \right] \tag{9.16}$$

For case 1, the solution can be explicitly determined, while an iterative numerical process is followed for the other two cases. Fig. 9.4 shows the evolution of crack density in cross-ply ($[0/90]_s$) laminate, with $\gamma = 193\ \mathrm{J/m^2}$, and $\beta = 0.9$. Case 3 with probability based on stress levels showed better agreement with experimental data.

9.2.1.2 Variational models

Variational methods provide two dimensional solutions to the BVP as shown in Fig. 9.1, and thus give more accurate stress fields for predicting multiplication of matrix cracks in cross-ply laminates. The first model for conducting stress analysis of a cross-ply laminate with periodic cracks in 90-plies was proposed by Hashin (1985), whose two-dimensional solution stress field is given by:

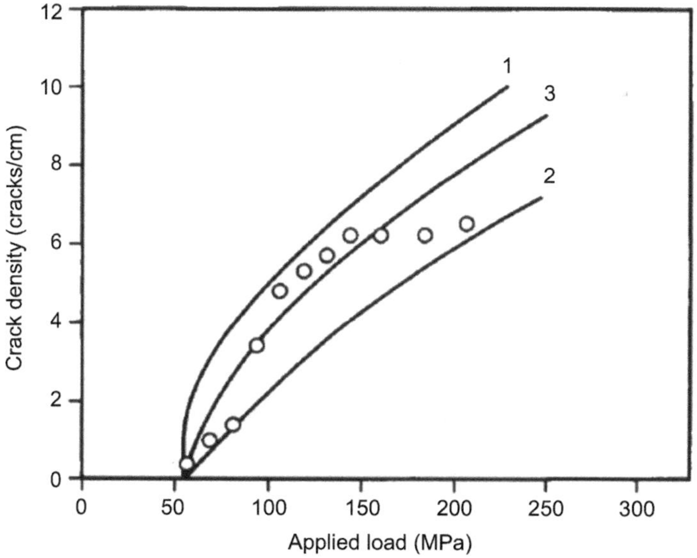

Figure 9.4 Prediction of transverse crack density evolution in $[0/90]_s$ laminates from energy-based shear lag model proposed by Laws and Dvorak (1988). Three different choices of probability distribution functions to describe failure location are utilized as shown. Reprinted with the kind permission, from Laws and Dvorak (1988), Copyright © 1988 by Sage Publications.

$$\begin{aligned}\sigma_{xx}^{90} &= \sigma_{xx0}^{90}[1-\phi(x)]\\ \sigma_{xz}^{90} &= \sigma_{xx0}^{90}\phi'(x)z\\ \sigma_{zz}^{90} &= \sigma_{xx0}^{90}\phi''(x)\frac{1}{2}\left[(1+\lambda)t_{90}^{2}-z^{2}\right]\end{aligned} \tag{9.17}$$

in 90°-ply, and

$$\begin{aligned}\sigma_{xx}^{0} &= \sigma_{xx0}^{0}+\sigma_{xx0}^{90}\frac{1}{\lambda}\phi(x)\\ \sigma_{xz}^{0} &= \sigma_{xx0}^{90}\phi'(x)\frac{1}{\lambda}[(1+\lambda)t_{90}-z]\\ \sigma_{zz}^{0} &= \sigma_{xx0}^{90}\phi''(x)\frac{1}{2\lambda}[(1+\lambda)t_{90}-z]^{2}\end{aligned} \tag{9.18}$$

in 0°-ply, where $\lambda = \frac{t_0}{t_{90}}$ is the ply thickness ratio, and ϕ denotes the perturbation field, which can be shown to be governed by the Euler–Lagrange equation:

$$\frac{d^4\phi}{d\xi^4}+p\frac{d^2\phi}{d\xi^2}+q\phi=0 \tag{9.19}$$

in terms of a nondimensional geometry parameter $\xi = \frac{x}{t_{90}}$ and the coefficients p and q defined by:

$$p=\frac{C_{02}-C_{11}}{C_{22}},\ q=\frac{C_{00}}{C_{22}} \tag{9.20}$$

The material constants in Eq. (9.20) are defined as:

$$\begin{aligned}C_{00} &= \frac{1}{E_2}+\frac{1}{\lambda E_1} C_{02} = \left(\lambda+\frac{2}{3}\right)\frac{\upsilon_{23}}{E_2}-\frac{\lambda}{3}\frac{\upsilon_{12}}{E_1}\\ C_{11} &= \frac{1}{3}\left(\frac{1}{G_{23}}+\frac{\lambda}{G_{12}}\right) C_{22} = (\lambda+1)\left(3\lambda^2+12\lambda+8\right)\frac{1}{60E_2}.\end{aligned} \tag{9.21}$$

where 1 is in fiber direction and 2, 3 are transverse directions, and E_i, G_{ij}, ν_{ij} $(i,j=1,2,3)$ denote the Young's moduli, shear moduli, and Poisson's ratio associated with respective directions.

The solution of Eq. (9.19) is obtained as:

$$\phi=A_1\cosh\alpha_1\xi\cos\alpha_2\xi+A_2\sinh\alpha_1\xi\sin\alpha_2\xi \tag{9.22}$$

where A_1 and A_2 are constants determined from the boundary conditions as:

$$A_1 = \frac{2(\alpha_1 \cosh \alpha_1\rho \sin \alpha_2\rho + \alpha_2 \sinh \alpha_1\rho \cos \alpha_2\rho)}{\alpha_1 \sin 2\alpha_1\rho + \alpha_2 \sinh 2\alpha_2\rho}$$
$$A_2 = \frac{2(\alpha_2 \cosh \alpha_1\rho \sin \alpha_2\rho - \alpha_1 \sinh \alpha_1\rho \cos \alpha_2\rho)}{\alpha_1 \sin 2\alpha_1\rho + \alpha_2 \sinh 2\alpha_2\rho} \tag{9.23}$$

where $\rho = \frac{l}{t_{90}}$ denotes the crack spacing normalized with the cracked ply thickness. When $4q/p^2 < 1$, $\phi(x)$ is given by Hashin (1985):

$$\phi = \frac{\alpha_2' \cosh \alpha_1'\xi}{\sinh \alpha_1'\rho(\alpha_2' \coth \alpha_1'\rho - \alpha_1' \coth \alpha_2'\rho)} + \frac{\alpha_1' \cosh \alpha_2'\xi}{\sinh \alpha_2'\rho(\alpha_1' \coth \alpha_2'\rho - \alpha_2' \coth \alpha_1'\rho)} \tag{9.24}$$

where $\alpha_1{}', \alpha_2{}'$ are defined as:

$$\alpha_1', \alpha_2' = \sqrt{\frac{-p}{2} \pm \sqrt{\frac{p^2}{4} - q}} \tag{9.25}$$

Nairn (1989) used Hashin's analysis to derive expressions for energy release rate during the formation of a new midway transverse crack in cross-ply laminates, given as:

$$G_m = \left(\sigma_c^2 \frac{E_2^2}{E_c^2} + \frac{\Delta\alpha T^2}{C_{00}^2}\right) t_{90} C_{22}[2\chi(\rho/2) - \chi(\rho)] \tag{9.26}$$

where T is temperature, $\Delta\alpha$ denotes the thermal expansion coefficient, and $\chi(\rho)$ is the energy release rate evaluated at normalized crack spacing ρ. Note here that in contrast to the shear lag analysis by Laws & Dvorak described above (Section 9.3), Nairn (1989); Nairn and Hu (1992) defined G_m as the matrix fracture toughness and determined it by fitting the ply cracking model predictions to experimental data. In the case where a new crack does not occur midway between existing adjacent cracks, expression for G_m in Eq. (9.26) should be rewritten as:

$$G_m = \left(\sigma_c^2 \frac{E_2^2}{E_c^2} + \frac{\Delta\alpha T^2}{C_{00}^2}\right) t_{90} C_{22}[\chi(\delta) + \chi(\rho - \delta) - \chi(\rho)] \tag{9.27}$$

where

$$[\chi(\delta)+\chi(\rho-\delta)-\chi(\rho)]=\frac{\int_0^{\rho/2}[\chi(\delta)+\chi(\rho-\delta)][1-\phi(\rho-2\delta)]\mathrm{d}\delta}{\int_0^{\rho/2}[1-\phi(\rho-2\delta)]\mathrm{d}\delta}-\chi(\delta) \tag{9.28}$$

So far, our discussion has been for $[0/90]_s$ type cross-ply laminates. However, when transverse ply cracking occurs in outer 90 plies, that is, in $[90_m/0_n]_s$ laminates, the constant C_{22} is given by $C_{22}=(\lambda+1)(3+12\lambda+8\lambda^2)\frac{1}{60E_2}$. Fig. 9.5 shows the model predictions for a $[0/90_3]_s$ glass/epoxy laminate with $G_m=330$ J/m^2 and a thermal residual stress of 13.6 MPa. Figs. 9.6 and 9.7 further demonstrate that the model predictions compare well with experimental data for a range of $[0/90_n]_s$ and $[90/0_n]_s$ laminates, respectively.

Vinogradov—Hashin analysis

More recently, Vinogradov and Hashin (2005) have derived a highly accurate damage evolution model based on Hashin's original variational analysis (Hashin, 1985) and combining it with finite fracture mechanics Hashin (1996). Following variational analysis for stress BVP, the change in complementary energy due to the presence of "*N*" transverse cracks can be written as:

$$\Delta\Pi^*=\sum_{n=1}^{N}\Delta\Pi_n^*=\left(\sigma_{xx0}^{90}\right)^2 t_{90}^2 C_{22}\sum_{n=1}^{N}\chi(\rho_n) \tag{9.29}$$

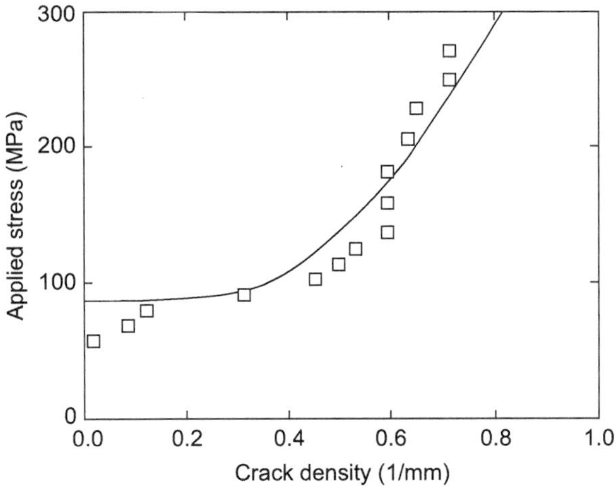

Figure 9.5 Evolution of crack density as a function of applied stress in $[0/90_3]_s$ glass-epoxy composites. The model fit is using the critical energy release rate equal to 330 J/m^2 and assuming that the initial level of the residual thermal stresses in the 90° plies was 13.6 MPa. Reprinted with the kind permission, Experimental data are from Nairn (1989), copyright © 1989 by Sage Publications.

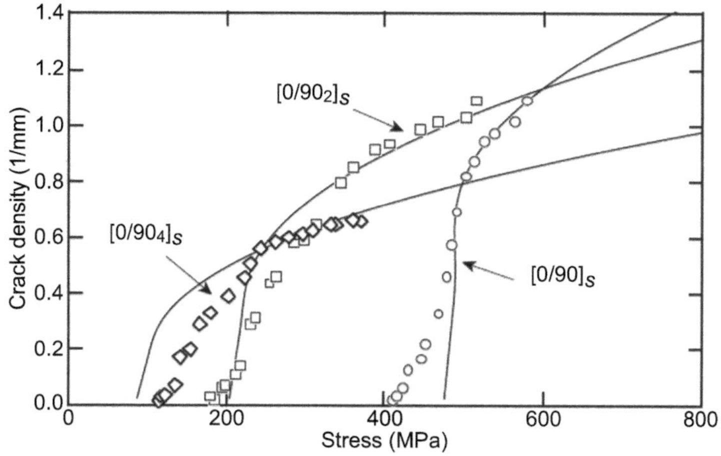

Figure 9.6 Evolution of ply crack density as a function of applied stress for AS4/3501−6 $[0/90_n]_s$ laminates. The experimental data are shown in symbols. The model fit is made with $G_{mc} = 250$ J/m^2.
Reprinted with the kind permission, from Nairn (2000), copyright Elsevier (2000).

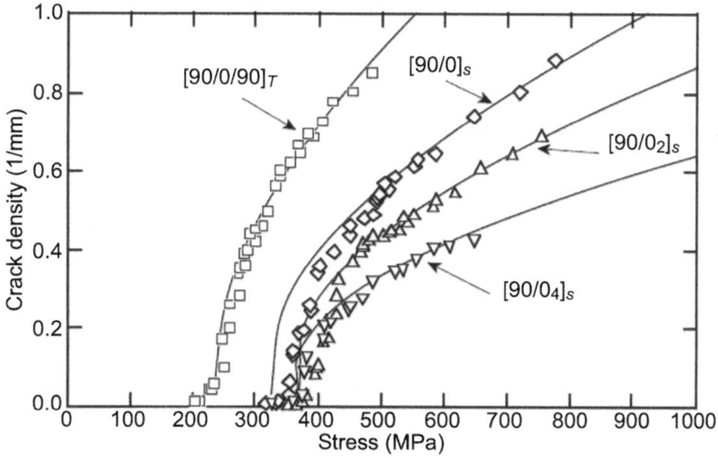

Figure 9.7 Evolution of ply crack density as a function of applied stress for AS4/3501−6 $[90/0_n]_s$ laminates. The experimental data are shown in symbols. The model fit is made with $G_{mc} = 230$ J/m^2.
Reprinted with the kind permission, from Nairn (2000), copyright Elsevier (2000).

where

$$\chi(\rho_n) = -\left.\frac{d^3\phi_n}{d\xi^3}\right|_{\rho_n} = 2\alpha_1\alpha_2\left(\alpha_1^2 + \alpha_2^2\right)\frac{\cosh(2\alpha_1\rho_n) - \cos(2\alpha_2\rho_n)}{\alpha_1 \sin(2\alpha_2\rho_n) + \alpha_2 \sinh(2\alpha_1\rho_n)} \tag{9.30}$$

In the above expressions, the summation is over all blocks bounded by adjacent cracks. Furthermore, it should be noted that $\rho_n = \frac{l_n}{t_{90}}$ is normalized crack spacing, rather than the crack density. The energy release rate when a new crack occurs from the state of laminate with "N" existing cracks is derived as:

$$\Delta\Gamma_n = \Pi^*\left(\boldsymbol{\sigma}^{N+1}\right) - \Pi^*\left(\boldsymbol{\sigma}^{N}\right) \tag{9.31}$$

where $\boldsymbol{\sigma}^N$ and $\boldsymbol{\sigma}^{N+1}$ are the stress fields before and after the formation of a new $(N+1)^{\text{th}}$ crack. For the formation of a new crack midway between existing cracks:

$$\Delta\Gamma_n = \left(\sigma_{xx0}^{90}\right)^2 t_{90}^2 C_{22} \sum_{i=1}^{N+1} \chi\left(\rho_i^{N+1}\right) - \left(\sigma_{xx0}^{90}\right)^2 t_{90}^2 C_{22} \sum_{i=1}^{N} \chi\left(\rho_i^{N}\right) \tag{9.32}$$

where σ_{xx0}^{90} is the stress in undamaged 90° ply and ρ_i^N is the nondimensional crack spacing for block i for $(N+1)$th crack formation. The criterion for new crack formation is given by:

$$\gamma = -\left(\sigma_{xx0}^{90}\right)^2 t_{90} C_{22} \frac{\mathrm{d}}{\mathrm{d}\overline{\rho}} \left(\frac{\overline{\chi}}{\overline{\rho}}\right) \overline{\rho}^2 \tag{9.33}$$

In terms of a lower bound to the laminate longitudinal modulus, the above expression can be written as:

$$\gamma = \frac{1}{2}\left(\sigma_{xx0}^{90}\right)^2 \frac{\mathrm{d}}{\mathrm{dA}} \left(\frac{1}{E_x^*}\right) V \tag{9.34}$$

where V is the laminate volume. The crack initiation stress for this model can be obtained in the limit $(\rho_n \rightarrow \infty)$, that is,

$$\chi(\infty) = 2\alpha_1\left(\alpha_1^2 + \alpha_2^2\right) \tag{9.35}$$

The model predictions as well as further details pertaining to statistical considerations while cracking are described in Section 9.5.

Recently, Vinogradov and Hashin (2010) have extended the analysis described above to the case of angle-ply laminates with $\left[\theta_m^{(1)}/\theta_n^{(2)}\right]_s$ layup containing ply cracks in middle layers. Following the approach for cross-ply laminates, the authors used the principle of minimum complementary energy to derive an optimal admissible stress field that satisfies equilibrium, boundary, and traction continuity conditions. Solving the resultant Euler–Lagrange equation, they obtained a lower bound on the stiffness of the cracked angle-ply laminates. The results for longitudinal stiffness of a cracked glass/epoxy $[0/45]_s$ laminate were reported to be in very good agreement with experimental data. Nonetheless, the study focused on the stiffness degradation owing to ply cracking and has not been extended for the case of damage evolution with respect to applied loading.

9.2.1.3 Fracture mechanics-based model

Based on the concepts of fracture mechanics, the transverse crack multiplication process can be studied using the virtual crack closure technique. Suppose a virtual crack is introduced between two existing cracks. If the work needed to close this crack is greater than the critical energy release rate, then this new crack is assumed to form. Following the analysis by Varna and coauthors (Joffe et al., 1999, 2001; Varna et al., 2001a,b), the work required to close N cracks from a state of $2N$ cracks (crack spacing, $s = 2l$) to N cracks (crack spacing, $s = l$) can be derived as:

$$W_{2N \to N} = 2N \frac{\left(\sigma_{xx0}^{90}\right)^2}{E_2} t_{90}^2 \left[2\widetilde{u}(l/2) - \widetilde{u}(l) \right] \tag{9.36}$$

where σ_{x0}^{90} is the far-field stress in 90° layer and $\widetilde{u}(s)$ represents the average crack opening displacement (COD) for cracks of spacing s, normalized with respect to the thickness of the cracked ply, and is defined as:

$$\widetilde{u}(s) = \frac{E_2}{\sigma_{xx0}^{90} (t_{90})^2} \int_0^{t_{90}} u(z,s) \mathrm{d}z \tag{9.37}$$

where $u(z,s)$ is the normal COD along the thickness (z) direction, evaluated at a crack spacing s. The formation of new cracks will occur when $W_{2N \to N}$ is greater than or equal to the total surface energy of the newly created surfaces, that is:

$$W_{2N \to N} \geq 2.N.2t_{90}.G_c \tag{9.38}$$

Substituting $W_{2N \to N}$ from Eq. (9.38) in Eq. (9.36), the criterion for crack multiplication is derived as:

$$\frac{\left(\sigma_{xx0}^{90}\right)^2}{2E_2} t_{90} \left[2\widetilde{u}(s/2) - \widetilde{u}(s) \right] \geq G_c \tag{9.39}$$

If a new crack does not form midway between two existing cracks, but at distance $l_1/2$ from one crack and $l_2/2$ from the second crack, then the cracking criterion is given by:

$$\frac{\left(\sigma_{xx0}^{90}\right)^2}{2E_2} t_{90} \left[2\widetilde{u}(l_1/2) - \widetilde{u}(l_1) + 2\widetilde{u}(l_2/2) - \widetilde{u}(l_2) \right] \geq G_c \tag{9.40}$$

The model predictions for crack density evolution in glass/epoxy $[\pm\theta/90_4]_s$ laminates are shown in Fig. 9.8 along with experimental data. In this analysis, CODs

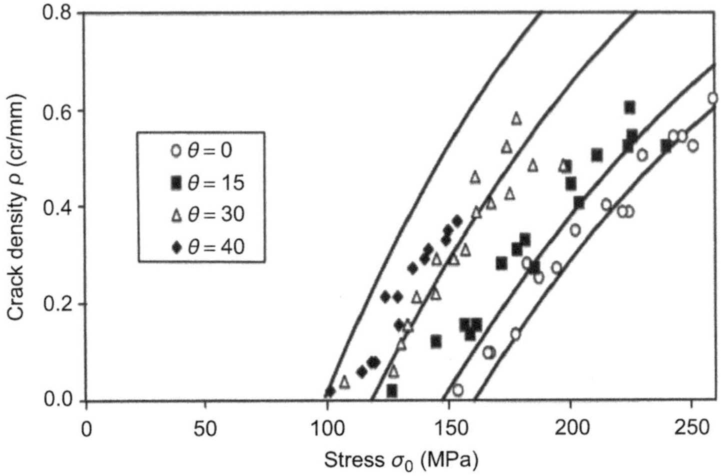

Figure 9.8 Predictions (*solid lines*) of crack density evolution in glass/epoxy $[\pm\theta/90_4]_s$ laminates using the fracture mechanics-based energy model developed by Joffe et al. (1999). Experimental data are shown in symbols.
Reprinted with the kind permission, from Joffe et al. (1999), copyright Elsevier (2001).

were calculated using finite element analysis, and G_c was assumed to be random with a Weibull distribution.

9.3 Damage evolution in multidirectional laminates

The ability to tailor laminate layup to obtain desired stiffness and strength properties without changing the constituent materials and the manufacturing process is a key advantage of using composites over metallic materials. To fully utilize capabilities of composite laminates, plies are placed in multiple orientations so as to yield multidirectional stiffness properties for a wide range of applications. For instance, laminates are constructed from a mix of plies in 0°, +45°, −45°, and 90° orientations such that their in-plane elastic properties are isotropic before damage. In these laminates loaded in uniaxial tension along the longitudinal direction, ply cracking would typically begin in transverse, that is, 90° plies, and multiply with increasing load. Damage analysis in such a laminate can be performed in a similar way as for cross-ply laminates. However, at higher loads, when cracks form in other off-axis, that is, ±45° plies, the local stress state becomes highly complex. In such cases, damage analysis that requires knowledge of this stress state is not effective. Instead, energy analysis is preferable and will be described next.

Singh and Talreja (2010) extended the fracture mechanics-based model described in Section 9.2.1.3 for the case of off-axis ply cracking in multidirectional laminates. Following their analysis, the work required to close N cracks in State 1 with crack spacing s (Fig. 9.2A), and $2N$ cracks in State 2 with crack spacing $s/2$ (Fig. 9.2B), respectively, can be derived as:

$$W_{N\to0} = N\frac{1}{\sin\theta}(t_\theta)^2\frac{1}{E_2}\left[\left(\sigma_{20}^\theta\right)^2\widetilde{u}_n^\theta(s) + \left(\sigma_{120}^\theta\right)^2\widetilde{u}_t^\theta(s)\right] \tag{9.41}$$

$$W_{2N\to0} = 2N\frac{1}{\sin\theta}(t_\theta)^2\frac{1}{E_2}\left[\left(\sigma_{20}^\theta\right)^2\widetilde{u}_n^\theta\left(\frac{s}{2}\right) + \left(\sigma_{120}^\theta\right)^2\widetilde{u}_t^\theta\left(\frac{s}{2}\right)\right] \tag{9.42}$$

where θ and t_θ denote orientation and total thickness, respectively, of cracked off-axis plies, and $\widetilde{u}_n^\theta, \widetilde{u}_t^\theta$ are the normalized average crack opening and sliding displacements (COD and CSD), respectively. They are evaluated as:

$$\widetilde{u}_n^\theta(s) = \frac{1}{(t_\theta)^2}\frac{E_2}{\sigma_{20}^\theta}\int_{-t_\theta/2}^{+t_\theta/2} u_n^\theta(z,s)\mathrm{d}z;\ \widetilde{u}_t^\theta(s) = \frac{1}{(t_\theta)^2}\frac{E_2}{\sigma_{120}^\theta}\int_{-t_\theta/2}^{+t_\theta/2} u_t^\theta(z,s)\ \mathrm{d}z \tag{9.43}$$

In above equations, u_n and u_t represent the relative opening and sliding displacement of the cracked surfaces, respectively.

Similar to Eq. (9.38) for the case of cracking in 90-plies of cross-ply laminates, the crack multiplication in off-axis plies would occur when the work required in going from State 1 to State 2 (which is the same as work needed to close N cracks in going from State 2 to State 1) exceeds or equals a critical value, that is, if:

$$W_{2N\to N} = W_{2N\to0} - W_{N\to0} \geq NG_c\frac{t_\theta}{\sin\theta} \tag{9.44}$$

where G_c is the critical value of energy required for multiple ply crack formation within the given laminate. It should be noted here that G_c should not be taken strictly as the critical energy rate defined using the usual linear fracture elastic mechanics, that is, as a fracture toughness value associated with growth of a single crack. Here, because of the multiple cracking process, G_c cannot be evaluated experimentally from a single crack test. Instead, it should be evaluated by fitting to experimental data model predictions for a representative laminate using the crack multiplication criterion described above. A convenient choice of a representative laminate is usually a cross-ply laminate, as discussed in Singh and Talreja (2013). Furthermore, for cracking in a general off-axis ply, a multimode criterion, such as the following, can be utilized.

$$\left(\frac{W_I}{G_{Ic}}\right)^M + \left(\frac{W_{II}}{G_{IIc}}\right)^N \geq 1 \tag{9.45}$$

where

$$w_I = \frac{\left(\sigma_{20}^\theta\right)^2 t_\theta}{E_2}\left[2\widetilde{u}_n^\theta\left(\frac{s}{2}\right) - \widetilde{u}_n^\theta(s)\right];\ w_{II} = \frac{\left(\sigma_{120}^\theta\right)^2 t_\theta}{E_2}\left[2\widetilde{u}_t^\theta\left(\frac{s}{2}\right) - \widetilde{u}_t^\theta(s)\right] \tag{9.46}$$

where G_{Ic} and G_{IIc} are the critical energy release rates in Mode I (opening mode) and Mode II (crack sliding mode), respectively. The multimode exponents M and N depend on the material system. For glass/epoxy based composite material, they are taken as $M = 1$, $N = 2$, following (Kashtalyan and Soutis, 2007). For the special case of cracking in 90° ply only, the sliding displacement is zero and hence the criterion for ply crack multiplication reverts back to Eq. (9.39) with $t_\theta = 2t_{90}$. The detailed analysis procedure for predicting damage evolution in off-axis plies of multidirectional laminates through this approach is described in Singh and Talreja (2010, 2013). As shown in Figs. 9.9 and 9.10 , the model predictions for damage evolution in glass/epoxy $[0/\pm\theta_4/0_{1/2}]_s$ and quasiisotropic ($[0/90/\mp 45]_s$) laminates (Singh and Talreja, 2010) are found to agree well with available experimental data.

In order to predict the overall stress–strain response of a composite laminate, the crack density evolution model needs to be combined with a damage mechanics model which predicts the stiffness property changes arising due to a given state of damage. One such approach developed is the synergistic damage mechanics (SDM) methodology; see Singh and Talreja (2008, 2009, 2013), Singh (2013), Talreja and Singh (2012), and Montesano and Singh (2015a) for further details. The recent World-Wide Failure Exercise (WWFE III) (Kaddour et al., 2013) in this field has been focused on damage mechanics models of predicting failure of composite laminates undergoing subcritical and critical damage. The model predictions for selected test cases from the exercise using the SDM methodology have been reported in Singh and Talreja (2013). The prediction results for two test cases are presented in Figs. 9.11 and 9.12. In these figures, both evolution of crack density in transverse and off-axis plies and overall stress–strain response of the multidirectional laminate are shown. As

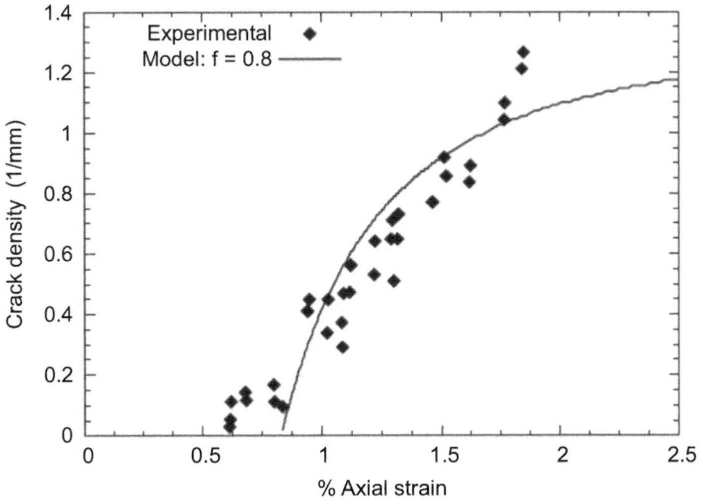

Figure 9.9 Damage evolution in $[0/\pm70_4/0_{1/2}]_s$ laminates. The experimental data are from Varna et al. (1999). The crack density is average of crack densities in +70° and −70°-plies. Reprinted, with the kind permission, from Singh and Talreja (2010), copyright Elsevier (2010).

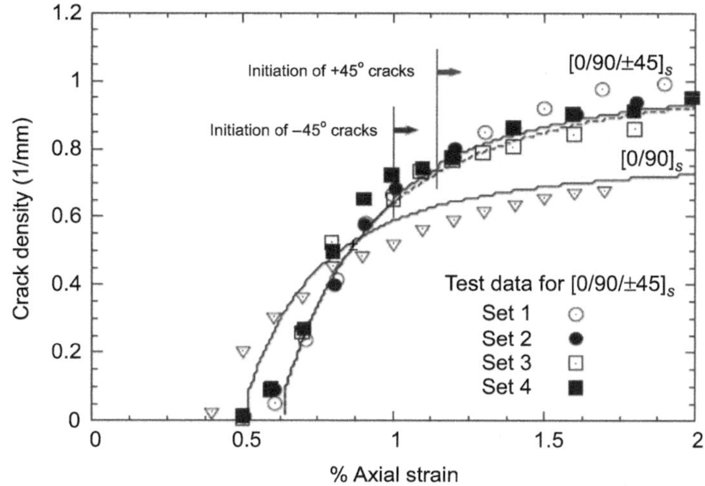

Figure 9.10 Evolution of 90°-crack density in $[0/90]_s$ and $[0/90/\pm45]_s$ laminates.
Reprinted, with the kind permission, from Singh and Talreja (2010), copyright Elsevier (2010); The experimental data are from Tong et al. (1997).

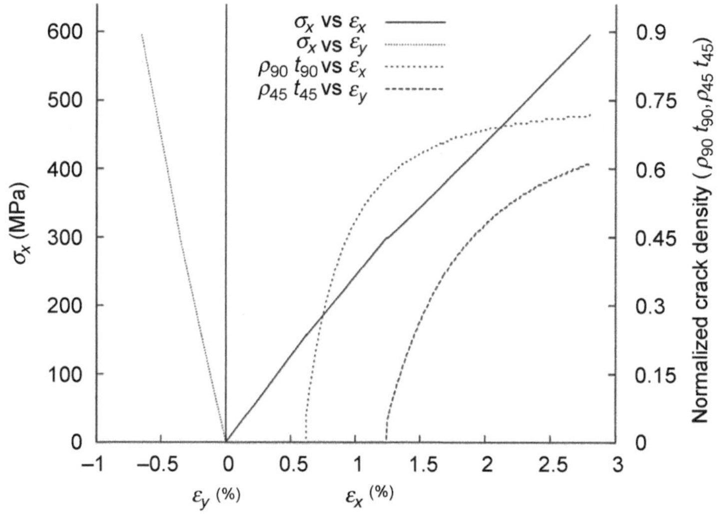

Figure 9.11 Evolution of crack density in 90 and 45 plies, and corresponding stress−strain response of $[0/90/\text{-}45/+45]_s$ glass/epoxy laminate with cracks in all off-axis modes for Test Case 6 of World-Wide Failure Exercise III. The crack initiation and the crack density evolution in both 45° plies were assumed to be identical.
Reproduced from Singh and Talreja (2013).

expected, cracks initiate first in 90° plies, and then appear in off-axis 45° plies. Based on previous experimental observations, 45° cracks do not develop fully in these quasiisotropic laminates (Test Case 6, Fig. 9.11), and hence their effective crack density

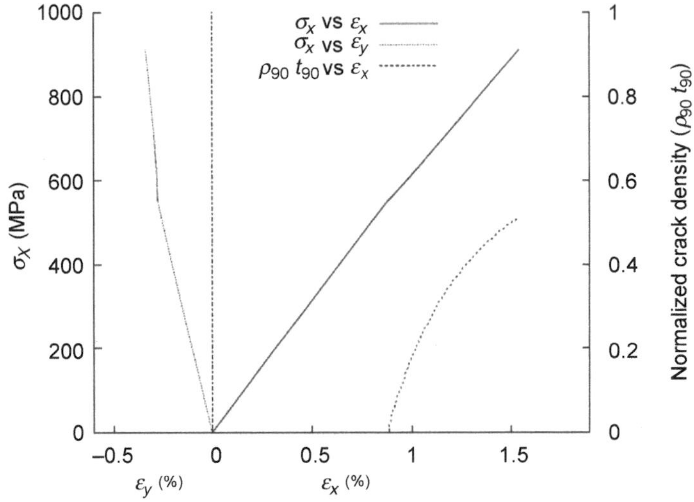

Figure 9.12 Evolution of crack density and corresponding stress−strain response and crack density evolution for Test Case 12 of World-Wide Failure Exercise III: $[45/90/\text{-}45/0]_s$ IM7/8552 laminate with cracks in 90°-plies only.
Reproduced from Singh and Talreja (2013).

was reduced to half to account for this effect. In the Test Case 12 (Fig. 9.12), the ply cracks were present only in the transverse plies, as 45° cracks did not grow until fiber fracture in these laminates. This test case also considered variation of different ply thicknesses, and the corresponding predictions for damage progression and stress−strain responses were made. These predictions showed that SDM model is capable of handling multimode damage scenarios. It was also observed that 90° damage contributes the most to the damage effects, both in terms of crack density evolution and for stiffness degradation.

9.4 Statistical aspects in multiple matrix cracking

The process of crack initiation is random because of variability in the matrix material's local resistance to fracture. This is mainly due to manufacturing-induced defects. These defects can be voids, incompletely cured regions and inclusions within the matrix, as well as weakly bonded regions of the fiber−matrix interfaces. In the early stages of damage development, the initiated cracks from defects therefore vary randomly in their location, sizes, shapes, and growth rates. Fig. 9.13 shows the progressive multiplication of transverse cracks in a IM7/977−3 $[0/90_4]_s$ laminates under uniaxial tensile loading. At 145 MPa applied stress, only few fully developed cracks and some partially developed cracks can be seen in the transverse plies, and they are randomly located through the laminate width. On further application of axial loads, the cracks typically form a roughly periodic array, thereby reducing the randomness in the cracking process. Based on this observation, it is essential for damage progression

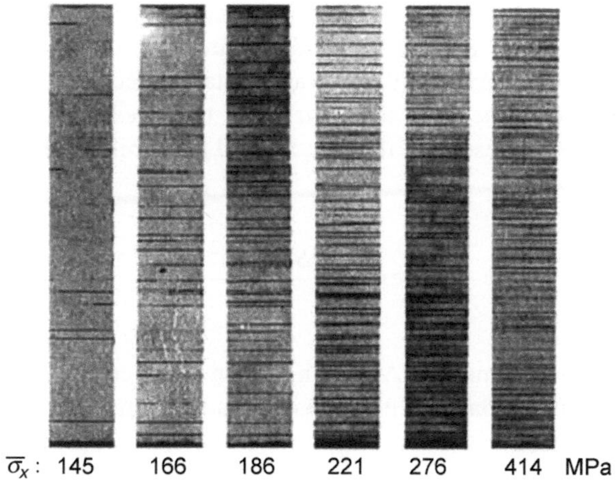

Figure 9.13 Evolution of transverse crack density with increasing loads in IM7/977−3 $[0/90_4]_s$ laminates under uniaxial tensile loading.
Source: Sun et al. (2003). Note: Need to add figure permission statement.

models to account for this randomness, especially in the early stages of ply cracking. Refs. Silberschmidt (2005); Silberschmidt (2006) provide a good overview of this topic. Fig. 9.14 shows crack distribution for $[0/90]_s$ glass/epoxy laminates loaded in quasistatic tensile loading. For comparison, the average axial stress distribution along laminate length, estimated by Berthelot and Le Corre (2000) is also shown in the figure.

Initial analysis of stochastic damage evolution was reported by Wang et al. (1980) for the case of cross-ply laminates. They postulated that the randomness in cracking process was due to a distribution of microcracks or inherent flaws in the material that

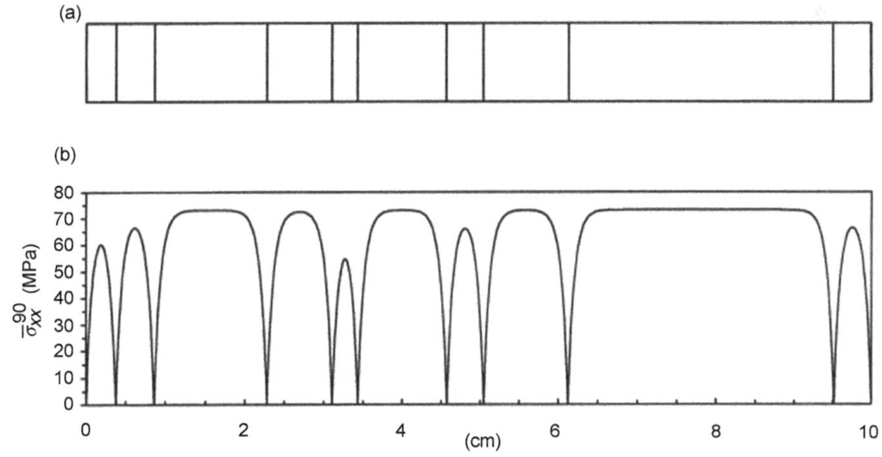

Figure 9.14 Crack distribution (A) and the corresponding variation (B) of the average longitudinal stress in the 90°-ply along the laminate length, estimated for [0/90]s glass/epoxy laminates.
Reprinted from Berthelot and Le Corre (2000), with permission from Elsevier.

are not visible until they propagate and grow to macroscopic dimensions and form fully developed ply cracks. The two key length parameters defining the distribution of these flaws are the average flaw size, $2a$, and the average distance between two adjacent flaws, S. The distributions of the flaw size, $f(a)$, and the crack spacing $f(S)$ along specimen length can then be described through normal probability distributions as:

$$f(a) = \frac{1}{a\sqrt{2\pi}} \exp\left[-\frac{(a-\mu_a)^2}{2\sigma_a^2}\right] f(S) = \frac{1}{S\sqrt{2\pi}} \exp\left[-\frac{(S-\mu_S)^2}{2\sigma_S^2}\right] \quad (9.47)$$

where, μ_a, μ_S, σ_a, and σ_S are fitting parameters. Accordingly, first ply cracking is assumed to initiate from the "worst" of the flaws. With increased loading, smaller flaws propagate to cause further transverse cracking. If σ_c is the longitudinal stress applied to the composite and $2a_0$ is the initial flaw size, transverse cracking would initiate when:

$$G(\sigma_c, a_0) = G_c \quad (9.48)$$

where G_c is the critical energy release rate, which is assumed to be constant along the laminate length.

The propagation of the flaw will be unstable if:

$$G(\sigma_c, a_0 + \Delta a) > G_c \quad (9.49)$$

where Δa denotes the increase in crack size upon loading. During applied loading, strain energy is stored in the composite material. When this energy is more than necessary for multiple flaws to propagate into fully grown transverse cracks, progressive cracking will take place. After first ply crack formation, stress field in the laminate is modified and hence the energy release rate for flaw propagation depends on its relative distance S from the existing crack given by:

$$G(\sigma_c, a) = R(S) G_0(\sigma_c, a) \quad (9.50)$$

where G_0 is the energy release rate for uncracked laminate, and $R(S)$ is the energy retention factor, with value between 0 and 1, that accounts for the presence of a neighboring crack. Similarly, for a flaw to propagate between two existing transverse cracks, energy release rate is defined as:

$$G(\sigma_c, a) = R(S_l) G_0(\sigma_c, a) R(S_r) \quad (9.51)$$

where S_l and S_r are distance of flaw from left and right crack, respectively.

The stochastic process described above was implemented by Wang et al. via a Monte Carlo scheme. The results showed a fair agreement with experimental data. However, the approach has not gained wide usage owing to multiple unknown parameters, which must be found by curve-fitting to experimental data.

9.5 Vinogradov–Hashin model

Probabilistic analysis has been performed for variational approach by Vinogradov and Hashin (2005). In their study, the authors delineated the stochastic nature of the cracking process through two probabilistic notions: "geometrical" and "physical." The "geometrical" uncertainty means that a crack can randomly appear at any location between two existing adjacent cracks, whereas the "physical" aspect accounts for the variation of material resistance to crack formation. The stochastic variation of distance between adjacent cracks is described by a distribution of crack spacing, that is:

$$\overline{\rho} = \int_0^\infty \rho p(\rho) \mathrm{d}\rho; \ \overline{\chi} = \int_0^\infty \chi(\rho) p(\rho) \, \mathrm{d}\rho \tag{9.52}$$

where $p(\rho)$ denotes the probability density function of distances between adjacent cracks. First ply cracking occurs in the limit $\rho \to \infty$ in the above equation. The energy release rate occurring during cracking process is derived as:

$$\gamma = \left(\sigma_{xx0}^{90}\right)^2 t_{90} C_{22} \chi(\infty) = 2\left(\sigma_{xx0}^{90}\right)^2 t_{90} C_{22} \alpha\left(\alpha^2 + \beta^2\right) \tag{9.53}$$

The above criterion is used to predict the initial stage of damage evolution. After a certain amount of crack density has developed in the laminate, any material block between two existing adjacent cracks will undergo further crack multiplication when:

$$\gamma = \left(\sigma_{xx0}^{90}\right)^2 t_{90} C_{22} \left[\chi\left(\frac{\rho + \xi}{2}\right) + \chi\left(\frac{\rho - \xi}{2}\right) - \chi(\rho)\right] \tag{9.54}$$

where ξ denotes the nondimensional coordinate of the new crack between the existing cracks. The "physical" nature of damage evolution, relating to local material resistance to cracking, is described through a probabilistic variation of the material property γ as a function of crack location by defining local fracture toughness. Accordingly,

$$\gamma = G(\xi) \tag{9.55}$$

A Weibull distribution is used to describe the variation of γ, that is:

$$p_r(\gamma) = \frac{\eta}{\gamma_0}\left(\frac{\gamma - \gamma_{\min}}{\gamma_0}\right)^{\eta - 1} \exp\left[-\left(\frac{\gamma - \gamma_{\min}}{\gamma_0}\right)\right], \gamma \geq \gamma_{\min} \tag{9.56}$$

where $\gamma_{\min}$ is the minimum possible value of γ, η, and γ_0 are parameters of the distribution, evaluated usually by fitting to experimental data. For predicting damage evolution for laminates with different thicknesses, Vinogradov and Hashin (2005) derived scaling relations for these damage model parameters. Consider two laminate configurations, $[0_{n1}/90_{m1}]_s$ and $[0_{n2}/90_{m2}]_s$. The parameters for the first laminate configuration, η_1, γ_{01} are evaluated by fitting the model with experimental data. The parameters η_2, γ_{02} for the second laminate configuration can then be computed using the following relations:

$$\frac{\Gamma\left(\frac{2+\eta_2}{\eta_2}\right)-\Gamma^2\left(\frac{1+\eta_2}{\eta_2}\right)}{\Gamma^2\left(\frac{1+\eta_2}{\eta_2}\right)}=\frac{m_2}{m_1}\frac{\Gamma\left(\frac{2+\eta_1}{\eta_1}\right)-\Gamma^2\left(\frac{1+\eta_1}{\eta_1}\right)}{\Gamma^2\left(\frac{1+\eta_1}{\eta_1}\right)}\gamma_{02}=\gamma_{01}\frac{\Gamma\left(\frac{1+\eta_1}{\eta_1}\right)}{\Gamma^2\left(\frac{1+\eta_2}{\eta_2}\right)} \tag{9.57}$$

where $\Gamma(x)$ represents the standard gamma function of the random variable x. For a detailed description of the iterative simulation procedure, the reader is referred to the original paper Vinogradov and Hashin (2005). Fig. 9.15 compares the model predictions with experimental data for cross-ply laminates made from two different ply materials. Clearly, model predictions fare well against the experimental data.

A strength-based model to account for randomness in ply cracking was developed by Berthelot and Le Corre (2000). This model utilized a probabilistic distribution to account for weakness areas in the laminate material and showed improvement in predictions for initial stages of damage evolution for $[0/90_2]_s$ carbon/epoxy laminate. As shown in Fig. 9.16, a clear divergence can be seen between model with and without consideration of weak areas at low crack densities. As explained before, this happens because, initially, cracking is preferred at areas where the resistance to fracture of the material is low (due to inherent defects) as compared to its average value in the whole laminate. Furthermore, the model deviates from experimental data at high crack densities because of possible delamination not considered in the model.

Defects arising from manufacturing processes are possible sites for nucleation of microcracks, which then can grow to form ply cracks, as noted above. In many cases, the size and spatial distribution of these defects is not known but can be accounted for by using appropriate probability distributions. Hence, strength-based damage initiation and evolution criteria are more suited. Huang et al. (2014) have derived a strength-based statistical model to predict ply cracking damage evolution in cross-ply laminates. In this model, the laminate length is divided into a number of elements, each possessing a single transverse strength. This elemental transverse strength σ_s is dependent on the local manufacturing defects and is assumed to be described by the Weibull distribution:

$$P_s(\sigma_s \leq \sigma) = 1 - \exp\left(-\left(\frac{\sigma}{\sigma_0}\right)^m\right) \tag{9.58}$$

where σ_0 and m are the two Weibull parameters called the characteristic strength and distribution shape constant, respectively. The authors used Hashin's variational analysis for estimating the axial stress (σ_{xx}) in the cracked plies of the cross-ply laminate. The progression of transverse cracking was subdivided into two regimes (see Fig. 9.17): (a) noninteractive regime, which occurs during the initial stages of the damage evolution such that the stress fields between cracks do not interact, and

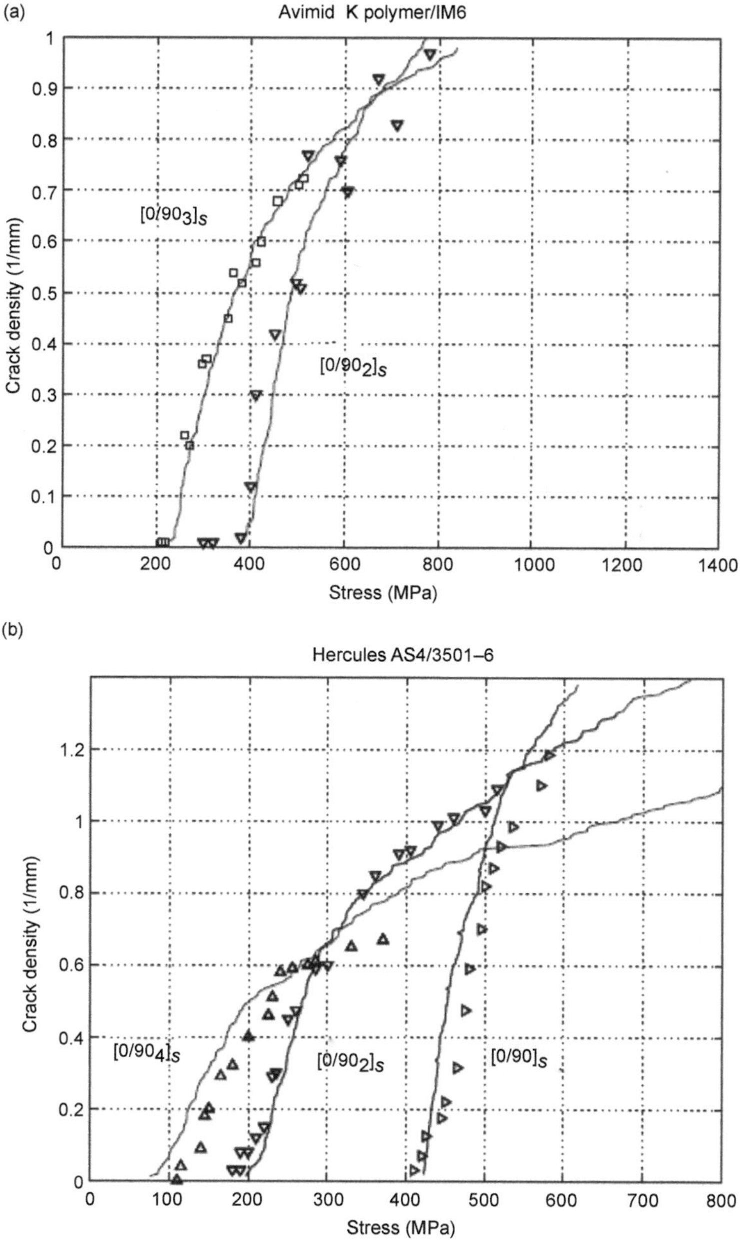

Figure 9.15 Prediction of crack density evolution from Vinogradov & Hashin model. The experimental data are from Liu and Nairn (1992).

Reprinted with the kind permission, from Vinogradov and Hashin (2005), Copyright Elsevier (2005).

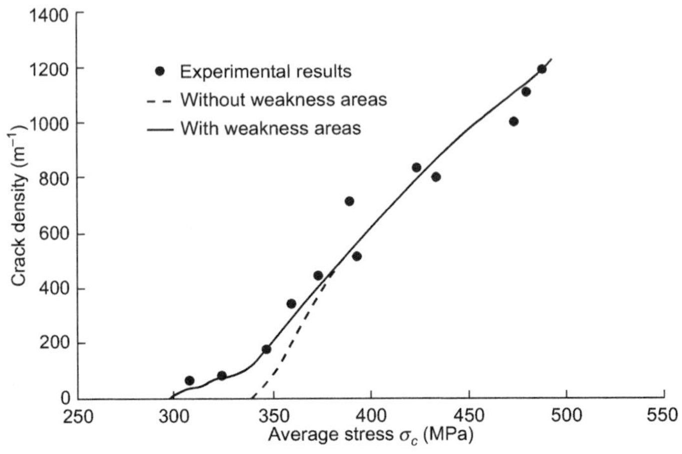

Figure 9.16 Effect of probabilistic model to predict evolution of crack density as a function of applied stress for $[0/90_2]_s$ carbon/epoxy laminate.
Reprinted, with the kind permission, from Berthelot and Le Corre (2000), copyright Elsevier (2000).

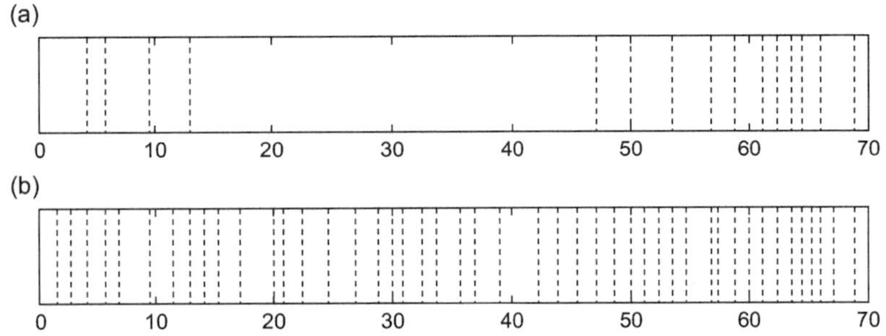

Figure 9.17 Distribution of cracks along the gauge length. Each vertical dash represents a transverse crack.
Reproduced from Huang et al. (2014). Note: Need to add permissions statement.

(b) interactive regime, where this interaction is significant. In the former case, the stress can be assumed to be equal to σ_{xx} in the uncracked 90° plies, whereas in the latter case, stress perturbations due to preexisting cracks need to be accounted for. The failure probability in the first stage is given as:

$$\rho = \frac{1}{l_0} \cdot P_s(\sigma_s \leq \sigma_{xx}) \tag{9.59}$$

This can be regarded as the Poisson's process (Manders, 1983; Huang and Talreja, 2014), and consequently the crack spacing is exponentially distributed with a cumulative probability of:

$$P_L(L \le L') = 1 - \exp(-\rho L') \tag{9.60}$$

For the interactive stage:

$$P_{L_i}^{(0)} = \begin{cases} P_L(L_i) - P_L(L_{i-1}) & i = 1, 2, ..., N-1 \\ 1 - P_L(L_{i-1}) & i = N \end{cases} \tag{9.61}$$

where $L_i = i/10$, and N is taken to be large enough so that $P_L(L_N)$ is close to 100%. As new cracks form under additional loading, the probability distribution of the block length changes since the long blocks are divided by new cracks into smaller ones. By tracking the fraction of block length at each step, the cracking evolution, that is, the average crack spacing and its variation, can be simulated. The failure probability for element $x_{i,j}$ is calculated from the given strength distribution as:

$$\lambda_{ij}^{(n+1)} = \frac{P_s\left(\sigma_{i,j}^{(n+1)}\right) - P_s\left(\sigma_{i,j}^{(n)}\right)}{1 - P_s\left(\sigma_{i,j}^{(n)}\right)} \tag{9.62}$$

Once the probability distribution of crack spacing is calculated, the crack density can be evaluated as:

$$\rho^{(n)} = \frac{1}{\sum_{i=1}^{N} L_i P_{L_i}^{(n)}} \tag{9.63}$$

The comparison of cumulative probability distributions of crack spacing from the model in the interactive regime and experimental data are presented in Fig. 9.18. It can be seen that the simulations agree very well with the experimental data.

The approach described above has been extended for off-axis ply crack evolution in multidirectional laminates as reported in Huang et al. (2014). The resultant predictions, shown in Fig. 9.19, are in close agreement with experimental data. Since no accurate analytical model exists to describe stress fields in a cracked multidirectional laminate, 3D finite element analysis was utilized to compute the stress variation. Note that this analysis is valid only for fully grown off-axis cracks, although partially grown cracks are often observed in off-axis plies. For instance, the experimental observations show that cracks do not grow completely in the +60° plies of $[0/60_2/0/\text{-}60_2]_s$ laminate before delamination or other critical mode of damage initiates.

9.6 Recent developments

Over the past decade, several studies have further advanced the above-described models in multiple ways. These advancements include modeling of multiaxial loading

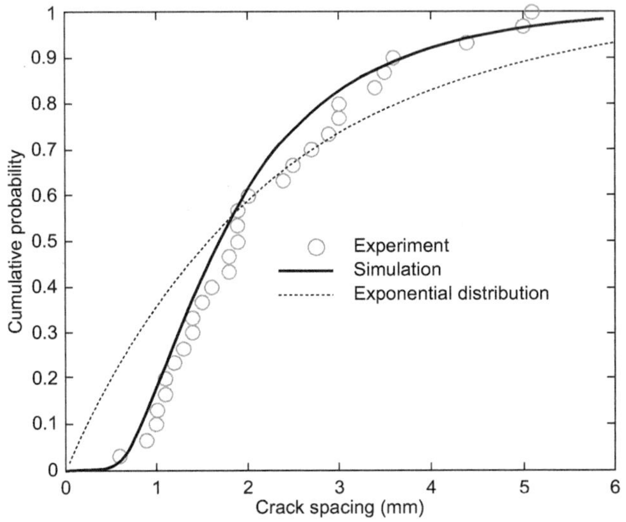

Figure 9.18 Cumulative probability distribution of crack spacing in a cross-ply laminate. Reproduced from Huang et al. (2014). Note: Need to add permissions statement.

Figure 9.19 Evolution of crack density in 60 ply of $[0/60_2/0/-60_2]_s$ laminate.
Reproduced from Huang and Talreja (2014). Note: need to add permissions statement.

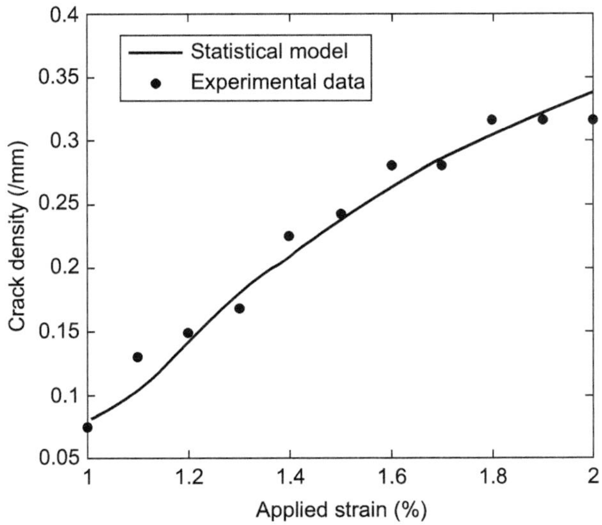

for multidirectional laminates, fatigue loading, and more integral use of stochastic models. In the works by Montesano and Singh (2015a,b, 2016), the energy-based damage progression model for multidirectional laminates described in Section 9.3 has been further improved by incorporating statistical aspects. Specifically, the stochastic aspects of ply cracking were accounted for by using a two-parameter Weibull distribution as follows:

$$G_{IC0} = G_0 \left[\ln\left(\frac{1}{1-F} \right) \right]^{1/m} \tag{9.64}$$

where G_0 and m are the Weibull distribution parameters and F is a random number in the interval [0,1]. The two Weibull parameters are evaluated from the mean and variance of G_{IC0}. These authors also noted that the resistance to ply crack multiplication tends to increase with increasing ply crack density due to crack shielding and adopted the following expression to define the critical energy release rate:

$$G_{IC} = G_{IC0} + G_{Icr}\{1 - \exp(-r\rho_\alpha)\} \tag{9.65}$$

where G_{IC0} is initial critical strain energy release rate and is evaluated numerically; G_{Icr} and r are material parameters; and ρ_α is the ply crack density. This damage progression model for multiple ply cracking has been extended to viscoelastic composite laminates (Berton et al., 2019), wherein the above equation was modified to incorporate the viscoelastic rate dependence of energy release rate. These models have been implemented in commercial finite element software packages such as ANSYS and Simulia (Abaqus) for complete durability analysis of structures made of composite laminate. For example, they assessed the durability of composite wind turbine blades under both quasistatic and cyclic loads (Montesano et al., 2016; Zuo et al., 2018). Similarly, a later study implemented this methodology for predicting matrix microcracking in composite structures subjected to low velocity impact (Berton et al., 2019). Stochastic models have also been developed by other authors. For instance, Jagannathan et al. (2016) improved the shear lag theory by incorporating probabilistic strength-based analysis to predict the evolution of matrix cracks in multidirectional laminates.

Computational micromechanics has also been utilized to investigate the mechanisms underpinning the initiation of matrix cracks from manufacturing induced defects. In one study, Sharifpour et al. (2020) investigated the role of ply constraints on the local stress states in cross-ply laminates containing manufacturing induced defects such as nonuniform fiber spatial distribution, microvoids, and microcracks. They observed that the emergence of the constraining effect highly depends on the presence and size of preexisting local defects, while different zones may exhibit varying degrees of constraining effects of the supporting plies. A local failure criterion is therefore required to be integrated within a multiscale modeling framework to predict the damage evolution in composite laminates.

An important advancement has been on analyzing the damage evolution during fatigue or cyclic loading. Carraro et al. (2017) developed a new multiscale strategy to predict the crack density evolution in multidirectional laminates subjected to cyclic loading. The model treated the off-axis crack initiation and propagation phases separately, wherein the initiation of ply cracks was by means of the S—N curve, whereas the propagation phase was treated by using the energy release rate concept in fracture mechanics. Another study by Glud et al. (2017) also utilized multiscale stress analysis to predict damage evolution in composite laminates under multiaxial fatigue loading. Both of these studies verified their modeling efforts with experimental observations.

Experimental techniques to assess crack density evolution in composite laminates have also improved substantially in past decades. One interesting study is by Finis et al. (2021) who have developed a procedure based on infrared thermography to observe crack density during constant amplitude fatigue tests.

References

Berthelot, J.M., Le Corre, J.F., 2000. Statistical analysis of the progression of transverse cracking and delamination in cross-ply laminates. Compos. Sci. Technol. 60 (14), 2659–2669. Comprehensive composite materials. Vol. 1 - 6. Amsterdam: Elsevier, 2000.

Berton, T., Najafi, F., Singh, C.V., 2019. Development and implementation of a multi-scale model for matrix micro-cracking prediction in composite structures subjected to low velocity impact. Compos. B Eng. 168, 140–151.

Carraro, P.A., Maragoni, L., Quaresimin, M., 2017. Prediction of the crack density evolution in multidirectional laminates under fatigue loadings. Compos. Sci. Technol. 145, 24–39.

De Finis, R., Palumbo, D., Galietti, U., 2021. An experimental procedure based on infrared thermography for the assessment of crack density in quasi-isotropic CFRP. Eng. Fract. Mech. 258, 108108.

Garrett, K.W., Bailey, J.A., 1977. Multiple transverse fracture in 90° cross-ply laminates of a glass fibre-reinforced polyester. J. Mater. Sci. 12 (1), 157–168.

Glud, J.A., Dulieu-Barton, J.M., Thomsen, O.T., Overgaard, L.C.T., 2017. A stochastic multiaxial fatigue model for off-axis cracking in FRP laminates. Int. J. Fatig. 103, 576–590.

Hashin, Z., 1985. Analysis of cracked laminates: a variational approach. Mech. Mater. 4 (2), 121–136.

Hashin, Z., 1996. Finite thermoelastic fracture criterion with application to laminate cracking analysis. J. Mech. Phys. Solid. 44 (7), 1129–1145.

Huang, Y., Talreja, R., 2014. Statistical analysis of oblique crack evolution in composite laminates. Compos. B Eng. 65, 34–39.

Huang, Y., Varna, J., Talreja, R., 2014. Statistical methodology for assessing manufacturing quality related to transverse cracking in cross ply laminates. Compos. Sci. Technol. 95, 100–106.

Jagannathan, N., Gururaja, S., Manjunatha, C.M., 2016. Probabilistic strength based matrix crack evolution in multi-directional composite laminates. Compos. B Eng. 87, 263–273.

Joffe, R., Varna, J., 1999. Damage evolution modeling in multidirectional laminates and the resulting nonlinear response. In: International Conference on Composite Materials -12. 5-9 July, Paris, France.

Joffe, R., Krasnikovs, A., Varna, J., 2001. COD-based simulation of transverse cracking and stiffness reduction in [s/90n]s laminates. Compos. Sci. Technol. 61 (5), 637–656.

Kaddour, A.S., Hinton, M.J., Li, S., et al., 2013. The background to part A of the third world-wide failure exercise (WWFE-III). J. Compos. Mater. 47 (20–21), 2417–2426.

Laws, N., Dvorak, G.J., 1988. Progressive transverse cracking in composite laminates. J. Compos. Mater. 22 (10), 900–916.

Liu, S.L., Nairn, J.A., 1992. The formation and propagation of matrix microcracks in cross-ply laminates during static loading. J. Reinforc. Plast. Compos. 11 (2), 158–178.

Manders, P.W., Chou, T.W., Jones, F.R., Rock, J.W., 1983. Statistical analysis of multiple fracture in [0/90/0] glass fiber/epoxy resin laminates. J. Mater. Sci. 19, 2876–2889.

Montesano, J., Singh, C.V., 2015a. A synergistic damage mechanics based multiscale model for composite laminates subjected to multiaxial strains. Mech. Mater. 83, 72–89, 2015.

Montesano, J., Singh, C.V., 2015b. Predicting evolution of ply cracks in composite laminates subjected to biaxial loading. Composites B 75 (2015), 264–273.

Montesano, J., Singh, C.V., 2016. Critical stiffness damage envelopes for multidirectional laminated structures under multiaxial loading conditions. Mater. Des. 91, 218–229.

Montesano, J., Chu, H., Singh, C.V., 2016. Development of a physics-based multi-scale progressive damage model for assessing the durability of wind turbine blades. Compos. Struct. 141, 50–62.

Nairn, J.A., Hu, S.F., 1992. The formation and effect of outer-ply microcracks in cross-ply laminates - a variational approach. Eng. Fract. Mech. 41 (2), 203–221.

Nairn, J.A., 1989. The strain energy release rate of composite microcracking: a variational approach. J. Mater. Sci. 23 (11), 1106–1129.

Nairn, J.A., 2000. Matrix microcracking in composites, 2000. In: Kelly, A., Zweben, C., Talreja, R., Manson, J.A. (Eds.), Polymer Matrix Composites. Comprehensive Composite Materials, vol. 2. Elsevier, pp. 403–432 (chapter 13).

Sharifpour, F., Montesano, J., Talreja, R., 2020. Assessing the effects of ply constraints on local stress states in cross-ply laminates containing manufacturing induced defects. Compos. B Eng. 199, 108227.

Silberschmidt, V.V., 2005. Matrix cracking in cross-ply laminates: effect of randomness. Compos Part A 36 (2), 129–135. SPEC. ISS.

Silberschmidt, V.V., 2006. Effect of micro-randomness on macroscopic properties and fracture of laminates. J. Mater. Sci. 41 (20), 6768–6776.

Singh, C.V., Talreja, R., 2008. Analysis of multiple off-axis ply cracks in composite laminates. Int. J. Solid Struct. 45 (16), 4574–4589.

Singh, C.V., Talreja, R., 2009. A synergistic damage mechanics approach for composite laminates with matrix cracks in multiple orientations. Mech. Mater. 41, 954–968.

Singh, C.V., Talreja, R, R., 2010. Evolution of ply cracks in multidirectional composite laminates. Int. J. Solid Struct. 47 (10), 1338–1349.

Singh, C.V., Talreja, R., 2013. A synergistic damage mechanics approach to mechanical response of composite laminates with ply cracks. J. Compos. Mater. 47 (20–21), 2475–2501.

Singh, C.V., 2013. A higher order synergistic damage model for prediction of stiffness changes due to ply cracking in composite laminates. Comput. Mater. Continua (CMC): Comput. Mater. Continua (CMC) 34 (3), 227–249.

Sun, Z., Daniel, I.M., Luo, J.J., 2003. Statistical damage analysis of transverse cracking in high temperature composite laminates. Mater. Sci. Eng.: A 341 1, 49–56.

Talreja, R., Singh, C.V., 2012. Damage and Failure of Composite Materials. Cambridge University Press, London, ISBN 9780521819428.

Tong, J., Guild, F.J., Ogin, S.L., Smith, P.A., 1997. On matrix crack growth in quasi-isotropic laminates - i. Experimental investigation. Compos. Sci. Technol. 57 (11), 1527–1535.

Varna, J., Joffe, R., Akshantala, N.V., Talreja, R, R., 1999. Damage in composite laminates with off-axis plies. Compos. Sci. Technol. 59 (14), 2139–2147.

Varna, J., Joffe, R., Talreja, R., 2001a. Mixed micromechanics and continuum damage mechanics approach to transverse cracking in [s,90(n)](s) laminates. Mech. Compos. Mater. 37 (2), 115–126.

Varna, J., Joffe, R., Talreja, R., 2001b. A synergistic damage-mechanics analysis of transverse cracking in [+/-theta/90(4)](s) laminates. Compos. Sci. Technol. 61 (5), 657–665.

Vinogradov, V., Hashin, Z., 2005. Probabilistic energy based model for prediction of transverse cracking in cross-ply laminates. Int. J. Solid Struct. 42 (2), 365–392.

Vinogradov, V., Hashin, Z., 2010. Variational analysis of cracked angle-ply laminates. Compos. Sci. Technol. 70 (4), 638–646.

Wang, A.S.D., Crossman, F.W., 1980. Initiation and growth of transverse cracks and edge delamination in composite laminates part 1. An energy method. J. Compos. Mater. 14, 71–87.

Zuo, Y., Montesano, J., Singh, C.V., 2018. Assessing progressive failure in long wind turbine blades under quasi-static and cyclic loads. Renew. Energy 119, 754–766.

Further reading

Aveston, J., Kelly, A., 1973. Theory of multiple fracture of fibrous composites. J. Mater. Sci. 8 (3), 352–362.

Berton, T., Montesano, J., Singh, C.V., 2016. Development of a synergistic damage mechanics model to predict evolution of ply cracking and stiffness changes in multidirectional composite laminates under creep. Int. J. Damage Mech. 25 (7), 1060–1078.

Crossman, F.W., Warren, W.J., Wang, A.S.D., Law, G.E., 1980. Initiation and growth of transverse cracks and edge delamination in composite laminates, part 2: experimental correlation. J. Compos. Mater. 14, 88–108.

Hashin, Z., 1988a. Micromechanics of cracking in composite materials. In: Inan, E., Markov, K. (Eds.), Continuum Models and Discrete Systems. Proceedings of the 9th International Symposium. World Scientific Publishing), Singapore, pp. 702–709.

Hashin, Z., 1988b. Thermal-expansion coefficients of cracked laminates. Compos. Sci. Technol. 31 (4), 247–260.

Heslehurst, R.B., 2014. Defects and Damage in Composite Materials and Structures. CRC Press.

Kashtalyan, M., Soutis, C., 2007. Stiffness and fracture analysis of laminated composites with off-axis ply matrix cracking. Compos. Part A 38 (4), 1262–1269.

Kwon, Y.W., Allen, D.H., Talreja, R., 2008. Multiscale Modeling and Simulation of Composite Materials and Structures, vol. 47. Springer, New York.

Montesano, J., McCleave, B., Singh, C.V., 2018. Prediction of ply crack evolution and stiffness degradation in multidirectional symmetric laminates under multiaxial stress states. Compos. B Eng. 133, 53–67.

Parvizi, Garrett, K.W., Bailey, J.E., 1978. Constrained cracking in glass fiber-reinforced epoxy cross-ply laminates. J. Mater. Sci. 13 (1), 195–201.

Reifsnider, K.L., 1982. Damage in Composite Materials. K. L. Reifsnider, Virginia Polytechnic Institute and State University.

Wang, A.S.D., 1984. Fracture mechanics of sublaminate cracks in composite materials. In: Composites Technology Review. ASTM, Philadelphia, pp. 45–62.

Compression failure of composite laminates

10

C. Soutis
Aerospace Research Institute, The University of Manchester, Manchester, United Kingdom

10.1 Introduction

The design of composite structures frequently includes discontinuities such as cut-outs and holes for joints, and they become critical regions under tensile or compressive loading. Failure in compression of fiber composite laminates with an open hole is by the initiation and growth of a microbuckle from the edge of the hole, Soutis *et* al (Soutis and Fleck, 1990; Soutis et al., 1991, 1993; Soutis, 1996; Soutis and Edge, 1997). They find that failure is governed by microbuckling in the 0° plies. The geometric inhomogeneity induces fiber rotation under increasing applied load; deformation localizes within a band and a microbuckle is initiated, Fig. 10.1. The microbuckle then propagates in a stable manner for 2–3 mm, and the component fails at a higher load than the initiation load. Soutis et al. (Soutis et al., 1991, 1993) compared this process in carbon-epoxy and carbon-PEEK laminates to an equivalent crack containing cohesive stresses. It is a crack bridging analysis that predicts the size of the buckled region as a function of the applied load, with the local stress supported by the buckled fibers decreasing linearly with the closing displacement of the microbuckle (linear softening cohesive zone law). The model is able to predict successfully the effects of hole size and lay-up upon the compressive strength and has recently been incorporated into a user-friendly computer program by Xin et al. (Xin et al., 1995). In the past, Soutis et al. (Soutis et al., 1991) performed tests to obtain both the laminate unnotched strength and the compressive energy release rate associated with fiber microbuckling, which are required as the model's input. From a design point of view, it is desirable to predict these laminate properties from the mechanical properties of the fibers and the matrix and from the lay-up geometry.

In this work, the unnotched strength for an IM7/8552 unidirectional (UD) laminate is estimated by a fiber microbuckling model and compared to experimental measurements. Then this value together with the laminate theory and the maximum stress failure criterion is used to predict the unnotched compressive strength of multidirectional quasiisotropic plates with emphasis on a ply-scaled $(+45_m/90_m/-45_m/0_m)_s$ laminate ($m \geq 1$). The critical fracture energy (intralaminar toughness associated with fiber breakage) in compression is calculated from a micromechanics model for fiber microbuckling and employed with the Soutis–Fleck cohesive zone model (Soutis et al., 1991, 1993) to predict the open-hole compressive (OHC) and the ply blocking effect; comparison with experimental data is made. The environment (hot/cold, wet/dry) has also an effect on notched strength and needs to be considered.

Modeling Damage, Fatigue and Failure of Composite Materials. https://doi.org/10.1016/B978-0-443-18489-5.00003-5

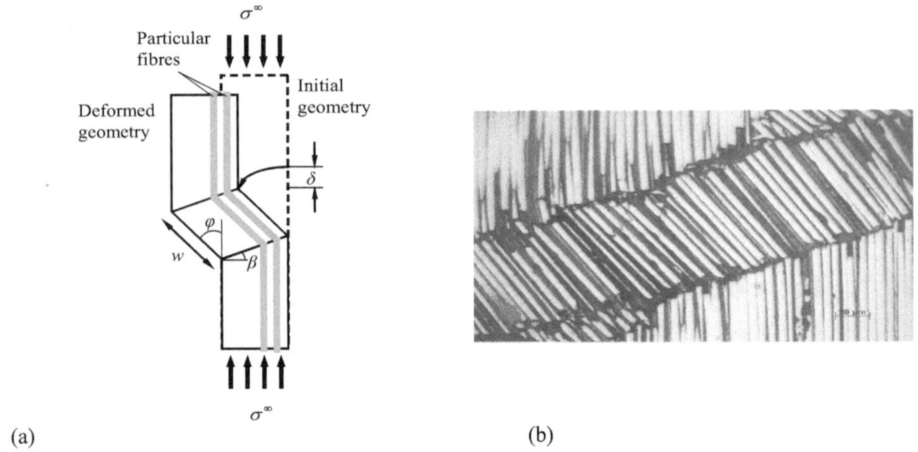

Figure 10.1 (a) A schematic showing the fiber kink band geometry. (b) An optical micrograph illustrating fiber microbuckling in a carbon fiber/epoxy unidirectional laminate (Lee and Soutis, 2007). The kink band width is 70–80 μm.

10.2 Modeling

10.2.1 Unidirectional strength

The unnotched compressive strength of long, aligned carbon fiber-reinforced plastics (CFRP) is significantly lower (30%–40%) than the tensile strength of the material due to kink band formation introduced by fiber instability (fiber microbuckling) (Rosen, 1965; Budiansky, 1983; Soutis, 1991, 1997; Berbinau et al., 1999a, 1999b; Soutis and Guz, 2006). The earliest attempt to model this behavior was given by Rosen (Rosen, 1965). Rosen predicted compressive strength based on an elastic fiber microbuckling and related buckling strength to the in-plane shear modulus of the composite (G_{12}). However, failure of modern filamentary composites occurs because of local nonlinear matrix deformation at composite strains well below the yield strain of the matrix and elastic analysis substantially over predicts compressive strength. Current models attribute the low compression strength and the mechanism of kink band formation to initial fiber misalignment (waviness), but fiber and fiber–matrix interface properties may also play an important role. For an elastic-perfectly plastic body, Budiansky (Budiansky, 1983) showed that

$$\sigma = \frac{\tau_y \left[1 + \left(\frac{\sigma_{T_y}}{\tau_y} \right)^2 \tan^2 \beta \right]^{\frac{1}{2}}}{\phi_0 + \phi} \tag{10.1}$$

where τ_y and σ_{Ty} are the in-plane shear and transverse yield stresses of the composite, respectively. ϕ_0 is the assumed fiber misalignment angle in the kink band, ϕ is the additional fiber rotation in the kink band under a remote stress σ, and β is the band orientation angle, as shown schematically in Fig. 10.1a; actual fiber microbuckling that occurs in an UD carbon fiber-epoxy laminate is shown by the optical micrograph in Fig. 10.1b (Soutis and Guz, 2006).

The critical stress $\sigma = \sigma_c$ is achieved at $\phi = 0$ in Eq. (10.1). The message here is that the compressive strength σ_c of UD laminates is very sensitive to fiber misalignment and even small values of $\phi_0 = 2{-}3$ degrees can cause a strength reduction of up to 40%. Of course, in addition to fiber waviness, void content (that may increase with thickness), fiber volume fraction, specimen geometry, test fixture design, and specimen misalignment in the fixture can contribute to premature failures and reduced measured compressive strength; many of these effects become difficult to be accounted in analytical work, but can be avoided by careful manufacturing and specimen design.

10.2.2 Multidirectional strength

Once the failure stress of the 0°-ply is known, the compressive strength of any multidirectional (MD) 0°-dominated lay-up, σ_{un}, can be estimated simply by the stiffness ratio method,

$$\sigma_{un} = \frac{\sigma_c}{NE_1}\sum_{k=1}^{N} n^{(k)} E_{x\theta}^{(k)} \tag{10.2}$$

where σ_{un} is the unnotched laminate strength, σ_c is the strength of the 0° lamina, N is the total number of the laminate in the laminate, E_1 is the 0° ply stiffness in the fiber direction, n is the number of plies of a given orientation θ, and $E_{x\theta}$ is the modulus of a ply of orientation θ in the loading direction (x). Alternatively, a ply-by-ply failure analysis using the classical laminate theory and the maximum stress failure criterion could be performed for more accurate predictions. Of course, in an MD laminate $(+45_m/90_m/\text{-}45_m/0_m)_s$, the stability of the 0° plies is affected by the fiber orientation and constraining effect provided by the neighboring off-axis layers and this depends very much on the ply thickness (or m value), especially near the free edge where a three-dimensional stress field is developed. The interlaminar shear stresses developed at the laminate free edge, between the 0° plies and the adjacent off-axis plies, were determined and incorporated into a fiber kinking model reported previously (Berbinau et al., 1999a, 1999b). The model is based on an initially misaligned fiber bundle on a nonlinear foundation (resin) that is capable of finite deflections.

$$E_f I\,\frac{d^4(v-v_0)}{dx^4} + \frac{A_f\sigma_{0^\circ-ply}}{V_f}\cdot\frac{d^2v}{dx^2} - 4r_f\left\{\left[\frac{d\tau_{zy}}{dy}\right]_{\frac{w}{2}}\right\}\cdot v - A_f G(\gamma)\cdot\frac{d^2(v-v_0)}{dx^2} = 0 \tag{10.3}$$

Eq. (10.3) gives a relationship for the compressive stress σ_0 developed in the 0° ply in terms of the maximum amplitude v of the 0° buckled fiber, initial fiber waviness v_0, fiber modulus E_f, fiber second moment of inertia I (or radius, r_f), fiber volume fraction V_f, shear modulus G(γ) as a function of applied load (that accounts for resin nonlinearity), and interlaminar shear stress τ_{zy} developed near the free edge during uniaxial compression; the other through-the-thickness tress components σ_{zz} (normal stress) and τ_{xz} (shear stress) can have an effect but are not included in Eq. (10.3). Analytical compressive strength predictions (Berbinau et al., 1999b) were compared with experimental measurements and found in a good agreement. It was also observed that the critical stress at which fiber microbuckling is triggered is affected by the interlaminar stresses but is more sensitive to the initial fiber waviness assumed. Residual thermal stresses trapped during the cooling process of the MD laminate fabrication can introduce matrix cracking and edge delamination and can lead to premature failure, especially in $(+45_m/90_m/\text{-}45_m/0_m)_s$ laminates where $m \geq 4$ (Soutis et al., 2002; Lee and Soutis, 2005, 2007). However, in the present chapter, since the emphasis is given on the prediction of the notched strength, the open hole is expected to dominate the failure of the multiply laminate (rather than edge effects), so the unnotched strength predicted by the maximum stress failure criterion, described above, is assumed for all *m* values, neglecting thermal and edge effects that can be introduced due to ply blocking.

10.2.3 Open-hole compressive strength: Cohesive zone model

The compressive strength is further reduced by the presence of fastener holes and access cut-outs. Previous work by Soutis and coworkers (Soutis and Fleck, 1990; Soutis et al., 1991, 1993; Soutis, 1996; Soutis and Edge, 1997) have found that open holes cause more than 40% reduction in the strength of carbon fiber-epoxy and carbon fiber-PEEK laminates and that damage was initiated by fiber microbuckling in the 0° plies at the edge of the hole. This process has been modeled with varying degrees of sophistication. Early models assumed that failure occurred when the maximum stress in the structure equals the unnotched strength of the material (maximum stress criterion) underestimating considerably the residual strength of the composite. To account for the local "ductility" of the material, researchers applied the average stress (ASFC) or point stress failure criteria (PSFC) (Tan, 1994). They introduced a characteristic length by assuming that fracture depends on attaining a critical stress (= unnotched strength) at a characteristic distance d_0 ahead of the notch or a critical average stress along a characteristic length a_0 ahead of the cutout. The characteristic distance is used as *a* free parameter to be fixed by best fitting the experimental data. Strength results obtained by the PSFC and ASFC for the IM7/8552 quasiisotropic laminate with a hole diameter/width ratio (*d/W*) equal to 0.2 are presented in Tables 10.2 and 10.3. Soutis et al. (Soutis et al., 1991, 1993) compared the damage zone (microbuckling surrounded by delamination) at the edge of the hole to a through-thickness line crack containing cohesive stresses, similar to Dugdale analysis (Dugdale, 1960), but for anisotropic layered plates. This equivalent crack is loaded on its faces by a normal traction, *T*, Fig. 10.2a, which decreases linearly with the crack closing displacement (CCD), 2*v*; Fig. 10.2b shows an optical micrograph of a T800/924C CFRP laminate where the fiber microbuckling, resembling a through-thickness crack, propagates across the plate width (Soutis et al., 1991).

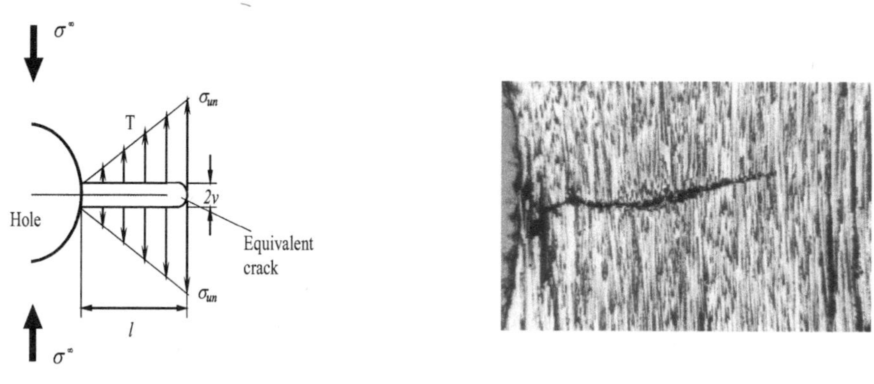

a) b)

Figure 10.2 (a) Damage zone is modeled as a line crack, loaded on its faces by a normal traction, *T*. (b) An optical micrograph showing 0°fiber microbuckling in a T800/924C multidirectional laminate containing a 5 mm hole; the length of the microbuckled zone is in the region of 2–3 mm (Soutis et al., 1991).

It is assumed that the length of the equivalent crack λ represents the length of the microbuckle. When the remote load is increased, the equivalent crack grows in length, thus representing microbuckle growth. The evolution of microbuckling is determined by requiring that the total stress intensity factor at the tip of the equivalent crack equals zero. When this condition is satisfied, stresses remain finite everywhere (Dugdale, 1960).

The equivalent crack length from the circular hole is deduced as a function of remote stress, σ^{∞}, that is,

$$\sigma^{\infty} = \sum_{i=1}^{n} \beta_i T_i = f(\lambda, \sigma_{un}, v_c, E, R, W) \tag{10.4}$$

detailed expressions for the functions β_i and T_i are given in Soutis et al. (1991). Eq. (10.4) gives an expression for the applied compressive stress as a function of microbuckling length, λ, unnotched strength, σ_{un}, critical CCD, v_c, laminate elastic properties, E, and geometry (plate width, W and hole radius, R). At a critical length of equivalent crack, λ_{cr}, the remote stress σ^{∞} attains a maximum value, designated σ_{cr}, where catastrophic failure occurs. The model contains two unknown parameters, which can be measured independently or predicted analytically: the unnotched strength σ_{un} and the critical CCD v_c, which is related to the area G_c (critical fracture energy) under the assumed linear traction-crack displacement curve.

For a linear softening cohesive zone law, Fig. 10.2, the critical strain energy release rate G_C is given by:

$$G_c = 2\int_0^{v_c} \sigma(v)\, dv = \sigma_{un}\, v_c \tag{10.5}$$

where v_c is the critical CCD on the crack traction-crack displacement curve, which is analogous to the crack opening displacement in tension. It is assumed that the fracture energy G_c represents the total energy per unit projected area dissipated by fiber microbuckling, matrix plasticity in the off-axis plies, and delamination.

In previous work (Soutis et al., 1991), G_C was obtained from a separate compressive kink propagation test, wherein the fracture toughness ($K_c = Y\sigma\sqrt{\pi a}$) of a laminate containing a sharpened long slit (=2a) is measured. The alternative method (Soutis and Curtis, 2000) is to analytically estimate the critical CCD v_c that appears in Eq. (10.5). Budiansky (Budiansky, 1983) in his microbuckling analysis related the crack overlap displacement 2 v_c that represents end-shortening δ in Fig. 10.1a explicitly to fiber diameter and fiber volume fraction by:

$$2v_c = w = \frac{\pi d_f}{4}\left(\frac{V_f E_f}{2\,\tau_y}\right)^{\frac{1}{3}} \approx 0.62\, d_f\left(\frac{V_f E_f}{\tau_y}\right)^{0.33} \tag{10.6}$$

where d_f is the fiber diameter, E_f is the fiber elastic modulus, and τ_y is the in-plane shear yield stress of the composite. Jelf and Fleck (Jelf and Fleck, 1994) examined the results of six experimental studies that appeared in the composites literature for continuous fibrous systems and found that the kink band width w satisfies the following empirical expression:

$$w = 2v_c = 0.68\, d_f\left(\frac{V_f E_f}{\tau_y}\right)^{0.37} \tag{10.7}$$

The correction factor in Eq. (10.7) accounts for material imperfections (fiber waviness, voids, resin-rich regions) and other damage modes in the form of matrix cracking and splitting that accompany the formation of fiber kinking/microbuckling, which are not accounted by the Budiansky model (Budiansky, 1983). Once the CCD and unnotched strength are known, the fracture energy associated with fiber microbuckling can be obtained from Eq. (10.5).

In the following sections, the linear softening cohesive zone model is applied to predict the compressive strength and critical microbuckling length of IM7/8552 ($+45_m$/90_m/-45_m/0_m)$_s$ notched carbon fiber-epoxy laminates loaded statically in uniaxial compression ($W = 32$ mm and $d/W = 0.2$). In addition, the unnotched strength of the UD and MD laminates is obtained.

10.3 Strength data and predictions

10.3.1 Compressive strength

The measured unnotched compressive strength of the unidirectional laminate (σ_{UD}) was 1590 MPa. Assuming that the failure mode is in the form of fiber microbuckling, Fig. 10.1, then the strength can be estimated from Eq. (10.1), provided that the initial fiber misalignment or waviness ϕ_0, the shear yield stress τ_y ($\approx$40 MPa), and the kink band orientation angle β (typically 10°−15°) are known. Table 10.2 presents the theoretical compressive strength of the IM7/8552 UD laminate as a function of fiber waviness and shear yield stress; it can be seen that a good correlation is obtained (1647 MPa) when $\phi_0 = 1.5$ degrees, $\tau_y = 40$ MPa, and $\beta = 15$ degrees. The difference between analysis and experiment is less than 4% and these "material" parameters can now be used to evaluate the unnotched and notched strength of the MD laminate $(+45_m/90_m/\text{-}45_m/0_m)_s$ for any specimen size, m value, and hole diameter. Assuming $\phi_0 = 1.5$ degrees, $\tau_y = 40$ MPa, and $\beta = 10$ degrees, a better match can be achieved for the UD strength, that is, 1580 MPa compared to 1590 MPa measured experimentally; this implies that less applied load is required to cause failure due to fiber microbuckling when the β value is reduced.

It is important to note that the fiber waviness, shear yield stress, and kink band orientation angle can be measured experimentally (Soutis, 1991, 1997). For the IM7/8552 system, the shear stress-strain response, Fig. 10.3, starts to divert from linearity at around 40 MPa, and here, this is taken as the yield point. Using the laminate plate theory and the maximum stress failure criterion with $\sigma_{UD} = 1647$ MPa ($=X_C$), a compressive strength of 635.6 MPa is predicted for the MD $(+45_m/90_m/\text{-}45_m/0_m)_s$ unnotched laminate, irrespectively of the m value (or $\sigma_{un} = 606.4$ MPa for $\beta = 10$ degrees in Eq. (10.1) which results to $X_C = 1580$ MPa); the simple stiffness ratio method of Eq. (10.2) results to a value of 534 MPa (about 16% lower than the ply-by-ply failure analysis). The shear stress-strain curve that has been experimentally determined and shown in Fig. 10.3 could be alternatively approximated by Eq. (10.8):

$$\tau(\gamma) = \tau_{ult}\left(1 - Exp\left(-\frac{G_{12}\gamma}{\tau_{ult}}\right)\right) \tag{10.8}$$

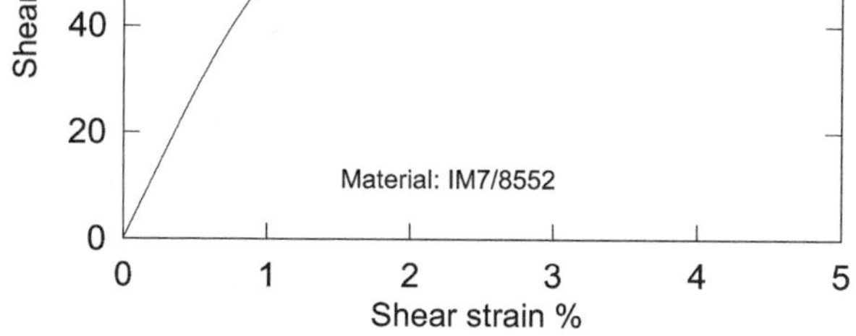

Figure 10.3 In-plane shear response of the IM7/8552 carbon fiber epoxy system. In the analysis of fiber microbuckling, see Section 10.2, the shear yield stress is taken at 40 MPa.

Table 10.1 Prediction of the IM7/8552 UD compressive strength.

τ_y	ϕ_0	β	σ_{11c} (X_c)	$\sigma_{11c}{}^{exp}$	Diff
MPa	°	°	MPa	MPa	%
	1.0	15	2470		+35.6
40	1.5	15	1647		+3.6
	1.5	10	1580	1590	<−1
	2.0	15	1235		−22.3
	1.0		3087		+48.5
50	2.5	15	2058	1590	+22.7
	2.0		1544		−2.9
		5	2890		+45
50	2.0	10	2963	1590	+46
		15	1544		−2.9

where τ_{ult} is the ultimate shear stress and G_{12} (secant modulus) is the elastic shear modulus, usually measured at 0.5% applied shear strain, γ (Berbinau et al., 1999a).

Identifying the τ_y value becomes a difficult task that can affect the predicted UD compressive strength, see Table 10.1; it also affects the outcome of Eq. (10.6) and hence the estimated critical fracture energy G_C value, which is needed for the OHC calculation.

10.3.2 Open-hole compressive strength

The notched compressive strength of the IM7/8522 ($+45_m/90_m/-45_m/0_m)_s$ laminate ($m \leq 3$), 32 mm wide with an open hole of 6.35 mm ($d/w = 0.2$) predicted by the ASFC assuming a characteristic length a_0 of 2.5 mm is 345 MPa, Table 10.2. This is 40% higher than that predicted by the maximum stress failure criterion (207 MPa) where an ideally brittle fracture is assumed. The characteristic length at the edge of the hole represents the development of damage (process zone) in the

Table 10.2 Open-hole compressive strength of IM7/8552 MD laminate ($d/W = 0.2$). by the Average Stress Failure Criterion (ASFC).

a_0	σ_n^{∞}	σ_n	
mm	MPa	MPa	σ_n/σ_{un}
2.0	331	324	0.510
2.5	353	345	0.543
3.0	370	363	0.571
4.0	401	392	0.618
5.0	425	416	0.655

Table 10.3 Open-hole compressive strength of IM7/8552 MD laminate ($d/W = 0.2$). Predicted by the Point Stress Failure Criterion (PSFC).

d_0	σ_n^{∞}	σ_n	
mm	MPa	MPa	σ_n/σ_{un}
0.5	299	293	0.461
1.0	355	347	0.547
1.5	410	401	0.632
2.0	453	444	0.699

form of matrix cracking, delamination, and fiber microbuckling that occurs in the 0° plies. This local damage reduces the stress concentration factor (elastic SCF, $k_t = 3.064$) at the edge of the hole and delays final failure to higher applied stresses. The results obtained from the PSFC for different assumed characteristic lengths d_0 are presented in Table 10.3. As explained earlier, the characteristic distance a_0 (or d_0) is used as a free parameter to be fixed by best fitting the experimental data; it varies with laminate stacking sequence and hole diameter, hence not a material property.

The advantage of employing the Soutis–Fleck cohesive zone model (Soutis et al., 1991) is that the critical damage length (fiber microbuckling length) that leads to catastrophic failure of the plate with the open hole is predicted by the model rather than assumed. The input parameters required are the laminate unnotched strength and the critical fracture energy associated with fiber microbuckling, which can be predicted from material (fiber and resin) properties or measured experimentally. For the $(+45/90/-45/0)_s$ laminate tested at room temperature, $V_f = 0.6$, $d_f = 4.5$ μm, $E_f = 276$ GPa, and $\tau_y = 40$ MPa, Eq. (10.6) results to a $2v_c = 45$ μm and hence $G_C = 14.3$ kJ/m^2 (or fracture toughness $K_C = 30$ MPa$\sqrt{\text{m}}$). Eqs. (10.5) and (10.7) result to a CCD value of $2v_c = 52$ μm and hence a higher fracture toughness ($K_C = 32$ MPa$\sqrt{\text{m}}$). It should be noted that Eqs. (10.6) and (10.7) are quite sensitive to the assumed fiber diameter, so careful measurements need to be made since this affects the calculated value of the critical fracture energy. The observed kink band width w ($= 2v_c$) from several optical micrographs that were obtained from sectioning studies in previous work by the author (Soutis, 1991, 1997; Lee and Soutis, 2007) was in the region of 70–90 μm that corresponds to a critical fracture energy $G_C = 22-29$ kJ/m^2 ($K_C = 37-42$ MPa$\sqrt{\text{m}}$).

In Table 10.4, the OHC strength of the quasiisotropic IM7/8522 laminate predicted by the cohesive zone model is presented together with the critical microbuckling length λ_{cr}^{A} as a function of the critical fracture energy G_C or fracture toughness K_C. The model at present cannot account for energy dissipated in damage modes as matrix cracking/splitting and delamination that may appear in $(+45_m/90_m/-45_m/0_m)_s$ as m becomes greater than 3. In previous work (Soutis and Lee, 2008), axial splitting appeared in laminates with blocked plies ($m > 3$), which blunt the hole, reduces the stress concentration factor that delays the triggering of fiber microbuckling, and results to enhanced notched strength (laminate becomes less notch sensitive). The fracture toughness value required in the model to accurately predict the measured strength

Table 10.4 Open-hole compressive strength of IM7/8552 MD laminate ($d/W = 0.2$). Predicted by the Soutis–Fleck cohesive zone model (Soutis et al., 1991).

K_C	G_C	σ_n	λ_{cr}^A	
MPa$\sqrt{m}$	kJ/m^2	MPa	mm	σ_n/σ_{un}
30	14.3	289	1.33	0.455
35	19.7	315	1.9	0.496
40	25.8	341	2.5	0.536
42	29.0	349	2.7	0.549
45	32.7	364	3.2	0.573
50	40.3	383	3.8	0.604
55	48.8	400	4.5	0.629
60	58.1	413	5.1	0.650

Notes: Unnotched strength $\sigma_{un} = 635.6$ MPa. The OHC predicted by the maximum stress criterion (ideally brittle) is $\sigma_n = 207$ MPa and that predicted by assuming ideally ductile behavior is $\sigma_n = 508$ MPa.

of the ply-level scaled $[45_4/90_4/\text{-}45_4/0_4]$s laminate examined in (Soutis and Lee, 2008) was $K_C = 55-60$ MPa$\sqrt{m}$ compared to 42 MPa$\sqrt{m}$ needed for the equivalent sublaminate level scaled $[45/90/\text{-}45/0]_{4s}$ laminate. This implies that the blocked lay-up ($m > 3$) is less notch sensitive than the laminate with the distributed plies; further analytical work is required to develop solutions that account for splitting and delamination, see Fig. 10.4 (Soutis and Spearing, 2002), that may develop near the hole edge in thicker (blocked) plies.

This may suggest that there isn't a laminate thickness effect but rather a ply-thickness effect (ply-level scaling), which in the case of the open-hole compression is rather beneficial. Fig. 10.5 presents experimental data that demonstrates the ply-scaled 4 mm thick laminate has a higher OHC strength when compared to sublaminate-level scaled one (Wisnom et al., 2010; Soutis, 2013).

In Table 10.5, the effect of the unnotched compressive strength on the open-hole compression is presented. It can be seen as the unnotched strength is reduced, the OHC is also reduced, but the critical microbuckling length is increased suggesting a more damage-tolerant laminate. These results are expected to help the composites designer to optimize strength and notch sensitivity by selecting appropriate fiber/matrix properties and lay-up configurations (Soutis and Spearing, 2002; Edgren et al., 2008). In order also to make comparisons with experimental data, the specimen geometry of the unnotched and notched scaled specimens should be carefully selected (Lee and Soutis, 2007) since bad specimen or end-tab design can lead to premature or invalid (near or within the end tab) failures. The overall specimen stability also needs to be sufficient to ensure that compressive failure occurs significantly before potential buckling, implying correct combination of specimen length and flexural stiffness. The maximum allowable specimen gauge section length can be determined on the basis of an Euler column buckling analysis assuming a pinned end strut with a rectangular

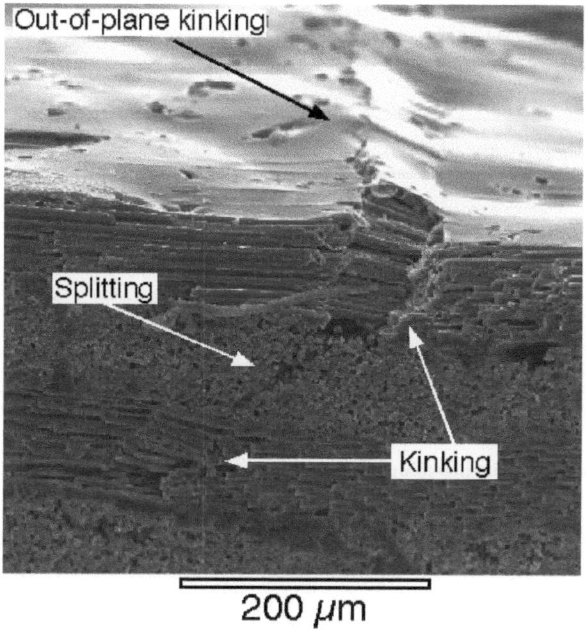

Figure 10.4 A scanning electron micrograph (SEM) showing damage in a notched carbon fiber/epoxy laminate loaded in axial compression. The cross-section is several millimeters behind the damage zone tip. Fiber kinking and matrix cracking that can affect the triggering of micro-buckling and hence fracture toughness of the laminate are clearly visible (Soutis and Spearing, 2002).

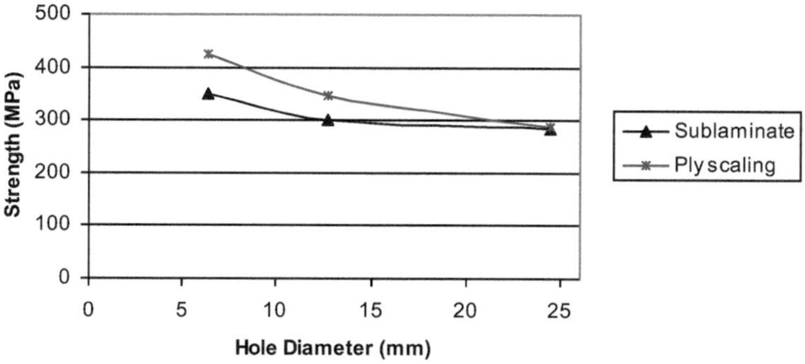

Figure 10.5 Effect of in-plane scaling on open-hole compressive strength of IM7/8552 4 mm thick laminates ($d/W = 0.2$). Axial splitting at the edge of the hole in the ply-scaled laminate delays fiber microbuckling (Wisnom et al., 2010).

Table 10.5 Effect of unnotched strength on the open-hole compressive strength of IM7/8552 MD $(+45_m/90_m/-45_m/0_m)_s$ laminate ($W = 32$ mm, $d/W = 0.2$).

σ_{un} MPa	606.4				635.6			
K_C MPa$\sqrt{m}$	30	40	50	60	30	40	50	60
σ_n MPa	283	334	374	400	289	341	383	413
λ^A_{cr} Mm	1.49	2.76	4.12	5.5	1.33	2.51	3.80	5.14

Note: The ply-level scaling (ply blocking) may affect the fracture toughness value, K_C and hence notch sensitivity.

cross-section and corrected to account for the influence of shear deflection due to transverse shear forces developed in anisotropic materials. Hence, for compressive failure to occur before Euler instability, the following condition must be satisfied:

$$\sigma_{(ult)} \leq \frac{0.67k\pi^2(EI)}{AL^2} \tag{10.9}$$

where E = longitudinal Young's modulus, I = second moment of area, A = cross-section area, and k is a constant taking into account the end support condition of the strut, that is, the manner in which it is held. For a rectangular cross-section, $I=At^2/12$, and Eq. (10.9) can be expressed in terms of critical length (max allowable), L, as:

$$L_{\max} \leq \pi t\sqrt{\frac{0.67kE}{12\sigma_{(ult)}}} \tag{10.10}$$

For a simply supported column, $k = 1$ will result to a conservative value.

Additionally, environment (hot/cold, wet/dry) will have an impact on notched strength. It has been shown (Soutis and Turkmen, 1997) that the strength properties of specimens tested in hot-wet conditions were substantially reduced and the final failure always occurred due to out-of-plane (through-the-thickness) fiber microbuckling. This is attributed to the reduction in matrix strength properties and weakening of the ply interface with increasing temperature and environmental conditioning. Thus, in compression, the matrix and interface play a key role in providing side support to the fibers and consequently resistance to fiber buckling. The Soutis—Fleck cohesive zone fracture model was successfully applied to predict the OHC strength for two different MD orthotropic laminates tested in hot-dry and hot-wet environments (Soutis, 2008, 2009). Compressive strength results for T800/924C carbon fiber-epoxy laminates with a 5 mm hole diameter ($d/W = 0.1$) are presented in Table 10.6. The unnotched compressive strength, σ_{un} and $G_C(\tau_y)$ are the input data for these predicted OHC results.

Table 10.6 Compressive strength results for T800/924C laminates with a 5 mm hole diameter ($d/W = 0.1$).

Test conditions °C, 95%RH,	OHC strength MPa	Predicted OHC strength* MPa	OHC strength MPa	Predicted OHC strength* MPa
$M_\infty = 1.42\%$	$[(\pm 45, 0_2)_3]_s$	$[(\pm 45, 0_2)_3]_s$	$[(90_2, 0_2)_3]_s$	$[(90_2, 0_2)_3]_s$
20-Dry	451.75	405.96	351.65	335.45
20-Wet	402.00	358.73	—	273.60
50-Dry	421.50	371.50	324.67	302.40
50-Wet	357.50	316.61	—	252.00
80-Dry	371.50	355.11	291.37	286.42
80-Wet	325.00	295.56	—	235.19
100-Dry	—	323.81		258.21
100-Wet	282.00	254.24	—	204.91

Notes: The unnotched compressive strength, σ_{un} and $G_C(\tau_y)$ presented in Table 10.3 are the input data for the predicted OHC results (Soutis, 2009).

10.4 Discussion and conclusions

The compressive strength of a carbon fiber-epoxy UD laminate can be successfully predicted from constituent properties and geometry of the failure mode, provided that the critical damage mode is that of fiber microbuckling, a fiber instability damage mode. The theoretical value of the IM7/8552 UD strength X_C (σ_{UD}) was 1580 MPa compared to 1590 measured experimentally assuming an initial fiber waviness $\phi_0 = 1.5$ degrees, shear yield stress $\tau_y = 40$ MPa, and kink band inclination angle $\beta = 10$ degrees. X_C is very sensitive to these three parameters (that relate to material and fabrication process) and results are summarised in Table 10.1.

Once the failure mode of the UD laminate has been established and the laminate constants for the examined system have been evaluated, the unnotched compressive strength of any MD laminate can be estimated by performing a ply-by-ply stress analysis and using the maximum stress failure criterion. For the IM7/8552 ($+45_m/90_m/-45_m/0_m)_s$, the strength was found equal to 606–635 MPa, irrespectively of the m value. The residual thermal stresses and edge effects expected to develop especially for $m > 3$ could introduce matrix cracking and edge delamination that could alter the critical failure mode, but are not considered in the present paper. They can be incorporated in our microbuckling model, Eq. (10.3), but since the emphasis is given in the open-hole compression, it is expected the hole to dominate the fracture and hence edge effects are neglected.

The predicted unnotched compressive strength σ_{un} together with the fracture toughness K_C (or critical energy release rate associated with fiber microbuckling, G_C) are used in the Soutis—Fleck cohesive zone model to predict the notched strength of the $(+45_m/90_m/\text{-}45_m/0_m)_s$ laminate. The K_C value for $m < 4$ is expected to be in the region of 40—42 MPa$\sqrt{\text{m}}$, while that of $m \geq 4$ could be 55—60 MPa$\sqrt{\text{m}}$ due to axial splitting that is expected as ply thickness increases (a local volume effect) (Soutis and Lee, 2008). The notched results as a function of K_C and σ_{un} are presented in Tables 10.4 and 10.5, respectively; the gray shaded row in Table 10.4 gives the expected notched strength for specimens $\leq$3 mm thick, while those of thickness $\geq$4 mm, provided that they develop axial splitting at the edge of the hole before the appearance of 0° fiber microbuckling, are predicted to show notched strength of up to 413 MPa; this is in a good agreement with measured data shown in Fig. 10.6 (Wisnom et al., 2010).

The predictions of the ASFC and PSFC for the $(+45_m/90_m/\text{-}45_m/0_m)_s$ plate with a $d/W = 0.2$ are shown in Tables 10.2 and 10.3 for comparison purposes. It should be noted that the characteristic length required as input in the point stress or average stress failure analysis needs to be determined by fitting the experimental data, and it varies with hole size and laminate lay-up; the expected notched strength for the thinner specimens ($\leq$3 mm) is also highlighted.

The advantage of the Soutis—Fleck cohesive zone model is that the fracture energy associated with fiber microbuckling can be estimated analytically from material properties. However, additional analytical work is required to account for the effect of axial splits that may develop at the edge of the hole, which lead to reduced stress concentration factor that delays the triggering of microbuckling and hence enhanced OHC strength; it is not so much the energy absorbed by splitting, since this is relatively small (typically 200—300 J/m^2), but its effect on the critical stress of fiber microbuckling initiation. The maximum stress failure criterion is expected to underestimate the notched failure stress by 30%—50% that could lead to a more conservative design and therefore a heavier structure.

The strength results obtained by the cohesive zone model can be compared to existing experimental data obtained for IM7/8552 $(+45_m/90_m/\text{-}45_m/0_m)_s$ and $[45/90/\text{-}45/0]_{ms}$ notched laminates where m takes different values resulting to thicknesses of 3—20 mm (24—160 plies), see work by Wisnom et al. (Wisnom et al., 2010; Soutis, 2013); Fig. 10.6 presents correlation of OHC strength predictions for m = 4 (4 mm thick laminates). Information required is the value of the shear yield stress and the geometry of the kink band, especially the kink band width, provided that fiber microbuckling/fiber kinking is the critical failure mode in these thicker panels. Penetrant-enhanced X-ray radiography or other nondestructive evaluation methods (Diaz Valdes and Soutis, 2002; Kessler et al., 2002; Diamanti et al., 2004, 2007; Qiu et al., 2006) can confirm if axial splitting develops in the thicker ($m \geq 4$) notched laminates that may have an effect on the selected K_C (or G_C) value. In addition, environment (hot/cold, wet/dry) has an impact on the compressive notched strength as shown for the T800/924 composite system in Table 10.6 and needs to be accounted in any structural design.

Research work by Davidson and Waas (Davidson and Waas, 2021) has also concluded that compressive failure in laminated composite bolted joints is controlled by the fiber kink failure developed in the axial plies, where is analyzed in the presence of transverse compressive loading. A micromechanics model is used to understand the

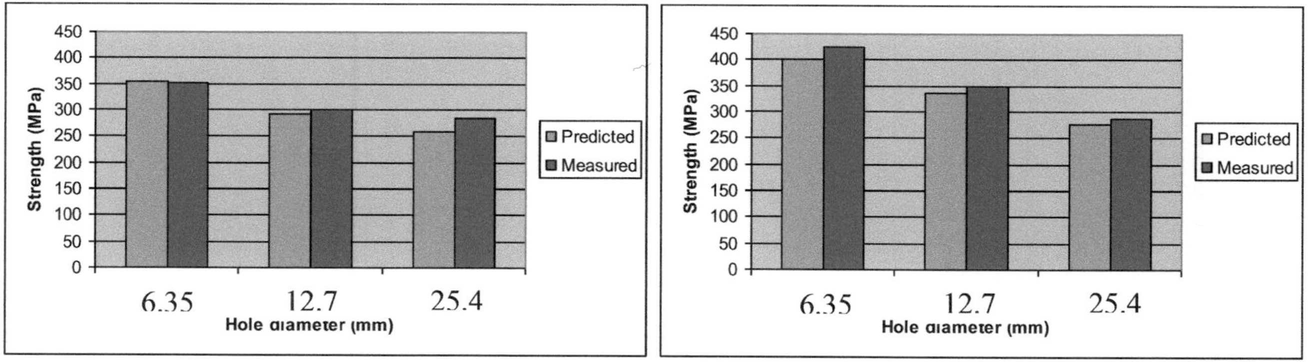

a) Sublaminate level scaling b) Ply level scaling

Figure 10.6 Correlation of open-hole compressive strength predictions for IM7/8552 4 mm thick laminates (Wisnom et al., 2010). (a) Sublaminate-level scaling. (b) Ply-level scaling.

influence of microscale and mesoscale fiber misalignment and their influence on compressive stiffness, strength, and postkinked response. Fabrication-induced defects in the form of resin rich regions, voids, and initial fiber misalignment affect the fiber stability of the axial plies (fiber orientation parallel to the applied compressive load) and need to be accounted in the analysis.

References

Berbinau, P., Soutis, C., Guz, I.A., 1999. Compressive failure of 0° unidirectional CFRP laminates by fibre microbuckling. Compos. Sci. Technol. 59 (9), 1451–1455.

Berbinau, P., Soutis, C., Goutas, P., Curtis, P.T., 1999. Effect of off-axis ply orientation on 0° fibre microbuckling. Composites, Part A 30 (10), 1197–1207.

Budiansky, B., 1983. Micromechanics. Comput. Struct. 16 (1), 3–12.

Davidson, P., Waas, A.M., 2021. Compressive failure due to kink band formation in the presence of transverse loading, and accounting for mesoscale and microscale misalignment. Compos. Struct. 265, 113760. https://doi.org/10.1016/j.compstruct.2021.113760.

Diamanti, K., Hodgkinson, J.M., Soutis, C., 2004. Detection of low-velocity impact damage in composite plates using Lamb waves. Struct. Health Monit. J. 3 (1), 33–41.

Diamanti, K., Soutis, C., Hodgkinson, J.M., 2007. Piezoelectric transducer arrangement for the inspection of large composite structures. Composites 38 (4), 1121–1130.

Diaz Valdes, S.H., Soutis, C., 2002. Real-time non-destructive evaluation of fibre composite laminates using low-frequency Lamb waves. J Acc. Soc. America 111 (5), 2026–2033.

Dugdale, D.S., 1960. Yielding of steel sheets containing slits. J. Mech. Phys. Solid. 8, 100–104.

Edgren, F., Soutis, C., Asp, L.E., 2008. Damage tolerance analysis of NCF composite sandwich panels. Compos. Sci. Technol. 68 (13), 2635–2645.

Jelf, P.M., Fleck, N.A., 1994. Compression failure mechanisms in unidirectional composites. J. Mat. Sci. 29, 3080–3085.

Kessler, S.S., Spearing, S.M., Soutis, C., 2002. Structural health monitoring in composite materials using lamb wave methods. Smart Mater. Struct. 11 (2), 269–278.

Lee, J., Soutis, C., 2005. Thickness effect on the compressive strength of T800/924C carbon fibre-epoxy laminates. Composites 36 (2), 213–227.

Lee, J., Soutis, C., 2007. A study on the compressive strength of thick carbon fibre/epoxy laminates. Compos. Sci. Technol. 67 (10), 2015–2026.

Qiu, Z.-X., Yao, X.-T., Yuan, J., Soutis, C., 2006. Experimental research on strain monitoring in composite plates using embedded SMA wires. Smart Mater. Struct. 15 (4), 1047–1053.

Rosen, B.W., 1965. Mechanics of composites strengthening. In: Fiber Composite Materials. American Society of Metals Seminar. ASM, Ohio, pp. 37–75, 1965.

Soutis, C., 1991. Measurements of the Static compressive strength of carbon fibre-epoxy laminates. Compos. Sci. Technol. 42 (4), 373–392.

Soutis, C., 1996. Failure of notched CFRP laminates due to fibre microbuckling: a topical review. J. Mech. Behav. Mater. 6 (4), 309–330.

Soutis, C., 1997. Compressive strength of unidirectional composites: measurement and prediction. ASTM-STP1242 13, 168–176.

Soutis, C., Analysis of hygrothermal effects on the compressive strength of CFRP laminates. 13th European Conference of Composite Materials (ECCM13), 2-5 June 2008, Stockholm, Sweden, Paper No ECCM-ID-0703.

Soutis, C., 2009. Modelling the open hole compressive strength of composite laminates tested in hot-wet conditions. Plast., Rubber Compos. 38 (2/3/4), 55–60.
Soutis, C., 2013. Compressive strength of composite laminates with an open hole: effect of ply blocking. J. Compos. Mater. 47 (20–21), 2503–2512.
Soutis, C., Curtis, P.T., 2000. A method for predicting the fracture toughness of CFRP laminates failing by fire microbuckling. Composites Part A 31 (7), 733–740.
Soutis, C., Edge, E.C., 1997. A method for the production of carpet plots for notched compressive strength of CFRP laminates. J. Aero. Eng. 211, 251–261.
Soutis, C., Fleck, N.A., 1990. Static compression failure of carbon fibre T800/924C composite plate with single hole. J. Compos. Mater. 24 (5), 536–558.
Soutis, C., Guz, I.A., 2006. Fracture of layered composites by internal fibre instability: effect of interlaminar adhesion. Aeronaut. J. 110 (1105), 185–195.
Soutis, C., Lee, J., 2008. Measuring the notched compressive strength of composite laminates: specimen size effects. Compos. Sci. Technol. 68 (12), 2359–2366.
Soutis, C., Spearing, S.M., 2002. Compressive response of notched woven fabric face-sheet honeycomb sandwich panels. J. Poly. Rubber Compos. 31 (9), 392–397.
Soutis, C., Turkmen, D., 1997. Moisture and temperature effects on the compressive failure of CFRP unidirectional laminates. J. Compos. Mater. 31 (8), 832–849.
Soutis, C., Fleck, N.A., Smith, P.A., 1991. Failure prediction technique for compression loaded carbon fibre-epoxy laminate with an open hole. J. Comp. Mat. 25, 1476–1498.
Soutis, C., Curtis, P.T., Fleck, N.A., 1993. Compressive failure of notched carbon fibre composites. Proc. Roy. Soc. Lond. A 440, 241–256 (1993).
Soutis, C., Lee, J., Kong, C., 2002. Size effect on the compressive strength of T300/924C carbon fibre-epoxy laminates. J. Poly. Rubber. Compos. 31 (8), 364–370.
Tan, S.C., 1994. "Stress Concentrations in Laminated Composites". Technomic Publishing Co.
Wisnom, M.R., Hallett, S.R., Soutis, C., 2010. Scaling effects in notched composites. J. Compos. Mater. 44 (2), 195–210.
Xin, X.J., Sutcliffe, P.P.F., Fleck, N.A., Curtis, P.T., 1995. Composites Compressive Strength Modeller. Cambridge University Engineering Department CUED MAT/TR139.

Delamination fracture in composite materials

11

Bent F. Sørensen
Composites Analysis and Mechanics Section, Department of Wind and Energy Systems, Technical University of Denmark, Roskilde, Denmark

11.1 Introduction

Over the latest decades, there has been a continuous increase in the use of composite materials in load-carrying structures where weight savings are of importance. The dominating class of composite materials in engineering structures are composites consisting of continuous, aligned fibers in a polymer matrix, typically made as laminates consisting of plies (laminas or fabrics of unidirectional fibers) stacked on top of each other, each ply having its own fiber orientation. A unidirectional ply is strong in the fiber direction but weak perpendicular to the fiber direction, so that cracks running parallel to the fibers such as transverse cracking of 90° plies and splitting of off-axis plies are common failure modes. Laminates can also fail by delamination, viz., cracking along the interface between adjacent plies. These failure modes can interact. It is difficult to predict the progressive damage evolution both under static and cyclic loading (fatigue).

Preferably, engineering structures should be designed using criteria formulated by the use of materials properties, that is, parameters that are independent of structural/component geometry and size. It is then possible to measure the relevant material properties by testing (typically small) test specimens in the laboratory and then use these to predict the strength or fatigue lifetime of much larger structures subjected to complicated loads using rigorous models or scaling laws. Stress-based criteria are widely used as design criteria for in-plane failure, that is, failure of individual plies. Fracture mechanics methods are often used for the characterization of delamination, that is, crack growth along interfaces between individual plies.

This chapter is devoted to delamination. In comparison with crack growth in isotropic materials, delamination is complicated by orthotropic elastic properties, mixed mode cracking, and sometimes the formation of a large-scale crack bridging zone. From an application point of view, crack bridging is beneficial, since it leads to a rising fracture resistance with crack extension (R-curve behavior) and creates an extra “build-in” safety margin, thus increasing the overall reliability and safety. However, it is more difficult to formulate engineering design laws and thus utilize crack bridging. In the modeling of large-sale bridging, the mechanics of crack bridging can be described in terms of traction-separation laws. It is still a challenge to measure and use traction-separation laws in structural design. Therefore, we will analyze delamination under two different situations: Small-scale fracture process zone

Modeling Damage, Fatigue and Failure of Composite Materials. https://doi.org/10.1016/B978-0-443-18489-5.00001-1

(FPZ), where fracture is confined to a small volume at a crack tip and large-scale FPZ, where fracture involves damage at a confined crack tip and a long bridging zone.

The main parts of this chapter concern delamination under static loading. Delamination crack growth under cyclic loading (fatigue crack growth) is of course also of great practical importance but will be covered in less detail. It is quite natural to consider composite structures and materials at several length scales. In this chapter, we will consider fracture at both the macroscale and the microscale. Analyzing test specimens at the macroscale, a ply is considered as being a homogeneous orthotropic material. Understanding and analysis of toughening mechanisms operating within the FPZ are accomplished at the microscale where the composite is considered as a heterogeneous material consisting of discrete fibers embedded in a matrix material.

The chapter is organized as follows. First, in Section 11.2, we introduce some basic fracture mechanics concepts for small-scale FPZ. Then, in Section 11.3 follows a brief summary of linear elastic fracture mechanics (LEFM), while advanced nonlinear fracture mechanics for large-scale fracture, based on traction-separation laws, is discussed in Section 11.4. Section 11.5 deals with delamination growth under cyclic loading, while Section 11.6 gives some perspectives for future developments. Major conclusions are summarized in Section 11.7.

11.2 Fracture mechanics concepts

A key feature of fracture mechanics is the viewpoint that materials and structures always have some manufacturing defects and flaws that will cause the initiation and growth of cracks leading to structural failure. Fracture mechanics is the discipline within solid mechanics that predicts the load-carrying capacity of structures while accounting for the presence of notches and cracks and taking the size of the crack into account. More precisely, fracture mechanics concerns growth of a crack, not the initiation of a crack from a manufacturing defect.

Physically, an FPZ is the damage zone in front of a crack tip within which the material degrades due to high stresses. Ahead of the FPZ, the material remains intact (i.e., it behaves elastically and reversibly), while the material is completely separated behind the FPZ. Typically, for a delamination crack in a laminated composite, the FPZ can include both a crack tip fracture zone and a bridging zone behind the crack tip in which single fibers or ligaments of multiple fibers connect the crack faces. The development of the microscale failure mechanisms can depend on the stress state acting at the FPZ. Understanding of the FPZ thus lies at the microscale.

Another microscale issue that deserves a comment is the size of the FPZ. In this chapter, we will distinguish between the case of a small-scale FPZ and a large-scale FPZ. The importance of this distinction will be clear in the next section. Here, we mention that the situation of delamination accompanied by large-scale bridging (LSB) (which is frequently seen in some types of composites such as glass fiber/polyester) must be analyzed differently than delamination of materials that show a small-scale FPZ (e.g., carbon fiber/epoxy composite).

Traditional LEFM concerns small-sale fracture, that is, the situation that the FPZ is much smaller in size than any other length of an elastic body. The approach is then to ignore the FPZ altogether so that the problem can be treated as a linear elastic boundary value problem, which will have a unique solution. Disregarding the FPZ, the problem will then be modeled as a sharp crack tip. A detailed linear elastic stress analysis will give a singular stress field at the crack tip. The stresses ahead of the crack tip follow a square root singularity. It is common to consider three so-called modes of cracking: pure normal opening (Δu_2), denoted Mode I, in-plane shearing (Mode II), Δu_1, and tearing or out-of-plane shearing, Δu_3, (Mode III). The crack tip deformations for the three modes are shown schematically in Fig. 11.1. For Mode I, there will be a singular normal stress ahead of the crack tip, σ_{22}, while the shear stresses, σ_{12} and σ_{23}, vanish ahead of the crack tip. For Mode II, there is a singular in-plane shear stress ahead of the crack tip, σ_{12}, while for Mode III, there is a singularity in the out-of-plane shear stress, σ_{23}. Mode I is particularly relevant for isotropic materials, since experience shows that for these materials, crack tends to grow in pure Mode I, along a path perpendicular to the largest principal tensile stress (Thouless and Evans, 1990). In contrast, for laminated composites made of plies consisting of long, aligned fibers that possess weak interfaces between the plies, cracks often remain "trapped" in the interface. The interface crack propagation along a weak plane occurs under mixed mode, that is, generally in a combination of Mode I, Mode II, and Mode III. Therefore, for composite laminates, it is relevant to characterize crack growth under mixed mode conditions. The combination of modes depends on the crack length, specimen geometry, and loads.

We also want to introduce the concepts of cohesive law and the work of separation per unit area. The mechanics of the FPZ will be expressed in terms of a cohesive law, which is a traction-separation relationship. Cohesive laws are usually used in models at the macroscale to represent the mechanics of the microscale failure mechanisms. For Mode I, a traction-separation law is the relationship between the normal traction, σ_n, and the normal opening, δ_n, that is, $\sigma_n = \sigma_n(\delta_n)$, see Fig. 11.2. The cohesive law is taken to be a material property (i.e., being the same law at any position along a specific interface). It is characterized in terms of a peak traction value, $\widehat{\sigma}_n$, and a critical

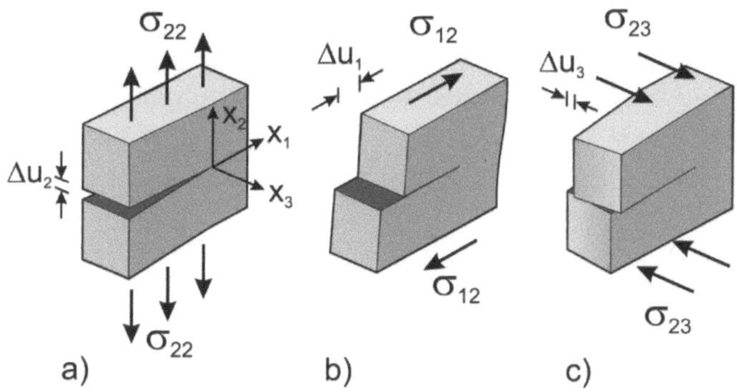

Figure 11.1 Definition of crack openings: (a) Mode I (normal opening), (b) Mode II (in-plane shearing), and (c) Mode III (out-of plane tearing).

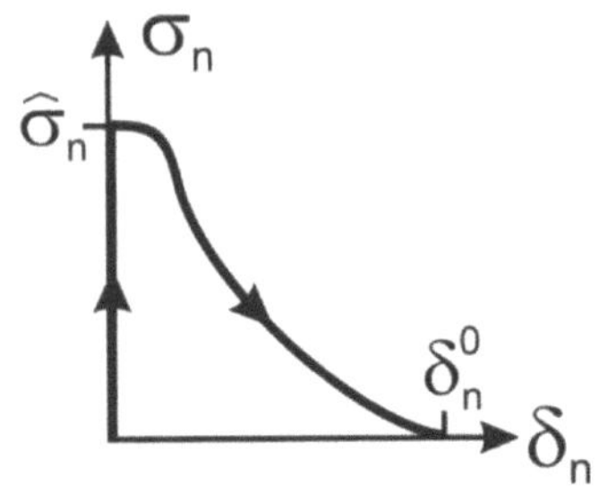

Figure 11.2 Definition of a cohesive law: the normal traction σ_n acting between crack faces is taken to be a function of the separation δ_n between crack faces.

separation, δ_n^0, which is the normal opening at which the cohesive traction vanishes—corresponding to complete breakage of the material within the FPZ. The area under the traction-separation law, that is, the work of the cohesive traction from intact (fully closed) to fully separated state, denotes the work of separation per unit area. The form of the cohesive law reflects the underlying microscale failure mechanisms. It can represent the crack tip only, the bridging zone only, or represent both the crack tip and the bridging zone.

A connection between cohesive laws and the overall fracture resistance, denoted J_R, can be established using the integration of the path-independent J integral (Rice, 1968). Consider two-dimensional problem in the $x_1 - x_2$ plane with a crack oriented in the $x_2 = 0$ plane. The J integral is defined as (using index notation and the Einstein summation rule):

$$J = \int_\Gamma \left(\Phi\, n_1 - \sigma_{ij} n_i \frac{\partial u_j}{\partial x_1} \right) dS \tag{11.1}$$

where $\Phi(\varepsilon_{ij})$ is the strain energy density which is a function of the strain tensor ε_{ij}, while σ_{ij} is the stress tensor, u_i the displacement vector, and n_i the outwards normal vector to an integration path S that goes from the lower traction free crack face to the upper traction-free crack face in a counterclockwise direction. The J integral is path independent, meaning that it takes the same value irrespective of the chosen integration path. When the J integral is evaluated along a crack face with cohesive tractions, the result becomes (Suo et al., 1992):

$$J_R = \int_0^{\delta_n^*} \sigma_n(\delta_n)\, d\delta_n \tag{11.2}$$

where δ_n^* is the end-opening of the active cohesive zone. This J integral result can be given a clear physical interpretation: the J integral is the area under the traction-separation relation from zero to the end-opening, that is, the work per unit area of the cohesive traction. With increasing applied load, δ_n^* increases, so that J_R increases. J_R attains is maximum value, denoted the work of separation, when the cohesive zone is fully developed, $\delta_n^* = \delta_n^0$. Note that J_R does not depend explicitly on the length of the active cohesive zone, L, nor on the opening shape of the crack, only the end-opening, δ_n^*. Therefore, we can write $J_R = J_R(\delta_n^*)$. Eq. (11.2) connects the traction-separation law to the so-called R-curve. An R-curve is the relationship between the fracture resistance and the crack extension, that is, J_R as a function of the crack extension, Δa.

For small-scale FPZ problems, the R-curve may be considered as being a material property, $J_R = J_R(\Delta a)$. But for large-scale PFZ problem, the R-curve behavior, $J_R(\Delta a)$, will show a dependence of specimen geometry (Suo et al., 1992). Then the relationship between J_R and Δa will not be a material property. Instead, the traction-separation law can be taken to be the relevant material property, $\sigma_n = \sigma_n(\delta_n)$. This issue will be discussed in more details later in Section 11.4.

11.3 Linear elastic fracture mechanics approach to delamination

11.3.1 Overview of LEFM

This section deals with fracture under the premise of a small-scale FPZ, traditionally referred to as LEFM. The assumption of LEFM can be summarized as follows: first, the material is taken to be linear elastic and small displacements and small strains are assumed. Second, it is assumed that the size of the FPZ is so small that it is embedded within a universal singular crack tip stress field (Rice, 1968). Within the field of fracture mechanics for metals, the assumption of a small-scale FPZ is sometime called small-scale yielding, since it is implicitly assumed that a plastic zone will form at the crack tip and that the size of the plastic zone is larger than the FPZ. The fulfillment of the small-scale FPZ assumption has the strong implication that an FPZ will be embedded in the same universal crack tip stress field for test specimens and components of different sizes and shapes (Rice, 1968). For all specimens, the FPZs will then be loaded in the same manner, and they will therefore all behave identically. The universal crack tip stress field can be seen as a stress field through which the external load and structure geometry communicate with the FPZ (and vice versa). Consequently, the critical condition, corresponding to the onset of crack propagation, can be expressed in terms of a critical magnitude of the universal stress field. The critical state (onset of cracking) can then be determined from a small test specimen and can subsequently be used as a criterion for predicting cracking and thus the load-carrying capacity of any large structure of that material. This transferability of the critical state due to similitude is one of the most powerful features of LEFM.

Within a LEFM analysis, the presence of the FPZ is disregarded. This is an important assumption that has two consequences: (1) A LEFM problem is linear, so that the principle of superposition applies. (2) The LEFM model does not include any energy dissipation. It follows that an energy balance must be set up as criterion for crack growth.

Introducing a coordinate system having its origin ($x_1 = x_2 = 0$) at the crack tip, LEFM gives the universal crack tip stress fields ahead of the crack tip (i.e., along the crack plane given by $x_2 = 0$, see Fig. 11.3a):

$$\sigma_{22} = \frac{K_I}{\sqrt{2\pi x_1}} \qquad \sigma_{12} = \frac{K_{II}}{\sqrt{2\pi x_1}} \qquad \sigma_{23} = \frac{K_{III}}{\sqrt{2\pi x_1}} \tag{11.3}$$

Here the parameters K_I, K_{II}, and K_{III} denote the Mode I, Mode II, and the Mode III stress intensity factors (SIFs) and x_1 is the coordinate ahead of the crack tip. The full crack-tip stress fields can be found in classical fracture mechanics treatments (Rice, 1968; Broek, 1986; Kanninen and Popelar, 1985). This crack-tip stress field given by Eq. (11.3) is not the complete stress field solution but is only the first (dominant) term in a Taylor series expansion. However, close to the crack tip, the singular terms will dominate. Retaining only the terms in Eq. (11.3), the problem is linear, since the stress intensity factors scale linearly with the applied tractions. Then, the crack-tip

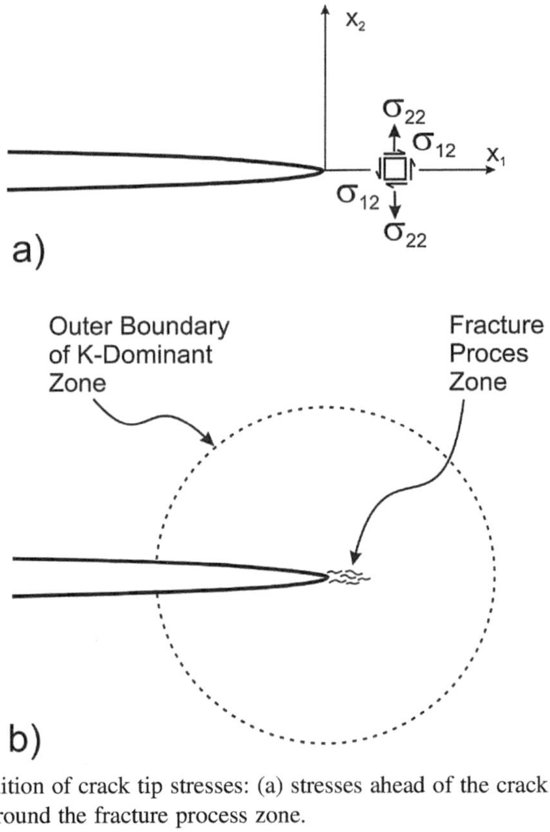

Figure 11.3 Definition of crack tip stresses: (a) stresses ahead of the crack tip and (b) the *K*-dominant region around the fracture process zone.

stresses will scale linearly to the applied tractions due to linearity of stress-strain behavior (linear elasticity)—therefore the name: LEFM. While the stress intensity factors depend linearly on the applied loads, they also depend on specimen geometry and dimensions, including (but not always) the crack length. In many cases, the stress intensity factors increase with crack length (with all other parameters held fixed) so that once crack growth begins, it becomes unstable under constant load. For pure Mode I, the critical value of K_I (corresponding to crack propagation), denoted K_{Ic}, is called the fracture toughness (SI units: N/m$^{3/2}$ or MPa m$^{1/2}$).

As discussed above, under the condition of small-scale FPZ, the singular stress field connects the external loads and geometry to the FPZ, and it is useful to think that the stress intensity factors control the magnitude of the stress field in an annulus around the FPZ as illustrated in Fig. 11.3b. This annulus is sometimes referred to as the K-dominant zone (Rice, 1968; Kanninen and Popelar, 1985).

Since no FPZ is modeled in LEFM, the crack tip is well-defined, and the concept of energy release rate can be introduced. The energy release rate is defined as the decrease in the potential energy, Π, per unit crack extension of the elastic body and applied forces and can be written as:

$$G = \frac{-1}{B}\frac{d\Pi}{da} \tag{11.4}$$

where B is the specimen thickness (in the x_3-direction) and a is the crack length. The potential energy release rate is a "global" parameter that involves the elastic energy of the body and the potential of the applied tractions. The energy release rate can be related to the stress intensity factors as:

$$G = \frac{\left(K_I^2 + K_{II}^2\right)}{E'} + \frac{(1+\nu)}{E}K_{III}^2 \tag{11.5}$$

where $E' = E$ for plane stress and $E' = E/\left(1-\nu^2\right)$ for plane strain with E being the Young's modulus and ν is the Poisson's ratio. Details of the derivation can be found in textbooks (Broek, 1986; Kanninen and Popelar, 1985). In this chapter, we will treat only plane stress cases since the plane stress/plane strain issue is considered of secondary importance. The implication is that a criterion for crack growth can be expressed both in terms of stress intensity factors and energy release rate. The onset of crack growth can be expressed in terms of a critical value of stress intensity factors or a critical value of energy release rate.

From an energy point of view, the criterion for crack propagation can be understood as an energy balance (Griffith, 1921): crack growth can occur only when the decrease in potential energy (for a small crack extension) equals (or exceeds) the energy consumed, G_{Ic}, called the fracture energy (SI units J/m^2). So while the elastic body seeks to minimize its potential energy (and extending the crack length decreases the potential energy), crack growth only occurs if the potential energy decrease is equal to or higher than the energy consumption associated with the crack growth.

For pure Mode I problems, the criterion for crack growth can be expressed in terms of a critical value of the Mode I stress intensity factor, K_{Ic}, or alternatively, in terms of a critical value of the energy release rate (Irwin, 1957):

$$G_{Ic} = \frac{K_{Ic}^2}{E'} \tag{11.6}$$

For mixed mode crack problems involving only K_I and K_{II}, a phenomenological mixed mode failure criterion can be written of the form (Jensen et al., 1990):

$$\left(\frac{K_I}{K_{Ic}}\right)^{\alpha} + \left(\frac{K_{II}}{K_{IIc}}\right)^{\beta} = 1 \tag{11.7}$$

where α and β are nondimensional parameter that are determined by curve fits to experimental data and K_{IIc} is the critical Mode II stress intensity factor, that is, the critical value of K_{II} under pure Mode II fracture. Alternatively, the crack grow criterion can be formulated in terms of the energy release rate and the mode mixity defined as (Hutchinson and Suo, 1992):

$$\psi = \tan^{-1}\left(\frac{K_{II}}{K_I}\right) = \tan^{-1}\left(\sqrt{\frac{G_{II}}{G_I}}\right). \tag{11.8}$$

In Eq. (11.8), G_I and G_{II} denote the Mode I and Mode II energy release rate, that is, the first two terms in the right-hand side of Eq. (11.5).

For instance, Hutchinson and Suo (1992) have proposed a criterion of the form:

$$G(\psi) = G_{Ic}^0\{1 + \tan^2[(1 - \Lambda)\,\psi]\} \tag{11.9}$$

where G_{Ic}^0 and Λ are curve fitting parameters. Such criteria are essential phenomenological, that is, they are fits to macroscale experimental mixed mode fracture data.

As mentioned above, LEFM was developed for metals in the 1950–1960s and has reached a mature stage where now catalogs are available for stress intensity factors and energy release rates for a large number of useful crack problems (Sih, 1973; Tada et al., 2000) and LEFM is being used for the safe design of the maintenance of a wide range of structures, such as aircrafts.

LEFM design involves determining a critical value of the stress intensity facture, K_{Ic}, (or alternatively, a critical value of the energy release rate, G_{Ic}) from a fracture mechanics test specimen. K_{Ic} (or G_{Ic}) is determined from catalog values of K_I (or G_I) of the particular test specimen and the critical load (the magnitude of the applied load at the onset of crack growth of a sharp, preexisting crack). Knowing G_{Ic}, components of other size and geometry can be designed such that G_I at design loads is always less than the critical value, G_{Ic}, for crack sizes that can be detected by nondestructive techniques, such as ultrasound scanning.

As mentioned earlier, fracture mechanics analyses for orthotropic materials are more complicated than for homogenous, isotropic materials. For orthotropic materials, the crack tip stress field still has a square-root singularity (Sih et al., 1965; Williams, 1989), but the elastic properties of the orthotropic material enter the equations for the energy release rate. For 2-D problems in the $x_1 - x_2$ plane, the linear elastic orthotropy parameters are:

$$E_1, E_2, G_{12}, \nu_{12}, \nu_{21}, \tag{11.10}$$

where E_1 and E_2 denote the Young's moduli in the x_1- and x_2-directions, respectively, G_{12} is the in-plane shear modulus, ν_{12} and ν_{21} are the major and minor Poisson's ratios, respectively. The so-called Maxwell–Betti reciprocal relationship connecting the Young's moduli and the Poisson's ratios reduces the number of independent constants to four ($\nu_{12}/E_1 = \nu_{21}/E_2$). It has been shown, however, that for traction prescribed boundary condition, the dependence can be expressed in terms of only two nondimensional parameters defined as (Suo et al., 1991):

$$\lambda = \frac{E_2}{E_1} \quad \rho = \frac{\sqrt{E_1 E_2}}{2G_{12}} - \sqrt{\nu_{12}\nu_{21}} \tag{11.11}$$

The λ-dependence can be handled analytically by a rescaling technique (Suo et al., 1991), while the dependence on ρ requires a complete (often numerical) analysis.

For cracks along interfaces between elastic dissimilar materials, such as, delaminations between plies of different orientations, the crack tip has oscillating singularity (Williams, 1959; Rice, 1988; Qu and Bassani, 1989; Suo, 1990), but this will not be discussed in this chapter.

11.3.2 Some common LEFM test specimens

The practical determination of G_{Ic} requires care and practical skills. First, appropriate fracture mechanics test specimens should be prepared, for example, from a laminate. In order to obtain an initial notch that acts as a sharp crack, a thin slip foil (maximum thickness 12 microns according to the ASTM standard, ASTM D5528-01, 2001) should be used as a starter crack. The slip foil is inserted during the manufacturing of the laminate. After consolidation of the laminate, fracture mechanics test specimens are cut out from the laminate and their dimensions are measured, since they are needed for calculation of the energy release rate. Experiments are typically conduced at a fixed displacement rate so that the load increases monotonically until crack propagation initiates. When crack growth is detected, the specimen is unloaded, and the crack length is recorded (this procedure is required since for most test specimen configurations, the crack length must be known to calculate the critical energy release rate). Alternatively, the crack length can be obtained from images or video recordings that link the load level and crack length. The specimen is then reloaded to determine a new onset of cracking. In this manner, a number of critical loads (onset of crack growth) and the associated crack length are recorded so that a number of fracture energy values can be determined for each specimen. A number (typically around five) of nominal identical specimens are tested to explore the variation from specimen to specimen.

The symmetric DCB (double cantilever beam) specimen loaded with wedge forces, P, shown in Fig. 11.4, is a commonly used Mode I test specimen. The energy release rate is given by (plane stress):

$$G_I = 12\frac{P^2a^2}{B^2H^3E_1}\left[1 + Y_I(\rho)\lambda^{-1/4}H / a\right]^2 \tag{11.12}$$

where $Y_I(\rho)$ is a nondimensional function that depends only on the second orthotropy parameter, ρ. The function $Y_I(\rho)$ is given by (Hutchinson and Suo, 1992):

$$Y_I(\rho) = 0.677 + 0.149(\rho - 1) - 0.013(\rho - 1)^2 \tag{11.13}$$

The dependency on the other orthotropy parameter λ is included explicitly in Eq. (11.12) through the term $\lambda^{-1/4}$. The first term inside the square bracket of Eq. (11.12) represents the standard beam solution, while the second term is a contribution for the shear deformation (root rotation). With increasing a, the second term vanishes relative to the first, so the terms inside the square brackets approach unity.

The ENF (end notch flexure) specimen, Fig. 11.5a, and the ELS (end-loaded split), Fig. 11.5b, are pure Mode II specimens. The plane stress energy release rate is given by (Hutchinson and Suo, 1992):

$$G_{II} = \frac{9P^2a^2}{4B^2H^3E_1}\left[1 + Y_{II}(\rho)\lambda^{-1/4}H / a\right]^2, \tag{11.14}$$

where P is the applied force and the nondimensional function $Y_{II}(\rho)$ is given by (Hutchinson and Suo, 1992):

$$Y_{II}(\rho) = 0.206 + 0.078(\rho - 1) - 0.008(\rho - 1)^2 \tag{11.15}$$

For both Eqs. (11.12) and (11.14), the terms in front of the square bracket include the crack length, a, in second power. This shows that for these specimens, the energy release increases with crack length under fixed load. As a result, crack growth under fixed load is unstable if the fracture resistance is constant. Under fixed grips, crack

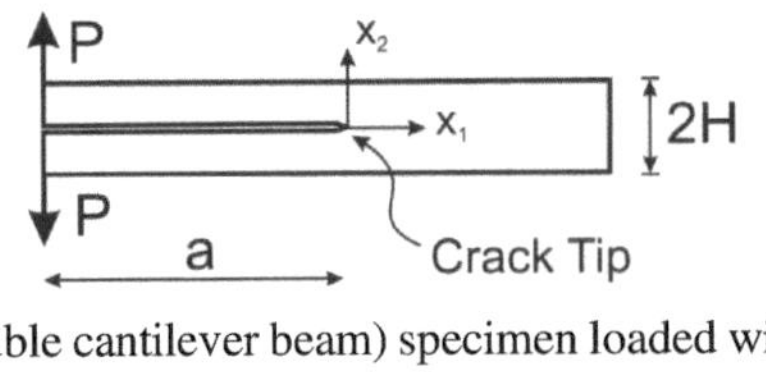

Figure 11.4 The DCB (double cantilever beam) specimen loaded with wedge forces is a Mode I specimen.

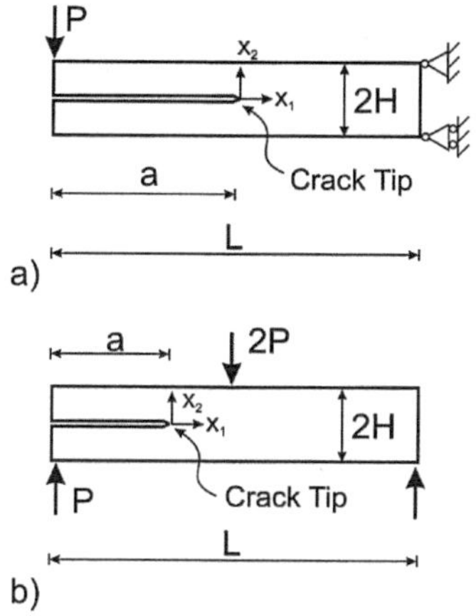

Figure 11.5 The ELS (end-loaded split) specimen and the ENF (end-notched-flexure) specimen are Mode II specimens. Due to contact between the crack faces, the applied force P is assumed to be shared equally between the two beams.

growth can be stable for the ENF specimen providing the crack length is sufficiently large (Carlsson et al., 1986).

The mixed mode end-notch split specimen, Fig. 11.6, can be obtained by superposition of the DCB and the ENF specimens. However, since both G_I and G_{II} change as a function of crack length, a, the mode mixity, which can be calculated from the definition (11.8) using Eqs. (11.12)–(11.15), is not constant as the crack advances:

$$\psi = \tan^{-1}\left(\frac{P_{II}(1 + Y_{II}(\rho)\lambda^{-1/4}H/a)}{P_I(1 + Y_I(\rho)\lambda^{-1/4}H/a)}\right) \tag{11.16}$$

where P_I and P_{II} are the associated Mode I and Mode II forces.

The MMB (mixed-mode *b* ending) specimen of Reeder and Crews, which is now an ASTM standard test method, is in principle also a superposition of the Mode I and Mode II specimen. The analysis considers nonlinear effects due to large displacements (Reeder and Crews, 1992).

11.3.3 Design with LEFM

The design process of a laminated structure by the use of LEFM can be as follows. First, the delamination resistance is characterized by the use of pure Mode I, Mode II, and mixed mode fracture mechanics test specimens for 4–6 different mode mixities. Then, a fracture criterion, the fracture energy as a function of the mode mixity, $G_c = G_c(\psi)$, for example, Eq. (11.9), is fitted to the fracture data. Next, a stress analysis of the component is conducted, for example, by the finite element method. The energy release rate G and mode mixity ψ are calculated for the component, for

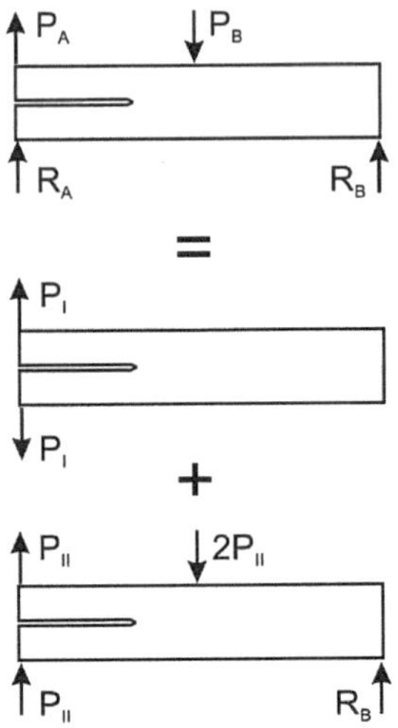

Figure 11.6 The MMB (mixed-mode bending) specimen can be analyzed as a superposition of a DCB and an ENF loadings.

example, for different crack lengths and crack locations. The design is considered safe under static loading if:

$$G(\psi) < G_c(\psi) \tag{11.17}$$

which should be understood as the energy release rate for the specific mode mixity being less than the critical energy release rate value (the fracture energy) for that particular mode mixity. In practice, safety factors are used to account, for example, variation in the measured material data.

It should be noted that for laminated composites, delamination can occur along a number of interfaces between different plies. It is thus important to characterize $G_c(\psi)$ for each interface in a laminate, for example, between plies having different fiber orientations and use the fracture criterion (11.17) for the relevant interface. In some case, it may not be known a priori along which interface the cracking will take place. Then, it can be necessary to consider cracking along several interfaces, that is, using Eq. (11.17) for several interfaces.

11.4 Advanced fracture mechanics

11.4.1 Overview of large-scale bridging

We now proceed to discuss delamination under LSB. LSB is when size of the bridging zone L is larger than the size of the *K*-dominant zone of a fracture mechanics test specimen (or model) with a sharp crack tip. Then, the end-opening of the FPZ will be located outside the *K*-dominant zone. It follows that the fracture depends not only on the displacement field inside the K-dominant zone, controlled by the SIF (or G), but also on the displacement field outside the K-dominant zone. K (or G) then loses its property as the sole parameter that controls the displacements at the FPZ. Consequently, LEFM does not apply. Instead, a traction-separation law, $\sigma_n = \sigma_n(\delta_n)$, can be used as the material law that describes the mechanics of the FPZ and the separation (the displacement jump within the active cohesive zone) must be calculated from stress analysis of the specific specimen under investigation.

It is instructive to consider two types of traction-separation laws, which we will call cohesive laws and bridging laws, respectively. A cohesive law was discussed in Section 11.2. A bridging law is another type of traction-separation law. The difference between a cohesive law and a bridging law can be understood by comparing Figs. 11.7 and 11.8. Using a cohesive law (Fig. 11.7), the mechanics of the entire FPZ is described in terms of the traction-separation law. Using the concepts of a bridging law (Fig. 11.8), the singular crack tip stress field at the crack tip is maintained and the traction-separation law represents only the tractions provided by bridging ligaments in the bridging zone.

A crack bridging problem can be analyzed by evaluation of the path-independent J integral locally along a path Γ_{loc} around the FPZ as shown schematically in Fig. 11.9.

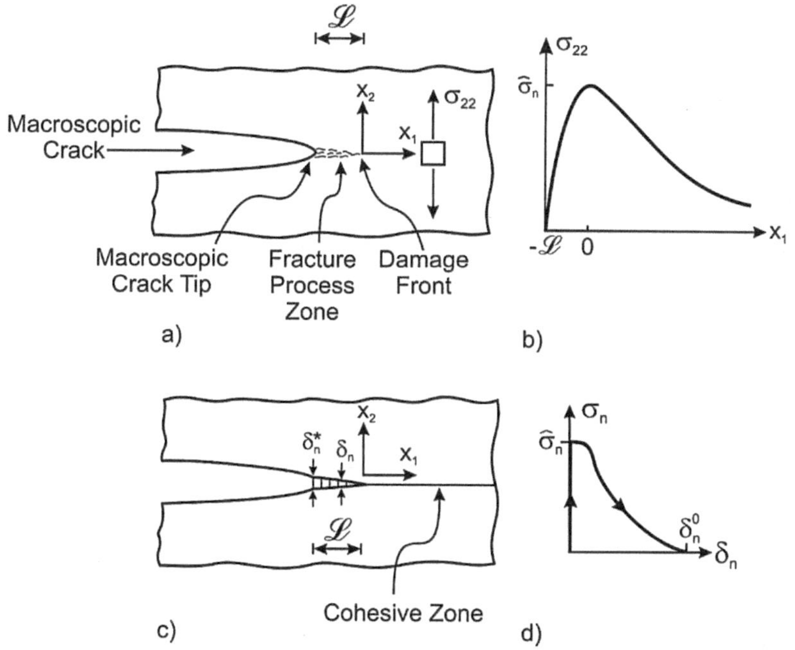

Figure 11.7 Schematic illustration of a cohesive zone, where there is no crack tip singular crack tip stress field: (a) a fracture process zone develops ahead of a crack tip, since (b) the material possesses a finite strength $\widehat{\sigma}_n$. (c) The cohesive zone opens ($\delta_n > 0$) when the stress σ_{22} reaches the cohesive strength, $\widehat{\sigma}_n$ of the cohesive law. (d) The cohesive traction σ_n is a function of separation, δ_n.

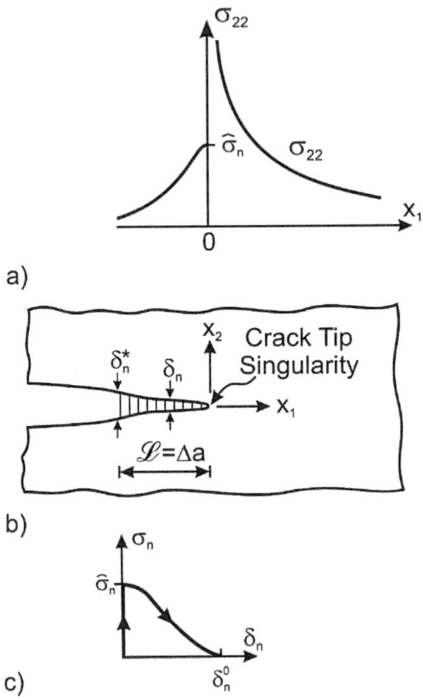

Figure 11.8 Schematics of a fracture problem analyzed with a bridging zone, where (a) there is a singular stress field at the crack tip, (b) a bridging zone develops, and (c) the traction-separat ion laws is called a bridging law.

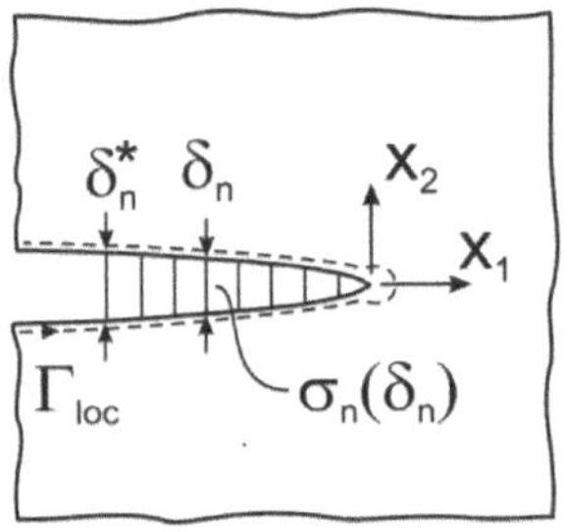

Figure 11.9 Integration path Γ_{loc} for the evaluation of the J integral locally around the fracture process zone.

The analysis can consider three stages as shown in Fig. 11.10. Consider a sharp, initially unbridged crack. In the unloaded state, the crack tip stress intensity (or equivalently the J integral evaluated around the crack tip, J_{tip}) is equal to zero. Increasing the load gradually increases J_{tip}. As long as J_{tip} is below a critical value, here denoted J_0, no cracking takes place; the crack tip maintains its position and the end-opening, δ_n^*, remains zero (Fig. 11.10a). Crack growth will start when the applied load is increased to a level that corresponds to $J_{\text{tip}} = J_0$. Then, with yet increasing load, the crack tip

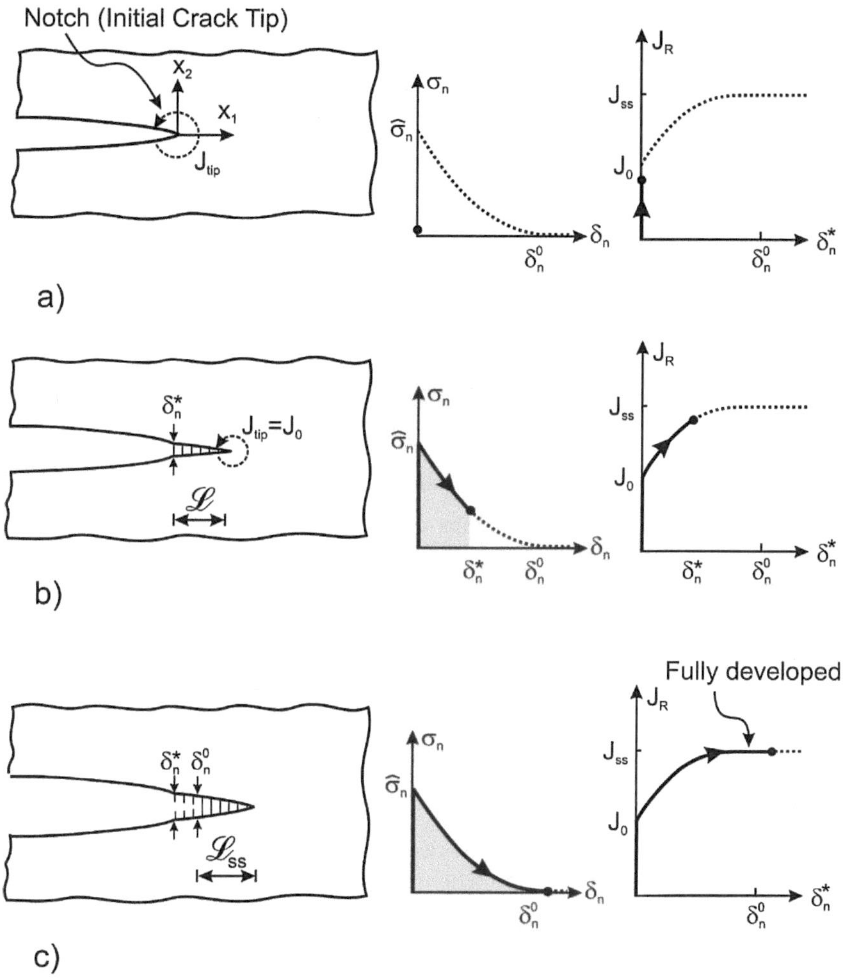

Figure 11.10 Schematics of an evolving bridging zone and its effect on the J integral: (a) initial configuration, (b) crack propagation, and (c) fully developed bridging zone.

advances, and a bridging zone now develops between the initial crack tip and the current crack tip position, see Fig. 11.8b. As a result, the end-opening at the original crack tip, δ_n^*, increases. Then the facture resistance increases according to:

$$J_R = J_0 + \int_0^{\delta_n^*} \sigma_n(\delta_n)\, d\delta_n, \tag{11.18}$$

where δ_n^* is the end-opening, that is, the displacement difference between the crack faces at the position of the initial crack tip. Since the presence of the bridging reduces

J_{tip}, the applied load must be increased to get more cracking so that J_{tip} remains at J_0. In most cases, this requires a higher applied load. Thus, the presence of a bridging zone leads to stable crack growth.

The fracture resistance rises with increasing crack extension, since a longer crack length leads to an increase in the end-opening as indicated in Eq. (11.18). Such a behavior is called R-curve behavior (rising crack growth resistance). When δ_n^* has reached the critical separation, δ_n^0, the fracture resistance attains its maximum value, J_{ss}, given by:

$$J_{ss} = J_0 + \int_0^{\delta_n^0} \sigma_n(\delta_n)\, d\delta_n. \tag{11.19}$$

When $\delta_n^* \geq \delta_n^0$, the bridging zone is said to be fully developed and the fracture resistance remains at J_{ss} for larger values of δ_n^*, as shown schematically in Fig. 11.10c. Although the fracture resistance remains at the steady-state value, the length of the bridging zone may change (Suo et al., 1992). For some specimens, the bridging zone length L maintains its size with increasing crack extension, so that the crack tip and the bridging zone translate in a self-similar manner along the specimen. Such specimens are called steady-state specimens. A family of steady-state specimens will be presented shortly.

For crack bridging problems, the increase in the macroscale fracture resistance results from the bridging tractions induced by the bridging ligaments at the microscale. Consider a fixed position within the bridging zone. With increasing opening, more ligaments are expected to break, and when δ_n^* reaches δ_n^0, all ligaments are broken so that the bridging tractions vanish at this position.

Note that for the special situation where $J_0 = 0$, the crack bridging problem looses the singular crack tip stress field and thus becomes a pure cohesive zone problem.

The determination of bridging laws (and/or cohesive laws) from experiments is challenging. One way is to first perform fracture mechanics tests during which the load, the crack length, and the displacement of the load-point are recording simultaneously. Then, the specimen is analyzed by numerical modeling, where the FPZ is modeled by a bridging law. The properties of the bridging law are initially unknown. By performing a number of simulations using different bridging law parameters (e.g., representing different shapes of the traction-separation relation) and comparing the predicted load–displacement relationships with the experimental data, the "true" bridging law is identified as the bridging law that gives the best agreement with the experimental data. This iterative guessing approach has the advantage that it requires no more experimental measurements than a regular LEFM test. The numerical modeling can involve the particular geometry under investigation, and no handbook solutions for J or G are required either. A drawback is that the approach requires the repetition of simulations for each new experimental data set that becomes available.

Another approach is based on the J integral. Differentiation of Eq. (11.18) gives (Suo et al., 1992):

$$\frac{\partial J_R}{\partial \delta_n^*} = \sigma_n\left(\delta_n^*\right) \tag{11.20}$$

The right-hand side can be understood as the normal traction at the position of the end-opening, that is, the separation at $x_1 = -L$. This relation then provides an approach to determine bridging/cohesive laws from experiments. The approach required the simultaneous measurements of J_R and δ_n^*, plotting these to perform differentiation according to Eq. (11.20). The fracture mechanics experiments must be conducted such that cracking occurs stably, so that the entire $J_R - \delta_n^*$ data can be recorded, from initiation to steady-state. Otherwise, only parts of the bridging law can be determined. This approach requires the determination of J_{ext} from load and geometry of the test specimens. A J integral analysis is thus required.

The relationship between the J integral value and the associated cohesive law is shown schematically in Fig. 11.11. The fact that the cohesive law is obtained by differentiation of the $J_R - \delta_n^*$ relationships provides a way get a feel for the shape of the $\sigma_n(\delta_n)$ law; high traction values will occur where the slope in a $J_R - \delta_n^*$ curve is high, and low traction values will be present where the slope is low.

For mixed-mode cracking, both normal and shear tractions will exist in the bridging/cohesive zone. These tractions are assumed to depend on both the local normal and tangential opening displacements, $\sigma_n = \sigma_n(\delta_n, \delta_t)$ and $\sigma_t = \sigma_t(\delta_n, \delta_t)$. Assume that the tractions are derived from a potential function, $\Phi(\delta_n, \delta_t)$ with $\Phi(0,0) = 0$:

$$\sigma_n(\delta_n, \delta_t) = \frac{\partial \Phi(\delta_n, \delta_t)}{\partial \delta_n} \text{ and } \sigma_t(\delta_n, \delta_t) = \frac{\partial \Phi(\delta_n, \delta_t)}{\partial \delta_t}. \tag{11.21}$$

These expressions should be understood as the tractions corresponding to a given set of opening values (δ_n, δ_t) are obtained by the partial derivate of the potential function evaluated at those opening values. Again, we assume that the traction-separation laws are the same at any position within the bridging zone. The potential function $\Phi(\delta_n, \delta_t)$ is unique for a given interface and therefore initially unknown and must be determined for a given case.

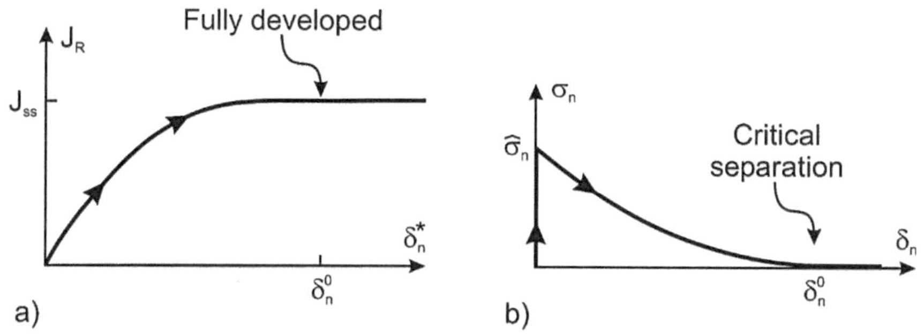

Figure 11.11 Relationship between (a) the fracture resistance and the cohesive law (b).

In order to determine the potential function for a specific material, we perform a J integral analysis along an integration path locally around the FPZ of a propagating crack. The result is (Sørensen and Kirkegaard, 2006):

$$J_R = J_0 + \int_0^{\delta_n^*} \sigma_n(\delta_n, \delta_t)d\delta_n + \int_0^{\delta_t^*} \sigma_t(\delta_n, \delta_t)d\delta_t = J_0 + \Phi\left(\delta_n^*, \delta_t^*\right) - \Phi(0, 0) \tag{11.22}$$

or (since $\Phi(0, 0) = 0$)

$$J_R = J_0 + \Phi\left(\delta_n^*, \delta_t^*\right) \Rightarrow \quad \Phi\left(\delta_n^*, \delta_t^*\right) = J_R - J_0. \tag{11.23}$$

Here, $\Phi\left(\delta_n^*, \delta_t^*\right)$ is the value of potential function for openings displacements corresponding to the end-opening and end-sliding. From Eq. (11.23), it follows that the increase in the fracture resistance J_R (beyond J_0) is equal to the potential function evaluated using the end-opening and end-sliding, δ_n^* and δ_t^*. In practice, any potential function $\Phi(\delta_n, \delta_t)$ can be fitted to the $J_R\left(\delta_n^*, \delta_t^*\right)$ data. A potential function with a sufficient high number of fitting parameters can fit the $J_R\left(\delta_n^*, \delta_t^*\right)$ data accurate.

Having obtained a potential function that fits the fracture resistance accurately, the bridging/cohesive laws are found by partial differentiation in accordance with Eq. (11.21).

The J integral equations described above can be used for the determination of mixed mode bridging laws to characterize mixed mode delamination crack growth. The procedure can be summarized as follows: first, a number of mixed mode experiments are conducted. During each experiment, data sufficient to calculate J_R as well as the end-opening and end-sliding are recorded, so that the fracture resistance can be calculated as a function of δ_n^* and δ_n^*, viz., $J_R\left(\delta_n^*, \delta_t^*\right)$. The initiation value, J_0, should be subtracted so that only data corresponding to the crack bridging is analyzed (alternatively, the entire fracture resistance can be built into the traction-separation laws, giving cohesive laws instead of bridging laws). Fitting a differentiable function (as the potential function, $\Phi(\delta_n, \delta_t)$) to the $J_R\left(\delta_n^*, \delta_t^*\right)$ data and performing the partial differentiation according to Eq. (11.21) now gives the cohesive/bridging laws. Examples of practical implementations of this approach—determination of mixed mode cohesive laws for a delamination problem—are given by Sørensen and Jacobsen (2009) and Erives et al. (2023). Sørensen and Jacobsen (2009) fitted a potential function given in terms of a sum of products of Chebyshev polynomials to the entire fracture resistance data, $J_R\left(\delta_n^*, \delta_t^*\right)$. Erives et al. (2023) used cylindrical coordinates and fitted one type of potential function to the crack tip fracture energy and another potential function to the bridging regime. The results of the two studies were fairly similar: The cohesive laws representing the crack tip region (typical openings up to 100 μm) attained high traction values (in the order of 30–50 MPa), while the bridging regime (openings in the order of several mm) gave low traction values (typically less than 1 MPa). Both studies found that the cohesive laws were coupled and the shapes of the traction-separation laws for the normal and shear tractions were different.

11.4.2 J integral specimens

The DCB loaded with wedge force (Fig. 11.12) is a Mode I specimen. An application of the J integral along the external boundaries gives (Paris and Paris, 1988; Olsson and Stigh, 1989):

$$J_{ext} = 2\frac{P\theta}{B}, \tag{11.24}$$

where θ is the rotation of the beam at the point where the force P is applied, see Fig. 11.12. In case of LSB, θ must be measured experimentally since the bridging tractions within the bridging zone influence the beam deflection and rotation. Note that the elastic properties of the laminate are not required for this approach. Also, crack length measurements are not required.

The ELS specimen (Fig. 11.13) is a Mode II specimen. Applying the J integral along the external boundaries of the specimen gives:

$$J_{\text{ext}} = \frac{P\theta}{B} - \frac{3P^2L^2}{4B^2H^3E_1}, \tag{11.25}$$

where L is the length of the specimen. The force P is assumed to be shared equally between the two beams, so that the forces applied to each beam are the same and $P_I = P_{II} = P/2$, see Fig. 11.13. Again, the beam rotation at the load-point, θ, must be measured experimentally.

For the MCB (mixed mode double cantilever beam) specimen, Fig. 11.14, the J-integral result can be written as (plane stress) (Högberg and Stigh, 2006):

$$J_{\text{ext}} = 12\frac{(Fa \sin \alpha)^2}{B^2H^3E_1} + \frac{F \sin \alpha}{B}(\theta_1 + \theta_2) + 4\frac{(F \cos \alpha)^2}{B^2HE_1} \tag{11.26}$$

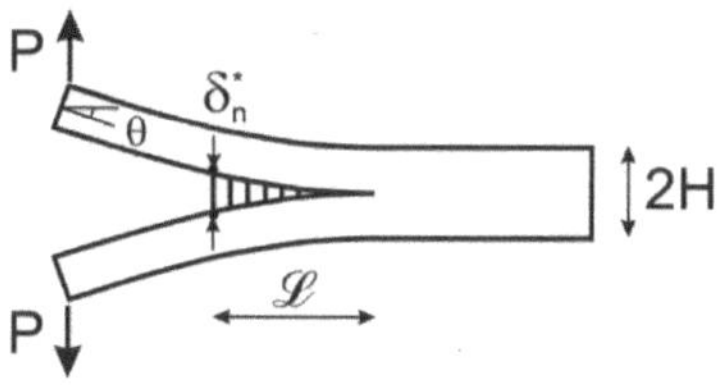

Figure 11.12 Sketch of the deformed DCB loaded with wedge force under large-scale bridging (LSB).

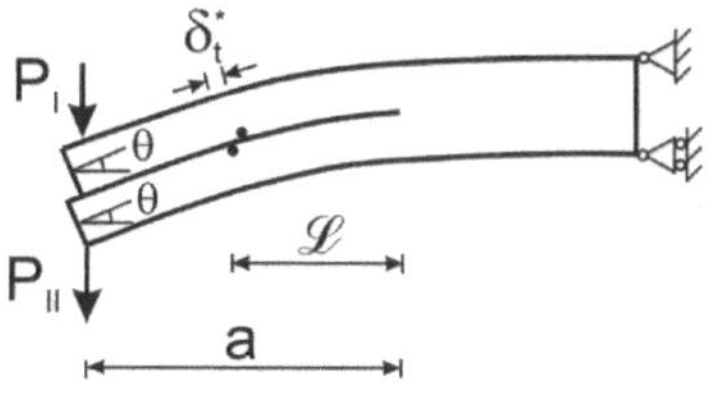

Figure 11.13 Sketch of the deformed ELS specimens experiencing LSB.

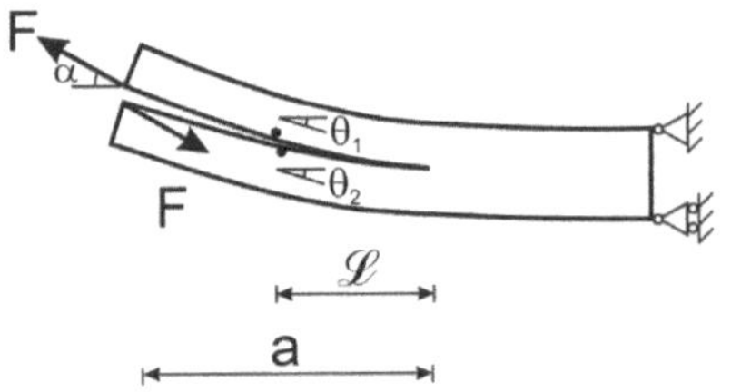

Figure 11.14 Schematics of the deformed MCB specimen experiencing LSB.

where α is the angle at which the force F apply, and θ_1 and θ_2 are the rotation of each of the beams at the end of the bridging/cohesive zone, see Fig. 11.14.

Sarrado et al. (2015) presented a J integral solution for the MMB specimen that generalizes the approach of calculating the J integral by measurements of four transverse loads and the associated beam rotations at the four points where the forces apply (analogous to Eq. 11.24). This result is quite general: No knowledge of elastic properties (e.g., lamina stiffness and layup) are required.

The beam specimens described so far are not steady-state specimens since the size of the fully developed bridging zone changes with increasing crack extension (Suo et al., 1992). They require the measurements of applied loads and beam rotations in order to determine J_{ext}.

A family of steady-state J integral fracture mechanics test specimens consists of DCB specimens loaded with axial forces and bending moments. For these test specimens, the only contributions to J_{ext} come from the loaded edges where the beams are subjected to a combination of uniform axial strain and pure bending, independent of the details of the bridging/cohesive law operating in the FPZ, provided that the beams are long and slender. Therefore, these configurations have closed-form analytical solutions to J_{ext} for LSB (Suo et al., 1992), and unlike the DCB loaded by transverse forces, measurement of beam rotation is not needed. Some of these specimens are shown in Fig. 11.15.

A symmetric DCB specimen loaded by equal moments (Fig. 11.15a) is a pure Mode I specimen. Under plane stress conditions, a J integral evaluation around the external boundaries gives:

$$J_{ext} = \frac{12M^2}{B^2H^3E_1}. \tag{11.27}$$

This specimen configuration was first suggested for LSB studies by Suo et al. (1992). This equation for J_{ext} is valid for orthotropic specimens. Note that the equation is independent of both λ and ρ.

A mixed-mode J integral specimen is the DCB specimen loaded with uneven bending moments (DCB-UBM). Taking the two bending moments of the two beam

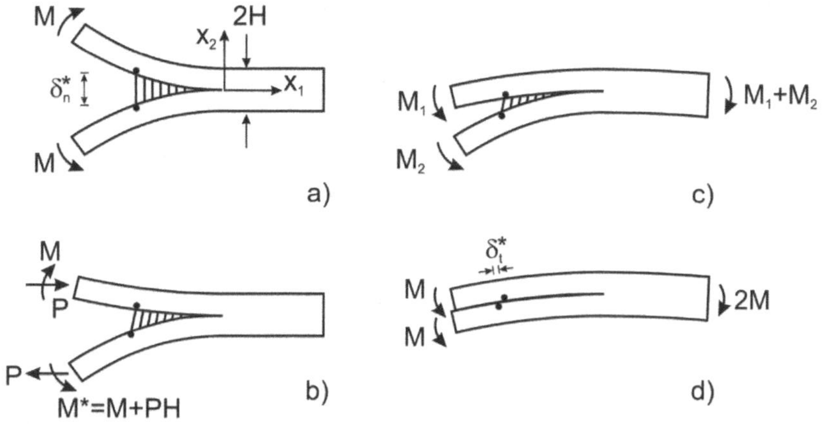

Figure 11.15 Family of steady-state J integral specimens (deformed state) for LSB: (a) DCB loaded with pure bending moments, (b) mixed mode specimen loaded with axial force and moments, (c) mixed mode DCB-UBM specimen, and (d) Mode II specimen loaded with equal moments.

arms, M_1 and M_2, positive in the anticlockwise direction as shown in Fig. 11.15c, the J integral result becomes (plane stress) (Sørensen et al., 2006):

$$J_{ext} = \frac{21\left(M_1^2 + M_2^2\right) - 6M_1M_2}{4B^2H^3E_1} \quad \text{for} \quad |M_1| \leq M_2. \tag{11.28}$$

For homogenous orthotropic specimens, the equation is independent of λ and ρ (Sørensen et al., 2006). Here, $M_1/M_2 = -1$ corresponds to pure Mode I, and Eq. (11.28) reduces to Eq. (11.27). For mixed mode, we have $-1 < M_1/M_2 < 1$. Setting $M_1/M_2 = 1$ gives pure Mode II (Fig. 11.14c), and the J integral can then be expressed as (plane stress):

$$J_{\text{ext}} = \frac{9M^2}{B^2H^3E_1}. \tag{11.29}$$

Toftegaard and Sørensen (2019) derived a J integral solution for another steady-state specimen, a beam-like multilayered or graded specimen loaded with axial force and bending moments (shown in Fig. 11.16). The solution also accounts for residual stresses. The J integral result can be written as:

$$\begin{aligned} J_{\text{ext}} = {} & \frac{\left(N_1 + N_1^{HT}\right)^2}{2B(EA)_1} + \frac{\left(N_2 + N_2^{HT}\right)^2}{2B(EA)_2} - \frac{\left(N_3 + N_3^{HT}\right)^2}{2B(EA)_3} \\ & + \frac{\left(M_1 + M_1^{HT}\right)^2}{2B(EI)_1} + \frac{\left(M_2 + M_2^{HT}\right)^2}{2B(EI)_2} - \frac{\left(M_3 + M_3^{HT}\right)^2}{2B(EI)_3} \end{aligned} \tag{11.30}$$

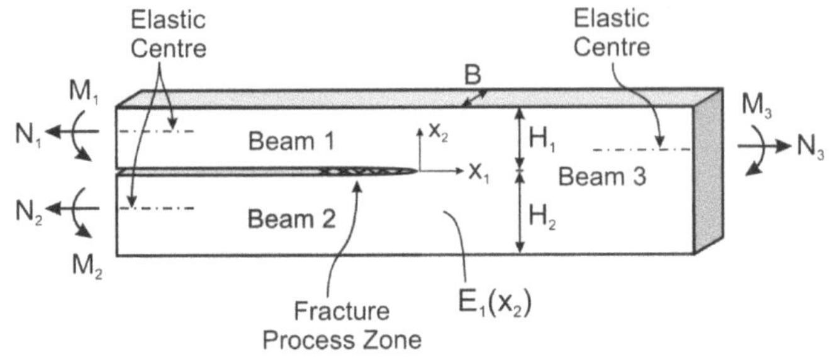

Figure 11.16 Beam-like specimen loaded with axial forces and bending moments. The elastic properties can vary as a function of position in the x_2-direction, layerwise or graded.

where N_i ($i = 1, 2, 3$) are the axial forces applied to beam section i and M_i are the applied moments. Furthermore, N_i^{HT} and M_i^{HT} are so-called hygrothermal forces and moments associated with residual stresses, while $(EA)_i$ is the extension stiffness for beam section i and $(EI)_i$ is the bending stiffness of beam section i. The J integral result (11.30) can be rewritten as:

$$J_{\text{ext}} = \frac{\varepsilon_1^0\left(N_1 + N_1^{HT}\right)}{2B} + \frac{\varepsilon_2^0\left(N_2 + N_2^{HT}\right)}{2B} - \frac{\varepsilon_3^0\left(N_3 + N_3^{HT}\right)}{2B} + \frac{\kappa_1^0\left(M_1 + M_1^{HT}\right)}{2B} + \frac{\kappa_2^0\left(M_2 + M_2^{HT}\right)}{2B} - \frac{\kappa_3^0\left(M_3 + M_3^{HT}\right)}{2B} \tag{11.31}$$

where ε_i^0 and κ_i^0 are the strain and curvature of the elastic center of beam i, respectively. Note that Eq. (11.31) is independent of the elastic properties. This implies that the J integral values can be calculated without the knowledge of the layup, the thickness, and elastic properties of the individual lamina; only values of forces, moments, strains, and curvatures must be measured. Eqs. (11.30) and (11.31) hold for multilayered and graded beams, providing the elastic properties are the same at the x_1-direction (i.e., only graded in the x_2-direction).

For the steady-state specimens described above, the J integral equations are all independent of the crack length. Using these test specimens, fracture tests will be stable under displacement (rotation) control since the forces and moments decrease when crack growth takes place under "fixed grip" conditions. Furthermore, these test specimens are "steady-state" specimens, that is, when the bridging zone is fully developed, cracking will occur in a steady-state fashion, in the sense that the FPZ retains its size and translates along with the crack tip in a self-similar manner.

11.4.3 Design with LSB

Design against delamination for laminates experiencing LSB is most conveniently done by the use of cohesive zone modeling. Then, a cohesive zone is specified along the expected crack path. For laminates, the cracking path will be along interfaces between different plies. Traction-separation laws in the form of cohesive laws are specified along the cohesive zone. No singular crack tip stress field zone will be modeled; the crack-tip fracture energy, J_0, must be included in the cohesive laws. In simulations using the finite element methods, cohesive laws can be implemented as zero-thickness or finite-thickness cohesive elements. The solution (the length and local opening of the active cohesive zone) must be obtained incrementally because the problem is nonlinear; the length of active cohesive zone and the distribution of cohesive traction depend on the structure. They are initially unknown for each load increment and must be determined iteratively by the finite element solver.

11.5 Delamination under cyclic loading

11.5.1 LEFM approach—the Paris–Erdogan law

Most load-carrying components are subjected to varying loads. For instance, the fuselage of an aircraft experiences internal pressure in high altitude so that each flight gives one major load cycle. A civil airplane will undergo in the order of 50–100,000 flights and thus the laminates of the fuselage will be subjected to about 10^5 load cycles. In comparison, a wind turbine rotor blade undergoes one rotation in, say, 3–10 s, inducing varying gravitational force directions about 10^8 times over a 20-year lifetime. Thus, delamination crack growth under cyclic loading is a major concern in many engineering composite structures. It is common to characterize the crack growth rate, da/dN, (where a denotes the crack length and N is the number of cycles) and explore whether a threshold value exists below which no crack growth occurs. For small-scale FPZ, the threshold can be expressed in terms of LEFM parameters, that is, in terms of a threshold value of the stress intensity factor K_{th} or energy release rate, G_{th}.

For metals, experience shows that within a broad range of crack growth rates, an empirical relation (Paris and Erdogan, 1963; Hertzberg, 1989)

$$\frac{da}{dN} = C_1(\Delta K)^{n_1} \tag{11.32}$$

can describe the relation between loading parameters and cyclic crack growth rate well. In Eq. (11.32), ΔK is the stress intensity range, $\Delta K = K_{max} - K_{min}$, where K_{max} and K_{min} are the maximum and minimum values of the stress intensity factors, respectively, while C_1 and n_1 are material constants that are determined by curve fitting of Eq. (11.32) to experimental data obtained from laboratory tests specimens. Recall that for isotropic materials, cracks propagate in pure Mode I. Thus, for isotropic metals, cyclic crack growth will be characterized in terms of the Mode I stress intensity factor range.

For composites undergoing delamination under small-scale FPZ, cyclic crack growth is often described in terms of a relation of similar form:

$$\frac{da}{dN} = C_2(\Delta G)^{n_2} \tag{11.33}$$

where $\Delta G = G_{max} - G_{min}$ with G_{max} and G_{min} are the maximum and minimum energy release rate values and while C_2 and n_2 are taken to be material constants. As noted by Rans et al. (2011), Eq. (11.33) is not quite equivalent to Eq. (11.32) since:

$$\Delta K = \frac{\sqrt{G_{max}} - \sqrt{G_{min}}}{E} \text{ and } \Delta G = \frac{(\Delta K)^2 - 2\Delta K K_{min}}{E}. \tag{11.34}$$

For small-scale FPZ problems for which the application of LEFM is valid, the crack growth rate determined from laboratory test specimens can be used to predict the cyclic crack growth rate for larger structures.

The philosophy of safe-life design relies on the use of fracture mechanics relations like Eq. (11.34) combined with inspection of structures by nondestructive techniques. The nondestructive methods are used for the detection of cracks larger than a certain size. Components having cracks larger than the detection size are then replaced with new components. The fracture mechanics Eqs. (11.32)–(11.34) are used to predict the number of cycles for the next inspections. The intervals between inspections are chosen such that a crack of the size of the detection limit cannot reach the critical length corresponding to unstable growth within the number of load cycles between the inspections.

Cyclic crack growth under LSB is more complicated. Consider a structure or a test specimen having an initially sharp, unbridged crack tip being subjected to cyclic loading with J_{ext} varying between fixed maximum and minimum values, J_{max} and J_{min}. In the first load cycles, the crack tip will advance in accordance to the Paris–Erdogan relation described above for a small-scale FPZ, if the fatigue threshold is exceeded. However, as the crack tip moves away from the initial crack tip, a bridging zone forms between the initial crack tip and the current crack tip position. The bridging traction within the bridging zone reduces the crack tip stress intensity range, ΔK_{tip} (or equivalently reduces both the maximum and minimum values of J_{tip}), so that the crack growth rate gradually decreases as the bridging zone evolves. Under some conditions, the crack tip stress intensity factor may fall below the threshold value. Then, crack arrest will occur. Otherwise, the bridging zone may evolve to become fully developed. For steady-state specimens, a steady-state situation can develop: the crack tip propagates in a self-similar fashion along the specimen at a constant crack growth rate, and the fully developed bridging zone maintains it size. Then, all bridging ligaments will undergo identical cyclic opening history. It seems reasonable that the growth rate of steady-state specimens will then take a constant value (this will not be the case for specimens and structures that are not steady-state specimens). Such a behavior has been quantified experimentally by Holmes et al. (2014).

Modeling (at the macroscale) of cyclic crack growth under LSB has not yet matured. One approach is incremental analysis of a loading-unloading cycle, to calculate the crack tip stress intensity range, ΔK_{tip} accounting for the bridging tractions. Then, using a Paris–Erdogan relationship, the crack growth rate can be estimated. Having estimated the crack growth rate, a subsequent cycle-by-cycle analysis can be conducted for a longer crack length ("cycle-jump" procedure).

Another approach is the approach of scaling a cyclic crack propagation problem (determination of ΔK_{tip}) to a static crack bridging problem (determination of K_{tip}) using an analytical scaling technique (McMeeking and Evans, 1990; Cox, 1993). This scaling can in some cases be performed analytically.

For both approaches, it is not clear, however, how the correct cyclic bridging law should be deduced from experiments and how they can be scaled correctly from one specimen geometry to another.

Some inspiration can be obtained by considering the crack bridging mechanisms at the microscale. Since the cyclic bridging law for fully developed bridging zone is likely to be a function of the crack growth rate or the number of cycles a bridging ligament spends from formation to breakage, a central parameter could be the number of load cycles that a bridging ligaments experience before failure, viz.,

$$N_{ss} = \frac{L_{ss}}{da/dN} \tag{11.35}$$

where L_{ss} is the fully developed steady-state bridging length of the particular specimen under a specific cyclic loading condition. It may be anticipated that two steady-state specimens having different bridging lengths and different cyclic load levels will have similar bridging stresses if they have identical N_{ss}. Thus, if the cyclic bridging stress distribution (for a given N_{ss}, denoted $(N_{ss})_1$) can be measured at a steady-state specimen (denoted specimen #1), it can be used for predicting (possibly by numerical modeling) the steady-state bridging length $(L_{ss})_2$ of another steady-state specimen (specimen #2). With $(L_{ss})_2$ calculated for specimen #2, the crack growth rate of specimen #2 can be estimated by (Sørensen and Jacobsen, 2002):

$$\left(\frac{da}{dN}\right)_2 = \frac{(L_{ss})_2}{(L_{ss})_1}\left(\frac{da}{dN}\right)_1 \tag{11.36}$$

where $(da/dN)_1$ is the crack growth rate of specimen #1 for the same N_{ss}.

11.6 Perspectives and trends

11.6.1 Mode mixity-dependent fracture resistance

Although the classical LEFM approach to fracture takes the outset at the potential energy of the whole body and the applied tractions (i.e., at the macroscale) and merely considers the energy absorbed by the FPZ, it is useful to recall that the fracture events actually occur at the microscale within the FPZ at the crack tip. A macroscopic Mode I crack has an FPZ that is largely associated with a single crack tip damage zone at the microscale. In contrast, for a macroscopic Mode II crack, the microscale damage within the FPZ is typically multiple microcracks oriented approximately 45° to the macroscopic cracking plane (the $x_1 - x_3$ plane in Fig. 11.1). These microcracks are in reality Mode I cracks at the microscale (Xia and Hutchinson, 1994) (pure shear corresponds to principal tensile and compressive stresses oriented at 45°). Since the failure events within the FPZ are quite different, it is not surprising that the macroscopic fracture energy changes with changing mode mixity.

Likewise, for LSB problems, for which the bridging zone extends out of the K-dominant region, the fracture resistance depends on the end-opening and end-sliding of the bridging zone in accordance with Eq. (11.23). As a result, the ratio between the shear and normal tractions may change with increasing crack extension, being controlled more by the global deformation of the specimen than by the displacements within the K-dominant zone.

11.6.2 Micromechanical models of crack bridging

An important point here is that an understanding of the failure mechanics through micromechanical models opens the possibility of microscale optimization. A beam theory-based micromechanical model of cross-over fiber bridging was developed by Spearing and Evans (1992), see Fig. 11.17. The model is based on the assumption that the bridging ligaments remain intact. The model includes displacements due to shear which make the result a bit complicated. Dropping the shear terms, the bridging law can be written as:

$$\sigma_n(\delta_n) = \left(\frac{2}{3}\right)^{3/4} \eta \, \frac{\left(G_c^i h\right)^{3/4} E^{1/4} \, b}{\sqrt{\delta_n}}, \tag{11.37}$$

where η is the number of bridging ligaments per unit cracked area, G_c^i is the interfacial fracture energy, E is the Young's modulus of the ligament, while h and b are the height and width of the ligaments, respectively. With other parameters held fixed, the model predicts that the bridging stress decreases with increasing opening as $\sigma_n \propto \delta_n^{-1/2}$. Invoking the J integral according to Eq. (11.18) and performing the integration leads to:

$$J_R(\delta_n^*) = \left(\frac{2}{3}\right)^{3/4} 2\eta \, \left(G_c^i h\right)^{3/4} E^{1/4} \, b\sqrt{\delta_n^*} \tag{11.38}$$

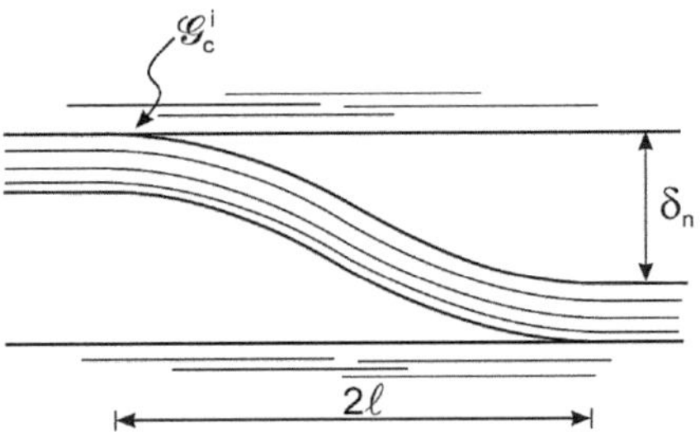

Figure 11.17 The Spearing–Evans micromechanical model of crack bridging under Mode I.

It is of interest to see how the fracture resistance depends on different parameters. It depends linearly on η and b, slightly weaker on G_c^i and h but even less on the Young's modulus, E. One way of enhancing the fracture resistance is thus to increase the interfacial fracture energy, G_c^i. Note that since the fracture resistance increases as the square root of δ_n^*, $J_R(\delta_n^*)$ approach infinity for $\delta_n^* \to \infty$. In other words, the micromechanical model predicts that the delamination resistance is unbounded provided that the ligaments remain intact. In practice, the toughening is bounded, since there are limitations on how large end-openings are obtained in real structures and because the bridging ligaments fail.

Using simple beam theory, a criterion for the prediction of fiber failure under pure normal opening (Mode I) can be constructed. The model predicts that fiber failure occurs when the maximum stress in the bridging ligament $\sigma_{\max}$ equals or exceeds the fiber strength σ_{fu} (for simplicity, we neglect fiber strength variation). The criterion for fiber failure can be written as (Sørensen et al., 2008):

$$G_c^i \geq \frac{\sigma_{fu}^2 h}{6E}. \tag{11.39}$$

where E is the Young's modulus of the ligament and h is the height of the ligament. This equation provides insight into the role of the interfacial fracture energy. For a sufficiently low value of G_c^i, the left-hand side of the equation will be lower than the right-hand side, so that the fiber failure criterion is not fulfilled. For a sufficiently high value of G_c^i, the inequality is satisfied, and fiber failure is predicted. It follows from Eq. (11.39) that a higher value of h increases the right-hand side of Eq. (11.39). Thus, bridging of ligaments consisting of several fibers (larger h) will enable fibers to survive for higher value of G_c^i than for ligaments consisting of a single fiber (h equal to one fiber diameter). Taking h equal to one fiber diameter will thus provide a lower estimate of the maximum value of G_c^i for which crack bridging by intact ligaments can occur. This estimate is conservative in the sense that if the criterion (11.39) is not satisfied for a ligament consisting of a single fiber, it will not be fulfilled for thicker ligaments consisting of multiple fibers. From the point of view of micromechanical optimization, G_c^i should be close to the critical value, so that the fracture resistance is maximized without the occurrence of fiber failure. Thus, knowing the fiber strength σ_{fu} and the ligament thickness, h (taken as the fiber diameter), Eq. (11.39) will predict the optimal value of the interfacial fracture energy.

A micromechanical model of cross-over bridging under mixed mode was developed by Sørensen et al. (2008), see Fig. 11.18. This model is based on the ordinary beam theory (ignoring shear deformation), and ligament failure is not included. Under mixed mode opening ($\delta_n \neq 0$ and $\delta_t \neq 0$), the model predicts the normal traction as:

$$\frac{\sigma_n}{\eta b h E} = \frac{\dfrac{\delta_n}{h}\left[\dfrac{G_c^i}{Eh}\right]^{3/2}}{\left[\left(\dfrac{\delta_t}{2h}\right)^2 + \sqrt{\left(\dfrac{\delta_t}{2h}\right)^4 + 6\dfrac{G_c^i}{Eh}\left(\dfrac{\delta_n}{2h}\right)^2}\right]^{3/2}} \tag{11.40}$$

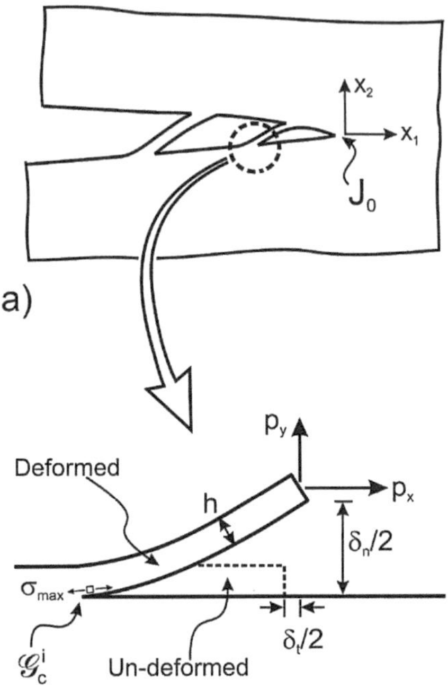

Figure 11.18 Schematics of a micromechanical model of mixed mode crack bridging.

and the shear traction is predicted to be:

$$\frac{\sigma_t}{\eta bhE} = \frac{\dfrac{\delta_t}{h}\left[\dfrac{G_c^i}{Eh}\right]^{1/2}}{\left[\left(\dfrac{\delta_t}{2h}\right)^2 + \sqrt{\left(\dfrac{\delta_t}{2h}\right)^4 + 6\dfrac{G_c^i}{Eh}\left(\dfrac{\delta_n}{2h}\right)^2}\,\right]^{1/2}}. \tag{11.41}$$

According to this model, the bridging laws are coupled: both the normal and the shear tractions depend on both the normal and tangential crack opening displacements.

For $\delta_t = 0$, Eq. (11.40) reduces to Eq. (11.37) (the pure Mode I result obtained from the Spearing–Evans model when the shear displacement is dropped) and Eq. (11.41) gives $\sigma_t = 0$. For pure Mode II ($\delta_n = 0$), Eqs. (11.40)–(11.41) give $\sigma_n = 0$ and

$$\sigma_t = \left[\frac{2G_c}{Eh}\right]^{1/2} \eta bhE, \tag{11.42}$$

that is, a constant value independent of the value of δ_t. Under mixed mode openings ($\delta_n \neq 0$ and $\delta_n \neq 0$), the normal traction, σ_n, decreases from a high value at small openings, and approaches zero asymptotically with increasing δ_n and δ_t. The shear traction, σ_t, increases to a near-constant value under mixed-mode opening, approaching a value similar to the bridging traction obtained under pure Mode II, Eq. (11.42).

It was shown by Sørensen et al. (2008) that a potential function exists for the mixed mode bridging tractions given by Eqs. (11.40) and (11.41). During cracking, the fracture resistance will be equal to the potential function evaluated at the end-openings plus the fracture energy of the crack tip (Eq. 11.23). For pure Mode I, the fracture resistance is predicted to increase as the square root of the end-opening, see Eq. (11.38), while for pure Mode II, the fracture resistance will increase linearly with the end-sliding. More details can be found in the cited paper.

A more accurate micromechanical model of fiber bridging (Grytten et al., 2021), based on large deflections, predicts that the mixed mode cohesive laws are coupled. Under pure normal opening, the normal traction decreases toward a constant (nonzero) value. Under pure tangential opening, the model gives the same result as the simple model, Eq. (11.42).

11.7 Summary

In this chapter, we have presented a brief overview of the current understanding of delamination from the perspective of fracture mechanics. Delamination is often encountered as a failure mode in laminated composites since these composites possess weak interfaces between differently oriented plies. Delamination involving a small-scale FPZ can be analyzed by LEFM, while delamination under LSB requires the use of a bridging/cohesive law. Suitable test specimens for mixed mode delamination under small-scale FPZ have been briefly described. Other fracture mechanics test specimens are suited for the determination of bridging laws by the use of a J integral approach. Delamination is also considered at the microscale. Understanding the toughening mechanisms, such as cross-over bridging by intact fibers, and the use of micromechanical models can lead to the development of more damage tolerant composite materials, potentially leading to an enhanced reliability of composite structures.

Acknowledgments

BFS was partially supported by the ReLife project (grant no. 64019-0543) from the Energy Technology Development and Demonstration Program (EUDP) of the Danish Energy Agency.

References

ASTM D5528-01, 2001. Standard Test for Mode I Interlaminar Fracture Toughness of Unidirectional Fiber-Reinforced Polymer Matrix Composites. American Society for Testing and Materials, West Conshohocken, Pennsylvania.

Broek, D., 1986. Elementary Engineering Fracture Mechanics, fourth ed. Mertinus Nijhoff Publishers, Dordrecht and Boston.

Carlsson, L.A., Gillespie Jr., J.W., Pipes, R.B., 1986. On the analysis and design of the end notch flexure (ENF) specimen for mode II testing. J. Compos. Mater. 20, 594–604.

Cox, B.N., 1993. Scaling for bridged cracks. Mech. Mater. 15, 87–98.
Erives, R., Sørensen, B.F., Goutianos, S., 2023. Extraction of mix-mode cohesive laws of a unidirectional composite undergoing delamination with large-scale fibre bridging. Compos. Part A 165, 107346.
Griffith, A.A., 1921. The Phenomena of Rupture and Flow in Solids, A221. Proc. Royal Soc. Lond., pp. 163–198
Grytten, F., Sørensen, B.F., Goutianos, S., Joki, R.K., Jørgensen, J.K., 2021. A micromechanical model of fiber bridging including effects of large deflections of the bridging fibers. Compos. Struct. 258, 113405.
Hertzberg, R.W., 1989. Deformation and Fracture of Engineering Materials, third ed. John Wiley and Sons, New York, pp. 517–562.
Holmes, J.,W., Liu, L., Sørensen, B.F., Wahlgren, S., 2014. Experimental approach for mixed-mode fatigue delamination crack growth with large-scale bridging in polymer composites. J. Compos. Mater. 48, 3111–3128.
Hutchinson, J.W., Suo, Z., 1992. Mixed mode cracking in layered materials. In: Hutchinson, J.W., Wu, T.Y. (Eds.), Advanced in Applied Mechanics, vol 29. Academic Press, Inc., Boston, pp. 63–191.
Högberg, J.L., Stigh, U., 2006. Specimen proposals for mixed mode testing of adhesive layer. Eng. Fract. Mech. 73, 2541–2587.
Irwin, G.R., 1957. Analysis of stress and strains near the end of a crack traversing a plate. J. Appl. Mech. 24, 361–364.
Jensen, H.M., Hutchinson, J.W., Kim, K.-S., 1990. Decohesion of a cut prestressed film on a substrate. Int. J. Solid Struct. 26, 1099–1114.
Kanninen, M.F., Popelar, C.H., 1985. Advanced Fracture Mechanics. Oxford University Press, New York.
McMeeking, R.M., Evans, A.G., 1990. Matrix fatigue cracking in fiber composites. Mech. Mater. 9, 217–227.
Olsson, P., Stigh, U., 1989. On the determination of the constitutive properties of thin interphase layers - an exact solution. Int. J. Fract. 41, R71–R76.
Paris, A.J., Paris, P.C., 1988. Instantaneous evaluation of J and C. Int. J. Fract. 38, R19–R21.
Paris, P.C., Erdogan, F., 1963. A critical analysis of crack propagation laws. J. Basic Eng. Ser. D 85, 528–534.
Qu, J., Bassani, J.L., 1989. Crack on bimaterial and bicrystal interfaces. J. Mech. Phys. Solid. 37, 417–433.
Rans, C., Alderliesten, R., Benedictus, R., 2011. Misinterpreting the results: how similitude can improve our understanding of fatigue delamination. Compos. Sci. Technol. 71, 230–238.
Reeder, J.R., Crews Jr., J.H., 1992. Redesign of the mixed-mode bending delamination testing. J. Compos. Technol. Res. 14, 12–19.
Rice, J.R., 1968. Mathematical analysis in the mechanics of fracture. In: Liebowitz, H. (Ed.), Fracture, an Advanced Treatice, vol. II. Academic Press, New York and London, pp. 191–311.
Rice, J.R., 1988. Elastic fracture mechanics concepts for interfacial cracks. J. Appl. Mech. 55, 98–103.
Sarrado, C., Turon, A., Renart, J., Costa, J., 2015. An experimentally data reduction method for mixed mode bending test based on the J integral approach. Compos. Sci. Technol. 117, 85–91.
Sih, G.C., 1973. Handbook of Stress-Intensity Factors. Lehigh University, Bethlehem, Pennsylvania.

Sih, G.C., Paris, P.C., Irwin, G.R., 1965. On cracks in anisotropic bodies. Int. J. Fract. Mech. 1, 189–203.

Spearing, S.M., Evans, A.G., 1992. The role of fiber bridging in the delamination resistance of fiber-reinforced composites. Acta Metall. Mater. 40, 2191–2199.

Suo, Z., 1990. Singularities, interfaces and cracks in dissimilar anisotropic media. Proc. Roy. Soc. Lond. A 427, 331–358.

Suo, Z., Bao, G., Fan, B., Wang, T.C., 1991. Orthotropy rescaling and implications for fracture in composites. Int. J. Solid Struct. 28, 235–248.

Suo, Z., Bao, G., Fan, B., 1992. Delamination R-curve phenomena due to damage. J. Mech. Phys. Solid. 40, 1–16.

Sørensen, B.F., Kirkegaard, P., 2006. Determination of mixed mode cohesive laws. Eng. Fract. Mech. 73, 2642–2661.

Sørensen, B.F., Jørgensen, K., Jacobsen, T.K., Østergaard, R.C., 2006. DCB-specimen loaded with uneven bending moments. Int. J. Fract. 141, 159–172.

Sørensen, B.F., Gamstedt, E.K., Østergaard, R.C., Goutianos, S., 2008. Micromechanical model of cross-over fibre bridging - prediction of mixed mode bridging laws. Mech. Mater. 40, 220–224.

Sørensen, B.F., Jacobsen, T.K., 2002. Crack bridging in composites: connecting mechanisms, micromechanics and macroscopic models. In: Pyrz, R., Schjødt-Thomsen, J., Rauhe, J.C., Thomsen, T., Jensen, L.R. (Eds.), Proceedings. Vol. 2. International Conference on New Challenges in Mesomechanics (Mesomechanics 2002), Aalborg (DK), August 26–30, 2002, vol 2002. Aalborg University, Aalborg, pp. 599–604.

Sørensen, B.F., Jacobsen, T.K., 2009. Delamination of fibre composites: determination of mixed mode cohesive laws. Compos. Sci. Technol. 69, 445–456.

Tada, H., Paris, P.C., Irwin, G.R., 2000. The Stress Analysis of Cracks Handbook. American Society of Mechanical Engineers.

Thouless, M.D., Evans, A.G., 1990. Comment on the spalling and edge-cracking of plates. Scripta Metall. Mater. 24, 1507–1510.

Toftegaard, H., Sørensen, B.F., 2019. General J integral solution for specimens loaded by moments, axial forces and residual stresses – a unifying stiffness formulation. Eng. Fract. Mech. 271, 106500.

Williams, M.L., 1959. The stresses around a fault or crack in dissimilar media. Bull. Seismol. Soc. Am. 49, 199–204.

Williams, J.G., 1989. Fracture mechanics of anisotropic materials. In: Friedrich, K. (Ed.), Application of Fracture Mechanics to Composite Materials. Elsevier Science Publishers, New York, pp. 3–38.

Xia, Z.C., Hutchinson, J.W., 1994. Mode II fracture toughness of a brittle adhesive layer. Int. J. Solid Struct. 31, 1133–1148.

Part Three

Modeling of damage and materials response in composite materials

The scale effect in composites, an explanation based on the mechanisms of damage

12

M.L. Velasco, E. Correa and F. París
School of Engineering, University of Seville, Seville, Spain

12.1 Introduction

The application of composite materials in all types of components and structures in different engineering branches has remarkably increased during the last decades. Their great versatility explains their use in many sectors, including the aeronautical, space, automotive (with special emphasis on competition), naval and wind power industries, and, of course, sport applications.

Just as an example, the progressive penetration of these materials in the aeronautical industry has been associated with a greater knowledge of them both in terms of manufacturing (with all the technological limitations that this entails), and design and calculation (which requires identification of failure mechanisms and their corresponding criteria). This increase in the use of composites has necessarily led to a shift from the use of composite materials in airfoils and secondary structures on early aircraft to their use on primary structures in the newest ones. The competition between the aircraft manufacturers to design lighter, safer and more ecological aircraft leads to a clear battle for a reduction in the weight of the different components of composite materials without forgetting the reduction of manufacturing costs, which leads to the need for a more precise selection of the thickness of the laminas to be used.

Since the appearance of composites and their application to the aeronautical industry in the 1960s, plies with a thickness of around 120 microns have been used, known as conventional thickness plies. The thickness of the weakest layers in a laminate, those oriented at 90 degrees with respect to the nominal direction of the load, has drawn particular attention because this is the layer that first experiences damage, and this damage may be either irrelevant or critical for the performance of the laminate in a structural component depending on several factors.

Thus, Parvizi et al. (1978) for glass—fiber composites, and Flaggs and Kural (1982) for carbon—fiber composites, among others, showed in the 70 and 80s that the damage in 90-degrees laminas appeared at a higher level of strain when its thickness decreased. This gave rise to the concept of *in situ strength*, taking the idea that the thinner the ply, the better its performance, which, without in-depth knowledge of the problem, was called *scale effect*.

However, it is obvious that the use of thin plies entails a higher workload, since this, in the classic autoclave manufacturing processes for example, is directly proportional

Modeling Damage, Fatigue and Failure of Composite Materials. https://doi.org/10.1016/B978-0-443-18489-5.00015-1

to the number of plies of the laminate. Therefore, it is logical to deduce that the thicker the plies of a laminate, the cheaper is to manufacture it. In this discussion about thickness, it must be said that it was not until very recently when technologically it became possible to have UltraThin Plies (UTPs), up to 20 microns thin.

The complexity of the internal structure of composite materials means that experimentation has been the key to guarantee the performance of a composite material structure, the pyramid of tests that is followed in the aeronautical industry being a good example of this. This procedure can give guarantees of satisfying safety conditions but it does not open the doors for an optimization of the material in order to improve its performance. París (2001) in a study for NASA on criteria to predict failure (using this word in generic terms for composite materials) established that the properties of a composite material cannot be improved without a deep knowledge of the damage mechanism at the level where the failure originates, i.e., the micromechanical level.

Since that publication, París et al. have been involved in a detailed study of the damage mechanisms of composite materials, particularly with regards to the intralaminar damage in the 90 degrees layer (transverse damage, Fig. 12.1) and its connection with interlaminar damage with the 0 degrees layer. In this sense, it is worth mentioning the identification of the beginning of intralaminar damage under tension and its propagation in the 90 degrees layer, París et al. (2003, 2007). The study was extended to compression loads, Correa et al. (2008a,b), cyclic loads, Correa et al. (2007), the effect of residual curing stresses, Correa et al. (2011), and biaxial loading, Correa et al. (2013), among others. All these studies have required a relevant amount of experimental work, collected among others in Correa et al. (2018). A great effort has been also devoted by other research groups to study the initial phase of the transverse failure, as is the case of Asp et al. (1996), and, more recently, Zhuang et al. (2018), specially focusing on numerical approaches.

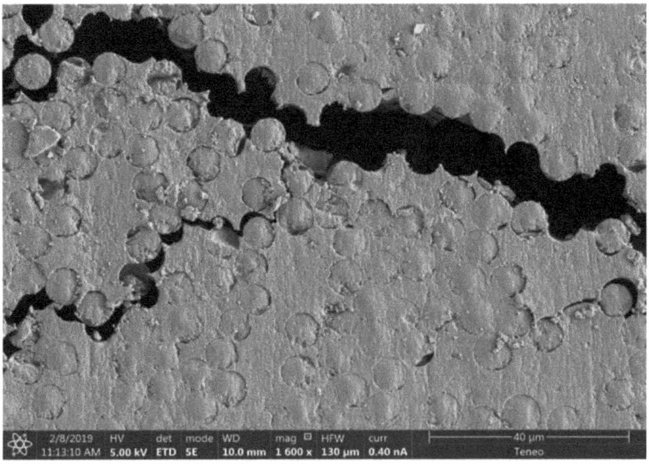

Figure 12.1 Microscopical image of transverse damage in a carbon epoxy 90 degrees layer.

The scale effect, for whose understanding the aforementioned investigations are required, has had different interpretations based on the possible presence of defects in the material; see, for instance, Dvorak and Laws (1987) or Li and Wisnom (1997) among others. The appearance of ultrathin plies, Tsai and Kawabe (2006), brought a new interest in the subject, producing a series of interesting works in the last decade. Thus, experimental works can be cited with some additional simulation by Saito et al. (2012) or Amacher et al. (2014), or purely of a numerical type, trying to reproduce the damage process in a micromechanical model, Saito et al. (2013) or Arteiro et al. (2014). A potential risk of developing numerical models assuming a damage process that might not be physically based may lead to a lack of correspondence with what is observed experimentally. An interesting contribution trying to connect the scale effect with modern approaches to the appearance of damage due to a double criterion based on stress and energy can be found in García et al. (2014).

Once the importance of micromechanics in the understanding of the developing of the mechanisms of damage was settled, and therefore its essential role in the prediction of failure at macromechanical level was established, a challenging task was undertaken in order to try to give a physically based explanation of the scale effect and therefore perform a critical evaluation of the in situ strength concept combining both numerical and experimental analyses. First, in París et al. (2018), numerical multiscale models of cross-ply laminates were developed in order to try to identify the effect of the 90 degrees layer thickness in the different micromechanical phases of the mechanism of damage. Velasco et al. (2018) provide details of the numerical modeling performed with boundary element method (BEM). Next, Velasco et al. (2020) presented numerical studies showing micromechanical new damage situations involving a higher number of fibers. And, finally, París et al. (2021) combined numerical and experimental results to propose a physically based explanation of the scale effect. In the present work, the main milestones of this study are explained, correlating predictions and experimental observations, leading to an entirely physically based explanation of the scale effect without fitting parameters.

This chapter is structured as follows. In Section 12.2, the cases under study are established, and the properties of the materials employed are detailed as well as the experimental and numerical tools. Section 12.3 presents the numerical study of the first phases of the mechanism of damage, i.e., the progression of a fiber–matrix debond under two possible scenarios of crack propagation. Section 12.4 deals with the next level of propagation, i.e., the generation of transverse damage in the 90 degrees lamina. Section 12.5 contains the results of the experimental studies performed, where the microscopical observation of the transverse failure is carried out on different cross-ply laminates as a function of the thickness of the 90-degree layer. Section 12.6 connects the numerical and experimental results to provide an explanation of the scale effect. Based on this explanation, Section 12.7 clarifies the concept of *in situ strength* and eventually shows the possibility of optimizing the thickness of the weakest lamina of a laminate. Finally, Section 12.8 summarizes the main conclusions of this study.

As clearly identified in the contents of this chapter described in the previous paragraph, the work developed has a heavy load of experimental work as well a significant development of modeling the behavior of the material at different levels. The connection between these two aspects of the work developed deserves a certain attention looking to describe the interaction between them. Bell and Truesdell (1973) distinguished between Theorists and Experimentists, using their own words: "The experimentist is a person who is engaged in the independent and systematic use of experiments to disclose patterns in nature which may or may not have been successfully considered by the theorist." This sentence highlights the different roles associated with different profiles involved in a research work which tends to develop a new knowledge. It is obvious that a comment like this, fully applicable in the 70s of previous century, has been altered in many senses mainly due to the appearance of computers.

The presentation in this chapter of the work developed in numerical and experimental Sections may at first sight reproduce the two different roles mentioned by Bell, although in this case performed by the same persons. The reality of the development of the knowledge presented in this chapter offers nevertheless a high level of interactivity between both roles. Thus, although the experimental and numerical parts are presented in different sections, there is a neat and beneficial interaction between them along the development of the work carried out. In this way, while numerical predictions producing conclusions not fully understood led to develop specific tests to contrast the predictions; also, on the contrary, experimental observations led to generate models to try to find the mechanical foundations of not easy to understand phenomena. All this fruitful interaction is not apparent when the entire scope of work developed is presented in a chapter, but the authors want to mention this particular feature of the work carried out, as is considered the key of having found a physically based explanation of the scale effect problem.

12.2 Methods, tools, and materials

12.2.1 Problem definition

The problem under study is the damage that appears in the 90 degrees layer of a cross-ply laminate when it is subjected to uniaxial tension. An example of one type of damage, involving a transverse crack in the 90 degrees ply, is shown in Fig. 12.2 at mesomechanical level. The damage may manifest in different forms depending on certain factors and therefore involving different mechanisms of damage that will be introduced and discussed in what follows. The work presented in this chapter mainly focuses on analyzing how the thickness of the 90 degrees layer affects the damage mechanisms involved in the failure process and the damage patterns that they will lead to.

12.2.2 Materials

Two prepregs with different thicknesses, conventional and ultrathin, have been used in this chapter. The conventional thickness material selected was AS4/8552 supplied by Hexcel, with an uncured lamina thickness of 190 microns. This material is widely used in the aircraft industry, the properties appearing in Table 12.1. Some of these

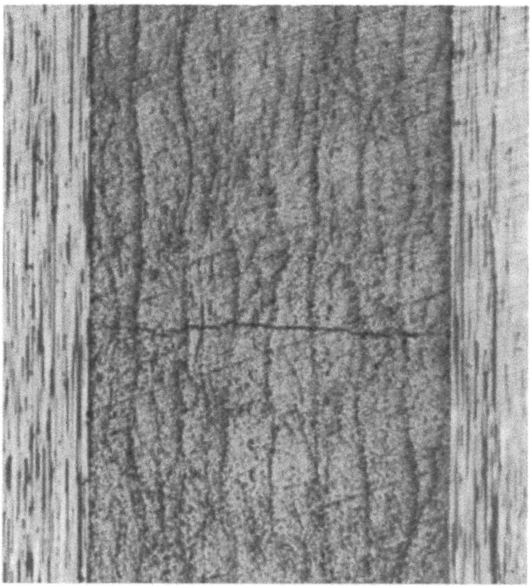

Figure 12.2 Transverse crack detected in the 90 degrees lamina of a $[0_4,90_8,0_4]$ laminate.

Table 12.1 Elastic properties of the carbon-epoxy material AS4/8552.

Material	Properties
0 degree layer (orthotropic)	$E_{11} = 135$ GPa, $E_{22} = 8.75$ GPa, $E_{33} = 8.75$ GPa, $\nu_{12} = 0.3$, $\nu_{13} = 0.3$, $\nu_{23} = 0.4$, $G_{12} = 4.75$ GPa
90 degrees layer (transversely isotropic)	E $(E_{22} = E_{33}) = 8.75$ GPa, $\nu = 0.4$
Matrix (epoxi, isotropic)	$E = 4.2$ GPa, $\nu = 0.32$
Fiber (carbon, transversely isotropic)	E $(E_{22} = E_{33}) = 15$ GPa, ν $(\nu_{23}) = 0.2$

properties have been obtained experimentally in-house (Graciani et al., 2020) and the others having been taken from the literature (Soden, 1998).

The ultrathin material used was SKY FLEX (SK) supplied by Triple H Composites with different grammages: 20, 50, 100, and 150 grams per square meter (gsm). The grammage of the lamina is approximately equivalent to the lamina thickness, the available lamina thicknesses then being 20, 50, 100, and 150 microns. Some properties of this material appear represented in Table 12.2 for the 150 gsm type.

The ultrathin material properties shown in Table 12.2 are very similar to those experimentally obtained by the authors (Sánchez-Carmona et al., 2023) from a material with the same characteristics but provided by another supplier (TP402/T700S −150 gsm of North Thin Ply Technology).

This chapter includes both an experimental campaign and modeling. The experimental tests will be performed with both materials, damage observation not requiring

Table 12.2 Elastic properties of the carbon-epoxy material of sky flex for 150 gsm prepreg provided by the supplier.

Material	Properties
150 gsm prepreg	E_{11} = 112,25 GPa, E_{22} = 7.92 GPa

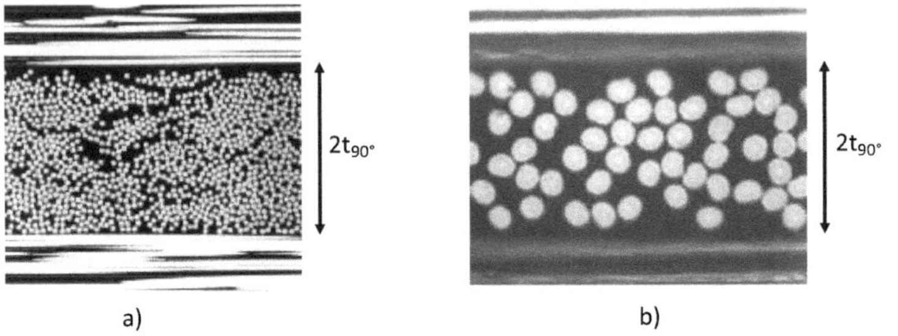

Figure 12.3 Lamina micrographs of the materials employed: (a) AS4/8552 with 190 μm nominal uncured thickness and (b) SK (UTP) with 50 μm nominal uncured thickness.

the precise knowledge of the properties of the material. However, with respect to the implementation of the numerical models employed in this work, all of them are referred to AS4/8552 material using the properties appearing in Table 12.1.

Fig. 12.3 shows micrographs of both materials. 50 gsm prepreg has been selected in the case of the UTPs, in order to show the particular "bean-like" transverse section of some fibers detected at low grammages plies. This section recovers its regular circular shape when the grammage increases. It has been checked that the fiber volume fraction, V_f, also depends in the UTP's case on the ply thickness. Specifically, Fig. 12.3b has associated a V_f value lower than that shown in Fig. 12.3a for the conventional thickness material. In any case, as occurs with the evolution of the shape of the transverse section, when the thickness of the ply increases, V_f tends to a regular value.

12.2.3 Experimental tools and methods

In order to obtain experimental evidence of the different damage morphologies that may appear in the 90 degrees layer, cross-ply laminates were manufactured with both materials by handly stacking. The laminates were cured in an autoclave (SCHOLZ BA-E44313-1) using the curing cycles recommended by the manufacturer. Subsequently, the laminates were reinforced on their ends with glass tabs to achieve a better grip with the clamps and the specimens were cut by a cutting machine with diamond disk. Prior to testing, the specimens were superficially treated by applying a thorough sanding and polishing process along the entire length of one of their free edges, to enable the accurate detection of damage using optical microscopic techniques. The steps followed for the surface preparation of the specimens together with some details of interest can be found in Velasco et al. (2023). The whole process of specimens preparation is illustrated in Fig. 12.4.

The dimensions of the specimens, following UNE-EN-2561:1995 (Spanish Standard UNE, 2561) technical standards, were 15 mm width by 250 mm length. The specimens were subjected to uniaxial tension with the objective of detecting the presence of damage in the 90 degrees layer. To this end, a universal testing machine INSTRON 4480 (load cell of 100 kN) and an INSTRON 2630-112 extensometer (gauge length of 50 mm) were used.

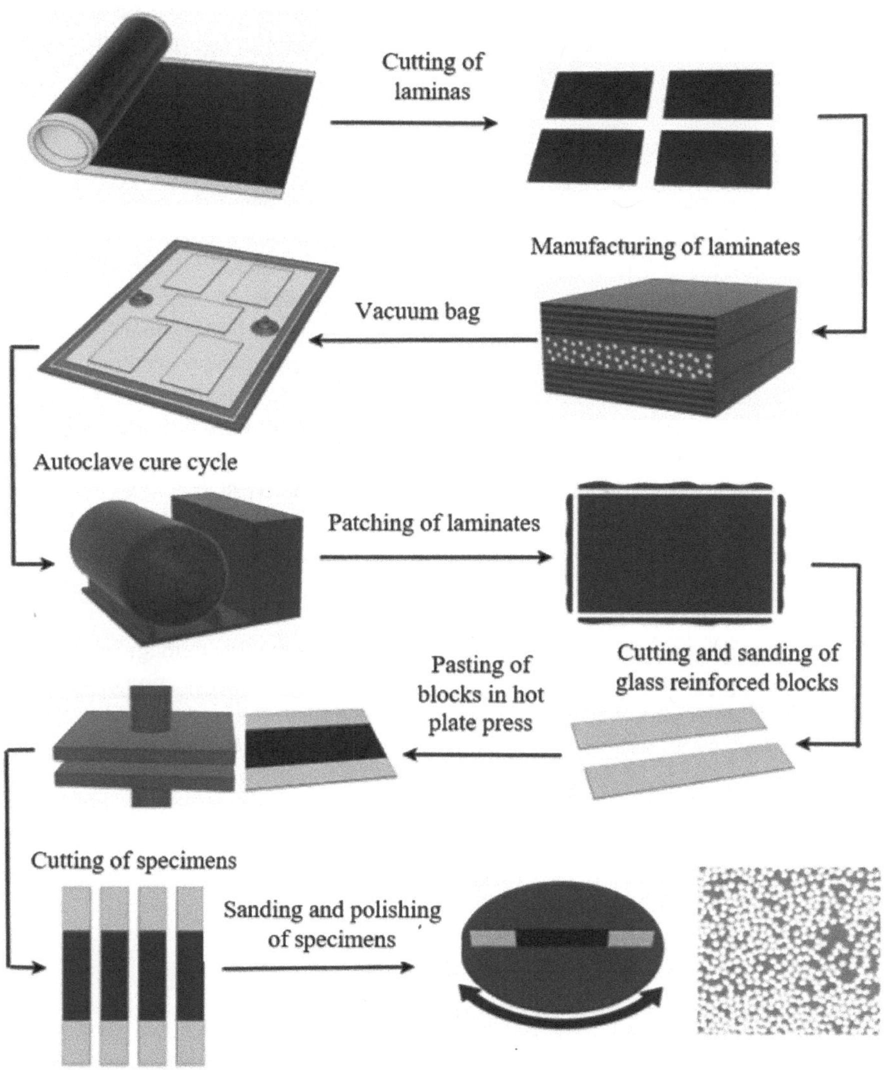

Figure 12.4 Schematic representation of the specimens manufacturing process.

The methodology applied for damage detection consisted of applying small load increments to the specimens. At each load increment, the test was stopped, the specimen was unloaded, extracted from the testing machine, and inspected thoroughly along its free polished edge. The load increments applied were not uniform because this methodology was combined with acoustic detection. Then, if during the application of a specific load increase, any sound emerging from the specimen was perceived, the test was also stopped, and the specimen was inspected.

Optical microscopical inspection was used for damage observation, employing a Nikon EPITHOT 200 optical binocular inverted microscope with an adapted INFINITY 2-5 camera from Lumera Corporation. The specimens were inspected in detail using different objectives with a magnification range from 10× to 1000×. In some cases where the thickness of the 90 degrees layer on the laminate was significant, the damage could be observed with the naked eye.

12.2.4 Numerical tools and methods

The numerical models carried out have been developed using the BEM, París and Cañas (1997). BEM is the ideal tool to study contact problems with presence of cracks, since displacements and stress vector components can be directly obtained along the boundaries of the solids involved in the model as independent variables. Both sets of variables are necessary to apply the foundations of fracture mechanics.

Damage in the 90 degrees layer is studied at micro- and mesomechanical level, it requiring to be able to represent both a debond between the fiber and the matrix (microlevel) and a transverse crack covering the total thickness of the 90 degrees layer (meso level). Consequently, different types of 2D models have been developed: micromechanical models (where damage is modeled at the fiber and matrix level), mesomechanical models (where damage is modeled at the layer level), and multiscale models, which are composed of a micromechanical cell (where the fiber and matrix are modeled) embedded inside the homogenized 90 degrees layer, Fig. 12.5.

The model represented in Fig. 12.5 is able to represent the first damage appearing in the 90 degrees ply, which is, according to the experimental observations and for the materials here employed, a debond between fiber and matrix. The details of the micromechanical cell with damage and features of the numerical modeling are shown in Fig. 12.6.

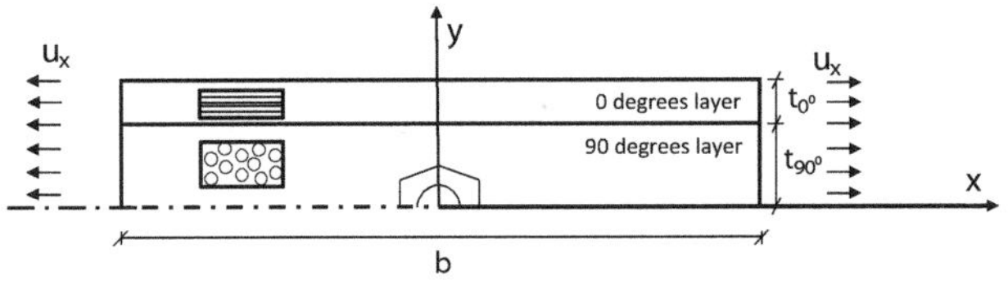

Figure 12.5 Schematic representation of a [0,90]s laminate multiscale plane model with a hexagonal micromechanical cell in the 90 degrees layer.

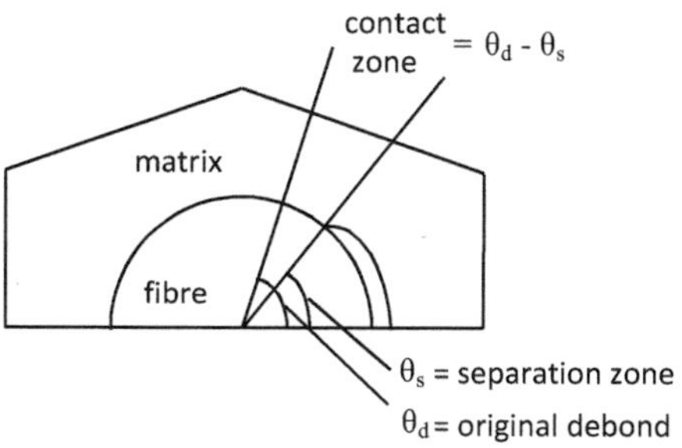

Figure 12.6 Detail of the micromechanical cell involving damage represented by a debond between fiber and matrix.

The progression of damage in the 90 degrees ply can be contemplated by means of different lengths of the debonds as well as considering the additional presence of kinked cracks emerging from the fiber–matrix interface and penetrating into the matrix.

The difficulty of performing the discretization of multiscale models such as that shown in Figs. 12.5 and 12.6 ought to be mentioned. When modeling the presence of debonds, very refined discretizations are needed at the tip of the interface crack, requiring elements smaller than 10^{-6} microns. As was previously mentioned, this is due to the fact that stresses and displacements are independent variables in the method. Nevertheless, the complexity of the discretizations that need to be performed is rewarded by the high accuracy achieved in the results just employing regular elements, with for instance linear or parabolic evolutions of the variables. Thus, BEM is particularly beneficial when the interest in the study carried out is referred, as in this case, to one point and its surrounding area and not to the whole domain of the bodies involved in the study. All details about the discretization of the multiscale models can be found in Velasco et al. (2018).

12.3 Damage onset: the appearance of a first debond

This section deals with the appearance of damage in the weakest lamina of a laminate, i.e., in our case the 90 degrees lamina of a generic $[0_m,90_n]_S$ laminate. The laminate considered in this section for the numerical study carried out is $[0_4,90_2,0_4]$ with $2t_{90^\circ} = 340$ μm, the total thickness of the 0 degree lamina being 680 μm. AS4/8552 properties are used in the modeling. Notice that the thicknesses of the 90 degrees laminas have been taken from the actual laminate and do not correspond exactly with those associated with the nominal thickness given by the supplier.

It will be assumed in this chapter that the damage in the 90 degrees lamina starts in the assumed weakest link of the microstructure of this lamina, i.e., along the interface between fiber and matrix. This is the evidence that the authors of this chapter have overwhelmingly found in the experimental observations of the onset of damage; other

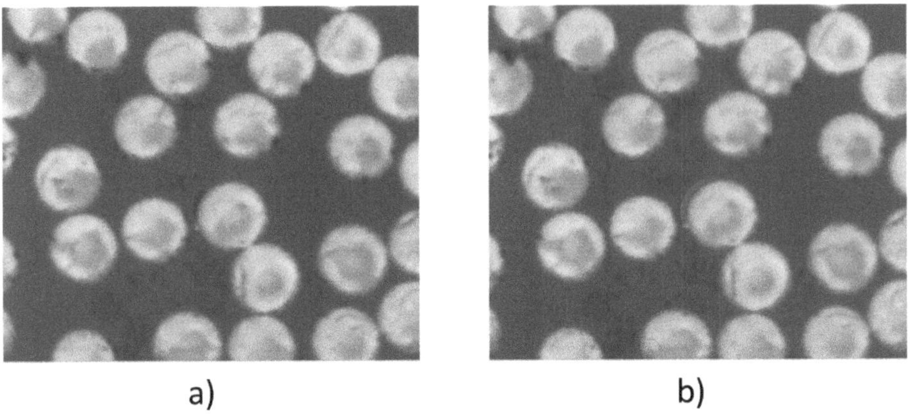

Figure 12.7 Presence of an isolated debond in the 90 degrees lamina.

authors (Asp et al., 1996) have associated the initial failure with matrix cavitation. The damage onset clearly depends on the relative properties of matrix and those associated with the interface between fiber and matrix.

Fig. 12.7 represents one image of a first debond as a manifestation of the abandoning of the pristine state. The conditions required to obtain this image will be explained later on in the chapter. This figure and all representing experimental observations correspond to horizontal loading.

To deal with the appearance of this debond and considering that we are going to employ arguments associated with Linear Elastic Fracture Mechanics (LEFM), there are two scenarios that can be taken into consideration when calculating values of the energy release rate and associated total energy released during the appearance of this debonding crack, Fig. 12.8. In the view corresponding to the explosive scenario, Fig. 12.8a, the interface crack would appear at once, whereas in the view corresponding to the progressive scenario, Fig. 12.8b, the debond appears sequentially, assuming that it starts at the point of maximum stress normal to the interface fiber—matrix. Both situations will be explored in what follows.

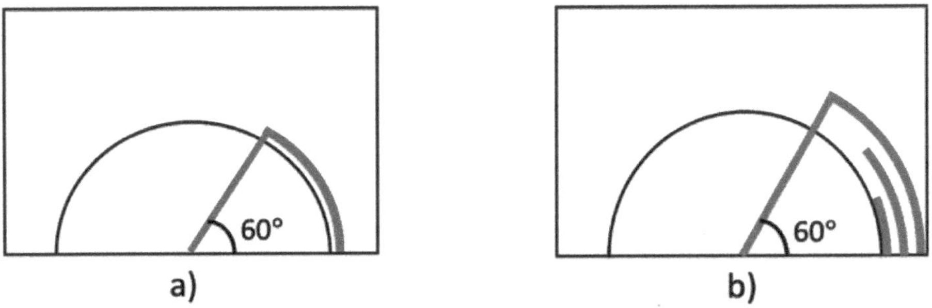

Figure 12.8 Representation of the two potential scenarios of damage: (a) explosive, (b) progressive, corresponding both to one debond associated, in the case of the figure, with the presence of a horizontal load.

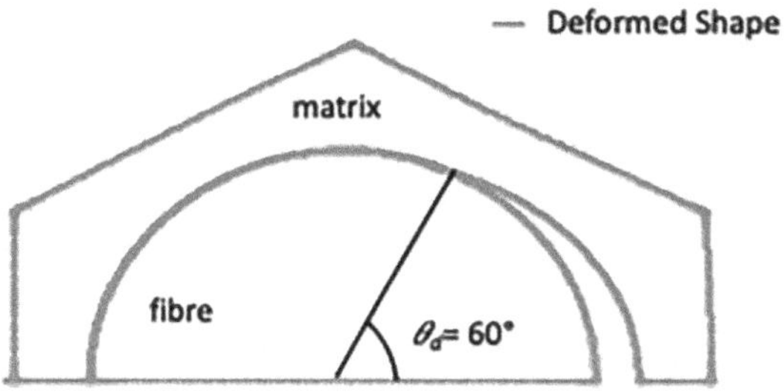

Figure 12.9 Detail of the cell of the integrated multiscale model of a $[0_4,90]_S$ laminate involving the presence of a debond (semidebond of 60 degrees in the Figure) between fiber and matrix.

12.3.1 Progressive scenario of the debonding process

The model used to study this situation is that represented in Figs. 12.5 and 12.6. It is a 2D multiscale integrated model in which the 0 degree oriented lamina is considered as homogeneous orthotropic and the 90 degrees oriented lamina as homogeneous isotropic. Inside the 90 degrees lamina, there is a cell with a nonhomogeneous constitution involving matrix and fiber, with the possibility of housing a debond between them, see Fig. 12.9, where the debond has been arbitrarily represented by 120 degrees; 60 degrees in the figure for the semidebond angle as a plane of symmetry at y = 0 (Fig. 12.5) has been assumed. Notice that the scheme represented in Fig. 12.9 corresponds to the deformed shape adopted by the cell under a horizontal loading. All details associated with this model can be found in París et al. (2018).

In the progressive view of the mechanism of damage, the total energy (one half using the symmetric model of the laminate) released by the laminate during the appearance of the debond, W^{pro}, can be obtained integrating the values of the energy release rates of the debonding crack along its whole path. The values of G, G_I, and G_{II} as a function of the semidebond angle are represented in Fig. 12.10 for $[0_4,90]_S$ laminate (n = 1).

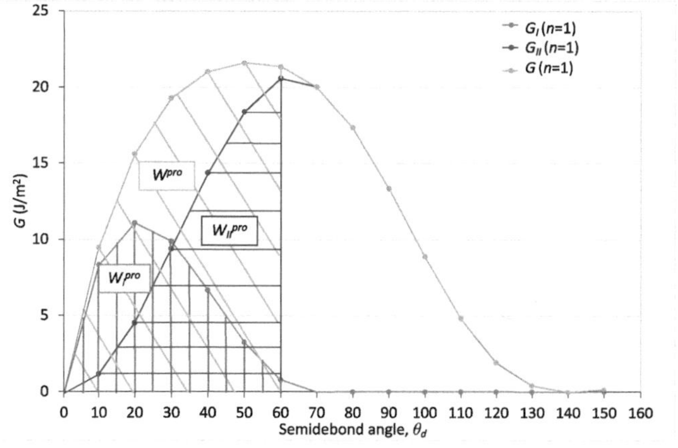

Figure 12.10 G, G_I, and G_{II} for different debonds. The dashed zones represent the total energy W and corresponding W_I and W_{II} released by the laminate for a semidebond angle of 60 degrees.

Table 12.3 W^{pro}, W_I^{pro}, and W_{II}^{pro} of a 60 degrees debond.

Progressive mechanism	θ_d = 60 degrees
$W_I{}^{pro}$ (10^{-3} J/m)	0.026644
W_{II}^{pro} (10^{-3} J/m)	0.038086
W^{pro} (10^{-3} J/m)	0.064730

The total values of the counterparts of half of the total energy released by the laminate (corresponding to the semidebonding) during the appearance of the debond in the progressive scenario are represented as W_I^{pro} and W_{II}^{pro} in Fig. 12.10. Notice the coherence of the evolution of G_I and G_{II} from a pure mode I when the debonding process starts, to a pure mode II, which initiates the arresting of the debonding crack.

The values of W^{pro} and its counterpart W_I^{pro} and W_{II}^{pro} for the case under study are represented in Table 12.3. These values will be compared in what follows with those coming from the explosive scenario.

12.3.2 Explosive scenario of the debonding process

In the explosive view of the appearance of the debond, the values of the stresses in the pristine state along the fiber–matrix interface as well as the opening of the lips of the debond in the final damaged state have to be calculated in order to obtain the values of the energy released by the laminate in this assumed mechanism of damage. Fig. 12.11 represents the values of σ_{rr} and $\sigma_{r\theta}$ along the fiber–matrix interface in the pristine state.

Fig. 12.12 represents the values of the relative normal displacements of the debond lips (opening), Δu_r, as well as the relative tangential displacements, Δu_θ.

The values of half of the total energy released by the laminate in this explosive scenario are obtained in accordance with Eqs. (12.1) and (12.2).

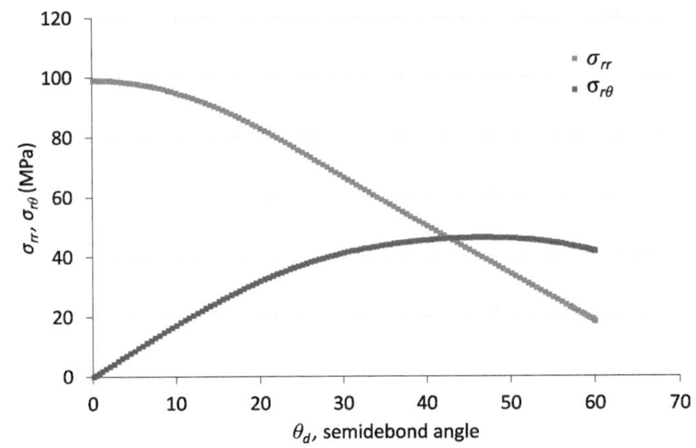

Figure 12.11 σ_r and $\sigma_{r\theta}$ along the fiber–matrix interface in the pristine state.

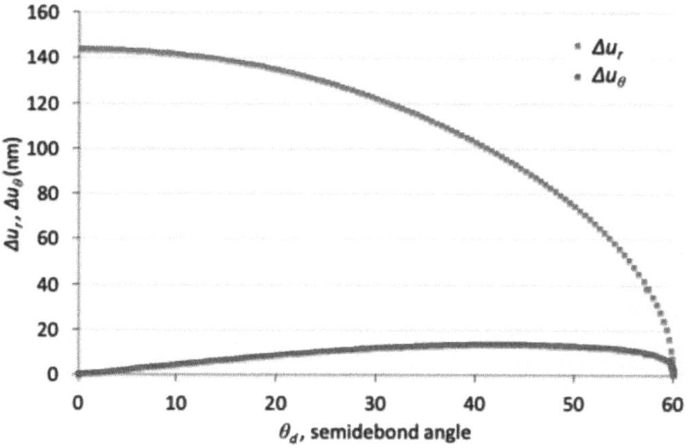

Figure 12.12 Δu_r and Δu_θ along the fiber—matrix interface in the pristine state.

$$W_I^{exp} = \left[\frac{1}{2}\int_0^{\theta_d} \sigma_{rr}\cdot\Delta u_r \, d\theta\right] \tag{12.1}$$

$$W_{II}^{exp} = \left[\frac{1}{2}\int_0^{\theta_d} \sigma_{r\theta}\cdot\Delta u_\theta \, d\theta\right] \tag{12.2}$$

The values obtained from previous formulas applied on stresses and displacements, represented in Figs. 12.11 and 12.12, respectively, are included in Table 12.4.

12.3.3 Comparison between the two considered scenarios of debond appearance

Fig. 12.13 represents the values of half of the total energy released by the laminate, W, and the corresponding counter parts, W_I and W_{II}, associated with modes I and II, respectively, for three different debonds (50, 60, and 70 degrees) and for the two scenarios under consideration, progressive and explosive. This figure includes the values represented in Tables 12.3 and 12.4 for $\theta_d = 60$ degrees.

The results represented in Fig. 12.13 are completely coherent. First, all values increase with the size of the debond. Second, the values of W obtained for each debond considered coincide in the explosive and progressive scenarios. This is coherent with

Table 12.4 W^{exp}, W_I^{exp}, and W_{II}^{exp} of a 60 degrees debond.

Explosive mechanism	$\theta_d = 60$ degrees
W_I^{exp} (10^{-3} J/m)	0.062540
W_{II}^{exp} (10^{-3} J/m)	0.002922
W^{exp} (10^{-3} J/m)	0.065462

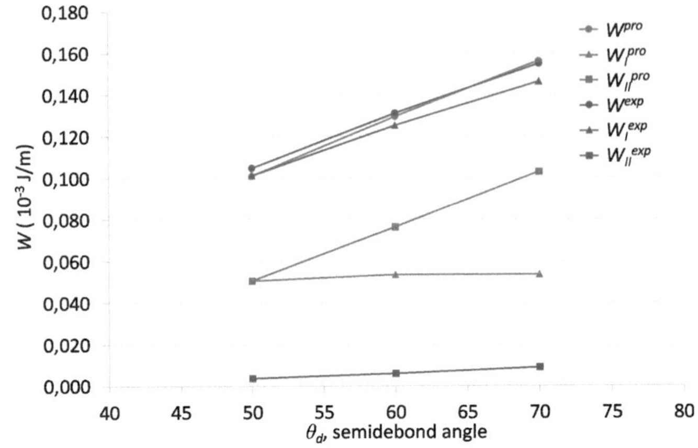

Figure 12.13 Comparison between W, W_I, and W_{II} for three values of the semidebonding angle (50, 60, and 70 degrees) under the progressive and explosive scenarios.

the concept of LEFM we are using, as the initial (pristine state) and final (debond) are the same for both the progressive and explosive scenarios.

Nevertheless, the distribution of W between modes I and II, W_I and W_{II}, is completely different between both approaches. Whereas in the case of the explosive scenario the failure is basically controlled by mode I as the whole debond is considered to appear at once, in the case of the progressive approach the mixity of the energy release rate evolves with the length of debond, as observed in Fig. 12.10. It is clear in this figure that starting from a pure mode I at the beginning of the damage, the tendency is to evolve to a pure mode II which makes the growth to stop when the semidebond approaches a value of 70 degrees. The involvement of mode II is clearly associated with this progressive scenario and it is obvious that the values of the fracture toughness involved in this mechanism are greater than those involved in the explosive approach.

Once the generation of an isolated debond under two different scenarios has been analyzed, it is instructive and necessary for the reasoning forwarded to explain the scale effect in composites, to study the effect of the 90 degrees ply thickness in the appearance of this debond. Fig. 12.14 represents the distribution of G for different values of the length of the debond for three different $[0,90_n,0]$ laminates with different thickness of the 90 degrees ply, nominally corresponding to $n = 3$, 1, and 0.25.

The message derived from these distributions of G for different thicknesses of the 90 degrees ply is clear: the appearance of a debond is independent of the thickness of the 90 degrees lamina. In other words, based on the analysis performed using Linear Elastic Interfacial Fracture Mechanics, there would be the same probability of a first debond appearance in the laminates under consideration using thin or thick laminas for the 90 degrees ply. The same probability, as we will see later on, does not necessarily imply the same possibility of observing them.

It is worth to mention that the results shown in Fig. 12.14 are obtained in all cases from a cell placed in the middle of the 90 degrees lamina. The cell was also placed in

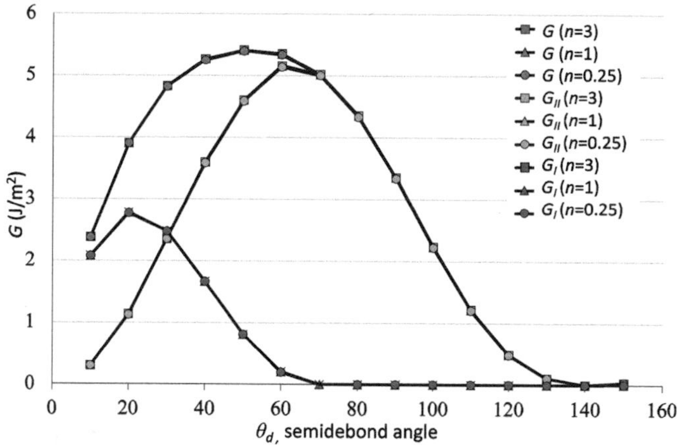

Figure 12.14 G distribution of a debond in the 90 degrees lamina as a function of the thickness of this lamina.

different positions not getting differences in the values of G represented in the figure, even when the cell is placed near to the interface with the 0 degrees lamina, a situation that in fact happens in the case of the thinnest lamina represented in Fig. 12.14.

The importance of having performed the two views of the appearance of the damage under consideration, explosive and progressive, can now be highlighted. The consequences derived from Fig. 12.14 will be the key to understand the scale effect, but they are obtained from a progressive view of the damage. The equivalence between the two approaches gives representativity to the results derived from Fig. 12.14.

12.4 Damage progression in the 90 degrees ply

Once the first damage in the 90 degrees ply (a fiber—matrix debond) has been studied and characterized, the generation and progression of subsequent transverse damage in the 90 degrees ply starting from the initial debond is going to be studied in this section. Fig. 12.15 covers the most plausible phases of damage consisting in a first debond whose extension is controlled by the evolution from a pure mode I to a pure mode

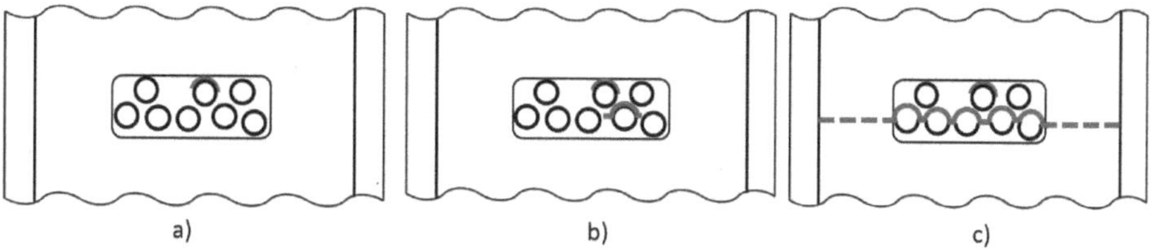

Figure 12.15 Tentative scheme of transverse damage generation in the 90 degrees ply: (a) Initial debond, (b) kinking of the debonding crack penetrating into the matrix, (c) transverse damage coalescence.

II, Fig. 12.15a, then the kinking of this debonding crack penetrating into the matrix, Fig. 12.15b, and, finally, the propagation of this transverse damage either through the matrix or along fiber–matrix interfaces as schematically suggested in Fig. 12.15c.

A similar scheme to that followed in the previous section for the study of the appearance of a debond can now be applied to model the subsequent stages of the described generation of a transverse damage in the 90 degrees ply and to characterize it from a fracture mechanics point of view. Fig. 12.16 schematically represents the two scenarios equivalent to those previously explored for the appearance of a debond, but now referred to a mesomechanical representation of the 90 degrees ply (modeling it as a homogeneous isotropic material in the plane represented in the figure).

The energy released by the appearance of the same damage, a transverse crack, in the 90 degrees ply, is now going to be calculated in accordance with the two scenarios described in Fig. 12.16, both on a homogenized view of the laminas.

The laminate selected in Sections 12.4.1 and 12.4.2 for the numerical analysis is $[0,90_4,0]$ with a thickness of each ply of 120 μm, the thickness of the 90 degrees lamina consequently being 480 μm.

12.4.1 Explosive scenario for the appearance of a transverse crack

In this scenario, the crack will be treated as if the appearance and reaching of its final extension occurred all at once. Then, the energy released by the laminate during the appearance of the transverse crack will be calculated making the stresses in the pristine state, σ_{90° in Fig. 12.17a, to work on the normal displacements of the lips of the transverse crack, u_{90° in Fig. 12.17b.

The energy released by the laminate during the appearance of the first transverse crack can be calculated by means of:

$$W = 2 \cdot 2 \cdot \frac{1}{2} \sigma_{90^\circ} \int_0^{t_{90^\circ}} u_{90^\circ} ds \tag{12.3}$$

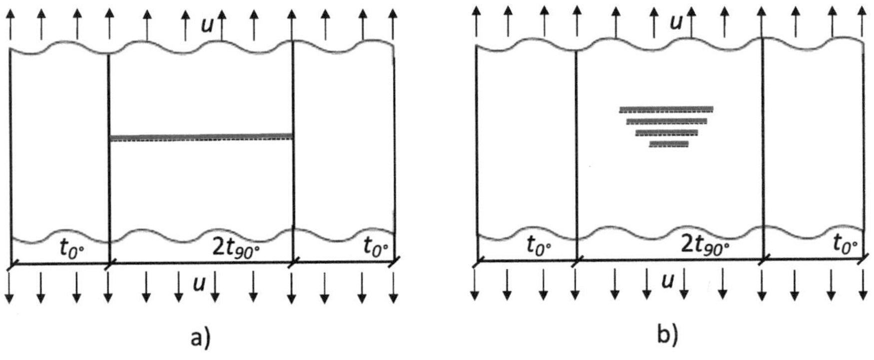

Figure 12.16 Two scenarios of damage representation: (a) explosive, (b) progressive.

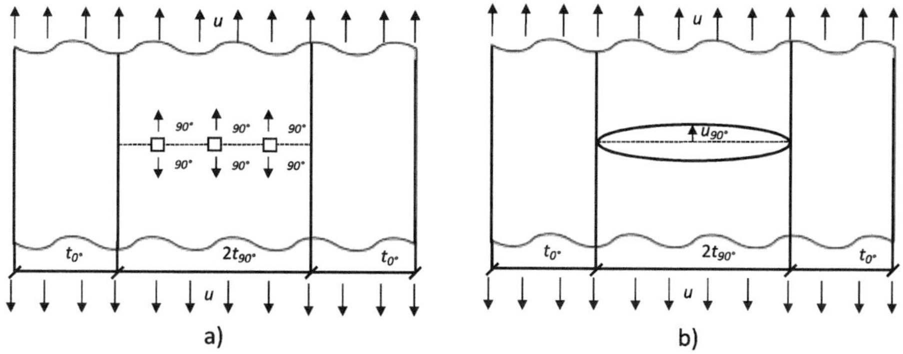

Figure 12.17 Schemes of the generation of a transverse crack in the 90 degrees ply considered as a homogeneous isotropic media: (a) pristine state, (b) cracked specimen.

where σ_{90° is directly calculated from Classical Laminate Theory (constant across 90 degrees lamina) and u_{90° is calculated from the BEM modeling of the problem París and Cañas (1997), Fig. 12.18.

The application of Eq. (12.3), using the values of σ_{90° and u_{90° described, leads to a value of the total work released by the laminate of $W = 0.41\ \text{J/m}^2$. This energy obviously corresponds with a pure mode I, in accordance with the homogeneous character of the 90 degrees ply assumed in this model.

12.4.2 Progressive scenario for the appearance of a transverse crack

In this case the energy released by the laminate, W, is going to be calculated by means of the values of the energy released rate, G, associated with cracks of different lengths (represented in Fig. 12.19).

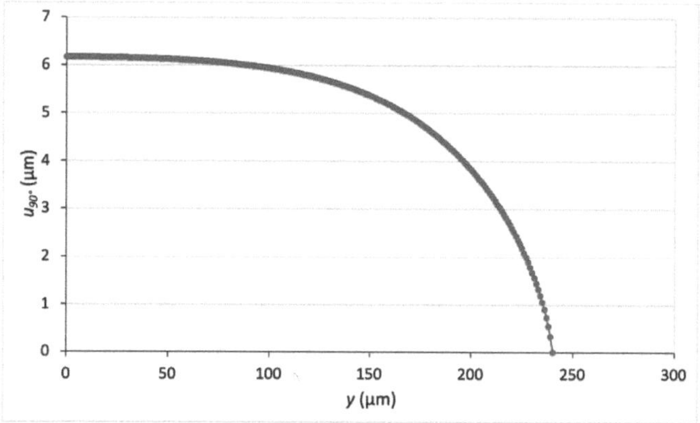

Figure 12.18 Normal displacements u_{90° of one of the lips of the transverse crack.

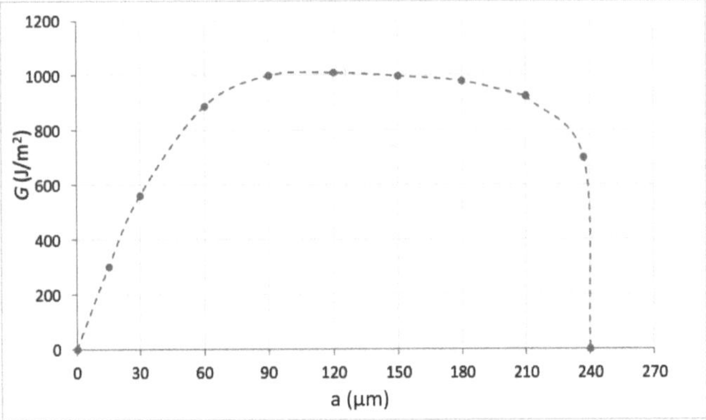

Figure 12.19 Evolution of the energy release rate with the length of the transverse crack corresponding to $t_{90^\circ} = 240$ μm.

Obviously, the values represented in Fig. 12.19 correspond to G_I and its integration along t_{90° leads to $W = 0.40$ J/m^2, a similar value to that obtained in the previous section with an explosive scenario. The small difference between both results, as both must be the same, is due to numerical errors in the two procedures to evaluate the work released.

12.4.3 Other considerations and conclusions

First of all, the difference between the two scenarios considered, both in the debond case and in the transverse crack case, has to be emphasized. In both cases, since we are applying LEFM theory, the value obtained for the total energy released by the crack under consideration in each case is the same independently of the scenario considered. However, whereas in the case of the transverse crack, both scenarios lead to the same conclusion (not only in the total work but also in the modes of fracture), in the case of the debond the total values coincide but the composition of modes is different in the two scenarios considered. The reason is obviously that whereas in the transverse crack case the study is on the homogeneous isotropic basis and the mode of fracture does not change along the length of the crack, in the case of the debond, and due to its circular shape, the mode changes along the length of the debonding crack. This fact is of extraordinary importance when trying to obtain fracture mechanical properties associated with the crack under consideration.

Once this question has been clarified, it is of main interest to elucidate if the scale effect can be supported by the balance of energy in the mesomechanical model of the cracked laminate. The equation that controls the energy balance corresponding to one half of the laminate is:

$$2\left[\frac{1}{2}\sigma_{90^\circ}\int_0^{t_{90^\circ}} u_{90^\circ}ds\right] = 2\gamma t_{90^\circ} \tag{12.4}$$

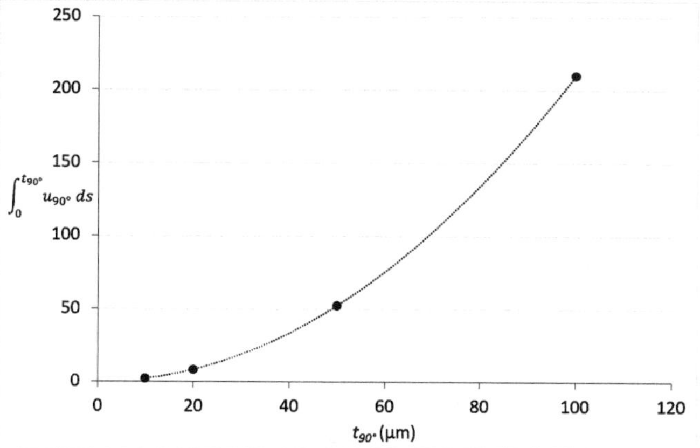

Figure 12.20 Evolution of the integral of the displacements as a function of t_{90°.

where γ represents the fracture toughness of the 90 degrees lamina considered as a homogeneous isotropic material. This balance corresponds to one half of the crack.

Fig. 12.20 represents the evolution of the integral of the displacements appearing in Eq. (12.4) for different thicknesses of the 90 degrees lamina. In this case, the laminate has the same scheme of that previously used ([0, 90, 0]), the 0 degree lamina having now a thickness of 100 μm and the values of the thickness of the 90 degrees lamina ($2t_{90^\circ}$) being varied as shown in Fig. 12.20.

It is noticeable how the value of this integral, which is directly responsible for the energy available to break the internal links of the material, increases in a nonlinear way with the value of t_{90°.

Now, if the balance represented by Eq. (12.4) is applied, employing for instance the values of γ obtained by Saito et al. (2012), the values of external displacements u acting on the laminate (see Fig. 12.17) and required for the fulfillment of the equation can be calculated as a function of t_{90°. Fig. 12.21 represents this external boundary condition but in the form of strain, ε, therefore meaning the strain value at which the damage appearance is assumed to take place.

The representation of the evolution of the strain required to produce the appearance of a transverse crack in the 90 degrees lamina for different thicknesses and for three values of γ clearly indicates the presence of the "scale effect" when the thickness of the 90 degrees lamina diminishes but not when the thickness increases. This fact will be discussed and explained in Section 12.6 once the experimental evidence is visited in the next section.

12.5 The experimental evidence

The detection of damage in the 90 degrees layer of cross-ply laminates subjected to uniaxial loading entails great difficulty, mainly because a first damage could consist

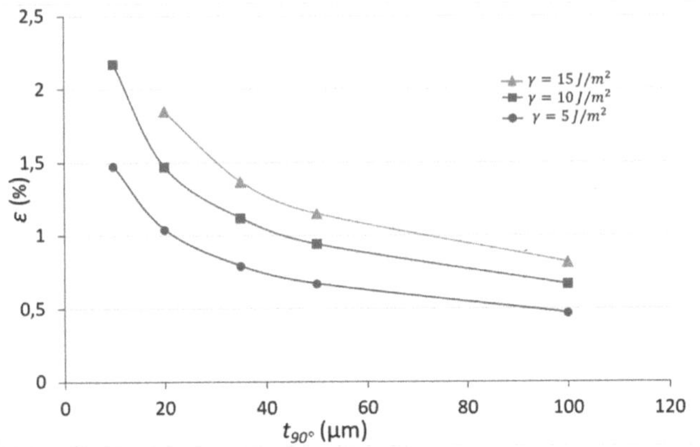

Figure 12.21 Evolution of the strain ε required to produce a transverse crack versus t_{90} for different materials characterized by γ.

of small fiber—matrix debonds that may go unnoticed. This experimental work involves tedious, repetitive, and time-consuming tasks.

In order to emulate the observations of Parvizi et al. (1978) and Flaggs and Kural (1982), two experimental campaigns were carried out, in one case using a material of conventional thickness and in the other case using a material with ultrathin thickness. In each case, cross-ply laminates were manufactured, their stacking sequences being detailed in Sections 12.5.1 and 12.5.2.

It is necessary to highlight the fact that the damage referred to by the previous authors, Parvizi et al. (1978) and Flaggs and Kural (1982), was in all cases a complete transverse crack. For this reason, the initial objective of these campaigns was to try to detect the value of the strain at which the appearance of a transverse crack occurred in the 90 degrees layer of each laminate. This first objective was crucial for the latter achievement of the physically based explanation of the scale effect that will be presented in Section 12.6.

As will be detailed in what follows, a complete transverse crack was not in all cases detected as the first damage. Thus, for this reason, before showing the damage observed in each manufactured laminate, it is useful to establish a schematic classification of the different damage morphologies that may appear in this type of materials. Fig. 12.22 shows a representation of the patterns observed, ordered from least to most advanced damage: isolated debonds, connected debonds or generalized transverse damage, transverse crack with slight indications of damage that indicates the presence of delamination, and, finally, complete transverse crack with associated delamination.

Five specimens have been tested for each laminate in order to obtain a representative average value, because, as is well known, a significant scatter may affect the results derived from tests involving transverse failure. For the sake of brevity, only a selection of micrographs is shown.

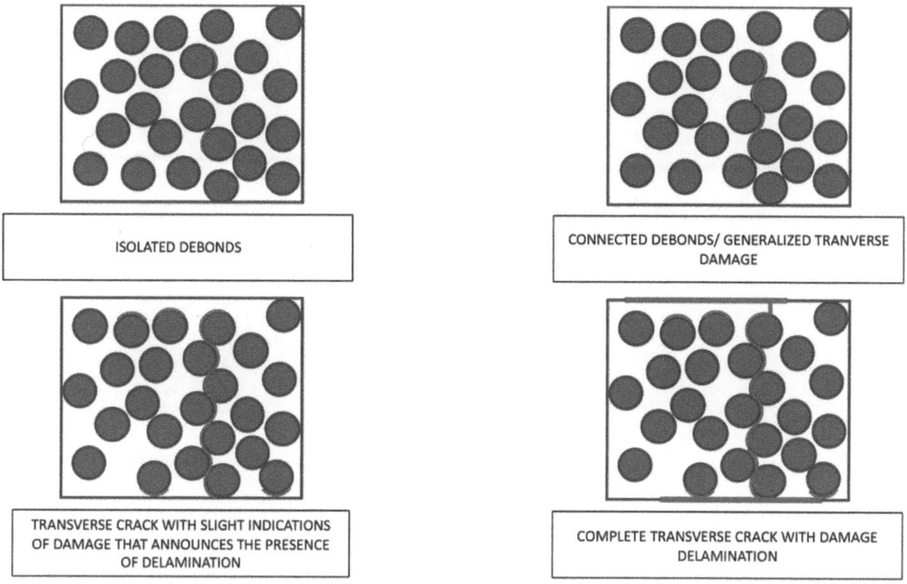

Figure 12.22 Schematic classification of the different damage morphologies observed in the 90 degrees layer of cross-ply laminates.

12.5.1 Laminates manufactured with conventional thickness plies

In this experimental campaign, cross-ply laminates were fabricated using AS4/8552 prepregs; details about this material and the methodology used to detect and observe damage were included in Section 12.2.2. The stacking sequences of these laminates were $[0_4,90_n,0_4]$ for $n = 1, 2, 4, 8$, and 16.

The different damage patterns found for each laminate will be shown in what follows.

12.5.1.1 Laminate $[0_4,90,0_4]$

Generalized transverse damage, i.e., connected and unconnected debonds following a crack path that mostly did not reach the interface between the layers (Fig. 12.23a), was directly detected as the first damage in the 90 degrees ply of this laminate. This damage was observed for a strain range between 0.45% and 0.49%.

As observed in Fig. 12.23a, the damage was moderately advanced, but it was not a complete transverse crack (defined as the initial object of the study). For this reason, the specimens were subjected to new load increments in order to try to observe the generation of a damage that was as similar as possible to a complete transverse crack; this damage can be observed in Fig. 12.23b and was detected for strain values of around 0.7%.

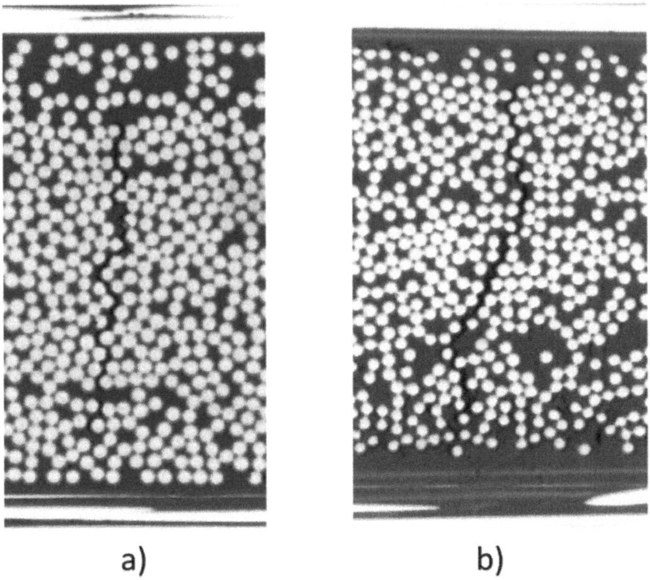

Figure 12.23 Damage found in the $[0_4,90,0_4]$ laminate for: (a) $\varepsilon = 0.46\%$, (b) $\varepsilon = 0.70\%$.

12.5.1.2 Laminate $[0_4,90_2,0_4]$

The damage found in the 90 degrees layer of this laminate is slightly more advanced than in the previous case ($[0_4,90,0_4]$). In some specimens, the first damage detected coincided with the presence of the expected complete transverse crack, as shown in Fig. 12.24. In this figure, it can be seen how the debonds between fiber and matrix are fully connected, either between them or by kinking through the matrix.

12.5.1.3 Laminate $[0_4,90_4,0_4]$

In all $[0_4,90_4,0_4]$ specimens, the first damage detected was a complete transverse crack, represented in Fig. 12.25a. In some cases, a small number of unconnected debonds with orientation parallel to the fibers of the 0 degree layer were observed at the ends of the transverse cracks, Fig. 12.25b.

12.5.1.4 Laminate $[0_4,90_8,0_4]$

The direct appearance of a complete transverse crack was detected in the 90 degrees layer of this laminate, Fig. 12.26a. In addition, at the ends of this crack (at the interface between the layers) now a significant number of unconnected and connected debonds parallel to the fibers of the 0 degree layer can be observed, Fig. 12.26b.

12.5.1.5 Laminate $[0_4,90_{16},0_4]$

The detection of the first damage in the 90 degrees layer of the $[0_4,90_{16},0_4]$ laminate was much more immediate than in the other laminates, as the damage was visible to a

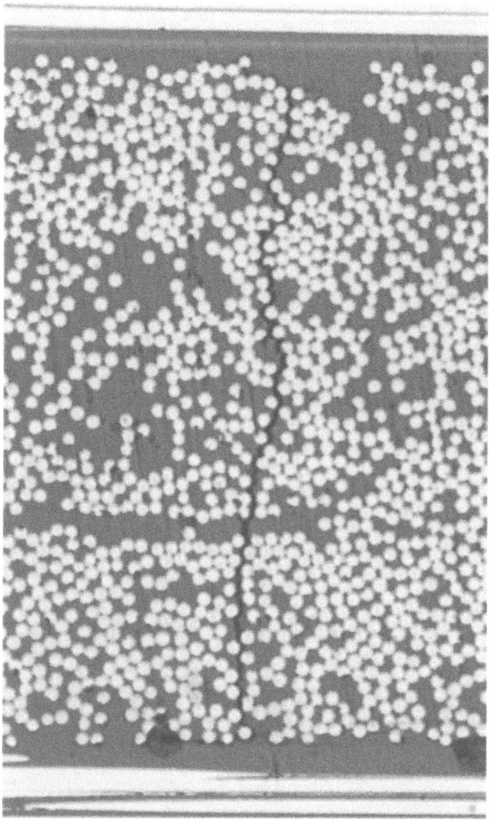

Figure 12.24 First damage observed in the 90 degrees layer of $[0_4,90_2,0_4]$ laminates for $\varepsilon = 0.72\%$.

naked eye. The morphology of the damage found consisted of a complete transverse crack, Fig. 12.27a, with extensive delamination between the 90 and 0 degree layers. The extensive area occupied by delamination damage can be observed in Fig. 12.27b.

Once the expected damage was identified in the 90 degrees layer of all selected laminates manufactured using the material with conventional thickness, the graphical representation shown in Fig. 12.28 was carried out, where the strain value responsible for the appearance of a complete transverse crack is represented versus the thickness of the 90 degrees layer. The strain associated with the nominal transverse strength of the material is also plotted in Fig. 12.28 as a reference for the values found, whose average values are highlighted in the figure. It can be deduced from this figure that for small values of the 90 degrees layer thickness, a higher strain value is necessary to generate a transverse crack.

12.5.2 Laminates manufactured with ultrathin thickness plies

An experimental campaign similar to that carried out for the material of conventional thickness was also developed using the SKY FLEX material fabricated by Triple H

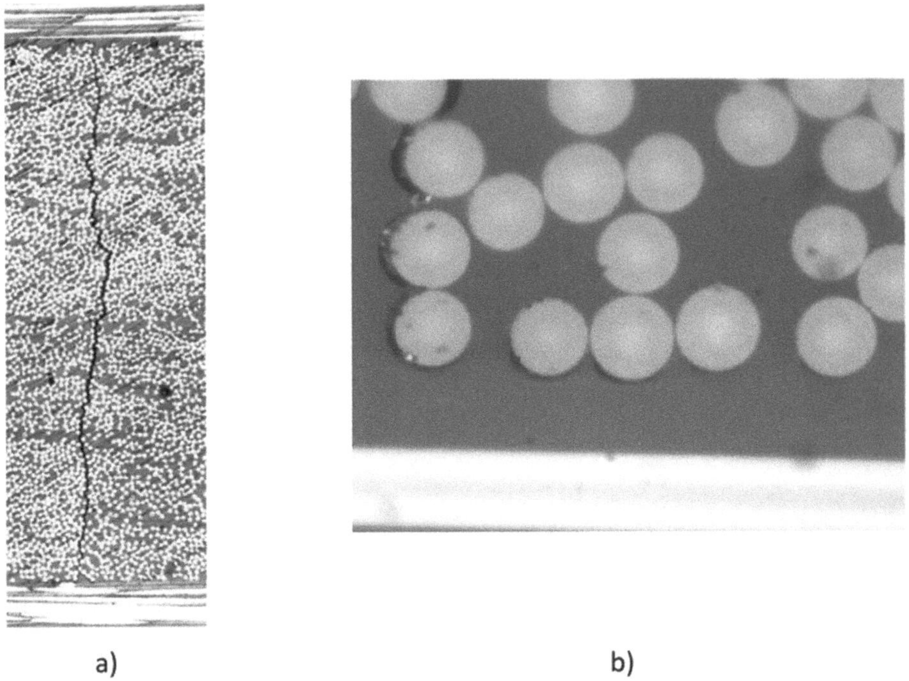

Figure 12.25 (a) First damage detected in a $[0_4{,}90_4{,}0_4]$ specimen at $\varepsilon = 0.57\%$, (b) micrograph showing the presence of debonds parallel to the fibers of the 0 degree layer.

Composites; the specifications of this material and the methodology used to detect and observe damage were provided in Section 12.2.2. The stacking sequence of these laminates was $[0_4{,}90{,}0_4]$, where the thickness of the 0 degree layer was always composed of four plies of 150 gsm, whereas the thickness of the 90 degrees layer was modified with the following thicknesses: 20, 50, 100, 150, and 450 gsm. In the four first cases, only one ply was used for the 90 degrees lamina, whereas in the case corresponding to 450 gsm, three plies of 150 gsm were used to create the 90 degrees lamina. The different damages found for each laminate will be shown in what follows.

12.5.2.1 Laminate $[0_4{,}90{,}0_4]$, $2t_{90^\circ} = 20$ gsm

This is the laminate with the thinnest 90 degrees layer thickness used by the authors. It is tremendously difficult to manipulate these plies and there is a certain uncertainty about the uniformity of distribution of fibers in a layer built with a ply of such a small thickness. Additionally, two significant observations can be made about this particular material, both being noticed in Fig. 12.29. First, the shape of the transverse section of the fibers is not circular, adopting a kind of bean-like section (as previously mentioned in Section 12.2.2). Second, the fibers content is noticeably smaller than the usually found in composite materials. The fiber volume fraction increases until a regular value when the material thickness of the ply increases.

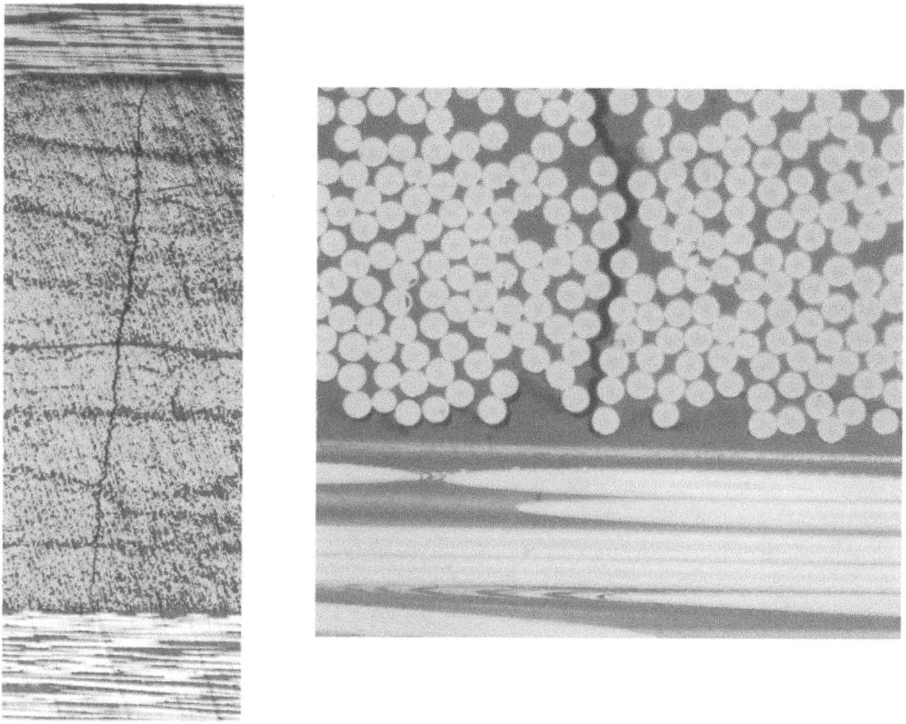

a) b)

Figure 12.26 (a) First damage detected (transverse crack) in a $[0_4, 90_4, 0_4]$ specimen at $\varepsilon = 0.48\%$, (b) detail of the damage morphology found at one end of a complete transverse crack, in the area near the interface between the layers.

The first damage detected in these laminates consisted only of isolated, or some connected, debonds, as can be seen in Fig. 12.29a together with a schematical representation with damage remarked in red (dark gray in print version). The observed debonds have a special morphology due to the irregular shape of the already mentioned fibers section. Because the first damage found did not correspond to a complete transverse crack, the specimens were subjected to new load increments until laminate failure. An outstanding result came up from these tests: laminate failure was reached without detecting a transverse crack. Fig. 12.29b shows a micrograph of the most advanced damage detected in the specimen before the total laminate failure, showing a group of unconnected and connected debonds that define a crack path but do not form a complete transverse crack.

12.5.2.2 Laminate $[0_4, 90, 0_4]$, $2t_{90^\circ} = 50$ gsm

The detection of a complete transverse crack was also not possible for the case of the laminate with a 90 degrees layer thickness of 50 gsm. The first damage observed

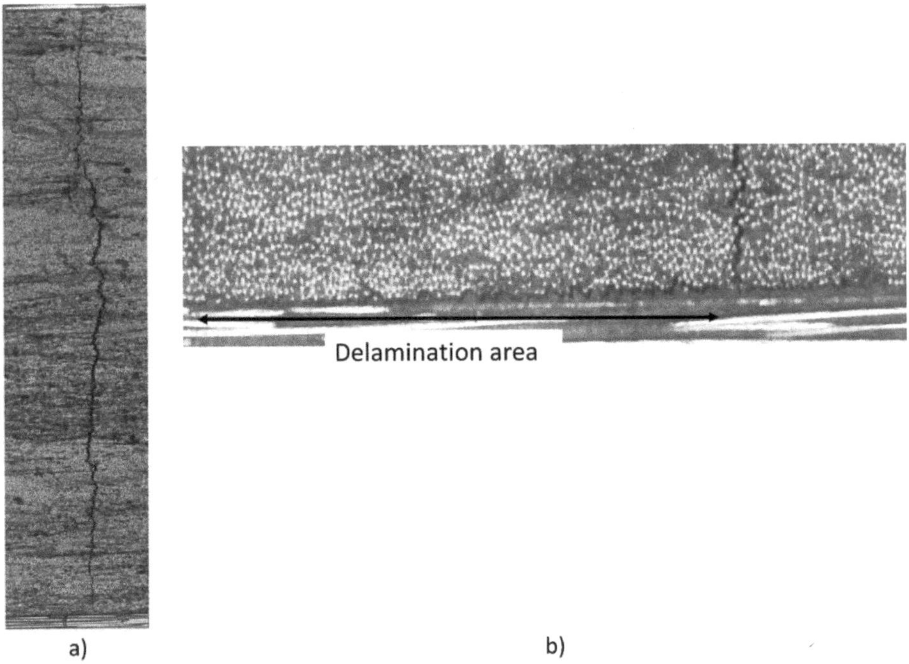

Figure 12.27 (a) First damage detected in a $[0_4,90_{16},0_4]$ specimen at $\varepsilon = 0.50\%$, (b) extension of the damage delamination found at the end of the transverse crack and close to the interface between the layers.

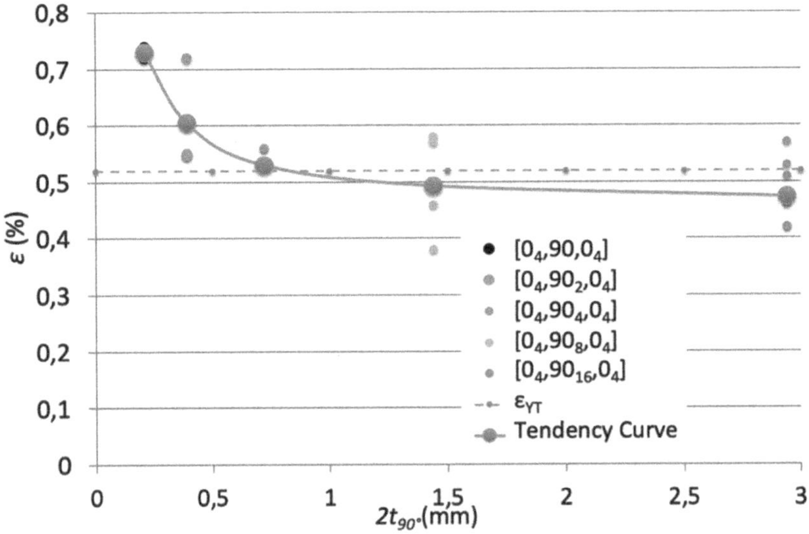

Figure 12.28 Graphical representation of the experimental values of the strain producing a complete transverse crack in $[0_4,90_n,0_4]$ specimens manufactured with a material of conventional thickness, AS4/8552.

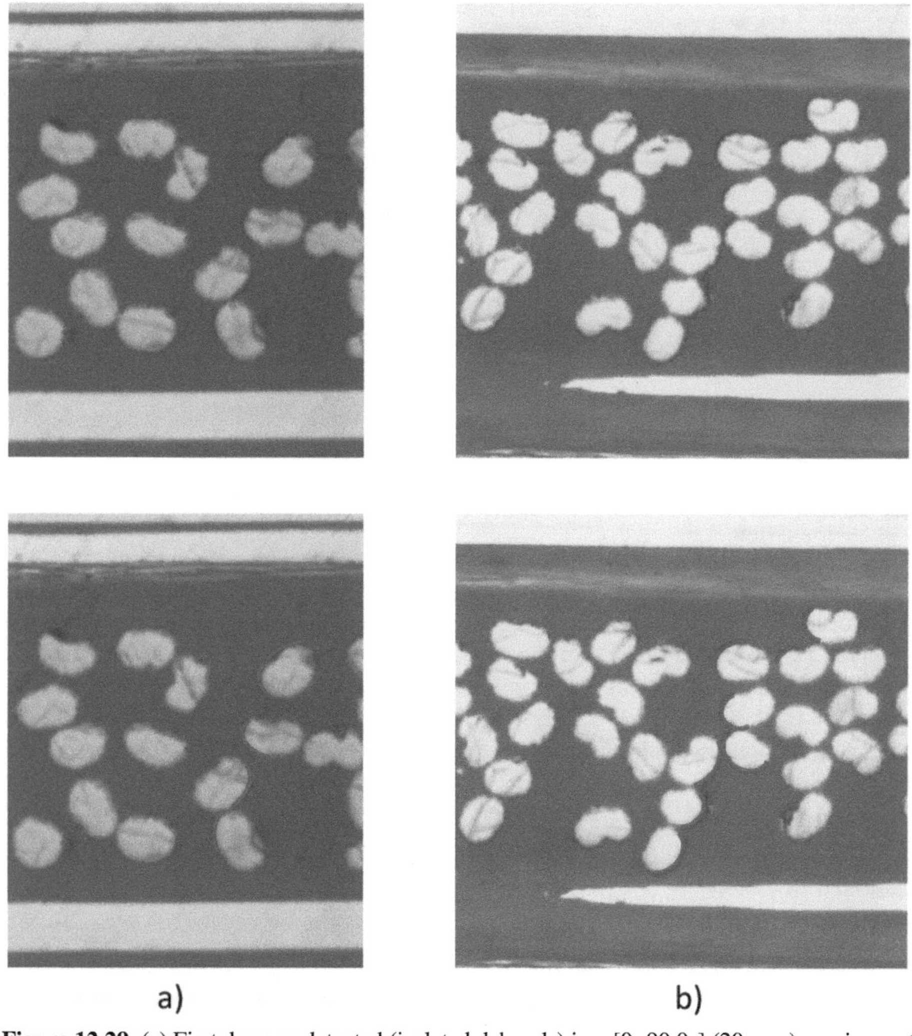

Figure 12.29 (a) First damage detected (isolated debonds) in a $[0_4, 90, 0_4]$ (20 gsm) specimen at $\varepsilon = 0.75\%$, (b) last damage observed on the 90 degrees layer for a strain value of 1.24%, before the laminate failure occurred.

consisted of isolated, and some connected, debonds and, consequently, the specimens were subjected to new load increments until the ultimate load of the laminate was reached. Fig. 12.30 presents the evolution of the first damage detected in the 90 degrees layer of a specific specimen, showing the morphology of the damage together with the strain value at which it was detected, since its appearance just before the total laminate failure.

Notice that for the ply thickness employed here, the shape of the transverse section of the fibers is almost circular; besides, the fiber volume fraction is significantly higher than that observed in the 20 gsm case.

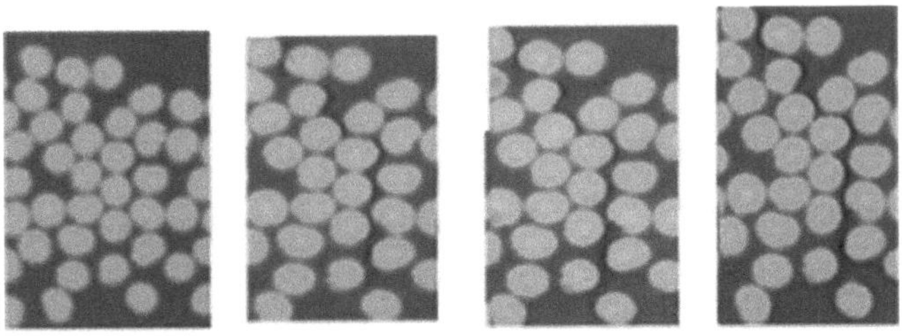

Figure 12.30 Damage evolution of the first damage detected in the 90 degrees layer with thickness of 50 gsm in $[0_4,90,0_4]$ laminate together with the strain value at which the damage was observed.

12.5.2.3 Laminate $[0_4,90,0_4]$, $2t_{90°} = 100$ gsm

For the 100 gsm case, the first damage detected consisted of isolated and connected debonds (similar to the first damage detected for the 50 gsm case). Aiming to detect the appearance of the full transverse crack, new load increments were applied, reaching the ultimate load of the laminate without being able to observe it. For this reason, the value of the strain selected for this case corresponds to the strain value from which no progress was produced in the morphology of the first damage detected under new load increments. The evolution of the morphology of the first damage detected in a single specimen is shown in Fig. 12.31, together with the strain value at which it was observed; the value of the strain from which no more damage advance occurred was

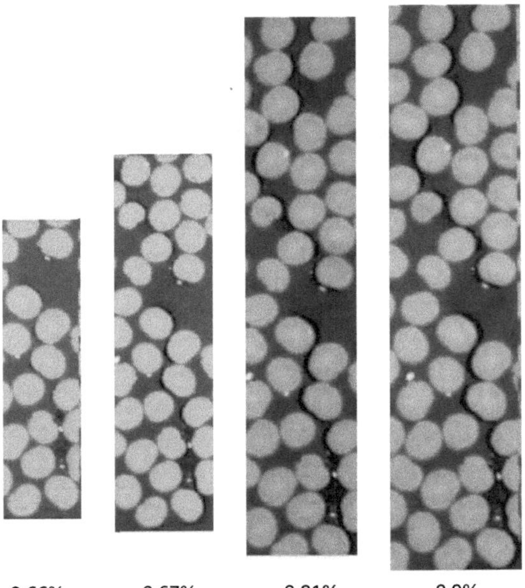

Figure 12.31 Damage evolution of the first damage detected in the 90 degrees layer (100 gsm) in the $[0_4,90,0_4]$ laminate together with the strain value at which the damage was observed.

0.9%. Higher values of this strain produced a wider opening of the already existent damage but no new breakages.

12.5.2.4 Laminate $[0_4,90,0_4]$, $2t_{90^\circ} = 150$ gsm

The first damage observed in the laminate with a layer thickness of 90 degrees equal to 150 gsm consisted of a generalized transverse damage formed by a number of significant debonds that are connected, Fig. 12.32a. As the detected damage did not consist of the expected complete transverse crack, as in the previous cases, new load increments were applied until a transverse crack appeared, Fig. 12.32b.

12.5.2.5 Laminate $[0_4,90,0_4]$, $2t_{90^\circ} = 450$ gsm

This laminate has the thickest 90 degrees layer made of Skyflex material, the 90 degrees lamina being obtained with three 150 gsm plies. In this case, the first damage detected was a complete transverse crack, it being the unique laminate of those selected in the experimental campaign with ultrathin material where this phenomenon was observed. In some of the specimens, at the ends of the transverse crack (at the interface between the layers), one or several debonds parallel to the fibers of the 0 degree layer could also be observed.

Fig. 12.33 shows the complete transverse crack as the first damage detected in the 90 degrees layer for a strain value equal to 0.75%, where an incipient delamination-like damage can also be observed.

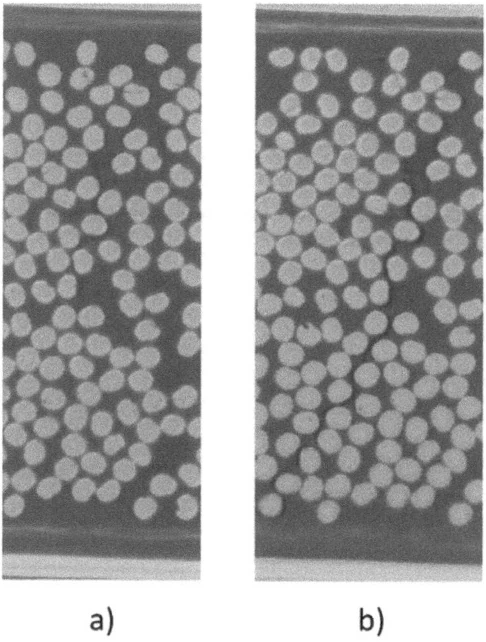

Figure 12.32 Damage found in the 90 degrees layer thickness (150 gsm) in the $[0_4,90,0_4]$ laminate: (a) first damage detected at a $\varepsilon = 0.66\%$, (b) complete transverse crack observed at a $\varepsilon = 0.76\%$.

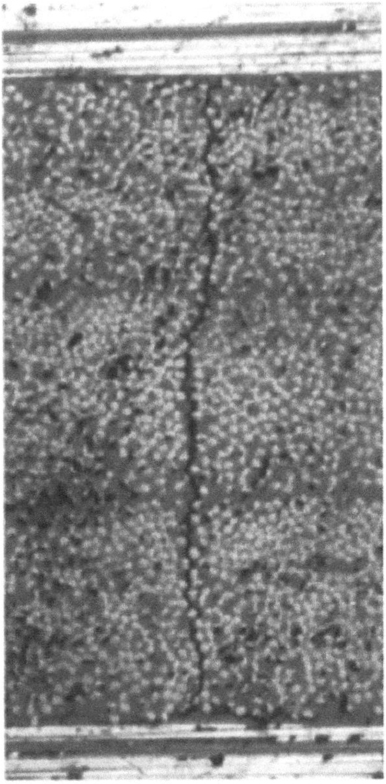

Figure 12.33 First damage detected consisting of a complete transverse crack in the 90 degrees layer (450 gsm) in the $[0_4,90,0_4]$ laminate for $\varepsilon = 0.75\%$.

As has been previously done with the experimental results of the conventional thickness laminates, the values of the thickness of the 90 degrees layer have been plotted versus the strain value for which the observed damage is assimilable to a complete transverse crack, see Fig. 12.34. For the graphical representation it has been assumed in this case, after experimental checking, that 1 gsm is equivalent approximately to 1 micron of thickness.

The values of the strain for which the maximum damage detected in the 90 degrees layer have been represented in Fig. 12.34. They can be assimilated to a complete transverse crack only in the case of having 150 μm or 450 μm of thickness in the 90 degrees layer. However, in the rest of laminates with ultrathin plies (20, 50, and 100 μm), a complete transverse crack is not observed since the debonds are not fully connected. A tendency line with the mean strain values at which the maximum observable damage is detected is plotted in Fig. 12.34. The strain associated with the nominal transverse strength of the material is also plotted in Fig. 12.34, in order to settle a reference for the values found. An increase of the laminate strain is necessary in the case of this material, to detect approximately the expected damage (transverse crack) as the thickness of the 90 degrees layer is decreased, the presence

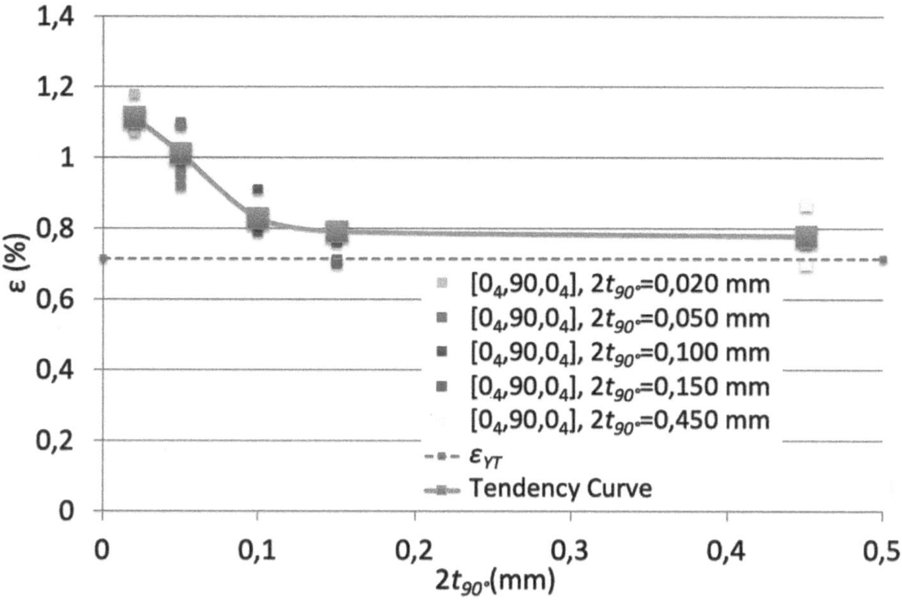

Figure 12.34 Graphical representation of the experimental values of the strain producing the most advanced detected damage in $[0_4,90,0_4]$ laminates manufactured with ultrathin plies.

of the scale effect being, in some way, also captured when the thickness of the 90 degrees lamina tends to zero.

12.6 A physically based explanation of the scale effect

All the ingredients to present a physically based explanation of the scale effect have already been generated and presented, both from a numerical point of view, Sections 12.3 and 12.4, and from an experimental point of view, Section 12.5.

The explanation will be built on a scheme of the evolution of the strain applied to the laminate and the value of the thickness of the 90 degrees ply, in a scheme similar to the presentation of the scale effect in the aforementioned works by Parvizi et al. (1978) or Flaggs and Kural (1982), among others. The successive steps of forming the explanation will be successfully entered into the scheme represented in Fig. 12.35.

First of all, Fig. 12.35 reflects the independence of the first damage appearance, i.e., a fiber–matrix debond, of the thickness of the 90 degrees ply. Obviously, this debond must necessarily correspond to the strain associated with the breakage of a 90 degrees specimen tested in the direction normal to the fibers, nominally denoted as $\varepsilon(Y_T)$. It is well known that this value, in fact all the values associated with the behavior normal to the fibers, exhibits a certain scatter. For this reason, the range of values where such event would appear has been represented by a band, ACBD in Fig. 12.35, where a first debond should appear in accordance with the numerical predictions developed in Section 12.3.

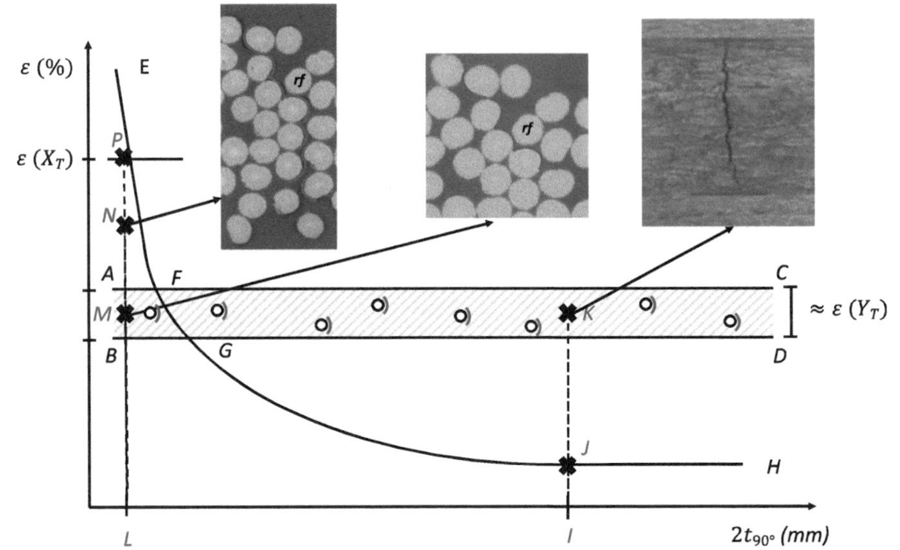

Figure 12.35 A physically based explanation of the scale effect: generation of damage for a certain value of the strain applied to the laminate for different values of the thickness of the 90 degrees lamina.

Now, regardless of the previous fact, the study carried out in Section 12.4 based on LEFM shows an evolution of the strain required to have energy enough to produce a transverse crack in the 90 degrees ply, as is shown schematically by the EFGH curve in Fig. 12.35. It is very important to note that this curve implies the assumption of having a defect identifiable as a crack. This curve clearly indicates that in the presence of a presumable defect in the 90 degrees lamina, the strain required to transform such initial defect into a transverse crack is very high for very a thin 90 degrees lamina and tends to zero when the thickness of the lamina tends to a high value. Whereas the former agrees with experimental observations, the latter does not agree at all, as it seemed that a strain plateau value was required for increasing values of the thickness of the 90 degrees ply to produce a transverse crack in the 90 degrees ply.

Combining the information coming from the two analyses carried out in Sections 12.3 and 12.4, a full physically based explanation can be given taking as reference two extreme situations, one corresponding to a thick 90 degrees ply and another corresponding to a very thin 90 degrees ply, both situations are considered separately in what follows.

The case of a thick 90 degrees lamina is considered by means of Point I in the figure. Now, when the strain corresponding to Point J is reached, there would be energy enough for an initial small crack to propagate until generation of a whole transverse crack, but, since we are still in the pristine state and, in accordance with the study carried out, the material has not reached yet the value of the strain able to produce a first debond, nothing will consequently happen from the damage appearance point of view. It is necessary to increase the strain until a level, Point K in Fig. 12.35, where a first debond will appear. Then, there will be energy enough to generate from this first debond a transverse crack, and due to the fact that there is an energy excess to produce this transverse crack, this excess will produce more damage in the laminate, it consisting of delamination between 0 and 90 laminas. Obviously, the greater the thickness of the 90 degrees ply, the greater the excess of energy available and, consequently, the greater the delamination damage produced; see images of damage included in Section 12.5 (see Figs. 12.26 and 12.27) corresponding to thick 90 degrees laminas.

The case of an ultrathin 90 degrees ply is considered by means of point L in Fig. 12.35. Nothing, in terms of damage, appears until reaching the strain indicated by point M, where a first debond arises. The problem now is that, due to the fact that point M is under the EFGH curve, there is not enough energy available to make this debond to evolve to a transverse crack. The damage associated with point M consists of one or several isolated debonds as represented in the figure. If the strain is again increased, point N can be reached increasing the number of observed debonds, but still without provoking a transverse crack. If the strain applied is increased again, the EFGH curve could be reached provoking the appearance of a transverse crack. Another possibility is, as suggested in Fig. 12.35, the reaching of the value of the strain at which the laminas with the fibers oriented 0 degrees fail, i.e., $\varepsilon(X_T)$, therefore originating the total failure of the laminate but with the absence of complete transverse crack in the 90 degrees layer.

Once the scale effect has been explained, it is adequate to turn the view to the figures included in Sections 12.5.1 and 12.5.2 that showed the evolutions of the damage with the strain value. One could think that the results represented there may be contradictory with the fact that the appearance of the damage, presumably isolated debonds, is associated with a fixed value of the strain. The explanation is that in the referenced figures in Sections 12.5.1 and 12.5.2, the checking of the scale effect was being forwarded and consequently the appearance of a transverse crack in the 90 degrees ply was being looked for and this fact was correctly reflected by Figs. 12.28 and 12.34.

Now, once the scale effect has been investigated, it is pertinent to repeat the observations associated with the effect of thin and ultrathin plies focusing on the presence of the first actual damage that appears in the material, devoting efforts in making it observable. These results are summarized in Fig. 12.36 for the ultrathin thickness material.

The results clearly show now the consistency of the explanation proposed for the scale effect. The first damage consisting in isolated debonds in the interval of t_{90} studied, between 50 and 400 μm, appears at the same strain level, as predicted by the theory developed. However, the smaller the thickness of the 90 degrees ply, the lighter the intensity of the damage appeared, as shown in the two pictures represented corresponding to 50 and 200 μm thickness.

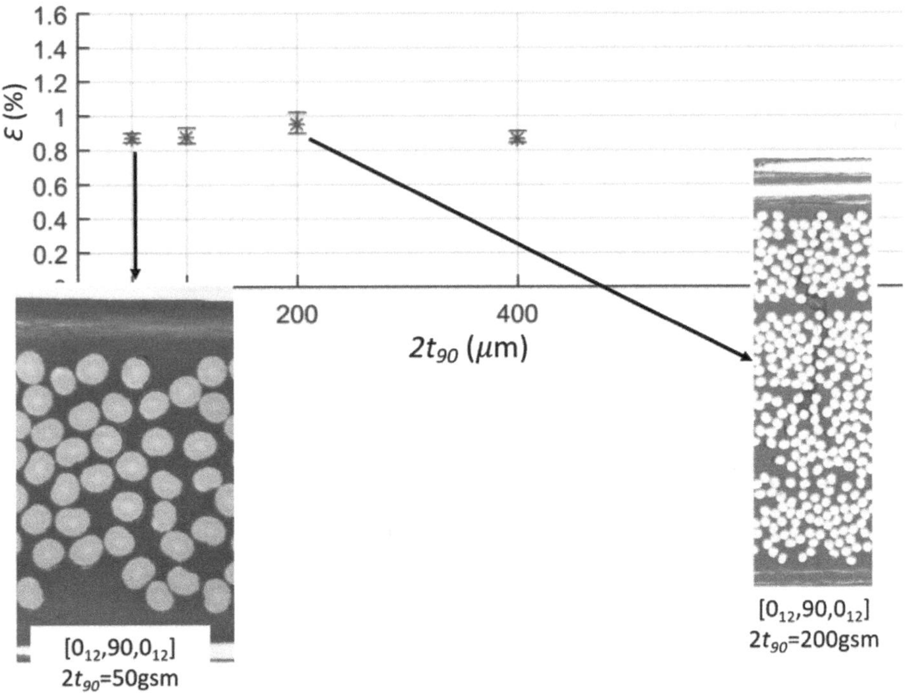

Figure 12.36 Evolution of the strain ε required to produce a first observable damage in the 90 degrees ply versus t_{90}.

12.7 Applications

Two applications of different kind derived from the generated knowledge are going to be presented in this section, to show the dimension of the understanding of the scale effect phenomenon. The first corresponds to a conceptual question and clarifies the concept of the in situ strength of a lamina of a laminate. The second has a more practical application due to the fact that it shows the possibility of optimizing the thickness of the weakest lamina of a laminate.

12.7.1 Remarks on the in situ strength concept

Fig. 12.37 shows a simplified schematic representation of the explanation about of the scale effect that the one presented on Fig. 12.35, where in red (dark gray in print version) appears a simplification of the failure of the 90 degrees lamina as a function of the strain applied. When the thickness of the 90 degrees lamina is relevant the failure is apparently controlled by a value of the strain, $\varepsilon(Y_T)$ associated with a nominal stress value, Y_T, whereas when the thickness is small the failure is apparently controlled by a value of ε associated with the energy required to make an existing crack to propagate across the 90 degrees lamina of the laminate. One would then be tempted to propose

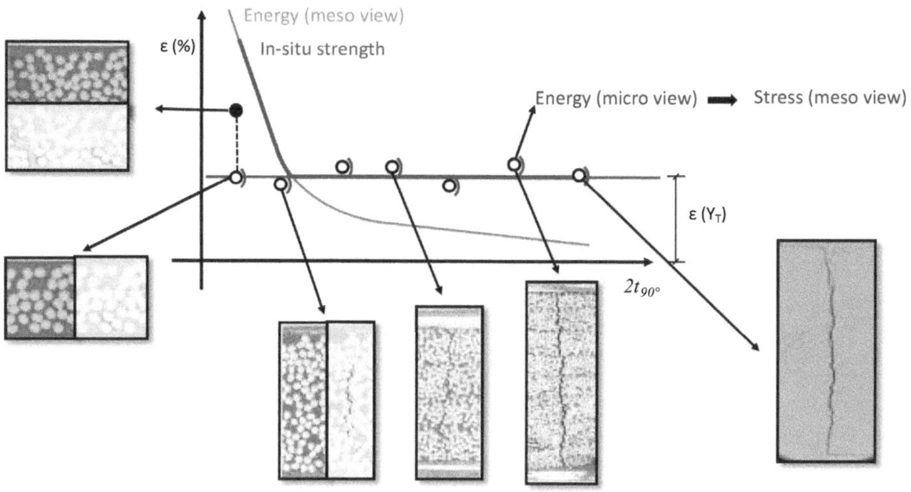

Figure 12.37 The in situ strength concept.

that both a value of the stress and a value of energy are required for a transverse crack to appear in the 90 degrees ply. The phenomenon is nevertheless much more complicated.

First of all, an agreement on the term failure (a material state), or strength (a property typically associated with a failure of the material), has to be reached. If failure is understood as the abandonment of the pristine state, then the scale effect does not exist, because as shown in Fig. 12.36, this abandonment appears always when the strain reaches $\varepsilon(Y_T)$. This abandonment of the pristine state can be either a total crack with delamination for thick laminas or a simple presence of one or two debonds when the lamina is very thin, both situations as well as intermediate cases appearing represented in Fig. 12.37.

On the contrary, if the term failure is associated with the presence of a total transverse crack of the weakest lamina of the laminate, then it might be adequate to use the term in situ strength. However, the problem is again more complicated as is possible to observe the final failure of the laminate (controlled by the breakage of the 0 degree lamina), with the damage in the 90 degrees ply consisting in isolated debonds or only generalized transverse failure, but not having reached the level of a complete transverse crack. Thus, even with this definition of strength, the presentation of failure by means of the simplified representation given by Fig. 12.37, with values associated with stress and energy, does not represent the events that take place as function of the thickness and the idea of a double criterion to explain the experimental observations is not necessary.

12.7.2 Optimizing the thickness of the weakest lamina

In accordance with previous ideas manifested in Section 12.7.1, the understanding of the scale effect based on experimental observations and on explanations physically

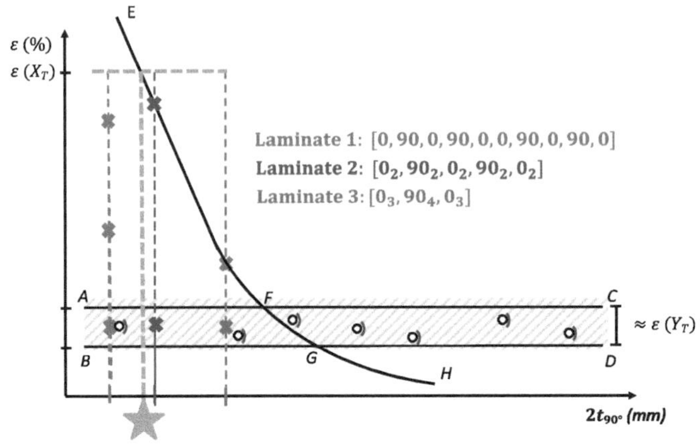

Figure 12.38 Optimization of the stacking sequence of a laminate.

based on mechanics and fracture mechanics concepts can lead us to selecting the most appropriate thickness of the 90 degrees layer of a laminate. Thus, if the lamina is too thick, a pattern of extensive damage (transverse crack plus delamination) will appear when the strain reaches the value $\varepsilon(Y_T)$. On the contrary, if the lamina is too thin it will never experience a complete transverse failure. Then, there should be a point, a value of the thickness of the 90 degrees ply, where the total breakage of the 0 degree lamina and the appearance of a transverse crack in the 90 degrees lamina occur at the same time, i.e., for the same value of the strain applied, higher than $\varepsilon(Y_T)$.

To prove this, the following testing campaign on three different laminates has been performed. The laminates are Laminate 1 $[0,90,0,90,0]_S$, Laminate 2 $[0_2,90_2,0]_S$, and Laminate 3 $[0_3,90_2]_S$, having all of them the same number of 0 (6) and 90 (4) degrees plies. Thus, they have the same total thickness, but different stacking sequences, giving rise in principle to different patterns of damage.

As shown schematically in Fig. 12.38, laminates 2 and 3 reach the curve controlled by energy before $\varepsilon(X_T)$ is reached, then allowing the appearance of complete transverse damage plus delamination. On the contrary in the case of laminate 1, the value of $\varepsilon(X_T)$ is reached without having touched the curve controlled by energy, which implies that the damage of the total laminate would be provoked by the failure of the 0 degree lamina without having a complete crack in the 90 degrees lamina.

It is obvious, as deduced from Fig. 12.38, that there must be a value of t_{90}, represented by a star in the figure, where the 0 degree plies reaches the allowable strain at the same time that the 90 degrees laminas present a complete transverse crack but without delamination; in this way, secondary negative effects (provoked by other loading cases as bending, torsion, or shear among others) are restricted in the laminate until the instant of the final failure.

12.8 Summary and conclusions

The scale effect has been revisited and investigated by means of experiments and numerical simulations. This study is pertinent due to the appearance of UTPs which offers the possibility of decreasing the weight of a composite component, without affecting its performance.

There are many possibilities of correlating observed results with variables of the structure under consideration, in this case a laminate, at the level at which the structure is calculated, in our case stresses at macro or mesoscale.

However, if one wants to get a deep knowledge of a phenomenon, in this case the scale effect derived from the damage in the 90 degrees ply of a $[0,90_n]_S$ laminate, it is absolutely necessary to go to the level at which the phenomenon under consideration occurs, in our case micro and meso-scale levels, integrating the information coming from experimental results and the prediction of behavior at both levels.

Thus, it has been considered that the weakest internal link of a composite is the interface between fiber and matrix, the damage in a 90 degrees lamina of a laminate starting by means of a debond between fiber and matrix. This assumption is supported overwhelmingly by the observations of all type of laminates and materials used in this study.

The predictions, based on a numerical model at a multiscale level, indicate that the appearance of a debond is independent of the thickness of the 90 degrees ply (this is an observation at micromechanical level), and all of them should appear at the same level of strain applied.

However, these debonds (isolated or connected) were not able to be observed when using plies of conventional thickness. On the contrary, it has been possible to observe them when using UTPs, verifying the numerical predictions obtained with the multi-scale model.

Why it is possible to observe isolated debonds using UTPs and not using plies of regular thickness?

To give an explanation to this question, we have to move now to the mesoscale, the appropriate scale at which the damage under consideration, in particular, transverse crack with potential delamination, takes place.

Based on fracture mechanics, the capacity to release energy per unit of length in the 90 degrees lamina, and therefore to break the internal links of the material (debonds and through the matrix cracks), increases with the thickness of the lamina. It implies that, in presence of a small crack, very small values of strain would release energy enough to break the internal links of the material in presence of thick 90 degrees laminae, but not in the case of UTPs laminae. However, in the pristine state, there are no cracks yet in the 90 degrees lamina. It is only when the strain reaches a specific value (at a level of the order of $\varepsilon(Y_T)$ of the material) that a first debond appears; at this point, there is a clear bifurcation in the behavior between regular and UTPs plies.

In the case of a ply of regular thickness, the energy balance detects that in presence of a small crack (the first debond) there is energy enough to propagate it generating a transverse crack and potentially delamination between 0 and 90 degrees laminas. This is the reason why in composites with laminae of regular thickness it was not possible to observe isolated debonds, the first observations being always generalized damage in the 90 degrees lamina and interface with the 0 degree lamina.

On the contrary, when UTPs are used, the appearance of a first debond generates a situation at which there is no energy enough to break the internal links of the material to produce a transverse crack but, instead, isolated debonds are finally observed. The strain must be increased in this case to sequentially observe new debonds that may or not coalesce to generate a transverse crack, it subjected to the possibility of reaching the failure of the 0 degree lamina before the total transverse crack arises in the 90 degrees lamina.

Once the phenomenon under consideration is fully understood, several related questions can be addressed.

- The explanation given requires no fitting parameters and it is based just on observations and predictions using the concepts of fracture mechanics and interfacial fracture mechanics.
- Models based on mechanisms of damage that do not reflect those actually taking place cannot detect and identify the scale effect. This is, for instance, the case of cohesive models since the advance of the damage at a microscale level with hundreds or thousands of fibers and with cohesive elements placed at all the interfaces matrix–fiber is not affected by the values of the thickness of the 90 degrees lamina.
- As previously explained, the unique quantity controlling the onset and propagation of damage is *energy*. A different question is that, in accordance with the explanation given, a variable at mesomechanical level, $\varepsilon(Y_T)$, or stress associate value, Y_T, controls the fact that there is *energy* enough to break at micromechanical level the internal links (in this case the adherence between fiber and matrix) of the material. Once this damage has arisen, the *energy* clearly, now at mesomechanical level, controls the generation and growth of a transverse crack in the 90 degrees lamina as well as the generation and growth of delamination-like damage along the 0–90 interfaces.
- A reflection about the in situ strength concept is pertinent. The idea of considering the in situ strength as something associated with the appearance of similar damages for different values of the strain depending on the value of the thickness of the 90 degrees lamina is too simplistic. The evolution of the intensity of the damage, going from a simple debond until a complete transverse crack plus delamination, is a continuous function of the 90 degrees lamina thickness. Values of the strain associated with different patterns of damage are not comparable at all. It does not prevent from accepting the presence of scale effect in all types of laminates having 90 degrees laminae of different thickness.

In any case, as highlighted in this chapter, the in situ strength concept additionally requires a clarification in terms of the damage observed. If the strength is associated with the abandon of the pristine state, the scale effect does not exist, because the damage, i.e., the first debond, appears at the same strain level, independently of the thickness of the 90 degrees lamina. If strength is understood as a property associated with a value of the strain originating generalized damage (a transverse crack in general terms), then it is acceptable to refer this fact as an in situ property, now understood with the total physically based explanation given.

- The knowledge of the mechanisms of damage involved in the generation and propagation of damage in the 90 degrees lamina of a laminate opens the possibility of measuring properties of the material at different scales. Specifically:
- G_{IC} of the fiber–matrix interphase
- G_{IC} of the 90 degrees lamina
- G_{IC}, G_{IIC} of the 0/90 degrees laminae interface can now be measured trying to reproduce the situations in which these properties play a significant role.
- A reflection on the use of thick/thin laminae is also pertinent. It is true that the designers of aeronautical components prefer to use thick laminas in order to reduce the manufacturing time and consequently the cost. However, in accordance with what has been shown in the

study carried out, the presence of laminas of relevant thickness may generate, at low strain level, significant damage in the 90 degrees ply and in the interfaces with the adjacent laminae. Then, experimental studies must be addressed to check the effect that the application of different types of loading, as shear, bending, or torsion, may have on the performance of laminates with different thicknesses of the 90 degrees lamina, the laminates consequently suffering different pattern of damages. A similar interest generates the different effect that the presence of fatigue may have in connection with different values of the 90 degrees laminas.

Acknowledgments

This work was supported by Consejería de Economía y Conocimiento Junta de Andalucía (Project P18-FR-3360) corresponding to the I+D+i Programa Operativo FEDER 2014–20, Spanish Ministry of Education, Science and Sports (Project MAT2016-80879-P) and Spanish Ministry of Science, Innovation and Universities (Project PID2021-126279OB-I00).

References

Amacher, R., Cugnoni, J., Botsis, J., Sorensen, L., Smith, W., Dransfeld, C., 2014. Thin ply composites: experimental characterization and modeling of size-effects. Compos. Sci. Technol. 101, 121–132.

Arteiro, A., Catalanotti, G., Melro, A.R., Linde, P., Camanho, P.P., 2014. Micro-mechanical analysis of the in situ effect in polymer composite laminates. Compos. Struct. 116, 827–840.

Asp, L., Berglund, L.A., Talreja, R., 1996. A criterion for crack initiation in glassy polymers subjected to a composite-like stress state. Compos. Sci. Technol. 56 (11), 1291–1301.

Bell, J.F., 1973. In: Truesdell, C. (Ed.), Encyclopedia of Physics, Mechanics of Solids I. Springer Verlag.

Correa, E., Gamstedt, E.K., París, F., Mantič, V., 2007. Effects of the presence of compression in transverse cyclic loading on fibre-matrix debonding in unidirectional composite plies. Comp.: Part A. 38, 2260–2269.

Correa, E., Mantič, V., París, F., 2008a. A micromechanical view of inter-fibre failure of composite materials under compression transverse to the fibres. Compos. Sci. Technol. 68, 2010–2021.

Correa, E., Mantič, V., París, F., 2008b. Numerical characterization of the fibre–matrix interface crack growth under transverse compression. Eng. Fract. Mech. 75, 4085–4103.

Correa, E., Mantič, V., París, F., 2011. Effect of thermal residual stresses on matrix failure under transverse tension at micromechanical level: a numerical and experimental analysis. Compos. Sci. Technol. 71, 622–629.

Correa, E., París, F., Mantič, V., 2013. Effect of the presence of a secondary transverse load on the inter-fibre failure under tension. Eng. Fract. Mech. 103, 174–189.

Correa, E., Valverde, M.I., Velasco, M.L., París, F., 2018. Microscopical observations of inter-fibre failure under tension. Compos. Sci. Technol. 155, 213–220.

Dvorak, G.J., Laws, N., 1987. Analysis of progressive matrix cracking in composites laminates II. First ply failure. J. Compos. Mater. 21, 309–329.

Flaggs, D.L., Kural, M.H., 1982. Experimental determination of the in situ transverse lamina strength in graphite/epoxy laminates. J. Compos. Mater. 16, 103–116.

García, I.G., Mantič, V., Blázquez, A., París, F., 2014. Transverse crack onset and growth in cross-ply [0/90]s laminates under tension. Application of a coupled stress and energy criterion. Int. J. Solid Struct. 51, 3844–3856.

Graciani, E., Justo, J., Zumaquero, P.L., 2020. Determination of in-plane and through-the-thickness coefficients of thermal expansion in composite angle brackets using digital image correlation. Compos. Struct. 238, 11139.

Li, D.S., Wisnom, M.R., 1997. Evaluating weibull parameters for transverse cracking in cross-ply laminates. J. Compos. Mater. 31, 935–951.

París, F., Cañas, J., 1997. Boundary Element Method: Fundamentals and Applications. Oxford University Press, USA.

París, F., Correa, E., Cañas, J., 2003. Micromechanical view of failure of the matrix in fibrous composite materials. Compos. Sci. Technol. 63, 1041–1052.

París, F., Correa, E., Mantič, V., 2007. Kinking of transverse interface cracks between fibre and matrix. J. Appl. Mech. 74, 703–716.

París, F., Velasco, M.L., Correa, E., 2018. Micromechanical study on the influence of scale effect in the first stage of damage in composites. Compos. Sci. Technol. 160, 1–8.

París, F., Velasco, M.L., Correa, E., 2021. The scale effect in composites: an explanation physically based on the different mechanisms of damage involved in failure. Compos. Struct. 257, 113089.

París, F., 2001. A Study of Failure Criteria of Fibrous Composites Materials. NASA/CR-2001-210661.

Parvizi, A., Garret, K.W., Bailey, J.E., 1978. Constrained cracking in glass fibre-reinforced epoxy cross-ply laminates. J. Mater. Sci. 13, 195–201.

Saito, H., Takeuchi, H., Kimpara, I., 2012. Experimental evaluation of the damage growth restraining in 90° layer of thin-ply CFRP cross-ply laminates. Adv. Compos. Mater. 66, 57–66.

Saito, H., Takeuchi, H., Kimpara, I., 2013. A study of crack suppression mechanism of thin-ply carbon-fiber-reinforced polymer laminate with mesoscopic numerical simulation. J. Compos. Mater. 48, 2085–2096.

Sánchez-Carmona, S., Velasco, M.L., Barroso, A., Correa, E., 2023. Thermomechanical Characterisation Data of 30 G/m2 and 150 G/m2 Cured Unidirectional Carbon/epoxy Tape Prepreg TP 402/T700S. Data in brief (Submitted for publication).

Soden, P., 1998. Lamina properties, lay-up configurations and loading conditions for a range of fibre-reinforced composite laminates. Compos. Sci. Technol. 58, 1011–1022.

Spanish Standard UNE-EN2561:1996 Material aeroespacial. Plásticos reforzados de fibra de carbono. Estratificados unidireccionales. Ensayo de tracción paralelamente a la dirección de la fibra.

Tsai, S.T., Kawabe, K., 2006. Thin Ply Laminates,US Patent (US 2006/0093802 A1).

Velasco, M.L., Graciani, E., Távara, L., Correa, E., París, F., 2018. BEM multiscale modelling involving micromechanical damage in fibrous composites. Eng. Anal. Bound. Elem. 93, 1–9.

Velasco, M.L., Correa, E., París, F., 2020. Interaction between fibres in the transverse damage in composites. Eng. Fract. Mech. 239, 107273.

Velasco, M.L., Correa, E., Sánchez-Carmona, S., París, F., 2023. Evolution of the Damage Onset and Morphology in [0/90n/0] Laminates when Increasing the Ply thickness., Comp. Part A (Submitted for publication).

Zhuang, L., Talreja, R., Varna, J., 2018. Transverse crack formation in unidirectional composites by linking of fibre/matrix debond cracks. Compos. Appl. Sci. Manuf. 107, 294–303.

Thermoelastic constants of damaged laminates: COD- and CSD-based methods

13

Janis Varna
Luleå University of Technology, Luleå, Sweden

13.1 Introduction

Long unidirectional (UD) fiber-reinforced composites are finding broad applications in "classical" fields such as aeronautics and marine industries, and also as structural elements in automotive applications. In most of the applications, structural elements contain laminates consisting of several layers with the fiber orientation in each layer selected to meet design requirements with respect to thermomechanical service loads and environmental exposure. Several microdamage modes appear in layers before the final macroscopic failure of the composite laminate. Owing to microdamage, the thermoelastic constants of the laminate are changed. Therefore, the composite structure is often designed using criteria in terms of limits for stiffness reduction and not with respect to final failure.

Among the five commonly used failure strain components of UD composites (longitudinal tensile and compressive strains, transverse tensile and compressive strains, and shear strain), the lowest is the transverse tensile strain to failure. Since all layers in an undamaged laminate are in iso-strain conditions (in the absence of bending, not considered here), the first observed damage mode usually is intralaminar cracking of layers with off-axis orientation with respect to the main load direction. Cracks are initiated as a result of combined action of transverse tensile stress and shear stress; see Parvizi and Bailey (1978), Jamison et al. (1984) for cracking due to transverse stresses.

Intralaminar cracks (Fig. 13.1) run in the direction of fibers in the layer (not breaking fibers) and the crack plane is perpendicular to the laminate midplane ($x-y$ plane in Fig. 13.1). The crack propagation is in mixed opening and sliding modes—Mode I and Mode II—and most researchers assume that Mode I is the dominant one. Usually, under axial (x-direction) tensile load, these cracks cover the whole thickness of the layer and propagate over the whole width (y-direction) of the specimen, except perhaps in very thin layers and under low-amplitude cyclic stress. In this chapter, we will call these cracks intralaminar cracks, although they are also called matrix cracks, tunneling cracks, transverse cracks, or sometimes, inclined cracks (the latter to denote intralaminar cracks in off-axis layers with a different orientation than 90 degrees).

The effect of each individual intralaminar crack on laminate stiffness is very small, but since the number of cracks increases during service life, they may significantly

Modeling Damage, Fatigue and Failure of Composite Materials. https://doi.org/10.1016/B978-0-443-18489-5.00005-9

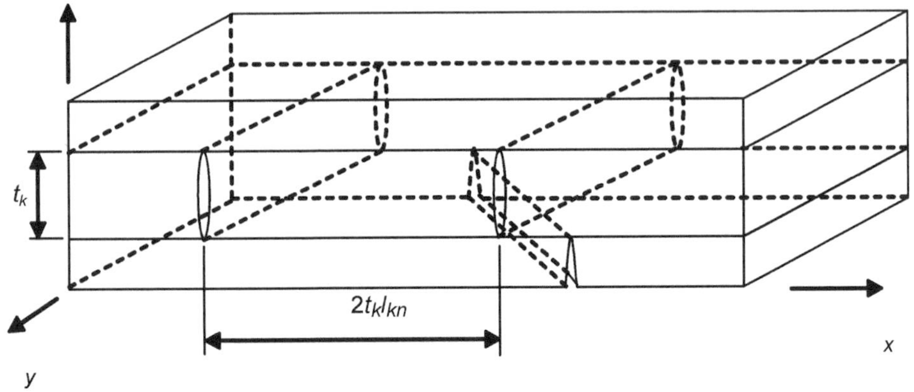

Figure 13.1 Schematic showing the upper part of a symmetric laminate with cracks in different layers.

impair the effective properties of the laminate (Ogin et al., 1985). They can also initiate other damage modes, such as local delamination at the intralaminar crack tip (Kashtalyan and Soutis, 2005; Takeda et al., 1995), "stitch cracks," and fiber breaks in neighboring plies (Fig. 13.2).

The severity of intralaminar damage in a layer is quantified with an average measure called crack density, which is defined as the number of cracks per unit distance normal to the crack plane calculated by counting the cracks in the layer over a certain distance. Often, in experiments, the cracks are counted on the free edge of a specimen over a distance. Since for calculating the crack density, this distance must be normal to the crack plane, which equals the distance along the specimen edge only for the 90 degrees layer, for other layer orientations, the proper distance is obtained by projection.

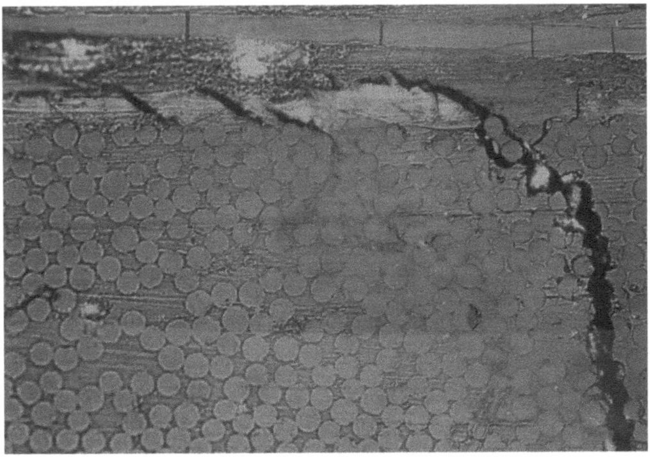

Figure 13.2 Intralaminar crack in a cross-ply laminate: Crack tip region at the 0/90 degrees interface with local interlayer delamination and fiber breaks in the 0 degrees layer.

The crack density thus calculated is linear density, which is different than the number of cracks per unit volume, and is generally different in different layers.

The crack density in the kth layer is denoted ρ_k (cracks/mm). The average distance between cracks, which is inverse to the crack density, is called average crack spacing $2l_k = 1/\rho_k$. More representatives for the stress state in a damaged layer are normalized spacing and normalized crack density. Normalized spacing is the average distance between cracks divided by the cracked ply thickness:

$$2l_{kn} = 2l_k/t_k, \tag{13.1}$$

whereas the normalized crack density is:

$$\rho_{kn} = \rho_k t_k \tag{13.2}$$

The meaning of normalized average spacing is similar to the fiber aspect ratio in short-fiber composites. For example, information that average spacing between cracks is 1.0 mm does not reveal the severity of damage because we do not know the ply thickness. If, for example, the ply thickness is $t_k = 0.125$ mm, the normalized crack density is low ($\rho_{kn} = 0.125$). In a region of very high crack density, often called a saturation region, due to large stress relief caused by interaction of stress perturbation, further cracking stops and the spacing between cracks approaches t_k. Hence, the maximum possible normalized crack density $\rho_{kn,max} \approx 1$.

Tensile transverse stress in the undamaged layer (called far-field stress) opens the crack, and it has two traction-free surfaces. Therefore, both the transverse normal stress and the in-plane shear stress on the crack surface are zero. With increasing distance from the crack, the transverse stress and the in-plane shear stresses start to recover, and when the distance is large, they may asymptotically approach the far-field values. The stress transfer from the undamaged layer to the damaged layer is accompanied by high out-of-plane shear stresses at the layer interface in the vicinity of the crack. The efficiency of the stress transfer (the distance needed to recover most of the far-field stress state) depends on the interface quality. If the interface is not damaged, the distance is much shorter than in the case of a partially delaminated interface as in Fig. 13.2. The stress recovery mechanism described is responsible for observed creation of multiple cracks in the same layer (multiple cracking) with increasing external load.

The stress transfer described explains how a new crack can be created between two existing cracks. However, it does not explain why only one crack appears first at a certain load level and the next crack is created only if the load is increased. The reason for a continuous increase of crack density is the statistical nature of the failure properties distribution in the layer. The failure properties (critical combination of transverse and shear stress) are not the same in all locations along the transverse direction of the layer: there are some weaker positions where the first cracks occur, and more strong positions requiring a larger applied load. For example, it can be a Weibull distribution for transverse tensile strength (Peters, 1984; Varna, 2006; Huang et al., 2011). Certainly, the position of the weakest element (or any element with given failure properties) is random but can depend on the manufacturing process. Only a few cracks are created at relatively low load because there are just a few weak locations. Then the cracking rate increases because load levels are reached where the material has the highest probability density of failure, and many positions have similar failure properties. After that, the cracking rate slows down because there are only a few positions with high strength (the probability density curve approaches zero).

Another phenomenon for slowing down the cracking rate in the high-crack density region is related to the stress state between cracks. When the normalized crack density is very high, there is not enough distance for stress recovery and the maximum values of the in-plane stresses between two cracks is lower than the far-field value corresponding to the applied load. Therefore, creation of a new crack requires applying a higher load to the laminate, and as cracks come closer, the additional load required to produce a new crack is increasing. This also explains why a saturation crack density is approached asymptotically.

Owing to the progression of microcracking, the macroscopic thermoelastic properties of the laminate are degraded with increasing crack density. The stiffness degradation mechanism can be illustrated considering laminate in a coordinate system where the x-direction coincides with the transverse direction of the damaged layer (this layer becomes the 90 degrees layer). Applying to this laminate macroscopic average tensile stress σ_x^{LAM} and comparing its average axial strain, $\varepsilon_x^{\mathrm{LAM}}$ before cracking and in the presence of cracks in the 90 degrees layer, we can compare the axial modulus of the laminate E_x^{LAM}. In an undamaged laminate, the stress in any layer does not depend on the coordinate (in other words, the stress distribution is uniform) and the classical laminate theory (CLT), which is based on an iso-strain assumption, can be used to calculate the strain $\varepsilon_{x0}^{\mathrm{LAM}}$ and stresses in layers, for example $\sigma_{x0}^{90^\circ}$ (lower index 0 is used to indicate the zero-damage case). If the 90 degrees layer has a crack, then $\sigma_x^{90^\circ}$ at the crack face is zero. Because of the stress transfer over the layer interface, this stress increases with the distance from the crack and somewhere far from the crack it could reach $\sigma_{x0}^{90^\circ}$. Thus,

$$\sigma_x^{90^\circ}(x) \le \sigma_{x0}^{90^\circ} \tag{13.3}$$

at any point in the damaged 90 degrees layer. Because of axial force balance, the axial force has to be the same in any cross-section of the laminate. This means that as seen from Eq. (13.3), in any cross-section, stress in the undamaged layers is, on average, larger than before damage. The larger axial stress results in larger local axial strains in these layers, leading to larger macroscopic deformation of the layer. Obviously, the result is larger laminate strain, $\varepsilon_x^{\mathrm{LAM}} \ge \varepsilon_{x0}^{\mathrm{LAM}}$, meaning that the damaged laminate axial modulus is lower, $E_x^{\mathrm{LAM}} \le E_{x0}^{\mathrm{LAM}}$. Quantitative estimation of the change requires knowledge of the stress distribution between cracks.

Realizing that not the local stress values but the average values of in-plane stresses in damaged layers govern the stiffness degradation, and that details of the stress distribution are less important, simplifies the quantification of stiffness reduction. Proof of that will be given in the following sections. The average stress between two cracks is always lower than $\sigma_{x0}^{90^\circ}$, as discussed above.

Using the divergence theorem (Allen and Yoon, 1998), one can show that the stress applied to the laminate is equal to the volume average stress in the laminate, which in turn has a rule of mixtures (RoM) relationship to the average stresses in layers. For the same applied force to undamaged and damaged laminates, from the RoM, it follows that the average stress in undamaged layers of the damaged laminate is high with the same consequences as described in the previous paragraph.

The simplest approach to account for reduced average stress in a damaged layer in a model is by replacing the cracked layer with an "effective layer," which has "effective"—read "reduced"—thermoelastic properties. How much the properties have to be reduced is not obvious. Physically, this approach is not correct; thermoelastic constants of the material in the damaged layer have not changed. Nevertheless, the reduction of elastic constants is a simple way to account for the effect of reduced average stress in the layer, still keeping the concept of iso-strain, which in the non-bending case is the basic assumption of CLT. An extreme case of this approach is the well-known ply-discount model, commonly used together with CLT. The common assumption in this approach that transverse and shear properties of a ply with cracks are zero is very conservative and does not reflect the real situation where the number of cracks is increasing in a stable manner during the service life. The ply-discount assumption corresponds to the case of an infinite number of cracks, or to the case with a totally delaminated damaged layer when the in-plane stresses between cracks approach zero. Therefore, the requirement that laminate stiffness with increasing crack density approaches the ply-discount model prediction must be satisfied by all models based on stress distribution.

More accurate value of the average stress change in the cracked layer for predicting laminate stiffness degradation can be obtained using analytical or numerical methods. In this chapter, we will use an alternative method to account for the average stress change. We know that under in-plane tension cracks open, under shear stress their surfaces slide, and under compression they are closed. Crack face opening and sliding displacements are characterized by average values called crack opening displacement (COD) and crack sliding displacement (CSD), respectively. The values of COD and CSD are proportional to the far-field stress in the layer and also to the ply thickness (size of the tunnel). Therefore, COD and CSD are normalized with respect to these quantities. These two rather robust parameters also depend on the normalized crack spacing (crack density).

It will be shown that the change of in-plane average stresses in the damaged layer due to cracking is proportional to normalized COD and CSD. It is easy to visualize this relationship imagining that we apply traction to the corresponding points on both crack faces to keep the crack closed, not allowing the crack faces to open or to slide (COD = CSD = 0). In this imaginary case, stresses between cracks would be the same as in the undamaged laminate and the laminate thermoelastic properties would not change (a closed crack with zero opening and sliding is not a crack from a stress-distribution point of view). In the tensile case, as soon as we remove the traction, points on crack faces will separate (by opening and/or sliding) and the stresses between cracks will be reduced. The larger the COD and CSD, the larger is the average stress reduction. The most extreme case is a fully delaminated unit between two cracks. Then the in-plane stresses in the cracked layer are zero. This corresponds to the maximum possible COD and CSD, which can be estimated easily by knowing that the 90 degrees-layer material is not deformed anymore and the whole load is carried by the rest of the layers. Thermal stresses must be included in this calculation.

As a consequence, the damaged laminate stiffness can be expressed in terms of change of average stress in damaged layers or, if it is more convenient, in terms of the normalized CODs and CSDs. The former method is more suitable when analytical stress distributions are used. The latter is preferable when crack face displacements COD and CSD have been calculated—for example, using the finite-element method (FEM).

In the GLOB-LOC (global–local relationships) model (Lundmark and Varna, 2005, 2006), which will be briefly repeated in this chapter, the damaged laminate stiffness was expressed in terms of density of cracks in all layers and two parameters: normalized COD and CSD.

The approach, called micromechanics modeling (see the review, for example, by Nairn et al., 1994), is based on stress perturbation analysis due to the presence of cracks. Most of the models are focused on approximate analytical determination of the local stress state in the repeating element between two cracks. The simplest calculation schemes used are based on shear lag assumptions or on variational principles (Nairn et al., 1994; Smith and Wood, 1990; Hashin, 1985; Varna and Berglund, 1991, 1994; Zhang et al., 1992). Most of the analytical solutions are applicable to the cross-ply type of laminates with cracks in 90 layers only. Increasing the accuracy requires models with increasing complexity and the calculation routines require sophisticated codes (Zhang et al., 1992; McCartney et al., 2000). To simplify determination of the effective properties of the damaged layer, an "equivalent constraint model" was introduced by Zhang et al. (1992), replacing the rest of damaged and undamaged layers with one layer, which would have the same effect on the damaged layer as the original lay-up. Zhang et al. (1992) used this approach together with a shear lag model. Using an improved stress model, this approach could be used to account for interaction of cracks belonging to different layers.

Average stresses calculated from the available analytical solutions will be used to calculate normalized COD and CSD to be used in the prediction of thermoelastic laminate constants.

To simplify analysis, analytical models often use the assumption that cracks are uniformly distributed in the layer (equidistant). It simplifies analysis and is expected to give sufficient accuracy. However, the crack distribution in the layer may be highly nonuniform, especially in the beginning of the cracking process. At low crack density, the stress distribution between two existing cracks has a large plateau region with constant high stress and at any position there is a site of possible failure. At high crack density, there is a distinct maximum in the stress distribution, and the probability is higher that a new crack will be created in the middle between the existing cracks.

The effect of the uniformity assumption on laminate stiffness prediction was addressed by Loukil et al. (2012), McCartney and Schoeppner (2000). Loukil et al. (2012) suggested the so-called "double-periodic" approach to calculate the COD of a crack in a nonuniform case as the average from two solutions for periodic crack systems representing the two different distances to neighboring cracks. It was shown that the elastic modulus reduction is higher for uniformly distributed cracks, and in this sense, periodic crack distribution models give a lower bound.

13.2 Stiffness of damaged laminates in terms of COD and CSD

13.2.1 Global–local relationships (GLOB-LOC model)

In this chapter, we analyze in-plane thermoelastic constants of symmetric laminates only. Part of an N-layer laminate is shown in Fig. 13.1. The kth layer of the laminate has thickness t_k, and the fiber orientation angle in this layer is θ_k. Direction 1 is the local fiber direction and direction 2 is the in-plane transverse direction. The total thickness of the laminate is $h = \sum t_k$. All layers of this laminate may contain intralaminar cracks. "Vector" (column) and matrix objects are denoted in the following by { } and [ ], respectively. A bar above the matrix and vector entities in the following text indicates layer characteristics in the global coordinate system x, y, z. According to CLT, the macroscopic elastic stress–strain relationship for the undamaged laminate, expressed through stiffness matrix $[Q]_0^{\mathrm{LAM}}$ and thermal expansion coefficient vector $\{\alpha\}_0^{\mathrm{LAM}}$, is:

$$\begin{Bmatrix} \sigma_{x0}^{\mathrm{LAM}} \\ \sigma_{y0}^{\mathrm{LAM}} \\ \sigma_{xy0}^{\mathrm{LAM}} \end{Bmatrix} = [Q]_0^{\mathrm{LAM}} \left(\begin{Bmatrix} \varepsilon_{x0}^{\mathrm{LAM}} \\ \varepsilon_{y0}^{\mathrm{LAM}} \\ \gamma_{xy0}^{\mathrm{LAM}} \end{Bmatrix} - \begin{Bmatrix} \alpha_{x0}^{\mathrm{LAM}} \\ \alpha_{y0}^{\mathrm{LAM}} \\ \alpha_{xy0}^{\mathrm{LAM}} \end{Bmatrix} \Delta T \right) \tag{13.4}$$

In a shortened notation, this constitutive law can be written as:

$$\{\sigma\}_0^{\mathrm{LAM}} = [Q]_0^{\mathrm{LAM}} \left(\{\varepsilon\}_0^{\mathrm{LAM}} - \{\alpha\}_0^{\mathrm{LAM}} \Delta T \right) \tag{13.5}$$

The undamaged laminate stiffness is calculated as:

$$[Q]_0^{\mathrm{LAM}} = \frac{1}{h} \sum_{k=1}^{N} [\overline{Q}]_k t_k \tag{13.6}$$

Thermal expansion coefficients $\{\alpha\}_0^{\mathrm{LAM}}$ of the undamaged laminate are obtained, solving the thermal problem for the laminate:

$$\frac{1}{h} \{N\}^{\mathrm{th}} = [Q]_0^{\mathrm{LAM}} \{\alpha\}_0^{\mathrm{LAM}} \tag{13.7}$$

where

$$\{N\}^{\text{th}} = \Delta T \sum_{k=1}^{N} \left[\overline{Q}\right]_k \{\overline{\alpha}\}_k t_k \tag{13.8}$$

and

$$\{\overline{\alpha}\}_k = [T]_k^T \begin{Bmatrix} \alpha_1 \\ \alpha_2 \\ 0 \end{Bmatrix} \tag{13.9}$$

In Eq. (13.9), $[T]_k^T$ is the transposed stress transformation matrix between local and global coordinates for the kth layer with fiber orientation angle θ_k. Using notation $=$ $cos\ \theta_k\ n\ = sin\ \theta_k$, this matrix is defined as:

$$[\mathrm{T}]_k = \begin{bmatrix} m^2 & n^2 & +2mn \\ n^2 & m^2 & -2mn \\ -mn & +mn & m^2 - n^2 \end{bmatrix} \tag{13.10}$$

The undamaged laminate compliance matrix is:

$$[S]_0^{\text{LAM}} = \left([Q]_0^{\text{LAM}}\right)^{-1} \tag{13.11}$$

The damaged laminate has an unknown stiffness matrix $[Q]^{\text{LAM}}$ and unknown vector of thermal expansion coefficients $\{\alpha\}^{\text{LAM}}$. Hence, the stress–strain relationship for the damaged laminate can be written as:

$$\{\sigma\}_0^{\text{LAM}} = [Q]^{\text{LAM}}\left(\{\varepsilon\}^{\text{LAM}} - \{\varepsilon\}^{\text{LAM}}\Delta T\right) \tag{13.12}$$

The objective of this section is to determine $[Q]^{\text{LAM}}$ and $\{\alpha\}^{\text{LAM}}$. Obviously, they depend on the UD composite and the undamaged laminate elastic constants and on crack densities in layers. Exact expressions linking macroscopic thermoelastic constants of damaged laminate with the local stress state in terms of COD and CSD were obtained in Lundmark and Varna (2005, 2006). These derivations for elastic constants of general symmetric laminate with intralaminar cracks in plies are repeated for completeness in Appendix 1. Thermal effects are analyzed in detail in Lundmark and Varna (2005). The expressions are as follows:

$$[Q]^{\mathrm{LAM}} = \left([I] + \sum_{k=1}^{N} \rho_{kn} \frac{t_k}{h} [K]_k [S]_0^{\mathrm{LAM}}\right)^{-1} [Q]_0^{\mathrm{LAM}} \tag{13.13}$$

$$[S]^{\mathrm{LAM}} = [S]_0^{\mathrm{LAM}} \left([I] + \sum_{k=1}^{N} \rho_{kn} \frac{t_k}{h} [K]_k [S]_0^{\mathrm{LAM}}\right) \tag{13.14}$$

$$\begin{aligned} \{\alpha\}^{\mathrm{LAM}} &= \left([I] + \sum_{k=1}^{N} \rho_{kn} \frac{t_k}{h} [S]_0^{\mathrm{LAM}} [K]_k \right) \{\alpha\}_0^{\mathrm{LAM}} \\ &\quad - \sum_{k=1}^{N} \rho_{kn} \frac{t_k}{h} [S]_0^{\mathrm{LAM}} [K]_k \left\{\overline{\alpha}\right\}_k \end{aligned} \tag{13.15}$$

The $[K]_k$ matrix function for a layer with index k is defined as:

$$[K]_k = \frac{2}{E_2} \left[\overline{Q}_k [T]_k^T\right] \begin{bmatrix} 0 & 0 & 0 \\ 0 & u_{2an}^{(k)} & 0 \\ 0 & 0 & \dfrac{E_2}{G_{12}} u_{1an}^{(k)} \end{bmatrix} [T]_k \left[\overline{Q}\right]_k \tag{13.16}$$

In Eqs. (13.13)–(13.15), $[I]$ is the identity matrix. As requested by thermodynamics considerations, the stiffness matrix of the damaged laminate $[Q]^{\mathrm{LAM}}$ is symmetric.

13.2.2 Thermoelastic constants of balanced laminates with cracks in 90 degrees layers

Analytical calculation of $[K]_k$ for cracked 90 degrees layers leads to a simple expression. Using the results in Eqs. (13.14) and (13.15) for a balanced laminate with central 90 degrees layer, the following expressions for laminate thermoelastic constants can be obtained:

$$\frac{E_x^{\mathrm{LAM}}}{E_{x0}^{\mathrm{LAM}}} = \frac{1}{1 + 2\rho_{90n} \frac{t_{90}}{h} u_{2an}^{90} c_2}, \quad \frac{E_y^{\mathrm{LAM}}}{E_{y0}^{\mathrm{LAM}}} = \frac{1}{1 + 2\rho_{90n} \frac{t_{90}}{h} u_{2an}^{90} c_4} \tag{13.17}$$

$$\frac{\nu_{xy}^{\mathrm{LAM}}}{\nu_{xy0}^{\mathrm{LAM}}} = \frac{1 + 2\rho_{90n} \frac{t_{90}}{h} u_{2an}^{90} c_1 \left(1 - \frac{\nu_{12}}{\nu_{yx0}^{\mathrm{LAM}}}\right)}{1 + 2\rho_{90n} \frac{t_{90}}{h} u_{2an}^{90} c_2}, \quad \frac{G_{xy}^{\mathrm{LAM}}}{G_{xy0}^{\mathrm{LAM}}} = \frac{1}{1 + 2\rho_{90n} \frac{t_{90}}{h} u_{1an}^{90} \frac{G_{12}}{G_{xy0}^{\mathrm{LAM}}}} \tag{13.18}$$

$$\frac{\alpha_x^{\text{LAM}}}{\alpha_{x0}^{\text{LAM}}} = 1 - 2\rho_{90n}\frac{t_{90}}{h}u_{2an}^{90}\frac{c_1}{\alpha_{x0}^{\text{LAM}}}\left(\alpha_2 - \alpha_{x0}^{\text{LAM}} - \nu_{12}\left(\alpha_{y0}^{\text{LAM}} - \alpha_1\right)\right) \tag{13.19}$$

$$\frac{\alpha_y^{\text{LAM}}}{\alpha_{y0}^{\text{LAM}}} = 1 - 2\rho_{90n}\frac{t_{90}}{h}u_{2an}^{90}\frac{c_3}{\alpha_{y0}^{\text{LAM}}}\left(\alpha_2 - \alpha_{x0}^{\text{LAM}} - \nu_{12}\left(\alpha_{y0}^{\text{LAM}} - \alpha_1\right)\right) \tag{13.20}$$

$$c_1 = \frac{E_2}{E_{x0}^{\text{LAM}}}\frac{1 - \nu_{12}\nu_{xy0}^{\text{LAM}}}{\left(1 - \nu_{12}\nu_{21}\right)^2}, \quad c_2 = c_1\left(1 - \nu_{12}\nu_{xy0}^{\text{LAM}}\right) \tag{13.21}$$

$$c_3 = \frac{E_2}{E_{y0}^{\text{LAM}}}\frac{\nu_{12} - \nu_{yx0}^{\text{LAM}}}{\left(1 - \nu_{12}\nu_{21}\right)^2}, \quad c_4 = c_3\left(\nu_{12} - \nu_{yx0}^{\text{LAM}}\right) \tag{13.22}$$

Index 90 is used for thickness, crack density, COD, and CSD in the 90 degrees layer. The quantities with upper index "LAM" are laminate constants; quantities with additional lower index 0 are undamaged laminate constants. If we simplify, assuming that the Poisson effects are neglected, $c_3 = c_4 = 0$ and E_y^{LAM} and α_y^{LAM} do not change because of damage in the 90 degrees ply. Shear modulus is not related to COD and depends on CSD only.

Thermoelastic property degradation of symmetric and balanced laminates with arbitrary location of damaged 90 degrees layers can be analyzed using Eqs. (13.17)–(13.22); however, even in a damaged state, the laminate has to remain symmetric: the same crack density has to be in both 90 degrees layers located symmetrically with respect to the midplane. Each of these two layers equally contributes to stiffness, and therefore terms with crack density in the layer, ρ_{90n}, must be multiplied by 2. If several couples of damaged 90 degrees layers of different thickness are present, a sum over all couple contributions has to be used. It must be noted that, as described in Appendix 2 to this chapter, the normalized COD and CSD of the 90 degrees layer depend on ply thickness and on the position of the damaged layer.

The class of laminates covered by Eqs. (13.17)–(13.20) is broader than just cross-ply laminates or laminates with 90 degrees layers. For example, any quasiisotropic laminate with an arbitrary cracked layer can be rotated to have the damaged layer as a 90 degrees layer. The only limitation is that the laminate after rotation is balanced with zero coupling terms in $[S]_0^{\text{LAM}}$. Quasiisotropic laminate is balanced in any system of coordinates $\left(S_{16}^{\text{LAM}} = S_{26}^{\text{LAM}} = 0\right)$.

13.2.3 Determination of normalized COD and CSD

From Eqs. (13.13)–(13.16), it follows that the thermoelastic constants of the damaged laminate depend on the elastic ply properties, details of laminate lay-up, and the dimensionless density of cracks in each layer. The influence of each crack is represented by the 3 × 3 displacement matrix in Eq. (13.16), which contains normalized COD (denoted u_{2an}^k) and normalized CSD (denoted u_{1an}^k) of the crack surface in the *k*th layer. In this chapter, we assume that all cracks in the same layer are equal: they have the same crack face displacements, and the crack distribution is uniform.

The average COD and CSD are defined as:

$$u_{2a}^{(k)} = \frac{1}{t_k}\int_{\frac{-t_k}{2}}^{\frac{+t_k}{2}} u_2^{(k)}(z)dz, \quad u_{1a}^{(k)} = \frac{1}{t_k}\int_{\frac{-t_k}{2}}^{\frac{+t_k}{2}} u_1^{(k)}(z)dz \tag{13.23}$$

In Eq. (13.23), $u_2^{(k)}, u_1^{(k)}$ is the distance in the transverse and longitudinal directions, respectively, between the corresponding points on the deformed crack surface and the surface connecting both crack tips (Fig. 13.3). If the crack spacing is nonuniform, each crack face has a different $u_2^{(k)}, u_1^{(k)}$.

Normalized COD and CSD used in this chapter are defined dividing the average displacements Eq. (13.23) by the layer thickness and the relevant far-field stress component in this layer (CLT at applied $\{\sigma\}_0^{LAM}$):

$$u_{2an}^{(k)} = \frac{u_{2a}^{(k)}}{\sigma_{T0}^{(k)} t_k} E_2 \quad u_{1an}^{(k)} = \frac{u_{1a}^{(k)}}{\sigma_{LT0}^{(k)} t_k} G_{12} \tag{13.24}$$

Elastic constants E_2 and G_{12} are introduced in Eq. (13.24) to have the normalized COD and CSD dimensionless.

Numerical values of $u_{2an}^{(k)}, u_{1an}^{(k)}$ are required to calculate, using Eqs. (13.13)–(13.16), the dependence of thermoelastic constants of damaged laminates on the density of cracks in different layers. FEM parametric analysis was performed by Lundmark and Varna (2005, 2006), Loukil et al. (2012), Joffe et al. (2001) and experimental measurements are presented in Varna et al. (1993, 1999); Loukil et al. (2014). The first studies were performed for so-called noninteractive cracks: a region

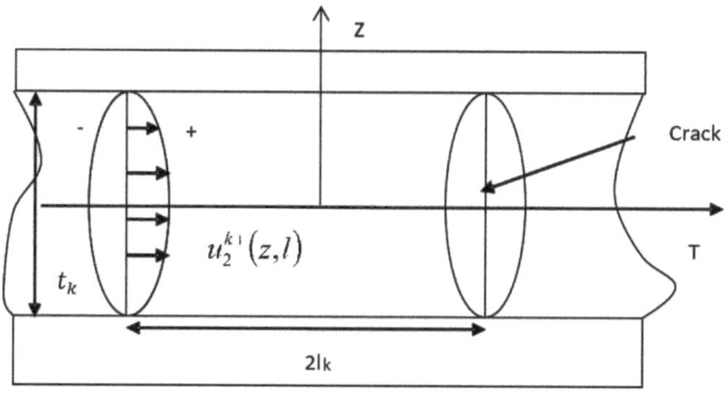

Figure 13.3 Fragment of the laminate RVE showing crack opening displacement of two equal cracks in a layer in its local coordinates.

where the distance between cracks in a layer is much larger than the crack (tunnel) size, which is equal to the ply thickness. In this region, the stress perturbations caused by two neighboring cracks do not overlap and the normalized COD and CSD do not depend on the value of the crack density. An upper index of 0 indicates normalized COD and CSD values in this region, $u_{1an}^{0(k)}$, $u_{2an}^{0(k)}$.

Parametric analysis of $u_{1an}^{0(k)}$ and $u_{2an}^{0(k)}$ was performed and the adjacent layer stiffness and thickness ratios were identified as the most significant material and geometrical parameters. Based on this analysis, simple fitting expressions were obtained to calculate $u_{1an}^{0(k)}$ and $u_{2an}^{0(k)}$ for cracks in different lay-ups. These expressions (see Appendix 2) are considered to be sufficiently general and accurate to be used for cracks in any laminate. Hence, there is no need to use FEM in any stiffness predictions presented in this chapter. Generalization of these fitting functions to the high crack density region (interactive cracks) is also given in Appendix 2.

13.3 Average stress state between cracks and average COD and CSD

An alternative approach to FEM for the normalized average COD and CSD determination is by expressing them through the average value of the in-plane stress perturbation caused by cracks. We will derive these relationships in this section. Employing these results, one can use any analytical model (shear lag, variational models, etc.) to find the in-plane stress distribution between two cracks and the requested average values.

We consider a laminate with a cracked kth layer shown in Fig. 13.4 in a coordinate system (x_1, x_2, z) where the x_1-axis is in the direction of the fibers in the kth layer and x_2 coincides with its transverse direction. The coordinate transverse to the laminate midplane is z. The origin is in the middle between two cracks with spacing $2l_k$, and $z = 0$ is in the middle between two surfaces of the kth layer. The kth layer in these coordinates is the 0 layer. We will use indexes 1, 2, and z to denote the corresponding stress, strain, and stiffness components in the 0 layer and x_1, x_2, and $z(3)$ for the same characteristics in the rest of the layers. In this system of coordinates, the average in-plane stresses applied to the laminate are σ_1^{LAM}, σ_2^{LAM} and the shear stress is σ_{12}^{LAM}.

The laminate is symmetric, and the damage state is also symmetric with respect to the midplane. The latter condition may have two simple realizations: (1) the middle layer has cracks or (2) a couple of symmetrically located layers on both sides of the midplane have the same crack density. In fact, even more complex cases of unsymmetrical damage state than described in this section can be treated, provided that due to geometrical and structural constraints the curvatures of the damaged laminate are equal to zero and that in average, iso-strain conditions for undamaged layers are applicable.

The cracked layer thickness is t_k and the size of the considered unit in direction x_1 is w. The set of layers with indexes $i = 1, 2, \ldots, N, i \neq k$, can formally be called "sublaminate" even if these layers in the laminate are separated by the damaged layer. Since

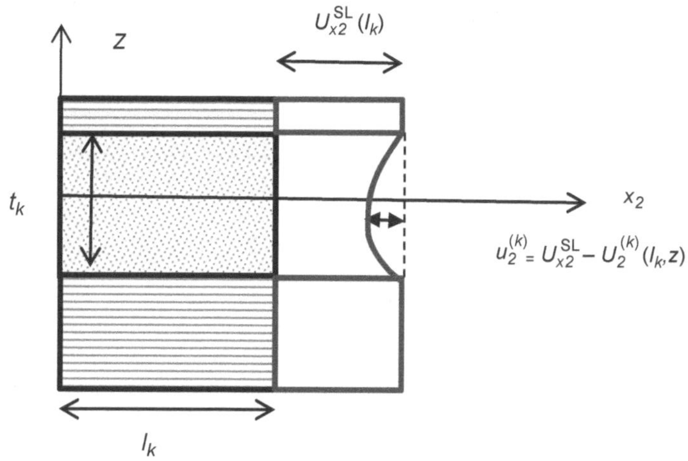

Figure 13.4 Half of the unit of the deformed laminate between two intralaminar cracks in the kth layer.

some of the layers in the sublaminate could be damaged, they are homogenized and characterized with effective stiffness (Varna, 2012). Index “SL” is used to denote average characteristics common for this group of layers.

In the following derivations, we first will consider the case when the damaged layer is the central layer. Expressions for the case of two symmetrical damaged layers will be given in Section 13.3.4.

13.3.1 Average stresses expressed via axial stress perturbation in the central damaged layer: normal loading

In the case of a cracked central layer, the sublaminate thickness is:

$$t_{SL} = h - t_k$$

The stress distributions in layers do not depend on x_1 and on the width of the laminate unit, w. First we consider case with normal loading $\sigma_1^{LAM}, \sigma_2^{LAM}$. We can assume that in the x_1 direction, generalized plane strain conditions are satisfied:

$$\varepsilon_{x_1}^{(i)} = \varepsilon_1 = const, i = 1, 2, \ldots, N \tag{13.25}$$

Displacements $U_{x_1}^{(i)}$ do not depend on x_2 and z, whereas $U_{x_2}^{(i)} = U_{x_2}^{(i)}(x_2, z), 1 = 1, 2, \ldots, N$ do. This displacement component at $x_2 = l_k$ is constant (z independent) and has the same value for all layers belonging to the sublaminate ($U_{x_2}^{SL}(l_k)$ is the notation used in Fig. 13.4). The displacement of the crack face is $U_2^{(k)}(l_k, z)$.

The crack with opening displacement $u_2^{(k)}$ causes stress perturbation in the kth layer and also in the rest of the layers. The transverse stress $\sigma_2^{(k)} = \sigma_T^{(k)}$ between cracks can be formally written in the following general form:

$$\sigma_2^{(k)} = \sigma_{20}^{(k)}(1 - f(x_2, z)) \tag{13.26}$$

Function f in Eq. (13.26) represents the unknown stress perturbation, and $\sigma_{20}^{(k)}$ is the transverse stress in the undamaged kth layer of the laminate calculated using CLT. The volume-averaged value of the stress in the layer in the half unit between two cracks is defined as:

$$\{\sigma\}^i_{(\mathrm{av})} = \frac{1}{V_i}\int_{V_i} \{\sigma\}^i dv i = 1, 2 \ldots N, V_i = t_i l_k w \tag{13.27}$$

From Eq. (13.26):

$$\sigma_{2(\mathrm{av})}^{(k)} = \sigma_{20}^{(k)}(1 - f_{(\mathrm{av})}) \tag{13.28}$$

$$f_{(\mathrm{av})} = \frac{1}{V_k}\int_{V_k} f(x_2, z) dv \tag{13.29}$$

We will show that the function $f_{(\mathrm{av})}$ also describes the summary change of the average stress components in the rest of the layers. For that purpose, we will use force balance and Eq. (13.25).

1. In any cross-section $x_2 = \mathrm{const}$, the force F_2 has to be equal to the applied force, F_{20}. We apply the same force to the undamaged laminate as to the damaged laminate. This equality formally can be integrated with respect to x_2:

$$\int_0^{l_k} F_2 dx_2 = \int_0^{l_k} F_{20} dx_2 \tag{13.30}$$

Using the definition of force as the stress integrated over the cross-section area, Eq. (13.30) can be rewritten as:

$$\int_{V_k} \sigma_2^{(k)} dv + \sum_{i \neq k} \int_{V_i} \sigma_{x_2}^{(i)} dv = \int_{V_k} \sigma_{20}^{(k)} dv + \sum_{i \neq k} \int_{V_i} \sigma_{x_2 0}^{(i)} dv \tag{13.31}$$

This expression can be rewritten as:

$$V_k\sigma_{2(av)}^{(k)} + \sum_{i \neq k} V_i\sigma_{x_2(av)}^{(i)} = V_k\sigma_{20}^{(k)} + \sum_{i \neq k} V_i\sigma_{x_20}^{(i)} \tag{13.32}$$

Next we multiply and divide by V_{SL} the terms in Eq. (13.32) containing the summation symbol, where $V_{SL} = t_{SL}l_kw$ is the volume of the sublaminate. Since the volume fraction in layered material is equal to the thickness fraction of the layers, Eq. (13.32) can be written as:

$$V_k\sigma_{2(av)}^{(k)} + V_{SL}\sum_{i \neq k} \frac{t_i}{t_{SL}}\sigma_{x_2(av)}^{(i)} = V_k\sigma_{20}^{(k)} + V_{SL}\sum_{i \neq k} \frac{t_i}{t_{SL}}\sigma_{x_20}^{(i)} \tag{13.33}$$

The average stress in the sublaminate can be written in the form of RoM of the average stresses in layers:

$$\sigma_{x_1(av)}^{SL} = \sum_{i \neq k} \frac{t_i}{t_{SL}}\sigma_{x_1(av)}^{(i)}, \ \sigma_{x_10}^{SL} = \sum_{i \neq k} \frac{t_i}{t_{SL}}\sigma_{x_10}^{(i)} \tag{13.34}$$

$$\sigma_{x_2(av)}^{SL} = \sum_{i \neq k} \frac{t_i}{t_{SL}}\sigma_{x_2(av)}^{(i)}, \ \sigma_{x_20}^{SL} = \sum_{i \neq k} \frac{t_i}{t_{SL}}\sigma_{x_20}^{(i)} \tag{13.35}$$

The average stress in the sublaminate reflects the summary effect of the crack on the average stress in all undamaged layers. Using Eq. (13.35), Eq. (13.33) can be written in terms of average stresses in the broken layer and in the sublaminate:

$$V_k\sigma_{2(av)}^{(k)} + V_{SL}\sigma_{x_2(av)}^{SL} = V_k\sigma_{20}^{(k)} + V_{SL}\sigma_{x_20}^{SL} \tag{13.36}$$

Substituting Eq. (13.28) in Eq. (13.36), we obtain:

$$\sigma_{x_2(av)}^{SL} = \sigma_{x_20}^{SL} + \sigma_{20}^{(k)} f_{(av)} \frac{t_k}{t_{SL}} \tag{13.37}$$

2. The force F_1 in any cross-section $x_1 = const$ is equal to the force applied to the undamaged laminate, F_{10}. Similarly, as analyzing force in the x_2 direction, we can integrate this equality with respect to x_1:

$$\int_{\frac{-w}{2}}^{\frac{+w}{2}} F_1 dx_1 = \int_{\frac{-w}{2}}^{\frac{+w}{2}} F_{10} dx_1 \tag{13.38}$$

From here, expressing the force through stresses in layers:

$$\int_{V_k} \sigma_1^{(k)} dv + \sum_{i \neq k} \int_{V_i} \sigma_{x_1}^{(i)} dv = \int_{V_k} \sigma_{10}^{(k)} dv + \sum_{i \neq k} \int_{V_i} \sigma_{x_1 0}^{(i)} dv \tag{13.39}$$

Using Eq. (13.38) in Eq. (13.37) leads to:

$$\sigma_{x_1(\text{av})}^{\text{SL}} = \sigma_{x_1 0}^{\text{SL}} + \sigma_{10}^{(k)} \frac{t_k}{t_{\text{SL}}} - \sigma_{1(\text{av})}^{(k)} \frac{t_k}{t_{\text{SL}}} \tag{13.40}$$

3. Averaged stress–strain relationships for the sublaminate. First we notice that the force component in the z-direction is zero and out-of-plane shear stresses are zero on the unit's boundaries, $x_2 = 0$; l_k and at $x_1 = \frac{\pm w}{2}$. Therefore, the integral of σ_z over $l_k \times w$ is zero at any value of z and:

$$\sigma_{z(\text{av})}^{(i)} = 0, i = 1, 2, \ldots, k, \ldots N \tag{13.41}$$

Hence, this stress component will not enter the averaged stress–strain relationships. The averaged stress–strain relationship for a layer belonging to the sublaminate can be written as:

$$\{\sigma\}_{(\text{av})}^{(i)} = [\overline{Q}]_i \{\varepsilon\}_{(\text{av})}^{(i)} \tag{13.42}$$

The effective stiffness should be used if a layer belonging to the sublaminate is damaged. Multiplying Eq. (13.42) by the ply thickness t_i and performing summation, we obtain:

$$\sum_{i \neq k} \{\sigma\}_{(\text{av})}^{(i)} t_i = \sum_{i \neq k} [\overline{Q}]_i t_i \{\varepsilon\}_{(\text{av})}^{(i)} \tag{13.43}$$

According to Eq. (13.25), $\varepsilon_{x_1(\text{av})}^{(i)} = \varepsilon_1$ is the same for all layers belonging to the sublaminate. Also, the average strains in the x_2 direction, $\varepsilon_{x_2(\text{av})}^{(i)}$, and even the shear strains, are the same:

$$\{\varepsilon\}_{(\text{av})}^{(i)} = \{\varepsilon\}_{(\text{av})}^{\text{SL}} \tag{13.44}$$

The left-side term in Eq. (13.43) can be expressed through the average sublaminate stress defined by Eqs. (13.34) and (13.35), leading to:

$$\{\sigma\}_{(\text{av})}^{\text{SL}} = [Q]^{\text{SL}} \{\varepsilon\}_{(\text{av})}^{\text{SL}} \tag{13.45}$$

$$[Q]^{\mathrm{SL}} = \sum_{i \neq k} \frac{t_i}{t_{\mathrm{SL}}} [\overline{Q}]_i \tag{13.46}$$

Eq. (13.45) can be inverted using the compliance matrix $[S]^S$:

$$\{\varepsilon\}^{\mathrm{SL}}_{(\mathrm{av})} = [S]^{\mathrm{SL}} \{\sigma\}^{\mathrm{SL}}_{(\mathrm{av})} \tag{13.47}$$

Assuming that the sublaminate is orthotropic, the first relationship of Eq. (13.47) is:

$$\varepsilon^{\mathrm{SL}}_{x_1(\mathrm{av})} = \frac{1}{E^{\mathrm{SL}}_{x_1}} \sigma^{\mathrm{SL}}_{x_1(\mathrm{av})} - \frac{\nu^{\mathrm{SL}}_{x_1x_2}}{E^{\mathrm{SL}}_{x_1}} \sigma^{\mathrm{SL}}_{x_2(\mathrm{av})} \tag{13.48}$$

The average value of the same strain component in the kth layer is:

$$\varepsilon^{(k)}_{1(\mathrm{av})} = \frac{1}{E_1} \sigma^{(k)}_{1(\mathrm{av})} - \frac{\nu_{12}}{E_1} \sigma^{(k)}_{2(\mathrm{av})} \tag{13.49}$$

Since

$$\varepsilon^{\mathrm{SL}}_{x_1(\mathrm{av})} = \varepsilon^{(k)}_{1(\mathrm{av})} = \varepsilon_1 \tag{13.50}$$

we obtain from Eqs. (13.43) and (13.48):

$$\sigma^{(k)}_{1(\mathrm{av})} = \nu_{12} \sigma^{(k)}_{2(\mathrm{av})} + \frac{E_1}{E^{\mathrm{SL}}_{x_1}} \sigma^{\mathrm{SL}}_{x_1(\mathrm{av})} - \nu^{\mathrm{SL}}_{x_1x_2} \frac{E_1}{E^{\mathrm{SL}}_{x_1}} \sigma^{\mathrm{SL}}_{x_2(\mathrm{av})} \tag{13.51}$$

Next, using Eqs. (13.28) and (13.37), we express in Eq. (13.51) $\sigma^{(k)}_{2(\mathrm{av})}$ and $\sigma^{SL}_{x_2(\mathrm{av})}$ via the stress perturbation function, $f_{(\mathrm{av})}$:

$$\sigma^{(k)}_{1(\mathrm{av})} = \nu_{12} \sigma^{(k)}_{20} - \nu_{12} \sigma^{(k)}_{20} f_{(\mathrm{av})} + \frac{E_1}{E^{\mathrm{SL}}_{x_1}} \sigma^{\mathrm{SL}}_{x_1(\mathrm{av})} - \nu^{\mathrm{SL}}_{x_1x_2} \frac{E_1}{E^{\mathrm{SL}}_{x_1}} \sigma^{\mathrm{SL}}_{x_20} - \nu^{\mathrm{SL}}_{x_1x_2} \frac{t_k}{t_{\mathrm{SL}}} \frac{E_1}{E^{\mathrm{SL}}_{x_1}} \sigma^{(k)}_{20} \tag{13.52}$$

Substituting Eq. (13.52) in Eq. (13.40), we obtain:

$$\sigma^{\mathrm{SL}}_{x_1(\mathrm{av})} = \sigma^{\mathrm{SL}}_{x_10} + b\sigma^{(k)}_{20} f_{(\mathrm{av})} \tag{13.53}$$

where

$$b = \frac{\left(\nu_{12} + \nu_{x_1x_2}^{\mathrm{SL}} \frac{E_1}{E_{x_1}^{\mathrm{SL}}} \frac{t_k}{t_{\mathrm{SL}}}\right) \frac{t_k}{t_{\mathrm{SL}}}}{1 + \frac{E_1}{E_{x_1}^{\mathrm{SL}}} \frac{t_k}{t_{\mathrm{SL}}}} \tag{13.54}$$

Substituting Eq. (13.53) in Eq. (13.40), we find expression for $\sigma_{1(\mathrm{av})}^{(k)}$

$$\sigma_{1(\mathrm{av})}^{(k)} = \sigma_{10}^{(k)} + c\sigma_{20}^{(k)} f_{(\mathrm{av})} \tag{13.55}$$

where

$$c = \frac{-\nu_{12} + \nu_{x_1x_2}^{\mathrm{SL}} \frac{E_1}{E_{x_1}^{\mathrm{SL}}} \frac{t_k}{t_{\mathrm{SL}}}}{1 + \frac{E_1}{E_{x_1}^{\mathrm{SL}}} \frac{t_k}{t_{\mathrm{SL}}}} \tag{13.56}$$

With this we have finished the expression of average stresses $\sigma_{1(\mathrm{av})}^{(k)}$, $\sigma_{2(\mathrm{av})}^{(k)}$, $\sigma_{x_1(\mathrm{av})}^{\mathrm{SL}}$, and $\sigma_{x_2(\mathrm{av})}^{\mathrm{SL}}$ through $f_{(\mathrm{av})}$.

13.3.2 Expression for COD for cracks in the central layer

Now we can link the average COD to average stresses found in Section 13.3.1.

We can calculate the average COD over the crack surface shown in Fig. 13.4 as:

$$u_{2a}^{(k)} = \frac{1}{wt_k} \int_{\frac{-w}{2}}^{\frac{+w}{2}} \int_{\frac{-t_k}{2}}^{\frac{t_k}{2}} \left[U_{x_2}^{\mathrm{SL}}(l_k) - U_2^{(k)}(l_k, z)\right] dx_1 dz \tag{13.57}$$

In fact, both functions under the sign of the integral do not depend on x_1 and the integration over x_1 is formal. Still, this integral form is necessary to reduce this expression to volume integrals and volume-averaged functions. Since $U_{x_2}^{SL}(l_k)$ is coordinate independent, the first term can be written as any integral divided by the integration area, for example, as:

$$u_{2a}^{(k)} = \frac{1}{wt_i} \int_{\frac{-w}{2}}^{\frac{+w}{2}} \int_{z_{i-1}}^{z_i} U_{x_2}^{\mathrm{SL}}(l_k) dx_1 dz - \frac{2}{wt_k} \int_{\frac{-w}{2}}^{\frac{+w}{2}} \int_0^{\frac{t_k}{2}} U_2^{(k)}(l_k, z) dx_1 dz \tag{13.58}$$

In Eq. (13.58), z_i and z_{i-1} are surface coordinates of the i-th layer. Next, we replace:

$$U_{x_2}^{\mathrm{SL}}(l_k) = \int_0^{l_k} \varepsilon_{x_2}^{(i)}(x_2, z)dx_2 \quad U_2^{(k)}(l_k, z) = \int_0^{l_k} \varepsilon_2^{(k)}(x_2, z)dx_2 \tag{13.59}$$

Substituting Eq. (13.59) in Eq. (13.58) and using Eq. (13.44), we obtain:

$$u_{2a}^{(k)} = l_k\left(\varepsilon_{x_2(\mathrm{av})}^{\mathrm{SL}} - \varepsilon_{2(\mathrm{av})}^{(k)}\right) = l_k\left(\frac{\sigma_{x_2(\mathrm{av})}^{\mathrm{SL}}}{E_{x_2}^{SL}} - \frac{\nu_{x_1x_2}^{\mathrm{SL}}}{E_{x_1}^{SL}}\sigma_{x_1(\mathrm{av})}^{\mathrm{SL}} - \frac{\sigma_{2(\mathrm{av})}^{(k)}}{E_2} + \frac{\nu_{12}}{E_1}\sigma_{1(\mathrm{av})}^{(k)}\right) \tag{13.60}$$

Hooke's law assuming an orthotropic sublaminate was used in Eq. (13.60). Substituting expressions for average stresses (Eqs. 13.28, 13.37, 13.53 and 13.55), in Eq. (13.60), after simple but tedious rearrangements we obtain:

$$u_{2a}^{(k)} = f_{(\mathrm{av})}\sigma_{20}^{(k)} l_k\left[\frac{Q_{22}t_k + Q_{x_2x_2}^{\mathrm{SL}}t_{\mathrm{SL}}}{t_{\mathrm{SL}}Q_{x_2x_2}^{\mathrm{SL}}Q_{22}} + \left(\nu_{12} - \nu_{x_1x_2}^{\mathrm{SL}}\right)^2\frac{t_k}{E_{x_1}^{\mathrm{SL}}t_{\mathrm{SL}} + E_1t_k}\right] \tag{13.61}$$

Normalizing as in Eq. (13.24):

$$u_{2an}^{(k)} = k_f f_{(\mathrm{av})} \tag{13.62}$$

$$k_f = \frac{E_2}{2\rho_{kn}}\left[\frac{Q_{22}t_k + Q_{x_2x_2}^{\mathrm{SL}}t_{\mathrm{SL}}}{t_{\mathrm{SL}}Q_{x_2x_2}^{\mathrm{SL}}Q_{22}} + \left(\nu_{12} - \nu_{x_1x_2}^{\mathrm{SL}}\right)^2\frac{t_k}{E_{x_1}^{SL}t_{\mathrm{SL}} + E_1t_k}\right] \tag{13.63}$$

13.3.3 CSD expression for cracks in the central layer

In this section, the macroscopic shear stress $\sigma_{12}^{\mathrm{LAM}}$ is applied to the laminate resulting in shear stress $\sigma_{12}^{(k)}$ in the damaged layer and shear stresses $\sigma_{x_1x_2}^{(i)}, i \neq k$ in layers belonging to the sublaminate. Shear stresses in layers of an undamaged laminate (CLT) at the same applied load have additional index 0; $\sigma_{120}^{(k)}$ and $\sigma_{x_1x_20}^{(i)}$.

Derivation of the relationship between shear stress perturbation and the average CSD is similar to the COD case. The in-plane shear stress between two cracks in the kth layer in the coordinate system defined in Fig. 13.4 can be written as:

$$\sigma_{12}^{(k)} = \sigma_{120}^{(k)}(1 - \varphi(x_2, z)) \tag{13.64}$$

For average values, we have:

$$\sigma_{12(\mathrm{av})}^{(k)} = \sigma_{120}^{(k)}(1 - \varphi_{(\mathrm{av})}) \tag{13.65}$$

The average values are defined according to Eq. (13.29). Integrating the tangential force equilibrium over $x_2 \in [0, l_k]$, we obtain:

$$\int_{V_k} \sigma_{12}^{(k)} dv + \sum_{i \neq k} \int_{V_i} \sigma_{x_1 x_2}^{(i)} dv = \int_{V_k} \sigma_{120}^{(k)} dv + \sum_{i \neq k} \int_{V_i} \sigma_{x_1 x_2 0}^{(i)} dv \tag{13.66}$$

This expression can be rewritten as:

$$V_k \sigma_{12(\text{av})}^{(k)} + \sum_{i \neq k} V_i \sigma_{x_1 x_2 (\text{av})}^{(i)} = V_k \sigma_{120}^{(k)} + \sum_{i \neq k} V_i \sigma_{x_1 x_2 0}^{(i)} \tag{13.67}$$

Next we multiply and divide by V_{SL} terms in Eq. (13.67) containing the summation symbol, where $V_{\text{SL}} = t_{\text{SL}} l_k w$ is the volume of the sublaminate. Since the volume fraction in a layered material is equal to the thickness fraction of layers, Eq. (13.67) can be written as:

$$V_k \sigma_{12(\text{av})}^{(k)} + V_{\text{SL}} \sum_{i \neq k} \frac{t_i}{t_{\text{SL}}} \sigma_{x_1 x_2 (\text{av})}^{(i)} = V_k \sigma_{120}^{(k)} + V_{\text{SL}} \sum_{i \neq k} \frac{t_i}{t_{\text{SL}}} \sigma_{x_1 x_2 0}^{(i)} \tag{13.68}$$

The average shear stress in a sublaminate has a RoM dependence on the average stresses in layers:

$$\sigma_{x_1 x_2 (av)}^{\text{SL}} = \sum_{i \neq k} \frac{t_i}{t_{\text{SL}}} \sigma_{x_1 x_2 (\text{av})}^{(i)}, \ \sigma_{x_1 x_2 0}^{\text{SL}} = \sum_{i \neq k} \frac{t_i}{t_{\text{SL}}} \sigma_{x_1 x_2 0}^{(i)} \tag{13.69}$$

Using Eq. (13.69), Eq. (13.68) can be written in terms of average stresses in the broken layer and in the sublaminate:

$$V_k \sigma_{12(\text{av})}^{(k)} + V_{\text{SL}} \sigma_{x_1 x_2 (\text{av})}^{\text{SL}} = V_k \sigma_{120}^{(k)} + V_{\text{SL}} \sigma_{x_1 x_2 0}^{\text{SL}} \tag{13.70}$$

Substituting Eq. (13.65) in Eq. (13.70), we obtain:

$$\sigma_{x_1 x_2 (\text{av})}^{\text{SL}} = \sigma_{x_1 x_2 0}^{\text{SL}} + \sigma_{120}^{(k)} \varphi_{(\text{av})} \frac{t_k}{t_{\text{SL}}} \tag{13.71}$$

The CSD is defined as the difference between the displacement in the fiber direction in the sublaminate at $x_2 = l_k$ (same value for all layers) and the corresponding point on the crack surface. Sliding displacements do not depend on the coordinate x_1. Since the sliding on both crack surfaces is of opposite sign, the displacement in the sublaminate $U_{x_1}^{\text{SL}}(l_k)$ at $x_2 = l_k$ cannot be z dependent. Hence, the crack face sliding displacement is:

$$u_1^{(k)}(z) = U_{x_1}^{\mathrm{SL}}(l_k) - U_1^{(k)}(l_k, z) \tag{13.72}$$

The average value of the CSD can be written as:

$$\begin{aligned} u_{1a}^{(k)} &= U_{x_1}^{\mathrm{SL}}(l_k) - \frac{1}{wt_k}\int_{\frac{-w}{2}}^{\frac{+w}{2}}\int_{\frac{-t_k}{2}}^{\frac{t_k}{2}} U_1^{(k)}(l_k, z)dx_1 dz \\ &= \frac{1}{wt_i}\int_{\frac{-w}{2}}^{\frac{+w}{2}}\int_{z_{i-1}}^{z_i} U_{x_1}^{\mathrm{SL}}(l_k)dx_1 dz - \frac{2}{wt_k}\int_{\frac{-w}{2}}^{\frac{+w}{2}}\int_{0}^{\frac{t_k}{2}} U_1^{(k)}(l_k, z)dx_1 dz \end{aligned} \tag{13.73}$$

Since

$$U_{x_1}^{\mathrm{SL}}(l_k) = U_{x_1}^{(i)}(l_k) = \int_0^{l_k} \gamma_{x_1x_2}^{(i)}(x_2, z)dx_2 \quad U_1^{(k)}(l_k, z) = \int_0^{l_k} \gamma_{12}^{(k)}(x_2, z)dx_2 \tag{13.74}$$

and

$$\gamma_{x_1x_2(\mathrm{av})}^{(i)} = \gamma_{x_1x_2(\mathrm{av})}^{\mathrm{SL}} \tag{13.75}$$

we obtain:

$$u_{1a}^{(k)} = l_k\left(\gamma_{x_1x_2(\mathrm{av})}^{\mathrm{SL}} - \gamma_{12(\mathrm{av})}^{(k)}\right) = l_k\left(\frac{\sigma_{x_1x_2(\mathrm{av})}^{\mathrm{SL}}}{G_{x_1x_2}^{\mathrm{SL}}} - \frac{\sigma_{12(\mathrm{av})}^{(k)}}{G_{12}}\right) \tag{13.76}$$

Substituting Eqs. (13.65) and (13.71) in Eq. (13.76), and normalizing according to Eq. (13.24), we obtain the relationship between the normalized CSD and the average shear stress perturbation $\varphi_{(\mathrm{av})}$:

$$u_{1an}^{(k)} = \varphi_{(\mathrm{av})}\frac{l_k}{t_k}\left(1 + \frac{G_{12}t_k}{G_{x_1x_2}^{\mathrm{SL}}t_{\mathrm{SL}}}\right) \tag{13.77}$$

Eq. (13.77) can be written as:

$$u_{1an}^{(k)} = k_\varphi \varphi_{(\mathrm{av})}, \quad k_\varphi = \frac{1}{2\rho_{kn}}\left(1 + \frac{G_{12}t_k}{G_{x_1x_2}^{\mathrm{SL}}t_{\mathrm{SL}}}\right) \tag{13.78}$$

13.3.4 COD and CSD relationship with average stresses in the case of two symmetrical damaged layers

Here, we consider the case of a symmetric laminate with two damaged layers of the same orientation and with the same crack density and stress perturbation function $f(x_2, z)$. The sublaminate thickness is:

$$t_{SL} = h - 2t_k \tag{13.79}$$

and factor 2 appears at integrals over V_k in Eq. (13.30), leading to:

$$\sigma^{SL}_{x_2(av)} = \sigma^{SL}_{x_2 0} + \sigma^{(k)}_{20} f_{(av)} \frac{2t_k}{t_{SL}} \tag{13.80}$$

Instead of Eq. (13.40), we obtain:

$$\sigma^{SL}_{x_1(av)} = \sigma^{SL}_{x_1 0} + \sigma^{(k)}_{10} \frac{2t_k}{t_{SL}} - \sigma^{(k)}_{1(av)} \frac{2t_k}{t_{SL}} \tag{13.81}$$

Eq. (13.51) is the same as in the case of a central cracked layer:

$$\sigma^{(k)}_{1(av)} = \nu_{12} \sigma^{(k)}_{2(av)} + \frac{E_1}{E^{SL}_{x_1}} \sigma^{SL}_{x_1(av)} - \nu^{SL}_{x_1x_2} \frac{E_1}{E^{SL}_{x_1}} \sigma^{SL}_{x_2(av)} \tag{13.82}$$

Expressing in Eq. (13.82) $\sigma^{(k)}_{2(av)}$ and $\sigma^{SL}_{x_2(av)}$ via stress perturbation function $f_{(av)}$ using Eqs. (13.28) and (13.80), we have two linear relationships linking $\sigma^{(k)}_{1(av)}$ and $\sigma^{SL}_{x_1(av)}$ with $f_{(av)}$. Solving this, we obtain:

$$\sigma^{SL}_{x_1(av)} = \sigma^{SL}_{x_1 0} + b \sigma^{(k)}_{20} f_{(av)} \tag{13.83}$$

where

$$b = \frac{\left(\nu_{12} + \nu^{SL}_{x_1x_2} \frac{E_1}{E^{SL}_{x_1}} 2 \frac{t_k}{t_{SL}} \right) 2 \frac{t_k}{t_{SL}}}{1 + \frac{E_1}{E^{SL}_{x_1}} 2 \frac{t_k}{t_{SL}}} \tag{13.84}$$

$$\sigma^{(k)}_{1(av)} = \sigma^{(k)}_{10} + c \sigma^{(k)}_{20} f_{(av)} \tag{13.85}$$

where

$$c = -\frac{\nu_{12} + \nu_{x_1x_2}^{\mathrm{SL}} \dfrac{E_1}{E_{x_1}^{\mathrm{SL}}} 2\dfrac{t_k}{t_{\mathrm{SL}}}}{1 + \dfrac{E_1}{E_{x_1}^{\mathrm{SL}}} 2\dfrac{t_k}{t_s}} \tag{13.86}$$

Expression for the average COD is:

$$u_{2a}^{(k)} = f_{(\mathrm{av})}\sigma_{20}^{(k)} l_k \left[\frac{Q_{22}2t_k + Q_{x_2x_2}^{\mathrm{SL}} t_{\mathrm{SL}}}{t_{\mathrm{SL}} Q_{x_2x_2}{}^{\mathrm{SL}} Q_{22}} + \left(\nu_{12} - \nu_{x_1x_2}^{\mathrm{SL}}\right)^2 \frac{2t_k}{E_{x_1}^{\mathrm{SL}} t_{\mathrm{SL}} + E_1 2t_k} \right] \tag{13.87}$$

Normalizing as in Eq. (13.24):

$$u_{2an}^{(k)} = k_f f_{(\mathrm{av})} \tag{13.88}$$

$$k_f = \frac{E_2}{2\rho_{kn}} \left[\frac{Q_{22}2t_k + Q_{x_2x_2}^{\mathrm{SL}} t_{\mathrm{SL}}}{t_{\mathrm{SL}} Q_{x_2x_2}^{\mathrm{SL}} Q_{22}} + \left(\nu_{12} - \nu_{x_1x_2}^{\mathrm{SL}}\right)^2 \frac{2t_k}{E_{x_1}^{\mathrm{SL}} t_{\mathrm{SL}} + E_1 2t_k} \right] \tag{13.89}$$

The average shear stress in the sublaminate in the case of two symmetric layers is:

$$\sigma_{x_1x_2(\mathrm{av})}^{\mathrm{SL}} = \sigma_{x_1x_20}^{\mathrm{SL}} + \sigma_{120}^{(k)} \varphi_{(\mathrm{av})} 2\frac{t_k}{t_{\mathrm{SL}}} \tag{13.90}$$

leading to:

$$u_{1an}^{(k)} = k_\varphi \varphi_{(\mathrm{av})}, \quad k_\varphi = \frac{1}{2\rho_{kn}} \left[1 + \frac{G_{12}2t_k}{G_{x_1x_2}^{\mathrm{SL}} t_{\mathrm{SL}}} \right] \tag{13.91}$$

Note that for this laminate configuration, factor 2 at the kth ply thickness appears in all terms inside brackets in Eqs. (13.89) and (13.91).

Expressions in this subsection can be generalized for laminates with 2 m, $m = 1$, 2,... cracked layers of the same orientation, the same damage state, and identical constraint from the rest of layers. This requires replacing in the above expressions the constant 2 by 2 m. It has to be kept in mind that t_{SL} is different in all of these cases.

13.3.5 Relationships for monoclinic sublaminates

In the preceding analysis, the sublaminate, constituted of all layers except the layer with cracks, was assumed macroscopically orthotropic in the coordinate system related to the material symmetry of the kth layer with cracks. In the general case, in the x_1, x_2, z coordinates this sublaminate can be monoclinic.

We will consider in this section the case with a central cracked layer. Expressions for case with two or more symmetrically located cracked layers can be obtained as described in Section 13.3.4.

The force balance-related relationships derived above Eqs. (13.28, 13.37, 13.40, 13.65 and 13.71) are independent of the form of Hooke's law.

Since for a monoclinic material/sublaminate S_{16}^{SL} and S_{26}^{SL} are not zero, the relationships in Eqs. (13.48) and (13.51) are different. In this case, it is more convenient to use compliances instead of engineering constants, obtaining instead of Eq. (13.51):

$$\sigma_{1(av)}^{(k)} = \frac{-S_{12}}{S_{11}}\sigma_{2(av)}^{(k)} + \frac{S_{11}^{SL}}{S_{11}}\sigma_{x_1(av)}^{SL} + \frac{S_{12}^{SL}}{S_{11}}\sigma_{x_2(av)}^{SL} + \frac{S_{16}^{SL}}{S_{11}}\sigma_{x_1x_2(av)}^{SL} \tag{13.92}$$

The expressions in Eqs. (13.64) and (13.76) are now slightly different:

$$u_{2a}^{(k)} = l_k\left(S_{22}^{SL}\sigma_{x_2(av)}^{SL} + S_{12}^{SL}\sigma_{x_1(av)}^{SL} + S_{26}^{SL}\sigma_{x_1x_2(av)}^{SL} - S_{22}\sigma_{2(av)}^{(k)} - S_{12}\sigma_{1(av)}^{(k)}\right) \tag{13.93}$$

$$u_{1a}^{(k)} = l_k\left(S_{16}^{SL}\sigma_{x_1(av)}^{SL} + S_{26}^{SL}\sigma_{x_2(av)}^{SL} + \frac{\sigma_{x_1x_2(av)}^{SL}}{G_{x_1x_2}^{SL}} - \frac{\sigma_{12(av)}^{(k)}}{G_{12}}\right) \tag{13.94}$$

Following the same procedure as in Sections 13.3.1−13.3.3, we obtain:

$$u_{2an}^{(k)} = f_{(av)}k_{11} + \varphi_{(av)}\frac{\sigma_{120}^{(k)}}{\sigma_{20}^{(k)}}\frac{E_2}{G_{12}}k_{12} \tag{13.95}$$

$$u_{1an}^{(k)} = f_{(av)}\frac{\sigma_{20}^{(k)}}{\sigma_{120}^{(k)}}\frac{G_{12}}{E_2}k_{21} + \varphi_{(av)}k_{22} \tag{13.96}$$

The coefficients k_{lm} in Eqs. (13.95) and (13.96) are defined as follows:

$$k_{11} = \frac{1}{2\rho_{kn}}\left[1 + \frac{S_{22}^{SL}}{S_{22}}\frac{t_k}{t_{SL}} - \frac{1}{S_{22}}\frac{\left(S_{12} + S_{12}^{SL}\frac{t_k}{t_{SL}}\right)^2}{S_{11} + S_{11}^{SL}\frac{t_k}{t_{SL}}}\right] \tag{13.97}$$

$$k_{12} = \frac{G_{12}}{2\rho_{kn}}\left[S_{26}^{SL} - S_{16}^{SL}\frac{S_{12}t_{SL} + S_{12}^{SL}t_k}{S_{11}t_{SL} + S_{11}^{SL}t_k}\right]\frac{t_k}{t_{SL}} \tag{13.98}$$

$$k_{21} = \frac{E_2}{2\rho_{kn}}\left[S_{16}^{SL}b + S_{26}^{SL}\frac{t_k}{t_{SL}}\right] \tag{13.99}$$

$$k_{22} = \frac{1}{2\rho_{kn}}\left[1 + \frac{G_{12}}{G_{x_1x_2}^{\mathrm{SL}}}\frac{t_k}{t_{\mathrm{SL}}} - \frac{G_{12}\left(S_{16}^{\mathrm{SL}}\right)^2 t_k}{S_{11}t_{\mathrm{SL}} + S_{11}^{\mathrm{SL}}t_k}\frac{t_k}{t_{\mathrm{SL}}}\right] \tag{13.100}$$

13.4 Analytical models for stress state between cracks

The notation for coordinate axes and parameters in this section is as in Fig. 13.4: the direction x_2 is also the transverse direction of the cracked layer. We consider a central cracked 90 degrees layer in a symmetric laminate. The cracked layer is embedded between two identical orthotropic support-laminates, S; the same constraint to the crack is assumed from both sides. The upper index s is used to denote elastic properties of each of the two constraining support-laminates. The thickness of the support-laminate is t_s.

We will consider two commonly used groups of micromechanics models (shear lag and variational models), each one of them leading to certain stress perturbation functions.

13.4.1 Uniaxial tensile loading in the x_2 direction

Expressions for stresses in all models used below are obtained assuming the plane stress conditions (analysis does not include stresses with index x_1). Expressions for $f_{(\mathrm{av})}$ are summarized next.

13.4.1.1 Shear lag models

The shear lag model is basically one-dimensional analysis, applicable to balanced symmetric [S/90]s laminates, containing cracked plies of 90 degrees orientation with respect to the tensile direction (Nairn et al., 1994). This model was first used for damaged laminates by Garrett and Bailey (1977). Many modifications of this model have been developed; however, the stress perturbation function is always given by expression:

$$f^{\mathrm{Sh}-L} = \frac{\cosh\left(\dfrac{\beta x_2}{t_k/2}\right)}{\cosh\left(\dfrac{\beta l_k}{t_k/2}\right)} \tag{13.101}$$

According to Eq. (13.29), the average is:

$$f_{(\mathrm{av})}^{\mathrm{Sh}-L} = \rho_{kn}\frac{1}{\beta}\tanh\left(\frac{\beta}{\rho_{kn}}\right) \tag{13.102}$$

The value of the shear lag parameter, β, governs the rate of the stress transfer from the support-laminate to the 90 degrees ply. The value of the shear lag parameter, β, depends on the assumptions made about the shear strain distribution.

Hypothesis A: Shear strain in the support-laminate is zero while the displacement in the 90 degrees ply varies linearly in z (Garrett and Bailey, 1977; Manders et al., 1983). In that case:

$$\beta^2 = G_{23}H \tag{13.103}$$

$$H = \frac{E^s_{x_2}t_s + E_2t_k/2}{E^s_{x_2}E_2t_s} \tag{13.104}$$

Hypothesis B: Zero shear strain in the S, whereas the displacement in the 90 degrees ply is parabolic in z (Smith and Wood, 1990), leading to:

$$\beta^2 = 3G_{23}H \tag{13.105}$$

Hypothesis C: Zero shear strain in the S and in the 90 degrees ply. All the shear strain is in a homogeneous layer at the S/90 interface, with the thickness and shear modulus, respectively, denoted as d_0 and G_m (Peters and Chout, 1987). In practice, the shear layer is often assumed to be a resin-rich region of very small thickness at the interface between the plies. G_m is then the shear modulus of the matrix:

$$\beta^2 = \frac{t_k/2}{d_0}G_mH \tag{13.106}$$

Obviously, with $G_m = G_{23}$ and $d_0 = t_k/2$, the *A* and *C* cases coincide. This is equal to the assumption that the shear layer coincides with the whole cracked layer. Equality of both modifications is also obtained, assuming $d_0 = t_k/2 \times G_m/G_{23}$.

In the *A* and *B* cases, the β values are significantly different, even if the *B* case can be considered as a refinement of the *A* case with a more realistic assumption for the displacement shape in the 90 degrees ply. In the *C* case, the shear layer thickness is the most unreliable decision: this region is assumed to be small compared to the ply thicknesses and, to be realistic, has to be smaller than the fiber diameter. Expressions for shear lag parameters for models where the intralaminar shear stress acts also in the support-laminate (entire or part of it) are given by Nuismer and Tan (1988).

13.4.1.2 Variational models

In Hashin's approach (Hashin, 1985), the stress perturbation function is obtained by minimizing the complementary energy. The assumption is made that the axial stress between two cracks does not depend on z but depends on x_2 only. Contrary to the shear lag model, in the case of Hashin's model, the perturbation function depends only on known elastic and geometrical parameters. We present the expression for $f^{\text{Hashin}}_{(\text{av})}$

(Joffe et al., 2001) generalized for the case when constraint layers are homogenized arbitrary orthotropic support-laminates:

$$f_{(av)}^{\text{Hashin}} = \rho_{kn} \frac{2AB}{A^2 + B^2} \frac{\cosh\left(\frac{2A}{\rho_{kn}}\right) - \cos\left(\frac{2B}{\rho_{kn}}\right)}{B \sinh\left(\frac{2A}{\rho_{kn}}\right) + A \sin\left(\frac{2B}{\rho_{kn}}\right)} \tag{13.107}$$

$$A = q^{1/4} \cos\frac{\theta}{2}, \quad B = q^{1/4} \sin\frac{\theta}{2}, \quad \tan\theta = \sqrt{\frac{4q}{p^2} - 1} \tag{13.108}$$

$$p = \frac{C_{02} - C_{11}}{C_{22}}, \quad q = C_{00}/C_{22} \tag{13.109}$$

$$\begin{aligned} C_{00} &= \frac{1}{E_2} + \frac{t_k}{2t_s E_{x_2}^s}, \; C_{02} = \frac{\nu_{23}}{E_2}\left(\frac{2t_s}{t_k} + \frac{2}{3}\right) - \frac{2t_s \nu_{x_2x_3}^s}{3t_k E_{x_2}^s}, \\ C_{11} &= \frac{1}{3G_{23}} + \frac{2t_s}{3t_k G_{x_2x_3}^s} \end{aligned} \tag{13.110}$$

$$C_{22} = \frac{1}{20E_3^s}\left(\frac{2t_s}{t_k}\right)^3 + \frac{1}{60E_2}\left[15\left(\frac{2t_s}{t_k}\right)^2 + 20\left(\frac{2t_s}{t_k}\right) + 8\right] \tag{13.111}$$

This solution is applicable if $4q/p^2 > 1$, which for typical polymeric composites is satisfied.

A more refined variational model (called the 2D-0 model in the following), in which the assumed shape of intralaminar shear stress distribution in the support-laminate is exponential with the shape parameter determined minimizing the complementary energy, was presented by Varna and Berglund (1991):

$$\sigma_2^{90} = \sigma_{20}^{90}(1 - \psi(x_2)) \tag{13.112}$$

$$\sigma_{x_2}^s = \sigma_{x_20}^s + \sigma_{20}^{90} \frac{t_k}{2t_s} \psi(x_2) \chi^{II}(z) \tag{13.113}$$

$$\chi(z) = \frac{1 - \cosh\Delta\left(\frac{z - \frac{h}{2}}{t_k/2}\right)}{\Delta \sinh\Delta \frac{t_s}{t_k/2}} \tag{13.114}$$

In Eq. (13.114), Δ is an unknown shape parameter. Roman *II* index denotes second derivative with respect to z.

Even in this model, the average stress perturbation is given by Eq. (13.107), with definitions of A, B, p, q, and θ given by Eqs. (13.108) and (13.109). However, expressions for C_{00}, C_{02}, C_{11}, C_{22} in this model are different:

$$C_{00} = \frac{1}{E_2} + \frac{1}{E^s_{x_2}} I_1(\Delta) \tag{13.115}$$

$$C_{02} = -2\frac{\nu_{23}}{E_2}\left(\frac{1}{6} + A^*\right) - 2\frac{\nu^s_{x_2x_3}}{E^s_{x_2}} I_4(\Delta) \tag{13.116}$$

$$C_{11} = \frac{1}{3G_{23}} + \frac{1}{G^s_{x_2x_3}} I_3(\Delta) \tag{13.117}$$

$$C_{22} = \frac{1}{E^s_3} I_2(\Delta) + \frac{1}{E_2}\left[A^{*2} + \frac{A^*}{3} + \frac{1}{20}\right] \tag{13.118}$$

where

$$A^* = \frac{-1}{2} + \frac{1 - \cosh 2\Delta\frac{t_s}{t_k}}{\Delta \sinh 2\Delta\frac{t_s}{t_k}} \tag{13.119}$$

$$I_1 = \frac{\Delta^2 \frac{t_s}{t_k}}{\sinh^2 2\Delta\frac{t_s}{t_k}}\left[1 + \frac{\sinh 4\Delta\frac{t_s}{t_k}}{4\Delta\frac{t_s}{t_k}}\right] \tag{13.120}$$

$$I_2 = \frac{\frac{t_s}{t_k}}{\Delta^2 \sinh^2 2\Delta\frac{t_s}{t_k}}\left[3 - 2\frac{\sinh 2\Delta\frac{t_s}{t_k}}{\Delta\frac{t_s}{t_k}} + \frac{\sinh 4\Delta\frac{t_s}{t_k}}{4\Delta\frac{t_s}{t_k}}\right] \tag{13.121}$$

$$I_3 = \frac{\frac{t_s}{t_k}}{\sinh^2 2\Delta\frac{t_s}{t_k}}\left[\frac{\sinh 4\Delta\frac{t_s}{t_k}}{4\Delta\frac{t_s}{t_k}} - 1\right] \tag{13.122}$$

$$I_4 = -I_3 - \frac{1}{2} - A^* \tag{13.123}$$

The shape parameter, Δ, is found to minimize the expression:

$$F(\Delta) = C_{00}(\Delta)\psi_{(\mathrm{av})}(\Delta) \tag{13.124}$$

13.4.2 In-plane shear loading

Three entirely different approaches were used by Hashin (1985), Tsai and Daniel (1992), Tan and Nuismer (1989), Nuismer and Tan (1988), leading to the same result. Tan and Nuismer (1989); Nuismer and Tan (1988) obtained expressions for shear modulus of the cross-ply laminate with 90 degrees cracks, by integrating the equilibrium and constitutive expressions over the ply thickness and obtaining a second-order differential equation for stress distribution. Tsai and Daniel (1992) reduced the 3D elasticity problem to a 2D problem in terms of displacements averaged over the ply thickness. In both cases, the obtained set of differential equations with constant coefficients was solved analytically. Due to the assumed linear through-the-thickness dependence of the out-of-plane shear stresses, the results of these models coincide with Hashin (1985) results, which used the principle of minimum complementary energy. The in-plane shear stress distribution between cracks according to these models is:

$$\sigma_{12}^{90} = \sigma_{12}^{\mathrm{LAM}}\left(1 - \frac{\cosh\left(\mu\dfrac{2x_2}{t_k}\right)}{\cosh\left(\mu\dfrac{2l_k}{t_k}\right)}\right) \tag{13.125}$$

The same form, but with different coefficients, follows also from more sophisticated models (Zhang et al., 1992; Giannadakis and Varna, 2014). According to Eq. (13.64), the stress perturbation function in all these models is:

$$\varphi(x_2) = \frac{\cosh\left(\mu\dfrac{2x_2}{t_k}\right)}{\cosh\left(\mu\dfrac{2l_k}{t_k}\right)}, \quad \varphi_{(\mathrm{av})} = \frac{t_k}{2\mu l_k}\,\frac{\sinh\left(\mu\dfrac{2l_k}{t_k}\right)}{\cosh\left(\mu\dfrac{2l_k}{t_k}\right)} \tag{13.126}$$

where

$$\mu^2 = \frac{1 + \dfrac{G_{12}}{G_{x_1x_2}^s}K_1}{\dfrac{1}{3} + \dfrac{G_{12}}{G_{x_1x_3}^s}K_2} \tag{13.127}$$

In Hashin's model and in models (Tsai and Daniel, 1992; Tan and Nuismer, 1989):

$$K_{1,\mathrm{Hashin}} = \frac{t_k}{2t_s}, \quad K_{2,\mathrm{Hashin}} = \frac{2t_s}{3t_k} \tag{13.128}$$

A bilinear shape function of the out-of-plane shear stress distribution across the constraint layer thickness was assumed by Zhang et al. (1992). The distance from the interface, where the stress perturbation becomes negligible (denoted by z_1), is not precisely defined:

$$\varphi(z) = \begin{cases} 0, & \frac{t_k}{2} + z_1 < z < \frac{h}{2} \\ \frac{z_1 + t_k/2 - z}{z_1}, & \frac{t_k}{2} \le z \le \frac{t_k}{2} + z_1 \end{cases} \tag{13.129}$$

In this model:

$$K_{1,\mathrm{Soutis}} = \frac{t_k/2}{z_1} \quad K_{2,\mathrm{Soutis}} = \frac{2z_1}{3t_k} \tag{13.130}$$

Giannadakis and Varna (2014) expressed that the shape function that describes the nonuniform out-of-plane shear stress distribution across the constraint layer thickness was selected in an exponential form with unknown parameter, Δ:

$$\varphi(z) = \frac{\exp\left(-2\Delta\frac{z - \frac{h}{2}}{t_k}\right) - 1}{\exp\left(2\Delta\frac{t_s}{t_k}\right) - 1} \tag{13.131}$$

This function increases exponentially, approaching the transverse crack tip. Parameters K_1 and K_2 in this model depend on Δ:

$$\begin{aligned} K_1 &= \frac{\Delta}{2}\frac{\exp\left(2\Delta\frac{t_s}{t_k}\right) + 1}{\exp\left(2\Delta\frac{t_s}{t_k}\right) - 1}, \\ K_2 &= \frac{\frac{2t_s}{t_k} - \frac{1}{2\Delta}\left(1 - \exp\left(4\Delta\frac{t_s}{t_k}\right)\right) + \frac{2}{\Delta}\left(1 - \exp\left(2\Delta\frac{t_s}{t_k}\right)\right)}{\left(\exp\left(2\Delta\frac{t_s}{t_k}\right) - 1\right)^2} \end{aligned} \tag{13.132}$$

Parameter Δ is found to minimize the following function:

$$U = \frac{\mu(\Delta)}{G_{12}}\left(\frac{1}{3} + \frac{G_{12}}{G^s_{x_1x_3}}K_2(\Delta)\right)\tanh\left(\mu(\Delta)\frac{2l_k}{t_k}\right) \tag{13.133}$$

13.5 Experimental data and simulation examples

13.5.1 Analytical formulas for axial modulus and shear modulus of cross-ply laminates

In this particular case, the global coordinates in Fig. 13.4 for laminates with cracked 90 degrees layers are $x = x_2$ and $y = x_1$. We assume that the two equal support-laminates applying constraint on cracks are orthotropic (no coupling in equations for normal and shear loading). Stress analysis based on the principle of minimum complementary energy in the case of monoclinic support-laminates (off-axis layer adjacent to cracked 90 degrees layer) was performed by Vinogradov and Hashin (2010). Unfortunately, the significance of the coupling between shear and normal stress perturbation functions and the error introduced by ignoring the coupling were not analyzed in the paper.

Assuming the plane stress state (in fact this assumption is not justified but still it is used in models described in Section 13.4.3), the undamaged modulus of the laminate can be calculated using the RoM:

$$E_{x0}^{\mathrm{LAM}} = \frac{E_2 t_k + 2E_x^s t_s}{t_k + 2t_s} \tag{13.134}$$

The axial modulus of the laminate with cracks in the 90 degrees layer can be found from the ratio of the applied stress to the average strain in the support-laminate (which is not damaged):

$$E_x^{\mathrm{LAM}} = \frac{\sigma_x^{\mathrm{LAM}}}{\varepsilon_x^{\mathrm{LAM}}} \tag{13.135}$$

$$\varepsilon_x^{\mathrm{LAM}} = \frac{\Delta l_k}{l_k} = \frac{U_x^s(l_k)}{l_k} = \varepsilon_{x(\mathrm{av})}^s = \frac{\sigma_{x(\mathrm{av})}^s}{E_x^s} \tag{13.136}$$

Since $t_{\mathrm{SL}} = 2t_s$, from Eq. (13.37) follows that:

$$\sigma_{x(\mathrm{av})}^s = \sigma_{x0}^s + \sigma_{20}^{(90)} f_{(\mathrm{av})} \frac{t_k}{2t_s} \tag{13.137}$$

In Eq. (13.137), $f_{(\mathrm{av})}$ is the average value of the stress perturbation defined by Eqs. (13.26)–(13.28). Then the CLT is used for the undamaged laminate to express stresses in layers via the applied stress (neglecting Poisson effects as required by plane stress conditions). Together with Eqs. (13.134)–(13.137), this leads to an expression for a damaged laminate modulus:

$$\frac{E_x^{\mathrm{LAM}}}{E_{x0}^{\mathrm{LAM}}} = \frac{1}{1 + \frac{t_k}{2t_s} \frac{E_2}{E_x^s} f_{(\mathrm{av})}} \tag{13.138}$$

The following expression for laminate shear modulus reduction is valid for all models described in Section 13.4.2:

$$\frac{G_{xy}^{\mathrm{LAM}}}{G_{xy0}^{\mathrm{LAM}}}=\frac{1}{1+\dfrac{G_{12}}{G_{xy0}^{\mathrm{LAM}}}\dfrac{t_k}{h}\dfrac{t_k}{2l_k}\mu\left(\dfrac{1}{3}+\dfrac{G_{12}}{G_{yz}^{s}}K_2\right)\tanh\left(\mu\dfrac{2l_k}{t_k}\right)} \tag{13.139}$$

For cross-ply laminates, $G_{xy}^{S}=G_{12}$, $G_{yz}^{S}=G_{23}$, and $G_{xy0}^{\mathrm{LAM}}=G_{12}$.

Using Eqs. (13.138) and (13.139), we obtain numerically identical results to using Eqs. (13.17) and (13.18) (with zero Poisson ratio).

13.5.2 Examples of calculation and experiments

Calculation results presented here are based on stiffness expressions in Section 13.2 with COD and CSD. When the GLOB-LOC model is used, the COD and CSD are calculated using expressions in Appendix 2. Predictions based on shear lag and variational models are obtained expressing the average COD and CSD via the stress perturbation functions $f_{(\mathrm{av})}$ and $\varphi_{(\mathrm{av})}$ of the corresponding model using Eqs. (13.62) and (13.77). First, we compare predictions with experimental data for glass fiber–epoxy cross-ply laminates (GF—EP-2). The thermoelastic constants of the UD layer are the longitudinal modulus $E_1=43.26$ GPa, the transverse modulus $E_2=13.3$ GPa, the in-plane shear modulus $G_{12}=3.4$ GPa, the in-plane Poisson ratio $\nu_{12}=0.29$, the out-of-plane Poisson's ratio $\nu_{23}=0.29$, and the thermal expansion coefficients $\alpha_1=4{\bullet}10^{-6}1/^{\circ}C$, $\alpha_2=17{\bullet}10^{-6}1/^{\circ}C$. The ply thickness is 0.144 mm. The axial modulus E_x^{LAM} and the Poisson ratio ν_{xy}^{LAM} reduction with increasing crack density in the 90 degrees layer is shown in Figs. 13.5–13.10.

The two shear lag models in Figs. 13.5 and 13.6 render very different predictions. Obviously, the modification C with a resin layer between the 0 degrees layer and the 90 degrees layer with assumed thickness of $d_0=7$ microns and resin shear modulus $G_m=1.1$ GPa does not lead to good predictions.

Modification A, which assumes that the whole 90 degrees layer is the shear layer, leads to good results, especially for the laminate with a thicker 90 degrees layer. The results presented confirm a well-known fact: since theoretical analysis cannot determine which is the best shear lag model, the shear lag parameter β in Eq. (13.101) can be used as a fitting parameter if experimental data or FEM results on some elastic properties reduction for the given composite system are known.

Predictions of Hashin's model and the 2D-0 model, which accounts for nonuniform axial stress distribution in the constraint layer, together with the same experimental data, are shown in Figs. 13.7 and 13.8. As expected, both models give lower bound to exact results of elasticity and the 2D-0 model predicts slightly higher laminate axial

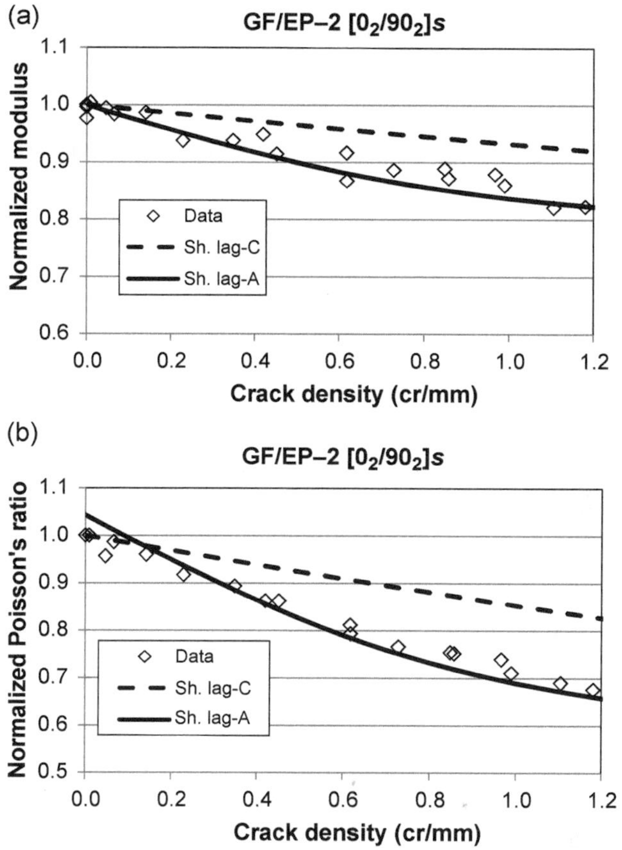

Figure 13.5 Elastic property reduction in a glass fiber—epoxy $[0_2/90_2]_s$ laminate according to two shear lag models. (a) Axial modulus, E_x^{LAM}. (b) Poisson ratio, ν_{xy}^{LAM}.

modulus E_x^{LAM} and the Poisson ratio ν_{xy}^{LAM}, which is due to its better accuracy in stress state description. However, the difference between predictions of both models is small, especially when the 90 degrees-layer thickness is larger (Fig. 13.8).

Despite variational models rendering the lower bound, we can see in Fig. 13.8b that at high crack density, the experimental data are lower than the predictions. The only explanation for this is that at high crack density, cracks in this laminate with a very thick 90 degrees layer are not straight and sharp: local delaminations propagate from the crack tip, increasing the COD and leading to more stiffness reduction than in any model that does not include delamination. At extremely high crack density, both variational models render indistinguishable results approaching the ply-discount value.

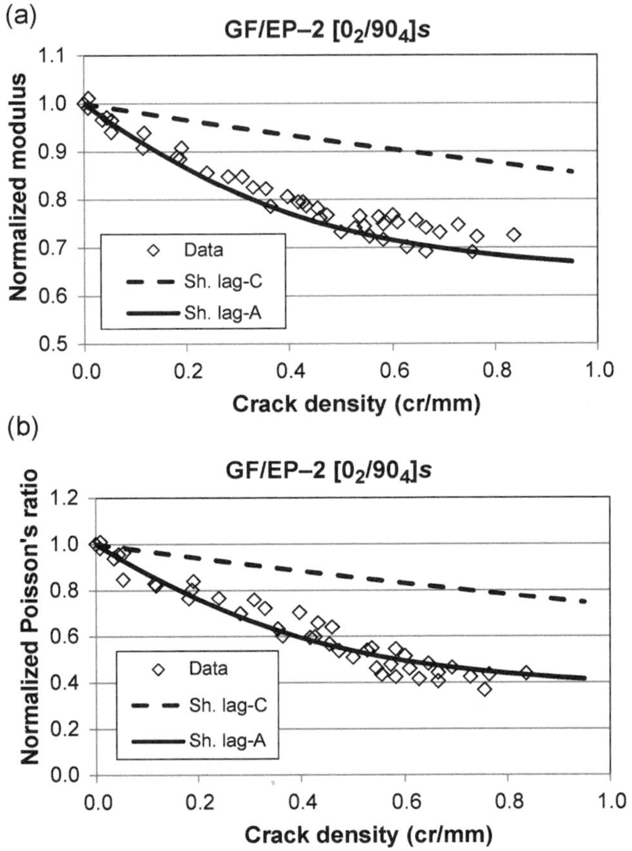

Figure 13.6 Elastic property reduction in a glass fiber–epoxy $[0_2/90_4]_s$ laminate according to two shear lag models. (a) Axial modulus, E_x^{LAM}. (b) Poisson ratio, ν_{xy}^{LAM}.

The GLOB-LOC model, where the COD and CSD values come from FEM-based fitting expressions (see Appendix 2), shows an excellent agreement with data for a $[0_2/90_2]_s$ laminate (Fig. 13.9). For a $[0_2/90_4]_s$ laminate with a thick 90 degrees layer, the agreement is excellent before the high crack density region is reached. At high crack density, experimental values are lower than predicted. The two possible explanations are (1) the simple crack interaction function Eq. (A2.7) gives too rough approximation of interaction; or (2) at high crack density, delaminations develop from intralaminar cracks, leading to larger stiffness reduction than according to the model, which does not include this phenomenon.

The GLOB-LOC expressions allow calculating change of any thermoelastic constant of any arbitrary symmetric laminate with crack densities that can be different

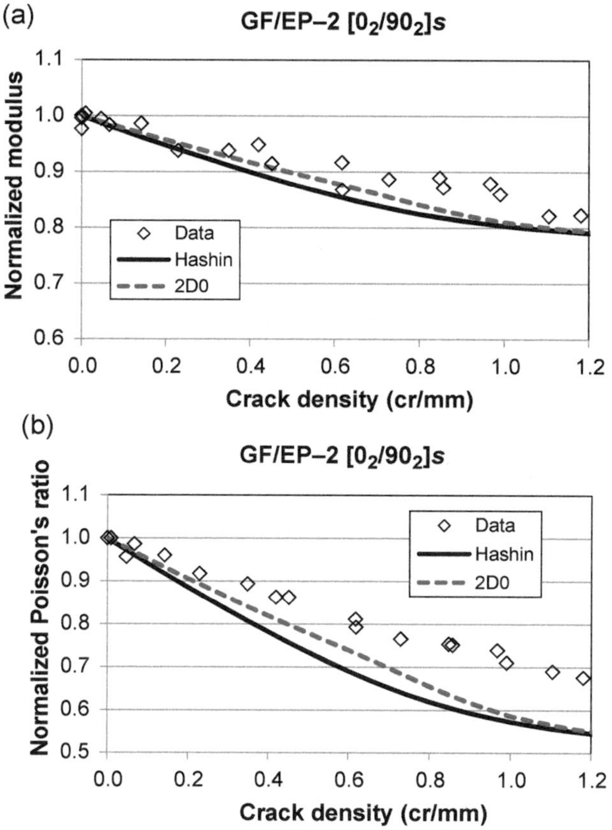

Figure 13.7 Elastic property reduction in a glass fiber–epoxy $[0_2/90_2]_s$ laminate according to two variational models. (a) Axial modulus, E_x^{LAM}. (b) Poisson ratio, ν_{xy}^{LAM}.

in all layers. The user can follow his or her preference in finding COD and CSD; a fitting expression from Appendix 2 can be used or stress perturbation functions according to any available analytical model.

In Fig. 13.11, reduction of the thermal expansion coefficient α_x^{LAM} of $[0_2/90_2]_s$ laminate with cracks in the 90 degrees layer is shown according to Hashin's model and the GLOB-LOC model with the power law for COD. Even in this case, Hashin's model gives the lower bound of the property change.

In Fig. 13.12, the GLOB-LOC model is applied to calculate the thermal expansion coefficient α_x^{LAM} of a carbon fiber–epoxy $[0/90]_s$ laminate with cracks in the 90 degrees layer. Elastic properties of this material are $E_1 = 138$ GPa, $E_2 = 10.3$ GPa, the in-plane shear modulus $G_{12} = 5.5$ GPa, and the in-plane Poisson ratio

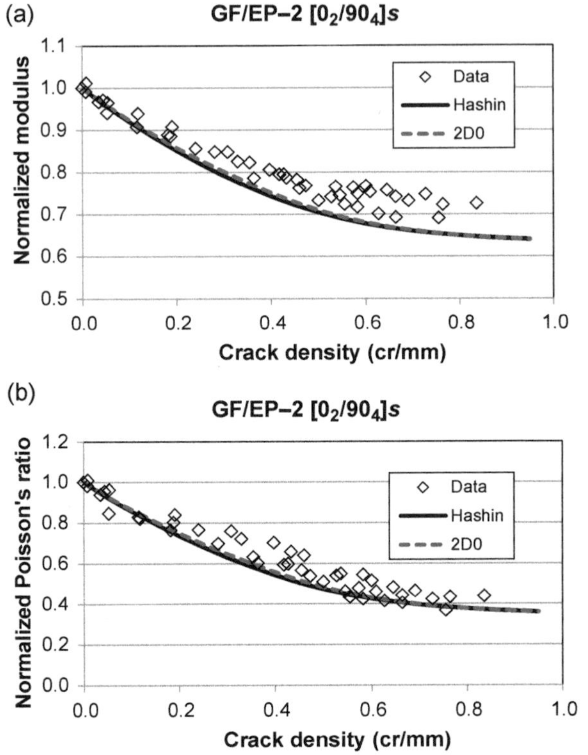

Figure 13.8 Elastic property reduction in a glass fiber−epoxy $[0_2/90_4]_s$ laminate according to two variational models. (a) Axial modulus, E_x^{LAM}. (b) Poisson ratio, ν_{xy}^{LAM}.

$\nu_{12} = 0.30$. The thermal expansion coefficients $\alpha_1 = 0.43\bullet10^{-6}$ 1/°C, $\alpha_2 = 25.87\bullet10^{-6}$ 1/°C. The ply thickness is 0.125 mm.

The calculation results and the experimental data from Kim et al. (2000) are also presented in Fig. 13.12. The GLOB-LOC simulations were performed assuming that cracks are noninteractive. The agreement is excellent before the interaction region. In the high crack density region, the expansion coefficient reduction experimentally is slightly smaller than predicted.

Predictions including interaction using the GLOB-LOC model can be found in Lundmark and Varna (2011). These results validate the GLOB-LOC model with power-law fitting functions for COD and CSD to be used as a reference to evaluate approximate analytical models. In Fig. 13.13, the degradation of the thermal expansion coefficient of GF/EP-2 quasiisotropic $[0/-45/+45/90]_s$ is shown. The same density of

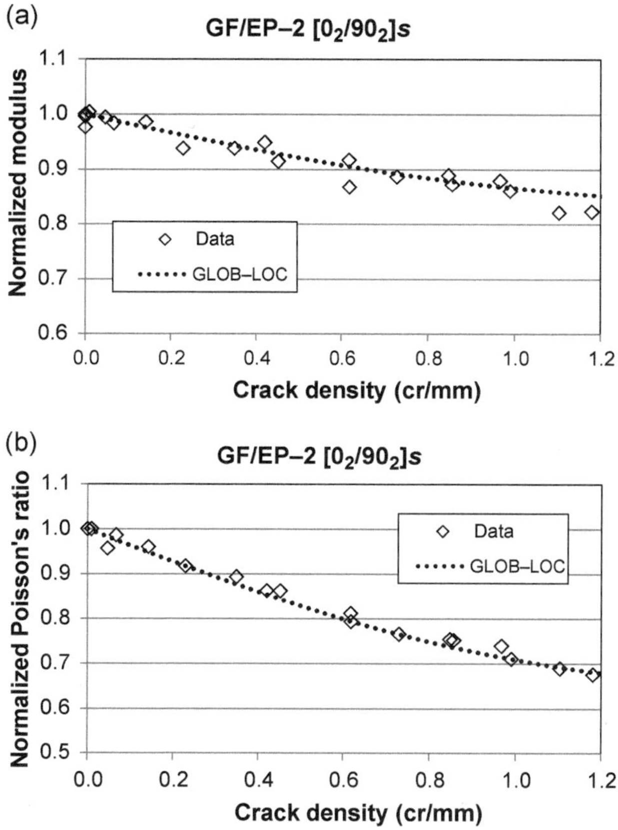

Figure 13.9 Elastic property reduction in a glass fiber–epoxy $[0_2/90_2]_s$ laminate according to GLOB-LOC model. (a) Axial modulus, E_x^{LAM}. (b) Poisson ratio, ν_{xy}^{LAM}.

intralaminar cracks in all layers is assumed. It may correspond to the case when cracks are introduced by thermal loading; subjecting the quasiisotropic specimen to very low temperature, transverse tensile thermal stresses in all layers would be the same. We can assume that it would introduce approximately the same number of cracks in all layers. Certainly, it is just an assumption used in simulation. In reality, differences in crack density may be caused by different layer thickness (the 90 degrees layer is two times thicker) and different location of layers (surface layers have different constraint conditions). In the crack density region considered, the reduction of α_x^{LAM} is rather linear, and as expected, Hashin's model predicts more reduction. However, the difference between both predictions is not very large.

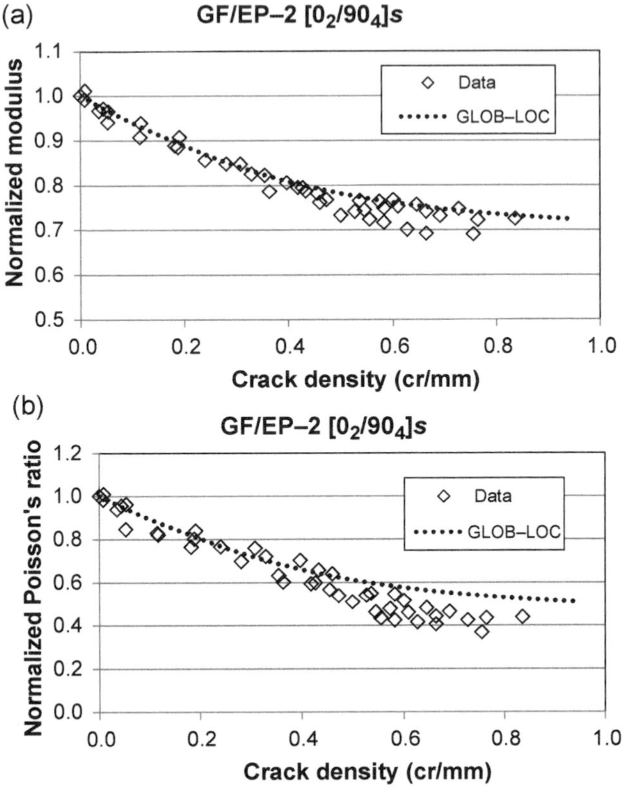

Figure 13.10 Elastic property reduction in glass fiber–epoxy $[0_2/90_4]_s$ laminate according to GLOB-LOC model. (a) Axial modulus, E_x^{LAM}. (b) Poisson ratio, ν_{xy}^{LAM}.

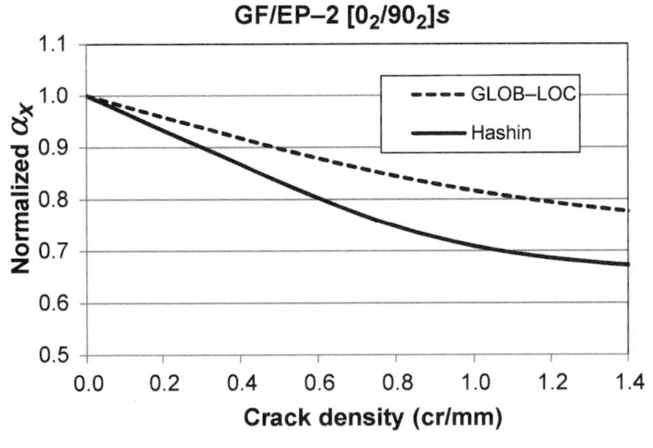

Figure 13.11 Reduction of thermal expansion coefficient α_x^{LAM} of a glass fiber–epoxy $[0_2/90_2]_s$ laminate according to the GLOB-LOC model and Hashin's model.

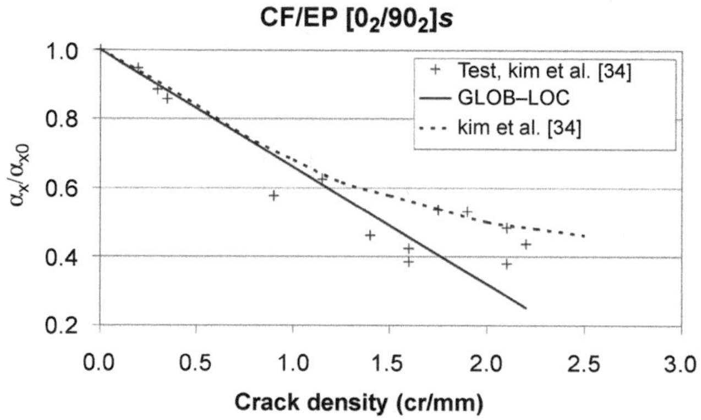

Figure 13.12 Reduction of thermal expansion coefficient α_x^{LAM} of a glass fiber–epoxy $[0_2/90_2]_s$ laminate according to the GLOB-LOC model. Test data and FEM predictions are also presented.

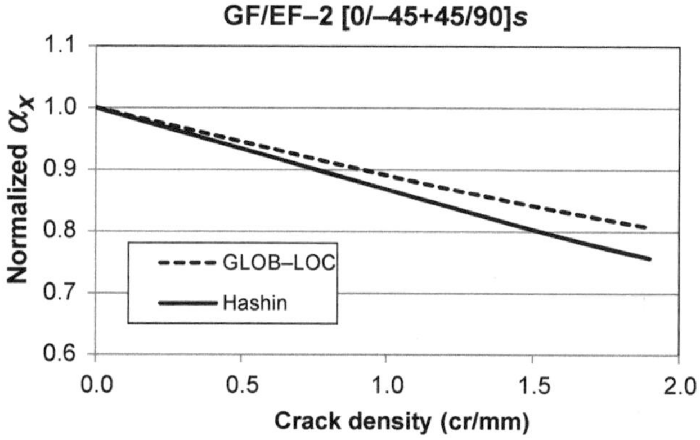

Figure 13.13 Reduction of thermal expansion coefficient α_x^{LAM} of a glass fiber–epoxy $[0/-45/+45/90]_s$ laminate according to the GLOB-LOC model and Hashin's model. The same crack density is assumed in all layers.

13.6 Conclusions

The methodology presented in this chapter allows predicting the thermoelastic properties of symmetric laminates with intralaminar cracks in layers. The derived exact relationships link the damaged laminate macroscopic thermoelastic constants with the number of cracks (crack densities) in different layers. In addition to the geometrical parameters of the laminate and the elastic properties of layers, these relationships have as input two local parameters characterizing the crack: average opening (COD) and sliding (CSD) displacements of its faces corresponding to unit stress. These rather robust parameters have been analyzed using FEM, and their dependence on layer properties and other parameters have been described by fitting functions. This simulation approach is called the GLOB-LOC model. It has been shown that alternatively any approximate analytical stress distribution model between cracks (shear lag, variational, etc.) can be used to calculate COD and CSD.

Models of different degrees of accuracy have been used to predict axial modulus, Poisson ratio, and thermal expansion coefficients for cross-ply and quasiisotropic laminates. Predictions have been compared with experimental data and the accuracy of models has been evaluated.

Appendix 1: Derivation of damaged laminate stiffness

Here, the derivation of Eqs. (13.13)–(13.16) using the GLOB-LOC approach is demonstrated for laminate stiffness only, not including thermal terms in Hook's law. The thermomechanical case analysis would follow the same path but with more complex expressions.

From the divergence theorem follows (Allen and Yoon, 1998) that the average stress applied to the laminate representative volume element (RVE) boundary is equal to the stress distribution averaged over the laminate volume:

$$\{\sigma\}_0^{\mathrm{LAM}} = \{\sigma\}^{(\mathrm{av})}, \tag{A1.1}$$

$$\{\sigma\}^{(\mathrm{av})} = \frac{1}{V}\int\int\int_V \{\overline{\sigma}\}dv \tag{A1.2}$$

In Eq. (A1.2), $V = hS$, where $S-$ is the in-plane surface area of the RVE. An integral over the laminate volume can be written as sum of integrals over volumes of all layers, leading to:

$$\{\sigma\}_0^{\mathrm{LAM}} = \sum_{k=1}^{N} \frac{t_k}{h}\{\overline{\sigma}\}_k^{(\mathrm{av})} \tag{A1.3}$$

Hook's law for the volume-averaged stresses in a layer is the same as in each point:

$$\{\overline{\sigma}\}_k^{(\mathrm{av})} = \left[\overline{Q}\right]_k\{\overline{\varepsilon}\}_k^{(\mathrm{av})} \tag{A1.4}$$

The volume-averaged strain in a damaged layer with volume $V_k = St_k$ is:

$$\{\overline{\varepsilon}\}_k^{(\mathrm{av})} = \frac{1}{St_k}\int\int\int_V \{\overline{\varepsilon}\}_k dv \tag{A1.5}$$

It can be expressed through the external boundary-averaged strains (they are equal to the laminate strain) and displacements on crack surfaces (Allen and Yoon, 1998):

$$\{\overline{\varepsilon}\}_k^{(\mathrm{av})} = \{\varepsilon\}^{\mathrm{LAM}} + \left\{\overline{\beta}\right\}_k \tag{A1.6}$$

where $\left\{\overline{\beta}\right\}_k$ is the Voigt vector representation of the Vakulenko–Kachanov tensor:

$$\overline{\beta}_{ij}^k = \frac{1}{V_k} \int\limits_{S_C} \frac{1}{2}\left(u_i n_j + u_j n_i\right) dS, i, j = x, y \tag{A1.7}$$

$$\overline{\beta}_1 = \overline{\beta}_{xx}, \quad \overline{\beta}_2 = \overline{\beta}_{yy} \quad \overline{\beta}_6 = 2\overline{\beta}_{xy} \tag{A1.8}$$

Integration in Eq. (A1.7) is over S_C, which is the total surface of all cracks in the layer; u_i are displacements of the points on the crack surface, and n_i is the outer normal to the crack surface.

Substituting Eq. (A1.4) in Eq. (A1.3) and using Eq. (A1.6), we obtain:

$$\{\sigma\}_0^{\mathrm{LAM}} = [Q]_0^{\mathrm{LAM}}\{\varepsilon\}^{\mathrm{LAM}} + \sum_{k=1}^{N} \frac{t_k}{h}\left[\overline{Q}\right]_k \left\{\overline{\beta}\right\}_k \tag{A1.9}$$

The second term in Eq. (A1.9) represents the effect on stiffness of the crack face displacements (opening and sliding). The applied strain, the volume average strain, and β_{ij} are all tensors with the same transformation expressions between local and global coordinates:

$$\left\{\overline{\beta}\right\}_k = [T]_k^T \{\beta\}_k \tag{A1.10}$$

To find the physical meaning of $\{\beta\}_k$, we consider the laminate in the local coordinates of the kth layer where indexes 1, 2, and 3 correspond to the longitudinal, transverse, and thickness directions as shown in Fig. 13.3. In these coordinates, layers supporting the cracked layer are monoclinic materials. The outer normal vector to the two faces of the crack surface has coordinates $n_1 = n_3 = 0$ and:

$$n_2^+ = -1, \quad n_2^- = +1 \tag{A1.11}$$

The "+" and "−" faces of the crack are indicated in Fig. 13.3.

Using definition Eq. (A1.7) for $\beta_{ij}^{(k)}$ in the kth layer of volume $V_k = Lwt_k$, with one crack surface equal to wt_k and with M cracks inside it, we see that the tensor contains only two nonzero elements, $\beta_{12}^{(k)}$ and $\beta_{22}^{(k)}$:

$$\beta_{12}^{(k)} = \frac{w}{2Lwt_k} \sum_{m=1}^{M} \int_{\frac{-t_k}{2}}^{\frac{t_k}{2}} \left[- u_1^{k+}(z) + u_1^{k-}(z)\right] dz \tag{A1.12}$$

=

$$\beta_{22}^{(k)} = \frac{w}{Lwt_k} \sum_{m=1}^{M} \int_{\frac{-t_k}{2}}^{\frac{+t_k}{2}} \left[- u_2^{k+}(z) + u_2^{k-}(z)\right] dz \tag{A1.13}$$

Since all cracks are equal and the crack distribution is uniform, $u_1^{k-} = -\, u_1^{k+}$ and

$u_2^{k-} = -\, u_2^{k+}$:

$$\beta_{12}^{(k)} = \frac{-M}{Lt_k} \int_{\frac{-t_k}{2}}^{\frac{+t_k}{2}} \left[u_1^{k+}(z)\right] dz, \qquad \beta_{22}^{(k)} = \frac{-2M}{Lt_k} \int_{\frac{-t_k}{2}}^{\frac{+t_k}{2}} \left[u_1^{k+}(z)\right] dz \tag{A1.14}$$

Using the definition of crack density and the definition of average COD and CSD (Eq. 13.23):

$$\beta_{12}^{(k)} = -\rho_k u_{1a}^{(k)}, \quad \beta_{22}^{(k)} = -2\rho_k u_{2a}^{(k)} \tag{A1.15}$$

In Eq. (A1.15), crack density may be replaced with the normalized crack density, ρ_{kn} (using Eq. 13.2), and the average COD and CSD may be replaced with normalized values (using Eq. 13.24). The obtained expressions can be written in the matrix form:

$$\{\beta\}_k = \begin{Bmatrix} 0 \\ \beta_{22} \\ 2\beta_{12} \end{Bmatrix}_k = -\,2\frac{\rho_{kn}}{E_2} \begin{bmatrix} 0 & 0 & 0 \\ 0 & u_{2an} & 0 \\ 0 & 0 & \dfrac{E_2}{G_{12}} u_{1an} \end{bmatrix} \begin{Bmatrix} \sigma_{L0} \\ \sigma_{T0} \\ \sigma_{LT0} \end{Bmatrix}_k \tag{A1.16}$$

According to CLT, the local stress $\{\sigma_0\}_k$ in the kth undamaged layer can be expressed through the applied laminate stress:

$$\begin{Bmatrix} \sigma_{L0} \\ \sigma_{T0} \\ \sigma_{LT0} \end{Bmatrix}_k = [T]_k \begin{Bmatrix} \sigma_{x0} \\ \sigma_{y0} \\ \sigma_{xy0} \end{Bmatrix}_k = [T]_k \left[\overline{Q}\right]_k \{\varepsilon\}_0^{\mathrm{LAM}} = [T]_k \left[\overline{Q}\right]_k [S]_0^{\mathrm{LAM}} \{\sigma\}_0^{\mathrm{LAM}} \tag{A1.17}$$

Substituting Eq. (A1.17) in Eq. (A1.16) and using the result in Eq. (A1.10) and Eq. (A1.9), we obtain:

$$\{\sigma\}_0^{LAM} = [Q]_0^{LAM}\{\varepsilon\}^{LAM} - 2 \times \sum_{k=1}^{N} \frac{t_k}{h}\frac{\rho_{kn}}{E_2}\left[\overline{Q}\right]_k [T]_k^T \begin{bmatrix} 0 & 0 & 0 \\ 0 & u_{2an} & 0 \\ 0 & 0 & \frac{E_2}{G_{12}} u_{1an} \end{bmatrix} [T]_k \left[\overline{Q}\right]_k [S]_0^{\mathrm{LAM}} \{\sigma\}_0^{\mathrm{LAM}} \tag{A1.18}$$

From here, after simple arrangements, we can express $\{\sigma\}_0^{\mathrm{LAM}}$ versus $\{\varepsilon\}^{LAM}$, which defines $[Q]^{\mathrm{LAM}}$. The expression obtained is Eq. (13.13).

Appendix 2: Expressions for COD and CSD

The COD, u_{2an}^0 of *noninteractive cracks,* is considered in a coordinate system where the cracked layer has 90 degrees orientation with respect to the x-axis (Fig. 13.14). In other words, the x-direction is direction 2 for the layer with the crack. The index k denoting the layer is omitted in the expressions below. A distinction must be made between cracks in surface layers and cracks in internal layers. Obviously, the normalized average COD of surface cracks is larger because the cracked layer is supported only from one side. The fitting expressions are presented for a symmetric case where the support layer S_1 has equal properties, orientation, and geometry to support layer S_2. The expression for u_{2an}^0 is:

$$u_{2an}^0 = A + B\left(\frac{E_2}{E_x^S}\right)^n \tag{A2.1}$$

In Eq. (A2.1), E_x^S is the Young's modulus of the support layer measured in the x-direction. For a crack in the *internal layer*:

$$A = 0.52 \quad B = 0.3075 + 0.1652\left(\frac{t_{90}}{2t_s} - 1\right)$$

$$n = 0.030667\left(\frac{t_{90}}{2t_s}\right)^2 - 0.0626\frac{t_{90}}{2t_s} + 0.7037 \tag{A2.2}$$

In Eq. (A2.2), t_s is the thickness of the adjacent support layer and t_{90} is the thickness of the cracked layer.

For a crack in the *surface layer*:

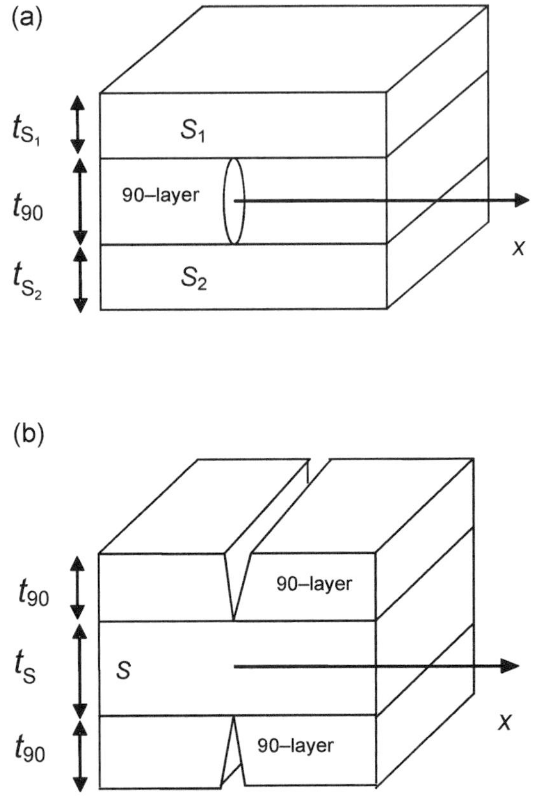

Figure 13.14 FEM models used to calculate COD of cracks (a) in the internal layer and (b) in surface layers.

$$A = 1.2, B = 0.5942 + 0.1901\left(2\frac{t_{90}}{t_s} - 1\right)$$

$$n = -0.52292\left(\frac{t_{90}}{t_s}\right)^2 + 0.8874\frac{t_{90}}{t_s} + 0.2576 \tag{A2.3}$$

The expressions presented correspond to a special case: the cracked layer is supported by a layer with certain E_x^s and thickness t_s followed by another damaged 90 degrees layer. Suggestions for calculations in more realistic cases when the support layers from different sides are different are given in Varna (2013).

CSD u_{1an}^0 (see Lundmark and Varna (2006) for details) also follows the power law:

$$u_{1an}^0 = A + B\left(\frac{G_{12}}{G_{xy}^S}\right)^n \tag{A2.4}$$

In Eq. (A2.4), G_{xy}^S is the in-plane shear modulus of the support layer.
For cracks in the *internal* layer:

$$A = 0.3, \quad B = 0.066 + 0.054\frac{t_{90}}{2t_s}, \quad n = 0.82 \tag{A2.5}$$

For cracks in the *surface layer*:

$$A = 0.6, \quad B = 0.134 + 0.105\frac{t_{90}}{t_s/2}, \quad n = 0.82 \tag{A2.6}$$

Eqs. (A2.1−A2.3) and Eqs. (A2.4−A2.6) show that the normalized average COD and CSD are larger for less stiff surrounding layers, and approach to a certain asymptotic value with an increasing support layer and cracked layer stiffness ratio. For thicker support layers, the COD and CSD are smaller. This effect of neighboring layers on the crack face displacements is called the "constraint effect."

Owing to a nonlinear shear stress−shear strain response, the secant shear modulus of the layer will change with increasing laminate strain and will affect the value of u_{1an}^0 calculated according to Eq. (A2.4).

When the distance between cracks decreases (high dimensionless crack density) the stress, perturbation regions of individual cracks overlap and the normalized average COD and CSD start to decrease. The u_{2an}^k has been related to COD of noninteractive cracks, u_{2an}^{0k}, by the relationship (Lundmark and Varna, 2011):

$$u_{2an}^k = \lambda_k(\rho_{kn})u_{2an}^{0k} \tag{A2.7}$$

The crack interaction function, λ, is a function of the crack density in the layer, and generally speaking it depends on material and geometrical parameters of the cracked layer and surrounding layers. For noninteractive cracks, $\lambda = 1$.

Detailed analysis of the effect of different parameters on interaction function was performed by Lundmark and Varna (2011) using FEM. Weak interaction (2%−5%) is observable at normalized spacing $2l_{90}/t_{90} = 2.5$. Further decrease of spacing (increase of crack density) may lead to a dramatic drop of the values of the interaction function to 0.3. The interaction of cracks in glass−epoxy laminates is stronger than in carbon−epoxy laminates. In the latter at a high stiffness ratio, the interaction function is not sensitive to the layer thickness ratio. In the former with lower layer stiffness ratio, the interaction is stronger if the support layer is thicker.

The calculated values of the interaction function were fitted by an empirical relationship with an origin in a simple shear lag model. The interaction function according to the shear lag model is:

$$\lambda_k = \tanh\left(\frac{k}{\rho_{kn}}\right) \tag{A2.8}$$

The shape function in Eq. (A2.8) was used to obtain the k value from the best fit. The best fit with this function to data corresponding to CF laminates is ($k_{CF} = 1.12$), and for GF laminates it is ($k_{GF} = 0.84$). The interaction effect on u_{2an} for cracks in the surface layer was analyzed by Loukil et al. (2013), where more accurate interaction functions also are presented. It has to be noted that this interaction function has to be used with the COD of noninteractive cracks calculated using FEM and not with the power law presented above. In this chapter, the approximate power law expressions for noninteractive cracks are used and, therefore, the interaction function from Loukil et al. (2013) is not the best option.

References

Allen, D.H., Yoon, C., 1998. Homogenization techniques for thermo-viscoelastic solids containing cracks. Int. J. Solid Struct. 35, 4035–4053.

Garrett, K.W., Bailey, J.E., 1977. Multiple transverse fracture in 90 degrees cross-ply laminates of a glass fibre-reinforced polyester. J. Mater. Sci. 12 (1), 157–168.

Giannadakis, K., Varna, J., 2014. Potential of a simple variational analysis in predicting shear modulus of laminates with cracks in 90-layers. J. Compos. Mater. 48 (15), 1843–1856.

Hashin, Z., 1985. Analysis of cracked laminates: a variational approach. Mech. Mater. 4, 121–136. North-Holland.

Huang, Y., Varna, J., Talreja, R., 2011. The effect of manufacturing quality on transverse cracking in cross ply laminates. In: Bhattacharyya, D., Lin, R., Srivatsan, T. (Eds.), Processing and Fabrication of Advanced Materials, vol. XIX. Centre for Advanced Composite Materials, University of Auckland, Auckland.

Jamison, R.D., Schulte, K., Reifsnider, K.L., Stinchcomb, W.W., 1984. Characterization and analysis of damage mechanisms in tension-tension fatigue of graphite/epoxy laminates. In: Effects of Defects in Composite Materials, vol. 836. ASTM STP, pp. 21–55.

Joffe, R., Krasnikovs, A., Varna, J., 2001. COD-based simulation of transverse cracking and stiffness reduction in $[S/90_n]_s$ laminates. Compos. Sci. Technol. 61, 637–656.

Kashtalyan, M., Soutis, C., 2005. Analysis of composite laminates with intra- and interlaminar damage. Prog. Aero. Sci. 41, 152–173.

Kim, R.Y., Crasto, A.S., Schoeppner, G.A., 2000. Dimensional stability of composite in space thermal environment. Compos. Sci. Technol. 60, 2601–2608.

Loukil, M.S., Varna, J., Ayadi, Z., 2012. Applicability of solutions for periodic intralaminar crack distributions to non-uniformly damaged laminates. J. Compos. Mater. 47 (3), 287–301.

Loukil, M.S., Varna, J., Ayadi, Z., 2013. Engineering expressions for thermo-elastic constants of laminates with high density of transverse cracks. Composites 48, 37–46.

Loukil, M., Ayadi, Z., Varna, J., 2014. ESPI analysis of crack face displacements in damaged laminates. J Compos Sci Technol 94 (9), 80–88.

Lundmark, P., Varna, J., 2005. Constitutive relationships for laminates with ply cracks in in-plane loading. Int. J. Damage Mech. 14 (3), 235–261.

Lundmark, P., Varna, J., 2006. Crack face sliding effect on stiffness of laminates with ply cracks. Compos. Sci. Technol. 66, 1444–1454.

Lundmark, P., Varna, J., 2011. Stiffness reduction in laminates at high intralaminar crack density: effect of crack interaction. Int. J. Damage Mech. 20, 279–297.

Manders, P.W., Chou, T.-W., Jones, R.J., Rock, J.W., 1983. Statistical analysis of multiple fracture in 0°/90°/0° glass fibre/epoxy resin laminates. J. Mater. Sci. 18 (10), 2876–2889.

McCartney, L.N., Schoeppner, G.A., 2000. Predicting the effect of non-uniform ply cracking on the thermo-elastic properties of cross-ply laminates. Compos. Sci. Technol. 62, 1841–1856.

McCartney, L.N., Schoeppner, G.A., Becker, W., 2000. Comparison of models for transverse ply cracks in composite laminates. Compos. Sci. Technol. 60, 2347–2359.

Nairn, J., Hu, S., 1994. Matrix microcracking. In: Pipes, R.B., Talreja, R. (Eds.), Damage Mechanics of Composite Materials, Composite Materials Series, vol. 9. Elsevier, Amsterdam, pp. 187–243.

Nuismer, R.J., Tan, S.C., 1988. Constitutive relations of a cracked composite lamina. J. Compos. Mater. 22, 306–321.

Ogin, S.L., Smith, P.A., Beaumont, P.W.R., 1985. Matrix cracking and stiffness reduction during the fatigue of a $[0/90]_s$ GFRP laminate. Compos. Sci. Technol. 22, 23–31.

Parvizi, A., Bailey, J.E., 1978. On multiple transverse cracking in glass fibre epoxy cross-ply laminates. J. Mater. Sci. 13, 2131–2136.

Peters, P.W.M., 1984. The strength distribution of 90-plies in 0/90/0 graphite-epoxy laminates. J. Compos. Mater. 18, 545–556.

Peters, P.W.M., Chout, T.-W., 1987. On cross-ply cracking in glass and carbon fibre reinforced epoxy laminates. Composites 18, 40–46.

Smith, P.A., Wood, J.R., 1990. Poisson's ratio as a damage parameter in the static tensile loading of simple cross-ply laminates. Compos. Sci. Technol. 38, 85–93.

Takeda, N., Ogihara, S., Kobayashi, A., 1995. Microscopic fatigue damage progress in CFRP cross-ply laminates. Composites 26, 859–867.

Tan, S.C., Nuismer, R.J., 1989. A theory for progressive matrix cracking in composite laminates. J. Compos. Mater. 23, 1029–1047.

Tsai, C.L., Daniel, I.M., 1992. The behaviour of cracked cross-ply composite laminates under shear loading. Int. J. Solid Struct. 29 (24), 3251–3267.

Varna, J., 2006. Quantification of damage and evolution modeling in multidirectional laminates. In: Proc. of the 2th RISÖ International Symposium on Material Science, Roskilde, Denmark, pp. 349–356.

Varna, J., 2012. On effective properties of layers in damaged composite laminates. Key Eng. Mater. 488–493. www.scientific.net/KEM.

Varna, J., 2013. Modeling mechanical performance of damaged laminates. J. Compos. Mater. 47 (20–21), 2443–2475.

Varna, J., Berglund, L.A., 1991. Multiple transverse cracking and stiffness reduction in cross-ply laminates. J. Compos. Technol. Res. 13 (2), 97–106.

Varna, J., Berglund, L.A., 1994. Thermo-elastic properties of composite laminates with transverse cracks. J. Compos. Technol. Res. 16 (1), 77–87.

Varna, J., Berglund, L.A., Talreja, R., Jakovics, A., 1993. A study of the crack opening displacement of transverse cracks in cross ply laminates. Int. J. Damage Mech. 2, 272–289.

Varna, J., Joffe, R., Akshantala, N.V., Talreja, R., 1999. Damage in composite laminates with off-axis plies. Compos. Sci. Technol. 59, 2139–2147.

Vinogradov, V., Hashin, Z., 2010. Variational analysis of angle-ply laminates. Compos. Sci. Technol. 70, 638–646.

Zhang, J., Fan, J., Soutis, C., 1992. Analysis of multiple matrix cracking in $[\pm\theta_m/90_n]_s$ composite laminates. Part 1. In-plane stiffness properties. Composites 23 (5), 291–304.

Microlevel approaches to modeling of damage in composite materials: Generalized plane strain analysis

14

L.N. McCartney
Department of Electromagnetic and Electrochemical Technologies, National Physical Laboratory, Teddington, United Kingdom

14.1 Introduction

Ensuring the structural integrity of composite components for service conditions is a very important design requirement relevant to all industrial sectors where composite materials are used in load-bearing situations. Reliable techniques for the prediction of the effective properties of composites based on the principles of mechanics can offer much scope for the reduction of development costs arising from make-and-test development procedures. Methodologies to achieve the design requirements must, however, take into account the complexities of composite material structures (e.g., fiber orientations), the geometry of the component, the nature of the loads applied, and the effects of temperature and moisture ingress, including residual stresses that arise during manufacturing processes.

There are three types of behavior that need to be considered when designing composite components. First of all, for relatively low applied loadings, the composite deforms as a linear elastic solid without the formation of any type of microstructural damage, that is, all interfaces between plies in laminates, and between fibers and matrix within plies, are perfectly bonded. Many models predicting the thermoelastic constants of unidirectional (UD) and laminated composites have been developed for this type of behavior. For larger applied loads, the composite experiences the second type of behavior, where microstructural damage formation occurs, leading to the progressive degradation of material properties and to nonlinear stress−strain behavior. Following the progressive stable growth of microstructural damage, the third type of behavior ensues, where instability and catastrophic failure of the structure occurs. Important design objectives are to delay the occurrence of microstructural damage, to extend the progressive microstructural damage growth phase of deformation, and to delay the onset of catastrophic failure. This chapter is concerned with the first two design objectives, with some emphasis on the determination of the dependence of many effective properties on the density of uniformly distributed arrays of ply cracks.

Modeling Damage, Fatigue and Failure of Composite Materials. https://doi.org/10.1016/B978-0-443-18489-5.00011-4

To achieve the overall design objectives, it is necessary to be able to understand and predict the nature of stress transfer between neighboring plies of a laminate when ply cracks form in response to loading. The immediate objective is to determine both the stress and displacement distributions within each ply of the laminate. In general, such distributions are three-dimensional and their determination using analytical methods is almost impossible due to both the presence of three-dimensional fields and singularities at both free surfaces (at interfaces) and at ply crack tips. The presence of the singularities can lead to difficulties, even if using numerical techniques such as finite-element or boundary element methods. To make progress using analytical methods, it is necessary to introduce approximations, and the purpose is to determine an approach that minimizes the loss of accuracy that must result. The objective of this chapter is to consider in some detail the approach to stress-transfer modeling known as the "generalized plane strain" approximation.

The generalized plane strain approach can be applied to any general symmetric laminate subject to general in-plane loading, although in achieving this objective the analysis becomes very complex, especially when considering laminates having many different ply orientations. Because of this, it is much more useful to describe the application of generalized plane strain conditions to simple laminates. This chapter will, therefore, consider in detail a simple cross-ply laminate in order to identify the types of extremely useful results that can be derived. Corresponding results for general symmetric laminates will be presented briefly, but readers are referred elsewhere for their derivation.

Section 14.2 introduces the fundamental equations and boundary conditions that should be satisfied, and defines conditions of generalized plane strain. Section 14.3 provides the solution for stress, strain, and the corresponding effective thermoelastic constants for a simple undamaged cross-ply laminate. The simplest type of stress-transfer model, known as shear lag analysis, is described in Section 14.4, and the generalized plane strain model for a cross-ply laminate is described in Sections 14.5−14.8, including the consideration of ply-crack closure. Analogous results for general symmetric laminates are given in Section 14.9 and prediction examples are given in Section 14.10.

14.2 Fundamental equations and conditions

Consider the model of a simple $[0/90]_s$ cross-ply laminate illustrated in Fig. 14.1, where two possible representative volume elements (RVEs) are shown. For the first shown in Fig. 14.1a, one ply crack in the inner 90 degrees ply is located on the plane $x_1 = 0$ and neighboring ply cracks (not shown) are on the planes $x_1 = \pm 2L$. For the second RVE shown in Fig. 14.1b, the plane $x_1 = 0$ is midway between two neighboring ply cracks in the inner 90 degrees ply on the planes $x_1 = \pm L$. The inner 90 degrees ply has total thickness $2a$ and the two outer 0 degree plies each have thickness denoted by b, so that the total thickness of the laminate is $2h$ where $h = a + b$. A set of Cartesian coordinates (x_1, x_2, x_3) is introduced such that the origin lies on the mid-plane

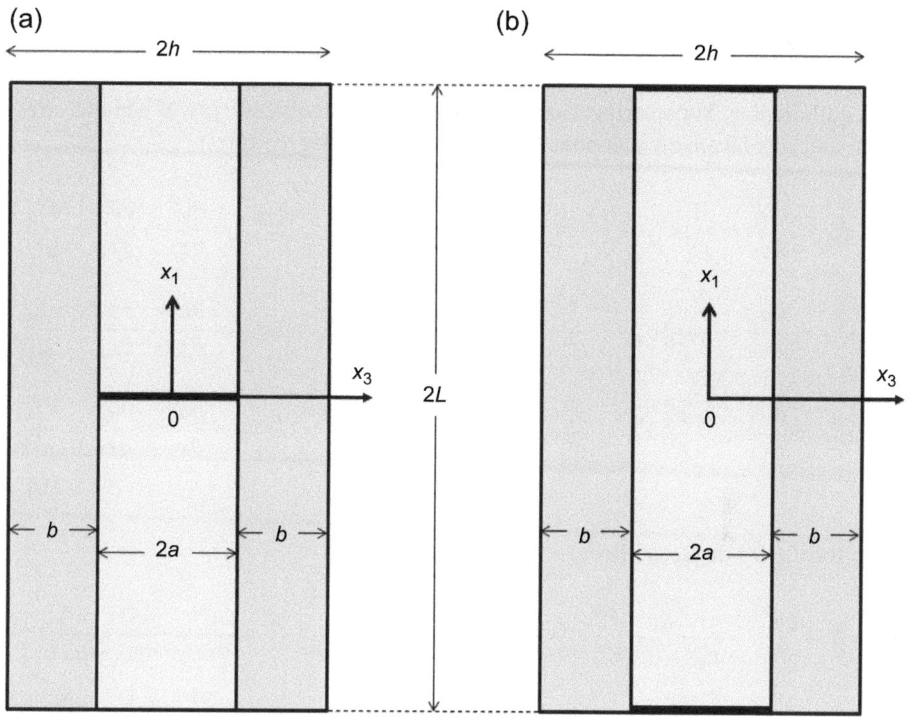

Figure 14.1 Representative volume elements for a cracked cross-ply laminate: (a) with crack on mid-plane, (b) with cracks on boundaries of representative region of laminate.

of the laminate at the midpoint between two neighboring cracks in the 90 degrees ply. The x_1-axis is directed along the principal loading direction and the x_3-axis is directed toward the through-thickness direction. The x_2-axis of the laminate, having width $2W$, is directed along the transverse in-plane loading direction.

14.2.1 Basic field equations

The following equilibrium equations must be satisfied for both the 0 degree and 90 degrees plies:

$$\frac{\partial \sigma_{11}}{\partial x_1} + \frac{\partial \sigma_{12}}{\partial x_2} + \frac{\partial \sigma_{13}}{\partial x_3} = 0, \tag{14.1}$$

$$\frac{\partial \sigma_{12}}{\partial x_1} + \frac{\partial \sigma_{22}}{\partial x_2} + \frac{\partial \sigma_{23}}{\partial x_3} = 0, \tag{14.2}$$

$$\frac{\partial \sigma_{13}}{\partial x_1} + \frac{\partial \sigma_{23}}{\partial x_2} + \frac{\partial \sigma_{33}}{\partial x_3} = 0, \tag{14.3}$$

where σ_{ij} refers to the stress components. The plies are regarded as transverse isotropic solids so that the stress–strain–temperature relations involve the axial and transverse values of Young's modulus E, Poisson ratio ν, shear modulus μ, and thermal expansion coefficient α. Superscripts 0 or 90 will be used to denote the ply to which a stress, strain, and displacement component refers. For the 0 degree plies:

$$\begin{aligned}
\varepsilon_{11}^{0} &= \frac{\partial u_1^0}{\partial x_1} = \frac{1}{E_A^0}\sigma_{11}^0 - \frac{\nu_A^0}{E_A^0}\sigma_{22}^0 - \frac{\nu_a^0}{E_A^0}\sigma_{33}^0 + \alpha_A^0 \Delta T, & 2\varepsilon_{12}^0 &= \frac{\partial u_1^0}{\partial x_2} + \frac{\partial u_2^0}{\partial x_1} = \frac{\sigma_{12}^0}{\mu_A^0}, \\
\varepsilon_{22}^{0} &= \frac{\partial u_2^0}{\partial x_2} = -\frac{\nu_A^0}{E_A^0}\sigma_{11}^0 + \frac{1}{E_T^0}\sigma_{22}^0 - \frac{\nu_t^0}{E_T^0}\sigma_{33}^0 + \alpha_T^0 \Delta T, & 2\varepsilon_{13}^0 &= \frac{\partial u_1^0}{\partial x_3} + \frac{\partial u_3^0}{\partial x_1} = \frac{\sigma_{13}^0}{\mu_a^0}, \\
\varepsilon_{33}^{0} &= \frac{\partial u_3^0}{\partial x_3} = -\frac{\nu_a^0}{E_A^0}\sigma_{11}^0 - \frac{\nu_t^0}{E_T^0}\sigma_{22}^0 + \frac{1}{E_t^0}\sigma_{33}^0 + \alpha_t^0 \Delta T, & 2\varepsilon_{23}^0 &= \frac{\partial u_2^0}{\partial x_3} + \frac{\partial u_3^0}{\partial x_2} = \frac{\sigma_{23}^0}{\mu_t^0},
\end{aligned} \tag{14.4}$$

while for the 90 degrees plies:

$$\begin{aligned}
\varepsilon_{11}^{90} &= \frac{\partial u_1^{90}}{\partial x_1} = \frac{1}{E_T^{90}}\sigma_{11}^{90} - \frac{\nu_A^{90}}{E_A^{90}}\sigma_{22}^{90} - \frac{\nu_t^{90}}{E_T^{90}}\sigma_{33}^{90} + \alpha_T^{90} \Delta T, & 2\varepsilon_{12}^{90} &= \frac{\partial u_1^{90}}{\partial x_2} + \frac{\partial u_2^{90}}{\partial x_1} = \frac{\sigma_{12}^{90}}{\mu_A^{90}}, \\
\varepsilon_{22}^{90} &= \frac{\partial u_2^{90}}{\partial x_2} = -\frac{\nu_A^{90}}{E_A^{90}}\sigma_{11}^{90} + \frac{1}{E_A^{90}}\sigma_{22}^{90} - \frac{\nu_a^{90}}{E_A^{90}}\sigma_{33}^{90} + \alpha_A^{90} \Delta T, & 2\varepsilon_{13}^{90} &= \frac{\partial u_1^{90}}{\partial x_3} + \frac{\partial u_3^{90}}{\partial x_1} = \frac{\sigma_{13}^{90}}{\mu_t^{90}}, \\
\varepsilon_{33}^{90} &= \frac{\partial u_3^{90}}{\partial x_3} = -\frac{\nu_t^{90}}{E_T^{90}}\sigma_{11}^{90} - \frac{\nu_a^{90}}{E_A^{90}}\sigma_{22}^{90} + \frac{1}{E_t^{90}}\sigma_{33}^{90} + \alpha_t^{90} \Delta T, & 2\varepsilon_{23}^{90} &= \frac{\partial u_2^{90}}{\partial x_3} + \frac{\partial u_3^{90}}{\partial x_2} = \frac{\sigma_{23}^{90}}{\mu_a^{90}},
\end{aligned} \tag{14.5}$$

where the strain and displacement components are denoted by ε_{ij} and u_i, respectively. The subscripts A, T, and t are attached to the properties to associate them, respectively, with the axial, in-plane transverse, and through-thickness directions of the lamina. It should be noted that the upper-case subscripts A and T are associated only with the in-plane directions, while the lower-case subscripts are associated with the through-thickness direction. The relations in Eq. (14.5) are either obtained by modifying directly the relations in Eq. (14.4) for the 0 degree plies or by using property transformations for rotations of the ply by an angle ±90 degrees. The thermoelastic constants of individual plies in a laminate are usually assumed to be transverse isotropic, so that:

$$E_t = E_T, \quad \nu_a = \nu_A, \quad \mu_a = \mu_A, \quad \alpha_t = \alpha_T \text{ and } E_T = 2\mu_t(1 + \nu_t).$$

14.2.2 Boundary and interface conditions

The loading of the laminate is achieved by applying uniform effective in-plane strains ε_A and ε_T together with a uniform through-thickness stress σ_t. Shear loading is not considered here. In order that the field equations can be solved uniquely, it is necessary to impose a sufficient number of boundary and interface conditions. The free surface ($x_3 = \pm h$) and interface ($x_3 = \pm a$) conditions will first be considered. On the free surfaces:

$$\sigma_{33}^{0} = \sigma_{t}, \quad \sigma_{13}^{0} = \sigma_{23}^{0} = 0, \quad \text{on } x_3 = \pm h, \tag{14.6}$$

and on the interfaces:

$$\begin{aligned} \sigma_{33}^{0} = \sigma_{33}^{90}, \quad \sigma_{13}^{0} = \sigma_{13}^{90}, \quad \sigma_{23}^{0} = \sigma_{23}^{90}, \quad \text{on } x_3 = \pm a. \\ u_1^{0} = u_1^{90}, \quad u_2^{0} = u_2^{90}, \quad u_3^{0} = u_3^{90}, \end{aligned} \tag{14.7}$$

The edges $x_2 = \pm W$ are such that in-plane transverse displacement is uniform, having the following values:

$$u_2^{0} = u_2^{90} = \pm W \varepsilon_{T}, \quad \text{on } x_2 = \pm W, \tag{14.8}$$

where ε_T is the in-plane transverse strain that is uniform everywhere in the laminate when generalized plane strain conditions are imposed. The edges $x_2 = \pm W$ are assumed to have zero shear stresses, so that:

$$\sigma_{12}^{0} = \sigma_{12}^{90} = 0, \quad \sigma_{23}^{0} = \sigma_{23}^{90} = 0, \quad \text{on } x_2 = \pm W. \tag{14.9}$$

For the above boundary conditions, and because of the symmetric nature of the laminate, there will be symmetry about $x_3 = 0$ of the stress, strain, and displacement distributions such that the following conditions are satisfied:

$$\sigma_{13}^{0} = \sigma_{13}^{90} = 0, \quad \sigma_{23}^{0} = \sigma_{23}^{90} = 0, \quad u_3^{0} = 0, \quad \text{on } x_3 = 0. \tag{14.10}$$

When applying laminate edge conditions on planes normal to the x_1-axis, two possible approaches can be made. Consider first of all the RVE shown in Fig. 14.1a, which can be used for undamaged laminates, and for damaged laminates where a ply crack in the 90 degrees ply is located at $x_1 = 0$. The edges $x_1 = \pm L$ are such that in-plane axial displacement is uniform, having the following values:

$$u_1^{0} = u_1^{90} = \pm L \varepsilon_{A}, \quad \text{on } x_1 = \pm L, \tag{14.11}$$

where ε_A is the effective axial applied strain. The edges $x_1 = \pm L$ are assumed to have zero shear stresses, so that:

$$\sigma_{12}^{0} = \sigma_{12}^{90} = 0, \quad \sigma_{13}^{0} = \sigma_{13}^{90} = 0, \quad \text{on } x_1 = \pm L. \tag{14.12}$$

These conditions imply that there is symmetry about the plane $x_1 = 0$, so that:

$$\sigma_{12}^{0} = \sigma_{12}^{90} = 0, \quad \sigma_{13}^{0} = \sigma_{13}^{90} = 0, \quad u_1^{0} = u_1^{90} = 0, \quad \text{on } x_1 = 0. \tag{14.13}$$

Consider now the RVE shown in Fig. 14.1b, where the 90 degrees ply cracks are located on the planes $x_1 = \pm L$. The boundary conditions applied are given as:

$$\sigma_{12}^{0} = \sigma_{12}^{90} = 0, \quad \sigma_{13}^{0} = \sigma_{13}^{90} = 0, \quad \sigma_{11}^{90} = 0, \quad u_1^{0} = \pm L\varepsilon_A, \quad \text{on } x_1 = \pm L. \tag{14.14}$$

It is clear from both the RVEs shown in Fig. 14.1, and the boundary conditions applied on planes normal to the x_1-axis, that for uniform arrays of ply cracks there is symmetry about the planes $x_1 = 0$ and $x_1 = \pm L$.

14.2.3 Generalized plane strain conditions

Generalized plane strain conditions are assumed so that the stress and strain distributions do not depend on the x_2-coordinate. This situation occurs when the displacement field is of the following form:

$$u_1 = u_1(x_1, x_3), \quad u_2 = \varepsilon_T x_2, \quad u_3 = u_3(x_1, x_3), \tag{14.15}$$

where $\varepsilon_{22} = \varepsilon_{\mathrm{T}}$ is the uniform in-plane transverse strain in both of the 0 degree and 90 degrees plies. In addition, it is assumed that on the laminate faces $x_3 = \pm h$ that $\sigma_{33}^{0} \equiv \sigma_{33}^{90} \equiv \sigma_{\mathrm{t}}$, where σ_{t} is a uniformly applied through-thickness stress and $\sigma_{13} \equiv \sigma_{23} \equiv 0$. If $\varepsilon_{\mathrm{T}} = 0$, then the laminate is highly constrained in the transverse direction, leading to the well-known conditions of plane strain deformation. It should be noted that the solution for undamaged laminates is automatically one of the generalized plane strains, as the transverse strain ε_{T} is uniform everywhere in those regions of the laminate that are sufficiently far away from the edges.

From Eq. (14.4), it then follows that for the 0 degree ply:

$$\sigma_{22}^{0} = \frac{\nu_A^0 E_T^0}{E_A^0}\sigma_{11}^{0} + \nu_t^0 \sigma_{33}^{0} - E_T^0 \alpha_T^0 \Delta T + E_T^0 \varepsilon_T, \tag{14.16}$$

and

$$\varepsilon_{11}^{0} = \frac{1}{\widetilde{E}_A^0}\sigma_{11}^{0} - \frac{\widetilde{\nu}_a^0}{\widetilde{E}_A^0}\sigma_{33}^{0} + \widetilde{\alpha}_A^0 \Delta T - \nu_A^0 \frac{E_T^0}{E_A^0}\varepsilon_T, \tag{14.17}$$

$$\varepsilon_{33}^{0} = -\frac{\widetilde{\nu}_a^{0}}{\widetilde{E}_A^{0}}\sigma_{11}^{0} + \frac{1}{\widetilde{E}_t^{0}}\sigma_{33}^{0} + \widetilde{\alpha}_t^{0}\Delta T - \nu_t^{0}\varepsilon_T, \tag{14.18}$$

where

$$\frac{1}{\widetilde{E}_t^{0}} = \frac{1}{E_t^{0}} - \frac{\left(\nu_t^{0}\right)^2}{E_T^{0}}, \qquad \widetilde{\alpha}_t^{0} = \alpha_t^{0} + \nu_t^{0}\alpha_T^{0}. \tag{14.19}$$

For the 90 degrees ply, it follows from Eq. (14.5) that:

$$\sigma_{22}^{90} = \nu_A^{90}\sigma_{11}^{90} + \nu_a^{90}\sigma_{33}^{90} - E_A^{90}\alpha_A^{90}\Delta T + E_A^{90}\varepsilon_T, \tag{14.20}$$

and

$$\varepsilon_{11}^{90} = \frac{1}{\widetilde{E}_{\mathrm{T}}^{90}}\sigma_{11}^{90} - \frac{\widetilde{\nu}_{\mathrm{t}}^{90}}{\widetilde{E}_{\mathrm{T}}^{90}}\sigma_{33}^{90} + \widetilde{\alpha}_{\mathrm{T}}^{90}\Delta T - \nu_{\mathrm{A}}^{90}\varepsilon_{\mathrm{T}}, \tag{14.21}$$

$$\varepsilon_{33}^{90} = -\frac{\widetilde{\nu}_{\mathrm{t}}^{90}}{\widetilde{E}_{\mathrm{T}}^{90}}\sigma_{11}^{90} + \frac{1}{\widetilde{E}_{\mathrm{t}}^{90}}\sigma_{33}^{90} + \widetilde{\alpha}_{\mathrm{t}}^{90}\Delta T - \frac{\nu_{\mathrm{a}}^{90}}{E_{\mathrm{A}}^{90}}E_{\mathrm{A}}^{90}\varepsilon_{\mathrm{T}}, \tag{14.22}$$

where

$$\frac{1}{\widetilde{E}_t^{90}} = \frac{1}{E_t^{90}} - \frac{\left(\nu_a^{90}\right)^2}{E_A^{90}}, \qquad \widetilde{\alpha}_t^{90} = \alpha_t^{90} + \nu_a^{90}\alpha_A^{90}. \tag{14.23}$$

It follows from Eq. (14.15) and the strain−displacement relations that $\varepsilon_{12} = \varepsilon_{23} = 0$ in both the 0 degree and 90 degrees plies and from the shear stress−strain relations in Eqs. (14.4) and (14.5) that $\sigma_{12} = \sigma_{23} = 0$. The equilibrium equation (14.2) is then automatically satisfied as, from Eqs. (14.16) and (14.20), and the stress−strain relations in Eqs. (14.4) and (14.5), σ_{22} is independent of x_2 as are both σ_{11} and σ_{13}. The remaining Eqs. (14.1) and (14.3) are then reduced to the following form:

$$\frac{\partial\sigma_{11}}{\partial x_1} + \frac{\partial\sigma_{13}}{\partial x_3} = 0, \tag{14.24}$$

$$\frac{\partial\sigma_{13}}{\partial x_1} + \frac{\partial\sigma_{33}}{\partial x_3} = 0. \tag{14.25}$$

When developing stress-transfer models for application under conditions of generalized plane strain, it is very useful to consider averages through the thickness of the plies of various physical quantities. Assuming symmetry about the mid-plane $x_3 = 0$ of the laminate, the average of any quantities $f_0(x_1, x_3)$ and $f_{90}(x_1, x_3)$ associated with the 0 degree and 90 degrees plies is defined, respectively, by:

$$\bar{f}_0(x_1) = \frac{1}{b}\int_a^h f_0(x_1, x_3)\mathrm{d}x_3, \qquad \bar{f}_{90}(x_1) = \frac{1}{a}\int_0^a f_{90}(x_1, x_3)\mathrm{d}x_3. \tag{14.26}$$

14.3 Solution for undamaged laminates

For undamaged laminates and in regions where edge effects are negligible, the axial and transverse strains in both the 0 degree and 90 degrees plies have the uniform values $\widehat{\varepsilon}_\mathrm{A}$ and $\widehat{\varepsilon}_\mathrm{T}$, respectively. The "hat" symbol is attached to the in-plane strains to distinguish them from the differing values ε_A and ε_T that will arise when the laminate is damaged. The through-thickness stress in both plies has the uniform value σ_t. For undamaged laminates, the uniform stress and strain components in the plies will be denoted by $\widehat{\sigma}_{ij}$ and $\widehat{\varepsilon}_{ij}$, respectively. It follows from the stress–strain relations in Eq. (14.4) that:

$$\begin{aligned}
\widehat{\varepsilon}_A &= \frac{1}{E_A^0}\widehat{\sigma}^0_{11} - \frac{\nu_A^0}{E_A^0}\widehat{\sigma}^0_{22} - \frac{\nu_a^0}{E_A^0}\sigma_t + \alpha_A^0\Delta T,\\
\widehat{\varepsilon}_T &= -\frac{\nu_A^0}{E_A^0}\widehat{\sigma}^0_{11} + \frac{1}{E_T^0}\widehat{\sigma}^0_{22} - \frac{\nu_t^0}{E_T^0}\sigma_t + \alpha_T^0\Delta T,\\
\widehat{\varepsilon}^0_{33} &= -\frac{\nu_a^0}{E_A^0}\widehat{\sigma}^0_{11} - \frac{\nu_t^0}{E_T^0}\widehat{\sigma}^0_{22} + \frac{1}{E_t^0}\sigma_t + \alpha_t^0\Delta T,
\end{aligned} \tag{14.27}$$

so that upon solving the first two equations for $\widehat{\sigma}^0_{11}$ and $\widehat{\sigma}^0_{22}$:

$$\frac{1}{\widetilde{E}_A^0}\widehat{\sigma}^0_{11} = \widehat{\varepsilon}_A + \nu_A^0\frac{E_T^0}{E_A^0}\widehat{\varepsilon}_T + \frac{\widetilde{\nu}_a^0}{\widetilde{E}_A^0}\sigma_t - \widetilde{\alpha}_A^0\Delta T, \frac{1}{\widetilde{E}_T^0}\widehat{\sigma}^0_{22} = \nu_A^0\widehat{\varepsilon}_A + \widehat{\varepsilon}_T + \frac{\widetilde{\nu}_t^0}{\widetilde{E}_T^0}\sigma_t - \widetilde{\alpha}_T^0\Delta T, \tag{14.28}$$

and on substituting in the expression for $\widehat{\varepsilon}^0_{33}$:

$$\begin{aligned}
\widehat{\varepsilon}^0_{33} = &-\frac{\nu_\mathrm{a}^0 + \nu_\mathrm{t}^0\nu_\mathrm{A}^0}{E_\mathrm{A}^0}\widetilde{E}^0_\mathrm{A}\widehat{\varepsilon}_\mathrm{A} - \left(\frac{\nu_\mathrm{t}^0}{E_\mathrm{T}^0} + \frac{\nu_\mathrm{a}^0\nu_\mathrm{A}^0}{E_\mathrm{A}^0}\right)\widetilde{E}^0_\mathrm{T}\widehat{\varepsilon}_\mathrm{T} + \left(\frac{1}{E_\mathrm{t}^0} - \frac{\nu_\mathrm{a}^0\widetilde{\nu}^0_\mathrm{a}}{E_\mathrm{A}^0} - \frac{\nu_\mathrm{t}^0\widetilde{\nu}^0_\mathrm{t}}{E_\mathrm{T}^0}\right)\sigma_\mathrm{t}\\
&+\left(\alpha_\mathrm{t}^0 + \frac{\nu_\mathrm{a}^0}{E_\mathrm{A}^0}\widetilde{E}^0_\mathrm{A}\widetilde{\alpha}^0_\mathrm{A} + \frac{\nu_\mathrm{t}^0}{E_\mathrm{T}^0}\widetilde{E}^0_\mathrm{T}\widetilde{\alpha}^0_\mathrm{T}\right)\Delta T,
\end{aligned} \tag{14.29}$$

where

$$\frac{1}{\widetilde{E}_{\rm A}^{0}}=\frac{1}{E_{\rm A}^{0}}\left(1-\left(\nu_{\rm A}^{0}\right)^{2}\frac{E_{\rm T}^{0}}{E_{\rm A}^{0}}\right),\quad \frac{1}{\widetilde{E}_{\rm T}^{0}}=\frac{1}{E_{\rm T}^{0}}\left(1-\left(\nu_{\rm A}^{0}\right)^{2}\frac{E_{\rm T}^{0}}{E_{\rm A}^{0}}\right),$$
$$\frac{\widetilde{\nu}_{\rm a}^{0}}{\widetilde{E}_{\rm A}^{0}}=\frac{\nu_{\rm a}^{0}+\nu_{\rm t}^{0}\nu_{\rm A}^{0}}{E_{\rm A}^{0}},\quad \frac{\widetilde{\nu}_{\rm t}^{0}}{\widetilde{E}_{\rm T}^{0}}=\frac{\nu_{\rm t}^{0}}{E_{\rm T}^{0}}+\frac{\nu_{\rm a}^{0}\nu_{\rm A}^{0}}{E_{\rm A}^{0}},$$
$$\widetilde{\alpha}_{\rm A}^{0}=\alpha_{\rm A}^{0}+\nu_{\rm A}^{0}\frac{E_{\rm T}^{0}}{E_{\rm A}^{0}}\alpha_{\rm T}^{0},\quad \widetilde{\alpha}_{\rm T}^{0}=\alpha_{\rm T}^{0}+\nu_{\rm A}^{0}\alpha_{\rm A}^{0}. \tag{14.30}$$

Similarly from Eq. (14.5):

$$\widehat{\varepsilon}_{A}=\frac{1}{E_{T}^{90}}\widehat{\sigma}_{11}^{90}-\frac{\nu_{A}^{90}}{E_{A}^{90}}\widehat{\sigma}_{22}^{90}-\frac{\nu_{t}^{90}}{E_{T}^{90}}\sigma_{t}+\alpha_{T}^{90}\Delta T,$$
$$\widehat{\varepsilon}_{T}=-\frac{\nu_{A}^{90}}{E_{A}^{90}}\widehat{\sigma}_{11}^{90}+\frac{1}{E_{A}^{90}}\widehat{\sigma}_{22}^{90}-\frac{\nu_{a}^{90}}{E_{A}^{90}}\sigma_{t}+\alpha_{A}^{90}\Delta T, \tag{14.31}$$
$$\widehat{\varepsilon}_{33}^{90}=-\frac{\nu_{t}^{90}}{E_{T}^{90}}\widehat{\sigma}_{11}^{90}-\frac{\nu_{a}^{90}}{E_{A}^{90}}\widehat{\sigma}_{22}^{90}+\frac{1}{E_{t}^{90}}\sigma_{t}+\alpha_{t}^{90}\Delta T,$$

so that upon solving the first two equations for $\widehat{\sigma}_{11}^{90}$ and $\widehat{\sigma}_{22}^{90}$:

$$\frac{1}{\widetilde{E}_{T}^{90}}\widehat{\sigma}_{11}^{90}=\widehat{\varepsilon}_{A}+\nu_{A}^{90}\widehat{\varepsilon}_{T}+\frac{\widetilde{\nu}_{t}^{90}}{\widetilde{E}_{T}^{90}}\sigma_{t}-\widetilde{\alpha}_{T}^{90}\Delta T,$$
$$\frac{1}{\widetilde{E}_{A}^{90}}\widehat{\sigma}_{22}^{90}=\nu_{A}^{90}\frac{E_{T}^{90}}{E_{A}^{90}}\widehat{\varepsilon}_{A}+\widehat{\varepsilon}_{T}+\frac{\widetilde{\nu}_{a}^{90}}{\widetilde{E}_{A}^{90}}\sigma_{t}-\widetilde{\alpha}_{A}^{90}\Delta T, \tag{14.32}$$

and on substituting in the expression for $\widehat{\varepsilon}_{33}^{90}$:

$$\widehat{\varepsilon}_{33}^{90}=-\left(\frac{\nu_{\rm t}^{90}}{E_{\rm T}^{90}}+\frac{\nu_{\rm a}^{90}\nu_{\rm A}^{90}}{E_{\rm A}^{90}}\right)\widetilde{E}_{\rm T}^{90}\widehat{\varepsilon}_{\rm A}-\frac{\nu_{\rm a}^{90}+\nu_{\rm t}^{90}\nu_{\rm A}^{90}}{E_{\rm A}^{90}}\widetilde{E}_{\rm A}^{90}\widehat{\varepsilon}_{\rm T}$$
$$+\left(\frac{1}{E_{\rm t}^{90}}-\frac{\nu_{\rm t}^{90}\widetilde{\nu}_{\rm t}^{90}}{E_{\rm T}^{90}}-\frac{\nu_{\rm a}^{90}\widetilde{\nu}_{\rm a}^{90}}{E_{\rm A}^{90}}\right)\sigma_{\rm t}+\left(\alpha_{\rm t}^{90}+\frac{\nu_{\rm a}^{90}}{E_{\rm A}^{90}}\widetilde{E}_{\rm A}^{90}\widetilde{\alpha}_{\rm A}^{90}+\frac{\nu_{\rm t}^{90}}{E_{\rm T}^{90}}\widetilde{E}_{\rm T}^{90}\widetilde{\alpha}_{\rm T}^{90}\right)\Delta T, \tag{14.33}$$

where

$$\frac{1}{\widetilde{E}_{\mathrm{A}}^{90}}=\frac{1}{E_{\mathrm{A}}^{90}}\left(1-\left(\nu_{\mathrm{A}}^{90}\right)^2\frac{E_{\mathrm{T}}^{90}}{E_{\mathrm{A}}^{90}}\right),\quad \frac{1}{\widetilde{E}_{\mathrm{T}}^{90}}=\frac{1}{E_{\mathrm{T}}^{90}}\left(1-\left(\nu_{\mathrm{A}}^{90}\right)^2\frac{E_{\mathrm{T}}^{90}}{E_{\mathrm{A}}^{90}}\right),$$

$$\frac{\widetilde{\nu}_{\mathrm{a}}^{90}}{\widetilde{E}_{\mathrm{A}}^{90}}=\frac{\nu_{\mathrm{a}}^{90}}{E_{\mathrm{A}}^{90}}+\frac{\nu_{\mathrm{t}}^{90}\nu_{\mathrm{A}}^{90}}{E_{\mathrm{A}}^{90}},\qquad \frac{\widetilde{\nu}_{\mathrm{t}}^{90}}{\widetilde{E}_{\mathrm{T}}^{90}}=\frac{\nu_{\mathrm{t}}^{90}}{E_{\mathrm{T}}^{90}}+\frac{\nu_{\mathrm{a}}^{90}\nu_{\mathrm{A}}^{90}}{E_{\mathrm{A}}^{90}}, \tag{14.34}$$

$$\widetilde{\alpha}_{\mathrm{A}}^{90}=\alpha_{\mathrm{A}}^{90}+\nu_{\mathrm{A}}^{90}\frac{E_{\mathrm{T}}^{90}}{E_{\mathrm{A}}^{90}}\alpha_{\mathrm{T}}^{90},\qquad \widetilde{\alpha}_{\mathrm{T}}^{90}=\alpha_{\mathrm{T}}^{90}+\nu_{\mathrm{A}}^{90}\alpha_{\mathrm{A}}^{90}.$$

From a consideration of mechanical equilibrium, the uniform ply stresses can be used to define, for an undamaged laminate, the effective axial stress $\widehat{\sigma}_{\mathrm{A}}$ and the effective in-plane transverse stress $\widehat{\sigma}_{\mathrm{T}}$ as follows:

$$h\widehat{\sigma}_{\mathrm{A}}=b\widehat{\sigma}_{11}^{0}+a\widehat{\sigma}_{11}^{90}, \tag{14.35}$$

$$h\widehat{\sigma}_{\mathrm{T}}=b\widehat{\sigma}_{22}^{0}+a\widehat{\sigma}_{22}^{90}. \tag{14.36}$$

The hat symbol is used to distinguish these effective stresses from those that will result when the laminate is damaged. Corresponding to the uniform through-thickness stress σ_{t}, an effective through-thickness strain, $\widehat{\varepsilon}_{\mathrm{t}}$, can be defined by the relation:

$$h\widehat{\varepsilon}_{\mathrm{t}}=b\widehat{\varepsilon}_{33}^{0}+a\widehat{\varepsilon}_{33}^{90}. \tag{14.37}$$

It should be noted that the value σ_{t} for the through-thickness stress of an undamaged laminate corresponds to the effective value when the laminate is damaged. On using the relations in Eqs. (14.35) and (14.36) together with the results in Eqs. (14.28) and (14.32), it can be shown that:

$$\begin{aligned}\widehat{\sigma}_{\mathrm{A}}&=A\widehat{\varepsilon}_{\mathrm{A}}+B\widehat{\varepsilon}_{\mathrm{T}}+C\sigma_{\mathrm{t}}-P\Delta T,\\ \widehat{\sigma}_{\mathrm{T}}&=B\widehat{\varepsilon}_{\mathrm{A}}+F\widehat{\varepsilon}_{\mathrm{T}}+G\sigma_{\mathrm{t}}-Q\Delta T,\end{aligned} \tag{14.38}$$

where

$$A=\frac{b}{h}\widetilde{E}_{\mathrm{A}}^{0}+\frac{a}{h}\widetilde{E}_{\mathrm{T}}^{90},\quad B=\frac{b}{h}\nu_{\mathrm{A}}^{0}\widetilde{E}_{\mathrm{T}}^{0}+\frac{a}{h}\nu_{\mathrm{A}}^{90}\widetilde{E}_{\mathrm{T}}^{90},\quad C=\frac{b}{h}\widetilde{\nu}_{\mathrm{a}}^{0}+\frac{a}{h}\widetilde{\nu}_{\mathrm{t}}^{90},$$

$$F=\frac{b}{h}\widetilde{E}_{\mathrm{T}}^{0}+\frac{a}{h}\widetilde{E}_{\mathrm{A}}^{90},\quad G=\frac{b}{h}\widetilde{\nu}_{\mathrm{t}}^{0}+\frac{a}{h}\widetilde{\nu}_{\mathrm{a}}^{90},$$

$$P=\frac{b}{h}\widetilde{E}_{\mathrm{A}}^{0}\widetilde{\alpha}_{\mathrm{A}}^{0}+\frac{a}{h}\widetilde{E}_{\mathrm{T}}^{90}\widetilde{\alpha}_{\mathrm{T}}^{90},\quad Q=\frac{b}{h}\widetilde{E}_{\mathrm{T}}^{0}\widetilde{\alpha}_{\mathrm{T}}^{0}+\frac{a}{h}\widetilde{E}_{\mathrm{A}}^{90}\widetilde{\alpha}_{\mathrm{A}}^{90}. \tag{14.39}$$

Upon solving Eq. (14.38) for the in-plane strains $\widehat{\varepsilon}_{\mathrm{A}}$ and $\widehat{\varepsilon}_{\mathrm{T}}$:

$$\widehat{\varepsilon}_{\mathrm{A}} = \frac{1}{E_{\mathrm{A}}^{(L)}}\widehat{\sigma}_{\mathrm{A}} - \frac{\nu_{\mathrm{A}}^{(L)}}{E_{\mathrm{A}}^{(L)}}\widehat{\sigma}_{\mathrm{T}} - \frac{\nu_{\mathrm{a}}^{(L)}}{E_{\mathrm{A}}^{(L)}}\sigma_{\mathrm{t}} + \alpha_{\mathrm{A}}^{(L)}\Delta T,$$
$$\widehat{\varepsilon}_{\mathrm{T}} = -\frac{\nu_{\mathrm{A}}^{(L)}}{E_{\mathrm{A}}^{(L)}}\widehat{\sigma}_{\mathrm{A}} + \frac{1}{E_{\mathrm{T}}^{(L)}}\widehat{\sigma}_{\mathrm{T}} - \frac{\nu_{\mathrm{t}}^{(L)}}{E_{\mathrm{T}}^{(L)}}\sigma_{\mathrm{t}} + \alpha_{\mathrm{T}}^{(L)}\Delta T, \tag{14.40}$$

where

$$E_{\mathrm{A}}^{(L)} = A - \frac{B^2}{F}, \qquad E_{\mathrm{T}}^{(L)} = F - \frac{B^2}{A}, \qquad \nu_{\mathrm{A}}^{(L)} = \frac{B}{F},$$
$$\nu_{\mathrm{a}}^{(L)} = C - \frac{BG}{F}, \qquad \nu_{\mathrm{t}}^{(L)} = G - \frac{BC}{A}, \tag{14.41}$$
$$\alpha_{\mathrm{A}}^{(L)} = \frac{1}{E_{\mathrm{A}}^{(L)}}\left(P - \nu_{\mathrm{A}}^{(L)}Q\right), \quad \alpha_{\mathrm{T}}^{(L)} = \frac{1}{E_{\mathrm{T}}^{(L)}}\left(Q - \nu_{\mathrm{A}}^{(L)}\frac{E_{\mathrm{T}}^{(L)}}{E_{\mathrm{A}}^{(L)}}P\right)$$

and where the superscript (L) denotes that the thermoelastic property refers to the undamaged laminate.

It should be noted that:

$$\tilde{E}_{\mathrm{A}}^{(L)} = A = \frac{b}{h}\tilde{E}_{\mathrm{A}}^{0} + \frac{a}{h}\tilde{E}_{\mathrm{T}}^{90}, \qquad \nu_{\mathrm{A}}^{(L)}\frac{E_{\mathrm{T}}^{(L)}}{E_{\mathrm{A}}^{(L)}} = \frac{B}{A} = b\nu_{\mathrm{A}}^{0}\frac{\tilde{E}_{\mathrm{T}}^{0} + a\nu_{\mathrm{A}}^{90}\tilde{E}_{\mathrm{T}}^{90}}{b\tilde{E}_{\mathrm{A}}^{0} + a\tilde{E}_{\mathrm{T}}^{90}},$$
$$\tilde{E}_{\mathrm{T}}^{(L)} = F = \frac{b}{h}\tilde{E}_{\mathrm{T}}^{0} + \frac{a}{h}\tilde{E}_{\mathrm{A}}^{90}, \qquad \nu_{\mathrm{A}}^{(L)} = \frac{B}{F} = b\nu_{\mathrm{A}}^{0}\frac{\tilde{E}_{\mathrm{T}}^{0} + a\nu_{\mathrm{A}}^{90}\tilde{E}_{\mathrm{T}}^{90}}{b\tilde{E}_{\mathrm{T}}^{0} + a\tilde{E}_{\mathrm{A}}^{90}}, \tag{14.42}$$

where

$$\tilde{E}_{\mathrm{A}}^{(L)} = \frac{E_{\mathrm{A}}^{(L)}}{1 - \left(\nu_{\mathrm{A}}^{(L)}\right)^2 E_{\mathrm{T}}^{(L)}/E_{\mathrm{A}}^{(L)}}, \qquad \tilde{E}_{\mathrm{T}}^{(L)} = \frac{E_{\mathrm{T}}^{(L)}}{1 - \left(\nu_{\mathrm{A}}^{(L)}\right)^2 E_{\mathrm{T}}^{(L)}/E_{\mathrm{A}}^{(L)}}. \tag{14.43}$$

Using Eq. (14.37) together with the results of Eqs. (14.29) and (14.33), it can be shown that:

$$\widehat{\varepsilon}_t = -A'\widehat{\varepsilon}_A - B'\widehat{\varepsilon}_T + C'\sigma_t + P'\Delta T, \tag{14.44}$$

where

$$A' = \frac{b}{h}\frac{\nu_a^0 + \nu_t^0\nu_A^0}{E_A^0}\tilde{E}_A^0 + \frac{a}{h}\left(\frac{\nu_t^{90}}{E_T^{90}} + \frac{\nu_a^{90}\nu_A^{90}}{E_A^{90}}\right)\tilde{E}_T^{90} = \frac{b}{h}\tilde{\nu}_a^0 + \frac{a}{h}\tilde{\nu}_t^{90},$$

$$B' = \frac{b}{h}\left(\frac{\nu_t^0}{E_T^0} + \frac{\nu_a^0\nu_A^0}{E_A^0}\right)\tilde{E}_T^0 + \frac{a}{h}\frac{\nu_a^{90} + \nu_t^{90}\nu_A^{90}}{E_A^{90}}\tilde{E}_A^{90} = \frac{b}{h}\tilde{\nu}_t^0 + \frac{a}{h}\tilde{\nu}_a^{90},$$

$$C' = \frac{b}{h}\left(\frac{1}{E_t^0} - \frac{\nu_a^0\tilde{\nu}_a^0}{E_A^0} - \frac{\nu_t^0\tilde{\nu}_t^0}{E_T^0}\right) + \frac{a}{h}\left(\frac{1}{E_t^{90}} - \frac{\nu_t^{90}\tilde{\nu}_t^{90}}{E_T^{90}} - \frac{\nu_a^{90}\tilde{\nu}_a^{90}}{E_A^{90}}\right),$$

$$P' = \frac{b}{h}\left(\alpha_t^0 + \frac{\nu_a^0}{E_A^0}\tilde{E}_A^0\tilde{\alpha}_A^0 + \frac{\nu_t^0}{E_T^0}\tilde{E}_T^0\tilde{\alpha}_T^0\right) + \frac{a}{h}\left(\alpha_t^{90} + \frac{\nu_a^{90}}{E_A^{90}}\tilde{E}_A^{90}\tilde{\alpha}_A^{90} + \frac{\nu_t^{90}}{E_T^{90}}\tilde{E}_T^{90}\tilde{\alpha}_T^{90}\right). \tag{14.45}$$

Upon substituting Eq. (14.40) in Eq. (14.44), it follows that:

$$\hat{\varepsilon}_t = -\frac{\nu_a^{(L)}}{E_A^{(L)}}\hat{\sigma}_A - \frac{\nu_t^{(L)}}{E_T^{(L)}}\hat{\sigma}_T + \frac{1}{E_t^{(L)}}\sigma_t + \alpha_t^{(L)}\Delta T, \tag{14.46}$$

where

$$\nu_a^{(L)} = A' - B'\nu_A^{(L)}, \quad \nu_t^{(L)} = B' - A'\nu_A^{(L)}\frac{E_T^{(L)}}{E_A^{(L)}},$$
$$\frac{1}{E_t^{(L)}} = C' + A'\frac{\nu_a^{(L)}}{E_A^{(L)}} + B'\frac{\nu_t^{(L)}}{E_T^{(L)}}, \quad \alpha_t^{(L)} = P' - A'\alpha_A^{(L)} - B'\alpha_T^{(L)}. \tag{14.47}$$

It can be shown that the same values result for the minor Poisson ratios $\nu_a^{(L)}$ and $\nu_t^{(L)}$ when using Eq. (14.41) or Eq. (14.47). It should be noted that the results (Eqs. 14.40 and 14.46 for the laminate) have exactly the same form as the stress–strain relations in Eq. (14.4) for the 0 degree ply.

14.4 Shear lag theory for cross-ply laminates

Stress-transfer phenomena in cross-ply laminates have been examined extensively using an approximate method of stress analysis, first developed for UD composites, that is known as shear lag theory. The objective now is to describe the nature of this methodology and to identify the various approximations that have to be made. It will be assumed in this section that the through-thickness stress σ_{33} is everywhere zero, i.e., $\sigma_t = 0$. The approach to be adopted follows the work of Nuismer and Tan (1988), as in later work by the author (McCartney, 1991). Nairn and Mendels (2001) have critically examined approaches to shear lag theory, including comparison of predictions with the results of finite-element analysis, and concluded that Nuismer and Tan's approach, which is now to be followed, is the most reliable shear lag methodology, although it is by no means a satisfactory solution to stress-transfer problems in cross-ply laminates.

The shear lag approach is based on averages defined over the regions occupied by the 0 degree and 90 degrees plies. Averaging Eq. (14.24) over the 0 degree and 90 degrees plies using Eq. (14.26) leads to the results:

$$\frac{d\overline{\sigma}_{11}^{0}}{dx_1} = \frac{\tau(x_1)}{b}, \qquad \frac{d\overline{\sigma}_{11}^{90}}{dx_1} = -\frac{\tau(x_1)}{a}, \tag{14.48}$$

where it has been assumed that $\sigma_{13}(x_1, h) = 0$ and $\sigma_{13}(x_1, 0) = 0$ because of symmetry about the plane $x_3 = 0$, and where $\tau\,(x_1) = \sigma_{13}(x_1, a)$ is the value of the shear stress on the interface $x_3 = a$. It has also been assumed that σ_{13} is continuous across the interface $x_3 = a$. It follows from Eq. (14.48) that:

$$b\overline{\sigma}_{11}^{0}(x_1) + a\overline{\sigma}_{11}^{90}(x_1) = h\sigma_{A}, \tag{14.49}$$

where σ_A is the effective stress applied in the axial direction. It is emphasized that Eq. (14.49) is valid for any value of x_1.

From Eqs. (14.4) and (14.5), the stress−strain relations involving the shear stress σ_{13} may be written for the 0 degree and 90 degrees plies as follows:

$$\frac{\partial u_1^0}{\partial x_3} + \frac{\partial u_3^0}{\partial x_1} = \frac{\sigma_{13}^0}{\mu_a^0}, \qquad \frac{\partial u_1^{90}}{\partial x_3} + \frac{\partial u_3^{90}}{\partial x_1} = \frac{\sigma_{13}^{90}}{\mu_t^{90}}. \tag{14.50}$$

To develop a shear lag model, these relations are now approximated by:

$$\frac{\partial u_1^0}{\partial x_3} = \frac{\sigma_{13}^0}{\mu_a^0}, \qquad \frac{\partial u_1^{90}}{\partial x_3} = \frac{\sigma_{13}^{90}}{\mu_t^{90}}. \tag{14.51}$$

Because $h \ll L$, the shear stress distribution is now assumed to have the following form:

$$\sigma_{13}^{0}(x_1, x_3) = \tau(x_1)\frac{h - x_3}{b}, \qquad \sigma_{13}^{90}(x_1, x_3) = \tau(x_1)\frac{x_3}{a}. \tag{14.52}$$

It is clear that $\sigma_{13}^{0}=\sigma_{13}^{90}$ on $x_3=a$, and that $\sigma_{13}^{0}(x_1,h)=0$. On multiplying Eq. (14.51) by $(h-x_3)$, on substituting the Eq. (14.52) for σ_{13}^{0}, and then integrating with respect to x_3 over the 0 degree ply, it can be shown that:

$$\frac{1}{b}\int_{a}^{h}(h-x_3)\frac{\partial u_1^{0}}{\partial x_3}\,\mathrm{d}x_3=\frac{b\tau(x_1)}{3\mu_{\mathrm{a}}^{0}}=\overline{u}_1^{0}(x_1)-u_1^{0}(x_1,a). \tag{14.53}$$

On multiplying Eq. (14.51) by x_3, on substituting the Eq. (14.52) for σ_{13}, and then integrating with respect to x_3 over the 90 degrees ply, it can be shown that:

$$\frac{1}{a}\int_{0}^{a}x_3\frac{\partial u_1^{90}}{\partial x_3}\,\mathrm{d}x_3=\frac{a\tau(x_1)}{3\mu_{\mathrm{t}}^{90}}=u_1^{90}(x_1,a)-\overline{u}_1^{90}(x_1). \tag{14.54}$$

Since there is perfect bonding at the interface $x_3=a$ so that $u_1^{0}(x_1,a)=u_1^{90}(x_1,a)$, the addition of Eqs. (14.53) and (14.54) leads to the important result:

$$\overline{u}_1^{0}(x_1)-\overline{u}_1^{90}(x_1)=\frac{1}{3}\left(\frac{b}{\mu_{\mathrm{a}}^{0}}+\frac{a}{\mu_{\mathrm{t}}^{90}}\right)\tau(x_1). \tag{14.55}$$

Clearly, the interfacial shear stress is proportional to the difference between the average axial displacements in the 0 degree and 90 degrees plies. It should be noted that a relationship of this form was derived by Nuismer and Tan (1988).

Since it has been assumed in this section that the through-thickness stress σ_{33} is zero, Eqs. (14.17) and (14.21) may be averaged using Eq. (14.26), so that:

$$\overline{\varepsilon}_{11}^{0}=\frac{\mathrm{d}\overline{u}_1^{0}}{\mathrm{d}x_1}=\frac{1}{\widetilde{E}_{\mathrm{A}}^{0}}\overline{\sigma}_{11}^{0}+\widetilde{\alpha}_{\mathrm{A}}^{0}\Delta T-\nu_{\mathrm{A}}^{0}\frac{E_{\mathrm{T}}^{0}}{E_{\mathrm{A}}^{0}}\varepsilon_{\mathrm{T}}, \tag{14.56}$$

$$\overline{\varepsilon}_{11}^{90}=\frac{\mathrm{d}\overline{u}_1^{90}}{\mathrm{d}x_1}=\frac{1}{\widetilde{E}_{\mathrm{T}}^{90}}\overline{\sigma}_{11}^{90}+\widetilde{\alpha}_{\mathrm{T}}^{90}\Delta T-\nu_{\mathrm{A}}^{90}\varepsilon_{\mathrm{T}}. \tag{14.57}$$

Differentiating these results with respect to x_1 and making using of Eq. (14.48):

$$\frac{\mathrm{d}^2\overline{u}_1^{0}}{\mathrm{d}x_1^2}=\frac{1}{\widetilde{E}_{\mathrm{A}}^{0}}\frac{\mathrm{d}\overline{\sigma}_{11}^{0}}{\mathrm{d}x_1}=\frac{1}{\widetilde{E}_{\mathrm{A}}^{0}}\frac{\tau(x_1)}{b}, \tag{14.58}$$

$$\frac{\mathrm{d}^2\overline{u}_1^{90}}{\mathrm{d}x_1^2}=\frac{1}{\widetilde{E}_{\mathrm{T}}^{90}}\frac{\mathrm{d}\overline{\sigma}_{11}^{90}}{\mathrm{d}x_1}=-\frac{1}{\widetilde{E}_{\mathrm{T}}^{90}}\frac{\tau(x_1)}{a}. \tag{14.59}$$

On subtracting these relations making use of the result Eq. (14.55), the following second-order ordinary differential equation for the interfacial shear stress distribution $\tau(x_1)$ may be derived:

$$\frac{d^2\tau(x_1)}{dx_1^2} = k^2\tau(x_1), \tag{14.60}$$

where

$$k^2 = \frac{3\mu_a^0\mu_t^{90}\left(b\widetilde{E}_A^0 + a\widetilde{E}_T^{90}\right)}{ab\widetilde{E}_A^0\widetilde{E}_T^{90}\left(b\mu_t^{90} + a\mu_a^0\right)}. \tag{14.61}$$

When considering the situation where $L \to \infty$, and as the plane $x_1 = 0$ is then remote from the ply cracks at $x_1 = \pm L$, the axial stresses σ_{11}^0 and σ_{11}^{90} on $x_1 = 0$ would have the uniform values $\widehat{\sigma}_{11}^0$ and $\widehat{\sigma}_{11}^{90}$ that occur in an undamaged laminate subject to an axial stress σ_A and transverse strain ε_T, implying that $\overline{\sigma}_{11}^0(0) = \widehat{\sigma}_{11}^0$ and $\overline{\sigma}_{11}^{90}(0) = \widehat{\sigma}_{11}^{90}$. The stress values $\widehat{\sigma}_{11}^0$ and $\widehat{\sigma}_{11}^{90}$ must be such that:

$$b\widehat{\sigma}_{11}^0 + a\widehat{\sigma}_{11}^{90} = h\sigma_A, \tag{14.62}$$

and it then follows from Eq. (14.48) that, since L is very large for this case, the condition $\overline{\sigma}_{11}^{90}(0) = \widehat{\sigma}_{11}^{90}$ and the use of Eq. (14.62) lead to the results:

$$\overline{\sigma}_{11}^0(x_1) = \frac{h\sigma_A}{b} - \frac{a}{b}\widehat{\sigma}_{11}^{90}\left(1 - \frac{\cosh(kx_1)}{\cosh(kL)}\right), \quad \overline{\sigma}_{11}^{90}(x_1) = \widehat{\sigma}_{11}^{90}\left(1 - \frac{\cosh(kx_1)}{\cosh(kL)}\right). \tag{14.63}$$

satisfying Eq. (14.49). The fact that the stress field in Eq. (14.63) can be specified only by imposing a boundary condition that is valid as $L \to \infty$ is regarded as a significant disadvantage of shear lag models. It is not clear to what extent ply interactions are being taken into account.

Since σ_{33} has been assumed to be zero, on averaging Eqs. (14.16) and (14.20) using Eq. (14.26), and on using Eq. (14.63), it can be shown that:

$$\overline{\sigma}_{22}^0(x_1) = -\frac{\nu_A^0 E_T^0}{E_A^0}\frac{a}{b}\widehat{\sigma}_{11}^{90}\left(1 - \frac{\cosh(kx_1)}{\cosh(kL)}\right) + \frac{\nu_A^0 E_T^0}{E_A^0}\frac{h\sigma_A}{b} - E_T^0\alpha_T^0\Delta T + E_T^0\varepsilon_T, \tag{14.64}$$

$$\overline{\sigma}_{22}^{90}(x_1) = \nu_A^{90}\widehat{\sigma}_{11}^{90}\left(1 - \frac{\cosh(kx_1)}{\cosh(kL)}\right) - E_A^{90}\alpha_A^{90}\Delta T + E_A^{90}\varepsilon_T. \tag{14.65}$$

For an undamaged laminate, the axial strains in both the 0 degree and 90 degrees plies have the same uniform value, $\widehat{\varepsilon}_{\mathrm{A}}$. It then follows from Eqs. (14.56) and (14.57) that:

$$\widehat{\varepsilon}_{\mathrm{A}} = \frac{1}{\widetilde{E}_{\mathrm{A}}^{0}}\widehat{\sigma}_{11}^{0} + \widetilde{\alpha}_{\mathrm{A}}^{0}\Delta T - \nu_{\mathrm{A}}^{0}\frac{E_{\mathrm{T}}^{0}}{E_{\mathrm{A}}^{0}}\varepsilon_{\mathrm{T}} = \frac{1}{\widetilde{E}_{\mathrm{T}}^{90}}\widehat{\sigma}_{11}^{90} + \widetilde{\alpha}_{\mathrm{T}}^{90}\Delta T - \nu_{\mathrm{A}}^{90}\varepsilon_{\mathrm{T}}. \tag{14.66}$$

Using Eq. (14.62), it then follows that:

$$\left(b\widetilde{E}_{\mathrm{A}}^{0} + a\widetilde{E}_{\mathrm{T}}^{90}\right)\frac{\widehat{\sigma}_{11}^{0}}{\widetilde{E}_{\mathrm{A}}^{0}} = h\sigma_{\mathrm{A}} - a\widetilde{E}_{\mathrm{T}}^{90}\left(\widetilde{\alpha}_{\mathrm{A}}^{0} - \widetilde{\alpha}_{\mathrm{T}}^{90}\right)\Delta T - a\widetilde{E}_{\mathrm{T}}^{90}\left(\nu_{\mathrm{A}}^{90} - \nu_{\mathrm{A}}^{0}\frac{E_{\mathrm{T}}^{0}}{E_{\mathrm{A}}^{0}}\right)\varepsilon_{\mathrm{T}}, \tag{14.67}$$

$$\left(b\widetilde{E}_{\mathrm{A}}^{0} + a\widetilde{E}_{\mathrm{T}}^{90}\right)\frac{\widehat{\sigma}_{11}^{90}}{\widetilde{E}_{\mathrm{T}}^{90}} = h\sigma_{\mathrm{A}} + b\widetilde{E}_{\mathrm{A}}^{0}\left(\widetilde{\alpha}_{\mathrm{A}}^{0} - \widetilde{\alpha}_{\mathrm{T}}^{90}\right)\Delta T + b\widetilde{E}_{\mathrm{A}}^{0}\left(\nu_{\mathrm{A}}^{90} - \nu_{\mathrm{A}}^{0}\frac{E_{\mathrm{T}}^{0}}{E_{\mathrm{A}}^{0}}\right)\varepsilon_{\mathrm{T}}. \tag{14.68}$$

The axial average displacement field, which satisfies the conditions $\overline{u}_1^0(0) = \overline{u}_1^{90}(0) = 0$, is derived on integrating Eqs. (14.56) and (14.57) using Eqs. (14.62) and (14.63), so that:

$$\overline{u}_1^0(x_1) = \frac{\widehat{\sigma}_{11}^{90}}{k\widetilde{E}_{\mathrm{A}}^{0}}\frac{a}{b}\frac{\sinh(kx_1)}{\cosh(kL)} + \widehat{\varepsilon}_{\mathrm{A}}x_1, \qquad \overline{u}_1^{90}(x_1) = -\frac{\widehat{\sigma}_{11}^{90}}{k\widetilde{E}_{\mathrm{T}}^{90}}\frac{\sinh(kx_1)}{\cosh(kL)} + \widehat{\varepsilon}_{\mathrm{A}}x_1. \tag{14.69}$$

Given that the 90 degrees ply is cracked, an effective axial applied strain for a cracked laminate ε_{A} is defined by:

$$\varepsilon_{\mathrm{A}} = \frac{\overline{u}_1^0(L)}{L} = \frac{a\Phi\widehat{\sigma}_{11}^{90}}{b\widetilde{E}_{\mathrm{A}}^{0}} + \widehat{\varepsilon}_{\mathrm{A}}, \qquad \text{where } \Phi = \frac{\tanh(kL)}{kL}. \tag{14.70}$$

So far the analysis has assumed generalized plane strain conditions where the transverse strain, ε_{T}, has the same prescribed uniform value for both the 0 degree and 90 degrees plies. In most applications, it is more convenient to prescribe the effective transverse stress, σ_{T}, which is defined by:

$$hL\sigma_{\mathrm{T}} = \int_0^h\int_0^L \sigma_{22}(x_1,x_3)\mathrm{d}x_1\mathrm{d}x_3 = \int_0^L\left(b\overline{\sigma}_{22}^0(x_1) + a\overline{\sigma}_{22}^{90}(x_1)\right)\mathrm{d}x_1. \tag{14.71}$$

Using Eqs. (14.64) and (14.65):

$$b\overline{\sigma}_{22}^{0}(x_1) + a\overline{\sigma}_{22}^{90}(x_1) = a\widehat{\sigma}_{11}^{90}\left(\nu_{\mathrm{A}}^{90} - \frac{\nu_{\mathrm{A}}^{0}E_{\mathrm{T}}^{0}}{E_{\mathrm{A}}^{0}}\right)\left(1 - \frac{\cosh(kx_1)}{\cosh(kL)}\right) + \frac{\nu_{\mathrm{A}}^{0}E_{\mathrm{T}}^{0}}{E_{\mathrm{A}}^{0}}h\sigma_{\mathrm{A}}$$
$$-\left(bE_{\mathrm{T}}^{0}\alpha_{\mathrm{T}}^{0} + aE_{\mathrm{A}}^{90}\alpha_{\mathrm{A}}^{90}\right)\Delta T + \left(bE_{\mathrm{T}}^{0} + aE_{\mathrm{A}}^{90}\right)\varepsilon_{\mathrm{T}}. \tag{14.72}$$

The substitution of Eq. (14.72) in Eq. (14.71) then leads to:

$$h\sigma_{\mathrm{T}} = \left(\nu_{\mathrm{A}}^{90} - \nu_{\mathrm{A}}^{0}\frac{E_{\mathrm{T}}^{0}}{E_{\mathrm{A}}^{0}}\right)\left(1 - \frac{\tanh(kL)}{kL}\right)a\widehat{\sigma}_{11}^{90} + \frac{\nu_{\mathrm{A}}^{0}E_{\mathrm{T}}^{0}}{E_{\mathrm{A}}^{0}}h\sigma_{\mathrm{A}}$$
$$-\left(bE_{\mathrm{T}}^{0}\alpha_{\mathrm{T}}^{0} + aE_{\mathrm{A}}^{90}\alpha_{\mathrm{A}}^{90}\right)\Delta T + \left(bE_{\mathrm{T}}^{0} + aE_{\mathrm{A}}^{90}\right)\varepsilon_{\mathrm{T}}. \tag{14.73}$$

For an undamaged laminate, the effective transverse stress is given as:

$$h\widehat{\sigma}_{\mathrm{T}} = \left(\nu_{\mathrm{A}}^{90} - \frac{\nu_{\mathrm{A}}^{0}E_{\mathrm{T}}^{0}}{E_{\mathrm{A}}^{0}}\right)a\widehat{\sigma}_{11}^{90} + \frac{\nu_{\mathrm{A}}^{0}E_{\mathrm{T}}^{0}}{E_{\mathrm{A}}^{0}}h\sigma_{\mathrm{A}} - \left(bE_{\mathrm{T}}^{0}\alpha_{\mathrm{T}}^{0} + aE_{\mathrm{A}}^{90}\alpha_{\mathrm{A}}^{90}\right)\Delta T$$
$$+\left(bE_{\mathrm{T}}^{0} + aE_{\mathrm{A}}^{90}\right)\varepsilon_{\mathrm{T}}, \tag{14.74}$$

with the result that Eq. (14.73) may be written as:

$$\sigma_{\mathrm{T}} = \left(\nu_{\mathrm{A}}^{0}\frac{E_{\mathrm{T}}^{0}}{E_{\mathrm{A}}^{0}} - \nu_{\mathrm{A}}^{90}\right)\frac{a}{h}\Phi\widehat{\sigma}_{11}^{90} + \widehat{\sigma}_{\mathrm{T}}, \qquad \text{where } \Phi = \frac{\tanh(kL)}{kL}. \tag{14.75}$$

It should be noted that $\sigma_{\mathrm{T}} \to \widehat{\sigma}_{\mathrm{T}}$, as $L \to \infty$.

This completes the development of relations for a shear lag model that can be applied to uniform distributions of ply cracks in a simple cross-ply laminate. In the next section, a more accurate stress-transfer model is developed, where an alternative, more accurate definition of the damage-dependent parameter Φ is given and used to predict the dependence of the thermoelastic constants of a cracked laminate on ply-crack density. Equivalent shear lag predictions can be made simply by using Eq. (14.75) for the parameter Φ.

The preferred shear lag approach described above is based on various approximations that are cause of concern from a mechanics point of view. It is useful now to identify clearly these approximations. The first approximation that is made concerns the neglect of the terms $\partial u_3/\partial x_1$ in the expression for the shear strain, ε_{13} (see Eqs. 14.50 and 14.51). This assumption is expected to lead to significant errors. It is observed that the equilibrium in Eq. (14.25) and the stress–strain relations in Eq. (14.4) and (14.5) have not been considered, indicating further sources of significant error. It is concluded, that while the preferred shear lag approach is relatively easy to develop and understand, it does suffer from having to make a number of assumptions that are expected to lead to unacceptable errors.

14.5 Generalized plane strain theory for cross-ply laminates

As mentioned in Section 14.4, stress-transfer phenomena in cross-ply laminates have been examined extensively using an approximate method of stress analysis, first developed for UD composites, that is known as the shear lag theory. The objective now is to improve the nature of this methodology by removing the various approximations that had to be made when developing shear lag solutions. The approach to be adopted extends previous work by McCartney (1992).

Consider a symmetric damaged laminate of length $2L$, width $2W$, and total thickness $2h$, for which the plies are uniformly thick and perfectly bonded together. In-plane loading is applied by imposing uniform axial and transverse displacements, denoted by $\pm U_A$ and $\pm U_T$, on the edges of the laminate, thus defining an effective axial strain $\varepsilon_A = U_A/L$ and an effective transverse strain $\varepsilon_T = U_T/W$. The faces of the laminate are subjected to a uniform applied stress denoted by σ_t. Corresponding to the effective in-plane strains ε_A and ε_T, axial and transverse effective applied stresses can be defined for damaged laminates by:

$$\sigma_A = \frac{1}{4Wh}\int_{-W}^{W}\int_{-h}^{h}\sigma_{11}\mathrm{d}x_2\mathrm{d}x_3, \qquad \sigma_T = \frac{1}{4Lh}\int_{-L}^{L}\int_{-h}^{h}\sigma_{22}\mathrm{d}x_1\mathrm{d}x_3. \tag{14.76}$$

Corresponding to the applied through-thickness stress, σ_t, an effective through-thickness strain, ε_t, can be defined for damaged and undamaged laminates by:

$$\varepsilon_t = \frac{1}{8LWh}\int_{-W}^{W}\int_{-L}^{L}[u_3(x_1,x_2,h) - u_3(x_1,x_2,-h)]\mathrm{d}x_1\mathrm{d}x_2. \tag{14.77}$$

For an undamaged laminate subject to these displacements, the axial and transverse strains will be uniform throughout all plies of the laminate, having the values ε_A and ε_T, respectively, and the through-thickness stress will be uniform throughout the laminate, having the value σ_t in each ply. It then follows from Eqs. (14.76) and (14.77) that $\sigma_A = \widehat{\sigma}_A$, $\sigma_T = \widehat{\sigma}_T$, and $\varepsilon_t = \widehat{\varepsilon}_t$, where $\widehat{\sigma}_A$, $\widehat{\sigma}_T$, and $\widehat{\varepsilon}_t$ are defined by Eqs. (14.35)–(14.37).

For generalized plane strain conditions, the shear stress distribution is assumed to have the following form:

$$\sigma_{13}^{0}(x_1,x_3) = C'(x_1)(h - x_3), \qquad \sigma_{13}^{90}(x_1,x_3) = \frac{b}{a}C'(x_1)x_3, \tag{14.78}$$

where $C(x_1)$ is a stress-transfer function that is to be determined and where the prime denotes that the derivative of the function C is taken. The relations in Eq. (14.78) ensure that the shear stress, σ_{13}, is continuous across the interface, $x_3 = a$. Upon substituting Eq. (14.78) in Eq. (14.24), and then integrating with respect to x_3, it can be shown that:

$$\sigma_{11}^{0}(x_1,x_3) = C(x_1) + \widehat{\sigma}_{11}^{0}, \qquad \sigma_{11}^{90}(x_1,x_3) = -\frac{b}{a}C(x_1) + \widehat{\sigma}_{11}^{90}, \tag{14.79}$$

where, from Eqs. (14.28) and (14.32), $\widehat{\sigma}_{11}^{0}$ and $\widehat{\sigma}_{11}^{90}$ are the values of the stress component σ_{11} in the 0 degree and 90 degrees plies, respectively, when the laminate is undamaged, so that $C(x_1) \equiv 0$. Using Eq. (14.35), it follows from Eq. (14.79) that the effective applied axial stress σ_{A} defined by Eq. (14.76) is given by:

$$h\sigma_{\mathrm{A}} = b\sigma_{11}^{0}(x_1,x_3) + a\sigma_{11}^{90}(x_1,x_3) = b\widehat{\sigma}_{11}^{0} + a\widehat{\sigma}_{11}^{90} = h\widehat{\sigma}_{\mathrm{A}}. \tag{14.80}$$

It follows that the effective applied axial stress for damaged laminate, σ_{A}, is equal to the effective axial stress, $\widehat{\sigma}_{\mathrm{A}}$, for the corresponding undamaged laminate.

On substituting Eq. (14.78) in Eq. (14.25) and then integrating with respect to x_3, it can be shown that:

$$\begin{aligned}\sigma_{33}^{0}(x_1,x_3) &= \frac{1}{2}C''(x_1)(h-x_3)^2 + \sigma_{\mathrm{t}}, \qquad \sigma_{33}^{90}(x_1,x_3)\\ &= \frac{b}{2a}C''(x_1)\left(ah - x_3^2\right) + \sigma_{\mathrm{t}},\end{aligned} \tag{14.81}$$

where the stress component σ_{33} is continuous across the interface $x = a$ and has zero value on the free surfaces $x_3 = \pm h$.

It follows from Eqs. (14.16) and (14.20) that the stress, σ_{22}, is given by:

$$\sigma_{22}^{0}(x_1,x_3) = \frac{\nu_{\mathrm{A}}^{0}E_{\mathrm{T}}^{0}}{E_{\mathrm{A}}^{0}}C(x_1) + \frac{1}{2}\nu_{\mathrm{t}}^{0}C''(x_1)(h-x_3)^2 + \widehat{\sigma}_{22}^{0}, \tag{14.82}$$

$$\sigma_{22}^{90}(x_1,x_3) = -\frac{b}{a}\nu_{\mathrm{A}}^{90}C(x_1) + \frac{b}{2a}\nu_{\mathrm{a}}^{90}C''(x_1)\left(ah - x_3^2\right) + \widehat{\sigma}_{22}^{90}, \tag{14.83}$$

where the values of $\widehat{\sigma}_{22}^{0}$ and $\widehat{\sigma}_{22}^{90}$ are given by Eqs. (14.28) and (14.32), respectively.

The stress representation in Eqs. (14.78)–(14.83) is now substituted in the stress–strain relation in Eq. (14.5) for the strain component ε_{33}^{90}, leading to the following result, on integration with respect to x_3 and on using Eq. (14.23):

$$u_3^{90}(x_1,x_3) = -\frac{b}{a}\frac{\widetilde{\nu}_{\mathrm{t}}^{90}}{\widetilde{E}_{\mathrm{T}}^{90}}C(x_1)x_3 + \frac{b}{6a}\frac{1}{\widetilde{E}_{\mathrm{t}}^{90}}C''(x_1)\left(3ah - x_3^2\right)x_3 + \widehat{\varepsilon}_{33}^{90}x_3, \tag{14.84}$$

where the value of $\widehat{\varepsilon}_{33}^{90}$ is given by Eq. (14.31) and where the symmetry condition $u_3^{90}(x_1,0) = 0$ has been imposed. The stress representation in Eqs. (14.78)–(14.83) is then substituted in the stress–strain relation Eq. (14.4) for the strain component ε_{33}^{0}, leading to, on integration with respect to x_3 and on using Eq. (14.19):

$$u_3^0(x_1,x_3) = \widehat{\varepsilon}_{33}^{0}(x_3-a) + \widehat{\varepsilon}_{33}^{90}a + \left(b\frac{\widetilde{\nu}_{\mathrm{t}}^{90}}{\widetilde{E}_{\mathrm{T}}^{90}} - \frac{\widetilde{\nu}_{\mathrm{a}}^{0}}{\widetilde{E}_{\mathrm{A}}^{0}}(x_3-a) \right) C(x_1)$$
$$+ \frac{1}{6}\left[\frac{1}{\widetilde{E}_{\mathrm{t}}^{0}}\left(b^3-(h-x_3)^3\right) + \frac{ab}{\widetilde{E}_{\mathrm{t}}^{90}}(2h+b) \right] C''(x_1), \tag{14.85}$$

where the value of $\widehat{\varepsilon}_{33}^{0}$ is given by Eq. (14.29) and where the continuity condition $u_3^0(x_1,a) = u_3^{90}(x_1,a)$ has been applied.

The next step is to use shear stress–strain relations to determine the form of the distributions of the displacement component u_1 in terms of the stress-transfer function $C(x_1)$. On substituting the stress representation in Eqs. (14.78)–(14.83) and (14.85) into the stress–strain relation (Eq. 14.4) for the strain component ε_{13}^{0}, this leads to the following result, on integration with respect to x_3:

$$u_1^0(x_1,x_3) = \left(\frac{\widetilde{\nu}_{\mathrm{a}}^{0}}{2\widetilde{E}_{\mathrm{A}}^{0}}(x_3-a)^2 - \frac{\widetilde{\nu}_{\mathrm{t}}^{90}}{\widetilde{E}_{\mathrm{T}}^{90}}b(x_3-a) + \frac{b^2-(h-x_3)^2}{2\mu_{\mathrm{a}}^{0}} \right) C'(x_1)$$
$$- \left(\frac{(h-x_3)^4 - b^4 + 4b^3(x_3-a)}{24\widetilde{E}_{\mathrm{t}}^{0}} + \frac{ab(2h+b)(x_3-a)}{6\widetilde{E}_{\mathrm{t}}^{90}} \right) C'''(x_1) + A(x_1). \tag{14.86}$$

where $A(x_1) \equiv u_1^0(x_1,a)$ is for the moment an arbitrary function of x_1. Similarly, on substituting the stress representation in Eqs. (14.78), (14.79), and (14.84) into the stress–strain relation in Eq. (14.5) for the strain component ε_{13}^{90}, and on integration with respect to x_3, it can be shown that:

$$u_1^{90}(x_1,x_3) = \frac{b}{2a}\left(\frac{1}{\mu_{\mathrm{t}}^{90}} - \frac{\widetilde{\nu}_{\mathrm{t}}^{90}}{\widetilde{E}_{\mathrm{T}}^{90}} \right)(x_3^2-a^2)C'(x_1) + \frac{b}{24a}\frac{1}{\widetilde{E}_{\mathrm{t}}^{90}}\left(x_3^4-a^4\right.$$
$$\left. -6ah(x_3^2-a^2)\right)C'''(x_1) + A(x_1), \tag{14.87}$$

where the continuity condition $u_1^0(x_1,a) = u_1^{90}(x_1,a)$ has been imposed.

The solution specified by Eqs. (14.78)–(14.87) automatically satisfies the equilibrium in Eqs. (14.24) and (14.25), and all the stress–strain relations in Eqs. (14.4) and (14.5), except for the two relations in Eqs. (14.4) and (14.5) involving the strains ε_{11}^{0} and ε_{11}^{90}, for any functions $C(x_1)$ and $A(x_1)$. It is possible, however, to satisfy these relations after they are averaged through the thickness of the 0 degree and 90 degrees plies, respectively, as will now be described.

On averaging Eqs. (14.4) and (14.5) using Eq. (14.26):

$$\frac{\partial \overline{u}_1^0}{\partial x_1} = \frac{1}{\widetilde{E}_{\rm A}^0}\overline{\sigma}_{11}^0 - \frac{\widetilde{\nu}_{\rm a}^0}{\widetilde{E}_{\rm A}^0}\overline{\sigma}_{33}^0 + \widetilde{\alpha}_{\rm A}^0 \Delta T - \nu_{\rm A}^0 \frac{E_{\rm T}^0}{E_{\rm A}^0}\varepsilon_{\rm T}, \tag{14.88}$$

$$\frac{\partial \overline{u}_1^{90}}{\partial x_1} = \frac{1}{\widetilde{E}_{\rm T}^{90}}\overline{\sigma}_{11}^{90} - \frac{\widetilde{\nu}_{\rm t}^{90}}{\widetilde{E}_{\rm T}^{90}}\overline{\sigma}_{33}^{90} + \widetilde{\alpha}_{\rm T}^{90} \Delta T - \nu_{\rm A}^{90}\varepsilon_{\rm T}. \tag{14.89}$$

It should be noted that Eqs. (14.88) and (14.89) are in fact the averages of Eqs. (14.17) and (14.21), respectively. On averaging Eqs. (14.79) and (14.81) over the 0 degree and 90 degrees plies, respectively, it can be shown that:

$$\overline{\sigma}_{11}^0(x_1) = C(x_1) + \widehat{\sigma}_{11}^0, \qquad \overline{\sigma}_{11}^{90}(x_1) = -\frac{b}{a}C(x_1) + \widehat{\sigma}_{11}^{90}, \tag{14.90}$$

$$\overline{\sigma}_{33}^0(x_1) = \frac{b^2}{6}C''(x_1) + \sigma_{\rm t}, \qquad \overline{\sigma}_{33}^{90}(x_1) = \frac{b(2a+3b)}{6}C''(x_1) + \sigma_{\rm t}. \tag{14.91}$$

On averaging Eqs. (14.86) and (14.87) over the 0 degree and 90 degrees plies, respectively, it can be shown that:

$$\overline{u}_1^0(x_1) = \left(\frac{\widetilde{\nu}_{\rm a}^0}{6\widetilde{E}_{\rm A}^0} - \frac{\widetilde{\nu}_{\rm t}^{90}}{2\widetilde{E}_{\rm T}^{90}} + \frac{1}{3\mu_{\rm a}^0}\right)b^2C'(x_1) - \left(\frac{1}{20\widetilde{E}_{\rm t}^0} + \frac{a(2a+3b)}{12b^2\widetilde{E}_{\rm t}^{90}}\right)b^4C''(x_1) + A(x_1), \tag{14.92}$$

$$\overline{u}_1^{90}(x_1) = \frac{a}{3b}\left(\frac{\widetilde{\nu}_{\rm t}^{90}}{\widetilde{E}_{\rm T}^{90}} - \frac{1}{\mu_{\rm t}^{90}}\right)b^2C'(x_1) + \frac{a^2(5b+4a)}{30b^3}\frac{1}{\widetilde{E}_{\rm t}^{90}}b^4C''(x_1) + A(x_1). \tag{14.93}$$

On substituting Eqs. (14.90)–(14.93) into Eqs. (14.88) and (14.89), on eliminating the terms $\widehat{\sigma}_{11}^{0}$ and $\widehat{\sigma}_{11}^{90}$ using Eqs. (14.28) and (14.32), it follows that:

$$\left(\frac{1}{20\widetilde{E}_{\mathrm{t}}^{0}}+\frac{a(2a+3b)}{12b^{2}\widetilde{E}_{\mathrm{t}}^{90}}\right)b^{4}C''(x_1)-\left(\frac{\widetilde{\nu}_{\mathrm{a}}^{0}}{3\widetilde{E}_{\mathrm{A}}^{0}}-\frac{\widetilde{\nu}_{\mathrm{t}}^{90}}{2\widetilde{E}_{\mathrm{T}}^{90}}+\frac{1}{3\mu_{\mathrm{a}}^{0}}\right)b^{2}C''(x_1)+\frac{1}{\widetilde{E}_{\mathrm{A}}^{0}}C(x_1)$$
$$=-\widehat{\varepsilon}_{\mathrm{A}}+A'(x_1), \tag{14.94}$$

$$\frac{a^{2}(5b+4a)}{30b^{3}\widetilde{E}_{\mathrm{t}}^{90}}b^{4}C''(x_1)+\frac{a}{3b}\left(\frac{4a+3b}{2a}\frac{\widetilde{\nu}_{\mathrm{t}}^{90}}{\widetilde{E}_{\mathrm{T}}^{90}}-\frac{1}{\mu_{\mathrm{t}}^{90}}\right)b^{2}C''(x_1)+\frac{b}{a}\frac{1}{\widetilde{E}_{\mathrm{T}}^{90}}C(x_1)$$
$$=\widehat{\varepsilon}_{\mathrm{A}}-A'(x_1). \tag{14.95}$$

On adding Eqs. (14.94) and (14.95), the term $\widehat{\varepsilon}_{\mathrm{A}}-A'(x_1)$ is eliminated and the following homogeneous fourth-order ordinary differential equation is obtained:

$$Fb^{4}C''(x_1)-Gb^{2}C''(x_1)+HC(x_1)=0, \tag{14.96}$$

where

$$F=\frac{1}{20\,\widetilde{E}_{\mathrm{t}}^{0}}+\frac{2}{15\,\widetilde{E}_{\mathrm{t}}^{90}}\frac{a}{b}\left(\frac{a^{2}}{b^{2}}+\frac{5}{2}\frac{a}{b}+\frac{15}{8}\right)>0,$$
$$G=\frac{1}{3}\left(\frac{1}{\mu_{\mathrm{a}}^{0}}+\frac{1}{\mu_{\mathrm{t}}^{90}}\frac{a}{b}+\frac{\widetilde{\nu}_{a}^{0}}{\widetilde{E}_{\mathrm{A}}^{0}}-\frac{\widetilde{\nu}_{\mathrm{t}}^{90}}{\widetilde{E}_{\mathrm{T}}^{90}}\frac{2a+3b}{b}\right),$$
$$H=\frac{1}{\widetilde{E}_{\mathrm{A}}^{0}}+\frac{1}{\widetilde{E}_{\mathrm{T}}^{90}}\frac{b}{a}>0. \tag{14.97}$$

From Eqs. (14.94) and (14.95), it is clear that the function $A(x_1)$ can be calculated using either of the following relations:

$$A(x_1)=\left(\frac{1}{20\widetilde{E}_{\mathrm{t}}^{0}}+\frac{a(2a+3b)}{12b^{2}\widetilde{E}_{\mathrm{t}}^{90}}\right)b^{4}C'''(x_1)-\left(\frac{\widetilde{\nu}_{a}^{0}}{3\widetilde{E}_{\mathrm{A}}^{0}}-\frac{\widetilde{\nu}_{\mathrm{t}}^{90}}{2\widetilde{E}_{\mathrm{T}}^{90}}+\frac{1}{3\mu_{\mathrm{a}}^{0}}\right)b^{2}C'(x_1)$$
$$+\frac{1}{\widetilde{E}_{\mathrm{A}}^{0}}\overline{C}(x_1)+\widehat{\varepsilon}_{\mathrm{A}}x_1, \tag{14.98}$$

$$A(x_1) = -\frac{a^2(5b+4a)}{30b^3\widetilde{E}_{\mathrm{t}}^{90}}b^4C'''(x_1) - \frac{a}{3b}\left(\frac{4a+3b}{2a}\frac{\widetilde{\nu}_{\mathrm{t}}^{90}}{\widetilde{E}_{\mathrm{T}}^{90}} - \frac{1}{\mu_{\mathrm{t}}^{90}}\right)b^2C'(x_1)$$
$$-\frac{b}{a}\frac{1}{\widetilde{E}_{\mathrm{T}}^{90}}\overline{C}(x_1) + \widehat{\varepsilon}_{\mathrm{A}}x_1, \tag{14.99}$$

where

$$\overline{C}(x_1) \equiv \int_0^x C(x)dx. \tag{14.100}$$

When there is symmetry about $x_1 = 0$, it follows that $C'(x_1) = 0$ and $C'''(x_1) = 0$, and the integration constant has been selected so that $A(x_1) = 0$. On substituting Eq. (14.98) in Eq. (14.92):

$$\overline{u}_1^0(x_1) = \frac{1}{\widetilde{E}_{\mathrm{A}}^0}\overline{C}(x_1) - \frac{\widetilde{\nu}_{\mathrm{a}}^0}{6\widetilde{E}_{\mathrm{A}}^0}b^2C'(x_1) + \widehat{\varepsilon}_{\mathrm{A}}x_1, \tag{14.101}$$

and on substituting Eq. (14.99) in Eq. (14.93):

$$\overline{u}_1^{90}(x_1) = -\frac{b}{a}\frac{1}{\widetilde{E}_{\mathrm{T}}^{90}}\overline{C}(x_1) - \frac{2a+3b}{6b}\frac{\widetilde{\nu}_{\mathrm{t}}^{90}}{\widetilde{E}_{\mathrm{T}}^{90}}b^2C'(x_1) + \widehat{\varepsilon}_{\mathrm{A}}x_1. \tag{14.102}$$

It should be noted that $\overline{u}_1^0(0) = \overline{u}_1^{90}(0) = 0$, consistent with the conditions in Eq. (14.13). Because the function $A(x_1)$ is now known in terms of $C(x_1)$, the u_1 displacement distributions in Eqs. (14.86) and (14.87) are fully specified in terms of the stress-transfer function, $C(x_1)$.

Dimensionless parameters r and s are defined as:

$$r = \frac{G}{2F}, \quad s = \sqrt{\frac{H}{F}}, \tag{14.103}$$

The most general solution of the differential Eq. (14.96) satisfying the symmetry condition $C(x_1) \equiv C(-x_1)$ is given by the following equation, if $s > r$:

$$C(x_1) = P\cosh\frac{px_1}{b}\cos\frac{qx_1}{b} + Q\sinh\frac{px_1}{b}\sin\frac{qx_1}{b}, \tag{14.104}$$

If $s = r$:

$$C(x_1) = P\cos\text{h}\frac{px_1}{b} + Q\frac{x_1}{b}\sin\text{h}\frac{px_1}{b}, \tag{14.105}$$

If $s < r$:

$$C(x_1) = P\cosh\frac{px_1}{b}\cosh\frac{qx_1}{b} + Q\sinh\frac{px_1}{b}\sinh\frac{qx_1}{b}, \tag{14.106}$$

where p and q are dimensionless parameters defined as:

$$p = \sqrt{\frac{1}{2}(r+s)}, \quad q = \sqrt{\frac{1}{2}|r-s|}. \tag{14.107}$$

Rather than considering the above three cases separately, it is more convenient to select Eq. (14.106) of the solution for the case $r > s$ and then allow the parameter q to be real zero or pure imaginary. This situation can be dealt with using software that includes the use of complex arithmetic. The solution in Eq. (14.106) is now expressed in the equivalent form:

$$C(x_1) = A\cosh\frac{(p+q))x_1}{b} + B\cosh\frac{(p-q)x_1}{b}, \tag{14.108}$$

where

$$\begin{gathered} P = A + B, \quad Q = A - B, \\ p = \sqrt{\frac{1}{2}(r+s)}, \quad q = \sqrt{\frac{1}{2}(r-s)}. \end{gathered} \tag{14.109}$$

This completes the development of the general solution for symmetric situations. Further progress depends upon the selection of appropriate boundary conditions for specific types of problem.

14.5.1 Solution for ply cracks

Consider now a uniform array of ply cracks, having density $\rho = 1/(2L)$, in the 90 degrees ply of the cross-ply laminate, as shown in Fig. 14.1b. The tractions on the ply crack surfaces must be zero so that from Eqs. (14.78) and (14.79):

$$\sigma_{13}^{90}(L, x_3) = \frac{b}{a}C'(L)x_3 = 0, \quad \text{implying } C'(L) = 0, \tag{14.110}$$

$$\sigma_{11}^{90}(L, x_3) = -\frac{b}{a}C(L) + \sigma_{11}^{90} = 0, \quad \text{implying } C(L) = \frac{a}{b}\hat{\sigma}_{11}^{90}. \tag{14.111}$$

On applying these conditions to the solution in Eq. (14.108), the parameters A and B must be selected so that:

$$A\cosh\frac{(p+q)L}{b}+B\cosh\frac{(p-q)L}{b}=\frac{a}{b}\widehat{\sigma}_{11}^{90}, \tag{14.112}$$

$$A\frac{p+q}{b}\sinh\frac{(p+q)L}{b}+B\frac{p-q}{b}\sinh\frac{(p-q)L}{b}=0. \tag{14.113}$$

On solving Eqs. (14.112) and (14.113):

$$A=-\frac{a}{b}\frac{(p-q)\tanh\dfrac{(p-q)L}{b}}{\cosh\dfrac{(p+q)L}{b}}\Lambda\widehat{\sigma}_{11}^{90},\qquad B=\frac{a}{b}\frac{(p+q)\tanh\dfrac{(p+q)L}{b}}{\cosh\dfrac{(p-q)L}{b}}\Lambda\widehat{\sigma}_{11}^{90}, \tag{14.114}$$

where

$$\frac{1}{\Lambda}=(q+p)\tanh\frac{(p+q)L}{b}+(q-p)\tanh\frac{(p-q)L}{b}. \tag{14.115}$$

The only boundary condition for a damaged laminate that has not been satisfied is given by Eq. (14.14). It is clear from Eqs. (14.86), (14.87), and (14.98) or Eq. (14.79) that it is not possible for these boundary conditions to be satisfied by the approximate solution derived. The boundary condition in Eq. (14.14) is now replaced by the following averaged condition:

$$\overline{u}_1^0=\pm L\varepsilon_{\mathrm{A}},\quad \text{on } x_1=\pm L. \tag{14.116}$$

It then follows from Eqs. (14.101) and (14.110) that:

$$\varepsilon_{\mathrm{A}}=\frac{1}{\widetilde{E}_{\mathrm{A}}^0}\frac{\overline{C}(L)}{L}+\widehat{\varepsilon}_{\mathrm{A}}. \tag{14.117}$$

It follows from Eqs. (14.100) and (14.108) that:

$$\overline{C}(x_1)=\frac{bA}{p+q}\sinh\frac{(p+q)x_1}{b}+\frac{bB}{p-q}\sinh\frac{(p-q)x_1}{b}, \tag{14.118}$$

so that on using Eq. (14.114):

$$\overline{C}(L)=\frac{a}{b}\widehat{\sigma}_{11}^{90}L\Phi, \tag{14.119}$$

where

$$\Phi = \frac{4\Lambda pq}{p^2 - q^2} \frac{b}{L} \tanh \frac{(p+q)L}{b} \tanh \frac{(p-q)L}{b}. \tag{14.120}$$

From Eq. (14.117) it follows that:

$$\varepsilon_{\mathrm{A}} = \frac{a\Phi \widehat{\sigma}_{11}^{90}}{b\widetilde{E}_{\mathrm{A}}^{0}} + \widehat{\varepsilon}_{\mathrm{A}}. \tag{14.121}$$

On using Eq. (14.36), it follows from Eqs. (14.82) and (14.83) that the effective applied transverse stress σ_{T} defined by Eq. (14.76) is given by:

$$\begin{aligned}\sigma_{\mathrm{T}} &= \frac{1}{4Lh} \int_{-L}^{L} \int_{-h}^{h} \sigma_{22} \mathrm{d}x_1 \mathrm{d}x_3 = \frac{1}{Lh} \int_{0}^{L} \int_{0}^{a} \sigma_{22}^{90} \mathrm{d}x_1 \mathrm{d}x_3 \\ &+ \frac{1}{Lh} \int_{0}^{L} \int_{a}^{h} \sigma_{22}^{0} \mathrm{d}x_1 \mathrm{d}x_3 = \frac{b}{Lh} \left(\nu_{\mathrm{A}}^{0} \frac{E_{\mathrm{T}}^{0}}{E_{\mathrm{A}}^{0}} - \nu_{\mathrm{A}}^{90} \right) \overline{C}(L) + \widehat{\sigma}_{\mathrm{T}},\end{aligned} \tag{14.122}$$

where use has been made of the relations $C'(0) = 0$ and $C'(L) = 0$. It follows that the effective applied transverse stress for a damaged laminate σ_{T} is not equal to the effective axial stress $\widehat{\sigma}_{\mathrm{T}}$ for the corresponding undamaged laminate. On using Eq. (14.119), it follows that:

$$\sigma_{\mathrm{T}} = \frac{a}{h} \left(\nu_{\mathrm{A}}^{0} \frac{E_{\mathrm{T}}^{0}}{E_{\mathrm{A}}^{0}} - \nu_{\mathrm{A}}^{90} \right) \Phi \widehat{\sigma}_{11}^{90} + \widehat{\sigma}_{\mathrm{T}}. \tag{14.123}$$

It is assumed that the values of $\sigma_{\mathrm{A}} = \widehat{\sigma}_{\mathrm{A}}$, $\varepsilon_{\mathrm{T}} = \widehat{\varepsilon}_{\mathrm{T}}$, σ_{t}, and ΔT are known, so that the stress–strain relations in Eq. (14.40) for an undamaged laminate may be written as:

$$\begin{aligned}\widehat{\varepsilon}_{\mathrm{A}} &= \frac{1}{E_{\mathrm{A}}^{(L)}} \sigma_{\mathrm{A}} - \frac{\nu_{\mathrm{A}}^{(L)}}{E_{\mathrm{A}}^{(L)}} \widehat{\sigma}_{\mathrm{T}} - \frac{\nu_{\mathrm{a}}^{(L)}}{E_{\mathrm{A}}^{(L)}} \sigma_{\mathrm{t}} + \alpha_{\mathrm{A}}^{(L)} \Delta T, \\ \varepsilon_{\mathrm{T}} &= -\frac{\nu_{\mathrm{A}}^{(L)}}{E_{\mathrm{A}}^{(L)}} \sigma_{\mathrm{A}} + \frac{1}{E_{\mathrm{T}}^{(L)}} \widehat{\sigma}_{\mathrm{T}} - \frac{\nu_{\mathrm{t}}^{(L)}}{E_{\mathrm{T}}^{(L)}} \sigma_{\mathrm{t}} + \alpha_{\mathrm{T}}^{(L)} \Delta T.\end{aligned} \tag{14.124}$$

14.6 Calculation of in-plane thermoelastic constants for damaged laminates

There are two approaches, involving complex algebra, that can be made to the estimation of the effective thermoelastic constants of a damaged laminate. In order for the readers to understand how the analysis is derived, the key steps of the complex algebra will be shown.

On using Eq. (14.123) to eliminate $\widehat{\sigma}_{\mathrm{T}}$, it follows that:

$$\widehat{\varepsilon}_{\mathrm{A}} = \frac{1}{E_{\mathrm{A}}^{(L)}}\sigma_{\mathrm{A}} - \frac{\nu_{\mathrm{A}}^{(L)}}{E_{\mathrm{A}}^{(L)}}\sigma_{\mathrm{T}} - \frac{\nu_{\mathrm{a}}^{(L)}}{E_{\mathrm{A}}^{(L)}}\sigma_{\mathrm{t}} + \frac{\nu_{\mathrm{A}}^{(L)}}{E_{\mathrm{A}}^{(L)}}\left(\nu_{\mathrm{A}}^{0}\frac{E_{\mathrm{T}}^{0}}{E_{\mathrm{A}}^{0}} - \nu_{\mathrm{A}}^{90}\right)\frac{a}{h}\Phi\widehat{\sigma}_{11}^{90} + \alpha_{\mathrm{A}}^{(L)}\Delta T, \tag{14.125}$$

$$\varepsilon_{\mathrm{T}} = -\frac{\nu_{\mathrm{A}}^{(L)}}{E_{\mathrm{A}}^{(L)}}\sigma_{\mathrm{A}} + \frac{1}{E_{\mathrm{T}}^{(L)}}\sigma_{\mathrm{T}} - \frac{\nu_{\mathrm{t}}^{(L)}}{E_{\mathrm{T}}^{(L)}}\sigma_{\mathrm{t}} - \frac{1}{E_{\mathrm{T}}^{(L)}}\left(\nu_{\mathrm{A}}^{0}\frac{E_{\mathrm{T}}^{0}}{E_{\mathrm{A}}^{0}} - \nu_{\mathrm{A}}^{90}\right)\frac{a}{h}\Phi\widehat{\sigma}_{11}^{90} + \alpha_{\mathrm{T}}^{(L)}\Delta T. \tag{14.126}$$

From Eq. (14.32), on eliminating ε_{T} using Eq. (14.126), and then on using Eqs. (14.34) and (14.125) to eliminate $\widehat{\varepsilon}_{\mathrm{A}}$:

$$\begin{aligned}
&\left[1 - \frac{\widetilde{E}_{\mathrm{T}}^{90}}{E_{\mathrm{T}}^{(L)}}\left(\nu_{\mathrm{A}}^{(L)}\frac{E_{T}^{(L)}}{E_{\mathrm{A}}^{(L)}} - \nu_{\mathrm{A}}^{90}\right)\left(\nu_{\mathrm{A}}^{0}\frac{E_{T}^{0}}{E_{\mathrm{A}}^{0}} - \nu_{\mathrm{A}}^{90}\right)\frac{a}{h}\Phi\right]\frac{\widehat{\sigma}_{11}^{90}}{\widetilde{E}_{\mathrm{T}}^{90}} \\
&= \left(1 - \nu_{\mathrm{A}}^{90}\nu_{\mathrm{A}}^{(L)}\right)\frac{\sigma_{\mathrm{A}}}{E_{\mathrm{A}}^{(L)}} - \left(\nu_{\mathrm{A}}^{(L)}\frac{E_{\mathrm{T}}^{(L)}}{E_{\mathrm{A}}^{(L)}} - \nu_{\mathrm{A}}^{90}\right)\frac{\sigma_{\mathrm{T}}}{E_{\mathrm{T}}^{(L)}} \\
&\quad - \left(\frac{\nu_{\mathrm{a}}^{(L)}}{E_{\mathrm{A}}^{(L)}} - \frac{\nu_{\mathrm{t}}^{90}}{E_{\mathrm{T}}^{90}} - \frac{\nu_{\mathrm{a}}^{90}\nu_{\mathrm{A}}^{90}}{E_{\mathrm{A}}^{90}} + \nu_{\mathrm{A}}^{90}\frac{\nu_{\mathrm{t}}^{(L)}}{E_{\mathrm{T}}^{(L)}}\right)\sigma_{\mathrm{t}} + \left(\alpha_{\mathrm{A}}^{(L)} + \nu_{\mathrm{A}}^{90}\alpha_{\mathrm{T}}^{(L)} - \widetilde{\alpha}_{\mathrm{T}}^{90}\right)\Delta T.
\end{aligned} \tag{14.127}$$

Using Eq. (14.42), it can be shown that:

$$\nu_{\mathrm{A}}^{(L)}\frac{E_{\mathrm{T}}^{(L)}}{E_{\mathrm{A}}^{(L)}} - \nu_{\mathrm{A}}^{90} = \left(\nu_{\mathrm{A}}^{0}\frac{E_{\mathrm{T}}^{0}}{E_{\mathrm{A}}^{0}} - \nu_{\mathrm{A}}^{90}\right)\frac{b}{h}\frac{\widetilde{E}_{\mathrm{A}}^{0}}{\widetilde{E}_{\mathrm{A}}^{(L)}}, \tag{14.128}$$

so that Eq. (14.127) may now be written as:

$$\xi\frac{\widehat{\sigma}_{11}^{90}}{E_{\mathrm{T}}^{90}}=\left(1-\nu_{\mathrm{A}}^{90}\nu_{\mathrm{A}}^{(L)}\right)\frac{\sigma_{\mathrm{A}}}{E_{\mathrm{A}}^{(L)}}-\left(\nu_{\mathrm{A}}^{0}\frac{E_{\mathrm{T}}^{0}}{E_{\mathrm{A}}^{0}}-\nu_{\mathrm{A}}^{90}\right)\frac{\widetilde{E}_{\mathrm{A}}^{0}}{\widetilde{E}_{\mathrm{A}}^{(L)}}\frac{b}{h}\frac{\sigma_{\mathrm{T}}}{E_{\mathrm{T}}^{(L)}}$$
$$-\left\{\frac{\nu_{\mathrm{a}}^{(L)}}{E_{\mathrm{A}}^{(L)}}-\frac{\nu_{\mathrm{t}}^{90}}{E_{\mathrm{T}}^{90}}+\nu_{\mathrm{A}}^{90}\left(\frac{\nu_{\mathrm{t}}^{(L)}}{E_{\mathrm{T}}^{(L)}}-\frac{\nu_{\mathrm{a}}^{90}}{E_{\mathrm{A}}^{90}}\right)\right\}\sigma_{\mathrm{t}}$$
$$+\left(\alpha_{\mathrm{A}}^{(L)}+\nu_{\mathrm{A}}^{90}\alpha_{\mathrm{T}}^{(L)}-\alpha_{\mathrm{T}}^{90}\right)\Delta T, \tag{14.129}$$

where

$$\xi=1-\left(\nu_{\mathrm{A}}^{0}\frac{E_{\mathrm{T}}^{0}}{E_{\mathrm{A}}^{0}}-\nu_{\mathrm{A}}^{90}\right)^{2}\frac{ab}{h^{2}}\frac{\widetilde{E}_{\mathrm{A}}^{0}\widetilde{E}_{\mathrm{T}}^{90}}{\widetilde{E}_{\mathrm{A}}^{(L)}\widetilde{E}_{\mathrm{T}}^{(L)}}\Phi. \tag{14.130}$$

The next step of the first approach is to eliminate $\widehat{\varepsilon}_{\mathrm{A}}$ and $\widehat{\sigma}_{11}^{90}$ in Eqs. (14.125) and (14.126) using Eqs. (14.121) and (14.129). On eliminating $\widehat{\varepsilon}_{\mathrm{A}}$ in Eq. (14.125):

$$\varepsilon_{\mathrm{A}}=\frac{1}{E_{\mathrm{A}}^{(L)}}\sigma_{\mathrm{A}}-\frac{\nu_{\mathrm{A}}^{(L)}}{E_{\mathrm{A}}^{(L)}}\sigma_{\mathrm{T}}-\frac{\nu_{\mathrm{a}}^{(L)}}{E_{\mathrm{A}}^{(L)}}\sigma_{\mathrm{t}}+\alpha_{\mathrm{A}}^{(L)}\Delta T+\left(1-\nu_{\mathrm{A}}^{(L)}\nu_{\mathrm{A}}^{90}\right)\frac{a}{b}\frac{\widetilde{E}_{\mathrm{A}}^{(L)}}{E_{\mathrm{A}}^{(L)}\widetilde{E}_{\mathrm{A}}^{0}}\Phi\widehat{\sigma}_{11}^{90}. \tag{14.131}$$

On eliminating $\widehat{\sigma}_{11}^{90}$ in Eqs. (14.126) and (14.131) using Eq. (14.129), the final form of the stress–strain relations for a damaged laminate is obtained, namely:

$$\varepsilon_{\mathrm{A}}=\frac{1}{E_{\mathrm{A}}}\sigma_{\mathrm{A}}-\frac{\nu_{\mathrm{A}}}{E_{\mathrm{A}}}\sigma_{\mathrm{T}}-\frac{\nu_{\mathrm{a}}}{E_{\mathrm{A}}}\sigma_{\mathrm{t}}+\alpha_{\mathrm{A}}\Delta T, \tag{14.132}$$

$$\varepsilon_{\mathrm{T}}=-\frac{\nu_{\mathrm{A}}}{E_{\mathrm{A}}}\sigma_{\mathrm{A}}+\frac{1}{E_{\mathrm{T}}}\sigma_{\mathrm{T}}-\frac{\nu_{\mathrm{t}}}{E_{\mathrm{T}}}\sigma_{\mathrm{t}}+\alpha_{\mathrm{T}}\Delta T, \tag{14.133}$$

where the thermoelastic constants of the damaged laminate are given by:

$$\frac{1}{E_{\mathrm{A}}}=\frac{1}{E_{\mathrm{A}}^{(L)}}+\left(1-\nu_{\mathrm{A}}^{(L)}\nu_{\mathrm{A}}^{90}\right)^{2}\frac{a}{b}\frac{\widetilde{E}_{\mathrm{A}}^{(L)}\widetilde{E}_{\mathrm{T}}^{90}}{\left(E_{\mathrm{A}}^{(L)}\right)^{2}\widetilde{E}_{\mathrm{A}}^{0}}\frac{\Phi}{\xi}, \tag{14.134}$$

$$\frac{1}{E_{\mathrm{T}}}=\frac{1}{E_{\mathrm{T}}^{(L)}}\left[1+\left(\nu_{\mathrm{A}}^{0}\frac{E_{\mathrm{T}}^{0}}{E_{\mathrm{A}}^{0}}-\nu_{\mathrm{A}}^{90}\right)^{2}\frac{ab}{h^{2}}\frac{\widetilde{E}_{\mathrm{A}}^{0}\widetilde{E}_{\mathrm{T}}^{90}}{\widetilde{E}_{\mathrm{A}}^{(L)}E_{\mathrm{T}}^{(L)}}\frac{\Phi}{\xi}\right],\quad\text{implying } E_{\mathrm{T}}=\xi E_{\mathrm{T}}^{(L)}, \tag{14.135}$$

$$\frac{\nu_{\mathrm{A}}}{E_{\mathrm{A}}}=\frac{\nu_{\mathrm{A}}^{(L)}}{E_{\mathrm{A}}^{(L)}}+\left(1-\nu_{\mathrm{A}}^{(L)}\nu_{\mathrm{A}}^{90}\right)\left(\nu_{\mathrm{A}}^{0}\frac{E_{\mathrm{T}}^{0}}{E_{\mathrm{A}}^{0}}-\nu_{\mathrm{A}}^{90}\right)\frac{a}{h}\frac{\widetilde{E}_{\mathrm{T}}^{90}}{E_{\mathrm{A}}^{(L)}E_{\mathrm{T}}^{(L)}}\frac{\Phi}{\xi},\tag{14.136}$$

$$\frac{\nu_{\mathrm{a}}}{E_{\mathrm{A}}}=\frac{\nu_{\mathrm{a}}^{(L)}}{E_{\mathrm{A}}^{(L)}}+\left\{\frac{\nu_{\mathrm{a}}^{(L)}}{E_{\mathrm{A}}^{(L)}}-\frac{\nu_{\mathrm{t}}^{90}}{E_{\mathrm{T}}^{90}}+\nu_{\mathrm{A}}^{90}\left(\frac{\nu_{\mathrm{t}}^{(L)}}{E_{\mathrm{T}}^{(L)}}-\frac{\nu_{\mathrm{a}}^{90}}{E_{\mathrm{A}}^{90}}\right)\right\}\left(1-\nu_{\mathrm{A}}^{(L)}\nu_{\mathrm{A}}^{90}\right)\frac{a}{b}\frac{\widetilde{E}_{\mathrm{A}}^{(L)}}{E_{\mathrm{A}}^{(L)}}\frac{\widetilde{E}_{\mathrm{T}}^{90}}{\widetilde{E}_{\mathrm{A}}^{0}}\frac{\Phi}{\xi},\tag{14.137}$$

$$\frac{\nu_{\mathrm{t}}}{E_{\mathrm{T}}}=\frac{\nu_{\mathrm{t}}^{(L)}}{E_{\mathrm{T}}^{(L)}}-\left\{\frac{\nu_{\mathrm{a}}^{(L)}}{E_{\mathrm{A}}^{(L)}}-\frac{\nu_{\mathrm{t}}^{90}}{E_{\mathrm{T}}^{90}}+\nu_{A}^{90}\left(\frac{\nu_{\mathrm{t}}^{(L)}}{E_{\mathrm{T}}^{(L)}}-\frac{\nu_{\mathrm{a}}^{90}}{E_{\mathrm{A}}^{90}}\right)\right\}\left(\nu_{A}^{0}\frac{E_{\mathrm{T}}^{0}}{E_{\mathrm{A}}^{0}}-\nu_{\mathrm{A}}^{90}\right)\frac{a}{h}\frac{\widetilde{E}_{\mathrm{T}}^{90}}{E_{\mathrm{T}}^{(L)}}\frac{\Phi}{\xi},\tag{14.138}$$

$$\alpha_{\mathrm{A}}=\alpha_{\mathrm{A}}^{(L)}+\left(1-\nu_{\mathrm{A}}^{(L)}\nu_{\mathrm{A}}^{90}\right)\left(\alpha_{\mathrm{A}}^{(L)}+\nu_{\mathrm{A}}^{90}\alpha_{\mathrm{T}}^{(L)}-\widetilde{\alpha}_{\mathrm{T}}^{90}\right)\frac{a}{b}\frac{\widetilde{E}_{\mathrm{A}}^{(L)}}{E_{\mathrm{A}}^{(L)}}\frac{\widetilde{E}_{\mathrm{T}}^{90}}{\widetilde{E}_{\mathrm{A}}^{0}}\frac{\Phi}{\xi},\tag{14.139}$$

$$\alpha_{\mathrm{T}}=\alpha_{\mathrm{T}}^{(L)}-\left(\alpha_{\mathrm{A}}^{(L)}+\nu_{\mathrm{A}}^{90}\alpha_{\mathrm{T}}^{(L)}-\widetilde{\alpha}_{\mathrm{T}}^{90}\right)\left(\nu_{\mathrm{A}}^{0}\frac{E_{\mathrm{T}}^{0}}{E_{\mathrm{A}}^{0}}-\nu_{\mathrm{A}}^{90}\right)\frac{a}{h}\frac{\widetilde{E}_{\mathrm{T}}^{90}}{E_{\mathrm{T}}^{(L)}}\frac{\Phi}{\xi}.\tag{14.140}$$

The results (Eqs. 14.132–14.140) show that the stress–strain relations of a damaged laminate are exactly the same as those for an undamaged laminate. The formation of damage affects only the values of the thermoelastic constants and not the form of the stress–strain relations.

On using Eq. (14.32) to substitute for $\widehat{\sigma}_{11}^{90}$ in both Eqs. (14.121) and (14.123), assume that the values of $\sigma_{\mathrm{A}}=\widehat{\sigma}_{\mathrm{A}}$, $\varepsilon_{\mathrm{T}}=\widehat{\varepsilon}_{\mathrm{T}}$, σ_{t}, and ΔT are known:

$$\varepsilon_{\mathrm{A}}-\widehat{\varepsilon}_{\mathrm{A}}=\left(\widehat{\varepsilon}_{\mathrm{A}}+\nu_{\mathrm{A}}^{90}\varepsilon_{\mathrm{T}}+\frac{\widetilde{\nu}_{\mathrm{t}}^{90}}{\widetilde{E}_{\mathrm{T}}^{90}}\sigma_{\mathrm{t}}-\widetilde{\alpha}_{\mathrm{T}}^{90}\Delta T\right)\frac{a}{b}\frac{\widetilde{E}_{\mathrm{T}}^{90}}{\widetilde{E}_{\mathrm{A}}^{0}}\Phi,\tag{14.141}$$

$$\sigma_{\mathrm{T}}-\widehat{\sigma}_{\mathrm{T}}=\left(\widehat{\varepsilon}_{\mathrm{A}}+\nu_{\mathrm{A}}^{90}\varepsilon_{\mathrm{T}}+\frac{\widetilde{\nu}_{\mathrm{t}}^{90}}{\widetilde{E}_{\mathrm{T}}^{90}}\sigma_{\mathrm{t}}-\widetilde{\alpha}_{\mathrm{T}}^{90}\Delta T\right)\left(\nu_{\mathrm{A}}^{0}\frac{E_{\mathrm{T}}^{0}}{E_{\mathrm{A}}^{0}}-\nu_{\mathrm{A}}^{90}\right)\frac{a}{h}\widetilde{E}_{\mathrm{T}}^{90}\Phi,\tag{14.142}$$

so that:

$$\sigma_{\mathrm{T}}-\widehat{\sigma}_{\mathrm{T}}=\lambda\widetilde{E}_{\mathrm{A}}^{(L)}\left(\varepsilon_{\mathrm{A}}-\widehat{\varepsilon}_{\mathrm{A}}\right),\tag{14.143}$$

where λ is a useful laminate constant defined by (see also Eq. 14.128):

$$\lambda = \nu_{\mathrm{A}}^{(L)} \frac{E_{\mathrm{T}}^{(L)}}{E_{\mathrm{A}}^{(L)}} - \nu_{\mathrm{A}}^{90} = \left(\nu_{\mathrm{A}}^{0} \frac{E_{\mathrm{T}}^{0}}{E_{\mathrm{A}}^{0}} - \nu_{\mathrm{A}}^{90} \right) \frac{b}{h} \frac{\widetilde{E}_{\mathrm{A}}^{0}}{\widetilde{E}_{\mathrm{A}}^{(L)}}. \tag{14.144}$$

14.7 Through-thickness properties of damaged laminates

On applying Eq. (14.77) to the ply-crack problem being considered:

$$\varepsilon_{\mathrm{t}} = \frac{1}{Lh} \int_{0}^{L} u_3^0(x_1, h) \mathrm{d}x_1. \tag{14.145}$$

On using Eq. (14.85) together with conditions $C'(0) = 0$ and $C'(L) = 0$:

$$\varepsilon_{\mathrm{t}} = \widehat{\varepsilon}_{\mathrm{t}} + \left(\frac{\widetilde{\nu}_{\mathrm{t}}^{90}}{\widetilde{E}_{\mathrm{T}}^{90}} - \frac{\widetilde{\nu}_{\mathrm{a}}^{0}}{\widetilde{E}_{\mathrm{A}}^{0}} \right) \frac{b}{h} \frac{\overline{C}(L)}{L}, \tag{14.146}$$

where the relation in Eq. (14.37) has been used. Since it has been assumed that $\sigma_{\mathrm{A}} = \widehat{\sigma}_{\mathrm{A}}$, it follows on using Eqs. (14.46), (14.119), and (14.123) that:

$$\varepsilon_{\mathrm{t}} = -\frac{\nu_{\mathrm{a}}^{(L)}}{E_{\mathrm{A}}^{(L)}} \sigma_{\mathrm{A}} - \frac{\nu_{\mathrm{t}}^{(L)}}{E_{\mathrm{T}}^{(L)}} \sigma_{\mathrm{T}} + \frac{1}{E_{\mathrm{t}}^{(L)}} \sigma_{\mathrm{t}} + \alpha_{\mathrm{t}}^{(L)} \Delta T - \Omega \frac{a}{h} \Phi \widehat{\sigma}_{11}^{90}, \tag{14.147}$$

where

$$\Omega = \frac{\widetilde{\nu}_{\mathrm{a}}^{0}}{\widetilde{E}_{\mathrm{A}}^{0}} - \frac{\widetilde{\nu}_{\mathrm{t}}^{90}}{\widetilde{E}_{\mathrm{T}}^{90}} - \frac{\nu_{\mathrm{t}}^{(L)}}{E_{\mathrm{T}}^{(L)}} \left(\nu_{\mathrm{A}}^{0} \frac{E_{\mathrm{T}}^{0}}{E_{\mathrm{A}}^{0}} - \nu_{\mathrm{A}}^{90} \right). \tag{14.148}$$

On substituting for $\widehat{\sigma}_{11}^{90}$ in Eq. (14.147) using Eq. (14.129), the following through-thickness stress–strain relation is obtained:

$$\varepsilon_{\mathrm{t}} = -\frac{\nu_{\mathrm{a}}}{E_{\mathrm{A}}} \sigma_{\mathrm{A}} - \frac{\nu_{\mathrm{t}}}{E_{\mathrm{T}}} \sigma_{\mathrm{T}} + \frac{1}{E_{\mathrm{t}}} \sigma_{\mathrm{t}} + \alpha_{\mathrm{t}} \Delta T, \tag{14.149}$$

where

$$\frac{\nu_\mathrm{a}}{E_\mathrm{A}}=\frac{\nu_\mathrm{a}^{(L)}}{E_\mathrm{A}^{(L)}}+\Omega\left(1-\nu_\mathrm{A}^{90}\nu_\mathrm{A}^{(L)}\right)\frac{a}{h}\frac{\widetilde{E}_\mathrm{T}^{90}}{E_\mathrm{A}^{(L)}}\frac{\Phi}{\xi}, \tag{14.150}$$

$$\frac{\nu_\mathrm{t}}{E_\mathrm{T}}=\frac{\nu_\mathrm{t}^{(L)}}{E_\mathrm{T}^{(L)}}-\Omega\left(\nu_\mathrm{A}^{0}\frac{E_\mathrm{T}^{0}}{E_\mathrm{A}^{0}}-\nu_\mathrm{A}^{90}\right)\frac{ab}{h^2}\frac{\widetilde{E}_\mathrm{A}^{0}\widetilde{E}_\mathrm{T}^{90}}{\widetilde{E}_\mathrm{A}^{(L)}E_\mathrm{T}^{(L)}}\frac{\Phi}{\xi}, \tag{14.151}$$

$$\frac{1}{E_\mathrm{t}}=\frac{1}{E_\mathrm{t}^{(L)}}+\Omega\left\{\frac{\nu_\mathrm{a}^{(L)}}{E_\mathrm{A}^{(L)}}-\frac{\nu_\mathrm{t}^{90}}{E_\mathrm{T}^{90}}+\nu_\mathrm{A}^{90}\left(\frac{\nu_\mathrm{t}^{(L)}}{E_\mathrm{T}^{(L)}}-\frac{\nu_\mathrm{a}^{90}}{E_\mathrm{A}^{90}}\right)\right\}\frac{a}{h}\widetilde{E}_\mathrm{T}^{90}\frac{\Phi}{\xi}, \tag{14.152}$$

$$\alpha_\mathrm{t}=\alpha_\mathrm{t}^{(L)}-\Omega\left(\alpha_\mathrm{A}^{(L)}+\nu_\mathrm{A}^{90}\alpha_\mathrm{T}^{(L)}-\widetilde{\alpha}_\mathrm{T}^{90}\right)\frac{a}{h}\widetilde{E}_\mathrm{T}^{90}\frac{\Phi}{\xi}. \tag{14.153}$$

The relations in Eqs. (14.137) and (14.138) are equivalent to the results in Eqs. (14.150) and (14.151) only if the following relation is satisfied:

$$\frac{\nu_\mathrm{a}^{(L)}}{E_\mathrm{A}^{(L)}}-\frac{\nu_\mathrm{t}^{90}}{E_\mathrm{T}^{90}}+\nu_\mathrm{A}^{90}\left(\frac{\nu_\mathrm{t}^{(L)}}{E_\mathrm{T}^{(L)}}-\frac{\nu_\mathrm{a}^{90}}{E_\mathrm{A}^{90}}\right)=\Omega\frac{b}{h}\frac{\widetilde{E}_\mathrm{A}^{0}}{\widetilde{E}_\mathrm{A}^{(L)}}. \tag{14.154}$$

The validity of the relation in Eq. (14.154) is established by making use of the relations in Eqs. (14.45) and (14.47) to show that:

$$\nu_\mathrm{a}^{(L)}+\nu_\mathrm{t}^{(L)}\nu_\mathrm{A}^{(L)}=\left(1-\left(\nu_\mathrm{A}^{(L)}\right)^2\frac{E_\mathrm{T}^{(L)}}{E_\mathrm{A}^{(L)}}\right)\left(\frac{b}{h}\widetilde{\nu}_\mathrm{a}^{0}+\frac{a}{h}\widetilde{\nu}_\mathrm{t}^{90}\right). \tag{14.155}$$

14.8 Consideration of ply-crack closure

A damaged laminate having cracks in the 90 degrees ply can always be loaded so that the ply cracks just close such that the stress normal to the ply cracks is zero everywhere in the 90 degrees ply. To understand this, it is necessary only to consider an undamaged laminate and to determine loading conditions for which the stress component $\widehat{\sigma}_{11}^{90}=0$. When an undamaged laminate is in this state, any number of ply cracks can be formed without changing in any way the stress or displacement distributions in the laminate. Let $\varepsilon_{\mathrm{A}}^{c}$, $\varepsilon_{\mathrm{T}}^{c}$, and σ_{t}^{c} be the in-plane laminate strains and through-thickness stress that generate a ply-crack closure condition.

From Eq. (14.31) at the point of ply-crack closure $\widehat{\sigma}_{11}^{90}=0$, so that:

$$\varepsilon_{\mathrm{A}}^{c}=-\frac{\nu_{\mathrm{A}}^{90}}{E_{\mathrm{A}}^{90}}\widehat{\sigma}_{22}^{90}-\frac{\nu_{\mathrm{t}}^{90}}{E_{\mathrm{T}}^{90}}\sigma_{\mathrm{t}}^{c}+\alpha_{\mathrm{T}}^{90}\Delta T,\quad \varepsilon_{\mathrm{T}}^{c}=\frac{1}{E_{\mathrm{A}}^{90}}\widehat{\sigma}_{22}^{90}-\frac{\nu_{\mathrm{a}}^{90}}{E_{\mathrm{A}}^{90}}\sigma_{\mathrm{t}}^{c}+\alpha_{\mathrm{A}}^{90}\Delta T. \tag{14.156}$$

On solving Eq. (14.156) for $\widehat{\sigma}_{22}^{90}$ and substituting the result in Eq. (14.156), it follows that:

$$\varepsilon_{\mathrm{A}}^{c}=-\nu_{\mathrm{A}}^{90}\varepsilon_{\mathrm{T}}^{c}-\left(\frac{\nu_{\mathrm{t}}^{90}}{E_{\mathrm{T}}^{90}}+\frac{\nu_{\mathrm{a}}^{90}\nu_{\mathrm{A}}^{90}}{E_{\mathrm{A}}^{90}}\right)\sigma_{\mathrm{t}}^{c}+\left(\alpha_{\mathrm{T}}^{90}+\nu_{\mathrm{A}}^{90}\alpha_{\mathrm{A}}^{90}\right)\Delta T. \tag{14.157}$$

14.8.1 Uniaxial loading in the axial direction

Consider the uniaxial loading of the laminate in the axial direction to the point where the ply cracks just close. Clearly $\widehat{\sigma}_{\mathrm{T}}=\widehat{\sigma}_{\mathrm{t}}=0$, so that from Eq. (14.40):

$$\begin{aligned}\overline{\varepsilon}_{\mathrm{A}}^{c}&=\frac{1}{E_{\mathrm{A}}^{(L)}}\overline{\sigma}_{\mathrm{A}}^{c}+\alpha_{\mathrm{A}}^{(L)}\Delta T,\\ \overline{\varepsilon}_{\mathrm{T}}^{c}&=\frac{-\nu_{\mathrm{A}}^{(L)}}{E_{\mathrm{A}}^{(L)}}\overline{\sigma}_{\mathrm{A}}^{c}+\alpha_{\mathrm{T}}^{(L)}\Delta T,\end{aligned} \tag{14.158}$$

where $\overline{\varepsilon}_{\mathrm{A}}^{c}$ and $\overline{\varepsilon}_{\mathrm{T}}^{c}$ are the in-plane laminate strains that arise when the uniaxially applied axial stress, $\overline{\sigma}_{\mathrm{A}}^{c}$, generates the ply-crack closure condition. For this uniaxial loading case, the closure relation in Eq. (14.157) reduces to:

$$\overline{\varepsilon}_{\mathrm{A}}^{c}=-\nu_{\mathrm{A}}^{90}\overline{\varepsilon}_{\mathrm{T}}^{c}+\left(\alpha_{\mathrm{T}}^{90}+\nu_{\mathrm{A}}^{90}\alpha_{\mathrm{A}}^{90}\right)\Delta T, \tag{14.159}$$

and on using Eq. (14.158):

$$\frac{\overline{\sigma}_{\mathrm{A}}^{c}}{\Delta T}=E_{\mathrm{A}}^{(L)}\frac{\alpha_{\mathrm{T}}^{90}-\alpha_{\mathrm{A}}^{(L)}+\nu_{\mathrm{A}}^{90}\left(\alpha_{\mathrm{A}}^{90}-\alpha_{\mathrm{T}}^{(L)}\right)}{1-\nu_{\mathrm{A}}^{90}\nu_{\mathrm{A}}^{(L)}}=-k_1. \tag{14.160}$$

14.8.2 Uniaxial loading in the in-plane transverse direction

Consider now the uniaxial loading of the laminate in the in-plane transverse direction to the point where the ply cracks just close. Clearly $\widehat{\sigma}_{\mathrm{A}}=\widehat{\sigma}_{\mathrm{t}}=0$ for this case, so that from Eq. (14.40):

$$\widetilde{\varepsilon}_{\mathrm{A}}^{c}=-\frac{\nu_{\mathrm{A}}^{(L)}}{E_{\mathrm{A}}^{(L)}}\widetilde{\sigma}_{\mathrm{T}}^{c}+\alpha_{\mathrm{A}}^{(L)}\Delta T,\quad \widetilde{\varepsilon}_{\mathrm{T}}^{c}=\frac{1}{E_{\mathrm{T}}^{(L)}}\widetilde{\sigma}_{\mathrm{T}}^{c}+\alpha_{\mathrm{T}}^{(L)}\Delta T, \tag{14.161}$$

where $\widetilde{\varepsilon}_{\mathrm{A}}^{c}$ and $\widetilde{\varepsilon}_{\mathrm{T}}^{c}$ are the in-plane laminate strains that arise when the uniaxially applied in-plane transverse stress, $\widetilde{\sigma}_{\mathrm{T}}^{c}$, generates the ply-crack closure condition. For this uniaxial loading case, the closure relation in Eq. (14.157) reduces to:

$$\widetilde{\varepsilon}_{\mathrm{A}}^{c} = -\nu_{\mathrm{A}}^{90} \varepsilon_{\mathrm{T}}^{c} + \left(\alpha_{\mathrm{T}}^{90} + \nu_{\mathrm{A}}^{90} \alpha_{\mathrm{A}}^{90}\right) \Delta T, \tag{14.162}$$

and on using Eq. (14.161):

$$\frac{\widetilde{\sigma}_{\mathrm{T}}^{c}}{\Delta T} = \frac{\alpha_{\mathrm{T}}^{90} - \alpha_{\mathrm{A}}^{(L)} + \nu_{\mathrm{A}}^{90}\left(\alpha_{\mathrm{A}}^{90} - \alpha_{\mathrm{T}}^{(L)}\right)}{\dfrac{\nu_{\mathrm{A}}^{90}}{E_{\mathrm{T}}^{(L)}} - \dfrac{\nu_{\mathrm{A}}^{(L)}}{E_{\mathrm{A}}^{(L)}}} = -k_2. \tag{14.163}$$

14.8.3 Uniaxial loading in the through-thickness direction

Consider now the uniaxial loading of the laminate in the through-thickness direction to the point where the ply cracks just close. Clearly $\widehat{\sigma}_{\mathrm{A}} = \widehat{\sigma}_{\mathrm{T}} = 0$ for this case, so that from Eq. (14.40):

$$\overset{\frown}{\varepsilon}{}_{\mathrm{A}}^{c} = -\frac{\nu_{\mathrm{a}}^{(L)}}{E_{\mathrm{A}}^{(L)}} \overset{\frown}{\sigma}{}_{\mathrm{t}}^{c} + \alpha_{\mathrm{A}}^{(L)} \Delta T, \qquad \overset{\frown}{\varepsilon}{}_{\mathrm{T}}^{c} = -\frac{\nu_{\mathrm{t}}^{(L)}}{E_{\mathrm{T}}^{(L)}} \overset{\frown}{\sigma}{}_{\mathrm{t}}^{c} + \alpha_{\mathrm{T}}^{(L)} \Delta T, \tag{14.164}$$

where $\overset{\frown}{\varepsilon}{}_{\mathrm{A}}^{c}$ and $\overset{\frown}{\varepsilon}{}_{\mathrm{T}}^{c}$ are the in-plane laminate strains that arise when the uniaxially applied through-thickness stress, $\overset{\frown}{\sigma}{}_{\mathrm{t}}^{c}$, generates the ply-crack closure condition. For this uniaxial loading case, the closure relation in Eq. (14.157) is written as:

$$\overset{\frown}{\varepsilon}{}_{\mathrm{A}}^{c} = -\nu_{\mathrm{A}}^{90} \overset{\frown}{\varepsilon}{}_{\mathrm{T}}^{c} - \left(\frac{\nu_{\mathrm{t}}^{90}}{E_{\mathrm{T}}^{90}} + \frac{\nu_{\mathrm{a}}^{90}\nu_{\mathrm{A}}^{90}}{E_{\mathrm{A}}^{90}}\right) \overset{\frown}{\sigma}{}_{\mathrm{t}}^{c} + \left(\alpha_{\mathrm{T}}^{90} + \nu_{\mathrm{A}}^{90} \alpha_{\mathrm{A}}^{90}\right) \Delta T, \tag{14.165}$$

and on using Eq. (14.164):

$$\frac{\overset{\frown}{\sigma}{}_{\mathrm{t}}^{c}}{\Delta T} = \frac{\alpha_{\mathrm{T}}^{90} - \alpha_{\mathrm{A}}^{(L)} + \nu_{\mathrm{A}}^{90}\left(\alpha_{\mathrm{A}}^{90} - \alpha_{\mathrm{T}}^{(L)}\right)}{\dfrac{\nu_{\mathrm{t}}^{90}}{E_{\mathrm{T}}^{90}} + \dfrac{\nu_{\mathrm{a}}^{90}\nu_{\mathrm{A}}^{90}}{E_{\mathrm{A}}^{90}} - \dfrac{\nu_{\mathrm{a}}^{(L)}}{E_{\mathrm{A}}^{(L)}} - \dfrac{\nu_{\mathrm{A}}^{90}\nu_{\mathrm{t}}^{(L)}}{E_{\mathrm{T}}^{(L)}}} = -k_3. \tag{14.166}$$

14.8.4 Derivation of important interrelationships

First of all, from Eqs. (14.160) and (14.163), it follows that:

$$\frac{\overline{\sigma}_{\mathrm{A}}^{c}}{\widetilde{\sigma}_{\mathrm{T}}^{c}} = \frac{E_{\mathrm{A}}^{(L)}}{E_{\mathrm{T}}^{(L)}} \frac{\nu_{\mathrm{A}}^{90} - \nu_{\mathrm{A}}^{(L)} E_{\mathrm{T}}^{(L)} / E_{\mathrm{A}}^{(L)}}{1 - \nu_{\mathrm{A}}^{90} \nu_{\mathrm{A}}^{(L)}} = \frac{k_1}{k_2} = k, \tag{14.167}$$

and from Eqs. (14.160) and (14.166):

$$\frac{\overline{\sigma}_{\mathrm{A}}^{c}}{\widehat{\sigma}_{\mathrm{t}}^{c}}=\frac{E_{\mathrm{A}}^{(L)}}{1-\nu_{\mathrm{A}}^{90}\nu_{\mathrm{A}}^{(L)}}\left\{\frac{\nu_{\mathrm{t}}^{90}}{E_{\mathrm{T}}^{90}}-\frac{\nu_{\mathrm{a}}^{(L)}}{E_{\mathrm{A}}^{(L)}}+\nu_{\mathrm{A}}^{90}\left(\frac{\nu_{\mathrm{a}}^{90}}{E_{\mathrm{A}}^{90}}-\frac{\nu_{\mathrm{t}}^{(L)}}{E_{\mathrm{T}}^{(L)}}\right)\right\}=\frac{k_1}{k_3}=k'. \tag{14.168}$$

It should be noted that the values of the constants k_1, k_2, k_3, k, and k' depend on the properties of the 90 degrees ply, and on the effective properties of the laminate when in an undamaged state. Clearly, their values are independent of any damage such as ply cracking. It should also be noted that the constants k and k' are independent of thermal expansion coefficients.

Consider now the relation in Eq. (14.134), which is written as:

$$\Psi=\frac{1}{E_{\mathrm{A}}}-\frac{1}{E_{\mathrm{A}}^{(L)}}=\left(1-\nu_{\mathrm{A}}^{(L)}\nu_{\mathrm{A}}^{90}\right)^2\frac{a}{b}\frac{\widetilde{E}_{\mathrm{A}}^{(L)}\widetilde{E}_{\mathrm{T}}^{90}}{\left(E_{\mathrm{A}}^{(L)}\right)^2\widetilde{E}_{\mathrm{A}}^{0}}\frac{\Phi}{\xi}, \tag{14.169}$$

so that:

$$\frac{\Phi}{\xi}=\frac{\left(E_{\mathrm{A}}^{(L)}\right)^2\widetilde{E}_{\mathrm{A}}^{0}b\Psi}{a\widetilde{E}_{\mathrm{A}}^{(L)}\widetilde{E}_{\mathrm{T}}^{90}\left(1-\nu_{\mathrm{A}}^{(L)}\nu_{\mathrm{A}}^{90}\right)^2}. \tag{14.170}$$

The ratio Φ/ξ is now eliminated in Eqs. (14.135)–(14.140), (14.152), and (14.153), so that on using Eqs. (14.34), (14.144), (14.154), (14.160), and (14.167)–(14.170):

$$\begin{gathered}\frac{1}{E_{\mathrm{A}}}-\frac{1}{E_{\mathrm{A}}^{(L)}}=\Psi,\quad \frac{1}{E_{\mathrm{T}}}-\frac{1}{E_{\mathrm{T}}^{(L)}}=k^2\Psi,\quad \frac{1}{E_{\mathrm{t}}}-\frac{1}{E_{\mathrm{t}}^{(L)}}=(k')^2\Psi,\\ \frac{\nu_{\mathrm{A}}^{(L)}}{E_{\mathrm{A}}^{(L)}}-\frac{\nu_{\mathrm{A}}}{E_{\mathrm{A}}}=k\Psi,\quad \frac{\nu_{\mathrm{a}}^{(L)}}{E_{\mathrm{A}}^{(L)}}-\frac{\nu_{\mathrm{a}}}{E_{\mathrm{A}}}=k'\Psi,\quad \frac{\nu_{\mathrm{t}}^{(L)}}{E_{\mathrm{T}}^{(L)}}-\frac{\nu_{\mathrm{t}}}{E_{\mathrm{T}}}=kk'\Psi,\\ \alpha_{\mathrm{A}}-\alpha_{\mathrm{A}}^{(L)}=k_1\Psi,\quad \alpha_{\mathrm{T}}-\alpha_{\mathrm{T}}^{(L)}=kk_1\Psi,\quad \alpha_{\mathrm{t}}-\alpha_{\mathrm{t}}^{(L)}=k'k_1\Psi.\end{gathered} \tag{14.171}$$

To summarize the results obtained so far, it is clear from Eq. (14.171) that the degradation of all the thermoelastic constants of a damaged laminate is governed by the three undamaged laminate constants k_1, k, and k' defined by Eqs. (14.160), (14.167), and (14.168), respectively, and a single function Ψ accounting for the damage state that is defined by Eq. (14.169).

On using Eqs. (14.160), (14.163), (14.166), and (14.171), it can be shown that:

$$\frac{\overline{\sigma}_{\mathrm{A}}^{c}}{\Delta T}=-\frac{\alpha_{\mathrm{A}}^{(L)}-\alpha_{\mathrm{A}}}{\dfrac{1}{E_{\mathrm{A}}^{(L)}}-\dfrac{1}{E_{\mathrm{A}}}}=-\frac{\alpha_{\mathrm{T}}^{(L)}-\alpha_{\mathrm{T}}}{\dfrac{\nu_{\mathrm{A}}}{E_{\mathrm{A}}}-\dfrac{\nu_{\mathrm{A}}^{(L)}}{E_{\mathrm{A}}^{(L)}}}=-\frac{\alpha_{\mathrm{t}}^{(L)}-\alpha_{\mathrm{t}}}{\dfrac{\nu_{\mathrm{a}}}{E_{\mathrm{A}}}-\dfrac{\nu_{\mathrm{a}}^{(L)}}{E_{\mathrm{A}}^{(L)}}}=-k_1, \tag{14.172}$$

$$\frac{\widetilde{\sigma}_{\mathrm{T}}^{c}}{\Delta T}=-\frac{\alpha_{\mathrm{A}}^{(L)}-\alpha_{\mathrm{A}}}{\dfrac{\nu_{\mathrm{A}}}{E_{\mathrm{A}}}-\dfrac{\nu_{\mathrm{A}}^{(L)}}{E_{\mathrm{A}}^{(L)}}}=-\frac{\alpha_{\mathrm{T}}^{(L)}-\alpha_{\mathrm{T}}}{\dfrac{1}{E_{\mathrm{T}}^{(L)}}-\dfrac{1}{E_{\mathrm{T}}}}=-\frac{\alpha_{\mathrm{t}}^{(L)}-\alpha_{\mathrm{t}}}{\dfrac{\nu_{\mathrm{t}}}{E_{\mathrm{T}}}-\dfrac{\nu_{\mathrm{t}}^{(L)}}{E_{\mathrm{T}}^{(L)}}}=-k_2, \tag{14.173}$$

$$\frac{\widehat{\sigma}_{\mathrm{t}}^{c}}{\Delta T}=-\frac{\alpha_{\mathrm{A}}^{(L)}-\alpha_{\mathrm{A}}}{\dfrac{\nu_{\mathrm{a}}}{E_{\mathrm{A}}}-\dfrac{\nu_{\mathrm{a}}^{(L)}}{E_{\mathrm{A}}^{(L)}}}=-\frac{\alpha_{\mathrm{T}}^{(L)}-\alpha_{\mathrm{T}}}{\dfrac{\nu_{\mathrm{t}}}{E_{\mathrm{T}}}-\dfrac{\nu_{\mathrm{t}}^{(L)}}{E_{\mathrm{T}}^{(L)}}}=-\frac{\alpha_{\mathrm{t}}^{(L)}-\alpha_{\mathrm{t}}}{\dfrac{1}{E_{\mathrm{t}}^{(L)}}-\dfrac{1}{E_{\mathrm{t}}}}=-k_3. \tag{14.174}$$

On using Eqs. (14.167), (14.168), and (14.171), it can also be shown that:

$$\frac{\dfrac{\nu_{\mathrm{t}}^{(L)}}{E_{\mathrm{T}}^{(L)}}-\dfrac{\nu_{\mathrm{t}}}{E_{\mathrm{T}}}}{\dfrac{\nu_{\mathrm{a}}^{(L)}}{E_{\mathrm{A}}^{(L)}}-\dfrac{\nu_{\mathrm{a}}}{E_{\mathrm{A}}}}=\frac{\dfrac{\nu_{\mathrm{A}}^{(L)}}{E_{\mathrm{A}}^{(L)}}-\dfrac{\nu_{\mathrm{A}}}{E_{\mathrm{A}}}}{\dfrac{1}{E_{\mathrm{A}}}-\dfrac{1}{E_{\mathrm{A}}^{(L)}}}=\frac{\dfrac{1}{E_{\mathrm{T}}}-\dfrac{1}{E_{\mathrm{T}}^{(L)}}}{\dfrac{\nu_{\mathrm{A}}^{(L)}}{E_{\mathrm{A}}^{(L)}}-\dfrac{\nu_{\mathrm{A}}}{E_{\mathrm{A}}}}=\frac{\alpha_{\mathrm{T}}-\alpha_{\mathrm{T}}^{(L)}}{\alpha_{\mathrm{A}}-\alpha_{\mathrm{A}}^{(L)}}=k, \tag{14.175}$$

$$\frac{\dfrac{1}{E_{\mathrm{t}}}-\dfrac{1}{E_{\mathrm{t}}^{(L)}}}{\dfrac{\nu_{\mathrm{a}}^{(L)}}{E_{\mathrm{A}}^{(L)}}-\dfrac{\nu_{\mathrm{a}}}{E_{\mathrm{A}}}}=\frac{\dfrac{\nu_{\mathrm{a}}^{(L)}}{E_{\mathrm{A}}^{(L)}}-\dfrac{\nu_{\mathrm{a}}}{E_{\mathrm{A}}}}{\dfrac{1}{E_{\mathrm{A}}}-\dfrac{1}{E_{\mathrm{A}}^{(L)}}}=\frac{\dfrac{\nu_{\mathrm{t}}^{(L)}}{E_{\mathrm{T}}^{(L)}}-\dfrac{\nu_{\mathrm{t}}}{E_{\mathrm{T}}}}{\dfrac{\nu_{\mathrm{A}}^{(L)}}{E_{\mathrm{A}}^{(L)}}-\dfrac{\nu_{\mathrm{A}}}{E_{\mathrm{A}}}}=\frac{\alpha_{\mathrm{t}}-\alpha_{\mathrm{t}}^{(L)}}{\alpha_{\mathrm{A}}-\alpha_{\mathrm{A}}^{(L)}}=k'. \tag{14.176}$$

14.9 Results for general symmetric laminates

The generalized plane strain model described in Sections 14.5−14.8, which applies only to a simple symmetric cross-ply laminate, has been extended to more complex multiple-ply symmetric cross-ply laminates and to general symmetric laminates. The analysis is rather complex and will not be included here as it has recently been published in detail as part of the author's contribution (McCartney, 2013a) to the third phase of the World-Wide Failure Exercise (Kaddour et al., 2013a) concerned with damage development and failure of composite laminates. The more general analysis dealing with multiple plies can be used together with ply refinement techniques to determine very accurately the stress and displacement distributions in laminates having uniform arrays of ply cracks of the same orientation in some of the plies of the laminate. The crack closure approach, used in Section 14.8 to determine interrelationships between thermoelastic constants, leads to the following important expressions, derived in reference (McCartney, 2013a), for the effective thermoelastic constants of a general symmetric laminate:

$$\frac{1}{\widehat{E}_{\mathrm{A}}} = \frac{1}{\widehat{E}_{\mathrm{A}}^{(L)}} + D, \tag{14.177}$$

$$\frac{1}{\widehat{E}_{\mathrm{T}}} = \frac{1}{\widehat{E}_{\mathrm{T}}^{(L)}} + k^2 D, \tag{14.178}$$

$$\frac{1}{\widehat{E}_{\mathrm{t}}} = \frac{1}{\widehat{E}_{\mathrm{t}}^{(L)}} + (k')^2 D, \tag{14.179}$$

$$\frac{\widehat{\nu}_{\mathrm{A}}}{\widehat{E}_{\mathrm{A}}} = \frac{\widehat{\nu}_{\mathrm{A}}^{(L)}}{\widehat{E}_{\mathrm{A}}^{(L)}} - kD, \tag{14.180}$$

$$\frac{\widehat{\nu}_{\mathrm{a}}}{\widehat{E}_{\mathrm{A}}} = \frac{\widehat{\nu}_{\mathrm{a}}^{(L)}}{\widehat{E}_{\mathrm{A}}^{(L)}} - k'D, \tag{14.181}$$

$$\frac{\widehat{\nu}_{\mathrm{t}}}{\widehat{E}_{\mathrm{T}}} = \frac{\widehat{\nu}_{\mathrm{t}}^{(L)}}{\widehat{E}_{\mathrm{T}}^{(L)}} - kk'D, \tag{14.182}$$

$$\widehat{\alpha}_{\mathrm{A}} = \widehat{\alpha}_{\mathrm{A}}^{(L)} + k_1 D, \tag{14.183}$$

$$\widehat{\alpha}_{\mathrm{T}} = \widehat{\alpha}_{\mathrm{T}}^{(L)} + kk_1 D, \tag{14.184}$$

$$\widehat{\alpha}_{\mathrm{t}} = \widehat{\alpha}_{\mathrm{t}}^{(L)} + k'k_1 D, \tag{14.185}$$

where

$$\frac{1}{\widehat{E}_{\mathrm{A}}} = \frac{1}{E_{\mathrm{A}}}\left(1 - \lambda_{\mathrm{A}}^2 \frac{\mu_{\mathrm{A}}}{E_{\mathrm{A}}}\right), \tag{14.186}$$

$$\frac{1}{\widehat{E}_{\mathrm{T}}} = \frac{1}{E_{\mathrm{T}}}\left(1 - \lambda_{\mathrm{T}}^2 \frac{E_{\mathrm{T}}\mu_{\mathrm{A}}}{E_{\mathrm{A}}^2}\right), \tag{14.187}$$

$$\frac{1}{\widehat{E}_{\mathrm{t}}} = \frac{1}{E_{\mathrm{t}}}\left(1 - \lambda_{\mathrm{t}}^2 \frac{E_{\mathrm{t}}\mu_{\mathrm{A}}}{E_{\mathrm{A}}^2}\right), \tag{14.188}$$

$$\frac{\widehat{\nu}_{\mathrm{A}}}{\widehat{E}_{\mathrm{A}}}=\frac{1}{E_{\mathrm{A}}}\left(\nu_{\mathrm{A}}+\lambda_{\mathrm{A}}\lambda_{\mathrm{T}}\frac{\mu_{\mathrm{A}}}{E_{\mathrm{A}}}\right), \tag{14.189}$$

$$\frac{\widehat{\nu}_{\mathrm{a}}}{\widehat{E}_{\mathrm{A}}}=\frac{1}{E_{\mathrm{A}}}\left(\nu_{\mathrm{a}}+\lambda_{\mathrm{t}}\lambda_{\mathrm{A}}\frac{\mu_{\mathrm{A}}}{E_{\mathrm{A}}}\right), \tag{14.190}$$

$$\frac{\widehat{\nu}_{\mathrm{t}}}{\widehat{E}_{\mathrm{T}}}=\frac{1}{E_{\mathrm{T}}}\left(\nu_{\mathrm{t}}+\lambda_{\mathrm{t}}\lambda_{\mathrm{T}}\frac{E_{\mathrm{T}}\mu_{\mathrm{A}}}{E_{\mathrm{A}}^{2}}\right), \tag{14.191}$$

$$\widehat{\alpha}_{\mathrm{A}}=\alpha_{\mathrm{A}}+\lambda_{\mathrm{A}}\frac{\mu_{\mathrm{A}}}{E_{\mathrm{A}}}\alpha_{\mathrm{S}}, \tag{14.192}$$

$$\widehat{\alpha}_{\mathrm{T}}=\alpha_{\mathrm{T}}+\lambda_{\mathrm{T}}\frac{\mu_{\mathrm{A}}}{E_{\mathrm{A}}}\alpha_{S}, \tag{14.193}$$

$$\widehat{\alpha}_{\mathrm{t}}=\alpha_{\mathrm{t}}+\lambda_{\mathrm{t}}\frac{\mu_{\mathrm{A}}}{E_{\mathrm{A}}}\alpha_{\mathrm{S}}, \tag{14.194}$$

where λ_{A}, λ_{T}, and λ_{t} are shear-coupling parameters, and where quantity D can be regarded as a definition of a macroscopic property degradation parameter arising in the field of continuum damage mechanics. The shear-coupling parameters are zero for undamaged balanced laminates and for balanced cross-ply laminates whether damaged or not. On using Eqs. (14.177)–(14.185), the effective thermoelastic constants of a damaged laminate can be calculated in terms of the damage-dependent parameter D, which can be defined using Eq. (14.177) as:

$$D=\frac{1}{\widehat{E}_{\mathrm{A}}}-\frac{1}{\widehat{E}_{\mathrm{A}}^{(L)}}. \tag{14.195}$$

In the results in Eqs. (14.180)–(14.185), the parameters k, k', and k_1 are constants for undamaged laminates defined by the relations:

$$k=\frac{\widehat{E}_{\mathrm{A}}^{(L)}}{\widehat{E}_{\mathrm{T}}^{(L)}}\,\frac{\nu_{\mathrm{A}}^{90}-\nu_{\mathrm{A}}^{(L)}\widehat{E}_{\mathrm{T}}^{(L)}/\widehat{E}_{\mathrm{A}}^{(L)}}{1-\nu_{\mathrm{A}}^{90}\widehat{\nu}_{\mathrm{A}}^{(L)}}, \tag{14.196}$$

$$k'=\frac{\widehat{E}_{\mathrm{A}}^{(L)}}{1-\nu_{\mathrm{A}}^{90}\widehat{\nu}_{\mathrm{A}}^{(L)}}\left[\frac{\nu_{\mathrm{t}}^{90}}{E_{\mathrm{T}}^{90}}-\frac{\widehat{\nu}_{\mathrm{a}}^{(L)}}{\widehat{E}_{\mathrm{A}}^{(L)}}+\nu_{\mathrm{A}}^{90}\left(\frac{\nu_{\mathrm{a}}^{90}}{E_{\mathrm{A}}^{90}}-\frac{\widehat{\nu}_{\mathrm{t}}^{(L)}}{\widehat{E}_{\mathrm{T}}^{(L)}}\right)\right], \tag{14.197}$$

$$k_1 = \frac{\widehat{E}_{\mathrm{A}}^{(L)} \left(\widehat{\alpha}_{\mathrm{A}}^{(L)} + \nu_{\mathrm{A}}^{90} \widehat{\alpha}_{\mathrm{T}}^{(L)} - \alpha_{\mathrm{T}}^{90} - \nu_{\mathrm{A}}^{90} \alpha_{\mathrm{A}}^{90} \right)}{1 - \nu_{\mathrm{A}}^{90} \widehat{\nu}_{\mathrm{A}}^{(L)}}. \tag{14.198}$$

14.10 Prediction examples for cross-ply laminates

The key results of this chapter are expressions for various effective thermoelastic properties of a simple cross-ply laminate having an array of uniformly spaced ply cracks. For the example to be considered, the ply properties shown in Table 14.1, typical of a transverse isotropic carbon fiber reinforced composite, are used.

When the ply properties given in Table 14.1 are used in conjunction with the formulas in Eqs. (14.134)−(14.140), and (14.150)−(14.153), for a set of ply-crack densities in the range of 0−4 cracks/mm, the results shown in Figs. 14.2−14.4 are obtained when $a = 0.6$ mm and $b = 0.3$ mm. The results shown assume the following identifications:

$$E_{\mathrm{A}} \equiv \mathrm{EA}, \quad E_{\mathrm{T}} \equiv \mathrm{ET}, \quad E_{\mathrm{t}} \equiv \mathrm{Est},$$

$$\nu_{\mathrm{A}} \equiv \mathrm{nuA}, \quad \nu_{\mathrm{a}} \equiv \mathrm{nusa}, \quad \nu_{\mathrm{t}} \equiv \mathrm{nust},$$

$$\mu_{\mathrm{A}} \equiv \mathrm{muA}, \quad \mu_{\mathrm{a}} \equiv \mathrm{musa}, \quad \mu_{\mathrm{t}} \equiv \mathrm{must},$$

$$\alpha_{\mathrm{A}} \equiv \mathrm{alA}, \quad \alpha_{\mathrm{T}} \equiv \mathrm{alT}, \quad \alpha_{\mathrm{t}} \equiv \mathrm{alst}.$$

Shear lag and generalized plane strain models for simple cross-ply laminates have both been used to obtain estimates for the effective thermoelastic constants arising from a uniformly spaced array of ply cracks in the 90 degrees ply. It is seen from the plots of effective property versus ply-crack density shown in Figs. 14.2−14.4 that the use of the shear lag approximation leads to predictions that are consistently greater than those resulting when using the generalized plane strain model. The ply discount estimates of effective properties (where the moduli of damaged plies are set to zero), as shown by the asymptotic limits as the ply crack density increases, are approached by the shear lag model much more slowly.

Table 14.1 Undamaged ply properties typical of a carbon fiber−reinforced polymer.

E_{A} = 140.77 GPa	E_{T} = 8.85 GPa	E_{t} = 8.85 GPa
$\nu_{\mathrm{A}} = 0.28$ $\mu_{\mathrm{A}} = 4.59$ GPa $\alpha_{\mathrm{A}} = 0.245 \times 10^{-6}\ \mathrm{K}^{-1}$	$\nu_{\mathrm{a}} = 0.28$ $\mu_{\mathrm{a}} = 4.59$ GPa $\alpha_{\mathrm{T}} = 45.6 \times 10^{-6}\ \mathrm{K}^{-1}$	$\nu_{\mathrm{t}} = 0.43$ $\mu_{\mathrm{t}} = 3.09441$ GPa $\alpha_{\mathrm{t}} = 45.6 \times 10^{-6}\ \mathrm{K}^{-1}$

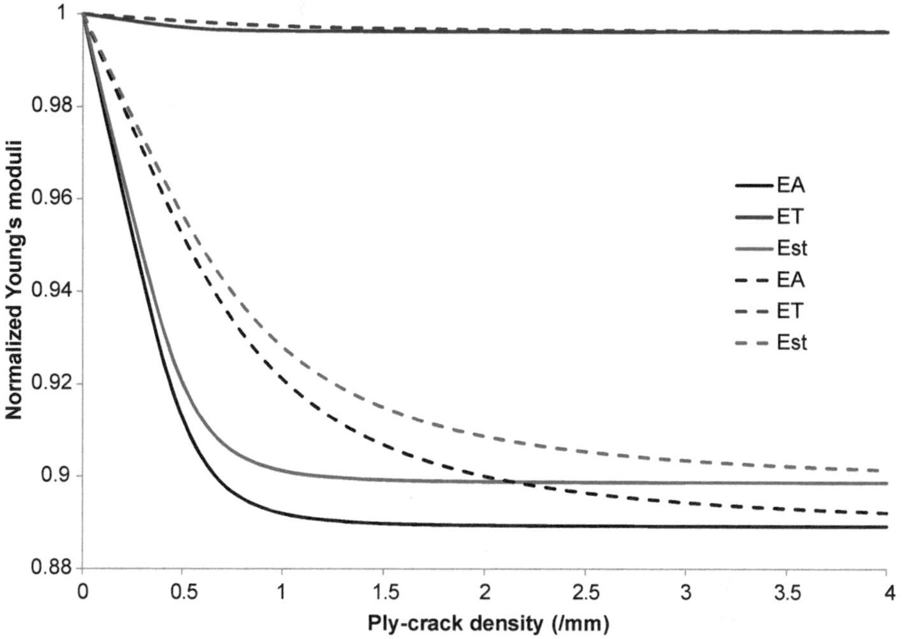

Figure 14.2 Predictions of the normalized effective Young's moduli of a $[0/90_2]_s$ laminate as a function of the density of a uniform distribution of ply cracks in the 90 degrees ply. (Continuous lines: generalized plane strain; dashed lines: shear lag approximation).

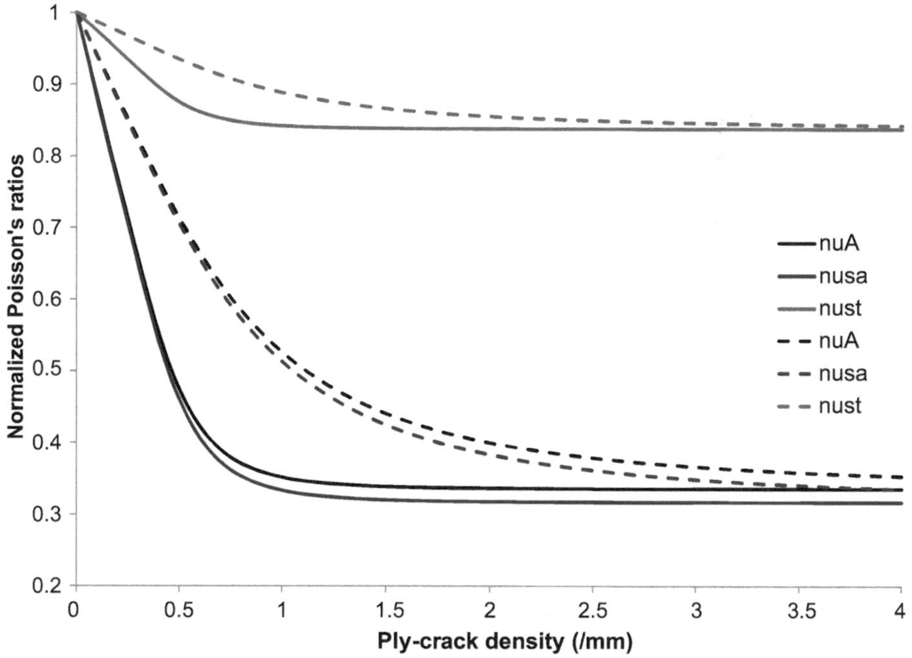

Figure 14.3 Predictions of the normalized effective Poisson ratios of a $[0/90_2]_s$ laminate as a function of the density of a uniform distribution of ply cracks in the 90 degrees ply. (continuous lines: generalized plane strain; dashed lines: shear lag approximation).

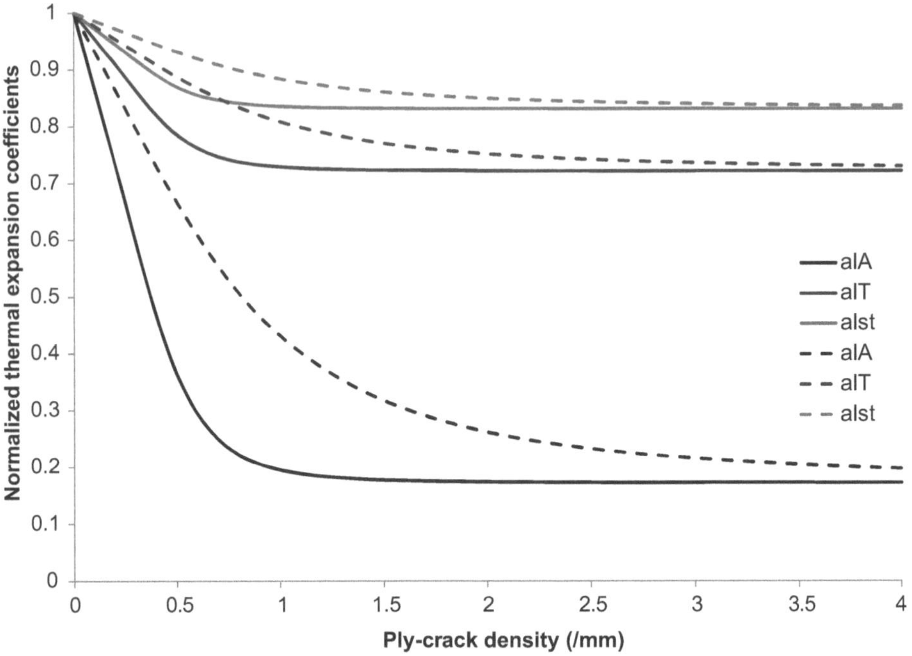

Figure 14.4 Predictions of the normalized effective thermal expansion coefficients of a $[0/90_2]_s$ laminate as a function of the density of a uniform distribution of ply cracks in the 90 degrees ply. (continuous lines: generalized plane strain; dashed lines: shear lag approximation).

By employing ply refinement techniques, more accurate predictions of the effective ply properties of damaged laminates can be made. The software required to achieve this has been written (PREDICT software) to apply the more general model in a user-friendly manner, and if applied to simple cross-ply laminates without ply refinement, it will predict results that correspond exactly to those that are obtained when using the analysis of this chapter.

Similar situations arise for ply cracking in the 90 degrees plies of general symmetric laminates for a variety of laminate configurations considered in the WWFE III International Exercise (Kaddour et al., 2013a) concerned with the assessment of damage models for composite laminates. Analogous results to those derived here for cross-ply laminates have been derived for general symmetric laminates (McCartney, 2013a,b) and assessed/discussed by the organizers of Exercise (Kaddour et al., 2013b). However, the author recommends that a great deal of caution be applied when considering the comparison of models, as the summary presented by the organizers (Kaddour et al., 2013b) relating to the generalized plane strain model developed by the author is very misleading, and some of their conclusions are not justified by the published information (McCartney, 2013b) presented to the Exercise by the author.

References

Kaddour, A.S., Hinton, M.J., Smith, P.A., Li, S., 2013a. The background to the third world-wide failure exercise. J. Compos. Mater. 47 (20–21), 2417–2426.

Kaddour, A.S., Hinton, M.J., Smith, P.A., Li, S., 2013b. A comparison between the predictive capability of matrix cracking, damage and failure criteria for fibre reinforced laminates: part A of the third world-wide failure exercise. J. Compos. Mater. 47 (20–21), 2749–2779.

McCartney, L.N., 1991. Analytical models of stress transfer in unidirectional composites and cross-ply laminates, and their application to the prediction of matrix/transverse cracking. In: Proc. IUTAM Symposium on 'Local Mechanics Concepts for Composite Materials Systems', Blacksburg, VA, pp. 251–282.

McCartney, L.N., 1992. Theory of stress transfer in a 0-90-0 cross-ply laminate containing a parallel array of transverse cracks. J. Mech. Phys. Solid. 40, 27–68.

McCartney, L.N., 2013a. Derivations of energy-based modelling for ply cracking in general symmetric laminates. J. Compos. Mater. 47 (20–21), 2641–2673.

McCartney, L.N., 2013b. Energy methods for modelling damage in laminates. J. Compos. Mater. 47 (20–21), 2613–2640.

Nairn, J.A., Mendels, D.A., 2001. On the use of planar shear-lag methods for stress-transfer analysis of multi-layered composites. Mech. Mater. 33, 335–362.

Nuismer, R.J., Tan, S.C., 1988. Constitutive relations of a cracked composite lamina. J. Compos. Mater. 22 (4), 306–321.

PREDICT Software. http://www.npl.co.uk/science-technology/composites-and-polymeric-materials/services/predict-software.

Semi-analytical techniques for analysis of cracked composite laminates

15

M. Hajikazemi[1] *and L.N. McCartney*[2]
[1]Department of Materials, Textiles and Chemical Engineering, Faculty of Engineering and Architecture, Ghent University, Ghent, Belgium; [2]Department of Engineering, Materials and Electrical Science, National Physical Laboratory, Teddington, Middlesex, United Kingdom

15.1 Introduction

Composite laminates, subject to quasi-static or fatigue thermomechanical loading, undergo progressive damage evolution involving different ply-level damage mechanisms, for example, ply cracking, delamination at the tips of ply cracks, and fiber breakage. The presence of geometrical discontinuities causes high stress concentrations that might locally trigger random initiations and interactions of these damage mechanisms. Away from these discontinuities, progressive ply-level damage evolution in laminates starts typically with ply cracks that typically traverse the thickness of the ply and run parallel to the fibers. Ply cracking degrades all the in-plane, out-of-plane, and bending laminate stiffnesses and changes its coefficients of thermal expansion, moisture absorption, and the natural vibration frequency. If laminate failure was defined to be the separation of the composite into two or more separate pieces, as in the case of metals, then ply cracking would not be normally a critical damage mechanism. However, in the design of composite laminates, failure should be defined as attainment of a certain damage state where its criticality depends on the application and on material performance. Take, for example, the use of glass/epoxy laminates in wind turbine blades where the presence of ply cracks with high crack density can significantly degrade the laminate stiffness. This stiffness reduction can largely increase the deformation of the blade that might even lead to impact with the ground. In addition, ply cracking triggers the development of other damage mechanisms that are critical from the final fracture point of view. Stress concentrations near the crack tips at the ply interfaces may cause either delamination, fiber breakage, further ply cracking, or sometimes all of these, in the adjacent plies. Thus, developing damage modeling techniques for the prediction of ply cracking initiation and propagation, and its effects on the effective properties of composite laminates, has been an important technological objective, offering much scope for the reduction of development costs associated with make-and-test development procedures. Considering numerous design possibilities for composite laminates, a reliable damage modeling approach must take into account the complexities of the laminate geometry (e.g., lay-up and thickness), the nature of the loads applied (e.g., multi-axial, bending, quasistatic, and fatigue), and the effects of temperature and moisture ingress,

Modeling Damage, Fatigue and Failure of Composite Materials. https://doi.org/10.1016/B978-0-443-18489-5.00014-X

including residual stresses that arise during manufacturing processes. Moreover, there are additional needs to apply damage models locally when designing laminated engineering components with complex shapes subjected to multiaxial local stress states. Such design requirements often require the use of finite element analysis (FEA). However, the capability of accurately degrading local properties to account for the growth of damage for such multi-axial stress states is crucial as it will lead to load transfer in structures and to the development of local "hot-spots" for failure.

The objective of this chapter is to introduce the contributions made by the authors toward the required modeling requirements namely "stress transfer models" and "damage evolution law" for the prediction of ply cracking initiation and propagation, and its impact on the effective properties of composite laminates. Section 15.2 introduces two semi-analytical stress transfer models known as the "generalized plane strain (GPS)/bending" and "variational" approximations and the range of stress transfer problems that they can solve. The extension of the models to include the important effects of bending deformations, out-of-plane shear stiffness reduction, and non-uniform distribution of ply cracks is also discussed. Section 15.3 introduces a damage evolution law which uses an energy balance approach to predict ply crack formation in general symmetric laminates in the presence of residual stresses under multiaxial quasistatic loading conditions. This damage evolution law is formulated in terms of effective properties of the damaged laminate and, thus, can be implemented using the effective properties obtained from any stress transfer technique. A brief discussion is included about the extension of the approach for the prediction of ply crack multiplication under fatigue loading conditions when progressive fatigue-induced growth of pre-existing defects leads eventually to fully developed ply cracks. Section 15.4 provides prediction examples where the results obtained using the GPS/bending and variational approach are compared with each other as well as with adequately refined FE results and available experiments. Section 15.5 gives a brief outlook on the application of the methods to delamination, and more complex damage scenarios that must be solved including the possibility of linking the approaches to FE methods in order to include the effects of damage mechanisms realistically at a structural level.

15.2 Stress transfer models

For the sake of simplicity, stress transfer approaches will be first explained in symmetric cracked laminates. A laminate made of $2N$ plies with general symmetric stacking sequence as shown in Fig. 15.1 is considered. Only the top half of the laminate (N plies) is shown as laminate symmetry is assumed. A Cartesian coordinate system where the x, y, and z directions, respectively, define the in-plane axial, in-plane transverse, and through-thickness directions, is located at the center of the laminate. There are $N-1$ interfaces in the top half of the symmetric laminate ($z > 0$) which are denoted by $z = z_i$, $i = 1, 2 \ldots N-1$. The plane of symmetry (midplane) is $z = z_0 = 0$, and the external top surface is $z = z_N = h$. The total thickness of the laminate is denoted by $2h$. The thickness of the ith ply is specified by $h_i = z_i - z_{i-1}$. The orientation of the ith ply is described by the angle θ_i (measured counterclockwise) between the loading and the

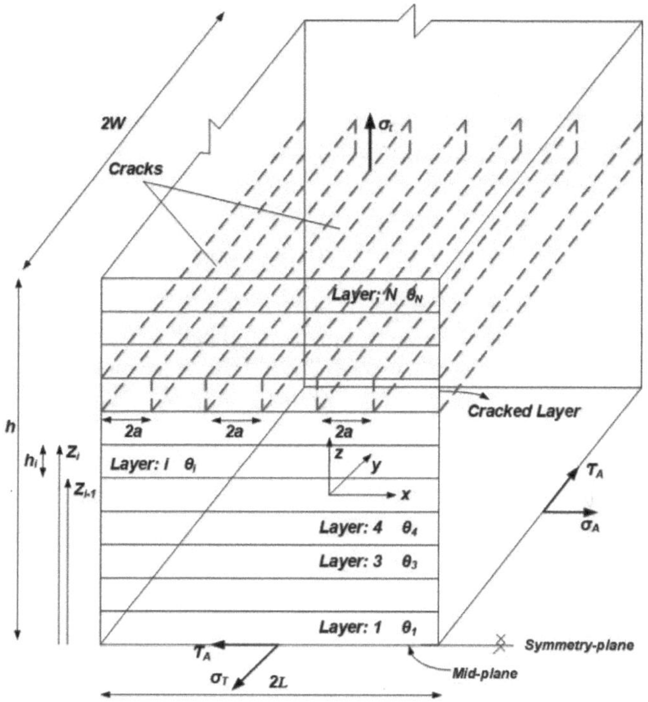

Figure 15.1 Schematic geometry of a general symmetric laminate containing uniformly distributed ply cracks in one of its plies (due to the symmetry with respect to the mid-plane only top half of the laminate is shown).

fiber direction of the layer. The stress and strain components and also material properties associated with the *i*th layer are denoted by a superscript or subscript i. It is now assumed that ply cracks are present in a single orientation (possibly in more than one ply); however, the symmetry with respect to the mid-plane is preserved after cracking. It is also assumed that the ply crack distribution in damaged plies is uniform, having a separation $2a$, and the cracks in each damaged ply of the laminate are in the same plane. By assuming a uniform distance between adjacent ply cracks ($2a$), a knowledge of the stress transfer in a unit cell between two adjacent ply cracks will be adequate. The aim of stress transfer models is to find the stress/displacement fields and subsequently the effective laminate properties as a function of crack density $\rho = 1/(2a)$. This problem has been the subject of many studies where several techniques with different accuracy and versatility are developed. Extensive literature reviews on stress transfer techniques can be found in McCartney (2022) and Hajikazemi and Van Paepegem (2021) with focus on earlier and recent techniques, respectively. Briefly, these approaches involve shear-lag models (Aveston and Kelly, 1973; Garrett and Bailey, 1977; Nairn and Mendels, 2001; Barbero and Cortes, 2010), the GPS model (McCartney, 1992a, 2000), the variational model (Hashin, 1985; Li and Hafeez, 2009; Vinogradov and Hashin, 2010; Hajikazemi and Sadr, 2014a, 2014b; Hajikazemi et al.,

2014a, 2018, 2019), models based on crack opening and sliding displacements (Lundmark and Varna, 2005, 2006; Varna, 2015; Loukil and Varna, 2019), and numerical methods (Herakovich et al., 1988; Yuan and Selek, 1993; Barbero et al., 2013; Garoz et al., 2019; Ahmadi et al., 2022).

15.2.1 GPS and variational stress transfer techniques for analysis of cracked symmetric laminates under triaxial and in-plane shear loading

Among the semi-analytical techniques that can be applied to arbitrary symmetric laminates, the GPS (McCartney, 2000) and variational (Hajikazemi and Sadr, 2014b; Hajikazemi et al., 2014a) models can provide highly accurate predictions for stress and displacement fields as well as effective laminate properties. The symmetric laminate in these models can be subject to general loading (Fig. 15.1), involving effective uniform applied stresses: in-plane axial σ_A, in-plane shear τ_A, in-plane transverse (either in terms of stress σ_T or strain ε_T), through-thickness (σ_t), and thermally induced residual stresses due to the uniform temperature difference ΔT. Let us assume that the symmetric laminate has ply cracks in a 90 degree ply (or any other single orientation). Having ply cracks in the 90 degree ply is not a limitation for these techniques as general loading is considered so that if cracks form first in another single orientation, the laminate can be rotated so that the crack planes are parallel to the y-axis (i.e., it becomes a 90 degree ply) and the applied stresses and effective properties transformed to appropriate values for the new orientation. Both methods assume that the stresses in the cracked material can be represented as a superposition of the stresses in the undamaged material and some as yet unknown stress transfer functions due to the presence of the cracks.

$$\sigma^i_{mn}(X) = \sigma^0_{mn}{}^{(i)} + \sigma^P_{mn}{}^{(i)}(X), \tag{15.1}$$

where $m, n = x, y, z$ and $X = x, z$. The first term in the right-hand side of Eq. (15.1) is the stress in the ith ply of an undamaged laminate. The last term in Eq. (15.1) is the perturbation stress in i^{th} ply, which in contrast to the stresses in the undamaged laminate, is a function of location within the ply.

GPS and variational techniques aim to find admissible stress fields that satisfy the stress equilibrium equations $\sigma^i_{mn,n} = 0$, and the required traction boundary and interface continuity conditions, namely:

- Zero shear traction condition on the external surfaces at $z = \pm h$: $\sigma^N_{xz} = \sigma^N_{yz} = 0$.
- Traction continuity condition at the interface between the plies at $z = z_i$, for $i = 1 \ldots N-1$: $\sigma^i_{zz} = \sigma^{i+1}_{zz}, \sigma^i_{xz} = \sigma^{i+1}_{xz}, \sigma^i_{yz} = \sigma^{i+1}_{yz}$.
- Zero traction condition on the crack surfaces at $x = \pm a$: $\sigma^c_{xx} = \sigma^c_{xy} = \sigma^c_{xz} = 0$, where superscript (c) indicates that the condition applies only to cracked plies. In addition, the stress field should be equivalent to the membrane forces/strains applied to the laminate.

The fundamental assumption of the analysis for the GPS model is that the transverse shear stresses ($\sigma^i_{xz}, \sigma^i_{yz}$) are linear functions of the through-thickness coordinate in each layer. Solving the equilibrium equations based on this assumption leads to the condition for which membrane stresses ($\sigma^i_{xx}, \sigma^i_{xy}$) and transverse normal stresses (σ^i_{zz}) are, respectively, independent of z and quadratic functions of z in each layer. The fundamental assumption of the variational model is that the membrane stresses ($\sigma^i_{xx}, \sigma^i_{xy}$) are uniform in the through-thickness direction (z-direction) of each layer and vary only in the x-direction. Similarly, this assumption leads to a condition in which transverse shear ($\sigma^i_{xz}, \sigma^i_{yz}$) and transverse normal stresses (σ^i_{zz}) are, respectively, linear and quadratic functions of z in each layer. Using the same form of variables in both approaches (see the details in (Hajikazemi and McCartney, 2018)), the following equations represent the admissible stress fields:

$$\sigma^i_{xx} = \sigma^{0\ (i)}_{xx} + \frac{p_{i-1}(x) - p_i(x)}{h_i}, \qquad \sigma^i_{xy} = \tau^{0\ (i)}_{xy} + \frac{q_{i-1}(x) - q_i(x)}{h_i}, \tag{15.2}$$

$$\sigma^i_{xz} = p'_i(x)\frac{z - z_{i-1}}{h_i} - p'_{i-1}(x)\frac{z - z_i}{h_i}, \qquad \sigma^i_{yz} = q'_i(x)\frac{z - z_{i-1}}{h_i} - q'_{i-1}(x)\frac{z - z_i}{h_i}, \tag{15.3}$$

$$\sigma^i_{zz} = \sigma_t + \frac{z - z_i}{2h_i}\left[(z - z_i)\, p''_{i-1}(x) - (z - z_i + 2h_i)\, p''_i(x)\right] + S_i(x),$$
$$\text{where} \quad S_{i-1}(x) = \frac{h_i}{2}\left[p''_{i-1}(x) + p''_i(x)\right] + S_i(x), \qquad S_N(x) = 0, \tag{15.4}$$

where $p_i(x), q_i(x), i = 1 \ldots N$ are unknown stress transfer functions of the x coordinate that are to be determined, and which have the property that they are identically zero for the case of undamaged laminates. It should be noted that $\sigma^i_{zz}(z = z_i) = \sigma_t + S_i(x)$. The fact that the stress components σ_{xz} and σ_{yz} are required to be zero on $z = 0$ (symmetry requirement) and $z = h$ (shear traction free) asserts that the functions $p_0(x), p_N(x), q_0(x)$ and $q_N(x)$ must be such that:

$$p_0(x) = p_N(x) \equiv 0, \qquad q_0(x) = q_N(x) \equiv 0. \tag{15.5}$$

It then follows that the value of the effective applied axial stress σ_A and in-plane shear stress τ_A has the same value for both cracked and uncracked laminates. One difference between the GPS and variational models is the loading condition for the in-plane transverse direction. In the GPS model, the in-plane transverse strain is one of the input loading parameters and thus has a uniform value ε_T in all layers of the cracked and uncracked laminates. Therefore, the in-plane transverse stress σ^i_{yy} can be defined in terms of the local values of the other stress components and the applied transverse strain ε_T:

$$\sigma^i_{yy} = \frac{1}{s^i_{22}}\left[\varepsilon_T - s^i_{12}\sigma^i_{xx} - s^i_{23}\sigma^i_{zz} - s^i_{26}\sigma^i_{xy} - \alpha^i_2 \Delta T\right], \tag{15.6}$$

where the thermoelastic constants s^i and α^i are, respectively, the components of the compliance matrix and coefficient of thermal expansion vector of the ith ply in the global coordinate system. Thus, σ_{yy}^i is a quadratic function of the through-thickness direction (z-direction) in each layer.

In the variational model, however, the effective in-plane transverse stress σ_T is an input loading parameter. Thus, similar to other membrane stresses, it is assumed that the transverse stress σ_{yy}^i is uniform in the z-direction of each layer, for the moment varying arbitrarily in the x-direction:

$$\sigma_{yy}^i = \sigma_{yy}^{0\ (i)} + \frac{r_{i-1}(x) - r_i(x)}{h_i}, \tag{15.7}$$

where $r_i(x)$, $i = 1 \ldots N$ are unknown functions of x coordinate that are to be determined. It is noted that Eq. (15.6) reduces by one the number of unknown stress components whose distribution needs to be found in the GPS model. Similarly, it has been shown in Hajikazemi and Sadr (2014a) that in the variational model, as a consequence of the minimum complementary energy principle, σ_{yy}^i is also not an independent stress component and indeed the perturbation functions $r_i(x)$ can be defined in terms of $p_i(x)$ and $q_i(x)$. Therefore, both methods involve the same number of unknown stress functions ($2N-2$) whose distributions need to be found. The representation for the admissible stress fields for both approaches satisfies all the stress equilibrium equations, traction boundary, and interface continuity conditions and balances the applied loads for any functions $p_i(x), q_i(x)$ $i = 1 \ldots N$, characterizing stress transfer in the cracked laminate.

For the GPS model (McCartney, 2000), it is assumed that the cracked laminate is under generalized in-plane strain conditions for which the displacement field in each layer (i) of laminate is of the following form:

$$u_i = f_{1,i}(x,z) + \varepsilon_A^0 x + \frac{1}{2}\gamma^0 y, \quad v_i = f_{2,i}(x,z) + \frac{1}{2}\gamma^0 x + \varepsilon_T y, \quad w_i = f_{3,i}(x,z), \tag{15.8}$$

where ε_A^0 and γ^0 are the effective axial and shear strains in the undamaged laminate when the stresses σ_A, σ_t, and τ_A are applied, when the in-plane transverse strain has the same value ε_T in all layers of the laminate, and when the temperature difference is uniform having the value ΔT. It should be noted that the functions $f_{k,i}$, $k = 1, 2, 3$, $i = 1 \ldots N$, appearing in Eq. (15.8) are all independent of y and have the property that they are identically zero for the case of undamaged laminates. The readers should refer to McCartney (2000), Hajikazemi and McCartney (2018) to find the explicit forms of functions $f_{k,i}$ in terms of unknown functions $p_i(x), q_i(x)$, $i = 1 \ldots N$. These explicit expressions are obtained by satisfying the stress–strain–displacement relationships (concerning out-of-plane strains) and the continuity of displacement fields at the interface between different layers.

Variational approach is known historically as a stress-based approach where in the process of finding unknown functions, there is no need to use displacement fields as the complementary energy is minimized and all the boundary conditions are all in terms of stress components. However, the existence of a complete elasticity solution for the GPS model involving both the stress and displacement fields (see Eq. 15.8), provides useful information about the form of displacement fields that should be considered for variational approach. It is shown in Hajikazemi and McCartney (2018) that taking a similar approach as for the GPS model, the displacement field can also be obtained using the variational approach as follows:

$$u_i = f_{1,i}(x,z) + \varepsilon_A^0 x + \frac{1}{2}\gamma^0 y, \quad v_i = f_{2,i}(x,z) + \frac{1}{2}\gamma^0 x + \bar{\varepsilon}_T y, \quad w_i = f_{3,i}(x,z). \tag{15.9}$$

It should be emphasized that in the variational model, unlike the GPS model, the in-plane transverse strain ε_T is not an input value and thus ε_{yy}^i is not uniform in the through-thickness direction of each layer. To overcome this issue and in order to have a strain field compatible with assumed admissible stress field, the average in-plane transverse strain through-thickness of the ith layer, which has the same value in all layers of the laminate, is considered in Eq. (15.9) and is defined by:

$$\bar{\varepsilon}_{yy}^i = \bar{\varepsilon}_T = \frac{1}{h_i}\int_{z_{i-1}}^{z_i}\left(s_{12}^i\sigma_{xx}^i + s_{22}^i\sigma_{yy}^i + s_{23}^i\sigma_{zz}^i + s_{26}^i\sigma_{yz}^i + \alpha_2^i\Delta T\right)dz, \quad h_i = z_i - z_{i-1}. \tag{15.10}$$

Finally, the expressions derived for the stress and displacement components in both approaches satisfy exactly the stress equilibrium equations together with the stress–strain–displacement relations involving the out-of-plane strains for *any* functions $p_i(x), q_i(x), i = 1 \ldots N-1$. In addition, the displacement and stress fields automatically satisfy the required interface continuity conditions. However, the unknown functions $p_i(x), q_i(x),\ i = 1 \ldots N-1$ are yet to be determined. The variational and GPS models implement different approaches to determine unknown functions. The GPS model uses an averaging procedure where two of stress–strain–displacement relations concerning the in-plane axial and in-plane shear strains are satisfied in an average sense through the thickness of each ply (or sub-layer of a ply when using ply refinement). The variational model uses the principle of minimum complementary energy to determine the solution. It is shown (Hajikazemi and McCartney, 2018) that both approaches will interestingly lead to a set of ordinary differential equations as follows that has the following very similar form:

$$\sum_{i=1}^{N-1}(F_1)_{ij}p_i''''(x)+\sum_{i=1}^{N-1}(F_2)_{ij}p_i''(x)+\sum_{i=1}^{N-1}(F_3)_{ij}p_i(x)+$$
$$\sum_{i=1}^{N-1}(F_4)_{ij}q_i''(x)+\sum_{i=1}^{N-1}(F_5)_{ij}q_i(x)=0,\ \ j=1...N-1, \tag{15.11}$$

$$\sum_{i=1}^{N-1}(F_6)_{ij}p_i''(x)+\sum_{i=1}^{N-1}(F_7)_{ij}p_i(x)+\sum_{i=1}^{N-1}(F_8)_{ij}q_i''(x)+$$
$$\sum_{i=1}^{N-1}(F_9)_{ij}q_i(x)=0,\ \ j=1...N-1, \tag{15.12}$$

where the coefficient matrices of F in Eqs. (15.11) and (15.12) may be determined explicitly in terms of ply elastic properties, see Hajikazemi and Sadr (2014a), McCartney (2013). Here, it should be noted that for a cross-ply laminate $[F_4]=[F_5]=[F_6]=[F_7]=0$, for $i,j=1\ldots N-1$, and Eqs. (15.11) and (15.12) become uncoupled. Two earlier works done by Hashin (Hashin, 1985) on the variational approach and McCartney (McCartney, 1992a; McCartney et al., 2016) on the GPS approach concern a two-layer cross-ply laminate (just two layers above the symmetry plane so that $N=2$) for which the form of governing equations in Eqs. (15.11) and (15.12) can be simplified as follows:

$$p_1''''(x)+\frac{F_2}{F_1}p_1''(x)+\frac{F_3}{F_1}p_1(x)=0, \tag{15.13}$$

$$q_1''(x)+\frac{F_9}{F_8}q_1(x)=0, \tag{15.14}$$

where the two single uncoupled ordinary differential equations can be solved in a closed form. In general, the governing equations in Eqs. (15.11) and (15.12) are a pair of coupled systems of simultaneous linear ordinary differential equations with constant coefficients. There are numerous standard treatments to solve these equations, which are not described here; however, for all these approaches, there is a need to solve an eigenvalue problem that is based on the size of the problem and is best solved numerically if $N>2$. That is the reason we have referred to the GPS and variational approaches as semi-analytical techniques.

Once the governing equations are solved and general solutions are obtained, the last step is to find values for the arbitrary constants of integration using boundary conditions that will lead to a unique solution. Considering the symmetry of the problem with respect to the $x=0$ plane, the boundary conditions in both approaches involve stress-free conditions for the cracked plies on the crack planes at $x=a$ and out-of-plane axial shear traction-free due to continuity considerations and periodic conditions for the uncracked plies at $x = a$. The physical construction of the problem itself does not provide any more boundary conditions without resorting to the use of displacement conditions. The required displacement boundary conditions in the GPS model are defined as follows:

$$u_i(a, y, z) = \varepsilon_A a + \frac{1}{2}\gamma\, y, \qquad v_i(a, y, z) = \frac{1}{2}\gamma a + \varepsilon_T y, \tag{15.15}$$

As these conditions cannot be satisfied exactly by the relations Eq. (15.8) because of the z dependence, the equivalent averages of above boundary conditions through the thickness of each layer are imposed on $x = a$ for uncracked layers. In a stress-based variational approach, displacements might not always be derived, and one has to find extra boundary conditions in terms of stresses for uncracked plies before the solution can be determined. It has been shown (Hajikazemi and Sadr, 2014a) that the second and third boundary conditions for each uncracked ply of a general symmetric laminate can be obtained mathematically from the variational calculus itself in terms of natural boundary conditions (Li and Hafeez, 2009). The natural boundary conditions obtained directly from the variational calculus which should be applied to uncracked plies are as follows (Hajikazemi and Sadr, 2014a):

$$[T_1](p'''(a)) + [T_4](q'(a)) = 0, \quad [T_6](q'(a)) = 0, \tag{15.16}$$

where $[T_1]$, $[T_4]$, and $[T_6]$ are constant matrices which are all obtained analytically in terms of ply properties (Hajikazemi and Sadr, 2014a). The mentioned boundary conditions for each approach provide exactly $3(N-1)$ boundary conditions in total, which is necessary and sufficient to ensure that the system of differential equations given by Eqs. (15.11) and (15.12) has a unique solution that is symmetric about $x = 0$.

Having the elastic fields for a cracked laminate as a function of the applied loads using the GPS or variational techniques, it is possible to obtain the effective laminate thermo-elastic constants. The method of estimating the elastic constants (McCartney, 2000; Hajikazemi et al., 2014a) is based on the fact that, when deriving the stress and displacement fields from the solution of the differential equations, the loading parameters $\sigma_A, \varepsilon_T, \sigma_t, \tau_A, \Delta T$ in the GPS model and the loading parameters $\sigma_A, \sigma_T, \sigma_t, \tau_A, \Delta T$ in the variational model are specified. Finally, provided that the effects of cracks can be effectively homogeneous at the macroscopic laminate level, the effective stress–strain–temperature relations of a cracked symmetric laminate can be written as follows

$$\varepsilon_A = \frac{1}{E_A}\sigma_A - \frac{\nu_A}{E_A}\sigma_T - \frac{\nu_a}{E_A}\sigma_t - \frac{\lambda_A}{\mu_A}\tau_A + \alpha_A \Delta T, \tag{15.17}$$

$$\varepsilon_T = -\frac{\nu_A}{E_A}\sigma_A + \frac{1}{E_T}\sigma_T - \frac{\nu_t}{E_T}\sigma_t - \frac{\lambda_T}{\mu_A}\tau_A + \alpha_T \Delta T, \tag{15.18}$$

$$\varepsilon_t = -\frac{\nu_a}{E_A}\sigma_A - \frac{\nu_t}{E_T}\sigma_T + \frac{1}{E_t}\sigma_t - \frac{\lambda_t}{\mu_A}\tau_A + \alpha_t \Delta T, \tag{15.19}$$

$$\gamma_A = -\frac{\lambda_A}{\mu_A}\sigma_A - \frac{\lambda_T}{\mu_A}\sigma_T - \frac{\lambda_t}{\mu_A}\sigma_t + \frac{1}{\mu_A}\tau_A + \alpha_S \Delta T, \tag{15.20}$$

where the parameters E, μ, ν, and α specify, respectively, Young's modulus, shear modulus, Poisson's ratio, and thermal expansion coefficients for the laminate. The parameters λ are ratios associated with the effect of shear stress on axial and transverse strains. The upper-case subscripts (A and T) are attached to axial and transverse laminate thermoelastic constants to specify that they refer to axial and transverse deformations, while the corresponding lower-case subscripts (a and t) denote laminate thermo-elastic constants that involve through-thickness stress and deformations. The upper-case subscript S denotes properties associated with shear deformations.

It has shown (McCartney, 2000; Hajikazemi and McCartney, 2018) that the stress fields obtained from the GPS and variational models are the optimal admissible stress field that can be developed based on the assumptions made for the admissible stress fields in Eqs. (15.2)–(15.7). The GPS and variational models satisfy, respectively, the Hellinger–Reissner (Reissner, 1950) and minimum complementary energy (Rosen, 1970) variational principles. In addition, both approaches are developed systematically so that they can deal with arbitrary number of plies N in the laminate. Thus, these approaches can benefit from the advantages a ply refinement technique (McCartney, 2000; Takeda et al., 2000) which can relax the effects of assumptions made in the model leading to accurate elasticity solutions. It is believed that the approaches taken in each model would lead to identical results if the models were under similar loading conditions in transverse direction.

In the subsequent sections, some extensions of GPS and variational models for dealing with more complex lay-ups and loading conditions will be very briefly introduced without discussing the details. The required references are though mentioned for interested readers to find the details.

15.2.2 Variational technique for cracked laminates under out-of-plane shear loading

In a laminate under triaxial $\sigma_A, \sigma_T, \sigma_t$ and in-plane shear τ_A loads as described in the previous section, the presence of cracks perturbs the uniform distribution of in-plane and out-of-plane normal stresses and induces high gradients of out-of-plane shear stresses leading to the triaxial and in-plane shear stiffness reductions. It is shown (Hajikazemi et al., 2019) that in a laminate under out-of-plane shear (τ_a, τ_t) loads (see Fig. 15.2a), the presence of cracks perturbs the uniform distribution of out-of-plane shear stresses and triggers high gradients of in-plane stresses leading to out-of-plane shear stiffness reductions. The study of this stress transfer mechanism in the cracked laminate is necessary for the modification of laminate out-of-plane shear stiffness and prediction of secondary damage modes, for example, ply cracking in neighboring plies and delamination at the interfaces. This problem is less studied in the literature due to difficulty of defining appropriate boundary conditions for the cracked unit cell. However, it was shown in the previous section that the variational approach does not

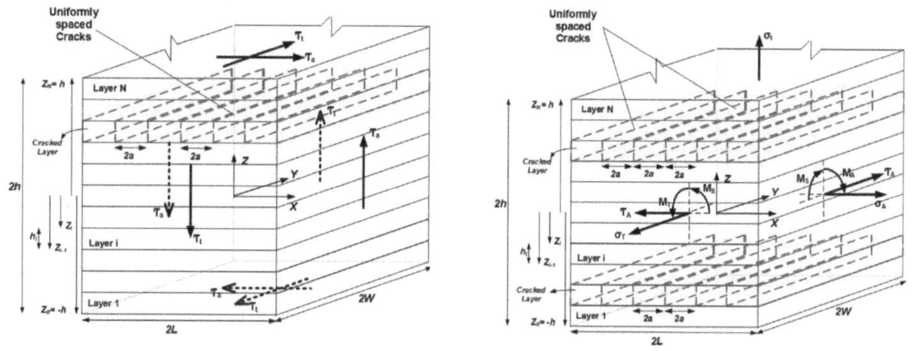

Figure 15.2 (a) Schematic geometry of a general laminate containing uniformly distributed ply cracks in one of its plies subject to out-of-plane shear (τ_a and τ_t) loads, (b) a general cracked laminate containing uniformly distributed ply cracks subject to combined triaxial (σ_A, σ_T, σ_t), in-plane shear (τ_A), and bending (M_A, M_T, M_S) loads.

inherently need the boundary conditions in terms of displacement fields and the appropriate boundary conditions can be obtained directly in the process of complementary energy minimization. Following a similar strategy as described in the previous section, a variational model is developed in Hajikazemi et al. (2019) for general laminates with ply cracks in 90 degree plies to obtain the stress fields under out-of-plane shear loads and the dependence of the out-of-plane shear stiffnesses on the crack density. It is shown that the effective out-of-plane shear stress—strain relations of a general laminate when having ply cracks in its 90 degree plies can be written as follows:

$$\gamma_t = \frac{1}{\mu_t}\tau_t - \frac{\lambda_S}{\mu_a}\tau_a, \tag{15.21}$$

$$\gamma_a = -\frac{\lambda_S}{\mu_a}\tau_t + \frac{1}{\mu_a}\tau_a, \tag{15.22}$$

where μ_t, μ_a, and λ_S and are, respectively, the out-of-plane transverse shear stiffness, axial shear stiffness, and shear coupling parameter for the cracked laminate. It is noteworthy that although the out-of-plane axial shear stiffness μ_a of a laminate can largely degrade as a function of ply cracks in the 90 degree plies, the transverse shear stiffness μ_t remains unchanged having the value for that of an undamaged laminate.

15.2.3 Generalized plane bending (GPB) and variational techniques for cracked laminates with general lay-ups under combined triaxial, shear, and bending loads

Many composite components encounter some sort of bending deformation, especially when considering the deformation of laminates in composite substructures. Moreover, nonsymmetric laminates are widely used in practice to achieve design requirements.

Due to the complexities involved, only a few stress transfer models have been developed to deal with general cracked laminates (possibly nonsymmetric) under combined general in-plane, out-of-plane, and bending loads (Fig. 15.2b). Briefly, some models that can predict stress transfer and/or stiffness reduction in cracked laminates under bending loads include: generalized plane bending (GPB) models (McCartney, 2001, 2008a), variational model (Hajikazemi et al., 2014b, 2017; Nairn, 2017), shear-lag model (Barbero and Cabrera Barbero, 2016), some approximate stiffness reduction models (Pupurs et al., 2016), finite element models (Barulich et al., 2018; Dinh et al., 2019; Schmitz and Horst, 2014), boundary element (BE) methods (McCartney and Blazquez, 2009), etc. Recently, Hajikazemi et al. (Hajikazemi et al., 2017) have reviewed almost all published bending stress transfer models focusing on the assumptions and limitations in each approach, where it is shown that GPB, variational, and FE models, in the framework of some common assumptions on the geometry of cracks, are more accurate despite different level of difficulties in the implementation of each approach. The current status of these models is as follows:

- GPB model: Applicable to non-symmetric cross-ply laminates containing a uniform distribution of ply cracks under combined biaxial in-plane (σ_A, σ_T), biaxial bending (M_A, M_T), and through-thickness (σ_t) loading conditions (McCartney, 2001, 2008a).
- FEM models: Applicable to general laminates containing distributions of ply cracks under combined general in-plane (σ_A, σ_T, τ_A) and bending (M_A, M_T, M_S) loading conditions (Dinh et al., 2019).
- Variational model: Applicable to general laminates containing a uniform distribution of ply cracks under combined general in-plane (σ_A, σ_T, τ_A), bending (M_A, M_T, M_S), and through-thickness (σ_t) loading conditions (Hajikazemi et al., 2014b, 2017, 2020a).

These analyses are too complex to be detailed here, but it is shown that the effective stress–strain–temperature relations of a cracked symmetric laminate with ply cracks in its 90 degree plies Eqs. (15.17)–(15.20) under triaxial and in-plane shear loading should be extended as follows to involve laminates with general lay-ups subject to combined triaxial, shear, and bending deformations:

$$\varepsilon_A = \frac{1}{E_A}\sigma_A - \frac{\nu_A}{E_A}\sigma_T - \frac{\nu_a}{E_A}\sigma_t - \frac{\lambda_A}{\mu_A}\tau_A - \frac{\widehat{\nu}_A}{\widehat{E}_A}M_A - \frac{\widehat{\eta}_A}{\widehat{E}_T}M_T - \frac{\widehat{\lambda}_A}{\widehat{\mu}_A}M_S + \alpha_A\Delta T, \tag{15.23}$$

$$\varepsilon_T = -\frac{\nu_A}{E_A}\sigma_A + \frac{1}{E_T}\sigma_T - \frac{\nu_t}{E_T}\sigma_t - \frac{\lambda_T}{\mu_A}\tau_A - \frac{\widehat{\nu}_T}{\widehat{E}_A}M_A - \frac{\widehat{\eta}_T}{\widehat{E}_T}M_T - \frac{\widehat{\lambda}_T}{\widehat{\mu}_A}M_S + \alpha_T\Delta T, \tag{15.24}$$

$$\varepsilon_t = -\frac{\nu_a}{E_A}\sigma_A - \frac{\nu_t}{E_T}\sigma_T + \frac{1}{E_t}\sigma_t - \frac{\lambda_t}{\mu_A}\tau_A - \frac{\widehat{\nu}_a}{\widehat{E}_A}M_A - \frac{\widehat{\eta}_a}{\widehat{E}_T}M_T - \frac{\widehat{\lambda}_t}{\widehat{\mu}_A}M_S + \alpha_t\Delta T, \tag{15.25}$$

$$\gamma_t = \frac{1}{\mu_t}\tau_t - \frac{\lambda_S}{\mu_a}\tau_a, \tag{15.26}$$

$$\gamma_a = -\frac{\lambda_S}{\mu_a}\tau_t + \frac{1}{\mu_a}\tau_a, \tag{15.27}$$

$$\gamma_A = -\frac{\lambda_A}{\mu_A}\sigma_A - \frac{\lambda_T}{\mu_A}\sigma_T - \frac{\lambda_t}{\mu_A}\sigma_t + \frac{1}{\mu_A}\tau_A - \frac{\widehat{\nu}_S}{\widehat{E}_A}M_A - \frac{\widehat{\eta}_S}{\widehat{E}_T}M_T - \frac{\widehat{\lambda}_S}{\widehat{\mu}_A}M_S + \alpha_S\Delta T, \tag{15.28}$$

$$\widehat{\varepsilon}_A = -\frac{\widehat{\nu}_A}{\widehat{E}_A}\sigma_A - \frac{\widehat{\nu}_T}{\widehat{E}_A}\sigma_T - \frac{\widehat{\nu}_a}{\widehat{E}_A}\sigma_t - \frac{\widehat{\nu}_S}{\widehat{E}_A}\tau_A + \frac{1}{\widehat{E}_A}M_A - \frac{\widehat{\delta}_A}{\widehat{E}_A}M_T - \frac{\widehat{\beta}_A}{\widehat{\mu}_A}M_S + \widehat{\alpha}_A\Delta T, \tag{15.29}$$

$$\widehat{\varepsilon}_T = -\frac{\widehat{\eta}_A}{\widehat{E}_T}\sigma_A - \frac{\widehat{\eta}_T}{\widehat{E}_T}\sigma_T - \frac{\widehat{\eta}_a}{\widehat{E}_T}\sigma_t - \frac{\widehat{\eta}_S}{\widehat{E}_T}\tau_A - \frac{\widehat{\delta}_A}{\widehat{E}_A}M_A + \frac{1}{\widehat{E}_T}M_T - \frac{\widehat{\beta}_T}{\widehat{\mu}_A}M_S + \widehat{\alpha}_T\Delta T, \tag{15.30}$$

$$\widehat{\gamma}_A = -\frac{\widehat{\lambda}_A}{\widehat{\mu}_A}\sigma_A - \frac{\widehat{\lambda}_T}{\widehat{\mu}_A}\sigma_T - \frac{\widehat{\lambda}_t}{\widehat{\mu}_A}\sigma_t - \frac{\widehat{\lambda}_S}{\widehat{\mu}_A}\tau_A - \frac{\widehat{\beta}_A}{\widehat{\mu}_A}M_A - \frac{\widehat{\beta}_T}{\widehat{\mu}_A}M_T + \frac{1}{\widehat{\mu}_A}M_S + \widehat{\alpha}_S\Delta T, \tag{15.31}$$

where all the effective laminate properties have the same meaning as they had for symmetric laminates, while the superscript "^" specifies bending laminate properties. Additionally, the parameters β and δ are the bending equivalent of Poisson's ratio, and they characterize the multiaxial bending deformation that occurs when a laminate is subject to bending about just one axis. The parameters $\widehat{\nu}$ and $\widehat{\eta}$ are also coupling equivalents of Poisson's ratio characterizing in-plane deformations when the laminate is under bending loads. The variational approach developed in Hajikazemi et al. (2020a) can be used to predict accurately all the effective laminate properties introduced in Eqs. (15.23)–(15.31). Similarly, GPB (McCartney, 2001, 2008a) can be used to predict the effective properties concerning the axial and transverse in-plane and bending deformations, and through-thickness loading.

15.2.4 Nonuniform distribution of ply cracking

Ply cracking is an inherently stochastic process due to the random variability of local material properties of the plies. Such randomness in microstructure and in failure evolution generally leads to nonuniform distributions of ply cracks (Fig. 15.3). There is enough experimental evidence (Silberschmidt, 2005, 2006; Andersons et al., 2008) that ply cracks in laminated composites are in general non-uniformly spaced. The stress analysis of a non-uniformly cracked laminate is much more difficult as an analysis must be carried out for the entire cracked laminate rather than a single unit cell. Therefore,

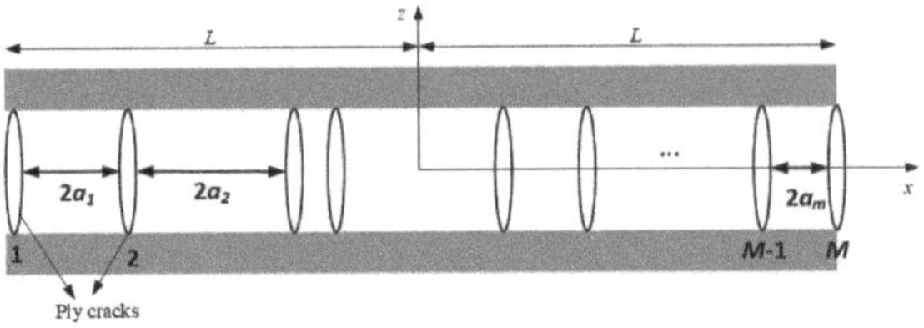

Figure 15.3 Schematic geometry of M nonuniformly distributed ply cracks.

due to the complexities involved, only a few approximate models (McCartney, 2000; Hajikazemi et al., 2019, 2020a; Hashin, 1986; McCartney and Schoeppner, 2002; Loukil et al., 2012) have contributed to the prediction of the effect of non-uniform ply cracking on the thermoelastic properties. Besides approximate approaches, Schoeppner and Pagano (Schoeppner and Pagano, 1999) have developed an exact axisymmetric model of stress transfer, using a variational technique based on the Reissner's energy functional, which is capable of estimating stiffness reduction of cross-ply laminates with non-uniformly spaced cracks. A variational model (Hajikazemi et al., 2018) is also developed to accurately address both the problems of stiffness reduction and stress evaluation in non-uniformly cracked general symmetric laminates. The analysis is carried out for the entire cracked laminate rather than for a unit cell. It is shown that modeling this non-uniformity is a crucial factor in predicting the initiation and propagation of ply cracks specially when dealing with the effects of manufacturing defects. This model is then used to examine the accuracy of the approximate approach developed by McCartney (McCartney, 2000; McCartney and Schoeppner, 2002) where use has been made of an averaging multiple periodic solutions. It is shown that this approximate approach can successfully consider the effects of non-uniform crack spacing on determination of the in-plane laminate thermoelastic constants leading to negligible errors when comparing with results of the accurate variational model (Hajikazemi et al., 2018). Very recently, the application of this approximate methodology has been enhanced (Hajikazemi et al., 2019) to predict the out-of-plane shear constants and bending properties (Hajikazemi et al., 2020a) of laminates with non-uniform ply cracking. The interesting feature of these approximate approaches is that they only need uniform solutions to predict the thermoelastic constants of a non-uniformly cracked laminate. The explicit formulations for prediction of thermo-elastic constants of a non-uniformly cracked laminate are written (McCartney, 2000; Hajikazemi et al., 2018, 2019) in terms of thermoelastic constants of uniformly cracked laminates, for which the methods explained in Sections (15.2.1)−(15.2.3) should be used.

15.2.5 Inter-relationships among thermo-elastic constants of cracked laminates

It is clear that in order to predict all the thermoelastic properties in Eqs. (15.17)−(15.31), several loading conditions must be applied in the stress transfer models. For example, to obtain the thermoelastic constants in Eqs. (15.17)−(15.20) concerning triaxial and in-plane shear deformation of a cracked symmetric laminate, five loading cases are required. To decrease the number of required loading cases for the calculation of laminate thermo-elastic constants and to have a more compact notation for the effective strain–stress–temperature relations of a cracked laminate, McCartney (McCartney, 2000) has used the concept of crack closure. It is based on the fact that a laminate containing ply cracks in the 90 degree ply can always be loaded in a way that the cracks just close so that the in-plane axial $\sigma_{xx}^{(90)}$ and shear stresses $\tau_{xy}^{(90)}$, $\tau_{xz}^{(90)}$ are zero for the entire length and thickness of the 90 degree ply. To better understand this, consider an uncracked symmetric laminate subject to a triaxial loading condition for which the in-plane axial and shear stresses are zero in the 90 degree ply, that is, $\sigma_{xx}^{0(90)} = \tau_{xy}^{0(90)} = 0$. Note that out-of-plane shear stresses are always zero in an uncracked laminate subject to triaxial loading. For such an uncracked laminate, any number of cracks can be formed without changing in any way the stress or displacement distributions in the laminate. This concept is implemented to derive a set of interrelationships among thermo-elastic constants of cracked and uncracked laminates with cross-ply, symmetric, and general lay-ups. As an example, we show such relationships for a symmetric cracked laminate Eqs. (15.17)–(15.20). To do so, the reduced form of relations Eqs. (15.17)–(15.20) is written as follows by replacing τ_A found from Eq. (15.20) in Eqs. (15.17)–(15.19):

$$\varepsilon_A^* \equiv \varepsilon_A + \lambda_A \gamma_A = \frac{1}{E_A^*}\sigma_A - \frac{\nu_A^*}{E_A^*}\sigma_T - \frac{\nu_a^*}{E_A^*}\sigma_t + \alpha_A^* \Delta T, \tag{15.32}$$

$$\varepsilon_T^* \equiv \varepsilon_T + \lambda_T \gamma_A = -\frac{\nu_A^*}{E_A^*}\sigma_A + \frac{1}{E_T^*}\sigma_T - \frac{\nu_t^*}{E_T^*}\sigma_t + \alpha_T^* \Delta T, \tag{15.33}$$

$$\varepsilon_t^* \equiv \varepsilon_t + \lambda_t \gamma_A = -\frac{\nu_a^*}{E_A^*}\sigma_A - \frac{\nu_t^*}{E_T^*}\sigma_T + \frac{1}{E_t^*}\sigma_t + \alpha_t^* \Delta T \tag{15.34}$$

where the reduced laminate properties are defined as follows:

$$\begin{gathered}
\frac{1}{E_A^*} = \frac{1}{E_A} - \frac{\lambda_A^2}{\mu_A}, \quad \frac{1}{E_T^*} = \frac{1}{E_T} - \frac{\lambda_T^2}{\mu_A}, \quad \frac{1}{E_t^*} = \frac{1}{E_t} - \frac{\lambda_t^2}{\mu_A}, \\
\frac{\nu_A^*}{E_A^*} = \frac{\nu_A}{E_A} + \frac{\lambda_A \lambda_T}{\mu_A}, \quad \frac{\nu_a^*}{E_A^*} = \frac{\nu_a}{E_A} + \frac{\lambda_A \lambda_t}{\mu_A}, \quad \frac{\nu_t^*}{E_T^*} = \frac{\nu_t}{E_T} + \frac{\lambda_T \lambda_t}{\mu_A}, \\
\alpha_A^* = \alpha_A + \lambda_A \alpha_S, \quad \alpha_T^* = \alpha_T + \lambda_T \alpha_S, \quad \alpha_t^* = \alpha_t + \lambda_t \alpha_S.
\end{gathered} \tag{15.35}$$

The above effective stress–strain relations concern triaxial deformation of a cracked general symmetric laminate where its in-plane shear deformation is restricted. It is shown (McCartney, 2000) that the thermoelastic constants defined in Eqs. (15.32)–(15.34) (or Eqs. (15.17)–(15.20)) are all dependent. To interpret these interrelationships, a macroscopic damage parameter is defined, as follows:

$$D_1 = \left(\frac{1}{E_A^*} - \frac{1}{E_A^{0*}} \right). \tag{15.36}$$

Then, the following interrelationships among the reduced thermoelastic constants of a general symmetric cracked laminate will be always valid:

$$\begin{aligned}
&-\left(\alpha_A^* - \alpha_A^{0*}\right) = D_1 k_1, \qquad -\left(\alpha_T^* - \alpha_T^{0*}\right) = D_1 \frac{k_1{}^2}{k_2}, \\
&-\left(\alpha_t^* - \alpha_t^{0*}\right) = D_1 \frac{k_1{}^2}{k_3}, \qquad \left(\frac{\nu_A^{0*}}{E_A^{0*}} - \frac{\nu_A^*}{E_A^*} \right) = D_1 \frac{k_1}{k_2}, \\
&\left(\frac{\nu_a^{0*}}{E_A^{0*}} - \frac{\nu_a^*}{E_A^*} \right) = D_1 \frac{k_1}{k_3}, \qquad \left(\frac{\nu_t^{0*}}{E_T^{0*}} - \frac{\nu_t^*}{E_T^*} \right) = D_1 \frac{k_1{}^2}{k_3 k_2}, \\
&\left(\frac{1}{E_T^*} - \frac{1}{E_T^{0*}} \right) = D_1 \frac{k_1{}^2}{k_2{}^2}, \qquad \left(\frac{1}{E_t^*} - \frac{1}{E_t^{0*}} \right) = D_1 \frac{k_1{}^2}{k_3{}^2}.
\end{aligned} \tag{15.37}$$

In the above inter-relationships, the three parameters k_1, k_2, and k_3 are constants for uncracked laminates all defined in McCartney (2000), Hajikazemi et al. (2020a). It can be seen that in order to obtain all damaged laminate properties defined in Eqs. (15.17)–(15.20), one needs only to apply two loading cases (concerning axial σ_A and in-plane shear τ_A) to calculate the macroscopic damage parameter D_1 and shear properties ($\mu_A, \lambda_A, \lambda_T, \lambda_t, \alpha_S$) as the remaining properties can all be defined in terms of these parameters and undamaged laminate properties. It is also noted that experimental measurement of some effective laminate constants (e.g., out-of-plane normal stiffness E_t) is very cumbersome. These inter-relationships can be used to estimate such laminate properties in terms of in-plane axial stiffness E_A, which are more easily measured. As these inter-relationships must always be satisfied, they can be used to benchmark the accuracy and quality of any stress transfer model in satisfying stress-free conditions at crack surfaces. A complete set of inter-relationships among thermoelastic constants of general laminates involving bending properties (Eqs. 15.23–15.31) are provided in Hajikazemi et al. (2020a) where it is shown that three independent macroscopic damage parameters are needed to characterize triaxial and bending thermo-mechanical deformation of a cracked general laminate. The number of independent macroscopic damage parameters and their variations due to additional damage resulting from increasing applied loads are very important when developing anisotropic continuum damage models (McCartney, 2003) for ply cracking analysis at a structural level.

15.3 Ply cracking damage evolution laws

Having sound models for the prediction of stress fields and effective thermo-elastic constants in undamaged and cracked laminates leads to the need for a damage evolution law to predict the load levels/cycles at which ply cracks initiate and grow. Clearly, such damage evolution laws depend on the nature of the applied load and, thus, are different for quasi-static, fatigue, or impact loadings. Focusing first on crack initiation and propagation under quasi-static loading conditions, early failure models for ply cracking assumed that ply cracks initiate when the stress in the off-axis plies reaches the transverse strength (Hahn and Tsai, 1974; Takeda and Ogihara, 1994) of the ply material. More advanced failure models are formulated based on the energy approaches in the framework of fracture mechanics (Barbero and Cortes, 2010; McCartney, 2000, 2005; Nairn et al., 1993; Nairn, 2000; Varna et al., 2001; Vinogradov and Hashin, 2005; Montesano and Singh, 2015; Montesano et al., 2018; Hajikazemi et al., 2020b). These works use the fracture mechanics in an unconventional form named as *finite fracture mechanics* (Hashin, 1996) which deals with fracture events characterized by the appearance of a fully formed new crack rather than the infinitesimal growth of one or more existing cracks. Finite fracture mechanics is accounting for the fact that in practice ply cracks form and cover the whole thickness of a ply instantaneously on the experimental time scale. Most of the energy-based models can only be applied to cross-ply laminates (Vinogradov and Hashin, 2005; Llobet et al., 2018) or very special lay-ups (Varna et al., 2001; Pakdel and Mohammadi, 2017) under uniaxial loading conditions with cracks in a single ply neglecting the effects of thermal residual stresses (Barbero and Cortes, 2010; Montesano and Singh, 2015; Montesano et al., 2018) and manufacturing defects. A general theoretical framework is now described (McCartney, 2000) for predicting ply crack formation in the 90 degree plies of general symmetric laminates subject to general triaxial loading conditions considering the effects of thermal residual stresses. The special feature of this energy-based criterion is that it is developed in terms of macroscopic laminate-level properties (effective laminate properties) even though the ply cracking occurs at the microscopic level. In this section, an extension of this energy-based failure criterion (Hajikazemi et al., 2020b) will be briefly presented using a compact notation for describing the prediction of ply crack formation in 90 degree plies of a symmetric laminate taking into account the effects of triaxial and shear loadings, thermal residual stresses, and possible mixed mode deformations. The compact notation will be expanded for a special case of ply crack simulation in a cross-ply laminate under uniaxial loading. To conclude this section, it will be shown how ply crack multiplication can be modeled in a cross-ply laminate under uniaxial fatigue loading.

15.3.1 Generalized framework for the prediction of ply cracking in the 90 degree ply of a symmetric laminate under combined general triaxial and shear loadings

Consider a general symmetric laminate, which already has some damage in the form of fully developed ply cracks (state 1) in the 90 degree ply. Consider then the simultaneous formation of new crack surfaces in the same ply, such that the resulting damage state is characterized as state 2. The locations of the cracks are assumed to be such that the overall deformation of the laminate is governed by the effective stress–strain–temperature relations Eqs. (15.17)–(15.20). The new crack surfaces are assumed to form quasi-statically under conditions of fixed effective multiaxial applied tractions and fixed temperature. From energy balance considerations and the fact that the kinetic energy is never negative, the criterion for crack formation under these conditions has the form:

$$\Delta G + \Delta\Gamma < 0, \quad \Delta G = V\left(g_{(2)} - g_{(1)}\right), \quad \Delta\Gamma = V\left(\Gamma_{(2)} - \Gamma_{(1)}\right), \tag{15.38}$$

where ΔG specifies the change in the total Gibbs free energy of the system (within a volume V) resulting from ply crack formation and $\Delta\Gamma$ is the energy absorbed in the laminate by the formation of the new ply cracks which characterizes the effective "fracture energy" of the ply, that is, its resistance to ply crack growth. The subscripts (1) and (2) correspond to the damage state 1 and the damage state 2, respectively. It can be shown (Hajikazemi et al., 2020b) that the average Gibbs free energy for unit volume of the laminate can be written in the following form:

$$\begin{aligned} g = & -\frac{\sigma_A^2}{2E_A} - \frac{\sigma_T^2}{2E_T} - \frac{\sigma_t^2}{2E_t} - \frac{\tau_a^2}{2\mu_a} - \frac{\tau_t^2}{2\mu_t} - \frac{\tau_A^2}{2\mu_A} + \frac{\nu_A}{E_A}\sigma_A\sigma_T + \frac{\nu_a}{E_A}\sigma_A\sigma_t + \\ & \frac{\nu_t}{E_T}\sigma_T\sigma_t + \frac{\lambda_A}{\mu_A}\sigma_A\tau_A + \frac{\lambda_T}{\mu_A}\sigma_T\tau_A + \frac{\lambda_t}{\mu_A}\sigma_t\tau_A + \frac{\lambda_S}{\mu_a}\tau_a\tau_t - \\ & [\sigma_A\alpha_A + \sigma_T\alpha_T + \sigma_t\alpha_t + \tau_A\alpha_S]\Delta T + f(\Delta T), \end{aligned} \tag{15.39}$$

where the function f is independent of the applied mechanical loads and where the stress terms with the superscript "o" are predefined crack closure stresses (McCartney, 2000). It can be clearly seen that the stress/strain relations given by Eqs. (15.17)–(15.20) can be recovered from the above equation using:

$$\begin{aligned} & \varepsilon_A = -\frac{\partial g}{\partial \sigma_A}, \quad \varepsilon_T = -\frac{\partial g}{\partial \sigma_T}, \quad \varepsilon_t = -\frac{\partial g}{\partial \sigma_t}, \\ & \gamma_t = -\frac{\partial g}{\partial \tau_t}, \quad \gamma_a = -\frac{\partial g}{\partial \tau_a}, \quad \gamma_A = -\frac{\partial g}{\partial \tau_A}. \end{aligned} \tag{15.40}$$

Eq. (15.38) can be written in the following more general form that distinguishes between Gibbs free energy terms (energy release rates) for three different deformation modes:

$$\chi_{\rm I}\,\Delta G_{\rm I}+\chi_{\rm II}\,\Delta G_{\rm II}+\chi_{\rm III}\,\Delta G_{\rm III}+\Delta\Gamma\left(G_{\rm I}^c,G_{\rm II}^c,G_{\rm III}^c\right)\;<\;0,\quad where$$
$$\Delta G\;=\;\chi_{\rm I}\,\Delta G_{\rm I}+\chi_{\rm II}\,\Delta G_{\rm II}+\chi_{\rm III}\,\Delta G_{\rm III},$$
$$\Delta G_{\rm I}=V\left(g_{I_{(2)}}-g_{I_{(1)}}\right),\ \Delta G_{\rm II}=V\left(g_{II_{(2)}}-g_{II_{(1)}}\right),\quad \Delta G_{\rm III}=V\left(g_{III_{(2)}}-g_{III_{(1)}}\right).$$
(15.41)

where subscripts I, II, and III show the different deformation modes. Moreover, $\Delta G_{\rm I}$, $\Delta G_{\rm II}$ and $\Delta G_{\rm III}$ specify, respectively, the change in the Gibbs free energy of the system due to mode I, II and III ply crack deformations. The parameters $\chi_{\rm I}, \chi_{\rm II}, \chi_{\rm III}$ are material-dependent parameters that control how mixed mode loading is modeled. In practice, it is the total energy release rate that is the physically important driving force for ply cracking. Eq. (15.41) asserts that mixed mode fracture is governed by the total energy release rate and the energy absorbed as a result of fracture that is a function of the fracture energy of the ply in mode I, II, and III, respectively, $G_{\rm I}^c$, $G_{\rm II}^c$, $G_{\rm III}^c$. So far, we have always balanced the *total* energy release rate ($\chi_{\rm I} = \chi_{\rm II} = \chi_{\rm III} \equiv 1$) with the *effective* fracture energy 2γ that is a function of the fracture energies for ply cracking in mode I, II, and III. However, it is possible to use any empirical mixed mode failure criterion.

The exact determination of the change in the Gibbs free energy of a cracked general symmetric laminate due to *pure* mode I, II and III ply crack deformations in terms of effective macroscopic laminate properties is impossible when using a finite fracture methodology which considers damage states rather than infinitesimal crack growths. However, applying the concept of crack closure and the inter-relationships among thermoelastic constants (McCartney, 2000; Hajikazemi and McCartney, 2018) to Eq. (15.39), it can be shown that the first two terms in criterion Eq. (15.41) for *ply crack formation in the 90 degree ply* of a general symmetric laminate, can be written in the following simple form:

$$g_{\rm I_{(2)}}-g_{\rm I_{(1)}}=-\frac{1}{2}\frac{\left[\varepsilon^*_{A_{(2)}}-\varepsilon^*_{A_{(1)}}\right]^2}{\dfrac{1}{E^*_{A_{(2)}}}-\dfrac{1}{E^*_{A_{(1)}}}},$$
$$g_{\rm II_{(2)}}-g_{\rm II_{(1)}}=-\frac{1}{2}\mu_{A_{(2)}}\gamma_{A_{(2)}}{}^2+\frac{1}{2}\mu_{A_{(1)}}\gamma_{A_{(1)}}{}^2, \qquad (15.42)$$

where ε_A^* is the in-plane axial strain for a symmetric laminate that is constrained so that the shear strains are zero, and E_A^* is the reduced axial stiffness modulus, defined already in the previous section. It should be emphasized that for a cross-ply symmetric laminate where there is no coupling between the triaxial and in-plane shear deformations, the expressions presented by Eq. (15.42) correspond exactly to the change of the Gibbs free energy due to pure mode I and II ply crack deformations. However, for a general symmetric laminate with coupling between triaxial and in-plane shear deformations, the expressions in Eq. (15.42) are just natural extensions of the definitions of pure mode I, II ply crack deformations but they do not correspond to exact *pure* mode deformations.

It can also be easily shown that the third term in criterion Eq. (15.41) for *ply crack formation in the 90 degree ply* of a general symmetric laminate can be written in the following simple form involving only macroscopic laminate properties:

$$g_{\mathrm{III}_{(2)}} - g_{\mathrm{III}_{(1)}} = -\frac{1}{2}\tau_a{}^2\left(\frac{1}{\mu_{a_{(2)}}} - \frac{1}{\mu_{a_{(1)}}}\right) - \frac{1}{2}\tau_t{}^2\left(\frac{1}{\mu_{t_{(2)}}} - \frac{1}{\mu_{t_{(1)}}}\right)$$
$$+ \tau_t\tau_a\left(\frac{\lambda_{S_{(2)}}}{\mu_{a_{(2)}}} - \frac{\lambda_{S_{(1)}}}{\mu_{a_{(1)}}}\right). \tag{15.43}$$

As discussed earlier, it is shown in Hajikazemi et al. (2019) that cracks in the 90 degree ply do not change the laminate out-of-plane transverse shear stiffness μ_t. Moreover, it is shown that the ratio λ_s/μ_a remains constant as cracks appear in the 90 degree ply of the laminate. Therefore, the second and third terms in relation Eq. (15.43) can be omitted leading to the following simple form:

$$g_{\mathrm{III}_{(2)}} - g_{\mathrm{III}_{(1)}} = -\frac{1}{2}\tau_a{}^2\left(\frac{1}{\mu_{a_{(2)}}} - \frac{1}{\mu_{a_{(1)}}}\right), \tag{15.44}$$

It is noted that the expressions in Eqs. (15.42) and (15.43) exactly partition the total average Gibbs free energy for unit volume of the laminate (Eq. (15.39)). It should also be mentioned that the change in Gibbs free energy defined by Eqs. (15.42) and (15.43) takes full account of anisotropy, thermally induced residual stresses, and can be applied for general mixed mode loading conditions when the effects of cracks in 90 degree plies are effectively homogeneous at the macroscopic level. This energy-based criterion is developed in terms of independent macroscopic laminate-level properties (E_A, λ_A, G_A and μ_a) even though the ply cracking occurs at a microscopic level. It is valid if the next damage state (state 2) is formed under fixed values of applied loads which means that the loading parameters (σ_A, σ_T, σ_t, τ_t, τ_a, and τ_A) and ΔT remain fixed during the formation of each cracking event, but implying that progressively increasing stresses will be needed to form new cracks.

15.3.2 Special case of ply crack growth in the 90 degree ply of a cross-ply laminate under quasistatic uniaxial loading

Consider a length $2L$ of a cross-ply laminate under a uniaxial loading condition (σ_A) where the thermally induced residual stresses characterized by the temperature difference ΔT is not negligible. It is assumed that on increasing the load, the average crack density will increase from ρ_1 (state (1)) to ρ_2 (state (2)). For a cross-ply laminate under uniaxial loading, Gibbs free energy terms concerning the deformations in mode II and III are omitted and Eq. (15.41) can be simplified as follows:

$$\Delta G_{\mathrm{I}} + \Delta\Gamma < 0\,. \tag{15.45}$$

For such a balanced laminates, $\lambda_A = 0$ as there is no coupling between axial and shear deformations, so that for uniaxial loading σ_A and in the presence of thermal residual stresses ΔT, we have:

$$g_{I_{(2)}} - g_{I_{(1)}} = -\frac{1}{2}\frac{\left(\left(\frac{1}{E_{A_{(2)}}} - \frac{1}{E_{A_{(1)}}}\right)\sigma_A + (\alpha_{A_{(2)}} - \alpha_{A_{(1)}})\Delta T\right)^2}{\left(\frac{1}{E_{A_{(2)}}} - \frac{1}{E_{A_{(1)}}}\right)}. \tag{15.46}$$

Moreover, the energy absorbed in a volume $V = 2L \times 2h$ of laminate by the formation of the new ply cracks is given by:

$$\Delta\Gamma = V\Gamma = V\frac{\rho h^{(90)} G_{\mathrm{I}}^c}{h}, \tag{15.47}$$

where h^{90} is the half thickness of 90 degree plies (only symmetric part) in the laminates, h is the half thickness of the laminate, and $\rho = \rho_2 - \rho_1$ is the average crack density of newly developed cracks. Then the condition for crack growth can be written as follows:

$$\frac{h}{2\rho h^{(90)}}\frac{\left(\left(\frac{1}{E_{A_{(2)}}} - \frac{1}{E_{A_{(1)}}}\right)\sigma_A + (\alpha_{A_{(2)}} - \alpha_{A_{(1)}})\Delta T\right)^2}{\left(\frac{1}{E_{A_{(2)}}} - \frac{1}{E_{A_{(1)}}}\right)} > G_{\mathrm{I}}^c. \tag{15.48}$$

The above equation can be further simplified by ignoring the effects of thermal residual stresses so that $\Delta T = 0$:

$$\frac{h}{2\rho h^{(90)}}\left(\frac{1}{E_{A_{(2)}}} - \frac{1}{E_{A_{(1)}}}\right)\sigma_A{}^2 > G_{\mathrm{I}}^c. \tag{15.49}$$

If the laminate is uncracked at the beginning of the loading $\rho_I = 0$, the crack initiation criterion can be simplified as follows:

$$\frac{h}{2\rho h^{(90)}}\left(\frac{1}{E_A(\rho)} - \frac{1}{E_A^0}\right)\sigma_A{}^2 > G_{\mathrm{I}}^c. \tag{15.50}$$

This result implies that the energy release rate (for the continuum model of a ply crack in the 90 degree ply of a laminate) $G^{(90)}$ for the ply cracking initiation is as follows:

$$G^{(90)} = \frac{h}{2\rho h^{(90)}}\left(\frac{1}{E_A(\rho)} - \frac{1}{E_A^0}\right)\sigma_A{}^2. \tag{15.51}$$

15.3.3 Outline of energy methods for fatigue damage modeling of laminates (the special case of cross-ply laminate under uniaxial fatigue loading)

In order to develop a practically useful fatigue life prediction design methodology, it is vital that the initiation and growth of fatigue damage preceding failure is capable of being predicted with confidence. As the prediction of the initiation of fatigue damage in composites is exceedingly difficult, even for metals, it is usual to consider fatigue damage growth assuming that defects of known size are already present in the material or structure. The overall objective of this section is, therefore, to briefly introduce some useful methods that can be recommended for use when predicting the fatigue growth of pre-existing ply defects. The reader must refer to McCartney (2008b) to find the details of developments.

The prediction of the number of cycles needed to initiate a ply crack, having a size that is measurable, is exceedingly difficult because of complex microscopic mechanisms that contribute to the formation of small ply cracks. Such mechanisms operate at the fiber/matrix interfaces and lead to debonded areas on fiber/matrix interfaces that can grow in the through-thickness and width directions, and between plies leading to ply cracks and delaminations. The variability of fiber/matrix interface quality and inhomogeneous fiber spacings leads to ply crack initiation times that are subject to significant scatter and to the need for a statistical approach. The pragmatic approach to this problem is to regard the ply crack initiation phase of fatigue damage as a known parameter, so that predictive modeling can concentrate on the growth of damage and the onset of catastrophic failure.

Let us consider fatigue ply cracking growth in the 90 degree plies of a cross-ply laminate under uniaxial fatigue load with maximum amplitude σ_A and R-ratio ($R = \sigma_{A,min}/\sigma_A$). For ply cracks in the 90 degree plies of cross-ply laminates, a fatigue crack growth law can be assumed of the form:

$$\frac{d\rho}{dN} = C\left[\Delta K^{(90)}\right]^m, \tag{15.52}$$

where C and m are material constants, and where $\Delta K^{(90)} = K^{(90)} - K_{\text{min}}^{(90)}$ is the energy-based effective stress range intensity factor for the fatigue crack growth of a ply crack in the 90 degree ply of the laminate. It is noted that the form of fatigue crack growth law in Eq. (15.52) implies incremental growth of the existing ply cracks as discussed in McCartney (2008b). However, in addition to the growth of pre-existing ply cracks, there is likely to be an incremental increase in the number of ply cracks present at any time due to the nucleation of new ply cracks of various sizes. This would require a modification of Eq. (15.52) which will not be discussed here. The methodology described in McCartney (1992b, 1992c) enables the calculation of the stress intensity factor K for the continuum model of a ply crack in a cross-ply laminate in terms of energy release rate Eq. (15.51), as follows:

$$K^{(90)} = \sqrt{\frac{E_T^{(90)} G^{(90)}}{\Omega^{(90)}}}, \tag{15.53}$$

where $E_T^{(90)}$ is the transverse stiffness of the 90 degree plies in the laminate and the parameter $\Omega^{(90)}$ characterizes the anisotropy of the material which is defined in McCartney (2008b). Using the energy release rate given in Eq. (15.51) for a cross-ply laminate under uniaxial loading, the relation (Eq. 15.53) can be written as follows:

$$K^{(90)} = \sqrt{\frac{E_T^{(90)} h \frac{1}{\rho}\left(\frac{1}{E_A(\rho)} - \frac{1}{E_A^0}\right)}{2h^{(90)}\Omega^{(90)}}}\sigma_A. \tag{15.54}$$

Using this value of $K^{(90)}$ in Eq. (15.52), the fatigue crack growth law for ply cracking growth in 90 degree plies of a cross-ply laminate under uniaxial fatigue loading can be simplified as follows:

$$\frac{d\rho}{dN} = C\left[\beta\sqrt{\frac{1}{\rho}\left(\frac{E_A^0}{E_A(\rho)} - 1\right)}\,\Delta\sigma_A\right]^m, \tag{15.55}$$

where

$$\beta = \sqrt{\frac{E_T^{(90)} h}{2h^{(90)} E_A \Omega^{(90)}}} \tag{15.56}$$

The stress transfer models introduced in the previous section can be used to provide the dependence of the in-plane axial stiffness E_A on crack density ρ to be used in the relation (Eq. 15.55) for the prediction of crack density as a function of applied cycles. It is noted that the stress transfer model described in the previous sections can deal with ply cracks which are fully developed through the thickness and across the width. Fatigue driven cracks, however, are not usually fully propagated across the width. To use the above equation for the prediction of fatigue crack growth using developed stress transfer models, one should consider that following manufacturing of the laminate or an impact during the service life, it is assumed that there are preexisting randomly distributed defects of the same size, and that the defects are located away from the edges, each occupying a different potential ply crack formation site, so that each ply crack formation site has at most just one small ply defect. The size of the defect is just large enough for the stress intensity factor to be independent of crack length. The ply defects are assumed to be distributed uniformly in the laminate (when viewed at the macroscopic level) corresponding to an effective initial crack density ρ_0. After N fatigue cycles having a fixed uniform amplitude $\Delta\sigma_A$ and peak stress σ_A, the initial defects will have grown in length so that the effective ply crack density is then given by ρ. The methodology implies that the gradual growth of the large number of small ply defects is equivalent (when viewed at the macroscopic level) to a uniform density ρ of fully developed ply cracks. Using the developed stress transfer models, there can be a continuous variation of the ply crack density by altering the separation of the representative fully developed cracks. This approach, the validity of which is very difficult to test, is a practical solution to the very complex problem involving the interaction of randomly distributed cracks that will have significant interaction as the effective crack density increases. It is also noted that for flat laminate coupons, edge effects (Hajikazemi and Van Paepegem, 2018) have important influence on the process of fatigue crack growth, especially on progressive ply crack initiation which is not considered in the stress transfer models considered above. However, this problem does not arise when considering the fatigue loading of laminated tubes.

15.4 Numerical examples

In this section, numerical results are presented for which the ply properties shown in Table 15.1, for different transversely isotropic glass/carbon fiber reinforced composites, are used.

In order to compare numerically the methodologies described in this paper for symmetric laminates (Section 15.2.1), an unbalanced symmetric laminate $[0/30/60/90]_s$ made of glass fiber/epoxy material (the material set #1) is considered. To compare stress and displacement fields, the laminate is assumed to be subject only to an effective applied axial stress of $\sigma_A = 100$ (MPa). Moreover, a system of equally spaced ply cracks in the 90^0 ply is selected so that the crack separation is $2a = 2$ mm (see Fig. 15.1), corresponding to crack density of 0.5 cracks/mm. Since both models have been designed for any number of plies, it is possible to improve the accuracy of predictions using ply refinement techniques by representing each ply in the laminate by a set of n elemental layers having the same thickness and properties. In this way, the effects of through-thickness variations in the stress and displacement distributions can be assessed and accurate elasticity solutions can be approached. Fig. 15.4a and b shows the through-thickness variation of, respectively, the in-plane axial stress σ_{xx} and in-plane axial displacement u, in the plane of ply crack ($x = a = 1$ mm) between the midplane at ($z = 0$) and the outer surface at ($z = h = 1$ mm). A perfect match is observed between the results obtained from the variational model and those obtained from GPS model for the in-plane axial stress and displacement. Indeed, it is not possible to distinguish the predictions of the two models for these stresses whether a ply refinement technique is implemented ($n = 10$) or not ($n = 1$). It can be clearly seen in Fig. 15.4a that in the region of cracked 90^0 plies, the in-plane axial stress is exactly zero as required. Moreover, it can be seen that both models can similarly predict the effect of the singular behavior near the transverse crack tip. To show the numerical stability of the approaches and to adequately capture stress concentrations near the crack tips, the results of variational approach are also added in Fig. 15.4a and b where additional ply refinements are implemented close to interfaces between the plies. To do so, each ply is first divided into seven elements of equal thickness. Further, elements adjacent to ply interfaces were successively subdivided into halves three times to ensure that the stress field near ply crack tips is adequately represented. It

Table 15.1 Unidirectional ([0 degree] laminate) material properties of plies (transversely isotropic).

Set of materials	E_A (GPa)	$E_T = E_t$ (GPa)	$G_A = G_a$ (GPa)	$\nu_A = \nu_a$	ν_t	α_A (10^{-6}/°C)	$\alpha_T = \alpha_t$ (10^{-6}/°C)	t_{ply} (*mm*)
#1	45.6	16.2	5.83	0.278	0.4	8.6	26.4	0.25
#2	44.7	12.7	5.8	0.297	0.411	8.43	18.44	0.144
#3	43	13	3.4	0.3	0.42	–	–	0.61
#4	41.7	13	3.4	0.3	0.42	6.72	29.3	0.203
#5	141.3	9.58	5	0.3	0.32	−1	26	0.25
#6	104	6.14	5	0.4	0.45	–	–	0.62
#7	28	8.2	2.8	0.28	0.4	–	–	0.32

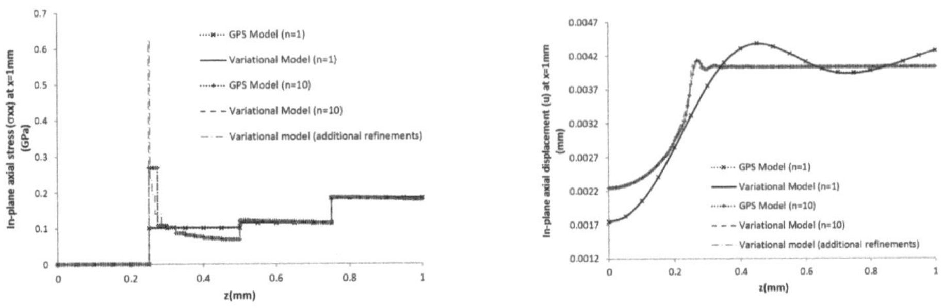

(a) (b)

Figure 15.4 Distribution of (a) in-plane axial stress σ_{xx} and (b) in-plane axial displacement (u) through the thickness of the $[0/30/60/90]_s$ laminate (material #1) in the plane containing cracks ($x = 1\ mm$) under applied uniaxial stress of $\sigma_A = 100\ (MPa)$.

is noted that the same results would be obtained if the GPS model was implemented using the same ply refinement. It is now useful to compare the accuracy of these approaches with highly refined finite element solutions for symmetric laminates. As both the GPS and variational models lead to similar results, only the variational approach is implemented for this comparison. A $[30/90_8/30_{0.5}]_s$ laminate made of the material set #2 (Garoz et al., 2019) with a uniform crack density of 1/mm has been loaded under combined in-plane shear $\tau_A = 30\ MPa$ and out-of-plane normal $\sigma_t = 10\ MPa$ loadings, and the results are compared with published finite element results in Garoz et al. (2019) wherever available. Fig. 15.5a shows the in-plane shear σ_{xy} and axial σ_{xx} stresses through the thickness at $x = a$, which is the plane that contains the cracks. Fig. 15.5b shows the interfacial stresses σ_{zz}, σ_{xz}, and σ_{yz} along the upper 30/90 interfaces of the 90 degree ply. There is perfect accordance between the FE and variational methods in Fig. 15.5, where the FE results are represented by symbols and variational (VA) results by lines.

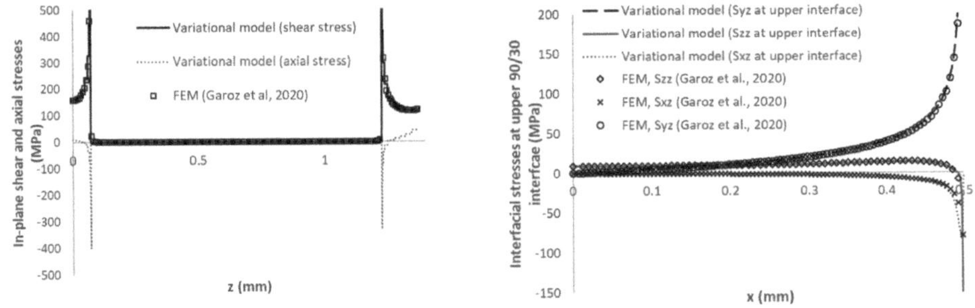

(a) (b)

Figure 15.5 Distribution of (a) in-plane axial σ_{xx} and shear σ_{xy} stresses through the thickness at the crack planes $x = 0.5\ mm$ (b) interfacial stresses at the 30/90 interface ($z = 8.5t_{ply}$) in the $[30/90_8/30_{0.5}]_s$ laminate (material #2) under combined in-plane shear $\tau_A = 30\ MPa$ and out-of-plane normal $\sigma_t = 10\ MPa$ loads.

The capability of both described stress transfer models in predicting stiffness reductions of symmetric laminates due to the presence of ply cracks is shown in Fig. 15.6a and b where the results of variational approach are compared with both FE (Garoz et al., 2019) and experimental results (Varna et al., 1999; Katerelos et al., 2006). Fig. 15.6a and b shows, respectively, the reduction of the normalized axial modulus E_A (normalized with respect to the uncracked modulus E_A^0) as a function of the crack density, ρ, for two different laminates: (i) the $[0/90_8/0_{0.5}]_s$ laminate made of material set #2 containing cracks in the 90^0 plies; (ii) and the $[0/45]_s$ laminate made of material set #3 containing cracks in the 45^0 plies. The good agreement between the results shows the mathematical accuracy of the variational (or GPS) methods as well as the realistic assumptions made in these approaches regarding the crack geometry. To show the effects of ply cracking on laminate properties other than the in-plane axial stiffness, a $[0/90/45/-45]_s$ laminate made of material set #4 is considered with uniformly spaced cracks in the 45 degree plies. Fig. 15.7 depicts the variation of the normalized (with respect to the uncracked properties) out-of-plane axial shear stiffness G_a, in-plane axial shear stiffness G_A, out-of-plane normal stiffness E_t, out-of-plane axial Poisson's ratio ν_a, and in-plane axial thermal expansion coefficients α_A, as a function of normalized crack density in the 45 plies ($\rho \times t_{ply}^{(45)}$). Besides good agreement with FEM solutions (Garoz et al., 2019), it is clear that out-of-plane shear stiffness can be largely degraded by the presence of ply cracks for which the models introduced in Section 15.2.2 are required to take this effect into account.

To study the effects of ply cracking on the performance of laminates under bending deformations, the models discussed in Section 15.2.3 should be implemented. Therefore, a $[0/90/45]_s$ symmetric laminate made of the material set #5 is first considered with uniformly spaced ply cracks with density of $\rho = 1/(2a) = 0.6$ (/mm) in only the upper 90 degree ply (nonsymmetrically cracked) under two different loading conditions $\tau_A = 10/1.5$ (MPa) and $M_S = 10/1.5$ $(MPa \times mm)$, the same as in Dinh et al. (2019). The laminate is unbalanced and nonsymmetrically cracked, thus all possible mechanical couplings are relevant for this analysis. The GPB model cannot be applied to this case as the laminate is unbalanced. The variational model is implemented using

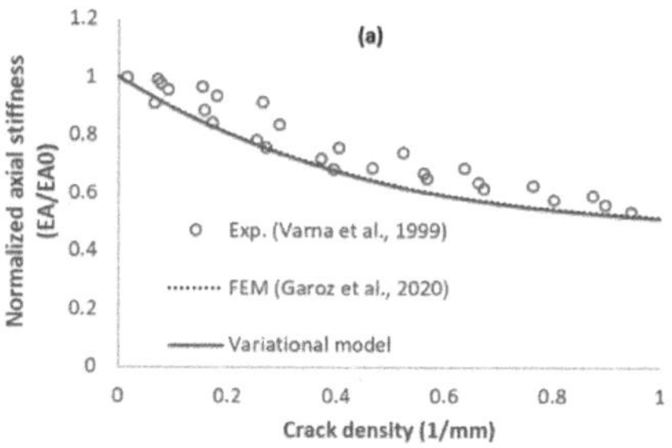

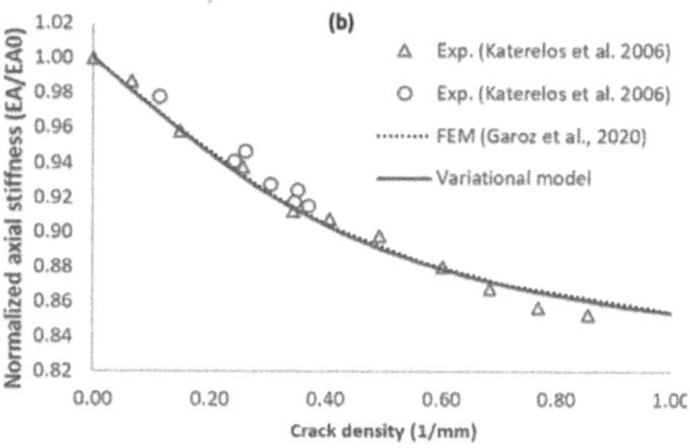

Figure 15.6 (a) Normalized axial stiffness as a function of the crack density ρ for the $[0/90_8/0_{0.5}]_s$ laminate made of material #2, cracks in the 90 degree plies, (b) normalized axial stiffness as a function of the crack density ρ (x-direction) for the $[0/45]_s$ laminate made of material #3, cracks in the 45 degree plies.

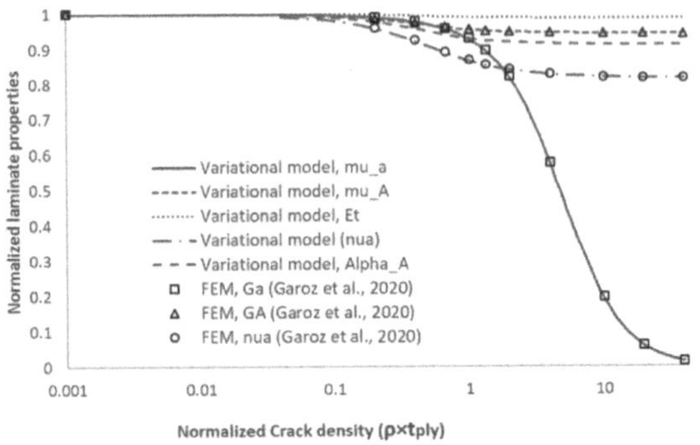

Figure 15.7 Normalized laminate properties, out-of-plane axial shear stiffness μ_a, in-plane axial shear stiffness μ_A, out-of-plane normal stiffness E_t, out-of-plane axial Poisson's ratio ν_a, and in-plane axial thermal expansion coefficients α_A, as a function of normalized crack density in 45 degree plies of a $[0/90/45/-45]_s$ laminate made of material #4.

the ply refinement technique where each ply is first divided into five elements of equal thickness. Further, elements adjacent to cracked interfaces were successively subdivided into halves three times to ensure that the stress field near ply crack tips is adequately represented. The results are compared with very refined FEM results of Dinh et al. (Dinh et al., 2019), wherever available. The through-thickness distribution of the in-plane axial σ_{xx} and shear τ_{xy} stresses at the plane containing cracks ($x = a$), respectively, for τ_A and M_S applied loading conditions is depicted in Fig. 15.8a and b. The general observation is that there is a perfect accordance between the results obtained from the variational approach and those of FEM. It can also be seen in Fig. 15.8 that stress-free conditions on crack surfaces ($x = \pm a$) are satisfied exactly by the variational approach, although FEM (like any other displacement-based

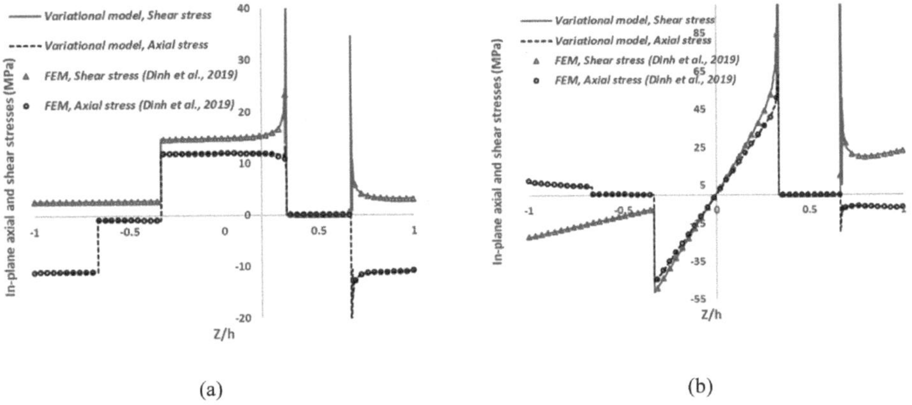

Figure 15.8 Variation of the in-plane axial (σ_{xx}) and shear (τ_{xy}) stresses at the plane ($x = a$) containing ply cracks, for a $[0/90/45]_s$ laminate (material #5) with uniformly spaced cracks in the upper 90 degree layer having density of $\rho = 1/(2a) = 0.6/mm$ under: (a) in-plane shear $\tau_A = 10/1.5$ (MPa) and (b) twisting $M_S = 10/1.5$ ($MPa.mm$) loading conditions.

approach) is unable to exactly satisfy this condition. To study the effects of ply cracking on thermoelastic constants of cracked laminates under bending deformations and to compare the performance of GPB and variational models, a $[90/0]_s$ laminate, made of material set#1, containing uniformly spaced ply cracks only in the upper 90 degree ply, is also considered. Fig. 15.9a shows the variation of normalized laminate elastic properties (with respect to the uncracked properties) versus crack density. Fig. 15.9b shows the variation of the normalized thermal expansion coefficients of the laminate versus crack density. Wherever possible, the results are compared with those of GPB model, showing perfect accordance. It is noted that in both models, the same amount of ply refinement is implemented ensuring that the results are converged. It is believed that the variational approach and the GPB model have the same quality and accuracy when applied to general cross-ply laminates. Finally, to check the quality of assumptions in the variational and GPB models when predicting the bending stiffness of cracked laminates, the results of bending stiffness reduction are compared with experiments. To do so, a carbon/epoxy $[90_2/0]_s$ laminate made of the material set #6 and a glass/epoxy $[90_3/0]_s$ laminate made of material set #7, the same as tested experimentally in Pupurs et al. (2022) are considered. The variational approach with an adequate level of ply refinement is implemented to calculate the normalized axial bending stiffness reduction $\widehat{E}_A$ (see Eq. 15.29) as a function of the normalized ply crack density ($\rho \times t_{ply}^{(90)}$) when ply cracks are present in one of the 90 degree plies (laminate is unsymmetrically cracked). Fig. 15.10a and b shows the variation of normalized axial bending stiffness as a function of normalized crack density in the cracked 90 degree ply where the variational results are compared with experimental results (Pupurs et al., 2022). The immediate observation is that the bending laminate properties can be significantly degraded even at rather small crack densities (<0.25/mm). It can be

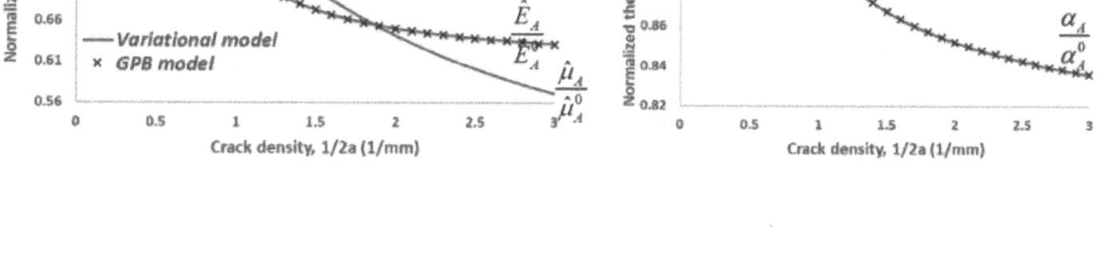

(a) (b)

Figure 15.9 Variation of normalized (with respect to uncracked properties), (a) effective elastic properties and (b) effective thermal expansion coefficients of a $[90/0]_s$ laminate (material #1) versus crack density in upper 90 degree ply (lower 90 degree ply is intact).

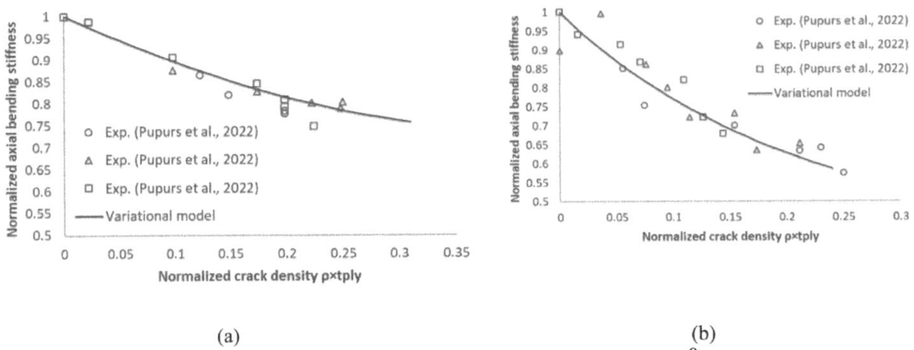

Figure 15.10 Variation of the normalized axial bending stiffness $\widehat{E}_A/\widehat{E}_A^0$ versus the normalized crack density $\rho \times t_{ply}^{(90)}$ in (a) a carbon/epoxy $[90_2/0]_s$ laminate (material #6) and (b) a glass/epoxy $[90_3/0]_s$ laminate (material #7) when ply cracks are present in one of the 90 degree plies.

also seen that there is very good agreement between the modeling and experimental results in both Fig. 15.10a and b, although more discrepancies (larger stiffness reduction in experiments) can be seen in Fig. 15.10a specially at higher crack densities. It should be noted that other forms of damage such as delaminations at the tips of ply cracks are very likely to form under bending loading conditions due to the high stress concentrations at the tips of ply cracks (see examples of such stress concentrations in Fig. 15.8a and b for a different lay-up). The presence of delaminations leads to additional stiffness reductions that can be crucial in some practical applications. Finally, it is noted that the same results would be obtained if the GPB model was employed instead of the variational model. The small differences between the assumptions of the GPB and the variational approaches, under bending deformations, are discussed in Hajikazemi et al. (2020a) where it is also shown that they lead to very similar results.

The results presented above were all concerned with uniform distributions of ply cracks; however, ply cracks are not always uniformly distributed. To investigate the effects of nonuniform distributions of ply crack on laminate properties, a $[45/90_3]_s$ laminate made of the material set#1 (with a different ply thickness $t_{ply} = 0.127$ mm) containing two cracks that might be uniformly or nonuniformly distributed is considered. The schematic diagram of this two-crack ($M = 2$) representative volume element (RVE) of length $2L$ is shown in Fig. 15.11. The average crack density (ρ) for this RVE is given by $1/L$ and is independent of the ratio of a_1/a_2. The variational model is first implemented to obtain the compliance matrices of each fragment assuming that ply cracks have a uniform distribution. These results are then used together with some approximate formulations (see references in Section 15.2.4) to predict the thermoelastic constants of the cracked $[45/90_3]_s$ laminate, having ply cracks in 90 degree plies with average crack density of $\rho = 1/L = 1$(/mm), with different degrees of nonuniformity (a_1/L). The variation of in-plane and bending axial stiffnesses is shown in Fig. 15.12a and b, respectively. The results are also compared with accurate FEM results presented in Hajikazemi et al. (2020a). One observation is that assuming a uniform distribution of ply cracks ($a_1/L = 0.5$) leads to a larger reduction of the in-plane and bending axial stiffnesses, leading to conservative predictions.

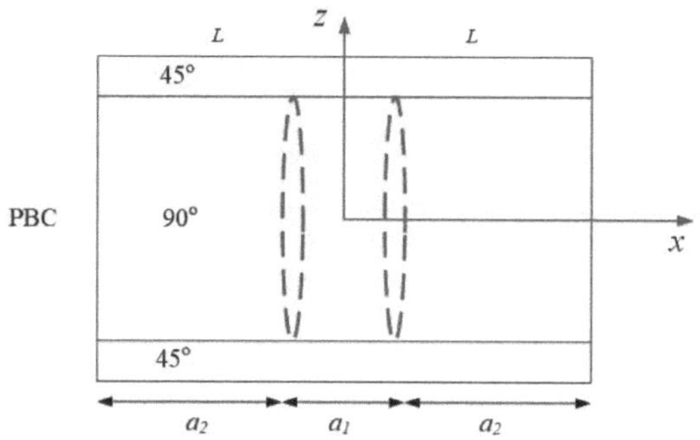

Figure 15.11 Schematic diagram of two nonuniformly spaced cracks in 90 degree ply of a $[45/90_3]_s$ laminate.

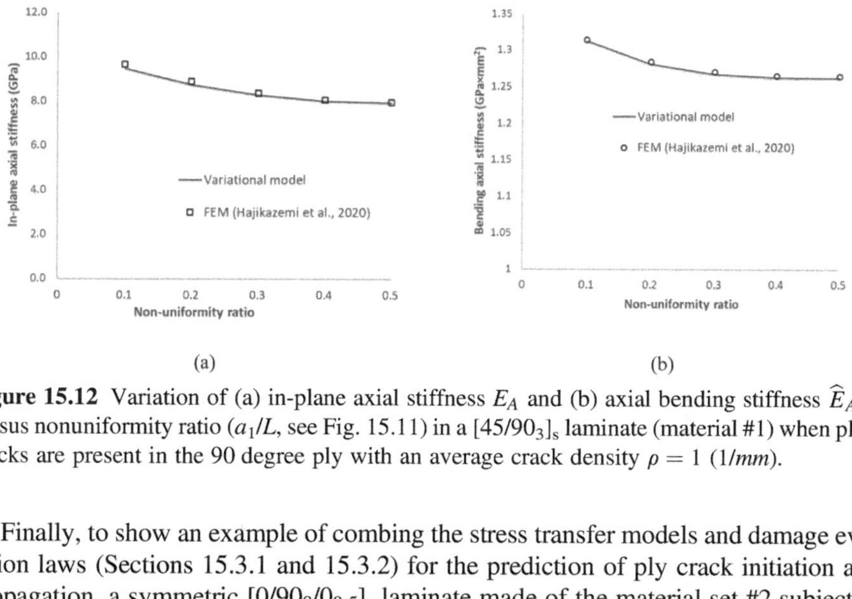

Figure 15.12 Variation of (a) in-plane axial stiffness E_A and (b) axial bending stiffness $\widehat{E}_A$ versus nonuniformity ratio (a_1/L, see Fig. 15.11) in a $[45/90_3]_s$ laminate (material #1) when ply cracks are present in the 90 degree ply with an average crack density $\rho = 1$ ($1/mm$).

Finally, to show an example of combing the stress transfer models and damage evolution laws (Sections 15.3.1 and 15.3.2) for the prediction of ply crack initiation and propagation, a symmetric $[0/90_8/0_{0.5}]_s$ laminate made of the material set #2 subject to uniaxial loading (σ_A), is considered. Ply crack growth in the 90 degree plies is concerned although 0 degree plies might potentially crack at high axial loads. To deal with ply cracking growth in more than one ply (orientation), the homogenization technique (McCartney, 2005; Hajikazemi et al., 2020b) can be used; however, the purpose is here to discuss a simple case. Interested readers can refer to Hajikazemi and Van Paepegem (2021), McCartney (2003, 2005) to find more details about ply crack growth simulation using GPS and variational models together with extensive validation results. It is assumed that the 90 degree ply has $M = 2^4+1 = 17$ equally spaced potential ply cracking sites in a $2L = 16$ mm length of laminate. The ply refinement technique is implemented in the variational stress transfer model to ensure converged results. The modeling results consider the effects of thermal residual stresses ($\Delta T = -120°C$) where the modeling input values are taken from Varna et al. (1999). It was first assumed that

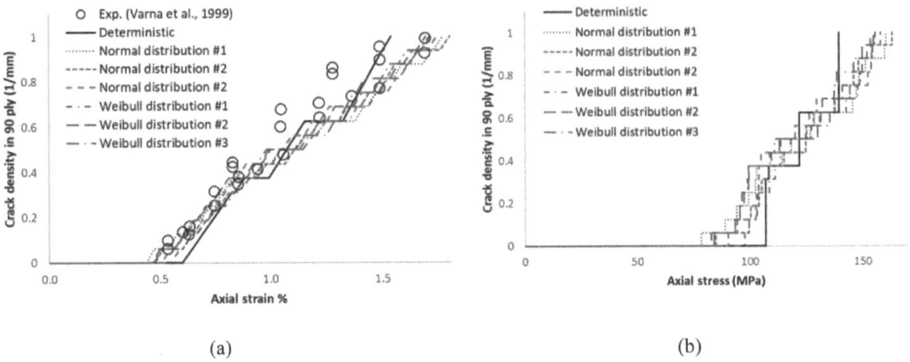

Figure 15.13 Variation of crack density in the 90 degree plies of a GFRP $[0/90_8/0_{0.5}]_s$ laminate (material #2) under uniaxial loading conditions versus (a) applied axial strain and (b) applied axial stress.

the same value of effective fracture energy ($G_I^c = 530\,J/m^2$) was used for all the potential ply cracking sites. Additionally, three probabilistic simulations are performed assuming that the effective fracture energy for each potential ply cracking site is taken at random from a normal distribution with mean value 530 J/m^2 and standard deviation (*SD*) 20%. Three more simulations are also performed where fracture energies are taken at random from a Weibull distribution with scale parameter 530 J/m^2 and shape parameter 6. The modeling results are compared with those of experiments (Varna et al., 1999) in Fig. 15.13a where the crack density in 90 degree plies is plotted versus the applied axial strain. Fig. 15.13b shows crack density in 90 degree plies versus the applied axial load. It should be mentioned that the accuracy of the variational stress transfer model (or GPS) for the prediction of thermoelastic properties of uniformly and nonuniformly cracked laminates has already been verified in previous sections. Therefore, the accordance with experimental results in Fig. 15.13a verifies the applicability of the suggested energy-based framework. It is noted that energy balance is an important physical principle that must be obeyed in any sound model of material behavior, alongside mass, charge, momentum and entropy balance. When modelling the failure of composite components, strength concepts are often introduced in the literature together with many well-known stress- and strength-based failure criteria. There is no physical justification for such an approach other than it can enable solutions to practical problems using empiricism. The energy balance approach described here, is an attempt to adhere to sound physical principles while avoiding the use of such popular empirical approaches.

15.5 Closing remarks and outlook

In this chapter, the GPS/bending and variational stress transfer models were introduced for the prediction of stress/displacement fields and stiffness reductions in both symmetric and nonsymmetric composite laminates containing uniformly or nonuniformly distributed ply cracks. The extension of the models to deal with complex loading conditions involving out-of-plane shear and bending deformations was then discussed. Generalized energy-based frameworks were introduced for the prediction of ply cracking initiation and propagation in symmetric laminates under quasistatic multiaxial loading conditions. It was briefly explained how the approach can be used to deal with cyclic stress states when predicting the growth of preexisting ply defects under fatigue loading. Indicative numerical results are shown to justify the accuracy and validity of the approaches.

One important remaining problem is to extend the capability of the energy-based damage evolution laws for the prediction of ply cracking under bending loads in general laminates under multiaxial, in-plane/out-of-plane shear, and bending deformations when residual stresses are present. This extension requires experimental observation of crack growth under bending loads in different lay-ups (possibly with different ply thickness) made of the same material properties to justify whether the same set of material parameters (e.g., ply fracture energies) can predict ply cracking fracture events in laminates under both in-plane and out-of-plane bending loading conditions. Such developments need to use some recently derived interrelationships (Hajikazemi et al., 2020a) among the thermoelastic constants of general laminates (Eqs. 15.23−15.31) together with the treatments for nonuniform ply cracking to involve the effects of thermally induced residual stresses and possible manufacturing defects on crack propagation under bending loads.

Another important problem is to include other damage mechanisms in the stress transfer models. This involves modeling of delaminations at the tips of ply cracks and fiber breaks in both symmetric and nonsymmetric laminates under complex loading conditions. When such stress transfer models are available, the next step will be to develop damage evolutions laws for predicting progressive formation of combinations of these damage mechanisms. Such damage evolutions laws must take into account the complexity of the laminate geometry (ply thickness, lay-up) together with multiaxial loading conditions involving residual stresses. Formulating such damage evolution laws, based on intrinsic effective material properties that account for damage initiation and propagation in different lay-ups of the same material, is of crucial practical importance that needs extensive experimental investigation. Very recently, a variational model has been developed (Hajikazemi et al., 2022) that predicts accurately the stress transfer problem in general symmetric laminates in the presence of both ply cracking and delamination under triaxial and in-plane shear loading conditions. The application of this approach should though be enhanced to deal with bending deformation and out-of-plane shear loadings.

Predicting the initiation and propagation of different damage mechanisms in real complex geometries is another challenge. To deal with such complex geometries, such as a wind turbine blade as a whole, one needs to use finite element techniques. The developments in this chapter can then be used as test data at coupon level to provide inputs for anisotropic continuum damage models that can be linked to FEM analyses. The advantages of the methods described in this chapter over the actual coupon level experiments should not be limited to their cost and time efficiencies. Semianalytical approaches can easily provide information needed regarding the damage state in laminates under multiaxial loading conditions or loading cases which are very difficult to simulate experimentally. Having much data on the material behavior of a laminate under different loading conditions will make it possible to develop an appropriate constitutive law that can be used as an anisotropic continuum damage model for use in FEM analyses at structural level.

Another approach to deal with damage analysis at structural level is to call directly the semianalytical approaches described at each integration point of FEM to update the damage state and consequently the laminate compliance matrix (e.g., for FEM shell elements) based on the local stress and damage state. Although the semianalytical approaches have split-second run time, this would still cause a large computational penalty to call these models at each integration point during a nonlinear FEM simulation of a large structure. Another issue is that damage in a composite laminate is a characteristic affecting the performance of the laminate as a whole even though that it might appear in only one of the plies. It means that, for example, the prediction of ply cracking in a 90 degree ply of a simple $[0/90]_S$ laminate cannot be accurately done based solely on the stress state and material properties of the 90 degree ply independent of the laminate load level or laminate geometry (e.g., adjacent ply orientation, thickness, etc.). This actually questions the accuracy of ply-level (mesoscale) continuum damage models that can be linked to FEM solid elements. One thought to resolve these issues is to link the semianalytical approaches with a homogenization concept. The homogenization concept is based on the replacement of a cracked ply in a laminate by an undamaged ply having modified elastic properties in order that the effective properties of both the discretely cracked and homogenized laminates are identical. Preliminary studies by one of the authors (MH) show that it is possible to use the semianalytical approaches considered here to provide adequate input data to train neural networks for the prediction of the effects of ply cracking and delamination on ply-level properties taking account of the effects of adjacent ply properties. These trained algorithms can then be used in conjunction with energy-based evolution laws as form-free constitutive laws for the determination of damage state in each individual ply of the laminate. The advantage of these form-free constitutive laws over direct implementation of the introduced semianalytical approaches at FEM structural analysis is that they do not introduce run-time penalties.

References

Ahmadi, H., Hajikazemi, M., Van Paepegem, W., 2022. A computational study about the effects of ply cracking and delamination on the stiffness reduction of damaged lamina and laminate. Int. J. Damage Mech. 31, 325–347. https://doi.org/10.1177/10567895211039268.

Andersons, J., Joffe, R., Spārniņš, E., 2008. Statistical model of the transverse ply cracking in cross-ply laminates by strength and fracture toughness based failure criteria. Eng. Fract. Mech. 75, 2651–2665. https://doi.org/10.1016/j.engfracmech.2007.03.007.

Aveston, J., Kelly, A., 1973. Theory of multiple fracture of fibrous composites. J. Mater. Sci. 8, 352–362. https://doi.org/10.1007/BF00550155.

Barbero, E.J., Cabrera Barbero, J., 2016. Analytical solution for bending of laminated composites with matrix cracks. Compos. Struct. 135, 140–155. https://doi.org/10.1016/j.compstruct.2015.09.021.

Barbero, E.J., Cortes, D.H., 2010. A mechanistic model for transverse damage initiation, evolution, and stiffness reduction in laminated composites. Compos. B Eng. 41, 124–132. https://doi.org/10.1016/j.compositesb.2009.10.001.

Barbero, E.J., Cosso, F.A., Campo, F.A., 2013. Benchmark solution for degradation of elastic properties due to transverse matrix cracking in laminated composites. Compos. Struct. 98, 242–252. https://doi.org/10.1016/j.compstruct.2012.11.009.

Barulich, N.D., Godoy, L.A., Dardati, P.M., 2018. Evaluation of cross-ply laminate stiffness with a non-uniform distribution of transverse matrix cracks by means of a computational meso-mechanic model. Compos. Struct. 185, 561–572. https://doi.org/10.1016/j.compstruct.2017.11.063.

Dinh, T.D., Garoz, D., Hajikazemi, M., Van Paepegem, W., 2019. Mesoscale analysis of ply-cracked composite laminates under in-plane and flexural thermo-mechanical loading. Compos. Sci. Technol. 175, 111–121. https://doi.org/10.1016/j.compscitech.2019.03.010.

Garoz, D., Hajikazemi, M., Dinh, T.D., Van Paepegem, W., 2019. Mesoscale finite element analysis of cracked composite laminates under out-of-plane loads using 3D periodic boundary conditions. Compos. Struct. 111699. https://doi.org/10.1016/j.compstruct.2019.111699.

Garrett, K.W., Bailey, J.E., 1977. Multiple transverse fracture in 90° cross-ply laminates of a glass fibre-reinforced polyester. J. Mater. Sci. 12, 157–168. https://doi.org/10.1007/BF00738481.

Hahn, H.T., Tsai, S.W., 1974. On the behavior of composite laminates after initial failures. J. Compos. Mater. 8, 288–305. https://doi.org/10.1177/002199837400800306.

Hajikazemi, M., McCartney, L.N., 2018. Comparison of variational and generalized plane strain approaches for matrix cracking in general symmetric laminates. Int. J. Damage Mech. 27, 507–540. https://doi.org/10.1177/1056789516685381.

Hajikazemi, M., Van Paepegem, W., 2018. A variational model for free-edge interlaminar stress analysis in general symmetric and thin-ply composite laminates. Compos. Struct. 184, 443–451. https://doi.org/10.1016/j.compstruct.2017.10.012.

Hajikazemi, M., 2021. 17—physics-based methodology for predicting ply cracking and laminate failure in symmetric composite laminates under multiaxial loading condition. In: Van Paepegem, W. (Ed.), Multi-Scale Continuum Mechanics Modelling of Fibre-Reinforced Polymer Composites. Woodhead Publishing, pp. 509–553. https://doi.org/10.1016/B978-0-12-818984-9.00017-2.

Hajikazemi, M., Sadr, M.H., 2014. A variational model for stress analysis in cracked laminates with arbitrary symmetric lay-up under general in-plane loading. Int. J. Solid Struct. 51, 516–529. https://doi.org/10.1016/j.ijsolstr.2013.10.024.

Hajikazemi, M., Sadr, M.H., 2014. Stiffness reduction of cracked general symmetric laminates using a variational approach. Int. J. Solid Struct. 51, 1483–1493. https://doi.org/10.1016/j.ijsolstr.2013.12.040.

Hajikazemi, M., Sadr, M.H., Hosseini-Toudeshky, H., Mohammadi, B., 2014. Thermo-elastic constants of cracked symmetric laminates: a refined variational approach. Int. J. Mech. Sci. 89, 47–57. https://doi.org/10.1016/j.ijmecsci.2014.08.008.

Hajikazemi, M., Sadr, M.H., Talreja, R., 2014. Variational analysis of cracked general cross-ply laminates under bending and biaxial extension. Int. J. Damage Mech. 24, 582–624. https://doi.org/10.1177/1056789514546010.

Hajikazemi, M., Sadr, M., Varna, J., 2017. Analysis of cracked general cross-ply laminates under general bending loads: a variational approach. J. Compos. Mater. 51, 3089–3109. https://doi.org/10.1177/0021998316682364.

Hajikazemi, M., McCartney, L.N., Van Paepegem, W., Sadr, M.H., 2018. Theory of variational stress transfer in general symmetric composite laminates containing non-uniformly spaced ply cracks. Compos. Appl. Sci. Manuf. 107, 374–386. https://doi.org/10.1016/j.compositesa.2018.01.021.

Hajikazemi, M., Garoz, D., Van Paepegem, W., 2019. Model to accurately predict out-of-plane shear stiffness reduction in general cracked laminates. Compos. Sci. Technol. 179, 88–96. https://doi.org/10.1016/j.compscitech.2019.05.009.

Hajikazemi, M., McCartney, L.N., Ahmadi, H., Van Paepegem, W., 2020. Variational analysis of cracking in general composite laminates subject to triaxial and bending loads. Compos. Struct. 239, 111993. https://doi.org/10.1016/j.compstruct.2020.111993.

Hajikazemi, M., McCartney, L.N., Van Paepegem, W., 2020. Matrix cracking initiation, propagation and laminate failure in multiple plies of general symmetric composite laminates considering the effects of manufacturing defects. Compos. Appl. Sci. Manuf. 2222, 5555.

Hajikazemi, M., Ahmadi, H., McCartney, L.N., Van Paepegem, W., 2022. A variational approach for accurate prediction of stress and displacement fields and thermo-elastic constants in general symmetric laminates containing ply cracking and delamination under general triaxial loading. Int. J. Solid Struct. 254–255, 111917. https://doi.org/10.1016/j.ijsolstr.2022.111917.

Hashin, Z., 1985. Analysis of cracked laminates: a variational approach. Mech. Mater. 4, 121–136. https://doi.org/10.1016/0167-6636(85)90011-0.

Hashin, Z., 1986. Analysis of stiffness reduction of cracked cross-ply laminates. Eng. Fract. Mech. 25, 771–778. https://doi.org/10.1016/0013-7944(86)90040-8.

Hashin, Z., 1996. Finite thermoelastic fracture criterion with application to laminate cracking analysis. J. Mech. Phys. Solid. 44, 1129–1145. https://doi.org/10.1016/0022-5096(95)00080-1.

Herakovich, C.T., Aboudi, J., Lee, S.W., Strauss, E.A., 1988. Damage in composite laminates: effects of transverse cracks. Mech. Mater. 7, 91–107. https://doi.org/10.1016/0167-6636(88)90008-7.

Katerelos, D.G., McCartney, L.N., Galiotis, C., 2006. Effect of off—axis matrix cracking on stiffness of symmetric angle-ply composite laminates. Int. J. Fract. 139, 529–536. https://doi.org/10.1007/s10704-006-0100-9.

Li, S., Hafeez, F., 2009. Variation-based cracked laminate analysis revisited and fundamentally extended. Int. J. Solid Struct. 46, 3505–3515. https://doi.org/10.1016/j.ijsolstr.2009.03.031.

Llobet, J., Maimí, P., Essa, Y., de la Escalera, F.M., 2018. Progressive matrix cracking in carbon/epoxy cross-ply laminates under static and fatigue loading. Int. J. Fatig. 119, 330–337. https://doi.org/10.1016/j.ijfatigue.2018.10.008.

Loukil, M.S., Varna, J., 2019. Crack face sliding displacement (CSD) as an input in exact GLOB-LOC expressions for in-plane elastic constants of symmetric damaged laminates. Int. J. Damage Mech. https://doi.org/10.1177/1056789519866000, 105678951986600.

Loukil, M.S., Varna, J., Ayadi, Z., 2012. Applicability of solutions for periodic intralaminar crack distributions to non-uniformly damaged laminates. J. Compos. Mater. 47, 287–301. https://doi.org/10.1177/0021998312440126.

Lundmark, P., Varna, J., 2005. Constitutive relationships for laminates with ply cracks in in-plane loading. Int. J. Damage Mech. 14, 235–259. https://doi.org/10.1177/1056789505050355.

Lundmark, P., Varna, J., 2006. Crack face sliding effect on stiffness of laminates with ply cracks. Compos. Sci. Technol. 66, 1444–1454. https://doi.org/10.1016/j.compscitech.2005.08.016.

McCartney, L.N., 1992. Theory of stress transfer in a 0°—90°—0° cross-ply laminate containing a parallel array of transverse cracks. J. Mech. Phys. Solid. 40, 27–68. https://doi.org/10.1016/0022-5096(92)90226-R.

McCartney, L.N., 1992. Mechanics for the growth of bridged cracks in composite materials: Part I. Basic principles. J. Compos. Technol. Res. 14, 133–146.

McCartney, L.N., 1992. Mechanics for the growth of bridged cracks in composite materials: Part II. Applications. J. Compos. Technol. Res. 14, 147–154.

McCartney, L.N., 2000. Model to predict effects of triaxial loading on ply cracking in general symmetric laminates. Compos. Sci. Technol. 60, 2255–2279. https://doi.org/10.1016/S0266-3538(00)00086-5.

McCartney, L.N., 2001. Predicting ply crack formation in cross-ply laminates subject to generalised plane bending. In: Proceedings 6th International Conference on Deformation and Fracture of Composites, pp. 57–66.

McCartney, L.N., 2003. Physically based damage models for laminated composites. Proc. Inst. Mech. Eng., Part L 217, 163–199. https://doi.org/10.1177/146442070321700301.

McCartney, L.N., 2005. Energy-based prediction of progressive ply cracking and strength of general symmetric laminates using an homogenisation method. Compos. Appl. Sci. Manuf. 36, 119–128. https://doi.org/10.1016/S1359-835X(04)00151-4.

McCartney, L.N., 2008. Approximate method of predicting ply crack formation in general symmetric laminates subject to biaxial loading and bending. Electronic proceedings of 13th European conference on composite materials (ECCM-13) 2–5.

McCartney, L.N., 2008. Energy methods for fatigue damage modelling of laminates. Compos. Sci. Technol. 68, 2601–2615. https://doi.org/10.1016/j.compscitech.2008.04.044.

McCartney, L.N., 2013. Derivations of energy-based modelling for ply cracking in general symmetric laminates. J. Compos. Mater. 47, 2641–2673. https://doi.org/10.1177/0021998313492359.

McCartney, L.N., 2022. Properties for Design of Composite Structures: Theory and Implementation Using Software. Wiley.

McCartney, L.N., Blazquez, A., 2009. Validating, using BEM, a model for ply cracks in cross-ply laminates subject to combined in-plane loading and out-of-plane bending. In: 2nd ECCOMAS Thematic Conference on the Mechanical Response of Composites, pp. 1–3.

McCartney, L.N., Schoeppner, G.A., 2002. Predicting the effect of non-uniform ply cracking on the thermoelastic properties of cross-ply laminates. Compos. Sci. Technol. 62, 1841–1856. https://doi.org/10.1016/S0266-3538(02)00091-X.

McCartney, L.N., 2016. Chapter 13 - microlevel approaches to modeling of damage in composite materials: generalized plane strain analysis. In: Talreja, R., Varna, J. (Eds.), Modeling Damage, Fatigue and Failure of Composite Materials. Woodhead Publishing, pp. 289–327. https://doi.org/10.1016/B978-1-78242-286-0.00013-3.

Montesano, J., Singh, C.V., 2015. Predicting evolution of ply cracks in composite laminates subjected to biaxial loading. Compos. B Eng. 75, 264–273. https://doi.org/10.1016/j.compositesb.2015.01.039.

Montesano, J., McCleave, B., Singh, C.V., 2018. Prediction of ply crack evolution and stiffness degradation in multidirectional symmetric laminates under multiaxial stress states. Compos. B Eng. 133, 53–67. https://doi.org/10.1016/j.compositesb.2017.09.016.

Nairn, J.A., 2000. Matrix Microcracking in Composites. Comprehensive Composite Materials. Elsevier, pp. 403–432. https://doi.org/10.1016/B0-08-042993-9/00069-3.

Nairn, J.A., 2017. Cross laminated timber properties including effects of non-glued edges and additional cracks. Eur. J. Wood Wood Prod. 75, 973–983. https://doi.org/10.1007/s00107-017-1202-y.

Nairn, J.A., Mendels, D.A., 2001. On the use of planar shear-lag methods for stress-transfer analysis of multilayered composites. Mech. Mater. 33, 335–362. https://doi.org/10.1016/S0167-6636(01)00056-4.

Nairn, J.A., Hu, S., Bark, J.S., 1993. A critical evaluation of theories for predicting microcracking in composite laminates. J. Mater. Sci. 28, 5099–5111. https://doi.org/10.1007/BF00361186.

Pakdel, H., Mohammadi, B., 2017. Experimental observation and energy based analytical investigation of matrix cracking distribution pattern in angle-ply laminates. Theor. Appl. Fract. Mech. 92, 146–154. https://doi.org/10.1016/j.tafmec.2017.06.007.

Pupurs, A., Varna, J., Loukil, M., Ben Kahla, H., Mattsson, D., 2016. Effective stiffness concept in bending modeling of laminates with damage in surface 90-layers. Compos. Appl. Sci. Manuf. 82, 244–252. https://doi.org/10.1016/j.compositesa.2015.11.012.

Pupurs, A., Loukil, M., Varna, J., 2022. Digital image correlation (DIC) validation of engineering approaches for bending stiffness determination of damaged laminates. Appl. Compos. Mater. 29, 1937–1958. https://doi.org/10.1007/s10443-022-10045-0.

Reissner, E., 1950. On a variational theorem in elasticity. J. Math. Phys. 29, 90–95.

Rosen, B.W., 1970. Thermoelastic energy functions and minimum energy principles for composite materials. Int. J. Eng. Sci. 8, 5–18. https://doi.org/10.1016/0020-7225(70)90010-8.

Schmitz, A., Horst, P., 2014. A finite element unit-cell method for homogenised mechanical properties of heterogeneous plates. Compos. Appl. Sci. Manuf. 61, 23–32. https://doi.org/10.1016/j.compositesa.2014.01.014.

Schoeppner, G.A., Pagano, N.J., 1999. 3-D thermoelastic moduli and saturation crack density for cross-ply laminates with transverse cracks. Int. J. Damage Mech. 8, 273–309. https://doi.org/10.1177/105678959900800304.

Silberschmidt, V.V., 2005. Matrix cracking in cross-ply laminates: effect of randomness. Compos. Appl. Sci. Manuf. 36, 129–135. https://doi.org/10.1016/j.compositesa.2004.06.008.

Silberschmidt, V.V., 2006. Effect of micro-randomness on macroscopic properties and fracture of laminates. J. Mater. Sci. 41, 6768–6776. https://doi.org/10.1007/s10853-006-0205-6.

Takeda, N., Ogihara, S., 1994. In situ observation and probabilistic prediction of microscopic failure processes in CFRP cross-ply laminates. Compos. Sci. Technol. 52, 183–195. https://doi.org/10.1016/0266-3538(94)90204-6.

Takeda, N., McCartney, L.N., Ogihara, S., 2000. The application of a ply-refinement technique to the analysis of microscopic deformation in interlaminar-toughened laminates with transverse cracks. Compos. Sci. Technol. 60, 231–240. https://doi.org/10.1016/S0266-3538(99)00121-9.

Varna, J., 2015. Strategies for stiffness analysis of laminates with microdamage: combining average stress and crack face displacement based methods: strategies for stiffness analysis of laminates. Z. Angew. Math. Mech. 95, 1081–1097. https://doi.org/10.1002/zamm.201400296.

Varna, J., Joffe, R., Akshantala, N.V., Talreja, R., 1999. Damage in composite laminates with off-axis plies. Compos. Sci. Technol. 59, 2139–2147. https://doi.org/10.1016/S0266-3538(99)00070-6.

Varna, J., Joffe, R., Talreja, R., 2001. A synergistic damage-mechanics analysis of transverse cracking in [$\pm\theta$/904]s laminates. Compos. Sci. Technol. 61, 657–665. https://doi.org/10.1016/S0266-3538(01)00005-7.

Vinogradov, V., Hashin, Z., 2005. Probabilistic energy based model for prediction of transverse cracking in cross-ply laminates. Int. J. Solid Struct. 42, 365–392. https://doi.org/10.1016/j.ijsolstr.2004.06.043.

Vinogradov, V., Hashin, Z., 2010. Variational analysis of cracked angle-ply laminates. Compos. Sci. Technol. 70, 638–646. https://doi.org/10.1016/j.compscitech.2009.12.018.

Yuan, F.G., Selek, M.C., 1993. Transverse cracking and stiffness reduction in composite laminates. J. Reinforc. Plast. Compos. 12, 987–1015. https://doi.org/10.1177/073168449301200905.

A multiscale approach to modeling of composite damage

16

Chandra Veer Singh [1] *and Ramesh Talreja* [2]
[1]Department of Materials Science and Engineering, University of Toronto, Toronto, ON, Canada; [2]Department of Aerospace Engineering, Department of Materials Science and Engineering, Texas A&M University, College Station, TX, United States

16.1 Introduction

It has long been recognized that the response of structures made of heterogeneous materials, such as composites, measured at a representative structural scale, called the "macro" or "global" scale, is reflected by the behavior of the heterogeneities (particles, fibers, etc.) lying within the volume of the material. The scale of heterogeneities is characterized as "micro" or "local." A field of mechanics, called micromechanics, that addresses relating the thermomechanical responses at the macroscale to the stress/strain fields at the microscale has developed to a high level of maturity; see Mura (1987) and Nemat-Nasser and Hori (1999). Most early works in micromechanics developed averaging methods, homogenization procedures, etc., and provided a conceptual foundation to the field. More recently, as the observation and measurement techniques for the characterization of heterogeneities have advanced, it has become clear that analytical methods of micromechanics are inadequate in accounting for real microstructures. Computational micromechanics has instead emerged as a viable field (Zohdi and Wriggers, 2008) and is today a very active field of research. A central issue in accounting for real (disordered) microstructures with computational micromechanics is constructing a representative volume element (RVE), see for instance, Ostoja-Starzewski (1999). As described in that work, the scale to represent RVE is the so-called "meso" scale. Thus, the hierarchy of micro-meso-macro scales in computational micromechanics is needed to connect the internal details of a heterogeneous solid to its response at the structural scale.

When the internal structure of a composite solid contains not only heterogeneities such as fibers and particles but also has "evolving" entities such as cracks, the micro-meso-macro hierarchy needs particular attention. In Chapter 5 of this book, the initiation of failure events was discussed. Such events often lead to crack formation at the microscale and are commonly called microcracks. Evolution of the microcracks in a composite structure, for example, a laminate, can consist of growth of the size and change of the shape and orientation of the microcracks. Furthermore, linkages and multiplication of the microcracks can also occur. A multiscale damage analysis must properly account for these aspects.

Modeling Damage, Fatigue and Failure of Composite Materials. https://doi.org/10.1016/B978-0-443-18489-5.00006-0

This chapter will examine the basic concepts of multiscale modeling with respect to damage in composite materials. The observed physical mechanisms of failure progression will then be reviewed. An integrative process of failure assessment with proper regard to these mechanisms will be presented. Some of the discussion to follow will draw upon the authors' previous writings on the subject (Singh and Talreja, 2008, 2009, 2013; Talreja, 2006, 2010, 2014).

16.2 Basic concepts and considerations

As noted above, one fundamental issue in multiscale modeling is the delineation of the hierarchy of the scales. For an undamaged composite with "stationary" microstructure, that is, where the heterogeneities do not evolve under loading, the geometrical features (size, shape, orientation, etc.) and their distributions will determine the separation of length scales in a selected order. For the case of uniform size, shape, and distribution of the heterogeneities, identifying a repeating unit cell containing one entity or a cluster of entities will suffice to describe the microscale. When the microstructural entities are disordered in their placement within the volume of a given composite, it would require defining an RVE to generate a mesoscale larger than the (micro) length scale of the entities themselves. The precise (minimum) RVE size is not unique but depends on the macro-level response of interest. Thus, RVE size for average elastic properties is not the same as that for the initiation of a given failure event such as yielding. The criterion associated with the occurrence of a given failure event determines how large the RVE size needs to be to capture initiation of that failure event in a statistical sense. Bulsara et al. (1999) treated this problem for the specific case of failure initiation in unidirectional (UD) composites under transverse loading when the fibers are distributed in an arbitrary manner, that is, without following a specific probability distribution. Their proposal of appropriate RVE size that minimizes the coefficient of variation of a measure of the critical local condition attained in simulated realizations of RVE can in principle be used for other failure initiation events.

Once failure events take the form of cracks, their further evolution poses another problem concerning RVE size. This is illustrated in Fig. 16.1, which schematically depicts two types of damage (multiple cracks) resulting from failure events. As shown there, in Type 1 damage, the individual cracks are associated with heterogeneities, for example, cracks formed by debonding of the heterogeneities (partly or fully) from the matrix, as illustrated in Fig. 16.1b. In Type 2 damage, on the other hand, cracks formed in the matrix are not connected to the heterogeneities, as illustrated in Fig. 16.1c. The two types of damage will produce different response measured over an RVE. For instance, assuming that the RVE shown in Fig. 16.1a is representative of the undamaged state of the composite in a selected region, let $\mathbf{u}$ be the displacement response on the boundary of the RVE on applying surface traction $\mathbf{t}$. Assuming now that the two types of damage are of the same intensity (measured by a given metric), the displacement response to the same traction $\mathbf{t}$ will change to $\mathbf{u_1}$ for the same RVE with Type 1 damage, while it will change to another value $\mathbf{u_2}$ for Type 2

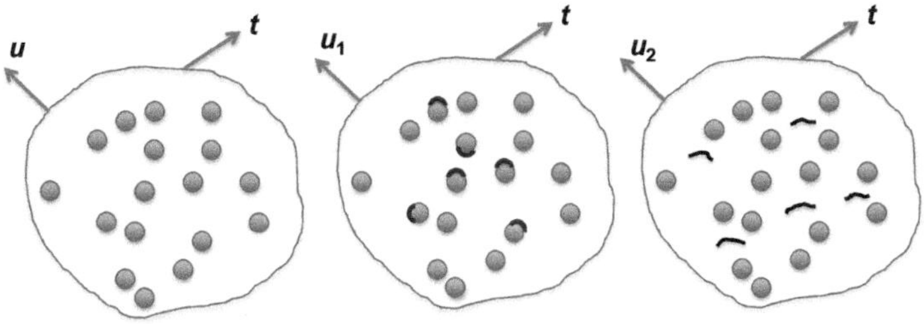

Figure 16.1 (a) A heterogeneous solid with disordered microstructure; (b) type 1 damage (debond cracks) associated with heterogeneities; and (c) type 2 damage (matrix cracks). The displacement response $\mathbf{u}$, $\mathbf{u_1}$, and $\mathbf{u_2}$ to the same traction $\mathbf{t}$ occurs in cases (a), (b), and (c), respectively.

damage. The difference in the displacement response for the two types of damage can be understood by realizing that the local perturbations in the stress (and strain) fields caused by cracks in the two types of damage will be different for the same prescribed boundary traction **t** due to differences in their interactions with the heterogeneities as well as among themselves. These considerations suggest that the RVE to be constructed for determining the changes in the composite response due to damage must be with respect to the particular damage induced in the composite by the external loading. This applies also to the evolution of damage, that is, the local conditions responsible for growth of the individual damage entities must be analyzed within the RVE appropriate for those entities.

16.3 Failure mechanisms

In Chapter 5 of this book, we examined the mechanisms of initiation of failure in a UD composite subjected to in-plane loading. Those mechanisms produce cracks at the fiber—matrix interface and in the matrix. Once such cracks have been produced, their further progression is of interest only in the context of a laminate, that is, when the considered UD composite is in the form of a ply in a laminate. Fig. 16.2 illustrates the triaxial stress state generated locally in the matrix of a ply, which in the homogenized state is in plane stress state. Note that the right part of Fig. 16.2 does not show the fiber-direction stress σ_x since it is assumed not to influence the crack formation in the matrix and at the fiber—matrix interfaces. If the UD composite is unconstrained, that is, is by itself and not bonded to other plies of a laminate, then the cracks in the matrix and at the fiber—matrix interfaces quickly link up and grow unstably to failure. Within a laminate, however, the crack growth is arrested at the ply interfaces and further growth then follows the fiber direction of the ply. Formation of the complete crack, spanning the ply thickness and extending along the fibers to the extent the ply stresses would allow, is of little practical interest. More consequential is the process of multiplication

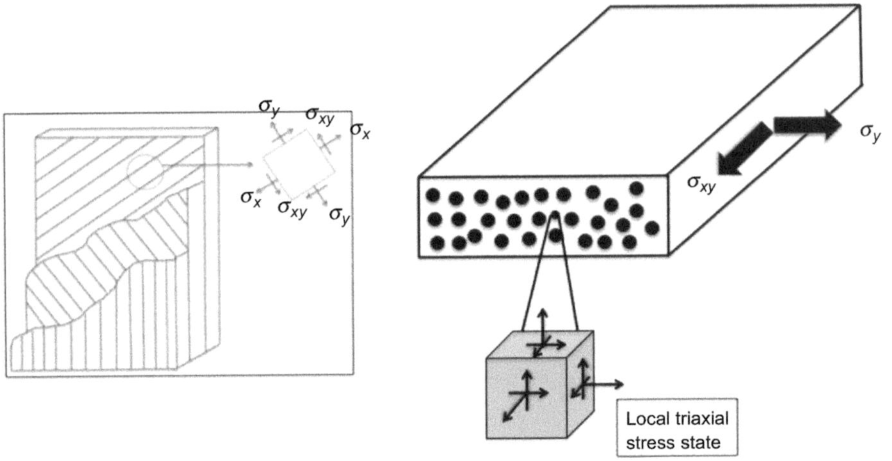

Figure 16.2 Illustration of the plane stress state of a homogenized ply within a laminate (*left*) and the local triaxial stress state in the matrix of the ply (*right*).

of such cracks within the ply in which they were formed. The field of study concerned with this process, known as transverse ply cracking (or multiple matrix cracking), is a mature field for cross-ply laminates and has been extensively reviewed by Berthelot (2003), who also discussed delamination caused by the transverse cracks by cracking the 0/90 ply interfaces. Multiple matrix cracking in transverse plies as well as in off-axis plies has been reviewed in Talreja and Singh (2012), and an updated version is in Chapter 8 of this book.

Fig. 16.3 gives a schematic description of the entire damage evolution process in a broad class of laminates subjected to in-plane loading based on Jamison et al. (1984) who conducted extensive experimental observations in tension–tension cyclic loading. The sequence of failure events, labeled as damage in the figure, is representative of the effects of ply constraint and interaction between cracks. The early stage in the development of damage consists of multiple matrix cracks in individual plies whose evolution rate increases at first and then slows down as increasing interaction between cracks relieves stresses within the plies. If all cracks reach a saturation state, then the collective pattern of cracks may be regarded as a state characteristic of the particular laminate. Further evolution of damage involves coupling of matrix cracks via ply interfaces, leading to extensive interface cracking (delamination). The final stage of damage consists of fiber failures.

16.4 Multiscale analysis

In devising a multiscale approach, it is useful first to define the ultimate purpose. If the purpose is to develop structural configuration of a given heterogeneous material, then starting at the scale of heterogeneities (microscale) and moving up-the-scale to the

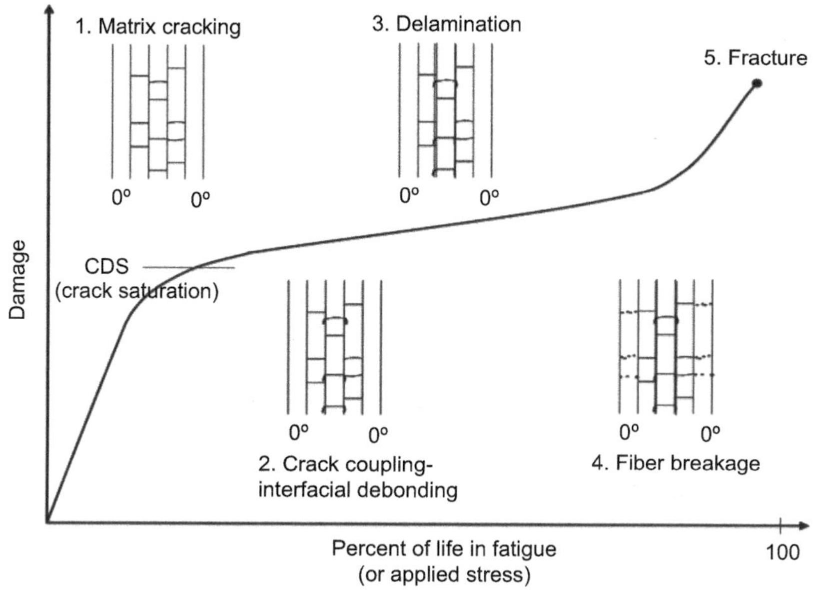

Figure 16.3 Schematic depiction of evolution of failure events (damage) in a broad class of laminates subjected to static or fatigue loading (Jamison et al., 1984).

structural scale would be an appropriate approach. This type of a hierarchical approach would allow parametric studies of the microstructure to optimize performance. If, however, a structural configuration (macro scale) has been determined based on some broader considerations of design, then a selected set of performance criteria, for example, durability, must be assessed. This is often the case. Fig. 16.4 depicts a flow chart of the durability assessment process for a composite structural part indicating the role of multiscale analysis. As shown there, one starts with (computational) stress analysis of a part of known geometry using as input the service environment imposed on the part and the material model (elasticity, viscoelasticity, etc.) describing the initial deformational behavior of the material. Based on the calculated stress fields and any prior experience with a similar part, one identifies critical regions often described as "hot spots" where issues of durability are likely to occur. Damage mechanisms are then analyzed in these regions to identify initiation and progression of failure events. Examples of the failure events were discussed above, referring to Fig. 16.3. The objective now would be to determine under what conditions would the given structural part reach criticality with respect to its design function such as retaining its deformational response adequately (stiffness-critical design) or retaining its load-bearing capability (strength-critical design). In each case, the multiscale analysis will use appropriate scales, in particular the RVE (mesoscale), to ascertain the respective criticality. For deformational analysis, the concepts and methods of damage mechanics incorporating multiscale approaches are the path to take. The outcome of this analysis provides stiffness degradation with damage and is used to update the

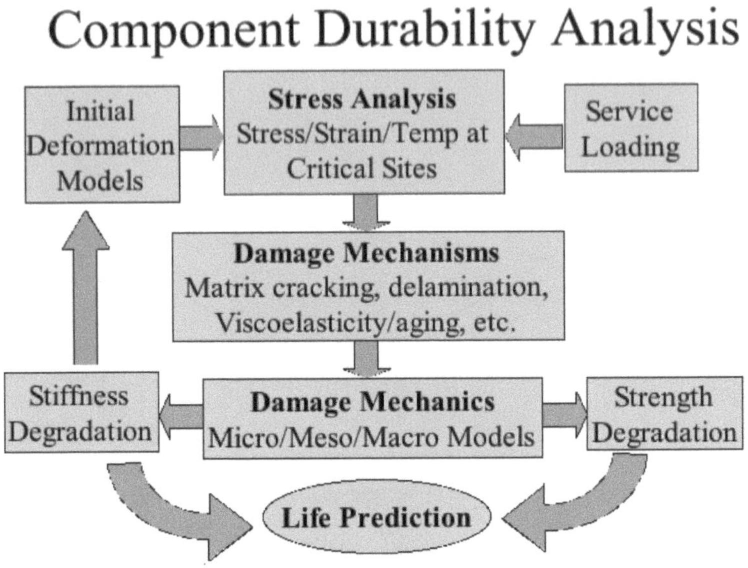

Figure 16.4 A flow chart describing the process of durability assessment of a composite part.

deformation models used for initial stress analysis. This is indicated in the left branch coming out of the damage mechanics "box" in Fig. 16.4. The right branch concerns strength degradation for which the analysis tools in damage mechanics focus on instability in progression of failure events. Further discussion in this chapter is restricted to multiscale analysis of deformational response, leaving failure theories directed at strength degradation for a later chapter (Chapter 15).

16.4.1 Continuum with internal state

A general approach for describing deformational response of materials with damage is the framework of continuum damage mechanics (CDM). In its classical form, CDM uses the concept of internal material state, first introduced by Kachanov (1958) for creep in metals to represent changes occurring within the volume of a solid that cause permanent changes in its deformational response. The internal changes conceived by Kachanov were not specific but can be attributed to energy dissipation processes and can be generalized to other material degradation processes such as distributed brittle microcracking. In any case, the incorporation of the changing internal material state in the classical CDM is by the so-called damage variables that have no length scale since they are simply nondimensional measures of the intensity of the internal microstructure changes. Thus, the classical CDM can be viewed as a single (macro) scale formulation. It is, however, possible to modify it to a multiscale formulation by associating a mesoscale with the internal state. As discussed in Section 16.2 above, damage in composites is an evolving microstructure whose proper description requires an RVE. Thus, in order for CDM to adequately treat composite materials with damage, it must be expanded from its classical single (macro)-scale form to a two-scale form, in which the mesoscale of an RVE is associated with the internal state. The RVE must then be constructed with appropriate representation of the microscale damage entities (microcracks). Incorporating the RVE as internal state in CDM will finally provide the micro-meso-macro scale bridging and thereby the desired multiscale framework for describing changes in deformational response due to damage. A framework of this nature is described next.

As reference, the classical CDM approach is illustrated in Fig. 16.5. Cast in the context of a boundary value problem, a heterogeneous solid is homogenized, that is, replaced by an equivalent homogeneous solid of properties that are averages of the heterogeneous solid. The boundary conditions prescribed as traction components t_i on the surface S_t and displacement components u_i on surface S_u initiate damage (distributed microcracks) within the volume. The distributed damage entities are homogenized and represented by internal state variables. The homogeneous continuum with damage with prescribed boundary conditions is now the new boundary value problem to be solved in the usual way. This is a single-scale model of composites with damage.

The multiscale version of the CDM is illustrated in Fig. 16.6. As depicted there, the internal state is not left unspecified as in Kachanov (1958) but is attributed an RVE of mesoscale at a generic material point P. The RVE is constructed to appropriately represent the microscale entities of damage (microcracks). A scheme is devised to associate field variables of micro- and mesoscales to the evolving internal state. These variables are described next.

A single damage entity, which is often a microcrack of arbitrary size and shape, can be represented by two vectors placed at a point on the surface S of the damage entity (Fig. 16.6). The vector **n** is a unit outward normal indicating the orientation of the damage entity, and the vector **a**, at the same point, represents an appropriate measure of the damage entity response to the external influences transmitted through the boundary

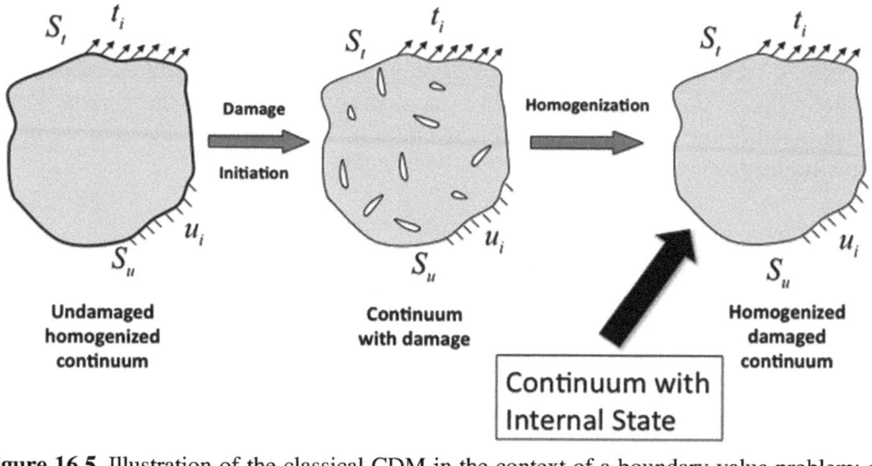

Figure 16.5 Illustration of the classical CDM in the context of a boundary value problem: a heterogeneous solid in pristine state is homogenized and subjected to boundary tractions t_i over the surface S_t and displacements u_i over the surface S_u; after initiation of damage, the solid is homogenized with damage regarded as internal state.

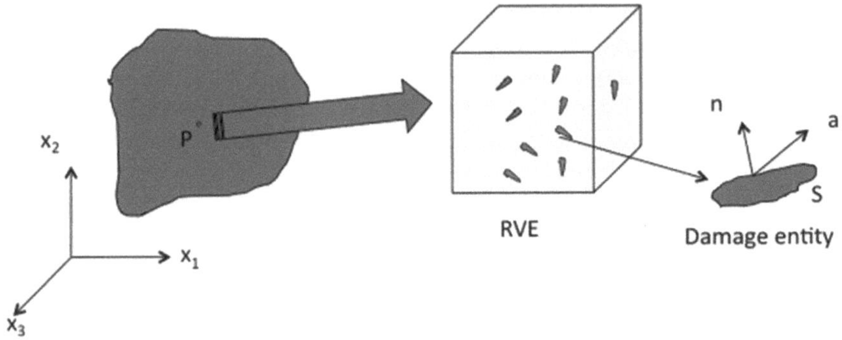

Figure 16.6 Schematic depiction of the multiscale version of CDM showing the internal state representation at a point P by an RVE of mesoscale in which damage entities of microscale are present.

(e.g., tractions and displacements illustrated in Fig. 16.5). For mechanical response, **a** is simply the displacement vector. For a microcrack, this displacement can be conveniently resolved in opening and sliding parts. The two vectors can be combined in a second order tensor by a dyadic product, which is named as a *damage entity tensor* given in component form by:

$$d_{ij} = \int_S a_i n_j dS \tag{16.1}$$

where the tensor components refer to the coordinate axes shown in Fig. 16.6 and integration is performed over the damage entity surface S. The size of this surface represents the *microscale*.

Assume that a volume V is found by certain appropriate criterion to represent the collective effect of the damage entities at point P such that the material response at that point due to damage can be given by a quantity defined over the volume V. Let such a quantity be defined as a simple average of the damage entity tensors in the following form:

$$D_{ij} = \frac{1}{V}\sum d_{ij} \tag{16.2}$$

where the summation is over all damage entities in the volume V. In a general case, not all damage entities respond similarly to external influences. It may be useful then to group damage entities in modes. Denoting the modes by α, and letting k_α denote the number of damage entities in the αth mode, a *damage mode tensor* is defined as:

$$D_{ij}^{(\alpha)} = \frac{1}{V}\sum_{k_\alpha} \left(d_{ij}\right)_{k_\alpha} \tag{16.3}$$

Each damage mode tensor is now the mesoscale representation of the evolving internal state of a composite with damage.

16.4.2 Microscale analysis

A microstructural entity that is initiated as a failure event, such as a microcrack, responds to external loading by inducing stress perturbation in the surrounding material. In the micromechanics approach, a suitable boundary value problem such as a unit cell subjected to uniform loading is solved to assess the desired perturbation. For cross-ply laminates with transverse cracks, Hashin (1985) took this approach by applying variational analysis to solve for the stress perturbations. For more general laminates where the cracks are oriented in planes not aligned with the planes of symmetry of the laminate, the stress analysis problem cannot be reduced to a two-dimensional one, as in Hashin (1985), and analytical solutions for local stress fields are not possible. Other approaches have been devised to estimate the elastic response of laminates with microcracks. Prior chapters on microlevel methods (Chapters 13 and 14) deal with such approaches. Here we describe the microscale analysis needed in the expanded CDM formulation described above.

The microscale descriptor in the expanded CDM formulation is the damage entity tensor d_{ij}, which, as defined by Eq. (16.1), is given by the orientation vector **n** and the microcrack "influence" vector **a**. The magnitude of this influence may be seen as a measure of the integrated stress perturbation effect caused by the microcrack on the surrounding material. This is clear when one notes that the stress perturbation is only possible when the microcrack surface conducts displacement, which is the quantity captured in the magnitude of the vector **a**. Separating this magnitude in the parts due to crack opening displacement (COD) and crack sliding displacement (CSD), and neglecting CSD, it may be expressed as:

$$a = \kappa c \tag{16.4}$$

where c is a characteristic length of the microcrack and κ is a coefficient representing the ability of the microcrack surfaces to conduct displacement under constraint imposed by the surrounding. This constraint comes from other cracks in the cracked ply and from the other plies of the laminate. The former is described as crack shielding effect and is caused by the mutual interaction between stress perturbations within the cracked ply. The latter is due to the displacement and stress continuity conditions imposed by the (assumed) perfect bonding between plies.

Evaluation of the coefficient κ, named as "constraint parameter," is essentially a microscale problem to be solved for a given array of cracks in a laminate. Since the stress perturbations for transverse cracks in cross-ply laminates can be evaluated analytically, it makes sense to use this case as the baseline (reference) case for one particular cross-ply laminate and cast the general problem in terms of the relative value of κ with respect to this value. This procedure has been adopted in Varna et al. (1999a,b, 2001) and Singh and Talreja (2008, 2009, 2013).

16.4.3 Mesoscale analysis

The mesoscale analysis consists of evaluating damage mode tensors, which are averages over the RVE volume, as defined by Eq. (16.3). In a general case of nonuniform distribution of damage entities, the volume V must be sufficiently large to capture a robust value of the damage mode tensor under consideration. This volume depends on the distribution of the damage entities. For interacting ply cracks, the nonuniformity of crack spacing is found to reduce as the degree of interaction between cracks increases. This is illustrated in Fig. 16.7 for transverse cracks in a cross-ply laminate. Based on observed data, the figure illustrates how the initial stage of transverse cracking is random in terms of the location of cracks along the length of a specimen. The randomness of the location of cracks reflects the random distribution of defects that initiate these cracks. As the crack multiplication process progresses, crack spacing reduces and interaction between cracks increases. The crack interaction produces the maximum axial stress midway between the interacting cracks (Hashin, 1985), leading to that location as the most favorable for new cracks to form unless the material defects elsewhere succeed in initiating cracks. The crack spacing becomes increasingly uniform with increasing applied load and eventually attains a constant value described as saturation spacing.

Based on the observed behavior of the transverse cracking process described above, the RVE size for this case is large in the initial stage of the process, and reduces as the cracking progresses, tending eventually to a unit (repeating) cell when the crack spacing attains the minimum (saturation) value. Huang et al. (2014) analyzed the statistical variation of transverse crack spacing in cross-ply laminates and proposed that the variation of this spacing in the early stage of the process can be used to evaluate the manufacturing quality. Huang and Talreja (2014) extended the analysis to oblique cracks in laminates reaching similar conclusions.

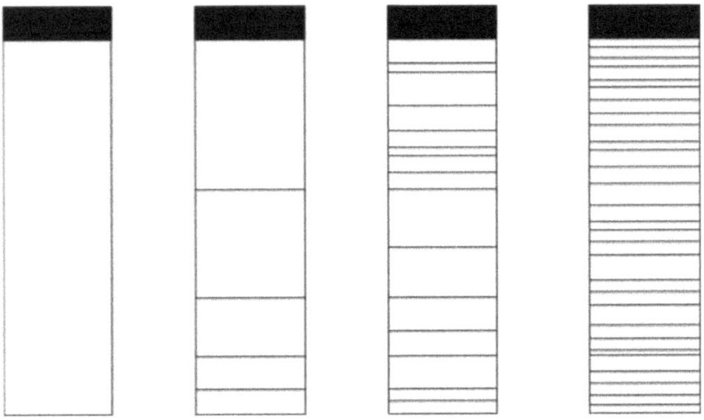

Figure 16.7 Development of crack spacing in a cross-ply laminate with increasing axial load (*left* to *right*), illustrating the reduction of crack spacing as well as increasing uniformity of crack spacing.

16.5 Application example for ply cracking in multidirectional laminates

In recent years, we have developed a comprehensive multiscale modeling approach for predicting the evolution of ply cracking and its effect on the stiffness properties in multidirectional laminates. Since multidirectional laminates consist of a mix of on-axis and off-axis plies, upon loading they develop ply cracks in multiple orientations. Hence, their analysis becomes extremely complex. Following our discussion in the preceding section, ply cracks in a specific orientation tend to be similar and hence can be grouped into a "damage mode." In this manner, the multiscale process can be utilized to effectively analyze the effect of multimode damage in multidirectional laminates. We have termed this approach as multiscale synergistic damage mechanics (SDM); its development for multiple types of multidirectional laminates with differing complexity of damage development can be found in Singh and Talreja (2009, 2010, 2013). Furthermore, an energy-based damage evolution model has been developed and incorporated along with the SDM methodology to yield full stress–strain response of the composite laminates undergoing subcritical damage. The details of this model are covered in Singh and Talreja (2013).

Let us first concentrate on ply cracking in off-axis plies of multidirectional laminates. Following the multiscale SDM approach (Singh and Talreja, 2009, 2010, 2013), the overall stiffness matrix for the whole laminate at a given state of ply cracking damage can be derived as:

$$\mathbf{C_{pq}} = \begin{bmatrix} \dfrac{E_x^0}{1-\nu_{xy}^0\nu_{yx}^0} & \dfrac{\nu_{xy}^0 E_y^0}{1-\nu_{xy}^0\nu_{yx}^0} & 0 \\ & \dfrac{E_y^0}{1-\nu_{xy}^0\nu_{yx}^0} & 0 \\ Symm & & G_{xy}^0 \end{bmatrix} + D_\theta \begin{bmatrix} 2a_1 & a_4 & 0 \\ & 2a_2 & 0 \\ Symm & & 2a_3 \end{bmatrix}. \qquad (16.5)$$

where the first term represents stiffness matrix for undamaged laminate, with $E_x^0, E_y^0, \nu_{xy}^0, G_{xy}^0$ representing the effective elastic moduli for the virgin laminate.

D_θ is an effective damage parameter, and a_i's are damage constants, obtained by fitting the SDM model to available stiffness data for a reference laminate at some given ply crack density. These data can be obtained using experimental measurements (see Singh and Talreja, 2008) or computational micromechanics (Singh and Talreja, 2009). For off-axis ply cracking, $D_\theta = \frac{\kappa_\theta t_c^2}{s_n^\theta t}$, where κ_θ is the constraint parameter, as previously defined in Eq. (16.4), for ply cracking in an off-axis ply oriented at an angle θ with respect to the longitudinal direction of the laminate; t_c and t, respectively, are the thicknesses of the cracked ply, and the whole laminate, and s_n^θ represents the normalized spacing between periodic cracks in the off-axis ply. Since experimental or numerical data can be obtained for stiffness degradation in cross-ply laminates more readily, the reference laminate is typically chosen as a cross-ply laminate. In that case, the constraint parameter for off-axis ply cracking relative to that for 90 degrees-cracking is given by Singh (2013) as:

$$k_{rel}(\theta) = \frac{k_\theta}{k_{90}} = \frac{\left(\overline{\Delta u_y}\right)_{\pm\theta_4}}{\left(\overline{\Delta u_y}\right)_{90_8}} \tag{16.6}$$

where the last term in Eq. (16.6) represents ratio of the average normalized COD in $\pm\theta$ to that in the 90 degrees-plies. The term in the denominator corresponds to 90 degrees-ply cracking in the reference cross-ply laminate. To illustrate the utility of the SDM model for a more complex damage scenario, consider damage development in a $[0_m/\pm\theta_n/90_r]_s$ laminate loaded in uniaxial tension in the longitudinal direction. Upon loading, the damage begins in the form of matrix cracks in the 90-plies. When load is increased further, these cracks multiply; and at certain load level, cracks initiate and multiply in $+\theta$ and $-\theta$ plies. The SDM model for this case was first developed in Singh and Talreja (2009). Following SDM methodology, this scenario can be analyzed by considering three damage modes, and the stiffness matrix for the damaged laminate is correspondingly given by:

$$\mathbf{C_{pq}} = \mathbf{C^0_{pq}} + \Delta\mathbf{C_{pq}} = \mathbf{C^0_{pq}} + \sum_{\alpha=1}^{N} \mathbf{C^{(\alpha)}_{pq}} \tag{16.7}$$

where $\Delta\mathbf{C_{pq}} = \mathbf{C^0_{pq}} + \sum_{\alpha=1}^{N} \mathbf{C^{(\alpha)}_{pq}}$ represents the stiffness changes brought by ply cracking in N damage modes and is derived as:

$$\Delta\mathbf{C_{pq}} = D\begin{bmatrix} 2a'_1 & a'_4 & 0 \\ & 2a'_2 & 0 \\ Symm & & 2a'_3 \end{bmatrix} \tag{16.8}$$

where, a'_i are the damage constants, and D represents the effective damage parameter for this case and is given as:

$$D = \frac{2t_0^2}{t}\left[\frac{1}{s_n^\theta}\frac{k_\theta}{k_\theta|_{\theta=90}}\left\{2(2n+r)^2 k_{90_{4n+2r}} - r^2 k_{90}\right\} + r^2\frac{k_{90}}{s^{90}}\right] \tag{16.9}$$

The terms in Eq. (16.9) involve three constraint parameters described as follows:

$$k_\theta = \frac{\left(\overline{\Delta u_y}\right)_{\pm\theta_{2n}}}{2nt_0}; \quad k_{90_{4n+2r}} = \frac{\left(\overline{\Delta u_y}\right)_{90_{4n+2r}}}{(4n+2r)t_0}; \quad k_{90} = \frac{\left(\overline{\Delta u_y}\right)_{90_{2r}}}{2rt_0} \tag{16.10}$$

Similar relations can be derived for the case where θ-plies are centrally located in the laminate, that is, a $[0_m/90_r/\pm\theta_n]_s$ laminate. For full expressions of the relevant constraint parameters, the reader is referred to Singh and Talreja (2009, 2013).

We have recently utilized this model to solve selected test cases in the World Wide Failure Exercise III, which provided an opportunity to examine initiation and evolution of subcritical damage and its effect on the overall mechanical response (Kaddour et al., 2013). The full report is presented in Singh and Talreja (2013), and only a limited set of results using the linear SDM model will be discussed here. Fig. 16.8 displays the stress−strain response of $[0/90_8/0]_T$ glass/epoxy laminate loaded in uniaxial tension along longitudinal direction. The response is highly nonlinear subsequent to damage initiation, depicting huge stiffness changes brought by ply cracking. Ply cracks are predicted to initiate approximately at 0.4% axial strain, and on increase in loading cracks are found to multiply quickly to a large crack density with a magnitude of 0.6 when normalized to cracked ply thickness (i.e., the crack size). The energy-based model to predict initiation and evolution of ply cracks in multidirectional laminates was initially developed in Singh and Talreja (2010) and has been covered in Chapter 8 in this book. From Fig. 16.8, it can also be observed that the thermal residual stresses incurred due to temperature mismatch between operating and manufacturing conditions change the initiation strain but not the overall trends in crack density evolution. Fig. 16.9 reports the SDM predictions for stress−strain response in both longitudinal and transverse directions for quasiisotropic $[0/90/-45/+45]_s$ glass/epoxy laminate

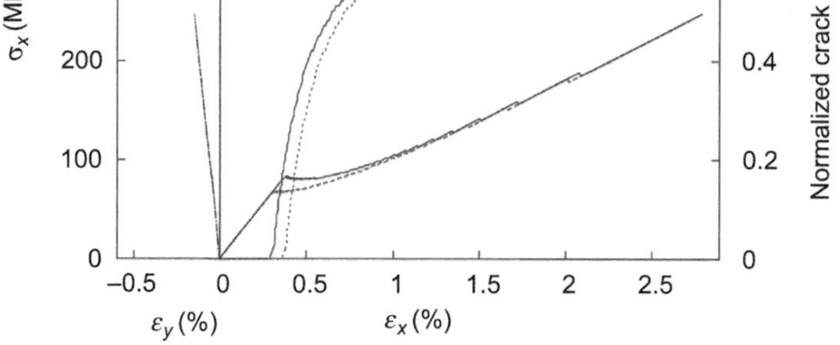

Figure 16.8 SDM model predictions for Test Case 4 of WWFE III depicting stress−strain response and evolution of crack density for $[0/90_8/0]_T$ glass/epoxy1 laminate. The dashed lines show the effect of strains generated due to thermal mismatch between 0 and 90 degrees plies during curing on damage initiation and progression.
Reprinted with kind permission from Singh and Talreja (2013), Copyright © 2013 by Sage publications.

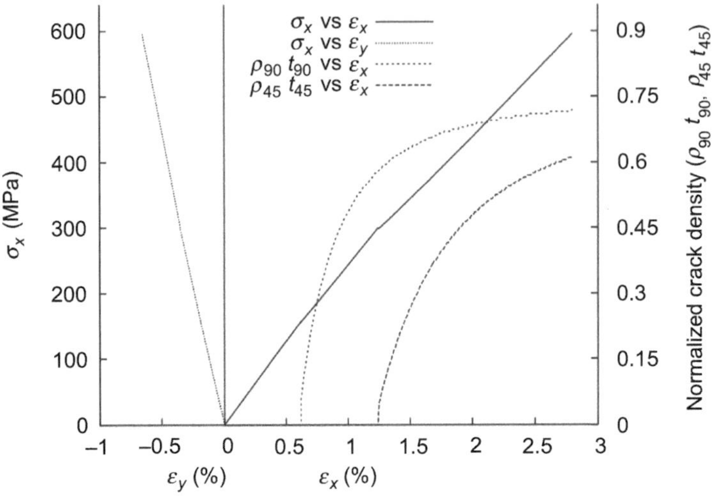

Figure 16.9 SDM model predictions for Test Case 6 of WWFE III depicting stress−strain response and evolution of ply crack density for $[0/90/-45/+45]_s$ glass/epoxy1 laminate with cracks in all off-axis modes. The crack initiation and the crack density evolution in both 45° plies are assumed to be identical.

with cracks in all off-axis modes. Crack density evolution plots suggest 90 degrees cracking initiates first, as expected. From the figure, it can be inferred that the 45 degrees cracking initiates at much higher strain levels and does not grow as quickly as the 90 degrees cracking.

16.6 Recent developments

Over the past few years, improvements have been made to the multiscale SDM approach described above. Some of the key developments are described as follows.

16.6.1 Nonlinear SDM model

Experimental observations of damage development in a variety of composite laminates have shown that the stiffness changes due to ply cracking are nonlinear in nature. In the beginning stages of damage initiation and progression, a quick and almost linear degradation in stiffness properties is observed. However, at high crack densities, the rate of change in the stiffness properties reduces, leading to almost negligible degradation rate at large crack density levels (>1.0 crack per unit thickness of the cracked ply). The SDM approach described above restricts the polynomial expansion of the free energy function to linear terms for simplification, and thus its stiffness predictions show a linear variation with crack density at all crack density levels. Hence, linear

SDM model predictions tend to show too large degradation in stiffness properties than observed experimentally. To overcome this limitation, a higher order SDM model has been developed for all damage scenarios described in preceding section. For details of the model development, the reader is referred to Singh (2013). Figs. 16.10 and 16.11 compare the two models for normalized longitudinal Young's modulus and Poisson's ratio of two multidirectional laminates undergoing multimode cracking. Fig. 16.10 considers the case with off-axis ply cracking in two damage modes, viz., $[0/\pm70_4/0_{1/2}]_s$ laminate, and Fig. 16.11 depicts model predictions for the case of three damage modes in a quasiisotropic laminate. From these figures, the nonlinear SDM model shows good agreement with test data.

16.6.2 Multiaxial SDM model

In practical applications, loading on a composite structure can be multiaxial. In such cases, multiple damage modes can become active simultaneously, complicating the

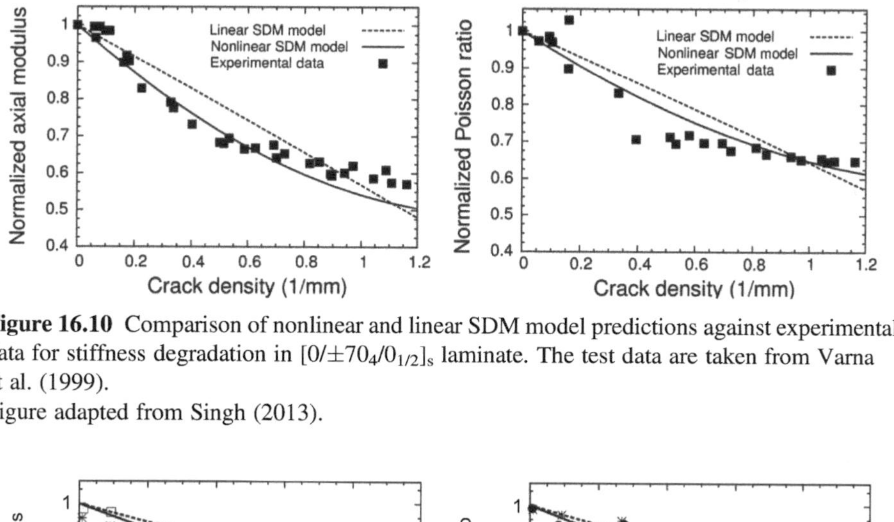

Figure 16.10 Comparison of nonlinear and linear SDM model predictions against experimental data for stiffness degradation in $[0/\pm70_4/0_{1/2}]_s$ laminate. The test data are taken from Varna et al. (1999).
Figure adapted from Singh (2013).

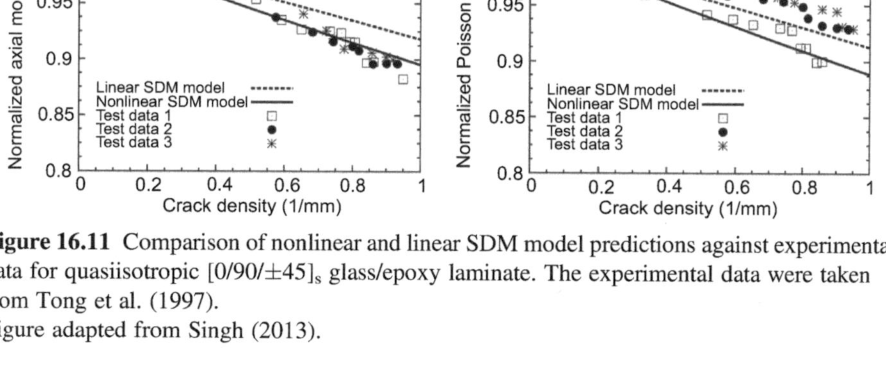

Figure 16.11 Comparison of nonlinear and linear SDM model predictions against experimental data for quasiisotropic $[0/90/\pm45]_s$ glass/epoxy laminate. The experimental data were taken from Tong et al. (1997).
Figure adapted from Singh (2013).

boundary value problem involved. Montesano and Singh (2015) have recently advanced SDM methodology to elastic response of damaged composite laminates subjected to multiaxial loading. In the cited paper, the newly developed multiaxial SDM model is shown to accurately capture the nonlinear stiffness degradation exhibited by cross-ply, quasiisotropic, and angle-ply laminates containing matrix cracks in multiple plies. The predicted stiffness results were shown to agree well with the experimental data as well as with independent finite element computations. Additionally, the new SDM model accurately captures shear deformation response, which was not possible before.

16.6.3 Multiscale structural analysis

Recently, we have implemented the multiscale SDM approach in commercial FE softwares such as ANSYS and ABAQUS for the purpose of full-scale stress and failure analysis of composite structures in practical engineering applications. For instance, Montesano et al. (2016) conducted multiscale progressive durability analysis for wind turbine rotor blade structures. They considered both quasistatic and fatigue loading cases and studied the role of blade geometry on damage evolution through the SDM approach. This analysis was later extended to account for debonding of the joint adhesive layer and predict service life of composite blade structures (Zuo et al., 2018). The multiscale SDM approach has also been utilized to investigate the damage behavior of a helicopter rotor blade made of CFRPs (Pao et al., 2020).

16.7 Concluding remarks

A multiscale constitutive framework for mechanical response of multidirectional composite laminates with intralaminar cracks in multiple plies now exists and has been described here. The framework is rooted in thermodynamics with internal variables. In contrast to traditional CDM treatments, where the internal variables are “hidden,” the current approach enters into the internal variables’ essential characteristics of microlevel entities via the so-called constraint parameters. These parameters are integral part of descriptors of damage entities, which are averages over appropriate RVEs. Thus, the approach has multiple scales, connecting the continuum (macro) level response characteristics to the microlevel damage entities via mesolevel descriptors. Also, since the constraint parameters can be determined by microlevel analyses, the constitutive framework enjoys synergism between micromechanics and CDM and is therefore called SDM.

The SDM approach has been applied to several specific multidirectional laminates, with up to three distinct microcrack orientations, supported by experimental data. Parametric studies have also been conducted to generate insight into the effects on mechanical response of various combinations of crack size and orientation. Recent extensions of the SDM approach to treat nonlinearities in stiffness–damage relationships and multiaxial loading have been cited. Finally, there are several case study examples of full multiscale structural analysis for practical engineering applications using the SDM multiscale modeling approach.

References

Berthelot, J.M., 2003. Transverse cracking and delamination in cross-ply glass-fiber and carbon-fiber reinforced plastic laminates: static and fatigue loading. Appl. Mech. Rev. 56 (1), 111–147.

Bulsara, V.N., Talreja, R., Qu, J., 1999. Damage initiation under transverse loading of unidirectional composites with arbitrarily distributed fibers. Compos. Sci. Technol. 59 (5), 673–682.

Hashin, Z., 1985. Analysis of cracked laminates: a variational approach. Mech. Mater. 4 (2), 121–136.

Huang, Y., Talreja, R., 2014. Statistical analysis of oblique crack evolution in composite laminates. Compos. B Eng. 65, 34–39.

Huang, Y., Varna, J., Talreja, R., 2014. Statistical methodology for assessing manufacturing quality related to transverse cracking in cross ply laminates. Compos. Sci. Technol. 95, 100–106.

Jamison, R.D., Schulte, K., Reifsnider, K.L., Stinchcomb, W.W., 1984. Characterization and analysis of damage mechanisms in tension-tension fatigue of graphite/epoxy laminates. In: Effects of Defects in Composite Materials. ASTM STP 836, American Society of Testing and Materials, Philadelphia, PA, pp. 21–55.

Kachanov, L.M., 1958. On the creep fracture time. Izv. ANSSR, Otd. Tekhn. Nauk 8, 26–31.

Kaddour, A.S., Hinton, M.J., Smith, P.A., Li, S., 2013. The background to the third world-wide failure exercise. J. Compos. Mater 47 (20-21), 2417–2426.

Montesano, J., Singh, C.V., 2015. A synergistic damage mechanics based multiscale model for composite laminates subjected to multiaxial strains. Mech. Mater. 83, 72–89.

Montesano, J., Chu, H., Singh, C.V., 2016. Development of a physics-based multi-scale progressive damage model for assessing the durability of wind turbine blades. Compos. Struct. 141, 50–62.

Mura, T., 1987. Micromechanics of Defects in Solids, vol. 3. Springer.

Nemat-Nasser, S., Hori, M., 1999. Micromechanics: Overall Properties of Heterogeneous Materials, vol. 2. Elsevier, Amsterdam.

Ostoja–Starzewski, M., 1999. Microstructural disorder, mesoscale finite elements and macroscopic response. Proc. Royal Soc. London Ser. A Math. Phys. Eng. Sci. 455, 3189–3199.

Pao, W.Y., Haldar, S., Singh, C.V., 2020. Performance analysis of composite helicopter blade using synergistic damage mechanics approach. AIAA J. 58 (2), 968–976.

Singh, C.V., Talreja, R., 2008. Analysis of multiple off-axis ply cracks in composite laminates. Int. J. Solid Struct. 45 (16), 4574–4589.

Singh, C.V., Talreja, R., 2009. A synergistic damage mechanics approach for composite laminates with matrix cracks in multiple orientations. Mech. Mater. 41 (8), 954–968.

Singh, C.V., Talreja, R., 2010. Evolution of ply cracks in multidirectional composite laminates. Int. J. Solids Struct. 47 (10), 1338–1349.

Singh, C.V., Talreja, R., 2013. A synergistic damage mechanics approach to mechanical response of composite laminates with ply cracks. J. Compos. Mater. 47 (20–21), 2475–2501.

Singh, C.V., 2013. A higher order synergistic damage model for prediction of stiffness changes due to ply cracking in composite laminates. CMC Comput. Mater. Continua 34 (3), 227–249.

Talreja, R., 2006. Multi-scale modeling in damage mechanics of composite materials. J. Mater. Sci. 41 (20), 6800–6812.

Talreja, R., 2010. On multi-scale approaches to composites and heterogeneous solids with damage. Phil. Mag. 90 (31–32), 4333–4348.
Talreja, R., Singh, C.V., 2012. Damage and Failure of Composite Materials. Cambridge University Press, London, ISBN 9780521819428.
Talreja, R., 2014. Multiscale modeling of damage in composite materials. In: Multiscale Modeling of Complex Materials. Springer Vienna, pp. 179–209.
Tong, J., Guild, F.J., Ogin, S.L., Smith, P.A., 1997. On matrix crack growth in quasi-isotropic laminates - I. Experimental investigation. Compos. Sci. Technol. 57 (11), 1527–1535.
Varna, J., Akshantala, N.V., Talreja, R., 1999a. Crack opening displacement and the associated response of laminates with varying constraints. Int. J. Damage Mech. 8 (2), 174–193.
Varna, J., Joffe, R., Talreja, R., 2001. Mixed micromechanics and continuum damage mechanics approach to transverse cracking in [S, 90n] s laminates. Mech. Compos. Mater. 37 (2), 115–126.
Varna, J., Joffe, R., Akshantala, N.V., Talreja, R., 1999b. Damage in composite laminates with off-axis plies. Compos. Sci. Technol. 59 (14), 2139–2147.
Zohdi, T.I., Wriggers, P. (Eds.), 2008. An Introduction to Computational Micromechanics, vol 20. Springer.
Zuo, Y., Montesano, J., Singh, C.V., 2018. Assessing progressive failure in long wind turbine blades under quasi-static and cyclic loads. Renew. Energy 119, 754–766.

Further reading

McCartney, L.N., 1998. Predicting transverse crack formation in cross-ply laminates. Compos. Sci. Technol. 58 (7), 1069–1081.
Nairn, J.A., Hu, S., 1994. Matrix microcracking. Compos. Mater. 187.
Varna, J., Berglund, L.A., 1994. Thermo-elastic properties of composite laminates with transverse cracks. J. Compos. Technol. Res. 16 (1), 77–87.

Part Four

Failure analysis of composite structures and joints

Computational modeling of damage in notched and unnotched composite laminates using a semi-discrete method

17

Anthony M. Waas and Minh Hoang Nguyen
Department of Aerospace Engineering, University of Michigan, Ann Arbor, MI, United States

17.1 Introduction

The scale and complexity of failure in structures made of composite materials is immensely important. Phenomena at different length scales interact with each other confounding the ability to cleanly separate the problem into subproblems whose solution when reconstructed and assembled would yield the whole response. This is not unlike in many interesting problems that arise in physics, which also displays a breakdown of the reductionist and constructionist hypotheses (Anderson, 1972). The hierarchical architecture displaying the many operational length scales during progressive failure is shown in Fig. 17.1. Damage and failure within the plies are referred to as intralaminar and in between the plies are referred to as interlaminar or delamination. When fibers rupture (Fuwa et al., 1975; Aroush et al., 2006; Phoenix, 1975; Curtin et al., 1998; Mortell et al., 2016), tensile failure occurs due to fibers breaking, whereas kink bands form under compressive loading due to a local instability (Jelf and Fleck, 1992; Basu et al., 2006; Khamseh and Waas, 1992; Ahn and Waas, 2002; Kyriakides et al., 1995; Waas and Schultheisz, 1996; Lee and Waas, 1999; Davidson et al., 2013). Unlike tensile failure, the driver for kink bands is the matrix nonlinear shear response and fiber misalignment. Matrix-dominated damage and failure comprise cracking due to transverse tension and/or shear, and in case of compression, frictional sliding of the crack faces is also important (Parvizi et al., 1978; McCartney, 1992; Manders et al., 1983; Takeda and Ogihara, 1994; Gamstedt and Sjogren, 1999; Talreja and Singh, 2012). Matrix failure in most cases is mixed, including shear loading (Liu and Piggott, 1995; Plumtree, 2002; Ng et al., 2010) and/or combined loads. Delamination is an important failure mechanism in laminated composites (Waas et al., 1990; Comiez et al., 1995; Shahwan and Waas, 1997; Reeder, 1992; Davidson et al., 2000; Camanho et al., 2003; Krueger, 2012; Wisnom, 2012) and may occur under different pure modes or mixed mode. The driver for delamination is the stress vectors on planes parallel to the laminae.

Modeling Damage, Fatigue and Failure of Composite Materials. https://doi.org/10.1016/B978-0-443-18489-5.00009-6

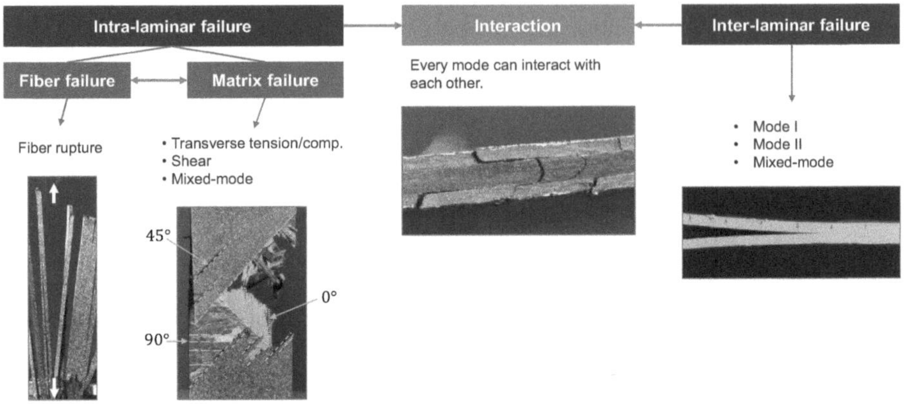

Figure 17.1 Possible failure modes in a composite laminate under tensile loading (Nguyen and Waas, 2022a).

While it is tempting to isolate and study failure modes in a lamina, it is the interaction between these failure modes that dictate the events in progressive failure. Therefore, the constructionist hypothesis fails in the problem of progressive failure of laminated composites (Anderson, 1972). It has been shown that intralaminar modes interact, e.g., fiber rupture and splitting or kinking and splitting (Prabhakar and Waas, 2013). Furthermore, matrix cracks may induce other cracks in adjacent plies. More importantly, intra and interlaminar cracks influence each other, e.g., a matrix crack can trigger delamination. Because of this, the problem of progressive failure to a high degree of fidelity is best resolved using a computational approach that subscribes to accurate models of the physics and mechanics of the operational mechanisms.

The issue of scale and complexity is displayed in the failure of a fiber reinforced solid. The material nonuniformity is due to the irregular fiber packing and orientation, the strength variability of the constituents, defects such as voids or ply waviness (e.g., automated fiber placement—manufacturing signature), and the nonuniform curing behavior within the composite due to temperature gradients at manufacturing. While it is nearly impossible to model all unintended deviations (imperfections) explicitly, one can attempt to model the accumulated effect of all imperfections, depending on the scale at which the constitutive model is formulated. For instance, a ply-level meso-scale model (homogenized approach) may lump the microscopic imperfections described via a few material parameters such as stiffness or strength variation.

It is known that material nonuniformity is reflected in the progression of the crack density in 90 degrees or off-axis plies, which has been a challenge to include in FE (finite element)-based progressive damage and failure models. In Kaddour et al. (2007), where results from the World Wide Failure Exercise are reported, the authors state that only a few models were able to capture the crack density progression, delamination, fiber splitting, and probabilistic aspects, which are among the most important and observable features in the progressive failure of a composite laminate.

This chapter contains two core sections: Section 17.2 describes the details of the computational framework and Section 17.3 presents predictions for different examples. The essence of semidiscrete meshing strategy is described in Section 17.2.2. The probabilistic modeling approach and the constitutive models are shown in Sections 17.2.3 and 17.2.4, respectively. Finally, prediction examples with unnotched and open-hole tensile laminates are presented in Section 17.3.

17.2 Computational modeling using the SD2M model

Continuum damage mechanics (Talreja, 1985) is a useful framework for describing the effects of microcracking damage and popular methods for efficiently predicting damage and failure of composites structures are described in Higuchi et al. (2017), Pineda and Waas (2013), Camanho et al. (2007), Abisset et al. (2011), Chen et al. (2013), Falco et al. (2018), Joseph et al. (2018a,b), Hyder et al. (2018), Cui and Lua (2020), Pham et al. (2019). However, continuum damage mechanics−based methods to capture sharp macroscopic matrix cracks are limited. Discrete crack methods usually provide a better way to represent strong discontinuities, i.e., sharp cracks (Higuchi et al., 2017; Hallett et al., 2009a,b; Jiang et al., 2007; Van der Meer et al., 2012; Swindeman et al., 2013; Chen et al., 2016; Achard et al., 2014; Serra et al., 2016; Joosten et al., 2018; Lu et al., 2019; Rudraraju et al., 2012). However, depending on the method, the numerical and modeling complexity may be quite challenging which can lead to prohibitively high computational costs (Oliver et al., 2006). Therefore, a gap exists where methods that are simultaneously of sufficient fidelity and yet are computationally efficient can be developed. A semidiscrete method that meets these needs is described in Nguyen and Waas, 2020a,b. An enhanced version of this method is utilized here for studying tensile failure of composite laminated structures.

17.2.1 Overview

The semidiscrete method is a compromise between continuum damage models and discrete models to achieve a good balance between fidelity and computational efficiency. It has been extended to the interlayer (enhanced semidiscrete damage modeling—eSD2M) (Nguyen and Waas, 2021, 2022b,c,d) model. The eSD2M framework has been utilized to model various problems, including open-hole tension (OHT) and scaling effects (Nguyen and Waas, 2021). As illustrated in Fig. 17.2, the proposed modeling framework consists of three main components that work seamlessly together. The first is a smart meshing strategy that creates a semidiscrete mesh. This mesh consists of strips of elements between bulk elements to represent discrete matrix (splitting) cracks that are parallel to the fibers (Nguyen and Waas, 2020a,b). This decomposition of the failure modes as a part of the meshing is novel. The bulk elements are not allowed to fail in matrix modes. The bulk elements fail in fiber-dominated modes. The splitting elements are continuum elements just like the bulk elements and the crack band (CB) method, as opposed to discrete numerical methods

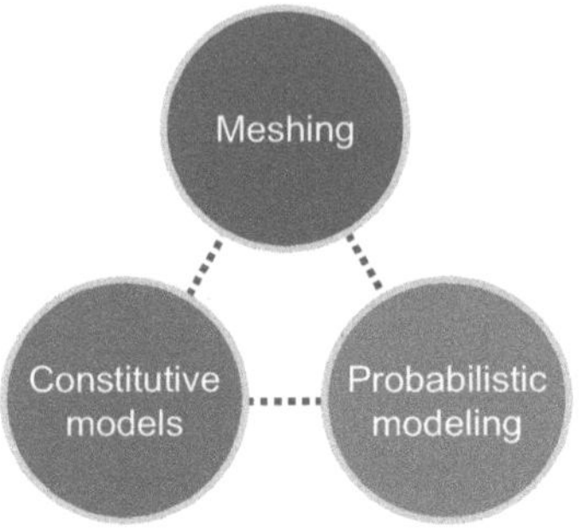

Figure 17.2 The semidiscrete damage modeling (SD2M) framework.

(like extended finite element method [xFEM] or variational multiscale cohesive method [VMCM]) is used to model the splitting mode. To improve the connectivity between layers, the semidiscrete meshing method was extended to the interlayers to create compatible meshes/nodes (Nguyen and Waas, 2022b). This is important to model intra and interlaminar failure mode *interactions*. The meshing technique is automated via Python scripting within the Abaqus CAE. Second, the semidiscrete mesh is enriched with randomized material properties. Each strip is assigned with different (randomly modeled) transverse and shear strength values (Nguyen and Waas, 2020a,b). Previous work has shown that this is needed to model the crack density progression accurately. A global—local randomization scheme is used to define global spatially varying properties such as fiber strength or interlaminar properties (Nguyen and Waas, 2022d). Finally, novel constitutive models are used to describe the material behavior. Schapery theory (ST) is used to model the prepeak nonlinear behavior due to microcracking, Hashin/Christensen failure criteria are used for the initiation of macroscopic failure, and a novel mixed-mode CB model is used to capture the postpeak softening (Nguyen and Waas, 2022d; Schapery, 1990; Hashin, 1980; Christensen, 2013). On-going work focuses on including residual stresses due to processing/curing in order to better predict the crack propagation.

17.2.2 Meshing concept

The meshing concept, which decomposes the bulk nonlinearity and localization zones and provides for a proper load transfer pathway, is adopted. The model consists of bulk material regions and thin strips of elements, placed along the fiber direction, as shown in Fig. 17.3. The failure modes are separated out into distinct domains. Matrix splitting failure is only active in the thin strips, while fiber failure is active in every element. By reducing the width of the matrix-splitting strips, the discreteness of the method is highly increased (approaching discrete methods), while the efficiency is maintained by using only continuum elements (CB) for the postpeak failure. Moreover, the present method aims to relax the requirements on the element shapes. Since the bulk material between the matrix-split strips does not fail in matrix modes, these elements can have arbitrary shapes, i.e., free mesh can be used for most of the elements. This helps to facilitate the meshing process significantly. Central to the success of the proposed

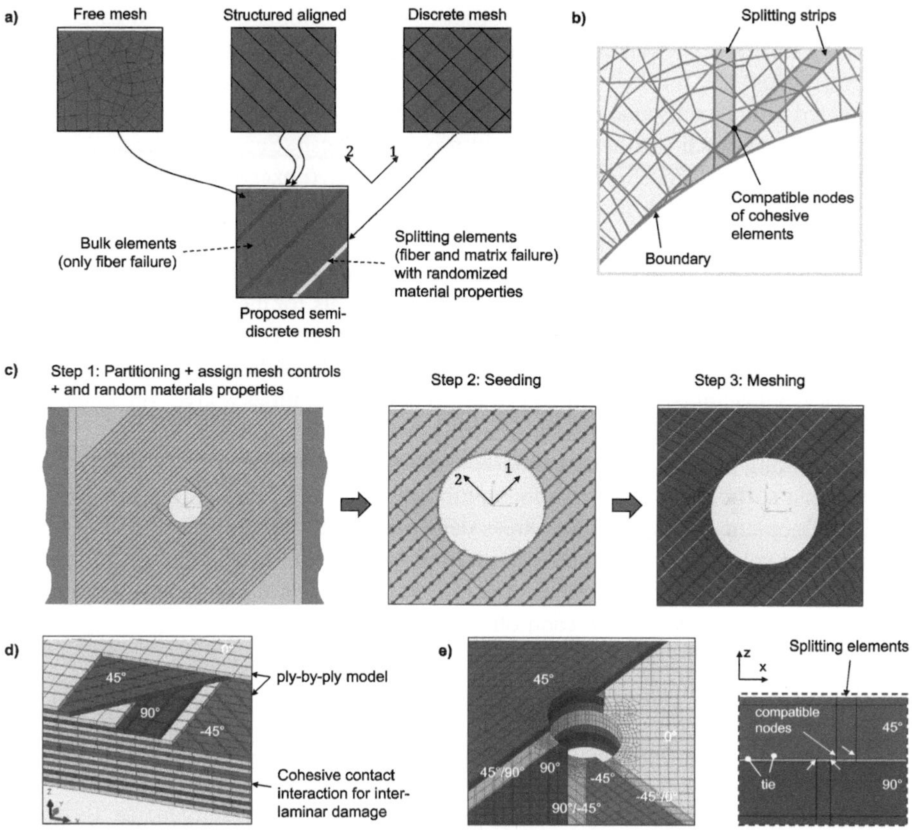

Figure 17.3 Meshing strategy: (a) Derivation of the semidiscrete mesh. (b) Meshing strategy for an interlayer for compatible mesh (eSD2M). (c) Meshing procedure for a ply. (d) Ply-by-ply model of a laminate. (e) Layer- by-layer model of a laminate using cohesive-element interlayers with compatible meshes (Nguyen and Waas, 2020a, 2022b).

method is the CB model (Bazant and Oh, 1983). More importantly, the element size of the bulk elements is allowed to exceed the Bazant limit (Pineda and Waas, 2013) required in the transverse direction; hence, only the Bazant limit for the fiber direction is required. As shown in Fig. 17.3, the proposed method takes advantage of the benefits of the free mesh, fiber-aligned mesh, and provides the discreteness associated with splitting cracks.

Clearly, the number of splitting elements used exceeds the number of physical matrix cracks.

Consider an open-hole specimen as an example, as shown in Fig. 17.3. The specimen is partitioned into elastic and fiber-aligned regions, between which a small mesh transition region is created. The matrix-split strips are created by partitioning along the fiber direction. The cells that are created by this process are then assigned with randomized material properties. Subsequently, a cut-out of any arbitrary shape can be added to the geometry, while maintaining the semidiscrete mesh. This is a significant benefit of this meshing method. No complicated algorithms are needed to generate the mesh. The bulk regions are assigned with a free mesh control, while the matrix-cracking strips are meshed with structured HEX elements to ensure a strip is represented with exactly one element along the transverse direction. A few additional partitions at the notch can be added to improve the mesh. Secondly, the specimen is seeded with a global mesh of the size that is allowed to exceed the Bazant limit for transverse cracks. Furthermore, the mesh can be locally refined in a flexible manner. For example, regions with expected high stress gradients can be seeded with a smaller size. The final mesh is shown in step 3. The refined mesh region near the hole can be seen. The colors represent different material assignments for the matrix-split elements. The bulk elements have an identical material assignment. However, fiber strength is also randomized during the simulation by the VUMAT, as described later.

The mesh spacing s in this model is set to about four times a layer thickness. It is then adjusted to the characteristic size of the cut-out, i.e., the hole diameter *d*—such that the splitting elements are tangential to the hole. It is important that a finite element model has a sufficiently fine mesh to capture possible stress concentrations. An inadequate refinement could give false stress data which are then used for the fiber failure criteria. The splitting at the hole—if captured by a model at all—would relieve the stress concentrations. However, the stress gradients are not completely eliminated, since the plies do interact with each other.

17.2.3 Probabilistic modeling

Two randomization schemes are used to describe the material nonuniformity. Transverse strength randomization is used for the splitting strips, and an elementwise global–local randomization scheme is applied to the entire domain, as shown in Fig. 17.4. While the transverse and shear strength values are closely connected to the mesh and the defined material sections, the fiber strength and interlayer properties are randomly assigned at every integration point in every element (within the VUMAT in the commercial software ABAQUS at its material points).

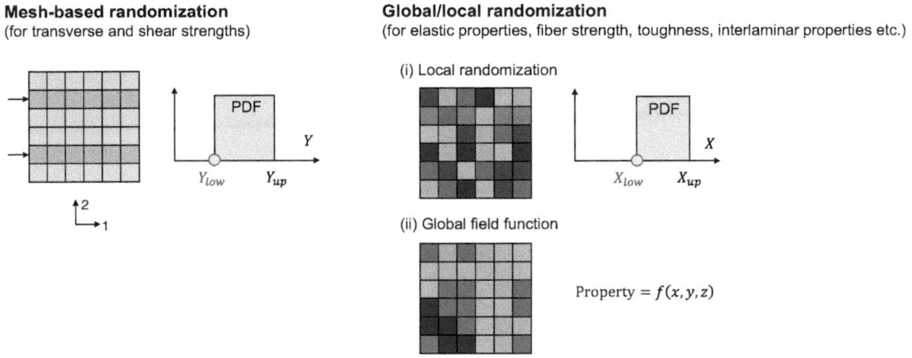

Figure 17.4 Proposed randomization methods for various properties (Nguyen and Waas, 2020a, 2022a).

Previous studies have shown that a uniform probability distribution for the transverse strength with lower and upper bounds can be calibrated to match the experimental data (Nguyen and Waas, 2020b). This was done using an unnotched cross-ply tensile laminate. The lower bound can be directly obtained from the first initiation strain and the upper bound can be calibrated to match the experiments. This distribution can then be used in the model to predict the behavior of other laminate configurations. With the inclusion of residual thermal stresses, the lower bound of the distribution is simply shifted. For future studies, the authors are interested in studying other types of distribution functions to improve the predictions even further.

For other properties (elastic properties, fiber strength, toughness, and interlaminar properties), there are two techniques for randomization: (i) Local randomization that samples a value from a probability distribution locally at every integration point. (ii) A mesh- independent global field function that is evaluated at every integration point. In the studies of this chapter, the local randomization scheme was used. In a recent paper (Nguyen and Waas, 2022a), the authors have proposed a modified Fourier series as a global field function to represent material nonuniformities of finite sizes. Future work will focus on further exploring this method.

17.2.4 Constitutive modeling

The constitutive modeling is implemented with two user subroutines **VUMAT** for Abaqus Explicit, while conveniently using available built-in element types. To determine the characteristic lengths for the CB method, the user subroutine **VUCHARLENGTH** is used. The main user subroutines used in the eSD2M model are associated with (a) **VUCHARLENGTH**, which calculates the characteristic lengths of an arbitrary 3D (continuum) element, (b) **VUMAT_LAMINA**, which uses ST for prepeak nonlinearity, and the CB method for mixed-mode postpeak softening, and applied to C3D8R and C3D6(R) elements, and (c) **VUMAT_COHEL**: A material model for the 3D cohesive elements (mixed-mode traction separation law) associated with the COH3D8 and COH3D6 cohesive elements.

Past studies have demonstrated the usefulness of the extended Schapery theory (EST) for a wide variety of predictive modeling (Pineda and Waas, 2013; Joseph et al., 2018b; Nguyen and Waas, 2020b; Thorsson et al., 2018; Ji and Waas, 2013; Lin et al., 2020). EST is able to model the prepeak nonlinearity due to microcracking (Schapery, 1990). The nonlinear shear response of a ply needed to implement ST can be obtained from ±45 degrees tensile tests. Since eSD2M employs transverse and shear strength variations, the experimentally obtained curves need to be extended using fitting functions. This is justified by the fact that the failure of such tests is usually triggered by the weakest material points, thus, representing lower bounds. The fitting curve was described in (Nguyen and Waas, 2022c). To model the transverse response in EST, a micromechanical FE model is employed. The goal is to quantify the effect of microcrack damage (quantified by the reduced Schapery parameter S_r) on the transverse response of the homogenized composite. Hence, the matrix is modeled using a nonlinear equivalent stress–strain behavior that can be inferred from the ±45 degrees tensile test, following the procedure by Ng et al., 2010. From the resulting transverse response, the transverse response function for the ST can be obtained using the same approach as for the shear response. The transition from the pre peak to the postpeak is modeled using initiation criteria. There are a number of different models, e.g., Hashin (1980), Christensen (2013), Puck and Schurmann (2002), or LaRC criteria (Davila et al., 2005). Fig. 17.5 shows the in-plane failure envelopes of a few selected failure criteria compared with micromechanical studies by Sun et al. (2019). The quadratic stress criterion (identical with Hashin for $a_{22} > 0$) has been used in previous EST versions. Christensen's failure criterion shows the best agreement with the micromechanical data, for transverse tension as well as compression. It should be noted that in the studies presented in this chapter, Hashin criteria were used. Results using Christensen's criterion were demonstrated in recent studies (Nguyen and Waas, 2022a,c). For most cases, the type of failure criteria used may not show significant differences since a strength randomness is implemented. Hence, effects due to material property uncertainties may diminish the differences associated with the type of failure criteria. On-going work examines how different failure criteria affect asymmetric crack propagation near cut-outs due to the presence of tensile, compressive and shear stresses.

Once either the fiber criterion (in either splitting or bulk elements) or matrix criterion (in only splitting elements) is met, the postpeak failure response in EST is modeled using the CB method. The CB method (Bazant and Oh, 1983) uses a characteristic length to regularize the energy dissipation, i.e., making the energy dissipation independent of the element size by scaling the stress—strain softening curve by a characteristic length. Corresponding to the failure criterion (fiber or matrix mode) that was

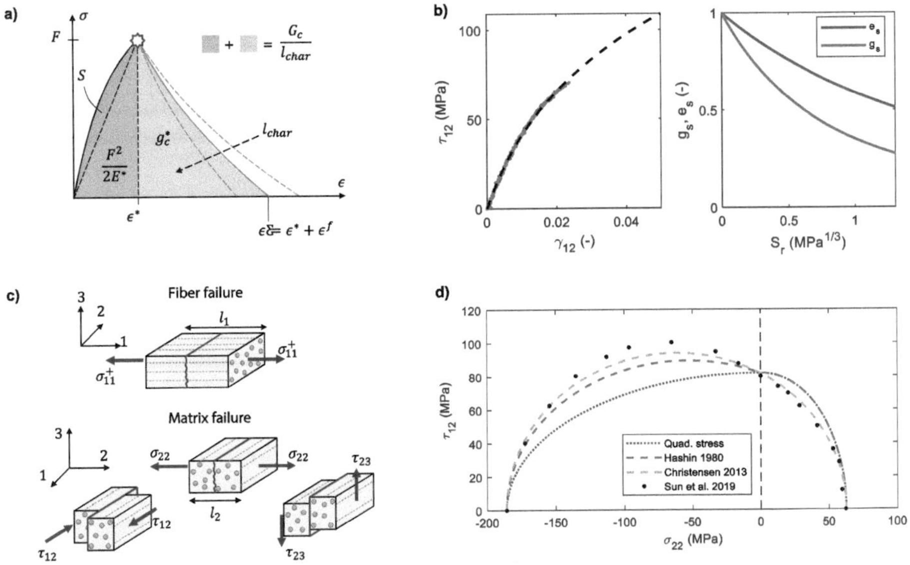

Figure 17.5 Constitutive modeling: (a) Generic nonlinear stress—strain curve captured using EST. (b) Nonlinear prepeak (shear) modeling using Schapery theory. (c) Failure planes considered in this chapter. (d) Failure criteria (Nguyen and Waas, 2020a, 2022a,c).

triggered, the CB routine uses the appropriate characteristic length that was determined and provided by the user subroutine VUCHARLENGTH. Using the material point coordinates and local material directions as well as nodal coordinates, the subroutine calculates the distance between the faces intersecting with corresponding material directional vector. Specifically, l_1 is used for the fiber mode and l_2 for the matrix mode. This estimated approach may be adequate and is more correct than using the default characteristic length provided by Abaqus, which may lead to wrong energy dissipation, also recently shown in Leone and Justusson (2021). Additionally, Xu and Waas (2016) showed that the input fracture toughness needs to be corrected when using logarithmic strain measures, as this strain measure is used by default in Abaqus Explicit. The authors demonstrated in Nguyen and Waas (2021) that mesh objectivity can be achieved using the G_c correction proposed in Xu and Waas (2016).

A novel mixed-mode law is described in Nguyen and Waas (2020a, 2022b). The mixed-mode law guarantees all traction components to vanish simultaneously when the material has failed completely. This was not the case in prior EST versions. Realizing this, Joseph et al. (2018b) proposed an incremental mixed-mode law that guarantees a simultaneous vanishing of all tractions at failure, with the assumption that the mode-mixity remains constant at the values at initiation. The present mixed-mode law implemented in the eSD2M framework adopts a nonincremental approach. One advantage of a total formulation is that unloading scenarios are easier to handle, especially when the mode-mixity can change. Furthermore, situations that call for one mode to unload but another keeps loading can also be handled conveniently. It is noted that the incremental approach (Joseph et al., 2018b) can also be formulated to do this.

The mixed-mode formulation uses the definition of an effective separation proposed by Nguyen and Waas (2016). It was then modified and generalized to use for the CB model as well, by using an effective strain measure (Fig. 17.6). Similar definitions of effective quantities have been proposed by other authors in the past (Camanho et al., 2003; Pham et al., 2019; Lu et al., 2019). The proposed mixed-mode law takes nondimensionalized softening laws for each pure mode and converts them into a mixed-mode behavior based on an effective strain. The advantage of this model is that any arbitrary softening shape can be used for each individual mode, and a simultaneous vanishing of all traction components can be guaranteed. The effective strain is defined by the three strain components associated with the failure plane. For detailed descriptions, the reader is referred to Nguyen and Waas (2020a, 2022b).

17.3 Predictions

To demonstrate the predictive capability of the eSD2M modeling framework, we consider four cases:

(i) Unnotched laminate with a "hard" (50% 0/40% ± 45/10% 90) layup: The stacking sequence of this 20-ply laminate is $[0/45/0/90/0/—45/0/45/0/—45]_s$ and the test was conducted at NIAR-NCAMP (Clarkson, 2012).

(ii) Open-hole laminate with the identical layup, also tested at NIAR-NCAMP (Clarkson, 2012).

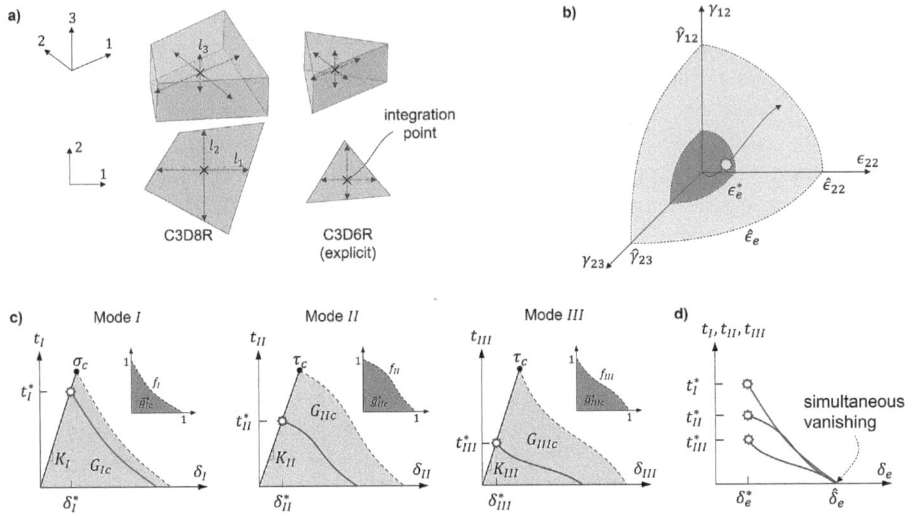

Figure 17.6 Constitutive modeling (continued): (a) Characteristic lengths for crack band. (b) Illustration of the effective strain. (c) Pure-mode traction-separation laws and example responses under mixed-mode. (d) Traction response governed by the effective strain (Nguyen and Waas, 2020a, 2022b).

Hole size effect: A numerical study on the effect of increasing hole diameter while the specimen width is kept constant. The layup is identical with the previous.

(iv) Delamination-prone laminate with an open hole: A laminate with scaled plies and the stacking sequence [454/904/—454/04]s. The specimen was tested by Green et al. (2007).

The material system for all specimens was IM7/8552, a carbon fiber—reinforced thermoset composite which has been widely studied in the literature (Krueger, 2012; Camanho et al., 2007; Chen et al., 2013; Hallett et al., 2009a; Lu et al., 2019; Clarkson, 2012; O'Brien and Krueger, 2001; O'Brien and Krueger, 2001; Boyd et al., 2014; Waas et al., 2018; Laffan et al., 2011). Since different studies were performed by different institutions under possibly varying processing conditions, there is an observable variation in the measured material properties. In addition, there is a difference in the cured ply thickness, which is 0.1829 mm for samples tested at NIAR-NCAMP (Clarkson, 2012) and 0.125 mm for specimens tested by Green et al. (2007). To be consistent, values of material properties will be used accordingly to the test series, often provided by the corresponding researchers. Thus, the material for cases (i), (ii), and (iii) will be denoted as batch A and the material for (iv) is denoted as batch B (Table 17.1). The material properties for batch A were taken from NIAR-NCAMP (Clarkson, 2012) when available. Additional properties were measured at the University of Washington by Waas et al. (2018). For batch B, the material values were obtained from studies by Hallett and coauthors. Additional properties were obtained by other numerical studies that attempted to predict the above-mentioned laminate.

The in-plane shear properties were measured by the authors of this chapter, since this test data were required to compute the Schapery parameters. Additionally, as

Table 17.1 Ply and interlayer properties of the IM7/8552 material system.

Property	Batch A	Batch B
Cured ply thickness $t_{p}i_{y}$ (mm)	0.1829 (Clarkson, 2012)	0.125 (Green et al., 2007)
Ply elastic properties		
$E11$ (GPa)	158 (Clarkson, 2012)	161 (O'Brien and Krueger, 2001)
$E22$ (GPa)	9.02 (Clarkson, 2012)	11.4 (O'Brien and Krueger, 2001)
$G12 = G13$ (GPa)	5.70[a]	5.70[a]
$V12 = V13$	0.316 (Clarkson, 2012)	0.32 (O'Brien and Krueger, 2001)
$V23$	0.436 (O'Brien and Krueger, 2001)	0.436 (O'Brien and Krueger, 2001)
Intralaminar strength values		
X_T,iow (MPa)	2500 (Clarkson, 2012)	2806 (Chen et al., 2013)
Yr,iow (MPa)	43.9[a]	43.9[a]
$S12,iow$ (MPa)	66.5[a]	66.5[a]
$S23,iow$ (MPa)	107[b] (Davila et al., 2005)	107[b] (Davila et al., 2005)
Strength distribution parameters		
$(X_{T,up}—X_{T,iow})/X_{T,iow}$	0.05[b]	0.05[b]
$(Y_{T,up}—Y_{T,iow})/Y_{T,iow}$	1.2[a]	1.2[a]
$(S_{12,up}—S_{12,iow})/S_{12,iow}$	0.43[b]	0.43[b]
Interlayer strengths		
σ_c (MPa)	55 (Waas et al., 2018)	40 (Lu et al., 2019)
τ_c (MPa)	72 (Waas et al., 2018)	60 (Lu et al., 2019)
Fracture energies		
$G^T{}_{fc}$ (kJ/m^2)	40 (Boyd et al., 2014)	40 (Boyd et al., 2014)
Gjc (kJ/m^2)	0.384 (Waas et al., 2018)	0.293 (Lu et al., 2019)
$G_{IIc} = G_{IIIc}$ (kJ/m^2)	1.335 (Waas et al., 2018)	0.631 (Lu et al., 2019)

[a]Measured.
[b]Assumed/estimated.

mentioned earlier, the transverse strength range with its lower and upper bounds was determined using cross-ply laminates. In this study, the intralaminar fracture toughness values were assumed to be identical with the interlaminar values.

17.3.1 Unnotched laminate

Results for progressive failure of an unnotched laminate are shown in Fig. 17.7. As expected for a "hard" laminate, the stress−strain response is stiff and the ultimate failure is abrupt. The experimental average value with ± one standard deviation (4.78%)

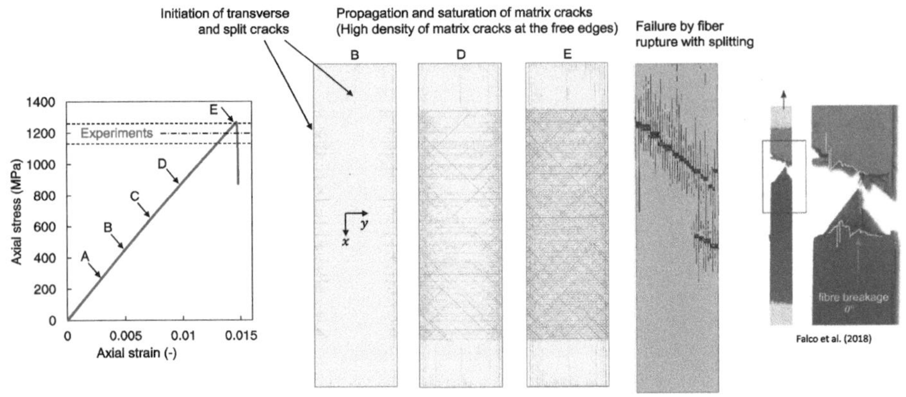

Figure 17.7 Predicted failure modes and progression for the unnotched laminate using SD2M (Nguyen and Waas, 2020b).
Image of a failed specimen from Falco et al. (2018).

is marked with the dashed lines. The predicted tensile strength is within one standard deviation, being 3.9% higher than the measured average value, which is a reasonable accuracy. More importantly, the progression of the failure modes (as observed in experiments) is modeled accurately, as shown in Fig. 17.7. The corresponding points on the stress–strain curve are displayed in this figure. The model predicts the initiation and progression of transverse and splitting cracks and the fiber failure localization that leads to final specimen failure. The first transverse cracks occur in the 90 degrees plies at the free edges. Splitting crack initiations were also observed in the 0 degree plies, which seem to arise from the ends of the specimen. With increasing loading (B–E), the damage accumulates at the free edges, including the angle plies. 0 degree splitting was also observed near the edges. Transverse cracks propagate toward the center (width) of the specimen. A high crack density was observed before the peak load, especially in the 90 degrees plies as expected. Final specimen failure is caused by fiber rupture at several locations: One near the elastic section and another around the center of the specimen, each with induced splitting. This shows the stochastic nature of the model.

17.3.2 Open-hole laminate

In this example, damage progression due to material nonuniformity, and damage localization due to stress concentrations near a geometric discontinuity, i.e., a notch or a hole, is studied. The layup is identical with the unnotched tension (UNT) layup (50/40/10), but the width is 1.5" (38.1 mm) and the hole diameter $d = 0.25$" (6.35 mm) (Clarkson, 2012).

The global (macroscopic) normalized stress–macroscopic strain response is plotted in Fig. 17.8. The predicted strength lies within a standard deviation of the test results (Clarkson, 2012) and differs from the mean value by 1.5%. The damage patterns are presented in Fig. 17.8, corresponding to the loaded states. Transverse

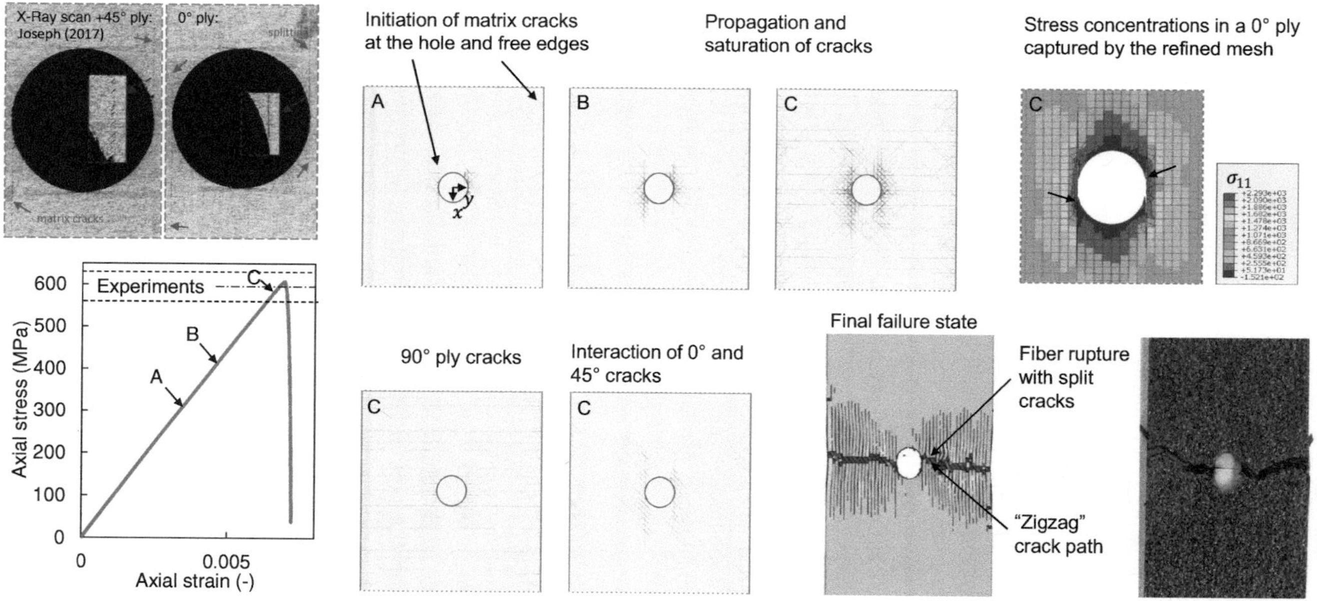

Figure 17.8 Predicted failure modes and progression for the open-hole laminate using SD2M (Nguyen and Waas, 2020b). X-ray scan of a tested specimen from Joseph (2017).
Image of a failed specimen from Falco et al. (2018).

cracks are seen to initiate at the hole and at the free edges. Clearly visible is the splitting in the 0° plies, extending from the open hole. With further loading (B—C), the crack density increases and existing cracks grow longer. Significant is the damage around the hole, especially along the 0° splits. Also noticeable are splitting cracks within the hole diameter and at the free edges. Fig. 17.8 reveals matrix cracks in a 45 degrees ply that were induced by the splitting failure in the neighboring 0 degree ply. The damage zone around the hole includes transverse cracking in the 90 degrees plies. Not insignificant is the crack initiation at the free edges, which propagates toward the center, but usually not further than $y = \frac{d}{2}$. This is due to the relaxed strain state behind and in front of the hole. Fig. 17.8 shows the axial stress distribution in a 0 degree ply. A clear discontinuity of the stress can be seen at the split cracks, tangential to the hole. The splitting has spread the zone of high stresses wider in the longitudinal direction, but there is still a stress concentration around the hole. However, it is not necessarily perfectly symmetric anymore, due to failure progression that breaks the symmetry. Here, the use of randomized material properties becomes important to obtain realistic predictions. Notice that fiber failure initiation might be slightly shifted. This is the case here, as shown in Fig. 17.8. The specimen fails due to fiber rupture, initiated at the hole. The failure propagates in a zigzag manner first following the 45 degrees plies underneath, rather than straight. It should be noted that the bent fibers are caused by dynamic effects during postfailure, as an artifact of the explicit method. A 'necking' near the center is visible, but not found in practice. However, since this is a snap shot, the image may change if the time at capture is altered.

17.3.3 Hole size effect

This study deals with the effect of hole diameter *d* while the specimen width is kept constant at 72 mm and was inspired by the study by Starnes and Williams (1983). The ultimate specimen strength values are compared with the baseline value from the unnotched sample ($d = 0$ mm). Three different hole sizes were studied, resulting in four data points on the plot in Fig. 17.9. Similarly to previous studies (Starnes and Williams, 1983), the open-hole strengths follow a curve underneath the "notch-insensitive" line. The structural strength is rapidly reduced from unnotched to the smallest hole size by over 50%.

The matrix cracks before significant fiber failure and the failure modes after the load drops are shown in Fig. 17.9. The unnotched specimen exhibited a lot more matrix cracks, especially at the free edges), compared to the open-hole specimens. Due to the stress concentrations near the hole, matrix cracks focused around the hole. Splitting in the 0 degree plies occurred tangential to the hole. The splitting is most apparent for the smaller hole size, which leads to a milder stress concentration compared to the other larger hole sizes. Therefore, the 12 mm data point does not lie on a straight line with the other two points. While the final fiber rupture occurred at random locations in a UNT specimen, it initiated near the hole for the OHT samples due to stress concentrations.

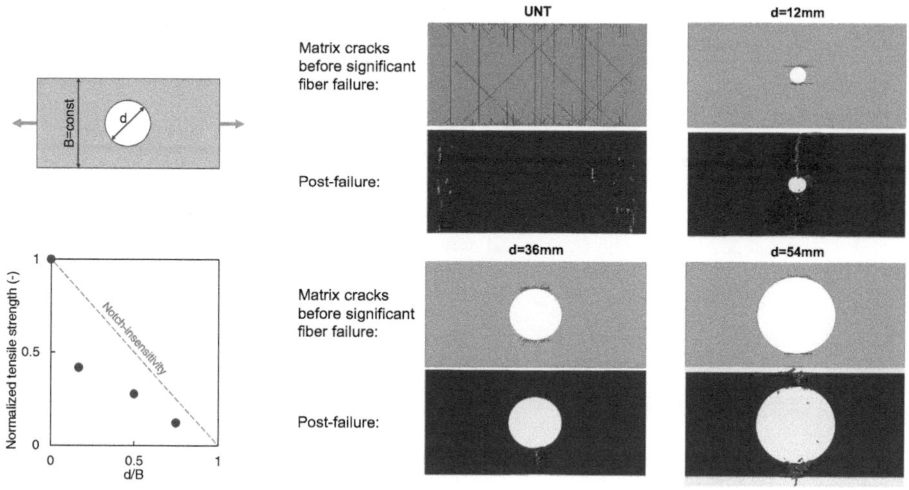

Figure 17.9 Predicted effects of the hole diameter on the failure modes and ultimate structural strength of open-hole laminates, using SD2M.

17.3.4 Delamination-prone laminate

Delamination is a dominant mode of failure for certain stacking sequences. Wisnom (2012), Green et al. (2007) showed that by scaling up plies and creating ply blocks, the failure mode shifts from "brittle" or "pull-out" type to a "delamination" type failure. The stacking sequence for the laminate studied in this chapter is $[45_4/90_4/{-}45_4/0_4]_s$. The total laminate thickness is 4 mm, the width is 16 mm, and the hole diameter is 3.175 mm.

A detailed presentation of the complete failure progression and mechanical response, which includes premature load drops by delamination and a final load drop due to fiber breakage, has not been presented in the literature to date, except for past results presented by us in Nguyen and Waas (2022b).

Prior to the first load drop, matrix cracking and splitting near the hole and the free edges are seen to occur. The first significant damage is observed at A* at about 50% of the first peak load. To highlight the prepeak damage pattern, a snap shot at point A is given in Fig. 17.10. Major split cracks in all plies near the hole are clearly visible. Between the matrix cracks, delamination has developed forming triangular shapes. Until the loading state B, the cracks continue to grow and a number of significant matrix cracks occurred at the free edges. The delamination continued to spread with the splitting due to a strong interaction at each interface in a stable manner. Some significant interface damage has been observed at the free edges as well in every interlayer except for the −45 degrees/0 degree interface. Finally, starting at the bottom outer 45 degrees ply, an abrupt growth of a major split crack through the width—localized on one side—was identified (see point C). It induced more cracking in the 90 degrees in its vicinity and a delamination spreading at the 45 degrees/90 degrees and 90 degrees/−45 degrees interfaces, extending from the hole to the free edge. This observed mechanism is the

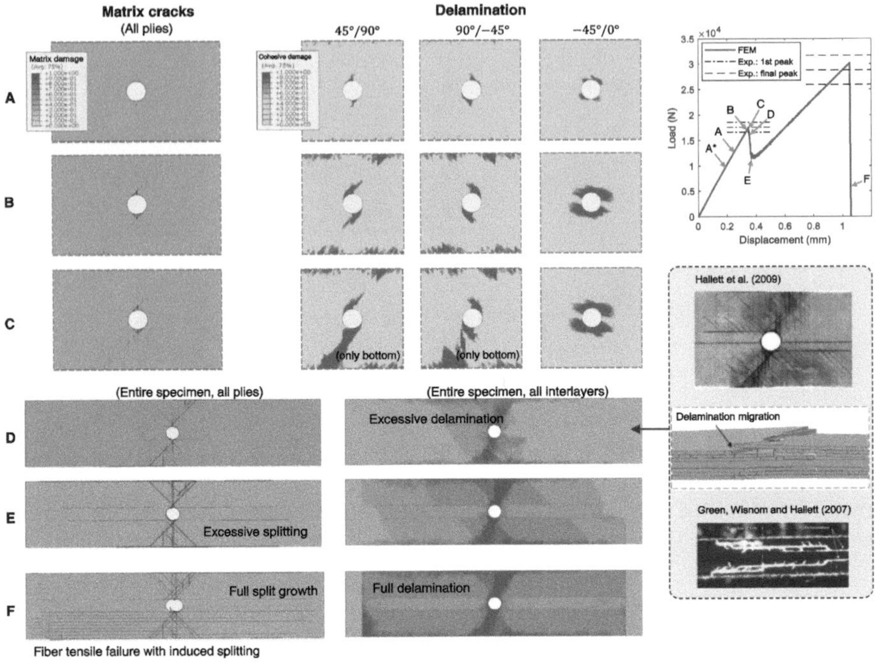

Figure 17.10 Predicted progression of matrix splitting and delamination in a ply-scaled open-hole laminate using eSD2M (Nguyen and Waas, 2022b). X-ray scan of a tested specimen from Hallett et al. (2009b). Side-view image showing delamination migration from Green et al. (2007).

start of the excessive *delamination migration and spreading* that caused the first load drop. Rapidly, all major matrix cracks extended to the free edges (point D). The 0 degree split cracks grew in an asymmetric manner. Due to a ply-to-ply interaction, subcritical matrix damage is triggered along the major split cracks. Delamination, as shown overlaid for all interlayers in Fig. 17.10, grew rapidly along with the intraply failure. Worth mentioning are the different delamination footprints in the two −45 degrees/0 degree interfaces, which continue to be true until full delamination was reached. Detachment between the middle 0 degree ply strip and the neighboring −45 degrees was observed at the upper −45 degrees/0 degree interface, where the delamination spreading was lagging behind the lower interface. It should be noted that the exact stages of delamination and 0 degree, i.e., the asymmetric damage and two subsequent peak loads due to delamination failure, were not yet captured by the current model. The authors believe that an improved material model may be the key to successfully predicting these features. As shown in Fig. 17.10e, the 0 degree spitting on both sides of the hole has almost reached to the grips. The unexplained 0 degree splitting that appeared to occur at the center of the hole (see Hallett et al. (2009b)) was also not captured.

In this problem, the loss of structural integrity occurs due to both splitting and delamination. With a reduced stiffness, the load kept increasing beyond the first load drop, with further growth of splits and the delamination footprint. The 0 degree split cracks have fully extended to the grips. Finally, fiber rupture was triggered near the grips, as shown in Fig. 17.10f. This is in agreement with the experimental observations in Green et al. (2007). The fiber rupture induced splitting at its vicinity, as observed in many experiments (Wisnom et al., 2008). Eventually, the 0 degree layer failed at every grip connection, except for the middle strips since they did not carry any load. It is worthwhile noting that the final peak load is predicted accurately within 5% and a few other simulations with a different randomized distribution of fiber strength showed an ultimate load closer to the experimental results, while the first peak load was unaffected.

17.4 Conclusions

This chapter has presented results for tensile failure of composite laminates using a novel semidiscrete damage model (eSD2M) to efficiently perform high-fidelity progressive failure analysis. In this method, all ply elements are modeled using a single-scale/meso-scale approach. Both bulk and splitting elements are modeled as a homogenized composite material with nonlinear stress—strain laws. Delamination was captured using cohesive zone models, which are realized using cohesive elements. For both intra and interlaminar failure, a novel mixed-mode law has been used.

The eSD2M model is able to capture the full failure progression and responses for unnotched and open-hole tensile laminates with high fidelity. The model captured the internal splitting and free-edge cracking in an unnotched laminate. By employing material strength randomness, the numerical framework was able to produce a steady increase in the crack density. The predicted final failure showed strong resemblance with the experimental observations, including fiber rupture and splitting. The unique meshing strategy made it possible to capture fine details of splitting and stress concentrations near a hole. The fact that the model framework can capture failure localizations both near cut-outs and at free edges is evidence of the versatility of the model. In a size effect study, the model was able to predict the strength knockdown due to the hole size. The predicted trend shows strong resemblance to data available in the literature. Even for the challenging case of a ply-scaled laminate, where delamination is the dominant failure mode while interacting with in-ply splitting, the model captured the full failure progression with fine details. This is possible due to the sophisticated meshing strategy, particularly the mesh connectivity between the plies and interlayers, as well as physically sound constitutive laws.

Recent and on-going work focuses on incorporating process-induced residual stresses, e.g., due to curing (Nguyen et al., 2023). Residual stresses have shown significant influence on the initiation and propagation of damage. When residual thermal stresses are included, the true in situ strength values need to be determined. Future work needs to be done to better characterize the strength variations. A well-thought-out material characterization plan is required to characterize these uncertainties. Incorporating residual stresses and a more careful material characterization plan have yielded improved predictions of the asymmetric damage progression in the OHT specimen (Nguyen and Waas, 2024). With more physics incorporated and the material properties better quantified, the model will become more universal and applicable to a wider range of problems.

Acknowledgments

The authors would like to acknowledge financial support through the Felix Pawlowski Collegiate Chair at the University of Michigan. This work is based on the PhD thesis of the second author at the University of Michigan, Ann Arbor.

References

Abisset, E., Daghia, F., Ladeveze, P., 2011. On the validation of a damage mesomodel for laminated composites by means of open-hole tensile tests on quasi-isotropic laminates. Compos. Appl. Sci. Manuf. 42 (10), 1515–1524. https://doi.org/10.1016/j.compositesa.2011.07.004.

Achard, V., Bouvet, C., Castanie, B., Chirol, C., 2014. Discrete ply modelling of open hole tensile tests. Compos. Struct. 113, 369–381. https://doi.org/10.1016/j.compstruct.2014.03.031.

Ahn, J., Waas, A.M., 2002. Prediction of compressive failure in laminated composites at room and elevated temperature. AIAA J. 40 (2), 346–358.

Anderson, P.W., 1972. More is different: broken symmetry and the nature of the hierarchical structure of science. Science 177 (4047), 393–396. URL. https://www.jstor.org/stable/1734697.

Aroush, D.R.-B., Maire, E., Gauthier, C., Youssef, S., Cloetens, P., Wagner, H., 2006. A study of fracture of unidirectional composites using in situ high-resolution synchrotron x- ray microtomography. Compos. Sci. Technol. 66 (10), 1348–1353. https://doi.org/10.1016/j.compscitech.2005.09.010.

Basu, S., Waas, A.M., Ambur, D.R., 2006. Compressive failure of fiber composites under multi-axial loading. J. Mech. Phys. Solid. 54 (3), 611–634. https://doi.org/10.1016/j.jmps.2005.09.004.

Bazant, Z.P., Oh, B.H., 1983. Crack band theory for fracture of concrete. Materiaux et Construction 16 (3), 155–177. https://doi.org/10.1007/BF02486267.

Boyd, S., Ji, W., Joseph, A., Waas, A.M., 2014. Single edge notch tension test on cross-ply laminated composites for intralaminar fracture properties. In: 55th AIAA/ASMe/ASCE/AHS/SC Structures, Structural Dynamics, and Materials Conference, p. 0852. https://doi.org/10.2514/6.2014-0852.

Camanho, P.P., Davila, C.G., De Moura, M., 2003. Numerical simulation of mixed-mode progressive delamination in composite materials. J. Compos. Mater. 37 (16), 1415–1438. https://doi.org/10.1177/0021998303034505.

Camanho, P.P., Maimi, P., Davila, C., 2007. Prediction of size effects in notched laminates using continuum damage mechanics. Compos. Sci. Technol. 67 (13), 2715–2727. https://doi.org/10.1016/j.compscitech.2007.02.005.

Chen, B., Tay, T., Baiz, P., Pinho, S., 2013. Numerical analysis of size effects on open-hole tensile composite laminates. Compos. Appl. Sci. Manuf. 47, 52–62. https://doi.org/10.1016/j.compositesa.2012.12.001.

Chen, B.Y., Tay, T., Pinho, S., Tan, V., 2016. Modelling the tensile failure of composites with the floating node method. Comput. Methods Appl. Mech. Eng. 308, 414–442. https://doi.org/10.1016/j.cma.2016.05.027.

Christensen, R.M., 2013. Timoshenko medal award paper—completion and closure on failure criteria for unidirectional fiber composite materials. J. Appl. Mech. 81 (1), 011011. https://doi.org/10.1115/1.4025177, 09 2013.

Clarkson, E., 2012. Hexcel 8552 im7 Unidirectional Prepreg 190 gsm & 35% RC Qualification Statistical Analysis Report. National Institute for Aviation Research.

Comiez, J.M., Waas, A.M., Shahwan, K.W., 1995. Delamination buckling; experiment and analysis. Int. J. Solid Struct. 32 (6), 767–782. https://doi.org/10.1016/0020-7683(94)00159-T time Dependent Problems in Mechanics.

Cui, X., Lua, J., 2020. Discrete crack informed continuum damage mechanics model in composite laminates. In: AIAA Scitech 2020 Forum, p. 0252. https://doi.org/10.2514/6.2020-0252.

Curtin, W., Ahn, B., Takeda, N., 1998. Modeling brittle and tough stress-strain behavior in unidirectional ceramic matrix composites. Acta Mater. 46 (10), 3409–3420. https://doi.org/10.1016/S1359-6454(98)00041-X.

Davidson, B.D., Gharibian, S.J., Yu, L., 2000. Evaluation of energy release rate-based approaches for predicting delamination growth in laminated composites. Int. J. Fract. 105 (4), 343–365. https://doi.org/10.1023/A:1007647226760.

Davidson, P., Pineda, E.J., Henrich, C., Waas, A.M., 2013. A unified model for predicting the open hole tensile and compressive strengths of composite laminates for aerospace applications. In: 54th AIAA/ASME/ASCE/AHS/ASC Structures, Structural Dynamics, and Materials Conference, p. 1613. https://doi.org/10.2514/6.2013-1613.

Davila, G., Camanho, P.P., Rose, C.A., 2005. Failure criteria for frp laminates. J. Compos. Mater. 39 (4), 323–345. https://doi.org/10.1177/0021998305046452.

Falco, O., Avila, R., Tijs, B., Lopes, C., 2018. Modelling and simulation methodology for unidirectional composite laminates in a virtual test lab framework. Compos. Struct. 190, 137–159. https://doi.org/10.1016/j.compstruct.2018.02.016.

Fuwa, M., Bunsell, A., Harris, B., 1975. Tensile failure mechanisms in carbon fibre reinforced plastics. J. Mater. Sci. 10 (12), 2062–2070. https://doi.org/10.1007/BF00557484.

Gamstedt, E., Sjogren, B., 1999. Micromechanisms in tension-compression fatigue of composite laminates containing transverse plies. Compos. Sci. Technol. 59 (2), 167–178. https://doi.org/10.1016/S0266-3538(98)00061-X.

Green, B.G., Wisnom, M., Hallett, S., 2007. An experimental investigation into the tensile strength scaling of notched composites. Compos. Appl. Sci. Manuf. 38 (3), 867–878. https://doi.org/10.1016/j.compositesa.2006.07.008.

Hallett, S.R., Green, B.G., Jiang, W.-G., Cheung, K.H., Wisnom, M.R., 2009a. The open hole tensile test: a challenge for virtual testing of composites. Int. J. Fract. 158 (2), 169–181. https://doi.org/10.1007/s10704-009-9333-8.

Hallett, S.R., Green, B.G., Jiang, W., Wisnom, M., 2009b. An experimental and numerical investigation into the damage mechanisms in notched composites. Compos. Appl. Sci. Manuf. 40 (5), 613–624. https://doi.org/10.1016/j.compositesa.2009.02.021.

Hashin, Z., 1980. Failure criteria for unidirectional fiber composites. J. Appl. Mech. 47 (2), 329–334. https://doi.org/10.1115/1.3153664.

Higuchi, R., Okabe, T., Nagashima, T., 2017. Numerical simulation of progressive damage and failure in composite laminates using xfem/czm coupled approach. Compos. Appl. Sci. Manuf. 95, 197–207. https://doi.org/10.1016/j.compositesa.2016.12.026.

Hyder, I., Leone, F.A., Justusson, B.P., Schaefer, J.D., Bergan, A., Wanthal, S., 2018. Implementation of a matrix crack spacing parameter in a continuum damage mechanics finite element model. In: American Society for Composites 33rd Technical Conference, Seattle, Washington. https://doi.org/10.12783/asc33/26052.

Jelf, P., Fleck, N., 1992. Compression failure mechanisms in unidirectional composites. J. Compos. Mater. 26 (18), 2706–2726. https://doi.org/10.1177/002199839202601804.

Ji, W., Waas, A., 2013. Progressive failure analysis for the interaction of interlaminar and intralaminar failure modes in composite structures with an initial delamination. Aeronaut. J. 117 (1187), 71–85.

Jiang, W.-G., Hallett, S.R., Green, B.G., Wisnom, M.R., 2007. A concise interface constitutive law for analysis of delamination and splitting in composite materials and its application to scaled notched tensile specimens. Int. J. Numer. Methods Eng. 69 (9), 1982–1995. https://doi.org/10.1002/nme.1842.

Joosten, M., Yang, Q., Blacklock, M., Wang, C., 2018. A cohesive network approach for modelling fibre and matrix damage in composite laminates. Compos. Struct. 206, 658–667. https://doi.org/10.1016/j.compstruct.2018.08.087.

Joseph, P., 2017. Progressive Damage and Failure Analysis of Composite Laminates, PhD Thesis, Aeronautics and Astronautics. University of Washington.

Joseph, P., Davidson, P., Waas, A.M., 2018a. Failure analysis of composite multi-bolt joints using intra-inter crack band model (i2cbm). In: 2018 AIAA/ASCE/AHS/ASC Structures, Structural Dynamics, and Materials Conference, p. 0977. https://doi.org/10.2514/6.2018-0977.

Joseph, P., Davidson, P., Waas, A.M., 2018b. Open hole and filled hole progressive damage and failure analysis of composite laminates with a countersunk hole. Compos. Struct. 203, 523–538. https://doi.org/10.1016/j.compstruct.2018.06.120.

Kaddour, A.S., Hinton, M., Li, S., Smith, P., 2007. Damage theories for fibre-reinforced polymer composites: the third world-wide failure exercise (wwfe-iii). In: Proceedings of the 16th International Conference on Composite Materials, pp. 8–13.

Khamseh, A.R., Waas, A., 1992. Failure mechanisms of uni-ply composite plates with a circular hole under static compressive loading. J. Eng. Mater. Technol. ASME 114 (3), 304–310.

Krueger, R., 2012. Development and Application of Benchmark Examples for Mixed-Mode i/ii Quasi-Static Delamination Propagation Predictions.

Kyriakides, S., Arseculeratne, R., Perry, E., Liechti, K., 1995. On the compressive failure of fiber reinforced composites. Int. J. Solid Struct. 32 (6), 689738. https://doi.org/10.1016/0020-7683(94)00157-R time Dependent Problems in Mechanics.

Laffan, M., Pinho, S., Robinson, P., McMillan, A., 2011. Translaminar fracture toughness: the critical notch tip radius of 0 plies in cfrp. Compos. Sci. Technol. 72 (1), 97–102. https://doi.org/10.1016/j.compscitech.2011.10.006.

Lee, S., Waas, A.M., 1999. Compressive response and failure of fiber reinforced unidirectional composites. Int. J. Fract. 100 (3), 275–306. https://doi.org/10.1023/A:1018779307931.

Leone, F.A., Justusson, B.P., 2021. Effects of characteristic element length on fracture energy dissipation in continuum damage mechanics models. J. Compos. Mater. https://doi.org/10.1177/00219983211023790, 00219983211023790.

Lin, S., Thorsson, S.I., Waas, A.M., 2020. Predicting the low velocity impact damage of a quasi-isotropic laminate using EST. Compos. Struct. 112530. https://doi.org/10.1016/j.compstruct.2020.112530.

Liu, K., Piggott, M., 1995. Shear strength of polymers and fibre composites: 2. Carbon/epoxy pultrusions. Composites 26 (12), 841–848. https://doi.org/10.1016/0010-4361(95)90877-3.

Lu, X., Ridha, M., Tan, V., Tay, T., 2019. Adaptive discrete-smeared crack (a-disc) model for multi-scale progressive damage in composites. Compos. Appl. Sci. Manuf. 125, 105513. https://doi.org/10.1016/j.compositesa.2019.105513.

Manders, P.W., Chou, T.-W., Jones, F.R., Rock, J.W., 1983. Statistical analysis of multiple fracture in 0/90/0 glass fibre/epoxy resin laminates. J. Mater. Sci. 18 (10), 2876–2889.

McCartney, L., 1992. Theory of stress transfer in a 0°—90°—0° cross-ply laminate containing a parallel array of transverse cracks. J. Mech. Phys. Solid. 40 (1), 27–68. https://doi.org/10.1016/0022-5096(92)90226-R.

Mortell, D., Tanner, D., McCarthy, C., 2016. An experimental investigation into multi-scale damage progression in laminated composites in bending. Compos. Struct. 149, 33–40. https://doi.org/10.1016/j.compstruct.2016.03.054.

Ng, W.H., Salvi, A.G., Waas, A.M., 2010. Characterization of the in-situ non-linear shear response of laminated fiber-reinforced composites. Compos. Sci. Technol. 70 (7), 1126–1134.

Nguyen, N., Waas, A.M., 2016. A novel mixed-mode cohesive formulation for crack growth analysis. Compos. Struct. 156, 253–262. https://doi.org/10.1016/j.compstruct.2015.11.015.

Nguyen, H., Waas, A.M., 2020a. A novel mode-dependent and probabilistic semi-discrete damage model for progressive failure analysis of composite laminates—part i: meshing strategy and mixed-mode law. Compos. C: Open Access 3, 100073. https://doi.org/10.1016/j.jcomc.2020.100073.

Nguyen, H., Waas, A.M., 2020b. A novel mode-dependent and probabilistic semi-discrete damage model for progressive failure analysis of composite laminates—part ii: applications to unnotched and open-hole tensile specimens. Compos. C: Open Access 3, 100071. https://doi.org/10.1016/j.jcomc.2020.100071.

Nguyen, M.H., Waas, A.M., 2021. High-fidelity progressive failure analysis of the open hole tensile (oht) behavior of composite laminates with a novel semi-discrete method. In: AIAA Scitech 2021 Forum, p. 0312. https://doi.org/10.2514/6.2021-0312.

Nguyen, M.H., Waas, A.M., 2022a. Efficient and validated framework for probabilistic progressive failure analysis of composite laminates. AIAA J. 1–21. https://doi.org/10.2514/1.J061499.

Nguyen, H., Waas, A.M., 2022b. Modeling delamination migration in composite laminates using an enhanced semi-discrete damage model (esd2m). Int. J. Solid Struct. 236–237, 111323. https://doi.org/10.1016/j.ijsolstr.2021.111323.

Nguyen, M.H., Waas, A.M., 2022c. Detailed experimental and numerical investigation of single- edge notched tensile cross-ply laminates. Compos. Struct. 279, 114731. https://doi.org/10.1016/j.compstruct.2021.114731.

Nguyen, M.H., Waas, A.M., 2022d. High-fidelity and computationally efficient digital tool for probabilistic progressive failure analysis of composite laminates: experiments and virtual testing. In: AIAA SCITECH 2022 Forum, p. 0663. https://doi.org/10.2514/62022-0663.

Nguyen, M.H., D'Mello, R.J., Waas, A.M., 2023. A multi-mechanism framework for cure-informed probabilistic progressive failure analysis of composite structures. AIAA SciTech 2023 Forum. https://doi.org/10.2514/6.2023-0593.

Nguyen, M.H., Waas, A.M., 2024. Numerical study on the effects of ply stacking and interlayer toughening on the failure modes and response of OHT laminates. Compos. Sci. Technol. https://doi.org/10.1016/j.compscitech.2023.109935.

O'Brien, T.K., Krueger, R., 2001. Analysis of Ninety Degree Flexure Tests for Characterization of Composite Transverse Tensile Strength.

Oliver, J., Huespe, A., Sanchez, P., 2006. A comparative study on finite elements for capturing strong discontinuities: E-fem vs x-fem. Comput. Methods Appl. Mech. Eng. 195 (37), 4732–4752. https://doi.org/10.1016/j.cma.2005.09.020 john H. Argyris Memorial Issue. Part I.

Parvizi, A., Garrett, K., Bailey, J., 1978. Constrained cracking in glass fibre-reinforced epoxy cross-ply laminates. J. Mater. Sci. 13 (1), 195–201. https://doi.org/10.1007/BF00739291.

Pham, D.C., Cui, X., Ren, X., Lua, J., 2019. A discrete crack informed 3d continuum damage model and its application for delamination migration in composite laminates. Compos. B Eng. 165, 554–562. https://doi.org/10.1016/j.compositesb.2019.02.045.

Phoenix, S., 1975. Probabilistic inter-fiber dependence and the asymptotic strength distribution of classic fiber bundles. Int. J. Eng. Sci. 13 (3), 287–304. https://doi.org/10.1016/0020-7225(75)90036-1.

Pineda, E.J., Waas, A.M., 2013. Numerical implementation of a multiple-isv thermodynamically-based work potential theory for modeling progressive damage and failure in fiber-reinforced laminates. Int. J. Fract. 182 (1), 93–122. https://doi.org/10.1007/s10704-013-9860-1.

Plumtree, L.S., 2002. Fatigue damage evolution in off-axis unidirectional cfrp. Int. J. Fatig. 24 (2–4), 155–159. https://doi.org/10.1016/S0142-1123(01)00068-8.

Prabhakar, P., Waas, A.M., 2013. Interaction between kinking and splitting in the compressive failure of unidirectional fiber reinforced laminated composites. Compos. Struct. 98, 85–92. https://doi.org/10.1016/j.compstruct.2012.11.005.

Puck, A., Schurmann, H., 2002. Failure analysis of frp laminates by means of physically based phenomenological models. Compos. Sci. Technol. 62 (12), 16331662. https://doi.org/10.1016/S0266-3538(01)00208-1.

Reeder, J., 1992. An Evaluation of Mixed-Mode Delamination Failure Criteria, National Aeronautics and Space Administration. Langley Research Center.

Rudraraju, S., Salvi, A., Garikipati, K., Waas, A.M., 2012. Predictions of crack propagation using a variational multiscale approach and its application to fracture in laminated fiber reinforced composites. Compos. Struct. 94 (11), 3336–3346. https://doi.org/10.1016/j.compstruct.2012.03.035.

Schapery, R., 1990. A theory of mechanical behavior of elastic media with growing damage and other changes in structure. J. Mech. Phys. Solid. 38 (2), 215–253. https://doi.org/10.1016/0022-5096(90)90035-3.

Serra, J., Bouvet, C., Castanie, B., Petiot, C., 2016. Scaling effect in notched composites: the discrete ply model approach. Compos. Struct. 148, 127–143. https://doi.org/10.1016/j.compstruct.2016.03.062.

Shahwan, K.W., Waas, A.M., 1997. Non-self-similar decohesion along a finite interface of unilaterally constrained delaminations. Proc. Math. Phys. Eng. Sci. 453 (1958), 515–550. https://doi.org/10.1098/rspa.1997.0029.

Starnes Jr., J.H., Williams, J.G., 1983. Failure characteristics of graphite-epoxy structural components loaded in compression. In: Mechanics of Composite Materials. Elsevier, pp. 283–306.

Sun, Q., Zhou, G., Meng, Z., Guo, H., Chen, Z., Liu, H., et al., 2019. Failure criteria of unidirectional carbon fiber reinforced polymer composites informed by a computational micromechanics model. Compos. Sci. Technol. 172, 81–95. https://doi.org/10.1016/j.compscitech.2019.01.012.

Swindeman, M.J., Iarve, E.V., Brockman, R.A., Mollenhauer, D.H., Hallett, S.R., 2013. Strength prediction in open hole composite laminates by using discrete damage modeling. AIAA J. 51 (4), 936–945. https://doi.org/10.2514/1.J051773.

Takeda, N., Ogihara, S., 1994. Initiation and growth of delamination from the tips of transverse cracks in cfrp cross-ply laminates. Compos. Sci. Technol. 52 (3), 309–318. https://doi.org/10.1016/0266-3538(94)90166-X.

Talreja, R., 1985. A continuum mechanics characterization of damage in composite materials. Proc. Math. Phys. Sci. 399 (1817), 195–216.

Talreja, R., Singh, C.V., 2012. Damage and Failure of Composite Materials. Cambridge University Press.

Thorsson, S.I., Waas, A.M., Rassaian, M., 2018. Low-velocity impact predictions of composite laminates using a continuum shell based modeling approach part a: impact study. Int. J. Solid Struct. 155, 185–200. https://doi.org/10.1016/j.ijsolstr.2018.07.020.

Van der Meer, F., Sluys, L., Hallett, S., Wisnom, M., 2012. Computational modeling of complex failure mechanisms in laminates. J. Compos. Mater. 46 (5), 603–623. https://doi.org/10.1177/0021998311410473.

Waas, M., Schultheisz, C.R., 1996. Compressive failure of composites, part ii: experimental studies. Prog. Aero. Sci. 32 (1), 43–78. https://doi.org/10.1016/0376-0421(94)00003-4.

Waas, A.M., Babcock, C., Knauss, W., 1990. An experimental study of compression failure of fibrous laminated composites in the presence of stress gradients. Int. J. Solid Struct. 26 (9), 1071–1098. https://doi.org/10.1016/0020-7683(90)90017-P.

Waas, M., Thorsson, S.I., Rassaian, M., 2018. Prediction of low-velocity face-on impact response and compression after impact (cai) of composite laminates using est and cohesive modeling (dczm). In: 2018 AIAA/ASCE/AHS/ASC Structures, Structural Dynamics, and Materials Conference, p. 1704. https://doi.org/10.2514/6.2018-1704.

Wisnom, M., 2012. The role of delamination in failure of fibre-reinforced composites. Phil. Trans. Math. Phys. Eng. Sci. 370 (1965), 1850–1870. https://doi.org/10.1098/rsta.2011.0441.

Wisnom, M., Khan, B., Hallett, S., 2008. Size effects in unnotched tensile strength of unidirectional and quasi-isotropic carbon/epoxy composites. Compos. Struct. 84 (1), 21–28. https://doi.org/10.1016/j.compstruct.2007.06.002.

Xu, W., Waas, A.M., 2016. Modeling damage growth using the crack band model; effect of different strain measures. Eng. Fract. Mech. 152, 126–138. https://doi.org/10.1016/j.engfracmech.2015.06.034.

Regularized eXtended Finite Element Modeling of fatigue response in laminated composites

18

Wei-Tsen Lu[1] *and Endel V. Iarve*[2]
[1]University of Texas at Arlington Research Institute, Institute for Predictive Performance Methodologies, Fort Wort, TX, United States; [2]University of Texas at Arlington, Mechanical and Aerospace Engineering, Arlington, TX, United States

18.1 Introduction

Damage in composite materials represents an evolution and interaction of a number of failure mechanisms, such as matrix cracks, delaminations, and fiber failure. These mechanisms tend to result in redistribution of internal stresses and can either increase or decrease the resistance of the composite structures to additional loads. Thus, the residual static strength in open-hole (OH) specimens after fatigue loading can exceed the static strength of pristine OH coupons both in tension and compression (Seneviratne et al., 2020). Progressive damage analysis (PDA) methodologies offer to capture interactions of different failure mechanisms and predict the strength and durability of composite structures. Two general classes of techniques are used for PDA: continuum damage mechanics (CDM) methods, where the effect of damage is represented by local stiffness degradation (Maimí et al., 2007a, b; Llobet et al., 2021) and discrete damage mechanics (DDM) methods, discussed below, where the effect of damage is directly modeled by introducing explicit representations of the displacement discontinuities associated with individual damage events. In laminated composites, the CDM methodology based on degrading element stiffness properties requires additional considerations to provide realistic results, namely, fiber-aligned meshing (Camanho et al., 2007; Nikishkov et al., 2010; Nikishkov and Makeev, 2011; Dalgarno et al., 2017). Introducing such meshes orients the matrix damage propagation with the transverse direction and allows to achieve pseudo-discrete damage patterns closely resembling the experiment. The cracking density is directly controlled by mesh refinement. The key advantage of these methodologies is their compatibility with the general framework of commercial finite element (FE) software and the ability to directly use both explicit and implicit of the shelf FE software.

The DDM approaches can be divided into two broad categories. In the first category, the anticipated damage propagation path is explicitly meshed in. Such special meshing is similar to the fiber-aligned meshes discussed above. The main difference between the CDM methods described above and the fixed path DDM (Nixon-Pearson et al., 2013; Tao et al., 2018; Sun and Hallett, 2018) is that the damage is

Modeling Damage, Fatigue and Failure of Composite Materials. https://doi.org/10.1016/B978-0-443-18489-5.00008-4

modeled along the element boundaries by implementing cohesive zone interfaces between each element row. The fixed path DDM like CDM can be implemented into the commercial software directly by using standard tools such as user-defined material property routines. However, FE meshing combining structural features and damage propagation path is in general complex and often prohibitive.

In the present work, we will focus on DDM where the displacement discontinuity is inserted on the fly independent of the FE mesh. Several of such techniques based on x-FEM have emerged in the last decade and have been developed for laminated composites to address various problems under static loading (van der Meer and Sluys, 2009; Duarte et al., 2017; Vellwock et al., 2018; Chen et al., 2016; Ma et al., 2021). These methods combine the kinematic part, which is responsible for displacement approximation enrichment allowing mesh-independent crack (MIC) modeling, and constitutive models, which control the displacement jump across the MIC faces as well as the delaminations. The distinct advantage of the DDM is the ability to accurately model the interaction between matrix cracks and delamination including delamination migration from one interface to another. A number of works (Adluru et al., 2019; Adluru et al., 2018; Rao et al., 2018) were devoted to modeling this important phenomenon under fatigue loading. However, only relatively simple matrix cracking patterns in quasi-2D problem statement were considered. In the present paper, we will focus on the problems resulting in dense 3D matrix cracking patterns. The phantom node method (PNM) (van der Meer and Sluys, 2010; Wang and Zhang, 2021), augmented finite element method (A-FEM) (Ma et al., 2021), and floating node method (FNM) in particular (Chen et al., 2016; Lu et al., 2018) have demonstrated such capability in the case of static analysis of OH tension problems. Fatigue loading generates significantly denser matrix cracking patterns before the structure loses its load-carrying capacity. The extension of the DDM framework to fatigue loading is conceptually straightforward and does not require any modification of the kinematic framework. Nevertheless, only limited work is devoted to the application of x-FEM to fatigue analysis. Fatigue evolution of a single delamination damage mode by using x-FEM is considered in Grogan et al. (2014), Teimouri et al. (2021), and in PNM (Wang and Xu, 2016) and FNM (de Carvalho et al., 2019). A more complex situation of matrix cracking and delamination interaction including delamination migration under fatigue loading is considered in Rao et al. (2018), Seshadri et al. (2018), and de Carvalho and Krueger (2016). While static analysis modeling of crack opening in almost all x-FEM models was performed by using cohesive zone model (CZM), the VCCT approach (Krueger, 2004) has been used under fatigue loading in several x-FEM approaches and in FNM in particular (Rao et al., 2018; Grogan et al., 2014; Seshadri et al., 2018). In this case, an additional constituent method was needed to insert a finite size initial open crack (Rao et al., 2018; Seshadri et al., 2018).

Only very limited number of works (Fang et al., 2017; Yang, 2017) (A-FEM) and (Iarve et al., 2017) (Regularized x-FEM) address complex matrix cracking and delamination frameworks such as those forming near OHs under fatigue loading. In these works, CZM methods were applied to compute the crack opening displacement without any assumptions regarding the initial crack dimension.

A variant of the x-FEM methodology termed as Regularized eXtended Finite Element Method (Rx-FEM) is further developed in this chapter to achieve high crack density with respect to a given mesh. At the same time, the method maintains the key feature of Rx-FEM where the elements are not partitioned but duplicated, which allows to maintain the Gauss integration scheme of the original element. The original Rx-FEM enrichment proposed in Iarve (2003) utilizes a phantom degree of freedom concept, which reduces to Handsbo and Handsbo (Hansbo and Hansbo, 2004) if the regularized step function is replaced by a discontinuous Heaviside function, as described in Ventura et al. (2009). Note that the nature of the regularized step function introduces a gradient zone over which the displacement enrichment is performed and thus generally requires a larger spacing between Rx-FEM cracks than the x-FEM cracks. In this chapter, we will extend the Rx-FEM framework by increasing phantom node multiplicity to eliminate the gap between gradient regions of neighboring cracks and to significantly increase the crack density for a given mesh.

Several types of fatigue CZM formulations have emerged in recent years. One type of formulation combines S–N relations for damage initiation and Paris law variant for damage growth (Tao et al., 2018; Iarve et al., 2017; Turon et al., 2018). Another type of CZM is based on assumed point local relationship between the displacement jump and the rate of growth of the damage variable (Dávila et al., 2020; Dávila, 2020; Liang et al., 2021; Joosten et al., 2022). The formulation (Dávila, 2020), in particular, significantly reduces the set of input parameters required for analysis, essentially requiring only Mode I crack growth rate characterization. CZM formulations derived via temporal homogenization have also been developed and applied within a multiscale analysis framework (Su and Oskay, 2021).

The enhanced Rx-FEM framework with increased crack density will be combined with CZM formulation (Lu et al., 2021, 2022; Gao et al., 2022), which provides consistent damage size within a range of the loading cycle increments. We will begin by describing the Rx-FEM development, followed by revisiting fatigue CZM formulation. The verification and validation of the DDM framework were performed for coupon-level laminated composites—double cantilever beam (DCB) and end-notched flexure (ENF). This was followed by two applications of DDM to subelement-level composite. First, a clamped tapered beam (CTB) specimen was under constant amplitude fatigue-loading conditions, where delamination migration was discussed. Second, a quasiisotropic OH laminate alongside two nontraditional laminates containing $\pm 30^{\circ}$ and $\pm 60^{\circ}$ plies considered in Iarve et al. (2017) under fatigue loading will be analyzed and compared to experimental data.

18.2 Computational methodology

In this section, the static DDM framework will be reviewed followed by a fatigue failure criterion. The failure criteria in the DDM framework are used to identify the location, crack orientation, and the number of cycles (in case of fatigue) at which matrix cracks are inserted. The discussion will continue with fatigue CZM development, including damage initiation, propagation, and finally describing the fatigue DDM framework.

18.2.1 Reduced crack spacing regularized extended finite element method

The Rx-FEM formulation (Iarve and Mollenhauer, 2015; Iarve et al., 2011) is extended for modeling more realistic and dense crack patterns in laminated composites. It is achieved by extending the original phantom degree of freedom concept to higher multiplicity phantom nodes. In Rx-FEM, the Heaviside step function used in typical x-FEM (Iarve, 2003) is replaced by a continuous function varying from 0 to 1 over a narrow volume few elements wide called gradient zone. This continuous function $\widetilde{H}(\xi^e)$ is approximated by the element shape functions as follows.

$$\widetilde{H}(\xi^e) = \sum_{i\in\Omega_e} N_i^e(\xi^e) h_i \tag{18.1}$$

where Ω_e is the set of all nodes of element e, ξ are the parametric coordinates, and N_i^e are the shape functions. The continuity of $\widetilde{H}$ function in Eq. (18.1) through the domain is assured by single field of nodal values computed for node i by summing over all elements sharing that node as

$$h_i = \frac{1}{2}\left\{1 + \left(\sum_{e\in\mho_i}\int_V N_i^e f\, \mathrm{d}V\right) \Big/ \left(\sum_{e\in\mho_i}\int_V N_i^e |f|\, \mathrm{d}V\right)\right\} \tag{18.2}$$

where $\mho_i$ is the set of all elements sharing the same node i, and f is the signed distance function of the crack surface. All integrals in Eq. (18.2) are taken over element domains in parametric coordinates. This definition involves only nodal values of continuous functions and thus the integrals can be calculated by standard Gauss quadrature. The coefficient h_i is equal to 0 or 1 if the signed distance function f does not change the sign within the domain of the shape functions N_i^e.

Multiple cracks in each ply will be considered. We will distinguish their sign distance function and step function by adding an additional index, that is, f_a or $\widetilde{H}_a$, where a is a crack index. For each crack, Eqs. (18.1) and (18.2) are used separately and the cracks are spaced sufficiently apart so that their gradient regions $|\nabla\widetilde{H}(f_a)| > 0$ are not allowed to overlap.

First, we construct the step function for each crack separately using Eqs. (18.1) and (18.2). Next, we assign each node a zone number.

0− if the node is outside a gradient region of any crack;

$(a1)$ if $\widetilde{H} > 0.5$;

$(a2)$ if $\widetilde{H} \leq 0.5$.

In the original formulation (Gao et al., 2022), the cracks were spaced so that the gradient zones were separated by one or more elements. In this case, each node could only belong to a gradient zone of a single crack, and if it did, then it was twinned so that the new phantom node would have a complimentary zone number, that is, $a2$ for $a1$ node and vice versa. Note that the phantom node is assigned a unique node number or label, thus increasing the model dimension. Recently, the original formulation was extended to allow higher crack density and permit the gradient zones not to have any separation (Lu et al., 2022). In this case, there will be nodes with multiple zone numbers. Consequently, in this case, we are creating higher multiplicity phantom nodes, so that several phantom nodes are created for a given node each with a complimentary node number for each zone of the original node. Typically, in laminated composites with parallel crack networks, each node will have no more than two zone numbers from neighboring cracks.

After all nodes are twinned, the element twins are created. We loop through all original elements and create element twins by scanning their nodes and collecting original and phantom nodes with the same zone number to form a new element twin. If any node has a zone 0, then the element is not twinned. Thus, the enriched FE model contains three types of elements: the original elements away from cracks; transition non-twinned elements, which contain nodes with zone 0 and either $a1$ or $a2$ but not both, which is intuitively clear in the case of parallel cracks such as in the composite laminates; and, finally, element twins of zone $a1$ and $a2$. The model assembly is standard, and the correct partitioning of the cracked space is assured by creating elements, which first inherit the original model spatial connectivity and then group the nodes with the same zone numbers. Present description is restricted to multiple node twinning but single element twinning only. The crack density enabled by this process will be suitable for fatigue analysis in laminated composites, as shown in the following numerical examples.

The described algorithm completely defines the element structure of the enriched FE model. The displacement and strain fields within each original or twinned element can be represented by using standard notations:

$$u = \mathbf{N}\mathbf{U}_{ak}, \varepsilon = \mathbf{B}\mathbf{U}_{ak} \tag{18.3}$$

where $\mathbf{N}$ is the shape function matrix, $\mathbf{B}$ is the strain approximation matrix of the original element, and $\mathbf{U}_{ak}$ is the nodal displacement vector, respectively. The index $k = 1, 2$ was introduced to denote the zone number. Eq. (18.3) clearly reflects that an element only contains degrees of freedom (DoF) with the same zone number. However, an element will mix the original and phantom DoF. The only exception are the transition elements which combine the zone=0 and the original DoF of the nodes at the "outside" of the gradient zone. It is the global field approximation where the Rx-FEM enrichment formulation employs essentially hybrid variational formulation. The displacement and strain fields in a ply are approximated independently:

$$U = \sum_{e+e_w} \widetilde{H}_{ak}\mathbf{N}\mathbf{U}_{ak}, E = \sum_{e+e_w} \widetilde{H}_{ak}\mathbf{B}\mathbf{U}_{ak} \tag{18.4}$$

The step function gradient is excluded from strain calculation, and the step function is used as a characteristic function of the volume partitioned by the cracks. The following notations have been introduced for the step function in twinned elements $\widetilde{H}_{a1} = \widetilde{H}(f_a)$ and $\widetilde{H}_{a2} = 1 - \widetilde{H}(f_a)$. To keep the notations concise, we include the original nontwinned elements in Eq. (18.4) by allowing $ak = 0$ and assuming $\widetilde{H}_{ak} = 1$. The stress is calculated based on the strain directly. The strain energy required for the variational formulation is:

$$W = \sum_{e+e_w} \int_{V_e} \widetilde{H}_{ak} \mathbf{U}_{ak}^T \mathbf{B}^\mathbf{T} \mathbf{D} \mathbf{B} \mathbf{U}_{ak} dv \tag{18.5}$$

where $\mathbf{D}$ is the orthotropic stiffness matrix.

CZM formulation (Turon et al., 2018), which provides the relationship between cohesive traction $\tau(\lambda)$ and the effective displacement jump $\lambda = \|u_{a1} - u_{a2}\|$, is used to control the MIC opening. To complete the Rx-FEM formulation, we consider the twin element pair and the respective displacement jump between the displacement fields in two zones:

$$u_{a1} - u_{a2} = \mathbf{N}(\mathbf{U}_{a1} - \mathbf{U}_{a2}) \tag{18.6}$$

This displacement jump results in cohesive tractions $\tau(\|u_{a1} - u_{a2}\|)$ which contribute to the Rx-FEM formulation. The potential energy variation of a cracked ply taking into account the cohesive forces is:

$$\delta \Pi = \delta W - \sum_{e_w} \int_{V_e} \left| \nabla \widetilde{H}(f_a) \right| \tau(\|u_{a1} - u_{a2}\|) \delta \|u_{a1} - u_{a2}\| dv \tag{18.7}$$

All functions in Eq. (18.7) are continuous, and standard Gauss quadratures are used. Note that the cohesive term in Eq. (18.7) is also integrated over a volume of each element. The step function gradient yields the surface area of the crack as a result of volume integration. It is more apparent if the regularized step function is replaced with discontinuous Heaviside function $H(f_a)$. In this case, the gradient turns into a delta function over the crack surface as discussed in Iarve et al. (2011). The nodal structure of the cohesive element (Liang et al., 2020; Swindeman et al., 2013) is different from a standard element and contains the DoF of both element twins, that is, the displacement fields u_{a1} and u_{a2}. To arrive at the laminate formulation, an additional term for interlaminar cohesive energy variation is needed. Similar to the intralaminar case, we construct this term based on cohesive force versus displacement jump relation $\tau(\lambda)$. However, the displacement jump in this case is the displacement difference between the plies n and $n+1$ at their interface, and the integration is performed over the interface area. Denoting the cohesive energy of that interface as Φ_n, its variation can be expressed as:

$$\delta \Phi_\text{n} = \int_{S_n} \tau(\|U^{n+1} - U^n\|) \delta \|U^{n+1} - U^n\| \tag{18.8}$$

The displacements in each ply are expressed according to Eq. (18.4), and the subsequent variation ties together u, u_{a1} and u_{a2} in both plies. Finally, we obtain the following variational expression for the laminate.

$$\sum_{n}^{N}\delta\Pi_{\mathrm{n}} - \sum_{n}^{N-1}\delta\Phi_{\mathrm{n}} - \delta A = 0 \tag{18.9}$$

where A is the work of external forces and N is the total number of plies. For additional details of the multiple phantom node twinning, the reader is referred to Lu et al. (2022). The multiple node twinning algorithm proposed above is implemented in BSAM20 (Hoos et al., 2023) software.

18.2.2 MIC insertion in fatigue regime and strength tracking method

Consider a material point with a stress state experiencing proportional cyclic stress between σ^{max} and $R\sigma^{max}$, where R is the stress ratio $\sigma^{min}/\sigma^{max}$. For clarity, the stress components are in the material coordinate system, so that the x_1 axis is oriented along the fiber direction and the x_3 axis is perpendicular to the ply interface. The LaRC failure criterion (Pinho et al., 2005) is used to calculate the static failure index f and the angle α of the matrix crack with respect to the ply interface. It is achieved by rotating the stress about the x_1 axis and computing quadratic failure index:

$$f = \max_{-\frac{\pi}{2}<\alpha\leq\frac{\pi}{2}}\left[\left(\frac{\sigma_n}{Y_t}\right)^2 + \left(\frac{\gamma_L}{S_L}\right)^2 + \left(\frac{\gamma_T}{S_T}\right)^2\right]^{1/2} \tag{18.10}$$

where $\sigma_n, \gamma_L, \gamma_T$ are the transverse normal, in-plane, and transverse shear stresses in the rotated coordinate frame, respectively, and Y_T, S_L, S_T are the transverse tensile strength, axial shear strength, and transverse shear strength, respectively. Eq. (18.10) is valid only if $\sigma_n \geq 0$, which is the case considered in the present work. In practice, these stresses are evaluated in various planes forming angle α with the ply interface and the maximum value of f as well as the corresponding angle α recorded as failure index and the crack angle, respectively. In the static analysis, the crack insertion corresponds to $f \geq 1$. During fatigue analysis, $f < 1$ and the number of cycles N until crack insertion needs to be calculated. This is accomplished by using the S–N relationship instead of the static failure criterion. The S–N relationship used below is a linear $\log_{10}$ expression:

$$f = 1 - sLog(N) \tag{18.11}$$

where the single coefficient s is dependent on the shear and normal stress combination. A mode mixity parameter B_σ is introduced in a way similar to the CZM but calculated based on stress ratio instead of the displacement jump, as follows.

$$B_\sigma = 1 - \frac{\left(\frac{\sigma_n}{Y_t}\right)^2}{\left(\frac{\sigma_n}{Y_t}\right)^2 + \left(\frac{\gamma_L}{S_L}\right)^2 + \left(\frac{\gamma_T}{S_T}\right)^2} \quad (18.12)$$

Thus $B_\sigma = 0$ corresponds to normal loading and $f = \frac{\sigma_n}{Y_t}$, and $B_\sigma = 1$ corresponds to shear loading and $f = \left(\left(\frac{\gamma_L}{S_L}\right)^2 + \left(\frac{\gamma_T}{S_T}\right)^2\right)^{1/2}$. Ideally, one should distinguish between the transverse shear and in-plane shear, but in practice, it depends on the availability of the respective static and fatigue input data. In the present work, we will only distinguish between normal loading and shear loading utilizing the corresponding parameter in Eq. (18.12). The parameter s will also have a separate value s_1 for normal loading and s_2 for shear loading. The so-called $B-K$ interpolation for intermediate values of B_σ will be performed as follows.

$$s = s_1 + (s_1 - s_2)B_\sigma^\eta \quad (18.13)$$

Eqs. (18.10)–(18.13) can be used to calculate the limit number of cycles N_{lim} to crack insertion under constant amplitude loading as:

$$Log(N_{lim}) = \frac{1-f}{s} \quad (18.14)$$

where parameter s is given by Eq. (18.13). An additional assumption regarding damage accumulation rules is required to generalize this approach to variable amplitude loading, The Palmgren-Miner linear damage accumulation rule is assumed. According to this rule, the failure after a sequence of cyclic loading with arbitrary amplitudes takes place if $d = 1$, where:

$$d = \sum_i \frac{\Delta N_i}{N_{\text{lim}}(f_i)} \quad (18.15)$$

And ΔN_i is the number of cycles at a given load amplitude f_i with $N_{\text{lim}}(f_i)$ being the limit number of cycles. The damage accumulation rule Eq. (18.15) allows the definition of residual strength as a value of cyclic load amplitude, which only requires one cycle to satisfy Eq. (18.15) for a given previous loading history $d < 1$, that is,

$$d + \frac{1}{N_{lim}(f_{r'})} = 1 \quad (18.16)$$

Eq. (18.16) establishes a unique relation between the damage history variable d and the residual strength. This allows us to use the residual strength $f_{r'}$ as a damage state identifier as opposed to the abstract variable d. This process was proposed in Seneviratne and Tomblin (2021) and can be applied to different fatigue methodologies, which don't include the linear damage accumulation rule (Sendeckyj, 1981). To complete the formulation, one needs to predict the residual strength $f_{r''}$ remaining after ΔN cycles of loading with arbitrary amplitude f provided that before the loading increment the residual strength was $f_{r'}$. The expression for accumulation rule can now be written as:

$$d+\frac{\Delta N}{N_{lim}(f)}+\frac{1}{N_{lim}(f_{r''})}=1 \tag{18.17}$$

where d is given by Eq. (18.16). After substitution, we arrive at the following relationship between residual strength $f_{r'}$ and strength $f_{r''}$

$$\frac{1}{N_{lim}(f_{r''})}=\frac{1}{N_{lim}(f_{r'})}+\frac{\Delta N}{N_{lim}(f)} \tag{18.18}$$

Eq. (18.18) provides a simple recursive relationship to compute new residual strength based on its previous value and new cycle increment and amplitude. This relationship only holds for assumptions in Eq. (18.9) and Eq. (18.11). In principle, the value of $f_{r''}$ can be obtained for any specific shape of the S−N curve which provides the number of cycles to failure or N_{lim}. The crack insertion will take place if the applied load amplitude exceeds the residual strength $f \geq f_{r''}$. The angle of the crack is that calculated in Eq. (18.10) while computing the value of f.

18.2.3 Fatigue cohesive law and strength tracking method

In the process of static deformation, material points under increasing external load move along the traction displacement jump curve continuously from undamaged to damaged portion and eventually to the fully separated state under increasing load or stored energy. In the fatigue regime, the external load amplitude is such that the cohesive traction reaches only a fraction of the static strength value Y and the transition from undamaged to damaged state of the fatigue CZM is one of the key parts of the formulation. The CZM formulation proposed in Iarve et al. (2017), which is further developed in the present chapter, proposes to modify the original CZM by replacing the static strength value with the residual strength after given fatigue loading history but maintaining the same value of the fracture toughness G_c. It is illustrated in Fig. 18.1. Denote the initiation and final displacement jump in the static regime as $\Delta_0 = Y/\text{K}$ and $\Delta_c = 2G_c/Y$, where K is the penalty stiffness. It is the displacement jump λ which defines the damage state and cohesive traction in CZM model. Assume that $f = \frac{\lambda}{\Delta_0} < 1$ is fatigue-loading amplitude and the material point is in the initiation phase. The idea of the damage initiation approach in Iarve et al. (2017) is that the initiation strength Y in CZM is continually modified by replacing static strength Y with residual strength $f_{r''}Y < f_{r'}Y < Y$, where the residual strength factor f_r is calculated according to Eq. (18.18). The CZM envelope is also modified for each new value of

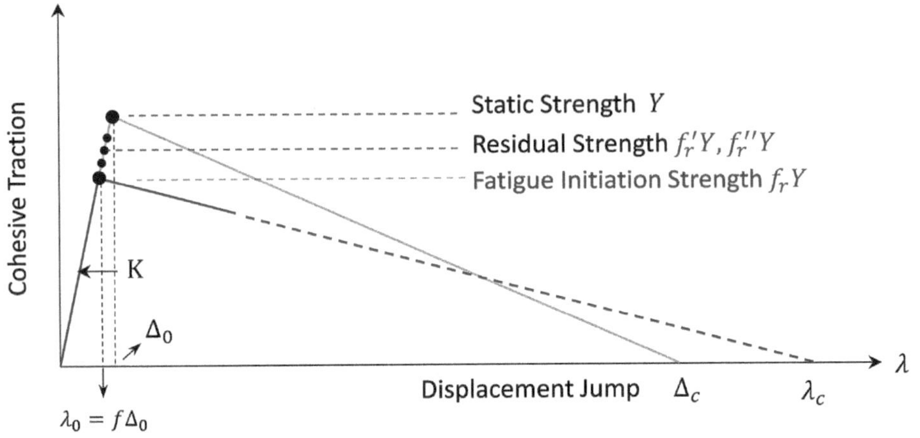

Figure 18.1 Schematic of a cohesive zone model for fatigue damage initiation.

residual strength by maintaining the G_c value as shown in Fig. 18.1. Note that the modification of the CZM envelope does not affect the material response since the penalty stiffness remains the same and equals to K. At some cycle increment $f \geq f_r$ and when the CZM transitions into the damage regime, the residual strength value corresponding to this transition will be denoted $Y_r = f_r Y$ and will remain unchanged. The initiation and final displacements are set to $\lambda_0 = Y_r/K$ and $\lambda_c = 2G_c/(Y_r)$ and are also kept unchanged through the propagation phase. As mentioned before, the proposed fatigue formulation is built upon consistent mode-dependent static CZM formulation (Turon et al., 2018). In this formulation, the penalty stiffness K and displacement jump λ are replaced with K_b and effective displacement jump λ_b which depend on the mode mixity B. We refer the reader to the original formulation for the detail of the static formulation. In the following, we will omit the subscript b and continue with K and displacement jump λ for the fatigue formulation assuming that the mode-dependent quantities are used. A damage norm RT is introduced as:

$$RT = \frac{\lambda - \lambda_0}{\lambda_c - \lambda_0}, \tag{18.19}$$

which defines the damage variable as:

$$D = \frac{RT\ \lambda_c}{RT\ \lambda_c + (1 - RT)\lambda_0} \tag{18.20}$$

and the cohesive traction magnitude as:

$$\tau = (1 - D)K\lambda \tag{18.21}$$

In addition, the energy release rate (ERR) is calculated as:

$$G = \int_{0}^{\lambda} \tau(\lambda) d\lambda \tag{18.22}$$

18.2.3.1 Event-based algorithm for two-dimensional application

In the event-based algorithm (EBA), one defines an increment of damage or damage event, such as new crack insertion or delamination extension for one element and computes the number of cycles required to advance to this event. This algorithm is extremely useful for 2D applications if damage events occur in a specific direction such as in a DCB test or ENF test. The Paris law is used in the propagation regime $\lambda > \lambda_0$ for an incremental increase of the damage variable. In the present work, an approach based on the original formulation of the Paris law (Deobald et al., 2017) in the power form of the change of the crack intensity factor during a loading cycle has been adopted as follows.

$$\frac{da}{dN} = C\left(\frac{\sqrt{G_{max}}}{\sqrt{G_c}} - \frac{\sqrt{G_{min}}}{\sqrt{G_c}}\right)^{2m} \tag{18.23}$$

where G_{max} and G_{min} are the maximum and minimum values of ERR during the cycle, and C and m are material fatigue constants, which depend on the mode mixity B. Paris law uses the ERR value to define the crack growth rate. The ERR, Eq. (18.22), for an integration point is calculated as the area under the red traction-displacement jump curve, as illustrated in Fig. 18.1. Eq. (18.24) is used to calculate the crack propagation rate for each interface element as:

$$E_p^{2D} = max\left(C\left(\frac{\sqrt{G_{max,i}}}{\sqrt{G_c}} - \frac{\sqrt{G_{min,i}}}{\sqrt{G_c}}\right)^{2m}\right) \tag{18.24}$$

where i is the number of integration points in the element. The element propagation rate could be used to calculate the number of cycles for the delamination to advance through this element if the delamination front was aligned with the element boundary. For each element, a value of effective length l_e is determined based on the gradient direction of the ERR inside the element. This procedure is very important if the surface elements have a large aspect ratio and are less critical if they have near square shape. The crack and/or delamination is propagated one element at a time so that the number of cycles ΔN_p^{2D} for propagation is:

$$\Delta N_p^{2D} = l_e / E_p^{2D} \tag{18.25}$$

where the length value for crack extension l_e is equal to the element length parallel to the crack propagation direction.

Note that we are not distinguishing between CZM evolution stages in matrix crack and delaminations: instead, in each case, the number of intervals is counted to the first occurrence of this event over the entire structure.

Specific to DDM is the MIC insertion, which can often happen during the static load ramp-up. If, however, no MIC was inserted in the static regime, then the process described in Section 18.2 is used to calculate the location and the cycle count for MIC insertion. Based on Eq. (18.15), we can compute the number of cycles until $d = 1$ at each load cycle step i, which in our context corresponds to the number of cycles to the insertion of a new matrix crack as:

$$\Delta N_c^{2D} = min\left[\frac{1-d}{N_{lim}(f_{r'})}\right] \tag{18.26}$$

A linear update of the damage norm, Eq. (18.19), in each integration point is performed proportionally to the ratio of the number of cycles at a given load amplitude ΔN during fatigue.

$$RT_f^{2D} = RT + (1-RT)\frac{\Delta N}{\Delta N_p^{2D}} \tag{18.27}$$

where the number of cycles at a given load amplitude ΔN is:

$$\Delta N = min\left(\Delta N_c^{2D}, \Delta N_p^{2D}\right) \tag{18.28}$$

In a complex DDM simulation, as the simulation progresses, the CZM propagation cycle count ΔN_p^{2D} becomes smaller and smaller. After it reaches a certain predefined minimum value ΔN_{min}, the propagation events determine the number of cycles per loading step $\Delta N = \Delta N_p^{2D}$, even if ΔN_c^{2D} is smaller and the cracks are inserted not exactly when the number of cycles is reached, rather the end of step is determined by delamination propagation. This may delay a crack insertion somewhat but allows the computation to proceed. This recursive procedure takes place until the total cycle $\sum \Delta N$ reaches the target value.

18.2.3.2 Cycle-based algorithm for three-dimensional application

As mentioned earlier, the EBA is extremely useful for two-dimensional (2D) applications if damage events occur in a specific direction such as in a DCB test or ENF test. However, if a three-dimensional (3D) structure with multiple elements meshed along the width direction is considered, the stagnation situation occurs as ΔN_p^{2D} at the crack tip element is getting smaller and smaller during damage evolution and it only allows one element to be released every loading step. Consequently, the proposed cycle-based algorithm overcomes this situation and multiple elements can be updated simultaneously. In the cycle-based algorithm, one predefined number of fatigue cycles on each solution step stimulates the damage which occurs during this amount of cycle. The modified Paris law will be used as stated in Eq. (18.23). The propagation stage of the CZM formulation is conceptually following that of Tao et al. (2018); however, the key ingredients such as the underlying static CZM, the form of the Paris law, and more importantly the element update are different. The comparison between element update criteria is shown in Fig. 18.2. The key difference is that in Tao et al. (2018) and Kawashita and Hallett (2012), only the crack tip elements were updated under fatigue loading requiring a global search over elements to determine their positions with respect to fully and partially damaged elements. In the present approach, the delamination front is very complex and defined by intersection with matrix cracks which would make such searches impractical. Instead, the damage norm is updated in all elements and realistic rates of propagation are achieved. The formulation is described below. The ERR Eq. (18.22) is calculated as the area under the traction-displacement jump curve, as illustrated in Fig. 18.3. Without restricting the generality, the point λ_1 with corresponding damage variable D_1 will be the first position of the material point in the damage phase after the initiation strength was sufficiently lowered. The corresponding ERR can be obtained as:

$$G_{max}^{[1]} = \frac{1}{2}\left[Y_r\lambda_1 + K(1-D_1)\lambda_1(\lambda_1-\lambda_0)\right] \tag{18.29}$$

Assume that λ_2 is the next position of the material point after a number of cycles and D_2 is the corresponding damage variable. We will calculate the ERR corresponding to λ_2 by using the trapezoidal rule as follows:

$$G_{max}^{[2]} = G_{max}^{[1]} + \frac{1}{2}\left[(K(1-D_1)\lambda_1 + K(1-D_2)\lambda_2)(\lambda_2-\lambda_1)\right] \tag{18.30}$$

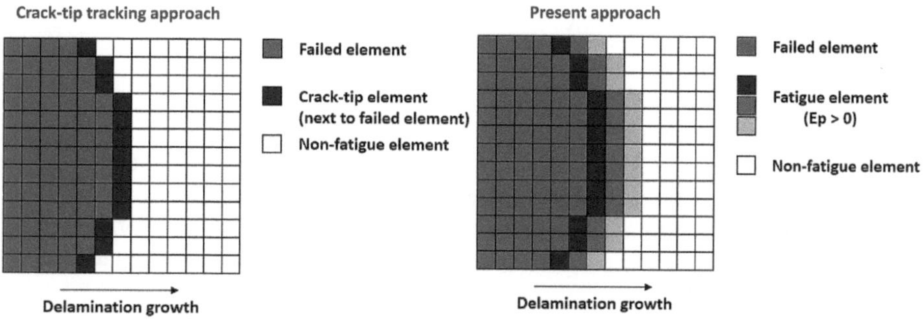

Figure 18.2 Comparison of element update criteria between crack-tip tracking approach (Tao et al., 2018; Kawashita and Hallett, 2012) and present approach.

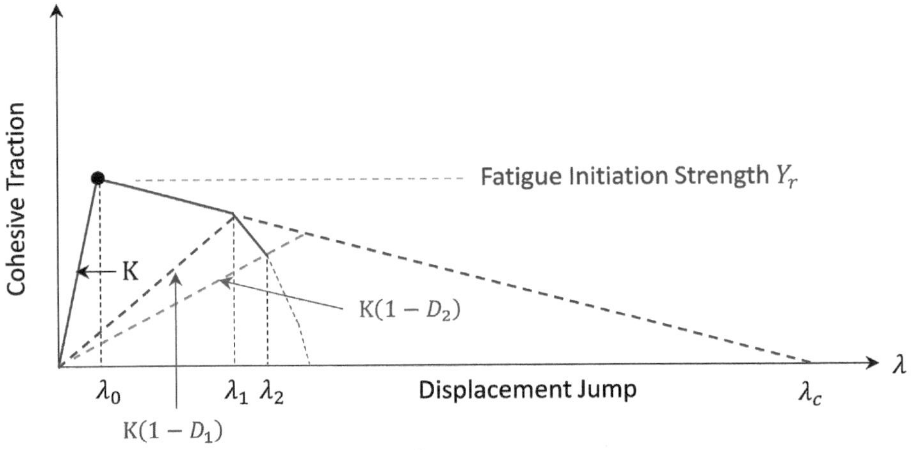

Figure 18.3 Schematic of a cohesive zone model for fatigue damage propagation.

and the $G_{min} = R^2 G_{max}$, where the R ratio is assumed to be constant through the model and equal to that of the applied load. Eq. (18.23) is used to calculate the maximum crack propagation rate for each interface element as:

$$E_p^{3D} = max\left(C\left(\frac{\sqrt{G_{max,i}}}{\sqrt{G_c}} - \frac{\sqrt{G_{min,i}}}{\sqrt{G_c}}\right)^{2m}\right) \tag{18.31}$$

where i is the number of integration points in the element. A linear update of the damage norm in each integration point is performed proportionally to the ratio of the cycle increment ΔN on the given step and the number of cycles required to propagate the entire element length:

$$\Delta N_p^{3D} = l_e / E_p^{3D} \tag{18.32}$$

so that:

$$RT_f^{3D} = RT + (1 - RT)\frac{\Delta N}{\Delta N_p^{3D}} \tag{18.33}$$

The above algorithm is implemented in an implicit analysis scheme. The updated damage norm in Eq. (18.33) is calculated at the beginning of a new step based on the converged and equilibrated displacement field at the end of the previous step. During the equilibrium iterations on the new fatigue step corresponding to ΔN cycles, the Newton–Raphson iterations are performed so that in each integration point we require that:

$$RT^u = \max\left(RT, RT_f^{3D}\right) \tag{18.34}$$

where the superscript u represents update. Eq. (18.34) allows for the damage norm update either in static or fatigue regime but does not add the two. It means that in each material point, the program selects either the damage norm due to static equilibrium or fatigue propagation. It should be noted that every element has individual ΔN_p^{3D} and depending on the value of ΔN a different fraction of the damage norm is updated in Eq. (18.33). In composite materials, the Paris law has extremely steep dependence on the ERR and this update has a significant effect only on the so-called crack front elements with the highest ERR or lowest $\Delta N_p^{front} = \min\left(\Delta N_p^{3D}\right)$. The update, Eq. (18.34), however, is performed in all elements. An apparent choice of ΔN would be that of the crack front element so that $\Delta N = \Delta N_p^{front}$. Such choice can be well reasoned when dealing with a single delamination since ΔN_p^{front} would be that of a group of elements forming the crack front. In the OH examples considered below, we will be dealing with multiple delaminations at each interface as well as multitude of matrix cracks. While the above reasoning for choosing $\Delta N = \Delta N_p^{front}$ is not applicable, we will, however, continue to use it as a default criterion for cycle jump selection, where the minima are taken over all elements $\Delta N_p^{3D} > 0$ involved in fatigue propagation. No topological considerations of element location with respect to the delamination front or matrix crack tip(s) are involved. As will be seen in numerical studies, $\Delta N = \Delta N_p^{front}$ is a conservative criterion and larger cycle jump values will provide accurate results as well. What is important is that within a meaningful cycle jump ΔN range, the predicted damage patterns appear consistent allowing to make conclusions upon convergence in the temporal domain.

18.3 Verification of Rx-FEM coupon-level analysis

In this section, the results of verification and partial validation of the DDM methodology are presented. The verification of these coupon models is accomplished by comparing them with experiments for DCB and ENF configurations.

18.3.1 Double cantilever beam analysis

The configuration of the DCB specimen and relevant experimental data, that is, the Paris law parameter, used for this study were described in Dávila et al. (2020). These parameters were then used in the present 2D and 3D simulations. The specimens were nominally 178 mm long and 12.7 mm wide. The initial crack length a_0 is 50.8 mm, the height of the arm h is 2.25 mm which contains 24 unidirectional IM7/8552 plies as shown in Fig. 18.4. The material properties of IM7/8552 (Dávila et al., 2020) are summarized in Table 18.1. The propagation zone (PZ) length is 10 mm, and the front of the propagation zone (FPZ) is 2.5 mm. To ensure accurate crack growth rates, the element edge size in the PZ is 0.1 mm, which is satisfied with the recommended mesh size provided by Dávila et al. (2020). The FPZ element edge size is 0.25 mm. The element mesh size

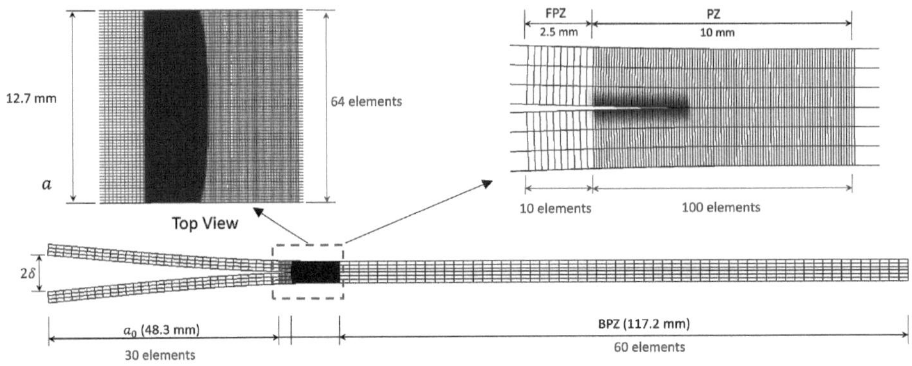

Figure 18.4 Double cantilever beam configuration.

Table 18.1 Material properties used in DCB, ENF, CTB, and OH tests.

Material properties	DCB (Dávila, 2020)	ENF (Krueger, 2011)	CTB (Adluru et al., 2018)	OHT (Iarve et al., 2017)
E_{11} (GPa)	146.67	161.00	157.20	164.0/137.4
E_{22}, E_{33} (GPa)	8.70	11.38	8.96	8.97
G_{12}, G_{13} (GPa)	5.16	5.20	5.08	5.02 (NLS)
G_{23} (GPa)	3.00	3.90	2.99	3.00
v_{12}, v_{13}	0.32	0.32	0.32	0.32
v_{23}	0.40	0.45	0.5	0.496
$\alpha_1 (10^{-6}/°C)$	0.00	0.00	0.00	0.00
$\alpha_2 (10^{-6}/°C)$	30.0	30.0	30.1	30.0
G_{IC} (N/mm)	0.24	0.21	0.24	0.256
G_{IIC} (N/mm)	0.739	0.78	0.739	0.649
Y_T (MPa)	64.1	98.6	127	100.0
Y_C (MPa)	285.7	199.95	200	247.0
S (MPa)	97.6	95.84	95.8	100.0
X_T (MPa)	2905	2501	2723	2900
X_C (MPa)	1580	1716	1200	1680

in the initial crack a_0 region is 1.61 mm. Three elements were used along the thickness direction of each arm. Note that the difference between 2D and 3D models is the total element number along the width direction. One element along the width direction is used to simulate 2D models. A total of 64 elements were used along the width direction for DCB with 3D models so crack growth rates along different width locations can be captured. Three displacements, 1.70, 1.92, and 2.25 mm, were studied, and crack growth rates were compared with the Paris law presented in Dávila et al. (2020).

The comparison between 2D and 3D crack growth rates of DCB simulations is shown in Fig. 18.5. The crack growth rates correlate well in 2D models. However, the crack growth rates observed from 3D models are not fully aligned with the given

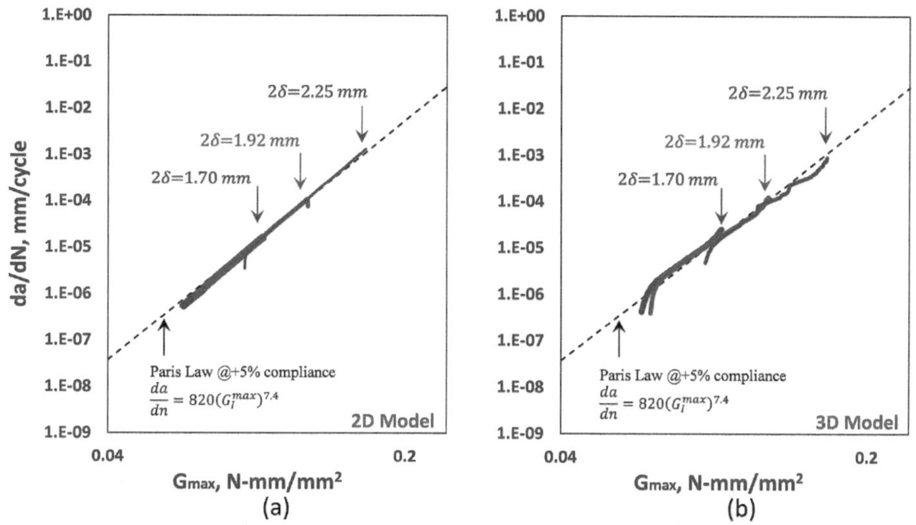

Figure 18.5 Crack growth rates versus G_{max} comparison for (A) two-dimensional (B) three-dimensional DCB models.

Paris law. According to Fig. 18.4, the crack front of a DCB model shows the toenail shape, which is expected in a DCB test. The crack front is symmetric with respect to the line along the longitudinal axis located at the mid-span width of the specimen. The larger crack growth rates are observed at this line. In contrast, smaller crack growth rates are located near the edges along the longitudinal axis. In the present work, the crack growth rates of 3D models shown in Fig. 18.5B are computed based on the symmetry line which has larger growth rates. Overall, both event-based fatigue algorithm for 2D applications and cycle-based fatigue algorithm for 3D applications provide excellent agreements in crack growth rates between numerical results and provided Paris law.

18.3.2 End-notched flexure analysis

ENF configuration (Krueger, 2011) is examined and is shown in Fig. 18.6, where 2L = 101.6 mm is the distance between supports and the active crack length is a_0 = 25.4 mm. Other dimensions are FPZ = 50.8 mm, and PZ = 101.6 mm, so the total precrack length is 76.2 mm, and the total length of the specimen is 177.8 mm. The width of the specimen is 25.4 mm, and the thickness of each arm is 2.25 mm. Along the length, all models were divided into different sections with different mesh refinements. In the FPZ section, 10 elements were used, and in the "a_0" section, the mesh size gradually reduces from 1 to 0.1 mm. The fine mesh was used within the PZ section except outside the support area. Through the specimen thickness direction, three elements were used in each arm. One element along the width direction is used to simulate 2D models. A total of 26 elements were used along the width direction for 3D ENF models so crack propagation rates along different width locations can be captured.

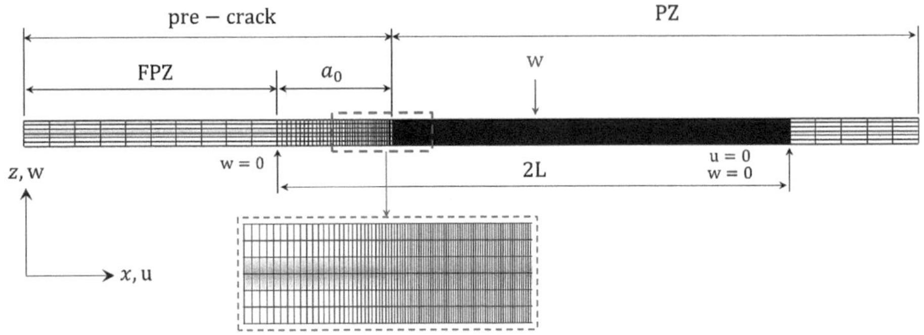

Figure 18.6 End-notched flexure configuration and boundary conditions.

The downward displacement is applied in the middle point of the length 2 L. The material properties used in the study are shown in Table 18.1.

Fig. 18.7 shows the predicted crack growth rates comparisons between 2D and 3D ENF models under applied displacement w of 1.0 and 0.78 mm as stated in Krueger (2011). Excellent agreement of crack growth rates between simulations and the Paris Law is observed in 2D models. However, identifying a crack growth rate in a 3D problem formulation is subject to some ambiguity. It is well known that the crack growth rate can be computed as dissipated energy divided by the incremental crack area as the crack grows. However, identifying the variation of crack area requires micromonitoring, which is not very practical. Therefore, the crack is assumed to uniformly grow in the width direction of the specimen. In addition, we used multiple points along the crack front to compute the crack growth rates. The scatter points shown in Fig. 18.7B were obtained by using crack growth rates for each element row through the width, a total of 27 rows in the width direction. The combined trendline for the points along the 3D crack front at three displacement amplitudes has the Paris law power coefficient $m = 5.63$, which is only 1.4% higher than the benchmark value.

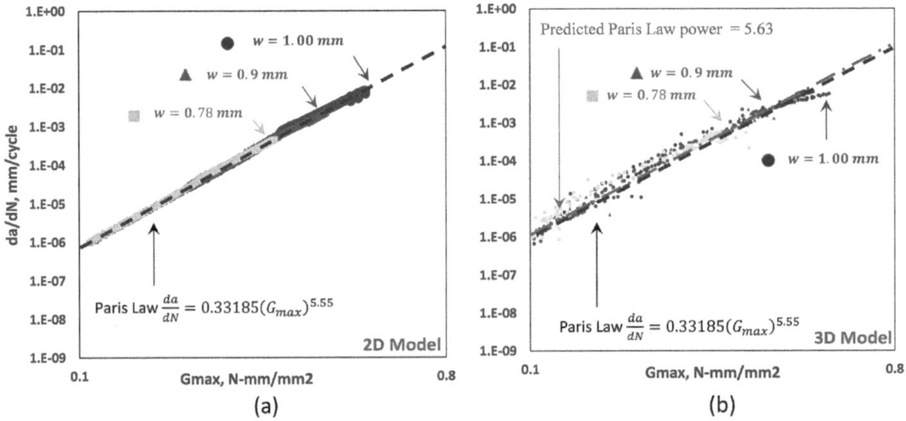

Figure 18.7 Crack growth rates versus G_{max} comparison for (A) two-dimensional (B) three-dimensional ENF models.

18.4 Validation of Rx-FEM subelement level analysis

In this section, the development and validation of Rx-FEM analysis tools to predict the durability and performance of structural elements at the subelement level are presented. The first representative subelement, the CTB specimen, which was designed and tested at NASA Langley Research Center, is considered to validate the 2D fatigue algorithm. This specimen was designed to study failure initiation, propagation, and delamination migration in both quasistatic and fatigue loading. The second representative subelement level analysis is OH laminates subjected to fatigue to assess the state of the art of progressive damage and the damage interaction. This OH specimen was designed and launched by the Air Force Research Laboratory (AFRL). The fatigue algorithm for 3D applications is selected and validated for the matrix crack, delamination, and interactions between them.

18.4.1 Clamped tapered beam

Laminated composites usually contain several stress concentration characteristics such as holes, stiffener attachments, and tapers. These characteristics often determine the load-carrying capacity of the structure. A CTB subelement was proposed by NASA to study the damage initiation and evolution from such stress concentration characteristics. Several damage patterns such as matrix crack initiation, delamination propagation, and migration from one interface to another are observed in the specimen. Previously, both static (Adluru et al., 2019) and fatigue (Adluru et al., 2018) simulations were studied using Rx-FEM. The essence of this methodology is the insertion of true displacement discontinuities independently of mesh orientation to represent matrix cracking. Multiple cracking in each ply is allowed. The present work will focus on the CTB validation using the 2D fatigue algorithm along with the strength tracking (ST) method.

CTB schematics, dimensions, boundary conditions, and load application points are shown in Fig. 18.8. The vertical displacement (u_z) on the left side is constrained. On the right side, both horizontal (u_x) and vertical displacement (u_z) are constrained. Only one element through the width direction was used taking advantage of the 2D natural

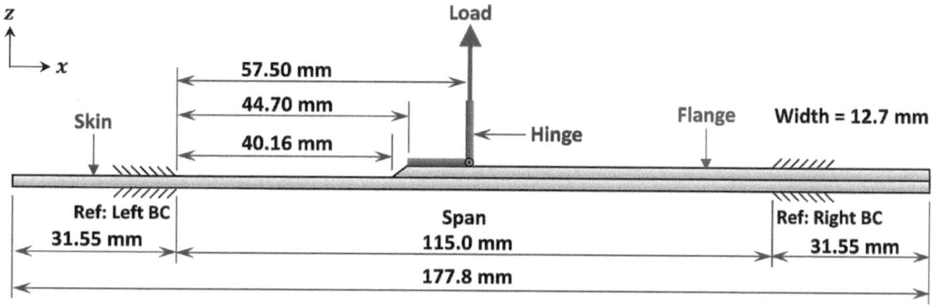

Figure 18.8 Clamped tapered beam configuration and boundary conditions.

problem with a symmetric condition $u_y = 0$ at plane where $y = 0$. In the current CTB fatigue test, the residual thermal stress is taken into consideration. It has a significant effect on the onset location of the crack insertion. A uniform temperature change $\Delta T = -165°C$ is applied for the difference between cure and room temperature. A three-stage load application procedure is utilized, where the first stage is under thermal loading, the second stage starts from static loading, and followed by the third stage under constant amplitude fatigue loading.

Both the flange and skin are made of 14 plies, 190 gm/m^2 IM7/8552. The material properties are listed in Table 18.2. The ply thickness is approximately 0.18 mm. The plies in the laminate layups are listed from the bottom to the top as $[0/90_3/0/90/0_2/90/0/90_2/0/90]$ in the skin and as $[90_3/0_2/90/0_2/90/0_2/90_3]$ in the flange, respectively. The fatigue analysis was conducted under displacement control with a stress ratio of R = 0.2. Two constant amplitude cycling loadings with a maximum arm displacement, 0.91 mm (high load) and 0.825 mm (low load) are investigated, which are approximately 80% and 70% of peak static load. The model was meshed in Abaqus CAE 2016. It was concluded that the fine mesh model, where mesh size is approximately equal to 1/2 ply thickness, is appropriate considering both accuracy and computational efficiency (Adluru et al., 2019). The characteristic distances and CTB failure path are illustrated in Fig. 18.9. The matrix crack forms on the tapered edge at the interface between 0° and 90° plies at point A. The crack is

Table 18.2 Material properties for clamped tapered beam and open-hole fatigue analyses.

Material properties	CTB (Adluru et al., 2018)	OH (Iarve et al., 2017)
S_I (normal)	0.0765	0.03
S_{II} (shear)	0.045	0.075
C_I (mm)	0.001	0.025
C_{II} (mm)	0.012	0.065
m_I	6.710	7.32
m_{II}	5.450	6.354
η	2.1	2.25

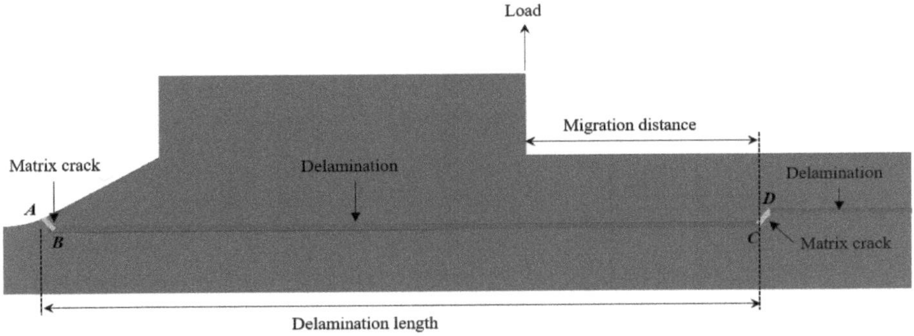

Figure 18.9 The damage events in the clamped tapered beam test as the crack grows.

inserted and propagated from point A to point B. Delamination initiation and propagation occurs between point B and point C. At point C, delamination migration from one interface to another occurs and the second delamination propagation continued after point D.

The dependence of crack growth rate in unidirectional composite tape laminates has been often observed as the effect of delamination resistance curve. This effect is called the R-curve effect, which is the effective increase in critical ERR as the crack propagates. The R-curve effect was also observed in generic interfaces between plies of different orientations as a result of delamination delving slightly into one of the bounding plies (Ratcliffe and Carvalho, 2014). Consequently, the fatigue predictions with the initiation value of the ERR ($G_{Ic}=0.24$ N/mm) and the steady-state value ($G_{Ic}=0.36$ N/mm) are performed. The delamination length versus corresponding fatigue cycles of CTB high load comparison using two different strain ERRs is shown in Fig. 18.10. It can be observed that the crack growth predictions appear somewhat conservative, especially for the predictions with the initiation ERR value, whereas the predictions with the steady-state value seem to provide more realistic results. The matrix crack was inserted at the location on the surface of the tapered region away from the interface with 0° ply, which correlates well with the observation stated by Adluru et al. (2018).

Unstable delamination propagations are observed in both simulations using the initiation ERR ($G_{Ic}=0.24$ N/mm) and the steady-state ERR ($G_{Ic}=0.36$ N/mm).

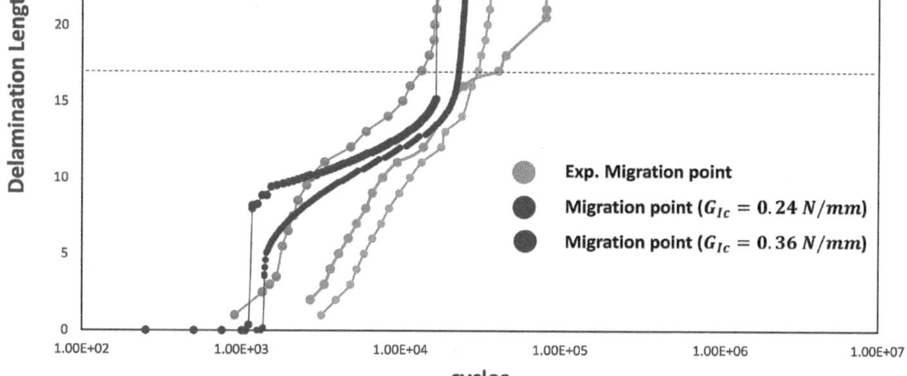

Figure 18.10 The delamination length versus cycles curves using $G_{Ic}=0.24$ N/mm, $G_{Ic}=0.36$ N/mm comparison between simulations and experiments. The CTB specimen is subjected to a high load where the arm displacement, 0.91 mm, is approximately 80% of the static peak load.

The delamination migration distances are slightly overpredicted due to unstable propagation. The delamination length versus cycles curve is almost a vertical straight line as shown in Fig. 18.10. It should be noted that the delamination initiation cycle of delamination using both ERR results provide excellent agreement with experiments.

The delamination length versus cycles curves using $G_{Ic} = 0.24$ N/mm, $G_{Ic} = 0.36$ N/mm subjected to low load (70% of static peak load) comparison between simulations and experiments is shown in Fig. 18.11. The matrix crack was inserted at the location on the surface of the tapered region away from the interface with 0° ply, which is identical to the result of CTB subjected to high load (80% of peak load). Unstable delamination propagation was observed only in the simulation using the initiation ERR ($G_{Ic} = 0.24$ N/mm). The stable propagation rate is predicted in the result using the steady-state ERR ($G_{Ic} = 0.36$ N/mm), which is much closer to the experimental data. The delamination propagation rate becomes larger near the location where the delamination migration occurs, but it is still within the scope of stable delamination propagation. The migration distance of steady-state ERR has an excellent agreement with experiments. Namely, the baseline predictions of the crack growth rate with $G_{Ic} = 0.24$ N/mm were similarly conservative. Much closer agreements with experiments were achieved by taking into account fiber bridging effects. Increasing the $G_{Ic} = 0.36$ N/mm motivated by the R-curve effect in the present work brought the results in very good agreement with the experiments, especially in the cycle of onset of delamination.

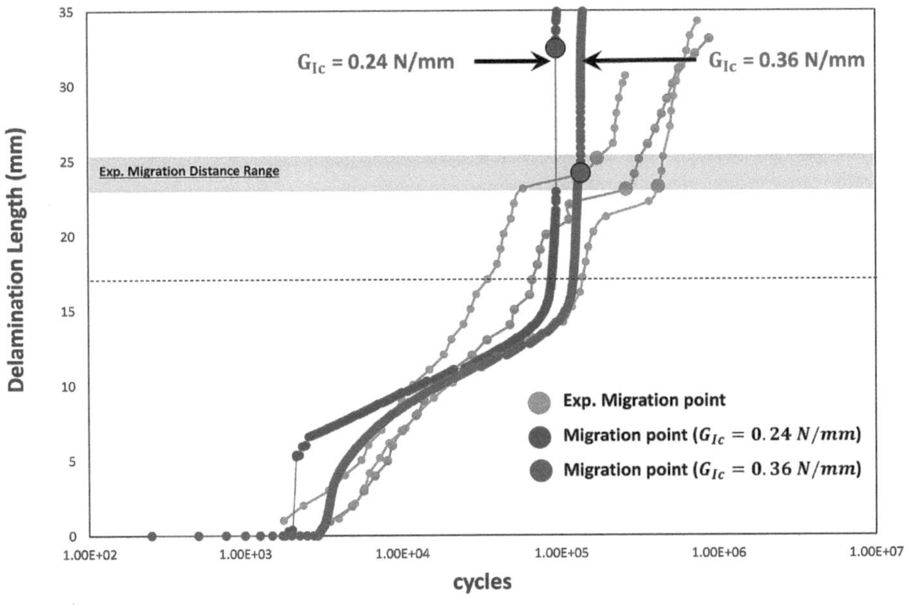

Figure 18.11 The delamination length versus cycles curves using $G_{Ic} = 0.24$ N/mm, $G_{Ic} = 0.36$ N/mm comparison between simulations and experiments. The CTB specimen is subjected to a low load where the arm displacement, 0.825 mm, is approximately 70% of the static peak load.

18.4.2 Open-hole tension–tension fatigue

Three IM7/977-3 laminates with stacking sequence $[30°/60°/90°/-60°/-30°]_{2s}$, $[0°/45°/90°/-45°]_{2s}$, and $[60°/0°/-60°]_{3s}$ previously considered in Iarve et al. (2017) will be examined. All specimens were 250 mm long and 38.1 wide with a 6.35 mm center hole and a 56 mm tab on the right. During the simulation, the left edge is constrained and fatigue loading in constant force amplitude regime was introduced through the tab as shown in Fig. 18.12. Only half of the symmetric laminate was modeled, and uniform tensile stress was applied at the right surface of the tab. Such load application was chosen to model the load control regime. The force resultant at the tab surface was calculated and divided by the specimen cross-section and reported as applied axial stress. Static and fatigue properties used in the simulation are shown in Tables 18.1 and 18.2, respectively. Nonlinear in-plane shear and thermal residual stress with $\Delta T = -159°C$ are included in the analysis. The specimens were subjected to fatigue loading under tension–tension cycles at $R = 0.1$ and were inspected by X-ray CT after a certain number of cycles. The loading amplitude and inspection cycle count for the three laminates are shown in Table 18.3.

Prior to discussing the results, we will illustrate the crack insertion and opening process. Fig. 18.13 displays the delamination at the 30°/60° interface of the $[30°/60°/90°/-60°/-30°]_{2s}$ laminate along with the matrix cracks in the 30° and 60° layers after 200,000 cycles. The left image shows the crack path, which has been inserted. The crack path was inserted at a given integration point where the failure index Eq. (18.10) is higher or equal to the residual strength coefficient f_r after a certain number of cycles.

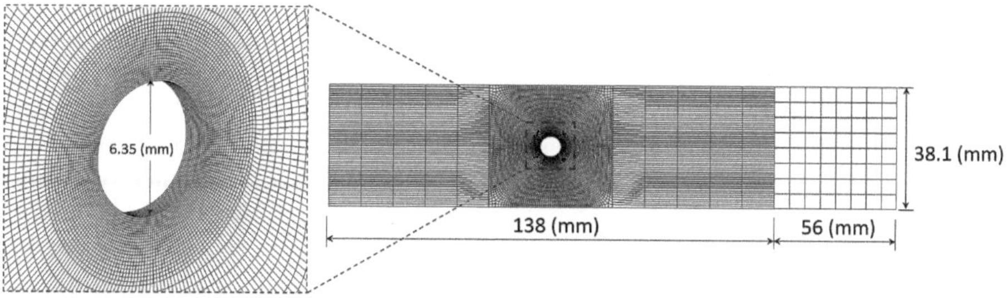

Figure 18.12 Open-hole specimen configuration.

Table 18.3 Description of selected open-hole specimens.

Stacking sequence	R-ratio	Fatigue loading at % static failure load	Cycles of X-ray CT inspection
$[30/60/90/-60/-30]_{2s}$	0.1	40%	200,000
$[0/45/90/-45]_{2s}$	0.1	50%	300,000
$[60/0/-60]_{3s}$	0.1	80%	200,000

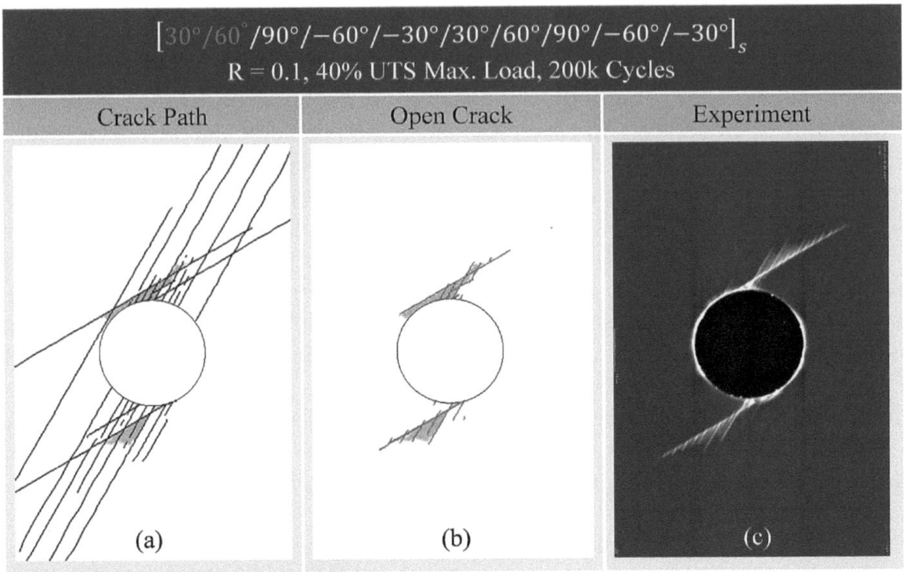

Figure 18.13 Delamination and matrix cracks at the outer 30°/60° interface after 200k cycles. (A) Crack path, (B) Open crack, (C) X-ray CT.

The crack path is inserted by means of Rx-FEM enrichment, and the lines in Fig. 18.13A are contours of the regularized step function $\widetilde{H}$ values of 0.5. The default length of the inserted crack is only limited by the domain boundary, that is, the OH contour and the outer rectangular contour. The finite crack path length is determined by positioning the previous cracks, so that the crack path is terminated when it gets as close as one element away from existing cracks, where the enrichment scheme described above is not valid. Fig. 18.13B displays open cracks selected as the damage variable in Eq. (18.20) greater than 0.995 (Iarve et al., 2017). While the open crack pattern is a subset of the overall crack pattern, only this pattern closely resembles the experimentally observed crack pattern. This illustrates the importance of each of the solution components, that is, the crack insertion as well as the crack opening methodologies.

We will proceed with comparing the delamination size and shape as well as the matrix cracking pattern observed experimentally with the open crack pattern and delamination size predicted for the three laminates in this work. Fig. 18.14 shows the delamination and matrix crack predictions of $[30°/60°/90°/-60°/-30°]_{2s}$ OH laminate in all interfaces after 200,000 cycles. At the outer 60°/90° and 90°/−60° interfaces, the X-ray CT image reveals short matrix crack patterns which appear to form approximately 30° bands. At the 60°/90° interface, this image is formed by 60° cracks, and at the 90°/−60°, it is formed by 90° cracks. Similar behavior is also shown by predictions. The matrix cracking pattern at the outer −60°/−30° interface is very different and represents a cloud of matrix cracks some of which are not attached to the hole. The same detail can be observed in the predicted patterns. Note that the damage patterns on

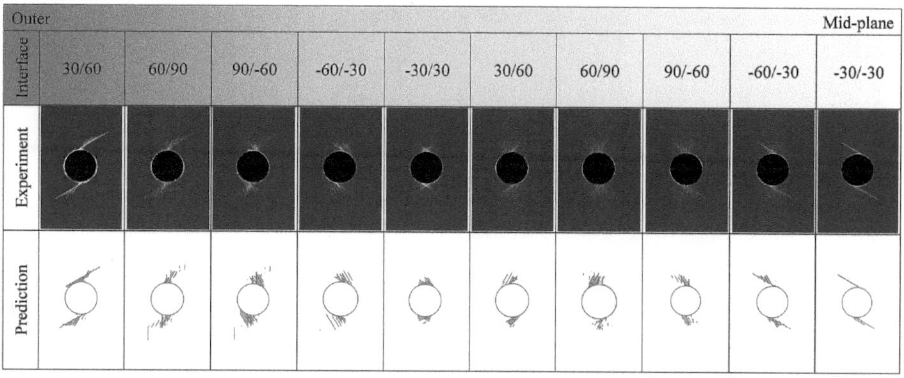

Figure 18.14 Delamination and matrix cracking distribution comparison between simulation and experimental results of OH $[30°/60°/90°/-60°/-30°]_{2s}$ specimen after 200k cycles.

the inner interfaces and outer interfaces between the same plies are also different. Compare, for example, the −60°/−30° interface in the outer and inner sublaminate. In the outer sublaminate, one observes a cloud-like pattern of matrix cracks, whereas, at the inner interface enlarged in Fig. 18.15, one observes long −30° cracks, and coincidentally both the CT image and the prediction show two closely spaced long cracks crossed by short −60° cracks.

Fig. 18.16 shows the delamination and matrix crack predictions for the quasiisotropic laminate $[0°/45°/90°/-45°]_{2s}$ after 300,000 cycles. Note that the top surface appears to be overtreated with die-penetrant and leaves shadows on the next surface as well. Nevertheless, the crack length and delamination agree well between the experiment and the prediction. The inner 45°/90° interface damage view is enlarged in Fig. 18.17. What appears to look like −45° cracks in the CT image is a sequence of short 90° cracks spun over the path of the −45° crack from the interface below. Similar pattern albeit less crisp is predicted as well. The number of matrix cracks predicted in the −45° ply is higher than observed in the experiment. However, the length of large cracks is approximately the same as that of the pattern of small 90° cracks around the hole and bounded by the −45° cracks. Fig. 18.18 shows the comparison between experimental data and predicted delamination and matrix cracking distribution for the $[60°/0°/-60°]_{3S}$ laminate. This laminate exhibits a significantly larger delamination size and extent of matrix cracking than the previously considered laminates. The predicted extent of both delamination and matrix cracking is also significantly larger than for previous laminates and close to experimental data, but it appears to be slightly overpredicted. Fig. 18.19 displays the inner −60°/60° interface at three different cycle counts of 50, 100, and 200K. A vast network of −60° and 60° matrix cracks develops at 50K cycles and further expands with increasing cycle count. While the predicted failure patterns follow the same trend, it appears that the predicted delamination size at 50K is larger than observed but grows slower. Overall, the damage pattern at 200K cycles is in good agreement with that observed in the experiment.

To evaluate the stability of the solution with respect to the cycle jump size, additional analysis with fixed $\Delta N = 100$, 500, and 1500 cycles was performed as shown

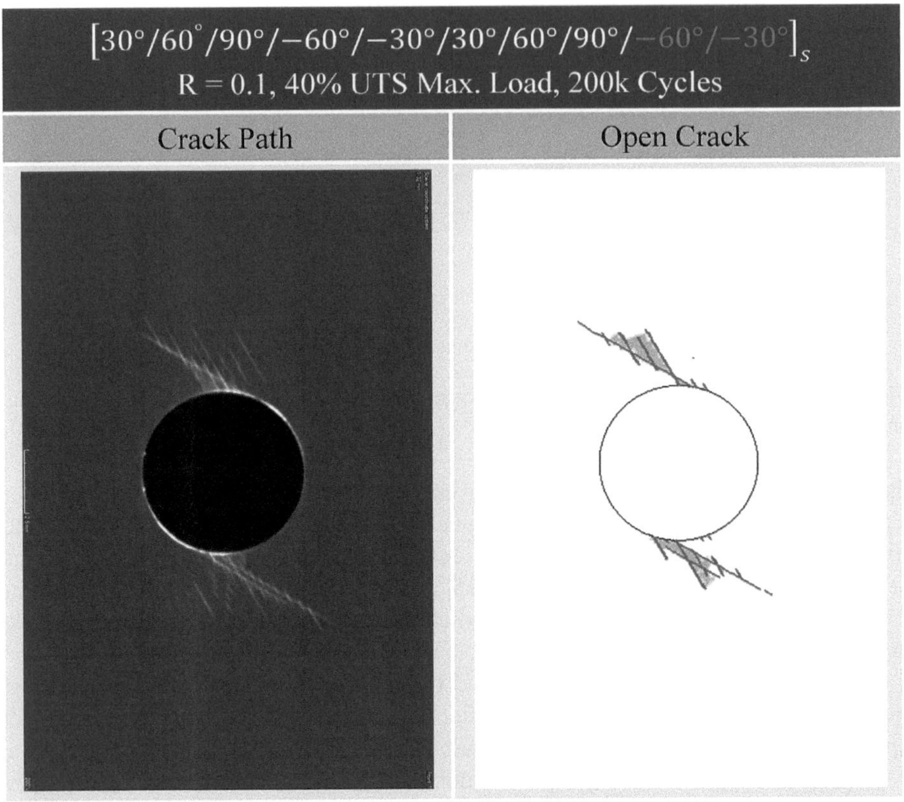

Figure 18.15 Comparison between experiment and simulation results at outer −60°/−30° interface of OH $[30°/60°/90°/-60°/-30°]_{2s}$ specimen after 200k cycle.

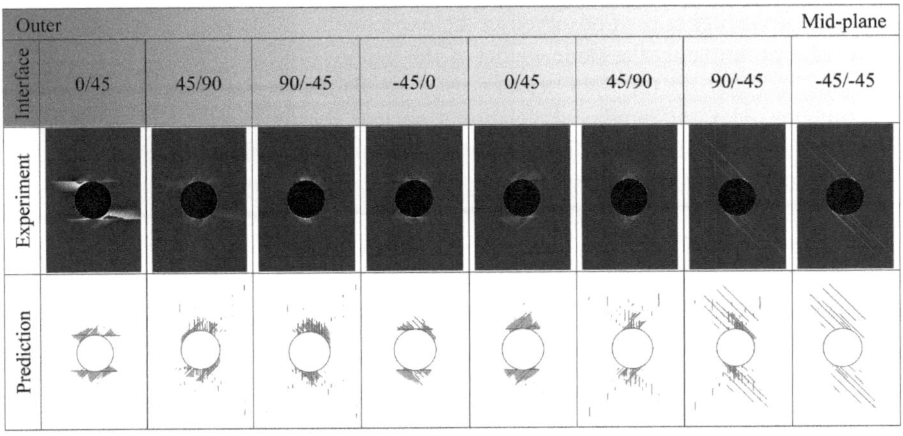

Figure 18.16 Delamination and matrix cracking distribution comparison between simulation and experimental results of OH $[0°/45°/90°/-45°]_{2s}$ specimen after 300k cycles.

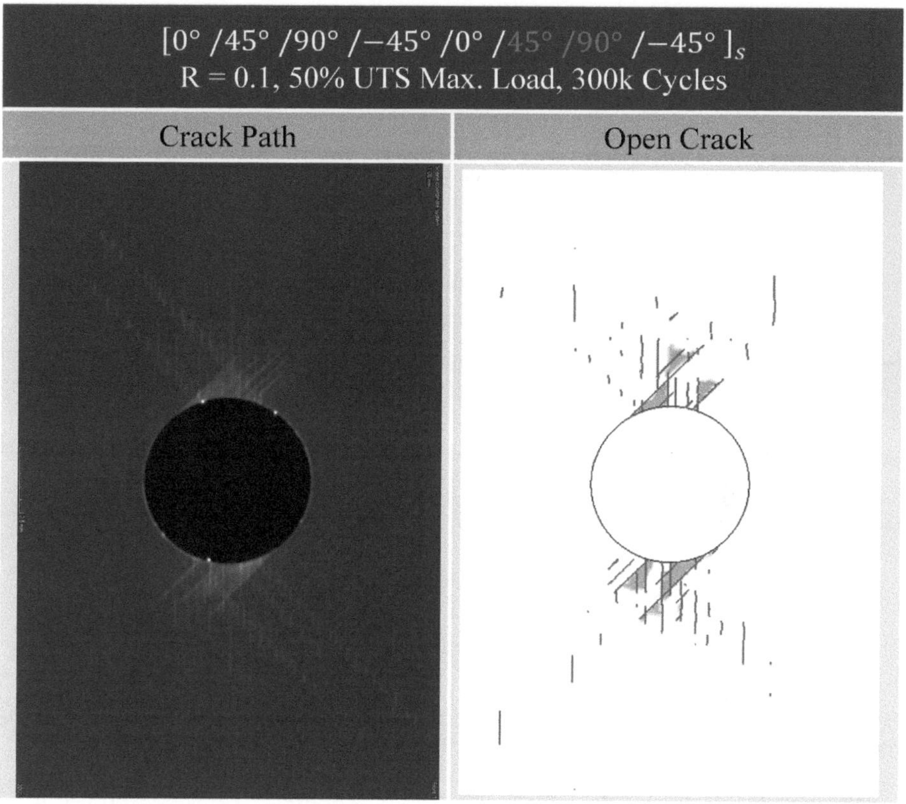

Figure 18.17 Delamination and matrix cracks at the inside 45°/90° interface after 300k cycles.

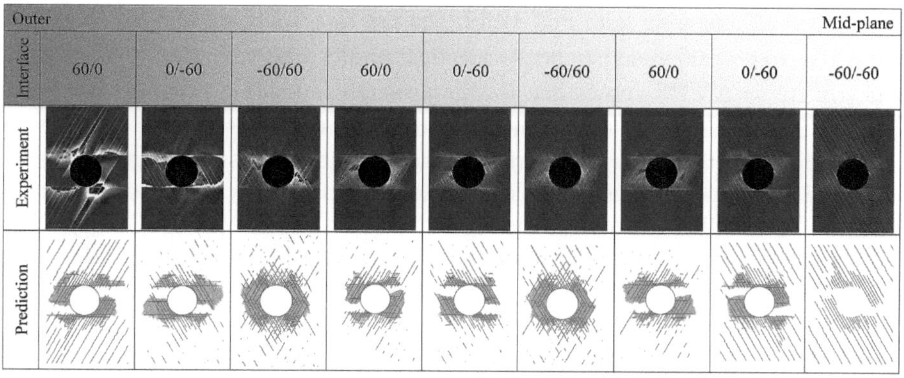

Figure 18.18 Delamination and matrix cracking distribution comparison between simulation and experimental results of OH $[60°/0°/-60°]_{3s}$ specimen after 200k cycles.

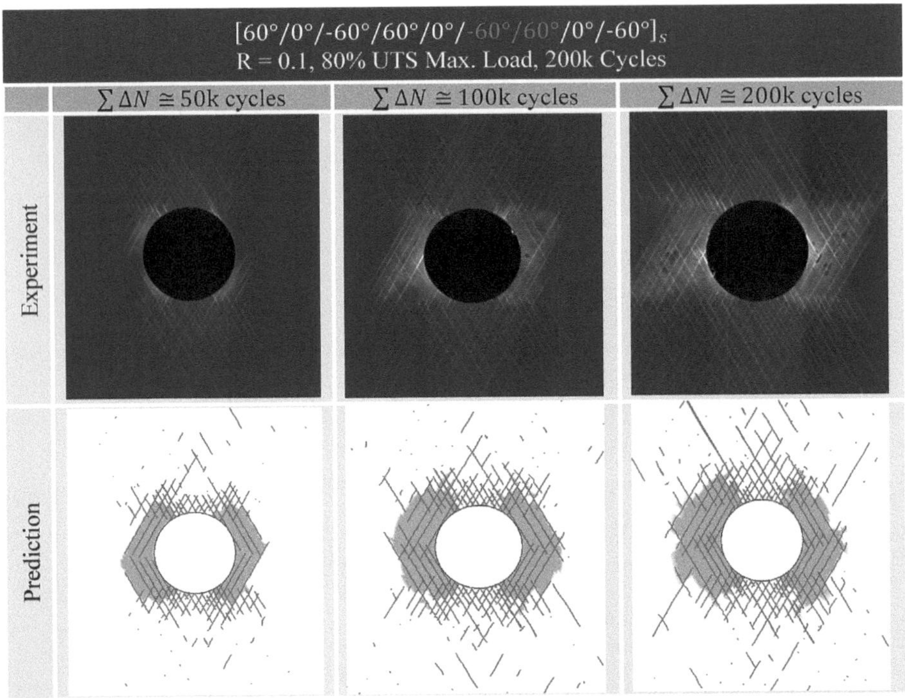

Figure 18.19 Damage evolution of delamination extent on inner −60°/60° interface after 50, 100, and 200k cycles.

in Fig. 18.20. The same inner −60°/60° interface at 200K cycles is considered. It can be observed that the damage patterns obtained with 100 and 500 cycle jump sizes are identical; however, the 1500 cycles per step appears too large and results in reduced damage size. Comparing with Fig. 18.19, one can also conclude that $\Delta N = \Delta N_p^{front}$ and $\Delta N = 100 - 500$ yield practically identical results. The analysis of the automatic step size $\Delta N = \Delta N_p^{front}$ shows that for the present example, the ΔN_p^{front} size varies from 1.5 cycles per step at the beginning of the analysis.

In addition to the damage pattern comparison, the stiffness loss predictions are compared with experimental data (Clay and Knoth, 2017). The comparison of residual stiffness prediction and experiments for $[0°/45°/90°/-45°]_{2s}$ and $[30°/60°/90°/-60°/-30°]_{2s}$ laminates after 1.5 million cycles is shown in Fig. 18.21. All six $[0°/45°/90°/-45°]_{2s}$ laminates experienced two-piece failure during experiments at very different cycle counts ranging from 0.4×10^6 to 2×10^6 cycles. The failure was sudden and without significant gradual stiffness loss. The specific cause of failure was not identified and is likely attributed to fiber failure, which is not accounted for in the analysis. It can be seen, however, that the predicted stiffness loss is in good agreement with the experimental data for surviving specimens in the 1.5×10^6 cycle range examined. Similarly, a good agreement can be seen for the $[30°/60°/90°/-60°/-30°]_{2s}$ laminates. A shorter cycle range of approximately

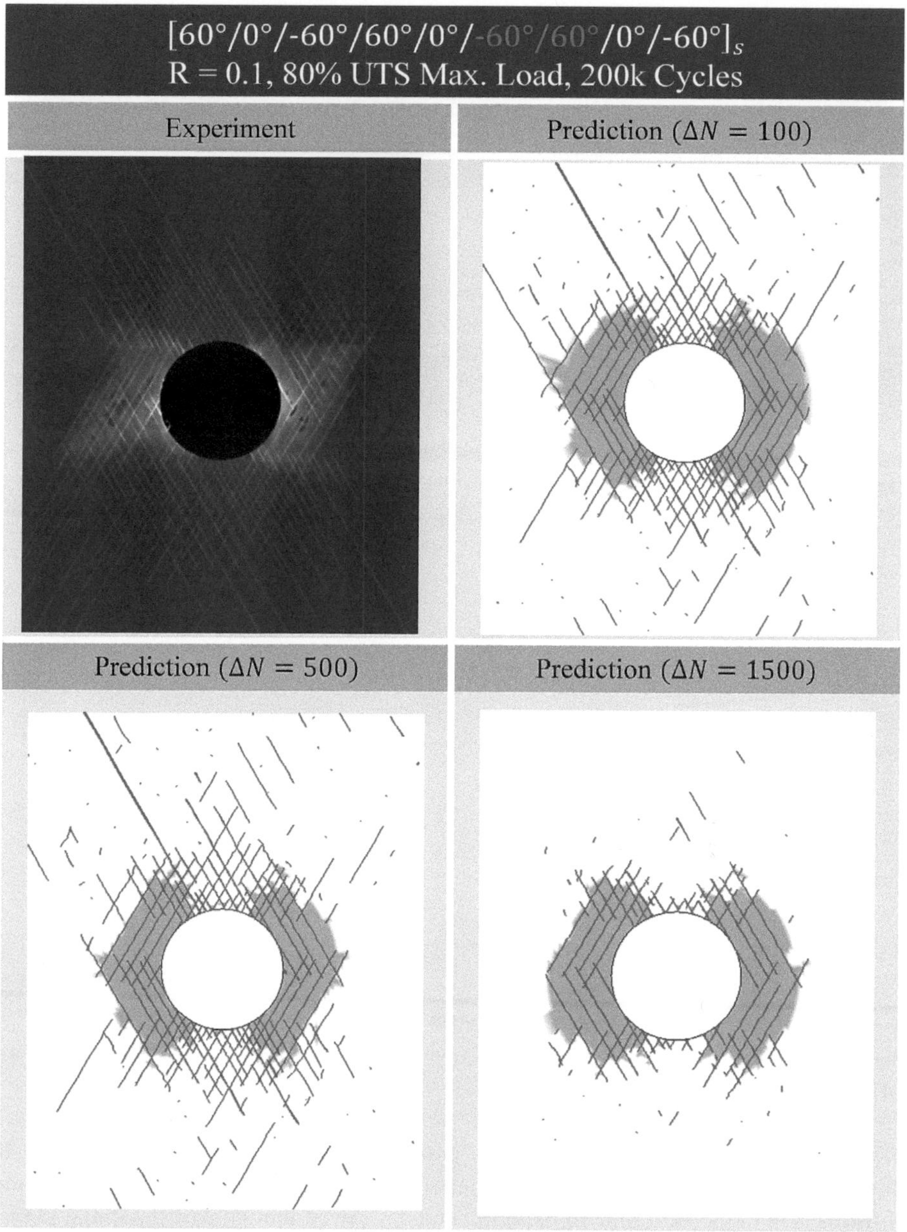

Figure 18.20 Delamination size and matric crack density comparison between experiment and simulation results using $\Delta N = 100$, 500, and 1500 cycles for OH $[60^\circ/0^\circ/-60^\circ]_{3s}$ specimen at $-60^\circ/60^\circ$ interface near the mid-plane.

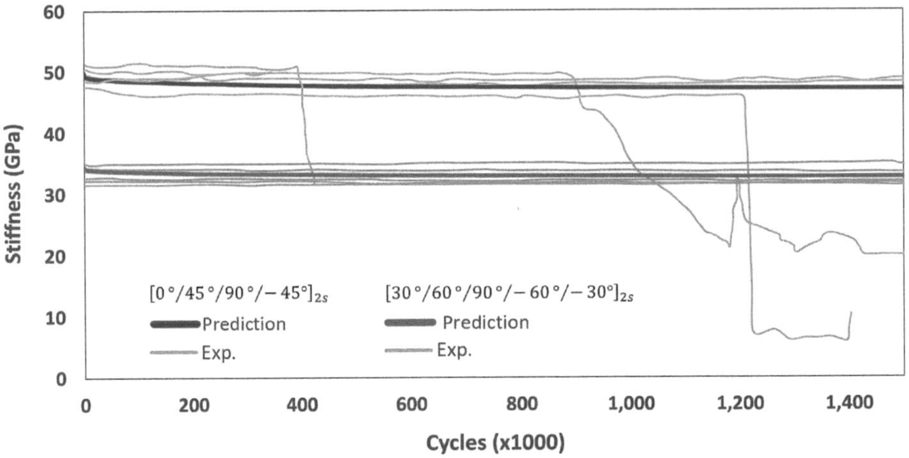

Figure 18.21 Residual stiffness versus cycles predictions and experiments for $[30°/60°/90°/-60°/-30°]_{2s}$ and $[0°/45°/90°/-45°]_{2s}$ OH specimens after 1.5 million cycles.

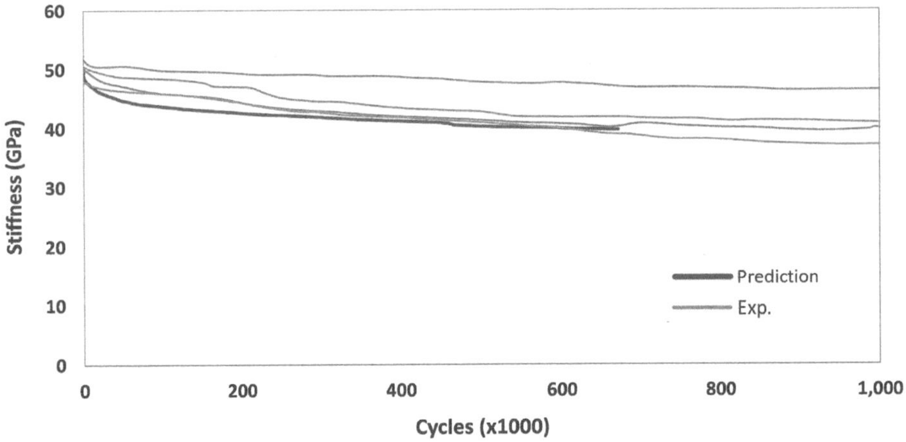

Figure 18.22 Residual stiffness versus cycles prediction and experiments for $[60°/0°/-60°]_{3s}$ OH specimen after 0.7 million cycles.

0.7×10^6 was examined for the $[60°/0°/-60°]_{3s}$ laminate as shown in Fig. 18.22. In this case, slightly faster stiffness degradation is predicted initially and the predicted stiffness is in the lower range of experimental data, even though past the initial range, the stiffness degradation rate is consistent with that reported in Clay and Knoth (2017).

18.5 Conclusions

In this work, the Rx-FEM framework for discrete modeling of damage evolution and interaction in laminated composite materials has been extended to increase discrete crack density for a given mesh resolution. This was achieved by extending the phantom node concept by introducing multiple phantom nodes corresponding to an original node in the vicinity of multiple close cracks. The algorithm preserves the Gauss integration within the cracked elements and allows to model matrix crack density comparable with experimentally observed in OH laminates under tension—tension fatigue.

The residual ST method, initially proposed for a semiempirical analysis of OH specimen test S—N data, has been implemented as a component of a high-fidelity PDA framework and has been combined with damage variable incremental update technique for the propagation phase within the CZM formulation. The CZM formulation directly uses the ST method to adjust the threshold load for damage initiation in every integration point according to the residual strength predicted locally by the ST method. In the propagation phase, an element maximum propagation rate has been used along with a damage variable update. Event-based cycle count has been used in 2D variant of the methodology, where critical event is delamination front propagation by 1 interval, whereas cycle increment-based method with incremental damage variable update in all elements regardless of their location with respect to the delamination front was used for 3D problems with complex crack fronts.

A CTB was considered under tension—tension fatigue loading for validation of 2D fatigue algorithm. The simulation results captured the main damage events including the first inserted matrix crack, delamination propagation, and delamination migration from one interface to another interface. The delamination onset cycles also exhibit excellent agreements with experiments.

Three different laminates containing OHs were considered under tension—tension fatigue loading for validation of 3D fatigue algorithm. A single unstructured FE mesh was used for discrete damage modeling in all laminates. Overall, the analysis results displayed excellent agreement with experimental data in terms of the size and location of damage including fine details of damage distribution such as small stitch cracking and delamination shapes.

Acknowledgments

This work was supported by Wichita State University, Wichita, KS, subcontract to the University of Texas at Arlington under AFRL contract FA8650-18-C-5289.

References

Adluru, H.K., Iarve, E.V., Hoos, K.H., 2018. Discrete damage modelling of clamped tapered beam specimen under fatigue loading. In: 33rd Technical Conference of the American Society for Composites 2018. https://doi.org/10.12783/asc33/26053.

Adluru, H.K., Hoos, K.H., Iarve, E.V., Ratcliffe, J.G., 2019. Delamination initiation and migration modeling in clamped tapered laminated beam specimens under static loading. Compos. Part A Appl. Sci. Manuf. https://doi.org/10.1016/j.compositesa.2018.12.020.

Camanho, P.P., Maimí, P., Dávila, C.G., 2007. Prediction of size effects in notched laminates using continuum damage mechanics. Compos. Sci. Technol. 67. https://doi.org/10.1016/j.compscitech.2007.02.005.

Chen, B.Y., Tay, T.E., Pinho, S.T., Tan, V.B.C., 2016. Modelling the tensile failure of composites with the floating node method. Comput. Methods Appl. Mech. Eng. 308. https://doi.org/10.1016/j.cma.2016.05.027.

Clay, S.B., Knoth, P.M., 2017. Experimental results of fatigue testing for calibration and validation of composite progressive damage analysis methods. J. Compos. Mater. 51, 2083–2100. https://doi.org/10.1177/0021998316670132.

Dalgarno, R.W., Action, J.E., Robbins, D.H., Engelstad, S.P., 2017. Failure simulations of open-hole IM7/977-3 coupons subjected to fatigue loading using Autodesk Helius PFA. J. Compos. Mater. 51. https://doi.org/10.1177/0021998316669579.

Dávila, C.G., 2020. From S-N to the Paris law with a new mixed-mode cohesive fatigue model for delamination in composites. Theor. Appl. Fract. Mech. 106. https://doi.org/10.1016/j.tafmec.2020.102499.

Dávila, C.G., Rose, C.A., Murri, G.B., Jackson, W.C., Johnston, W.M., 2020. Evaluation of Fatigue Damage Accumulation Functions for Delamination Initiation and Propagation. No NF1676L-36100.

de Carvalho, N.v., Krueger, R., 2016. Modeling fatigue damage onset and progression in composites using an element-based virtual crack closure technique combined with the floating node method. In: Proceedings of the American Society for Composites - 31st Technical Conference, ASC 2016.

de Carvalho, N.v., Mabson, G.E., Krueger, R., Deobald, L.R., 2019. A new approach to model delamination growth in fatigue using the Virtual Crack Closure Technique without remeshing. Eng. Fract. Mech. 222. https://doi.org/10.1016/j.engfracmech.2019.106614.

Deobald, L.R., Mabson, G.E., Engelstad, S., Prabhakar Rao, M., Gurvich, M.R., Seneviratne, W., et al., 2017. Guidelines for VCCT-Based Interlaminar Fatigue and Progressive Failure Finite Element Analysis. (No. NASA/TM-2017-219663).

Duarte, A.P.C., Díaz Sáez, A., Silvestre, N., 2017. Comparative study between XFEM and Hashin damage criterion applied to failure of composites. Thin-Walled Struct. 115. https://doi.org/10.1016/j.tws.2017.02.020.

Fang, E., Cui, X., Lua, J., 2017. A continuum damage and discrete crack-based approach for fatigue response and residual strength prediction of notched laminated composites. J. Compos. Mater. 51. https://doi.org/10.1177/0021998317705975.

Gao, Z., Adluru, H.K., Lu, W., Hoos, K.H., Iarve, E.V., 2022. Two-way global-local coupled failure modeling with Rx-FEM. In: Proceedings of The American Society for Composites-Thirty-Seventh Technical Conference.

Grogan, D.M., Leen, S.B., Brádaigh OCM, O., 2014. An XFEM-based methodology for fatigue delamination and permeability of composites. Compos. Struct. 107. https://doi.org/10.1016/j.compstruct.2013.07.050.

Hansbo, A., Hansbo, P., 2004. A finite element method for the simulation of strong and weak discontinuities in solid mechanics. Comput. Methods Appl. Mech. Eng. 193. https://doi.org/10.1016/j.cma.2003.12.041.

Hoos, K.H., Iarve, E.V., Zhenjia, G., Wei-Tsen, L., 2023. BSAM20 User Manual. University of Texas at Arlington, TX.

Iarve, E.V., 2003. Mesh independent modelling of cracks by using higher order shape functions. Int. J. Numer. Methods Eng. 56. https://doi.org/10.1002/nme.596.

Iarve, E.V., Mollenhauer, D.H., 2015. Mesh-independent matrix cracking and delamination modeling in advanced composite materials. Num. Modell. Fail. Advan. Compo. Mater. https://doi.org/10.1016/B978-0-08-100332-9.00009-8.

Iarve, E.V., Gurvich, M.R., Mollenhauer, D.H., Rose, C.A., Dávila, C.G., 2011. Mesh-independent matrix cracking and delamination modeling in laminated composites. Int. J. Numer. Methods Eng. 88. https://doi.org/10.1002/nme.3195.

Iarve, E.V., Hoos, K.H., Braginsky, M., Zhou, E., Mollenhauer, D.H., 2017. Progressive failure simulation in laminated composites under fatigue loading by using discrete damage modeling. J. Compos. Mater. 51, 2143–2161. https://doi.org/10.1177/0021998316681831.

Joosten, M.W., Dávila, C.G., Yang, Q., 2022. Predicting fatigue damage in composites subjected to general loading conditions. Compos. Part A Appl. Sci. Manuf. 156. https://doi.org/10.1016/j.compositesa.2022.106862.

Kawashita, L.F., Hallett, S.R., 2012. A crack tip tracking algorithm for cohesive interface element analysis of fatigue delamination propagation in composite materials. Int. J. Solid Struct. 49. https://doi.org/10.1016/j.ijsolstr.2012.03.034.

Krueger, R., 2004. Virtual crack closure technique: history, approach, and applications. Appl. Mech. Rev. 57. https://doi.org/10.1115/1.1595677.

Krueger, R., 2011. Development and application of benchmark examples for mode II static delamination propagation and fatigue growth predictions. In: National Aeronautics and Space Administration, (No NASA/CR-2011-217305), Langley Research Center.

Liang, Y.J., McQuien, J.S., Iarve, E.V., 2020. Implementation of the regularized extended finite element method in Abaqus framework for fracture modeling in laminated composites. Eng. Fract. Mech. 230. https://doi.org/10.1016/j.engfracmech.2020.106989.

Liang, Y.J., Dávila, C.G., Iarve, E.V., 2021. A reduced-input cohesive zone model with regularized extended finite element method for fatigue analysis of laminated composites in Abaqus. Compos. Struct. 275. https://doi.org/10.1016/j.compstruct.2021.114494.

Llobet, J., Maimí, P., Essa, Y., Martin de la Escalera, F., 2021. A continuum damage model for composite laminates: Part III - fatigue. Mech. Mater. 153. https://doi.org/10.1016/j.mechmat.2020.103659.

Lu, X., Chen, B.Y., Tan, V.B.C., Tay, T.E., 2018. Adaptive floating node method for modelling cohesive fracture of composite materials. Eng. Fract. Mech. 194. https://doi.org/10.1016/j.engfracmech.2018.03.011.

Lu, W.-T., Gao, Z., Adluru, H.K., Hoos, K.H., Seneviratne, W., Mollenhauer, D., et al., 2021. Fatigue damage modeling in laminated composites by using Rx-FEM and strength tracking method. In: Proceedings of the American Society for Composites—Thirty-Sixth Technical Conference on Composite Materials.

Lu, W.T., Gao, Z., Adluru, H.K., Hoos, K.H., Seneviratne, W.P., Mollenhauer, D.H., et al., 2022. Fatigue damage modeling in laminated composite by using Rx-FEM and strength tracking method. Compos. Part A Appl. Sci. Manuf. 163, 107199.

Ma, Z., Chen, J., Yang, Q., Li, Z., Su, X., 2021. Progressive fracture analysis of the open-hole composite laminates: experiment and simulation. Compos. Struct. 262. https://doi.org/10.1016/j.compstruct.2021.113628.

Maimí, P., Camanho, P.P., Mayugo, J.A., Dávila, C.G., 2007. A continuum damage model for composite laminates: Part I - constitutive model. Mech. Mater. 39. https://doi.org/10.1016/j.mechmat.2007.03.005.

Maimí, P., Camanho, P.P., Mayugo, J.A., Dávila, C.G., 2007. A continuum damage model for composite laminates: Part II - computational implementation and validation. Mech. Mater. 39. https://doi.org/10.1016/j.mechmat.2007.03.006.

Nikishkov, Y., Makeev, A., 2011. Fatigue Life Assessment for Composite Materials. ICCM International Conferences on Composite Materials.

Nikishkov, Y., Markeev, A., Seon, G., 2010. Simulation of damage in composites based on solid finite elements. J. Am. Helicopter Soc. 55. https://doi.org/10.4050/JAHS.55.042009.

Nixon-Pearson, O.J., Hallett, S.R., Harper, P.W., Kawashita, L.F., 2013. Damage development in open-hole composite specimens in fatigue. Part 2: numerical modelling. Compos. Struct. 106. https://doi.org/10.1016/j.compstruct.2013.05.019.

Pinho, S.T., Dávila, C.G., Camanho, P.P., Iannucci, L., Robinson, P., 2005. Failure Models and Criteria for FRP under In-Plane or Three-Dimensional Stress States Including Shear Non-linearity. Nasa/Tm-2005-213530.

Rao, P.M., Mordasky, M.D., Gurvich, M.R., de Carvalho, N., Seshadri, B.R., Ratcliffe, J.G., 2018. Validation of a discrete damage mechanics methodology using the static and fatigue behavior of carbon/epoxy tapered specimens. In: AIAA/ASCE/AHS/ASC Structures, Structural Dynamics, and Materials Conference, 2018, Vol. 0. https://doi.org/10.2514/6.2018-1228.

Ratcliffe J, Carvalho N. Investigating Delamination Migration in Composite Tape Laminates. Nasa/Tm−2014-218289 2014.

Sendeckyj, G.P., 1981. Fitting Models to Composite Materials Fatigue Data. ASTM International.

Seneviratne, W., Tomblin, J., 2021. Life of Composites under Variable-Amplitude Fatigue Using St Method. AIAA Scitech 2021 Forum.

Seneviratne, W., Tomblin, J., Walimunige, R., December 17, 2020. Advanced Material Characterization and Structural Certification (AMCSC) Volume 1 of 2. National Institute for Aviation Research (NIAR).

Seshadri, B.R., de Carvalho, N.v., Ratcliffe, J.G., Mabson, G.E., Deobald, L., 2018. Simulating the clamped tapered beam specimen under quasi-static and fatigue loading using floating node method. In: AIAA/ASCE/AHS/ASC Structures, Structural Dynamics, and Materials Conference, 2018, Vol. 0. https://doi.org/10.2514/6.2018-0971.

Su, Z., Oskay, C., 2021. Mesh size objective fatigue damage propagation in laminated composites using the multiscale discrete damage theory. Comput. Mech. 67. https://doi.org/10.1007/s00466-021-01978-7.

Sun, X.C., Hallett, S.R., 2018. Failure mechanisms and damage evolution of laminated composites under compression after impact (CAI): experimental and numerical study. Compos. Part A Appl. Sci. Manuf. 104. https://doi.org/10.1016/j.compositesa.2017.10.026.

Swindeman, M.J., Iarve, E.V., Brockman, R.A., Mollenhauer, D.H., Hallett, S.R., 2013. Strength prediction in open hole composite laminates by using discrete damage modeling. AIAA J. https://doi.org/10.2514/1.J051773.

Tao, C., Mukhopadhyay, S., Zhang, B., Kawashita, L.F., Qiu, J., Hallett, S.R., 2018. An improved delamination fatigue cohesive interface model for complex three-dimensional multi-interface cases. Compos. Part A Appl. Sci. Manuf. 107. https://doi.org/10.1016/j.compositesa.2018.02.008.

Teimouri, F., Heidari-Rarani, M., Haji Aboutalebi, F., 2021. An XFEM-VCCT coupled approach for modeling mode I fatigue delamination in composite laminates under high cycle loading. Eng. Fract. Mech. 249. https://doi.org/10.1016/j.engfracmech.2021.107760.

Turon, A., González, E.v., Sarrado, C., Guillamet, G., Maimí, P., 2018. Accurate simulation of delamination under mixed-mode loading using a cohesive model with a mode-dependent penalty stiffness. Compos. Struct. 184, 506−511. https://doi.org/10.1016/j.compstruct.2017.10.017.

van der Meer, F.P., Sluys, L.J., 2009. Continuum models for the analysis of progressive failure in composite laminates. J. Compos. Mater. 43. https://doi.org/10.1177/0021998309343054.

van der Meer, F.P., Sluys, L.J., 2010. Mesh-independent modeling of both distributed and discrete matrix cracking in interaction with delamination in composites. Eng. Fract. Mech. 77. https://doi.org/10.1016/j.engfracmech.2009.11.010.

Vellwock, A.E., Vergani, L., Libonati, F., 2018. A multiscale XFEM approach to investigate the fracture behavior of bio-inspired composite materials. Compos. B Eng. 141. https://doi.org/10.1016/j.compositesb.2017.12.062.

Ventura, G., Gracie, R., Belytschko, T., 2009. Fast integration and weight function blending in the extended finite element method. Int. J. Numer. Meth. Eng. 77. https://doi.org/10.1002/nme.2387.

Wang, C., Xu, X., 2016. An extended phantom node method study of crack propagation of composites under fatigue loading. Compos. Struct. 154. https://doi.org/10.1016/j.compstruct.2016.07.022.

Wang, C., Zhang, C., 2021. Discussions on extension of traditional cohesive element for delamination modeling of laminates used in combination with phantom node intraply elements. Compos. Struct. 261. https://doi.org/10.1016/j.compstruct.2021.113588.

Yang, Q., 2017. ARO 1.2: Solid Mechanics: Augmented Finite Element Method for High-Fidelity Analysis of Structural Composites. University of Miami-Coral Gables.

Modeling the crack initiation in unidirectional laminates under multiaxial fatigue loading

19

P.A. Carraro and M. Quaresimin
University of Padova, Vicenza, Italy

19.1 Introduction

[1]Multidirectional laminates under fatigue loadings undergo a progressive damage evolution from the beginning of fatigue life to the final failure. It is proven by several works in the literature that the first evidence of macroscopic damage, occurring since the early stages of fatigue life, is represented by the initiation and propagation of intralaminar cracks in the off-axis layers (Adden and Horst, 2006, 2010; Lafarie-Frenot and Hénaff-Gardin, 1991; Quaresimin et al., 2014; Tong, 2001, 2002; Tong et al., 1997; Wharmby and Ellyin, 2002; Yokozeki and Aoki, 2002). The accumulation of such cracks leads to the degradation of the elastic properties of a laminate, as well as to the final failure, because they act as stress concentrators causing the initiation of delaminations and fiber failure. Accordingly, design against fatigue failure and degradation requires the development of a criterion for predicting crack initiation in the unidirectional (UD) plies of a multidirectional laminate. It is important to say that the stress state, which a ply is subjected to within a multidirectional laminate, is in general multiaxial, it being due to external loads in different directions ('external' multiaxiality) or to the material anisotropy ('internal' multi-axiality). The equivalence of crack initiation and propagation phenomena between these two kinds of multiaxial conditions was proved in Quaresimin et al. (2014), provided that the local stress state in the plies is the same.

However, a crack initiation criterion has to be capable of dealing with multiaxial stress states leading to a matrix-dominated material response.

Some criteria can already be found in the literature, and they can be classified as follows:

- Macro-mechanical models.
- Micro-mechanical models.

Macro-mechanical models are the most common, and they are based on macroscopic quantities such as the stresses in the material coordinates system (σ_1, σ_2 and σ_6) or

[1] This chapter was first published as Chapter 7 "Modeling the crack initiation in UD laminates under multiaxial fatigue loading" by P. A. Carraro and M. Quaresimin in *Fatigue of Textile Composites*, ed. Carvelli and Lomov, Woodhead Publishing Limited, 2015, ISBN: 978-1-78242-281-5.

Modeling Damage, Fatigue and Failure of Composite Materials. https://doi.org/10.1016/B978-0-443-18489-5.00002-3

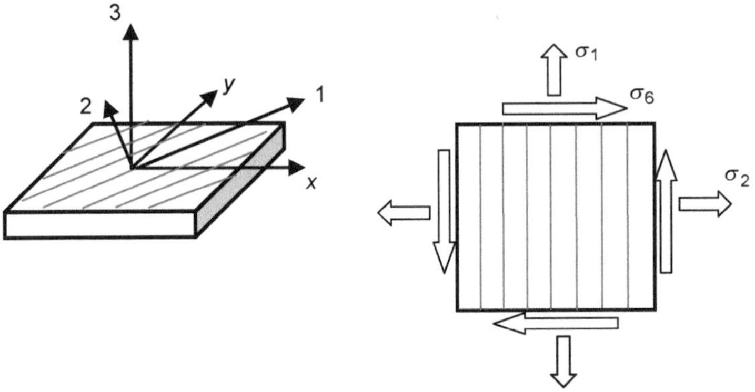

Figure 19.1 Definition of reference systems and stresses.

the associated strain energy density, often involved in a polynomial expression. See Fig. 19.1 for the definition of stress components.

Tsai–Hill and Tsai–Wu criteria are some classical examples of the static behavior. An improvement to them is achieved by Puck's polynomial criterion (Puck and Shurmann, 1998), which involves the macroscopic stresses acting on the fracture plane, meant as the plane in which the final fracture occurs.

With regard to the fatigue behavior, Kawai et al. (2001) extended the Tsai–Hill polynomial criterion to fatigue loading, combined with a continuum damage model to obtain a power law for the S–N curves.

El-Kadi and Ellyin (1994) proposed to use a polynomial expression involving the strain energy density components instead of the stresses, normalizing each contribution with respect to the static limit value. Both Kawai et al. and El-Kadi and Ellyin obtained reasonable agreement with their own experimental data.

However, it is worth mentioning that these criteria describe the fatigue behavior for both fiber and matrix-dominated response with one equation only, which seems not consistent from a physical point of view.

Before those attempts, two polynomial criteria were proposed (Hashin and Rotem, 1973), separating the fiber and the matrix-dominated behaviors, depending on which is the critical component in the composite lamina (fiber or matrix). For the matrix-dominated behavior, they proposed to involve in a polynomial expression only the transverse and in-plane shear stresses, weighted by means of fatigue functions for which an experimental calibration is needed.

The fiber and interfiber-related failure are treated separately also by the Puck's criterion based on the fracture plane concept. This criterion, initially conceived for the static behavior, was later extended to the case of cyclic loading (Sun et al., 2012).

Micromechanical models are based on the use of local stresses in the matrix or at the fiber–matrix interface for formulating criteria for fatigue crack initiation. Conceptually, this approach implicitly recognizes that the initiation of a macrocrack in a UD lamina is driven by damage evolution at the microscopic scale by means of irreversible mechanisms (Talreja, 1981, 2006).

One of the first attempts of considering the local stresses (or microstresses) acting in the matrix was published by Reifsneider and Gao (1991). They used the Mori—Tanaka theory in order to evaluate the average transverse stress and in-plane shear stress in the matrix. Then, they proposed to involve them in a polynomial criterion similar to that of Hashin, using fatigue functions of empirical derivation. Later, Plumtree and Cheng (1999) developed a model based on the local stresses acting on the fracture plane, defined as the plane normal to the transverse direction. Accordingly, the relevant stresses turned out to be the local transverse and in-plane shear stresses only, whose local peak values were calculated by means of finite element (FE) analyses of a fiber—matrix unit cell.

In spite of the use of the local stresses, these criteria are still of phenomenological nature because they are not developed on the basis of damage mechanisms taking place at the microscopic scale.

In the present chapter, a criterion developed by the authors (Carraro and Quaresimin, 2014) for fatigue crack initiation in UD plies under multiaxial fatigue loading leading to a matrix-dominated behavior is presented. The criterion is based on a micromechanical approach and on the damage mechanisms occurring at the microscopic scale.

19.2 Peculiarities of fatigue failure

Some preliminary considerations about the peculiarities of fatigue failure are reported here.

When a UD lamina is subjected to a cyclic loading condition leading to a matrix-dominated response, final failure occurs without a visible (macroscopic) progressive damage (Awerbuch and Hahn, 1981; El-Kadi and Ellyin, 1994; Quaresimin and Carraro, 2013). In fact, a stable propagation phase is never observed, and when a macrocrack nucleates, it propagates unstably leading the lamina to the complete fracture in just a few cycles. It can be concluded that the matrix-dominated fatigue failure is controlled by the initiation phase, if the word “initiation” refers to a macrocrack propagating in the direction of the fibers.

However, as already mentioned, at a microscopic level, a progressive and irreversible degradation process takes place during fatigue life and leads to the initiation of a macrocrack (Talreja, 1981, 2006). Thus, the number of cycles spent for the initiation of a macrocrack is controlled by the damage evolution occurring at the microscopic level. As a consequence, the definition of a crack initiation criterion consists in the identification of the driving force for damage evolution at the microscopic scale. In order to do that, the concept of the *local nucleation plane* is introduced in the following.

As already mentioned, Puck’s criterion is based on the hypothesis that the effective stresses to be considered as the cause of the static failure are the stresses lying on the fracture plane.

According to Puck, positive values of σ_2 and σ_6 lead to a fracture plane of which the normal is the transverse direction (2-axis). Therefore, the effective stresses are the global stress components σ_2 and σ_6. The same concept and definition of the fracture plane is adopted also by Plumtree and Cheng (1999), and also for them, in the case of an off-axis lamina, the stress components to be accounted for are σ_2 and σ_6, combined in a polynomial criterion.

It is important to point out that the fracture plane considered by Puck and by Plumtree-Cheng is actually the plane where final separation occurs. It can be defined as the *macroscopic fracture plane*. Observations of fracture surfaces (Awerbuch and Hahn, 1981; Quaresimin and Carraro, 2014; Shiino et al., 2012) show the wide presence of shear cusps in the matrix between the fibers, which can be reasonably thought to be the results of the initiation of multiple inclined microcracks in the matrix between the fibers. This seems to be also supported by the in situ observations on 45 degree off-axis specimens statically tested (Cox et al., 1994).

The *local nucleation plane* is, therefore, defined as the plane in which the microcrack initiates in the matrix, as shown in Fig. 19.2. The angle between that plane and the 1-axis is referred to as β_c.

According to these observations, the driving force for this kind of damage evolution should be searched in the stress components normal to the local nucleation plane. It is here reasonably assumed that the local nucleation plane is perpendicular to the local

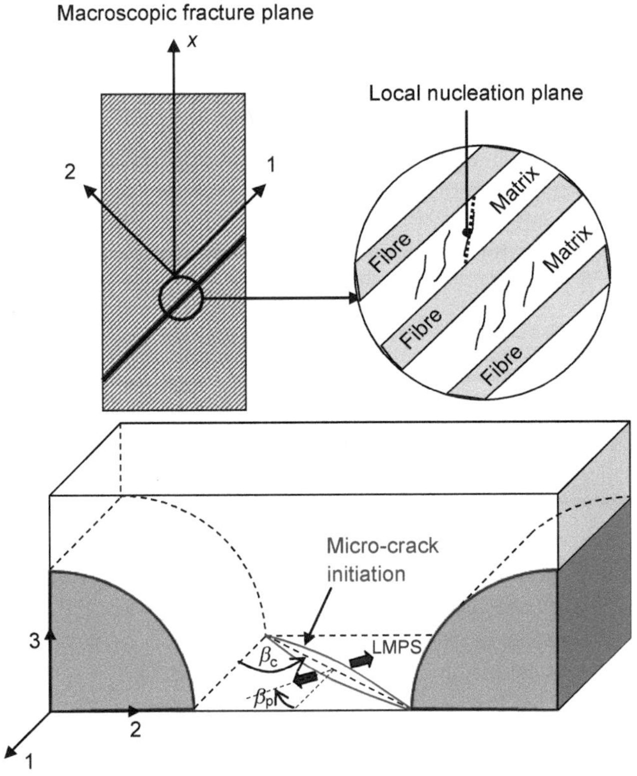

Figure 19.2 Concept of the local nucleation plane.

maximum principal stress (LMPS) in the matrix. As a consequence, the effective stress to be considered is the LMPS in the matrix, whose orientation with respect to the fiber direction is β_p, so that $\beta_c = \beta_p + \pi/2$.

It is important also to mention that the orientation of the local nucleation plane, normal to the LMPS, is representative of the degree of multi-axiality of the local stress state at the microscopic scale.

When a lamina is subjected to a pure transverse stress loading condition, the local nucleation plane is normal to the 2-direction. In fact, in this case, no shear cusps can be observed in the matrix between the fibers (Quaresimin and Carraro, 2014). In addition, it is proved that in the case of pure transverse tension, the local stress state at or close to the fiber–matrix interface is nearly hydrostatical (Asp et al., 1996a,b). As a consequence, the failure occurs in the form of cavitation-induced matrix cracking and debonding. Asp et al. showed that a suitable criterion for predicting the static failure in the case of pure transverse stress is based on reaching a critical value for the dilatational energy density expressed in Eq. (19.1):

$$U_v = \frac{1 - 2\nu}{6E} I_1^2 \tag{19.1}$$

where I_1 is the first local stress tensor invariant.

Extending this finding also to the case of cyclic loading, a change is expected in the leading damage mode moving from a loading condition near the pure transverse stress to another one, characterized by the presence of a high enough shear stress component. In the first case, microscale damage evolution is assumed to be driven by the local hydrostatic stress in the matrix (LHS $= I_1/3$), whereas in the latter by the LMPS. Accordingly, the following criterion for multiaxial fatigue is proposed: two different parameters, representative of two different driving forces, have to be used for the representation of the S–N curves for a UD lamina, according to the multiaxial condition:

- The peak of the local hydrostatic stress (LHS $= I_1/3$), for nearly pure transverse tension case.
- The peak of LMPS for enough high shear stress component.

19.3 Calculation of local stresses

As already discussed, the LHS and LMPS parameters, representative of the driving forces for the microscopic damage evolution, have to be expressed in terms of local stresses (or microstresses) acting in the matrix and at the fiber–matrix interface. A multiscale approach, to link the applied stresses σ_1, σ_2, and σ_6 to the local stress fields, is, therefore, adopted.

If a regular fibers array of the square or hexagonal type is assumed for a UD lamina, unit cells as those shown in Fig. 19.3 are representative of the homogenized lamina in terms of average mechanical properties and damage initiation and evolution at the microscale. According to micromechanics theory, the global stresses σ_1, σ_2, and σ_6 are the average stresses acting on the faces of the unit cell.

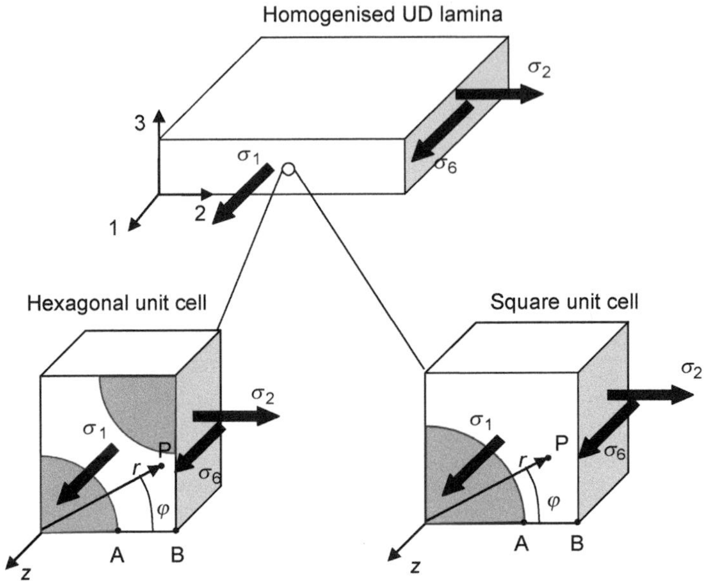

Figure 19.3 Definition of the fiber–matrix unit cells for micromechanical analysis.

Of course, regular fibers distributions are not representative of the real microstructure in composite laminae. In spite of this, these types of unit cells provide reliable results for the application of the proposed multiaxial criterion. This conclusion is based on the detailed analysis of the possible effects of local fiber volume fraction and local orientation of the fiber array presented by the authors (Carraro and Quaresimin, 2014), showing that the orientation of the local nucleation plane, and therefore the local degree of multiaxiality, depends only very weakly on these parameters.

The microstresses can be calculated by means of FE analyses of a fiber–matrix unit cell subjected to the average (or macroscopic) stresses σ_1, σ_2, and σ_6. Periodic boundary conditions have to be applied on the surfaces of the quarter of the unit cell, as explained in Zhang and Xia (2005). In addition, residual stresses due to the different thermal expansion of the two phases during the cooling process must be accounted for. In the FE code, thermal loads have been applied as a uniform temperature gap $\Delta T = T_c - T_r$, where T_c is the curing temperature, and T_r is the room temperature. Since tensile loading conditions are analyzed here, and the epoxy behavior is usually linear elastic in that range, linear elastic FE analyses have been carried out with the software ANSYS 11 using 20 nodes solid elements.

At each point P of the unit cell, the mechanical microstresses in polar coordinates (r, φ, z) can be defined in terms of stress concentration factors $k_{i,jl}$ relating the macrostress σ_i to the local stress σ_{jl}, as in Eq. (19.2). Finally, the thermal stresses are related to the temperature gap ΔT by means of thermal concentration factors h_{jl}, as in Eq. (19.2).

$$\begin{Bmatrix} \sigma_{rr} \\ \sigma_{\varphi\varphi} \\ \sigma_{zz} \\ \sigma_{r\varphi} \\ \sigma_{\varphi z} \\ \sigma_{rz} \end{Bmatrix}^P = \begin{bmatrix} k_{1,rr} & k_{2,rr} & 0 \\ k_{1,\varphi\varphi} & k_{2,\varphi\varphi} & 0 \\ k_{1,zz} & k_{2,zz} & 0 \\ k_{1,r\varphi} & k_{2,r\varphi} & 0 \\ 0 & 0 & k_{6,\varphi z} \\ 0 & 0 & k_{6,rz} \end{bmatrix}^P \begin{Bmatrix} \sigma_1 \\ \sigma_2 \\ \sigma_6 \end{Bmatrix} + \Delta T \begin{Bmatrix} h_{rr} \\ h_{\varphi} \\ h_{zz} \\ 0 \\ 0 \\ 0 \end{Bmatrix}^P \quad (19.2)$$

As shown in Carraro and Quaresimin (2014), the stress and thermal concentration factors are functions of the fiber volume fraction V_f and of the elastic properties of the fiber/matrix system (the dependence on the matrix modulus is actually very weak). The analyses reported later in the chapter are for a typical glass/epoxy system, of which the tensile modulus E, Poisson's ratio ν, and coefficients of thermal expansion (CTE) are listed in Table 19.1.

The peak values of the LMPS and LHS have been found to be always at the points A or B (mostly at point A) of Fig. 19.3, where, due to symmetry, some stress components vanish, allowing one to adopt the following simple expressions for *LMPS* and *LHS* (valid in the segment AB):

$$LMPS = \frac{1}{2}\left[\sigma_{rr} + \sigma_{zz} + \sqrt{\sigma_{rr}^2 + 4\sigma_{rz}^2 - 2\sigma_{rr}\sigma_{zz} + \sigma_{zz}^2}\right] \quad (19.3)$$

$$LHS = \frac{\sigma_{rr} + \sigma_{\varphi\varphi} + \sigma_{zz}}{3} \quad (19.4)$$

In Eqs. (19.3) and (19.4), the local stresses obtained from Eq. (19.2) must be substituted.

19.4 Validation

In this section, it will be shown that fatigue crack initiation data can be suitably described by two scatter bands and master curves relating the number of cycles to crack initiation to the peak values of the *LHS* or *LMPS* in the unit cell.

Table 19.1 Typical glass/epoxy material properties used in the FE analyses.

	E (MPa)	ν	CTE/°C
Glass	70,000	0.22	7×10^{-6}
Epoxy	3200	0.37	67.5×10^{-6}

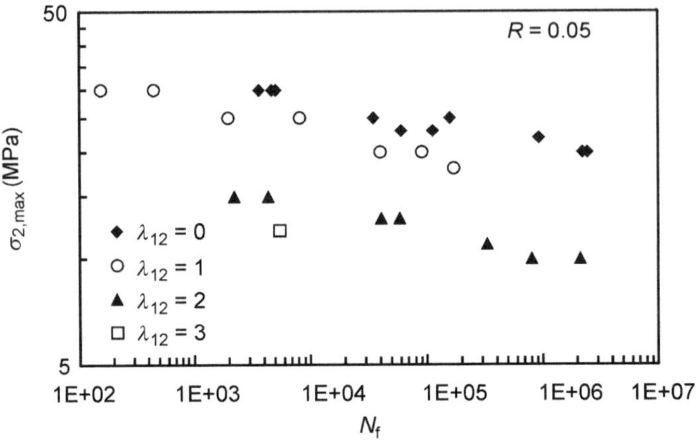

Figure 19.4 Crack initiation data for biaxially loaded tubes; R = 0.05 (Carraro and Quaresimin, 2014; Quaresimin and Carraro, 2013).

Quaresimin and Carraro carried out experimental tests on UD tubes under biaxial fatigue loadings characterized by the presence of the transverse stress σ_2 and the in-plane shear σ_6, combined with several values of the biaxiality ratio $\lambda_{12} = \sigma_6/\sigma_2$ from 0 to infinity (Quaresimin and Carraro, 2013). Crack initiation data, reported in Fig. 19.4, in terms of the maximum cyclic value of the transverse stress, highlight that different fatigue curves are obtained for different biaxiality ratios (pure torsion results cannot be shown because σ_2 is zero in that case).

In Fig. 19.5a and b, crack initiation data are reported in terms of the LMPS_{max} computed by means of the stress and thermal concentration factors for the square and hexagonal arrays, respectively. In both cases, it can be seen that data for λ_{12} equal to 1 to ∞ are collapsed in a single narrow scatter band. This indicates that the LMPS is a good representation of the driving force for the initiation and evolution of inclined microcracks between the fibers, leading to the initiation of a macrocrack.

The accuracy of the criterion is proved by the limited values of the scatter indexes T_{LMPS}, defined as the ratio of the LMPS values at 10^6 cycles related to the 10% and 90% probability of survivor curves.

The fact that both unit cell types provide the same degree of accuracy is due to the fact that the orientation of the local nucleation plane predicted by the square and hexagonal arrays is almost the same (the difference being of about 0.3 degree). The way the global stresses combine for the computation of the LMPS depends only on the angle β_c. Since it remains unvaried in these two conditions, the capability of the LMPS of collapsing data for different (shear-dominated) multiaxial conditions is the same for the two types of unit cell.

In both cases, the pure tension curve ($\lambda_{12} = 0$) is outside the scatter band, but this is not surprising because in that case, the LHS should be the driving force for damage evolution at the microscale. In fact, as shown in Fig. 19.6a and b for the two kinds of array, the pure tension curve in terms of LHS exhibits higher values of the LHS parameter with respect the other conditions.

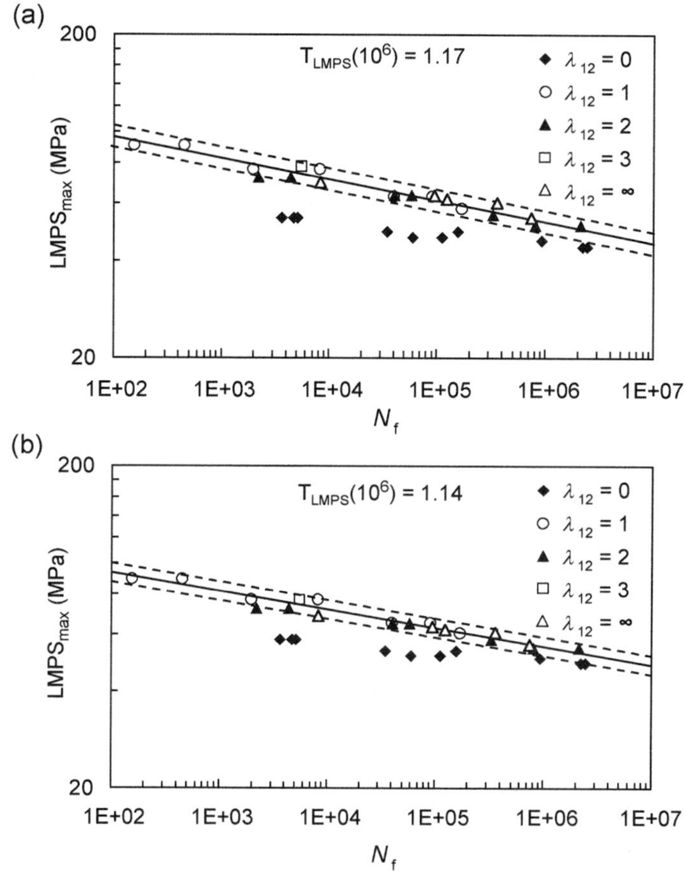

Figure 19.5 Data reanalyzed in terms of the LMPS computed with the (a) Square and (b) Hexagonal unit cell.

Since equivalent performances are provided by the two kinds of unit cells, only results for the square one will be shown in the remaining part of the chapter. In all the figures reported in this section, master curves and 90%−10% probability of survival scatter bands are shown. Data were fitted with a power law in the form $N_f \cdot \text{LMPS}^k = C$ or $N_f \cdot \text{LHS}^k = C$, assuming a log-normal distribution of the number of cycles to crack initiation. The results are listed in Table 19.2 for all the cases analyzed, related to the square unit cell.

Further crack initiation results were obtained by Quaresimin and Carraro on $[0_F/90_3/0_F]$ glass/epoxy tubes, where three layers of 90 degree UD plies were constrained by an external and internal thin fabric ply (Quaresimin and Carraro, 2014). For this kind of specimen, subjected to combined tension-torsion as well, the nucleation of the first crack on the 90 degree layers was followed by its stable propagation in the circumferential direction, and eventually by multiple cracking. However, in

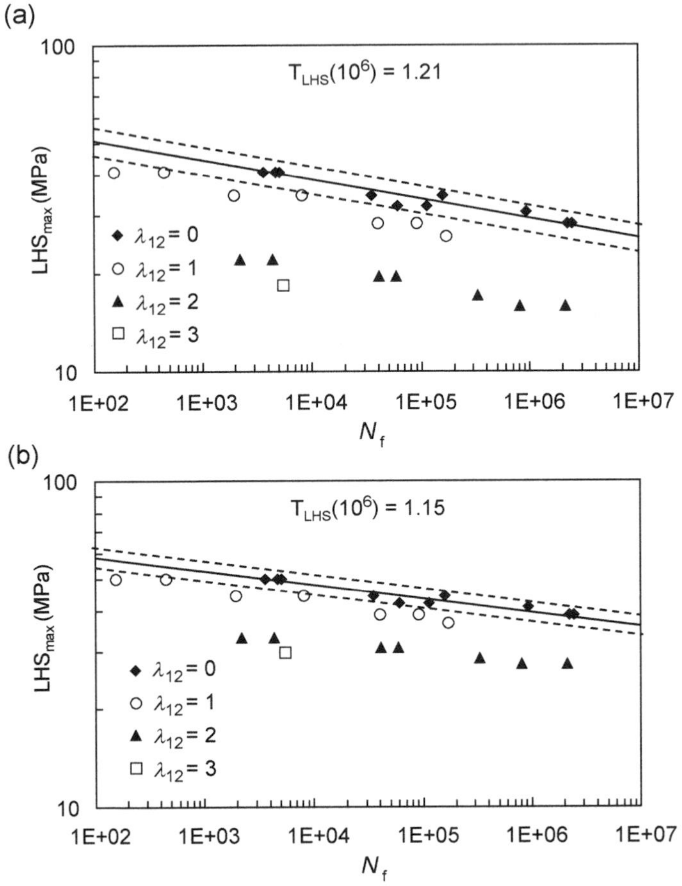

Figure 19.6 Data from Quaresimin and Carraro (2013) reanalyzed in terms of the LHS computed with the (a) Square and (b) Hexagonal unit cell.

Fig. 19.7a, the S–N curves for the case $R = 0.05$ are reported for the first transverse crack nucleation in terms of the maximum cyclic transverse stress in the 90 degree plies. Different curves are obtained for different values of the bi-axiality ratio λ_{12}, if expressed in terms of the transverse stress, but a single and narrow scatter band in terms of LMPS can be shown to describe fatigue data for $\lambda_{12} = 1$ and 2 (Fig. 19.7b). Data related to $\lambda_{12} = 0.5$ and $\lambda_{12} = 0$ can be, instead, described by a single scatter band in terms of LHS as shown in Fig. 19.7c.

Hashin and Rotem presented experimental results on flat off-axis specimens subjected to uniaxial cyclic loading with $R = 0.1$ (Hashin and Rotem, 1973). In the material reference system, a multiaxial stress state is originated, leading to a matrix-dominated fatigue failure that coincides, for these UD specimens, with the initiation of the first macrocrack. Fatigue data reported in terms of the maximum cyclic value of the global stress $\sigma_{x,max}$ are shown in Fig. 19.8a. Different S–N

Table 19.2 Results of the statistical analysis on the analyzed scatter bands.

Refs.	Specimen	Driving force	*k*	*C* 50%	*C* 90%	T_{LMPS} or T_{LHS} (10^6)
Quaresimin and Carraro (2013)	$[90_4]$ tubes	LMPS	15.13	131.03	121.09	1.17
Quaresimin and Carraro (2013)	$[90_4]$ tubes	LHS	16.67	67.2	61.03	1.21
Quaresimin and Carraro (2014)	$[0_F/90_3/0_F]$ tubes	LMPS	13.22	166.84	148.29	1.27
Quaresimin and Carraro (2014)	$[0_F/90_3/0_F]$ tubes	LHS	13.34	91.1	81.5	1.25
Hashin and Rotem (1973)	Off-axis laminae	LMPS	13.32	85.38	73.35	1.36
El-Kadi and Ellyin (1994)	Off-axis laminae	LMPS	11.13	158.4	131.17	1.46
El-Kadi and Ellyin (1994)	Off-axis laminae	LHS	14.47	67.94	52.91	1.65
Quaresimin et al. (2014)	Constrained off-axis laminae	LMPS	9.82	267.02	244.84	1.19

curves are obtained for different values of the off-axis angle. Fig. 19.8b shows, instead, that all data are collected in the same narrow scatter band when presented in terms of the LMPS.

Off-axis test results were also reported by El-Kadi and Ellyin with $R = 0$, 0.5, and -1 for different off-axis angles (El-Kadi and Ellyin, 1994). For the time being, as the criterion was developed on the basis of the damage mechanisms observed in the absence of compressive parts of the cycles, it can be considered applicable for nonnegative load ratios only. Results for $R = 0$ are reported in Fig. 19.9. Again the LMPS and LHS parameters are shown to collapse crack initiation data in two scatter bands to be used when the shear stress is high or very low, respectively.

In a work by Quaresimin et al. (2014), experimental data were presented on glass/epoxy laminates under uniaxial cyclic loading. The stacking sequence was $[0/\theta_2/0/-\theta_{2S}]$, with $\theta = 50$ and 60 degrees. S−N curves for the initiation of the first fatigue cracks in the off-axis layers were reported in terms of the maximum cyclic transverse stress in those plies. An example is reported in Fig. 19.10a for the cracks initiated in the θ_2 layers. In Fig. 19.10b, the data are presented in terms of the LMPS, showing a good compaction effect again.

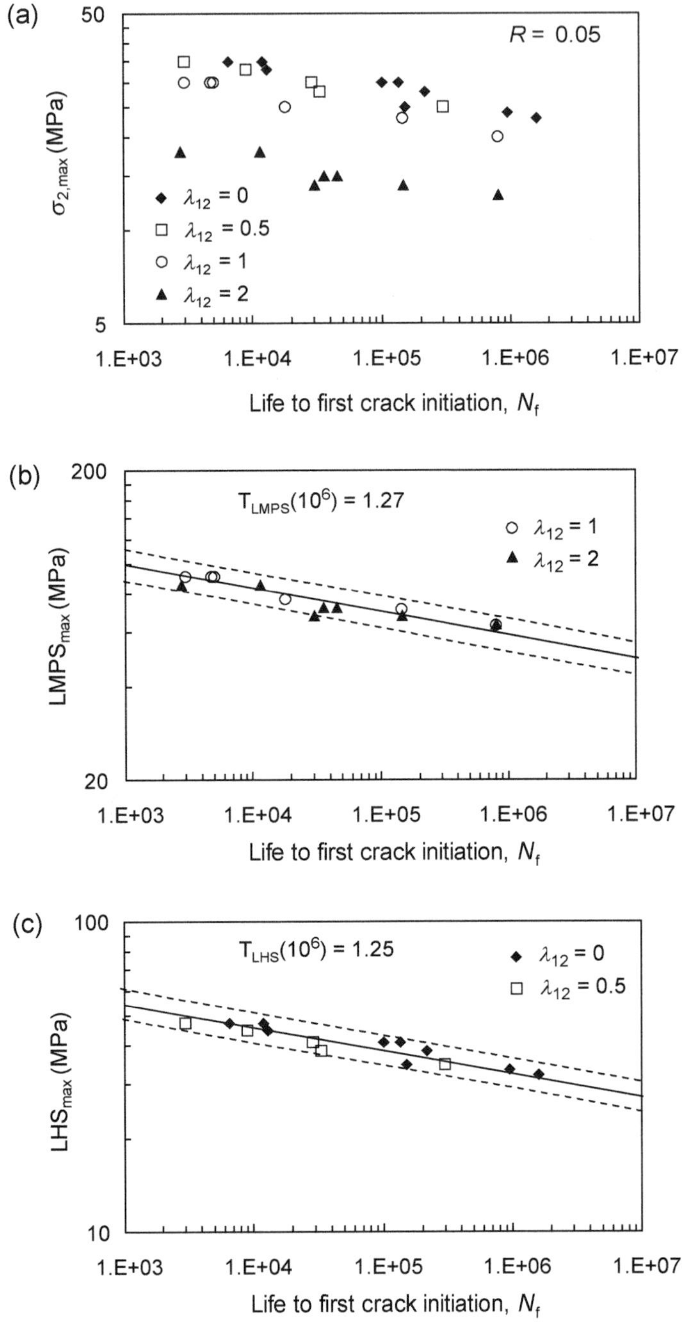

Figure 19.7 Fatigue results for $[0_F/90_3/0_F]$ tubes with $R = 0.05$ (Quaresimin and Carraro, 2014): maximum (a) Cyclic transverse stress, (b) LMPS, and (c) LHS against the number of cycles for first crack initiation N_f.

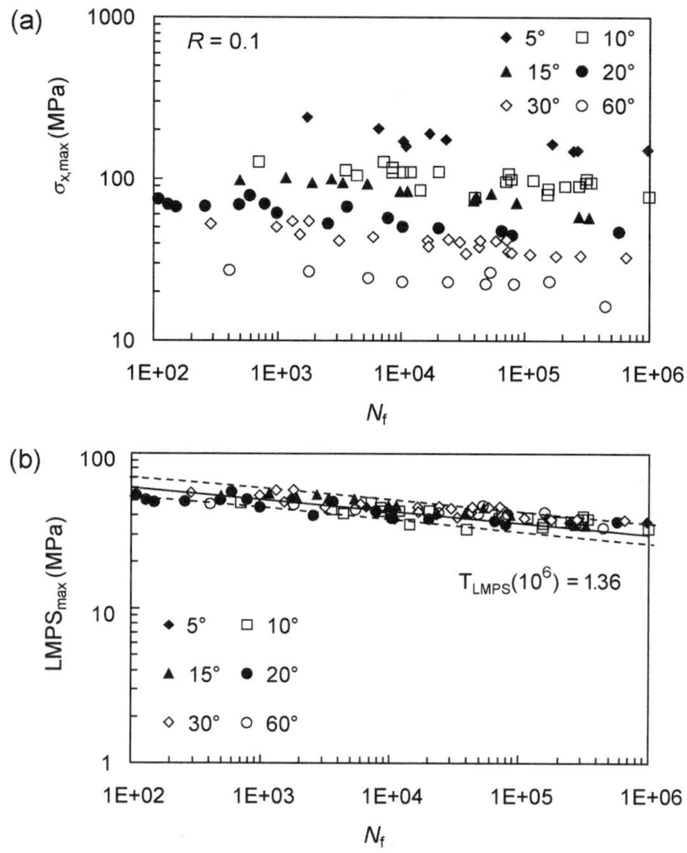

Figure 19.8 Fatigue results for flat specimens with R = 0.1 (Hashin and Rotem, 1973): maximum (a) Cyclic global stress and (b) LMPS against the number of cycles for first crack initiation N_f.

19.5 Constant–life diagrams

The two local stress parameters LHS and LMPS have been shown to describe the fatigue crack initiation behavior of UD composites by means of two scatter bands only, to be used whether the shear stress is very low or high enough, respectively. As a consequence, the complete characterization of the nonfiber-dominated fatigue behavior of a UD lamina requires only two experimental S–N curves to be obtained. The first one is, for example, related to a pure transverse stress condition, and once expressed in terms of LHS, it will be representative of the hydrostatic stress-driven fatigue failure. The second testing condition has to be characterized by an enough high shear stress contribution, achieved, for example, by means of a low off-axis angle (15–30 degree) or a high biaxiality ratio ($\lambda_{12} \geq 1$), and the related S–N curve, in terms of LMPS, will be representative of the LMPS-driven fatigue failure. The so-

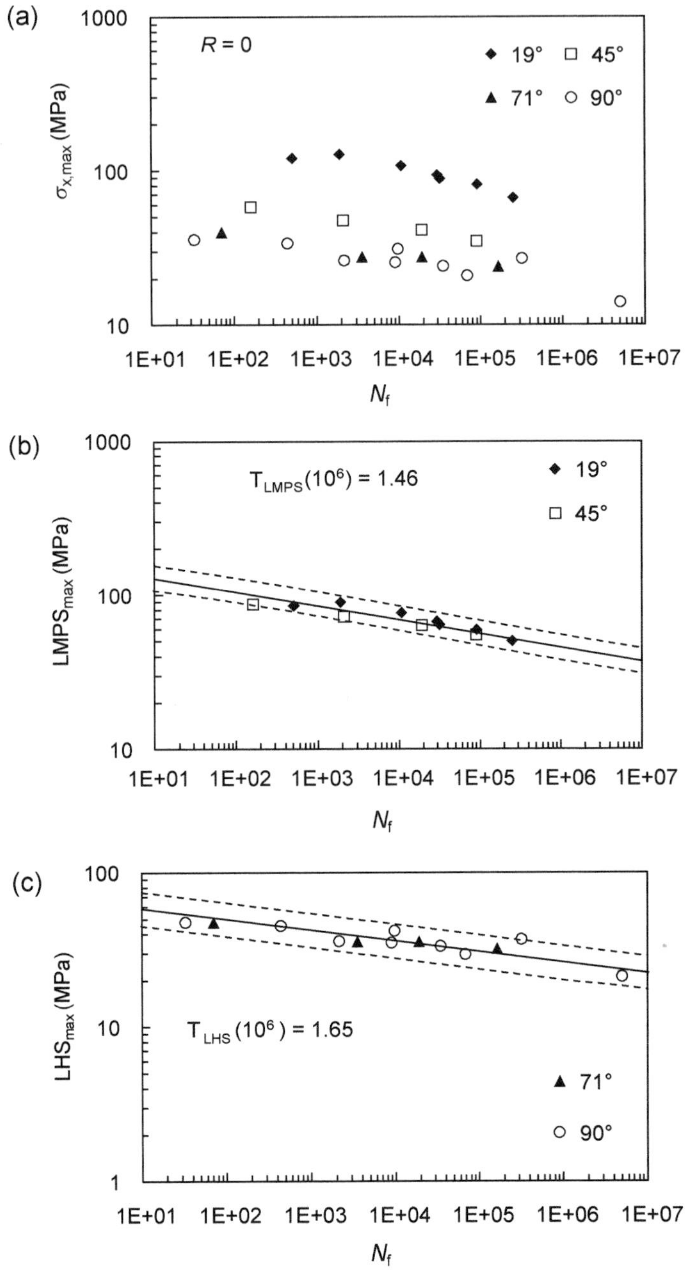

Figure 19.9 Fatigue results for flat specimens with R = 0 (El-Kadi and Ellyin, 1994): maximum (a) Cyclic global stress, (b) LMPS, and (c) LHS against the number of cycles for first crack initiation Nf.

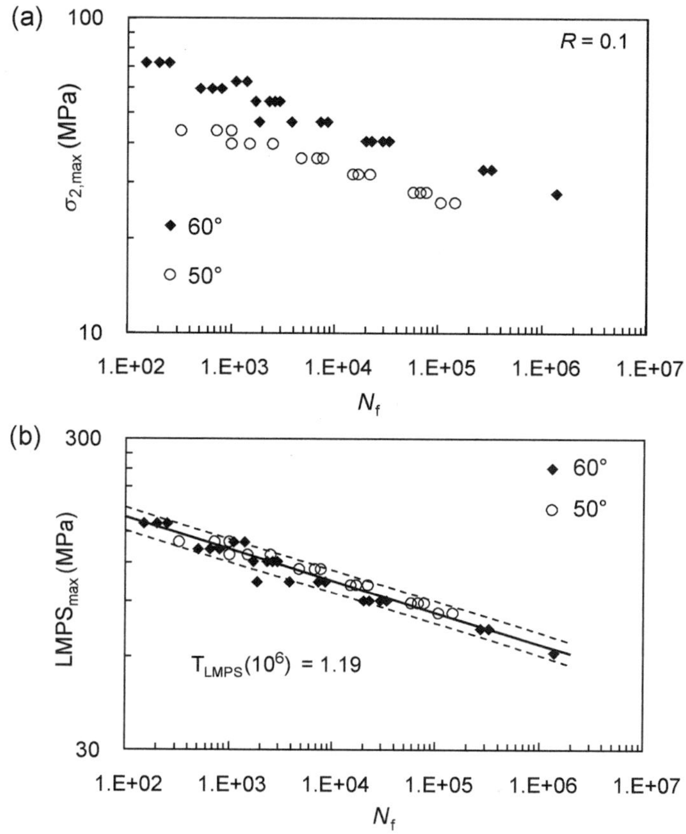

Figure 19.10 Fatigue results on flat specimens with $R = 0.1$ (Quaresimin et al., 2014): maximum (a) Cyclic transverse stress and (b) LMPS against the number of cycles to crack initiation N_f.

obtained master curves and scatter bands can be used to predict the cycles spent for crack initiation in a UD, also within a laminate, in multiaxial conditions different from those of the coducted tests.

In order to identify the transition between the two damage conditions (LHS of LMPS-controlled), constant–life diagrams can be predicted according to the proposed criterion. As an example, considering the $[90_4]$ tubular specimens subjected to combined tension/torsion analyzed in Section 19.4, the value of the transverse stress $\sigma_{2,max}$ related to a given crack initiation life N_f can be predicted, as a function of λ_{12}, by imposing the two criteria:

- LHS = LHS(N_f).
- LMPS = LMPS(N_f).

The values of LMPS(N_f) and LHS(N_f) can be calculated by means of the master curves in terms of LMPS and LHS, respectively. The two predictions for $\sigma_{2,max}(\lambda_{12})$

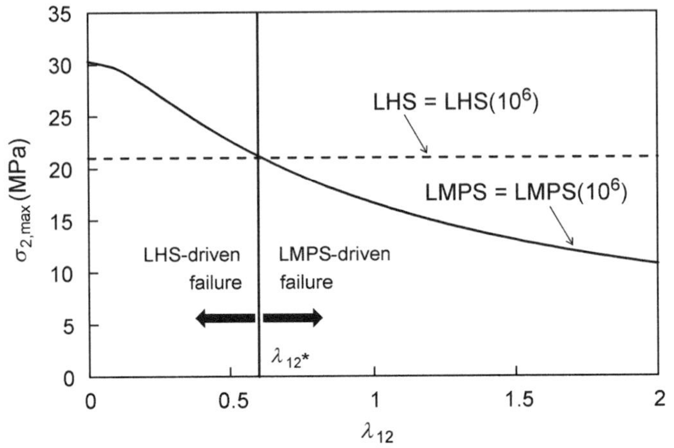

Figure 19.11 Constant–life diagram for unidirectional tubes subjected to tension/torsion.

are shown in Fig. 19.11 for $N_f = 10^6$ cycles. Obviously, the most critical criterion is that providing the lower value of $\sigma_{2,max}$ for a given biaxiality ratio. Therefore, a particular value λ_{12} can be identified as the transition between a hydrostatic stress and a maximum principal stress-driven fatigue failure. In real composites, there will not be a transition point. Instead, there will be a transition zone that is, however, expected to occur around the point corresponding to λ_{12}^*, which is a property of the composite system.

In the case of off-axis fatigue testing on the same material, constant–life diagrams can be predicted in terms of the stress in the loading direction σ_x as a function of the off-axis angle θ, by means of the same procedure already discussed. An example is shown in Fig. 19.12, for the same material properties of the previous application.

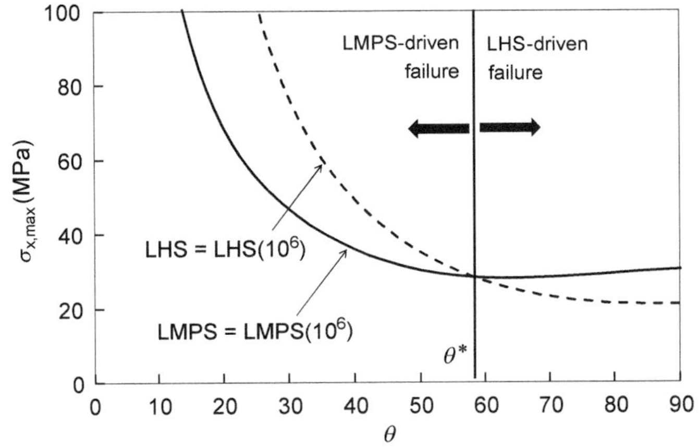

Figure 19.12 Constant–life diagram for off-axis-loaded unidirectional laminae.

19.6 Conclusions

A damage-based criterion for predicting crack initiation in UD laminae under multiaxial fatigue loading has been presented. On the basis of experimental observations, two local stress parameters, LHS and LMPS, have been identified as the driving forces for damage evolution at the microscale in the presence of a very low or high enough shear stress contribution, respectively. As proof, it has been shown that the LMPS parameter is suitable to collapse in one single scatter band crack initiation data related to multiaxial loading conditions, not in nearly pure transverse stress. In fact, it is proved that in last conditions, the local stress state in the matrix is highly hydrostatical, and it is reasonable to assume that this produces a change in the leading damage mode and, therefore, in the parameter to be used as representative of the driving force. In this work, the LHS has been shown to be a good parameter to collect fatigue data in the case of loading conditions with a low shear stress, compared to the transverse stress. Therefore, only two scatter bands, and related master curves, can be used for predicting fatigue crack initiation in composite laminae, depending on the multiaxial stress state.

References

Adden, S., Horst, P., 2006. Damage propagation in non-crimp fabrics under bi-axial static and fatigue loading. Compos. Sci. Technol. 66, 626–633. https://doi.org/10.1016/j.compscitech.2005.07.034.

Adden, S., Horst, P., 2010. Stiffness degradation under fatigue in multi-axially loaded non-crimped-fabrics. Int. J. Fatig. 32, 108–122. https://doi.org/10.1016/j.ijfatigue.2009.02.002.

Asp, L.E., Berglund, L.A., Talreja, R., 1996a. A criterion for crack initiation in glassy polymers subjected to a composite-like stress state. Compos. Sci. Technol. 56, 1291–1301. https://doi.org/10.1016/S0266-3538(96)00090-5.

Asp, L.E., Berglund, L.A., Talreja, R., 1996b. Prediction of matrix initiated transverse failure in polymer composites. Compos. Sci. Technol. 56, 1089–1097. https://doi.org/10.1016/0266-3538(96)00074-7.

Awerbuch, J., Hahn, H.T., 1981. Off-axis fatigue of graphite/epoxy composite. In: Fatigue of Fibrous Composite Materials, ASTM STP 723. American Society for Testing and Materials, pp. 243–273.

Carraro, P.A., Quaresimin, M., 2014. A damage based model for crack initiation in unidirectional composites under multiaxial cyclic loading. Compos. Sci. Technol. 99, 154–216. https://doi.org/10.1016/j.compscitech.2014.05.012.

Cox, B.N., Dadkhah, M.S., Morris, W.L., Flintoff, J.G., 1994. Failure mechanisms of 3D woven composites in tension, compression, and bending. Acta Metall. Mater. 42 (12), 3967–3984. https://doi.org/10.1016/0956-7151(94)90174-0.

El-Kadi, H., Ellyin, F., 1994. Effect of stress ratio on the fatigue of unidirectional fibre glass-epoxy composite laminae. Composites 25 (10), 917–924. https://doi.org/10.1016/0010-4361(94)90107-4.

Hashin, Z., Rotem, A., 1973. A fatigue failure criterion for fibre-reinforced materials. J. Compos. Mater. 7, 448–464.

Kawai, M., Yajima, S., Hachinohe, A., Takano, Y., 2001. Off-axis fatigue behaviour of unidirectional carbon fiber-reinforced composites at room and high temperatures. J. Compos. Mater. 35, 545–576. https://doi.org/10.1106/WQMQ-524H-6PKL-NGCY.

Lafarie-Frenot, M.C., Hénaff-Gardin, C., 1991. Formation and growth of 90° ply fatigue cracks in carbon/epoxy laminates. Compos. Sci. Technol. 40, 307–324. https://doi.org/10.1016/0266-3538(91)90087-6.

Plumtree, A., Cheng, G.X., 1999. A fatigue damage parameter for off-axis unidirectional fiber reinforced composites. Int. J. Fatig. 21, 849–856. https://doi.org/10.1016/S0142-1123(99)00026–2.

Puck, A., Shurmann, H., 1998. Failure analysis of FRP laminates by means of physically based phenomenological models. Compos. Sci. Technol. 58, 1045–1067. https://doi.org/10.1016/S0266-3538(01)00208-1.

Quaresimin, M., Carraro, P.A., 2013. On the investigation of the biaxial fatigue behaviour of unidirectional composites. Compos. B Eng. 54, 200–208. https://doi.org/10.1016/j.compositesb.2013.05.014.

Quaresimin, M., Carraro, P.A., 2014. Damage initiation and evolution in glass/epoxy tubes subjected to combined tension-torsion fatigue loading. Int. J. Fatig. 63, 25–35. https://doi.org/10.1016/j.ijfatigue.2014.01.002.

Quaresimin, M., Carraro, P.A., Pilgaard Mikkelsen, L., Lucato, N., Vivian, L., Brøndsted, P., et al., 2014. Damage evolution under internal and external multiaxial cyclic stress state: a comparative analysis. Compos. B Eng. 61, 282–290. https://doi.org/10.1016/j.compositesb.2014.01.056.

Reifsneider, K.L., Gao, Z., 1991. A micromechanics model for composites under fatigue loading. Int. J. Fatig. 13, 149–156. https://doi.org/10.1016/0142-1123(91)90007-L.

Shiino, M.Y., De Camargo, L.M., Cioffi, M.O.H., Voorwald, H.C.J., Ortiz, E.C., et al., 2012. Correlation of microcrack fracture size with fatigue cycling on non-crimp fabric/RTM6 composite in the uniaxial fatigue test. Compos. B Eng. 43, 2244–2248. https://doi.org/10.1016/j.compositesb.2012.01.074.

Sun, X.S., Haris, A., Tan, V.B.C., Tay, T.E., Narasimalu, S., Della, C.N., 2012. A multi-axial fatigue model for fiber-reinforced composite laminates based on Puck's criterion. J. Compos. Mater. 46, 449–469. https://doi.org/10.1177/0021998311418701.

Talreja, R., 1981. Fatigue of composite materials: damage mechanisms and fatigue-life diagrams. Proc. Roy. Soc. Lond.: Math. Phys. Eng. Sci. 378, 461–475.

Talreja, R., 2006. Multi-scale modeling in damage mechanics of composite materials. J. Mater. Sci. 41, 6800–6812. https://doi.org/10.1007/s10853-006-0210-9.

Tong, J., 2002. Characteristics of fatigue crack growth in GFRP laminates. Int. J. Fatig. 24, 291–297. https://doi.org/10.1016/S0142-1123(01)00084–6.

Tong, J., 2001. Three stages of fatigue crack growth in GFRP composite laminates. J. Eng. Mater. Technol.-Transac. ASME 123, 139–143. https://doi.org/10.1115/1.1286234.

Tong, J., Guild, F.J., Ogin, S.L., Smith, P.A., 1997. Off-axis fatigue crack growth and the associated energy release rate in composite laminates. Appl. Compos. Mater. 4, 349–359.

Wharmby, A.W., Ellyin, F., 2002. Damage growth in constrained angle-ply laminates under cyclic loading. Compos. Sci. Technol. 62, 1239–1247. https://doi.org/10.1016/S0266-3538(02)00075-1.

Yokozeki, T., Aoki, T., Ishikawa, T., 2002. Fatigue growth of matrix cracks in the transverse direction of CFRP laminates. Compos. Sci. Technol. 62, 1223–1229. https://doi.org/10.1016/S0266-3538(02)00068-4.

Zhang, Y., Xia, Z., 2005. Micromechanical analysis of interphase damage for fiber reinforced composite laminates. CMC-Computers Mater. Continua 2, 213–226. https://doi.org/10.3970/cmc.2005.002.213.

Incorporating manufacturing defects in damage and failure analysis

20

Ramesh Talreja
Department of Aerospace Engineering, Department of Materials Science and Engineering, Texas A&M University, College Station, TX, United States

20.1 Introduction

Composite materials have impacted the aeronautics and space industries by allowing structural designs in performance regimes that would not be feasible with monolithic materials. The lightweight advantages of composites have in recent years led to exponential rise in their usage in nonaerospace applications such as automobiles and wind turbines. While the cost of manufacturing in aerospace applications is an important consideration, it is a crucial limiting factor in nonaerospace applications. Reduction in manufacturing cost is possible if some defects can be tolerated, as approaching a defect-free state can make the product cost prohibitive. A strategy for cost-effective manufacturing is depicted in Fig. 20.1. As indicated there, a cost−performance trade-off must be conducted to reduce the manufacturing cost. In order to achieve this, one must quantify the manufacturing process in terms of parameters that can be related to the material state of the end product. The material state in turn must be characterized in terms of parameters that describe the microstructure as produced by

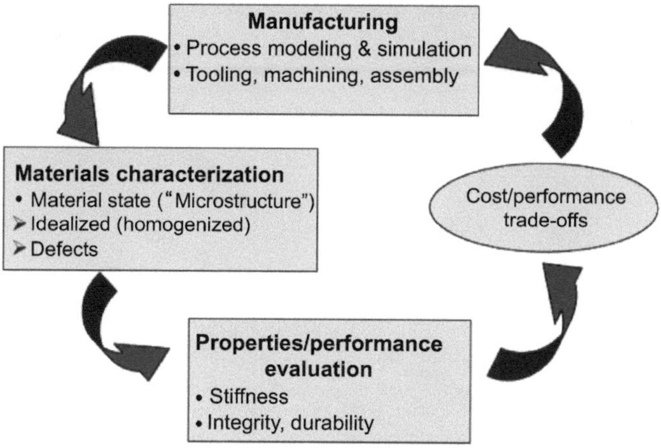

Figure 20.1 A schematic depiction of cost-effective design strategy for composite structures.

Modeling Damage, Fatigue and Failure of Composite Materials. https://doi.org/10.1016/B978-0-443-18489-5.00017-5

the manufacturing process. The current engineering practice is to use descriptors of the material state that are based on idealized (homogenized) microstructure. Examples of this are elastic properties, such as Young's modulus, and failure characteristics, for example, tensile and compressive strengths. When measured as experimental values, these properties reflect the composition of the material state but do not explicitly carry the information concerning manufacturing-induced defects. When a structural part is subjected to a service environment, the material state can undergo permanent changes that manifest themselves as degradation of the initially measured properties. Design requirements dictate limits to such degradation in order to have safe performance expressed as structural integrity and durability. The current design strategy is to manufacture a product aiming at the least defects, followed by tests to assure satisfaction of the performance limits. The certification authorities, in particular in the civilian aircraft industry, enforce this empirical trial-and-error approach, driven by the needs of safety. It would be unreasonable to compromise safety, but perhaps one can question whether this is the only way to assure safety. If the performance indicators (property degradation limits) can be assured at reduced cost of manufacturing, then a cost-effective design can be achieved without compromising safety. The means to achieve this is the cost–performance trade-off, which in turn requires quantitative iteration of the processing–microstructure–performance loop.

Fig. 20.2 further illustrates the nature of the cost-effective design process. As depicted there, the process consists of improving the design to better levels of cost-effectiveness by sequentially conducting the iteration of the processing–microstructure–performance loop under the imposed constraints of cost. The microstructure to be quantified is, however, of the real initial material state (RIMS), which includes defects induced by the manufacturing process along with the geometry and

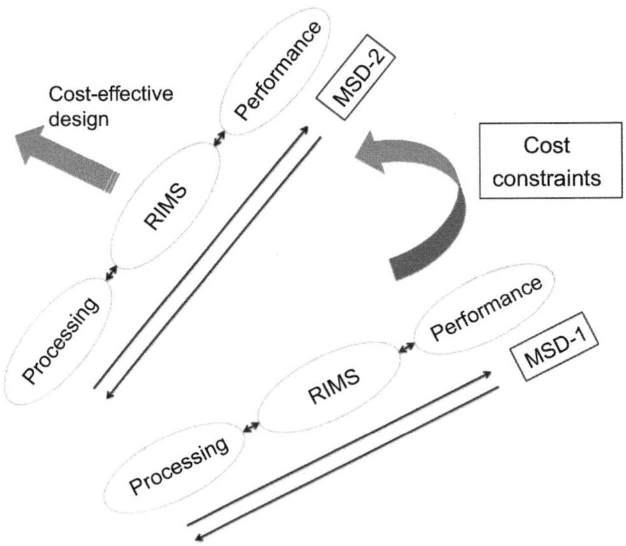

Figure 20.2 The sequential-iterative process of cost-effective design.

properties of the composite constituents. The three-way linkage connecting the manufacturing process, RIMS, and product performance is known as manufacturing-sensitive design (MSD), which constitutes the iterative part of the sequential design to increasing levels of cost-effectiveness. This chapter will focus on characterization of RIMS and connecting it to the performance indicators, which quantify the degradation incurred in RIMS. The discussion here will relate to the concepts of multiscale modeling such as a representative volume element (RVE) described in the earlier chapters (Chapters 14 and 15).

20.2 Real initial material state

RIMS is the material state induced in the end-product resulting from the completed manufacturing process, and it consists of the expected (desired) microstructure and all irregularities (undesired elements) in it. The irregularities, or defects, can be categorized systematically as fiber defects, for example, misalignment, waviness, and breakages of fibers as well as nonuniformity in their distribution within the matrix; matrix defects, for example, voids and incompletely cured regions; and interface defects, consisting of lacking adhesion between fibers and matrix and between layers. Some illustrative examples of defects are shown in Figs. 20.3–20.6. Fig. 20.3 shows large voids in the matrix (>100 μm) resulting from RTM (resin transfer molding). Voids comparable to the fiber diameter (≈ 10 μm) are seen in the image in Fig. 20.4, showing a cross-section of a composite made by a liquid compression molding process. The nonuniformity of fiber distribution is also evident in this image. Fig. 20.5 shows examples of voids produced by intentionally not following the recommended manufacturing process. Fig. 20.5a shows voids resulting from moisture dissolved in the matrix due to incorrect handling of prepregs (Grunenfelder and Nutt, 2010), and Fig. 20.5b gives an example of a void trapped in the matrix that did not get removed because of omitting vacuum from the autoclave molding process (Huang et al., 2011). Fig. 20.6 shows a severe case of fiber waviness in a wind turbine blade that develops delamination during service as a consequence of this defect.

Recently, the ability to observe internal defects in a composite has advanced to the extent that images of defects such as voids can be produced with great clarity. This has resulted in quantification of void size, shape, location, and distributions. Fig. 20.7 shows images obtained by microfocus X-ray computed tomography scans (Lambert et al., 2012).

20.3 RIMS–performance relationships

Referring to Fig. 20.2 depicting the cost-effective manufacturing process, the following discussion will focus on establishing RIMS–performance relationships. The other elements of the cost-effective design strategy, that is, the processing–RIMS relationships and cost–performance trade-off, are outside the scope of this exposition.

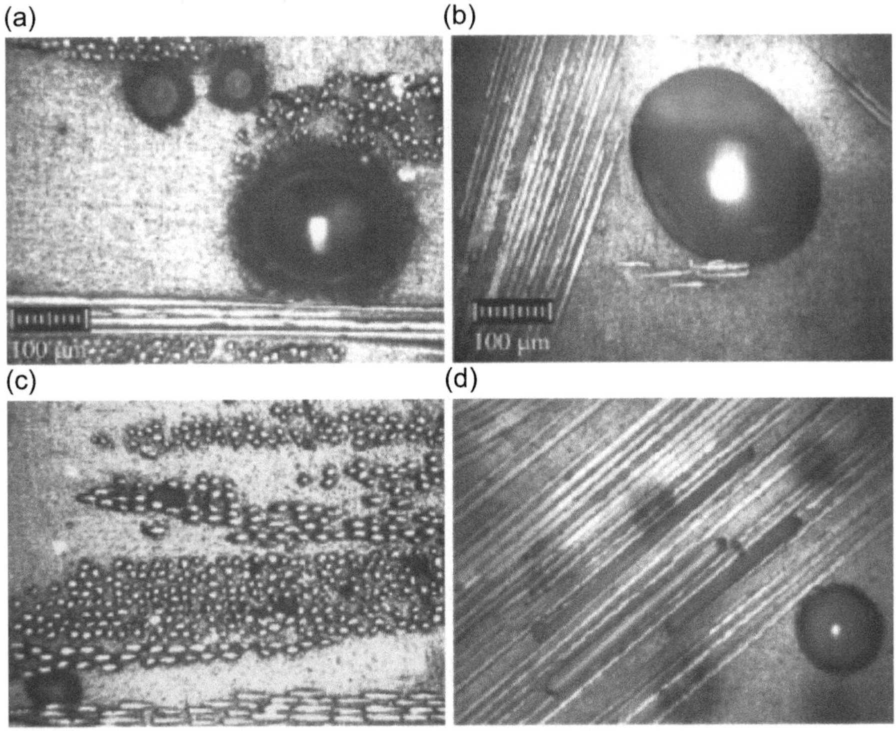

Figure 20.3 Large voids in matrix resulting from a resin transfer molding (RTM) process (Hamidi et al., 2005). (a)–(d) are images taken from different regions of the manufactured part.

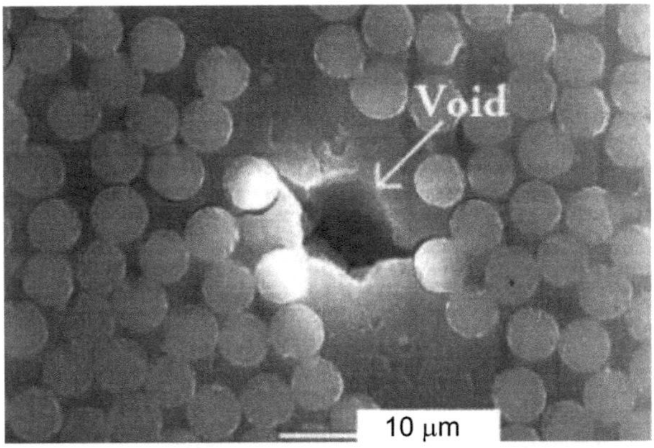

Figure 20.4 Cross-section of a composite showing irregular fiber distribution and a void in the matrix (Wood and Bradley, 1997).

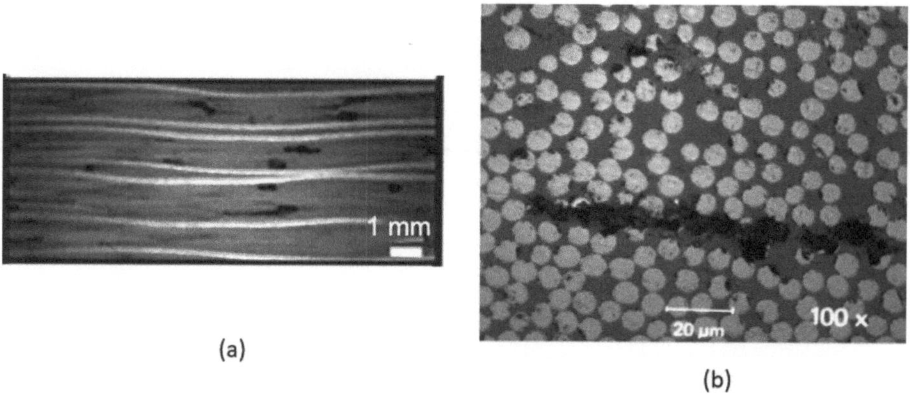

Figure 20.5 Voids in the matrix created by irregularities of the autoclave molding process. (a) Not removing moisture (Grunnenfelder and Nutt, 2010). (b) Not applying vacuum (Huang et al., 2011).

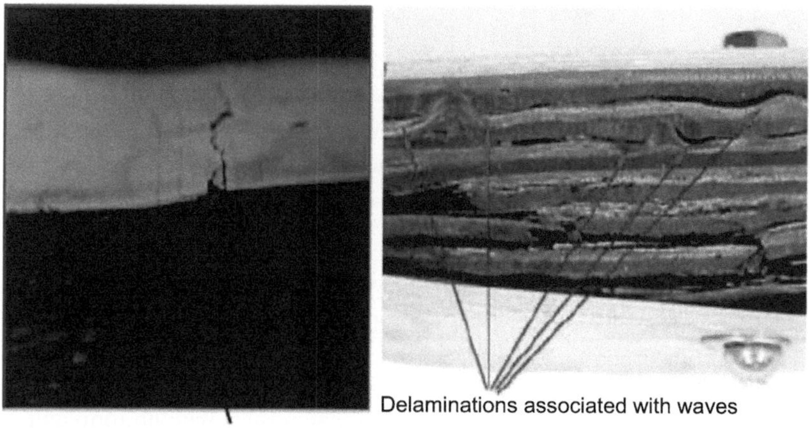

Figure 20.6 Fiber waviness in a wind turbine blade (left) and delamination caused by it in service loading (Cairns et al., 2011).

20.3.1 Describing RIMS

The principle guiding RIMS description is to capture sufficient details of the microstructure for a given purpose concerning evaluation of performance. For instance, if the performance requirements are specified as limiting values of structural deflections, then the microstructure description should target determining relevant deformation characteristics, for example, Young's modulus. Traditionally, defects have been neglected for this purpose, assuming small effects, or have been accounted for by embedding them in a homogenized solid. For failure assessment, the need to capture the details of the microstructure, including defects, is generally higher and depends on the failure mode involved. In fact, initiation of failure is governed by the local stress

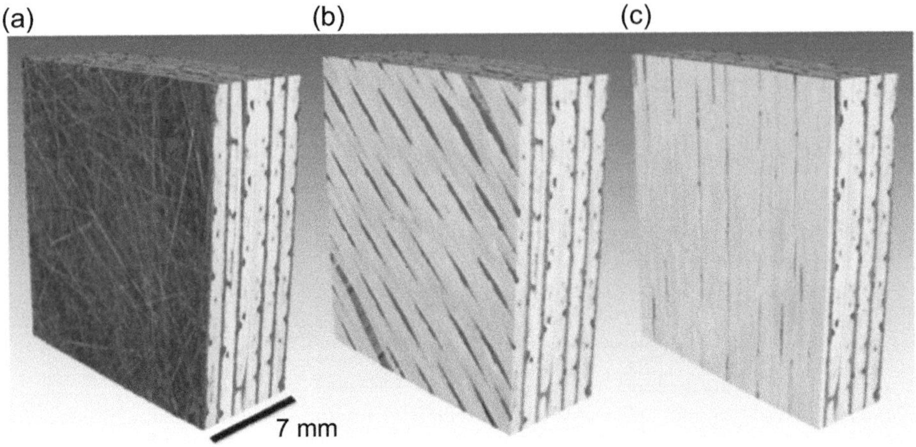

Figure 20.7 Images produced by microfocus X-ray computed tomography scans reported in Lambert et al. (2012) showing voids in a glass−epoxy [0/45/−45]3s laminate. Image (a) shows the entire laminate. Images (b) and (c) show voids in the 45° and 0° ply, respectively.

fields, which in turn depend on the microstructure, and defects in particular. A proper account of defects is not only necessary for the assessment of failure, it is a key element in MSD and cost reduction. The following will illustrate RIMS descriptions in the context of assessing different performance metrics.

20.3.2 Elastic properties

Estimating elastic properties of a composite from properties of matrix and fibers and geometrical details such as size, shape, and distribution of fibers is a well-developed field. However, incorporating manufacturing defects in property estimation is not as fully developed. Hsiao and Daniel (1996) developed an analytical model to account for axial compressive stiffness of unidirectional composites with uniform and graded fiber waviness. Karami and Garnich (2005) analyzed the effect of fiber waviness on this property by a finite element method. The most common defects in the matrix are voids, and there has been a significant attention in the literature to the effects of these defects on composite elastic properties. Most studies have extended the classical inclusion theories (Mura et al., 1996) to account for voids (Berryman, 1993) where voids are considered as inclusions having zero properties. Other efforts considered voids embedded in a homogeneous composite (Chao and Huang, 1999). Huang and Talreja (2005) examined actual voids resulting from an autoclave manufacturing process (Bowles and Frimpong, 1992) (Fig. 20.8), and incorporated the observed shape and size characteristics of those voids in a computational scheme for calculating the in-plane and out-of-plane elastic properties of a unidirectional composite. They found that while the in-plane properties were mildly affected by voids, the out-of-plane properties were significantly degraded. Fig. 20.9 shows the predicted and experimental values of the out-of-plane Young's modulus as a function of the void volume fraction.

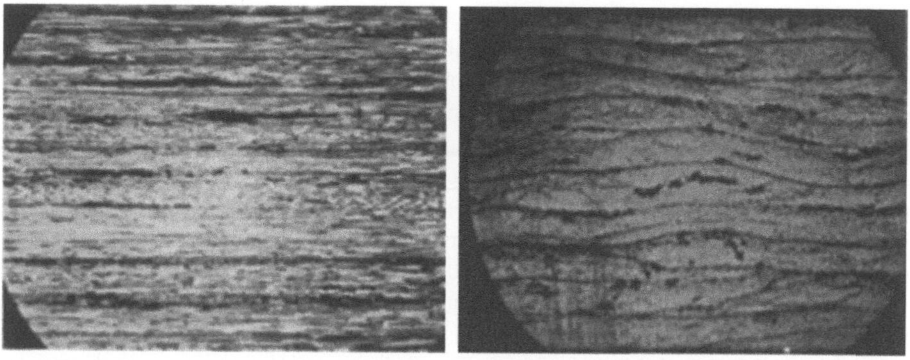

Figure 20.8 Cut sections parallel to fibers (left) and normal to fibers (right) showing voids in a unidirectional composite manufactured by the autoclave process (Bowles and Frimpong, 1992).

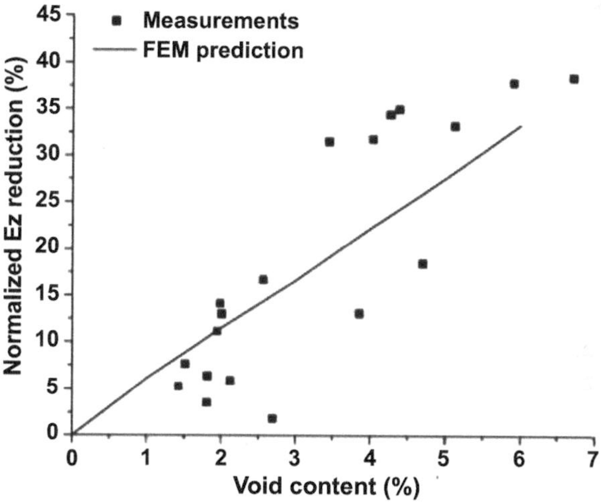

Figure 20.9 Experimental and predicted values of the normalized reduction in the through-the-thickness Young's modulus of a unidirectional composite with the void content.

A novel feature of the Huang−Talreja approach was to not embed voids in a homogenized composite, as is commonly done, but to place the voids as a part of the microstructure. Thus, the local displacement of fibers caused by the presence of voids was accounting for by adjusting the fiber volume fraction around the voids.

20.3.3 Failure initiation

Initiation of failure in the matrix and at fiber−matrix interfaces was discussed in Chapter 3 of this book. The mechanisms for these failure modes are governed by the local stress states, which in turn depend on the microstructure, including defects.

The defects of significance in this context are the irregularity of fiber distribution and the presence of voids in the matrix. Several methodologies have been put forward for characterizing nonuniform fiber distributions. Pyrz (1994) proposed that a real fiber distribution be described by random distributions of points in patterns categorized as hard-core and various clusters. Such distributions were generated from finite-size samples using random numbers. Certain preselected distribution features, for example, nearest neighbor spacing, captured from observed images using Dirichlet tessellations, were used to discriminate between the distributions. Bulsara et al. (1999) developed a simulation scheme for generating statistically similar realizations of the actual microstructure based on a radial distribution function. They also addressed the question of how large these realizations need to be in order to be representative with respect to the considered failure-initiation mechanism. A systematic study was conducted for the crack-initiation mechanisms under transverse tension of unidirectional composites.

Fig. 20.10 shows an example of a realization of fiber distribution in the cross-section of a unidirectional composite with 40 fibers using the scheme developed in Bulsara et al. (1999). The corresponding variations of the average values of the maximum local dilatational and distortional energy densities with imposed total lateral strain consisting of mechanical strain superposed on thermal residual strain are shown in Fig. 20.11. These results are used in assessing which of the two competing failure mechanisms, that is, cavitation-induced fiber—matrix debonding or matrix cracking, are likely to occur (Talreja, 2014).

The effect of voids on failure initiation under tensile loading normal to fibers was experimentally observed in a composite manufactured by the RTM process (Varna et al., 1995). Chowdhury et al. (2008) conducted an advanced failure analysis of a unit cell of a composite containing fibers, matrix, and voids, and loaded transverse to fibers. That analysis accounted for the effects of strain rate, temperature, and

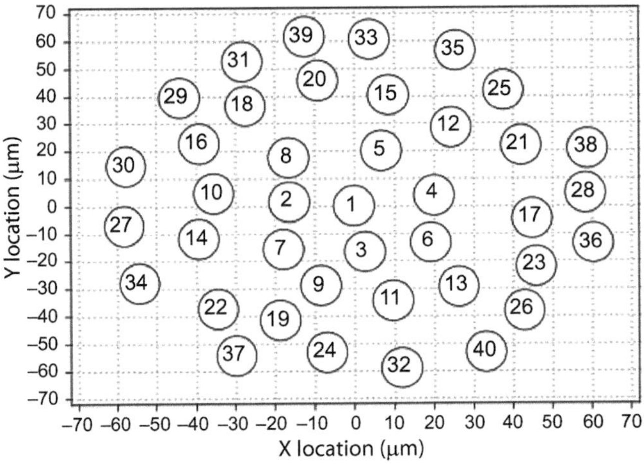

Figure 20.10 A representative cross-section of a UD composite constructed by randomly distributing 40 fibers in the matrix.

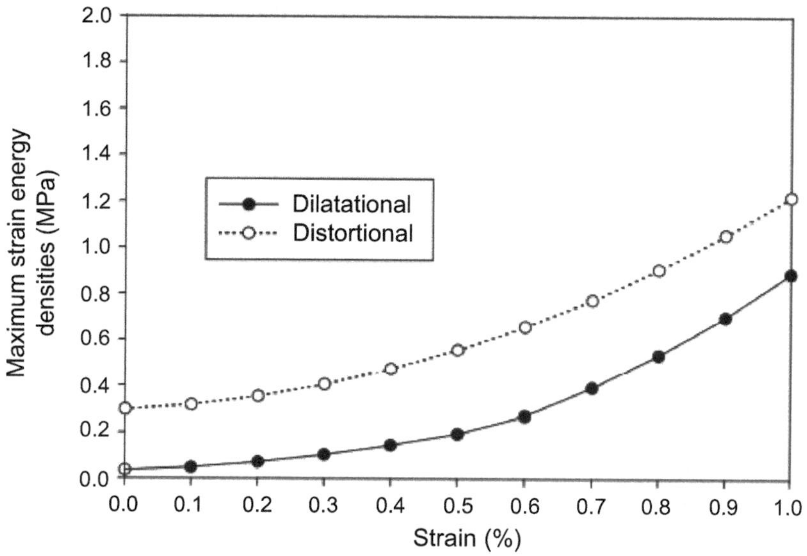

Figure 20.11 Variations of the maximum values of dilatational and distortional strain energy densities occurring locally within the RVE shown in Figure 20.9 with the applied transverse strain.

pressure on yielding as well as all observed nonlinearities in the postyield behavior of a glassy polymer. The results showed that in the presence of voids, the fiber—matrix debonding was likely at low temperatures, irrespective of the strain rate, while at moderately high temperatures, the failure initiation shifts to void-induced cracking.

20.3.4 Multiple cracking

Multiple cracking occurs in laminates within plies under the constraint of neighboring plies. This phenomenon has been studied extensively, and several chapters of this book review the evolution of cracking (Chapters 1 and 8) and the effects on composite response (Chapters 12—14). All of these studies do not consider manufacturing defects in keeping with the traditional approach. It is known that the significance of defects is largely in the early stage of the multiple-cracking process, where interaction between cracks is low or absent. As the stress perturbations caused by individual cracks interact, the mutual influence of cracks is to relax the stress field between two neighboring cracks (known as the shielding effect) and reduce the potential region in which next crack is formed on increasing the applied stress. This localization of potential cracking region reduces the effect of distributed defects in initiating cracks.

Huang et al. (2014) studied the effects of manufacturing defects on multiple-cracking evolution in a cross-ply laminate by intentionally producing defects under different autoclave-processing conditions. By using or eliminating vacuum and pressure in different combinations, they introduced different defect populations in the laminate. The resulting evolution of the number density of transverse cracks in the

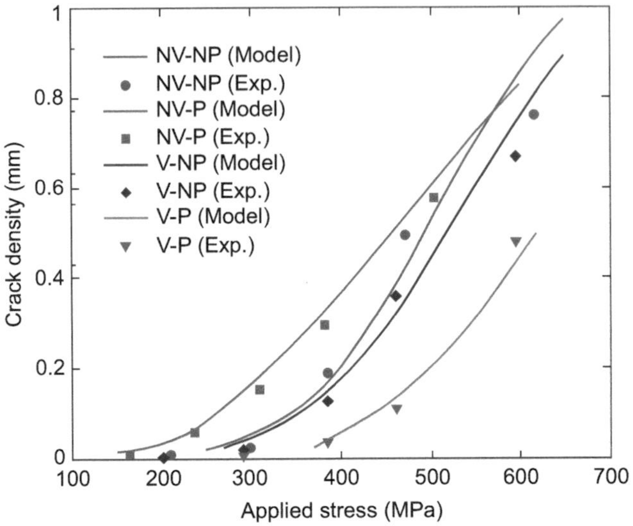

Figure 20.12 The initial stage of crack density evolution in a cross-ply laminate with different autoclave processing conditions: V-P, vacuum and pressure applied; V-NP, vacuum applied, pressure not applied; NV-P, vacuum not applied, pressure applied; and NV-NP, neither vacuum nor pressure applied.
From Huang et al. (2014).

90-degree plies was found to depend on the processing conditions. The effects were most prominent in the early stages of the multiple-cracking process and are shown in Fig. 20.12. Huang et al. (2014) developed a statistical model to capture the effect of defects on the crack density as a practical tool for assessing the manufacturing quality.

20.3.5 Conditions for crack growth

A large crack in a composite laminate can exist either as an unbonded region of the interface between plies or as a crack produced at the interface, for example, by a lateral impact. In either case, it is of interest to know how manufacturing defects would affect the conditions for the growth of such a crack. One scenario to consider in this context is the presence of manufacturing-induced voids that may lie ahead of the crack front in the matrix. Ricotta et al. (2008) studied the effects of voids on Mode I energy release rate by introducing voids ahead of the crack tip in a double cantilever beam. Fig. 20.13 illustrates the results by plotting the effect of the void shape and location ahead of the crack tip. As seen there, the incentive for the crack to advance is enhanced significantly by the presence of voids. Zhuang and Talreja (2014) presented another study of the effect of voids on a delamination crack growth under post-buckling (out-of-plane) deformation of the delaminated plate when a void lies ahead

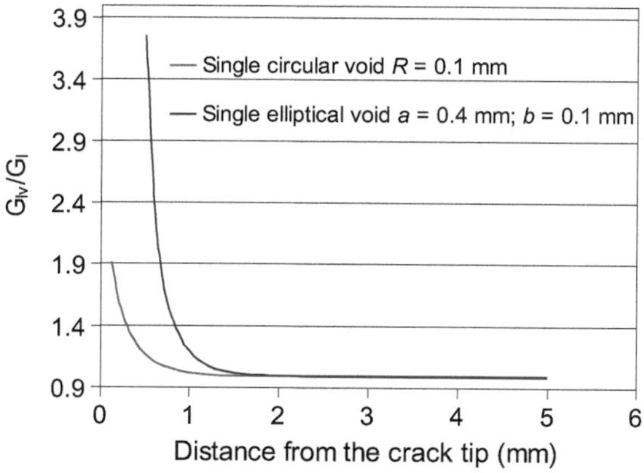

Figure 20.13 Energy release rate in the presence of a void (GIv), normalized by its value without a void (GI) for a circular void of radius R = 0.1 mm and for an elliptical void of major semiaxis a = 0.4 mm and minor semiaxis b = 0.1 mm, plotted against distance of the void center from the crack tip.

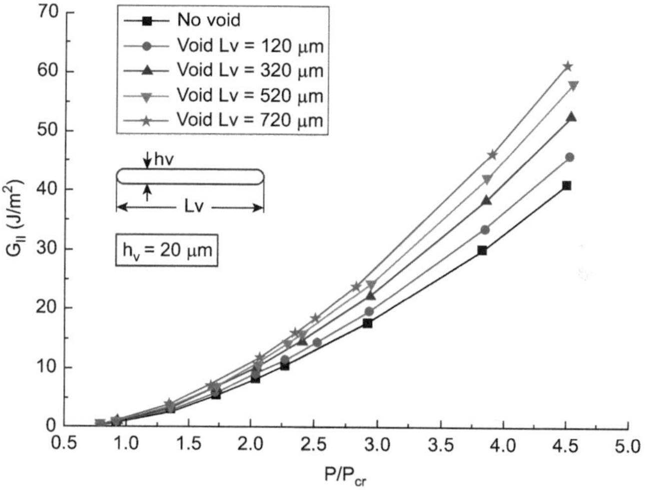

Figure 20.14 Effects on the mode II energy release rate (GII) of a delamination crack in postbuckling growth by voids of different shape lying at a distance of 0.14 mm ahead of the delamination front.
From Zhuang and Talreja (2014).

of the delamination front. Fig. 20.14 shows one illustrative result where the effect on the Mode II energy release rate (G_{II}) by voids of different shape lying at a fixed distance is depicted. The crack growth is in mixed mode with the Mode II being the dominant part.

20.3.6 *Final failure*

Final failure of composite laminates comes from breakage of fibers in the last load-bearing ply. Prior to this event, that ply often delaminates from the plies to which it was bonded. As described in Chapter 15, the depletion of load-bearing capacity of a laminate culminating in the final failure involves increasing localization of interconnected cracks. In the final stage of the evolution of damage, the complexity of the failing elements makes it difficult to delineate the role of manufacturing defects, except in the fiber breakage process, which is inherently statistical because of the fiber defects.

20.4 Comprehensive failure analysis with defects

The total failure process of a composite laminate involves initiation of the first failure event, progression of collective, increasingly interactive failure events, and localization of interconnected failure surfaces, leading to delamination and fiber breakage. The scales that characterize energy dissipation change from the first failure event to the final failure. A comprehensive failure analysis must therefore be at multiple scales. The role of manufacturing defects is accordingly different at each scale.

Fig. 20.15 provides an overview of the comprehensive multiscale failure assessment scheme depicting the incorporation of manufacturing defects. The descriptors associated with RIMS correspond to the operating scales of the failure events. In the lamina level failure analysis, the RVE in which the defects are incorporated, labeled as RVE-1, are the defects that play a role in initiating fiber—matrix debonding and matrix failure. Additionally, fiber failure, such as that resulting from kink bands, can also be treated in RVE-1. Further progression of failure involves multiple cracking under the mutual constraints of plies in a laminate, leading to localization, delamination, and final failure. All of these events have interactions that extend beyond a single-ply level. Failure analysis of these events must incorporate defects in a laminate-level RVE, labeled as RVE-2 in Fig. 20.15.

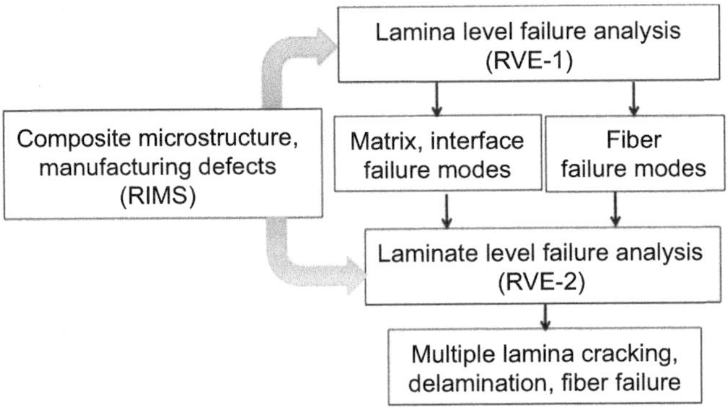

Figure 20.15 Overview of the scheme for incorporating manufacturing defects in damage and failure analysis of composite laminates.

20.5 Conclusions

Damage and failure analysis of composite materials has entered a new phase where manufacturing defects can be incorporated. This has been possible because of advances in capabilities to observe and quantify the defects as well as to conduct computational multiscale failure analyses. Although the effects of defects on various measured materials' response characteristics such as interlaminar shear strength and fatigue life have been known, a comprehensive failure assessment strategy incorporating manufacturing defects has not existed. Such a strategy has been proposed here and a methodology has been discussed that accounts for the defects as an integral part of the material state. The concept underlying the methodology is a characterization of the material state by appropriate descriptors that capture features of the microstructure at selected length scales using relevant RVEs. Defects of interest are thus represented in these descriptors. Realizations of statistical equivalency to the real material state are then produced and subjected to stress and failure analysis.

The new approach to failure analysis discussed here provides a rational alternative to current failure criteria that are formulated on homogenized composites and are therefore incapable of multiscale treatment of failure-incorporating defects. The approach also provides a systematic pathway to reducing manufacturing cost by conducting cost−performance trade-off.

References

Berryman, J.G., 1993. Role of porosity in estimates of composite elastic constants. In: Composite Material Technology Conference Proceedings, vol 53. American Society of Mechanical Engineers, pp. 223−235.

Bowles, K.J., Frimpong, S., 1992. Void effects on the interlaminar shear strength of unidirectional graphite-fiber-reinforced composites. J. Compos. Mater. 26 (10), 1487−1509.

Bulsara, V.N., Talreja, R., Qu, J., 1999. Damage initiation under transverse loading of unidirectional composites with arbitrarily distributed fibers. Compos. Sci. Technol. 59 (5), 673−682.

Cairns, D.S., Riddle, T., Nelson, J., 2011. Wind turbine composite blade manufacturing: the need for understanding defect origins, prevalence, implication and reliability. In: Report SAND2011−1094 Sandia National Laboratories: Albuquerque, New Mexico.

Chao, L.P., Huang, J.H., 1999. Prediction of elastic moduli of porous materials with equivalent inclusion method. J. Reinforc. Plast. Compos. 18 (7), 592−605.

Chowdhury, K.A., Talreja, R., Benzerga, A.A., 2008. Effects of manufacturing-induced voids on local failure in polymer-based composites. J. Eng. Mater. Technol. 130 (2), 021010.

Grunenfelder, L.K., Nutt, S.R., 2010. Void formation in composite prepregs−Effect of dissolved moisture. Compos. Sci. Technol. 70 (16), 2304−2309.

Hamidi, Y.K., Aktas, L., Altan, M.C., 2005. Three-dimensional features of void morphology in resin transfer molded composites. Compos. Sci. Technol. 65 (7), 1306−1320.

Hsiao, H.M., Daniel, I.M., 1996. Elastic properties of composites with fiber waviness. Compos. Appl. Sci. Manuf. 27 (10), 931−941.

Huang, H., Talreja, R., 2005. Effects of void geometry on elastic properties of unidirectional fiber reinforced composites. Compos. Sci. Technol. 65 (13), 1964–1981.

Huang, Y., Varna, J., Talreja, R., 2011. A statistical approach to evaluate the effect of manufacturing quality on transverse cracking in cross ply laminates. In: Proceedings of 18th International Conference on Composites Materials.

Huang, Y., Varna, J., Talreja, R., 2014. Statistical methodology for assessing manufacturing quality related to transverse cracking in cross ply laminates. Compos. Sci. Technol. 95, 100–106.

Karami, G., Garnich, M., 2005. Effective moduli and failure considerations for composites with periodic fiber waviness. Compos. Struct. 67 (4), 461–475.

Lambert, J., Chambers, A.R., Sinclair, I., Spearing, S.M., 2012. 3D damage characterisation and the role of voids in the fatigue of wind turbine blade materials. Compos. Sci. Technol. 72 (2), 337–343.

Mura, T., Shodja, H.M., Hirose, Y., 1996. Inclusion problems. Appl. Mech. Rev. 49 (10), 118–127.

Pyrz, R., 1994. Quantitative description of the microstructure of composites. Part I: morphology of unidirectional composite systems. Compos. Sci. Technol. 50, 197–208.

Ricotta, M., Quaresimin, M., Talreja, R., 2008. Mode I strain energy release rate in composite laminates in the presence of voids. Compos. Sci. Technol. 68 (13), 2616–2623.

Talreja, R., 2014. Assessment of the fundamentals of failure theories for composite materials. Compos. Sci. Technol. 105, 190–201.

Varna, J., Joffe, R., Berglund, L.A., Lundström, T.S., 1995. Effect of voids on failure mechanisms in RTM laminates. Compos. Sci. Technol. 53 (2), 241–249.

Wood, C.A., Bradley, W.L., 1997. Determination of the effect of seawather on the interfacial strength of an interlayer E-glass/graphite/epoxy composite by in situ observation of transverse cracking in an environment SEM. Compos. Sci. Technol. 57 (8), 1033–1043.

Zhuang, L., Talreja, R., 2014. Effects of voids on postbuckling delamination growth in unidirectional composites. Int. J. Solid Struct. 51 (5), 936–944.

Damage simulations in composite structures in the presence of stress gradients

21

J. Reinoso [1], *A. Blázquez* [1] *and F. París* [2]
[1]University of Seville, Seville, Spain; [2]School of Engineering, University of Seville, Seville, Spain

21.1 Introduction

The use of carbon fiber–reinforced composites (CFRCs) for lightweight structures has been gradually increased due to their superior material properties, especially in terms of their specific stiffness and strength ratios, when compared to the metallic materials traditionally used. This trend has been noteworthy in aerospace and aeronautical components. Additionally, these materials are also being incorporated into automotive applications as well as wind-energy systems, among other industrial applications. In this context, stiffened panels are extensively used in the construction of aircrafts, offering notable advantages to the mechanical performance in comparison with other configurations. This typology generally consists of a flat or curved primary surface (skin), which is usually reinforced with stringers in order to increase the structural stiffness of the set. The stringers are attached to the skin via diverse techniques, mainly using a series of rivets or adhesive layers (cobonded or cocured joints) or using hybrid fastening procedures that combine both of the previous methodologies, for example, bolted-bonded joints.

From the structural point of view, in practical applications, the load-carrying capacity of these panels can be notably affected by:

1. Specific geometric features due to design requirements or conflict between adjacent regions. For instance, this is the case with the variation of the transverse fuselage section and the stringer-frame intersections that provoke the termination of some of the stringers of the panel (areas known as runout regions). Notable stress concentrations appear around these locations due to the abrupt modification of the primary loading paths.
2. External loading conditions that can induce the appearance of structural instabilities from the geometric signature, entering into the so-called postbuckling range. Throughout the postbuckling evolution, these instabilities generally lead to the initiation and development of potential delaminations and skin–stringer debonding processes at different critical locations that cause the specimen to collapse.
3. The presence of manholes for inspection of lateral boxes, or the holes for the windows in the fuselage, is also an element carrying out stress concentrations.
4. Geometrical designs to solve interferences between structural elements (such as ribs or frames with stiffeners, among others) or to facilitate the location of auxiliary aircraft systems may imply the presence of reentrant corners that produce significant stress gradients.

Modeling Damage, Fatigue and Failure of Composite Materials. https://doi.org/10.1016/B978-0-443-18489-5.00012-6

Referring to the analysis of runout regions, comprehensive investigations of different geometric concepts have been accomplished in the last few decades (Falzon et al., 2001; Falzon and Davies, 2003; Falzon and Hitchings, 2003; Greenhalgh and García, 2004; Psarras et al., 2011; Reinoso et al., 2012a, 2013). In an effort to provide a better understanding of their mechanical performance, Greenhalgh and García (2004) conducted detailed experimental studies of runout specimens at coupon level, with special emphasis on the identification of the different failure processes. Falzon and co-workers (Falzon et al., 2001; Falzon and Davies, 2003; Falzon and Hitchings, 2003) performed experimental and numerical investigations of similar coupons under uniaxial compressive conditions, especially evaluating the role of different aspects such as the stacking sequence and geometric runout terminations, among others. Along these lines, Psarras et al. (2011) extended the investigation by designing optimized baseline definitions that substantially improved the crack growth stability. From the computational standpoint, based on the finite-element method (FEM), most of the studies devoted to runout panels employed the popular virtual crack closure technique (VCCT) (Falzon et al., 2001; Falzon and Davies, 2003; Falzon and Hitchings, 2003; Psarras et al., 2011) and cohesive zone model (CZM) (Reinoso et al., 2012a, 2013) strategies to trigger interlaminar failure events, which were those mainly responsible for structural collapse.

Regarding the postbuckling response of composite-stiffened panels, due to their potential applicability for the construction of aerospace structures, a large number of experiments evaluating different panel configurations and loading conditions have been recently carried out (Blázquez et al., 2012; Boni et al., 2012; Falzon et al., 2000; Hur et al., 2008; Lauterbach et al., 2010; McCarthy et al., 2002; Reinoso et al., 2012b; Zimmermann et al., 2006). In this setting, FEM has been the most popular numerical tool for predicting arbitrary complex postbuckling evolutions. Currently, most of these studies focus their attention on the development of numerical procedures that enable the accurate reproduction of the overall general response of the specimen (Blázquez et al., 2012; Falzon et al., 2000; Lauterbach et al., 2010; Degenhardt et al., 2006; Romeo and Frulla, 1997; Stevens et al., 1995). Additionally, some investigations have also incorporated computational techniques such as VCCT, CMZs, and continuum damage mechanics (Krueger et al., 2009; Orifici et al., 2008a,b; Vescovini and Bisagni, 2012), to model interlaminar and intralaminar failure events. However, for multistringer panels, the required computational costs render the use of such modeling strategies at the global level rather impractical. To overcome such limitations, several studies have proposed the use of single-panel configurations (Orifici et al., 2008a,b) and simplified models that can eventually predict the structural collapse in an efficient and reliable fashion (Orifici et al., 2008a; Mittelstedt and Beerhorst, 2009).

To extend the range of applicability of damage-prediction tools to geometrically complex structures, global/local methodologies represent potential alternatives. These techniques provide a competitive solution strategy due to the high fidelity level of local phenomena and, at the same time, preserve the manageability of the associated computational costs. In particular, models based on shell/3D modeling strategies have been exploited in both composite stiffened panels in postbuckling regime and in runout panels (Reinoso et al., 2012a; Krueger et al., 2009).

This chapter is concerned with the modeling of damage in composite structures in the presence of stress gradients. From the computational perspective, finite-element (FE) simulations of such structures are carried out through the use of global/local techniques using the commercial FE package ABAQUS (2009) and incorporating material degradation models into the local level of analysis. The components analyzed refer to two fundamental configurations: (1) specimens including runout regions with different geometric definitions subjected to uniaxial tensile loading, and (2) a multistiffened panel under uniaxial compressive loading along the longitudinal direction of the specimen undergoing postbuckling evolution prior to collapse. The chapter is organized as follows. A description of global/local techniques especially devoted to composites is given in Section 21.2. Section 21.3 presents the description of the numerical techniques herein employed to account for interlaminar damage. The application of these inelastic models into the two aforementioned typologies of composite-stiffened panels is addressed in Section 21.4. Finally, the main conclusions of the current research along with future developments are covered in Section 21.5.

21.2 Global/local FE techniques

With the aim of improving the level of reliability of composite structure simulations, techniques connecting several FE models of different size and/or fidelity have been widely exploited in the last 2 decades. These techniques differ in how the connection between the different models is carried out, all of them considering two models: a global model and a detailed model. Three main classifications attending to different features can be considered: (1) global/local and multiscale analysis; (2) loose and tight couplings; and (3) one-way and two-way transferences.

From the authors' point of view, the pivotal distinction between the global/local and multiscale denominations refers to the level of scale used to represent the material. On the one hand, the same scale of study (the same physical hypothesis with regard to continuum media, homogeneity, etc.) is used in global/local techniques. On the other hand, in multiscale approaches, the scale separation becomes effective (including some aspects of the internal structure of the material) through the use of the representative volume element concept (Feyel, 2003) and homogenization procedures (Ernst et al., 2010; Fish et al., 1999; Ghosh et al., 2001).

Each of these approaches (global/local and multiscale) can be classified with reference to the interrelation between both levels of analysis in loose coupling and tight coupling. The term "loose coupling" implies the use of spatially separated models (or at least of distinct simulations); the global model and the detailed model are solved sequentially (first the global and then the detailed model). In contrast, the term "tight coupling" refers to numerical techniques that simultaneously solve the systems of equations corresponding to the global and local models (Chrupalla et al., 2011; Labeas et al., 2012).

Additionally, attending to the information exchange between both models, the techniques used can be classified as one-way and two-way couplings. In a one-way coupling, the information transfer is only conducted in one direction, for example, from the global to the local model or vice versa. Conversely, in a two-way coupling technique, the transfer is accomplished in both directions (Chrupalla et al., 2011; Labeas et al., 2012). Although usually loose coupling is synonymous with one-way transference, and tight coupling with two ways, any combination can be imagined and implemented.

Focusing on global/local procedures for composites, in particular, the commercial software ABAQUS offers two possibilities for performing such analyses by means of the so-called submodeling and shell-to-solid coupling techniques (ABAQUS, 2009). Both methods are based on a mesh refinement scheme, but the nature of each of them differs substantially from the other. Thus, the submodeling technique is based on solving the global and local models sequentially: first the global model is solved, and subsequently these results are taken as prescribed boundary conditions on the delimiting surfaces of the local one. Therefore, both the models are treated in separated analyses (loose coupling and one-way technique) without the local model influencing the global one. In contrast, the shell-to-solid coupling technique allows the interaction between both regions throughout the solution process, since the refined region becomes part of the entire model right from the beginning of the computation. This latter strategy can be categorized as a tight coupling and two-way strategy. Fig. 21.1 illustrates a schematic overview of the main distinction between the global/local techniques already described, for the particular case of one of the composite stringer runout specimens studied in this chapter (see Section 21.4).

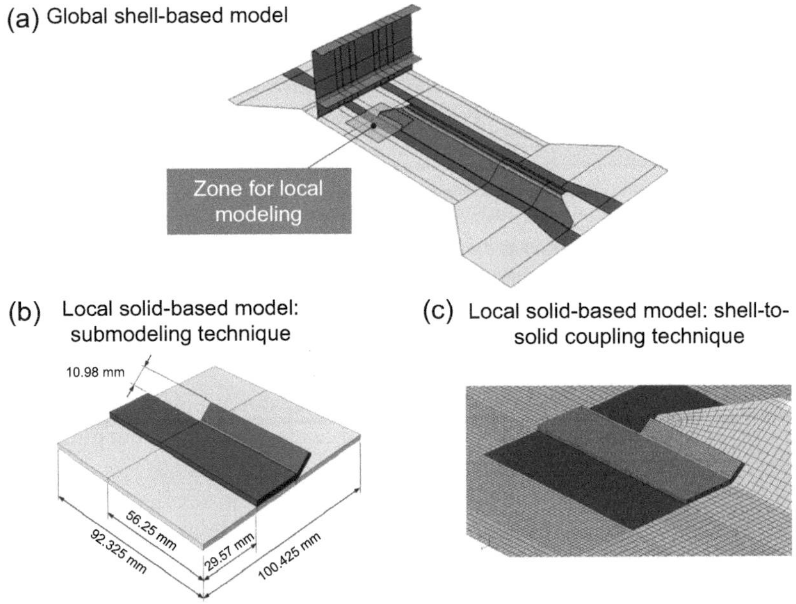

Figure 21.1 Overview of the submodeling and shell-to-solid global/local techniques: application for a composite runout specimen. (a) The complete runout component. (b) Submodeling technique: the geometric definition of the submodel. (c) Shell-to-solid coupling technique: the area of interest with a refined mesh is integrated into and solved with the whole structure.

21.3 Numerical techniques for interlaminar damage failure

In the present investigation, damage events are only taken into account through the use of interface elements in order to model the skin–stringer debonding processes, since this failure mechanism can be considered as the most relevant one for the analyzed panels (Section 21.4). Regarding the interface law, two main typologies are now considered: (1) the cohesive interface behavior with gradual deterioration of the virgin mechanical properties of the interface (ABAQUS, 2009; Dávila et al., 2007) and (2) the linear elastic brittle interface model (LEBIM), in which an abrupt failure of the interface is envisaged (Távara et al., 2011).

21.3.1 Cohesive zone models

CZMs have received a great deal of attention during the last 2 decades and have been employed for predictions of composite interlaminar failure. There exists an overwhelming number of decohesion laws in the literature, ranging from exponential or trapezoidal, among other typologies (see Reinoso et al., 2012a; Dávila et al., 2007), the references therein given.

A simple traction-separation law (TSL) governing the interface behavior can be illustrated (for simplicity for the 2D case), by means of a bilinear decohesion as follows (Fig. 21.2a), and whose simplified description is referred in the sequel to Mode I for the sake of brevity: (1) an initial stage (0–1) that is characterized by an initial stiffness (K_n, K_t corresponding to the normal and shear components, respectively); (2) once the combination of interfacial normal and shear tractions fulfill the damage initiation criterion, point 1, the stiffness is gradually reduced (following a linear degradation law in this case) to zero via points 2 and 3 (fully debonded). The irreversibility of the process after point 1 is accounted for through a damage variable. The area under the curve 0–1–2–3 identifies the associated fracture toughness, G_{Ic} (Mode I), and G_{IIc} (Mode II). The complete interface debonding version is attained using the 3D version of the Benzeggagh–Kenane failure criterion (Benzeggagh and Kenane, 1996), whose mathematical representation reads as follows:

$$G_c = G_{Ic} + (G_{IIc} - G_{Ic})\left(\frac{G_{II} + G_{III}}{G_I + G_{II} + G_{III}}\right)^{\eta} \tag{21.1}$$

In Eq. (21.1), G_c identifies the mixed fracture toughness; G_I, G_{II}, and G_{III} represent the energy release rate (ERR) associated with the fracture Modes I, II, and III, respectively; and G_{Ic} and G_{IIc} are the fracture toughness of Modes I and II, respectively. Finally, η stands for an experimental fitting parameter.

To elucidate the role of the B–K exponent for damage estimations, analyzing Eq. (21.1) from a mathematical perspective leads to the observation that a higher B–K exponent yields a lower value of the exponential term. Therefore, the mixed fracture toughness (G_c) tends to more closely maintain the value of the fracture toughness of

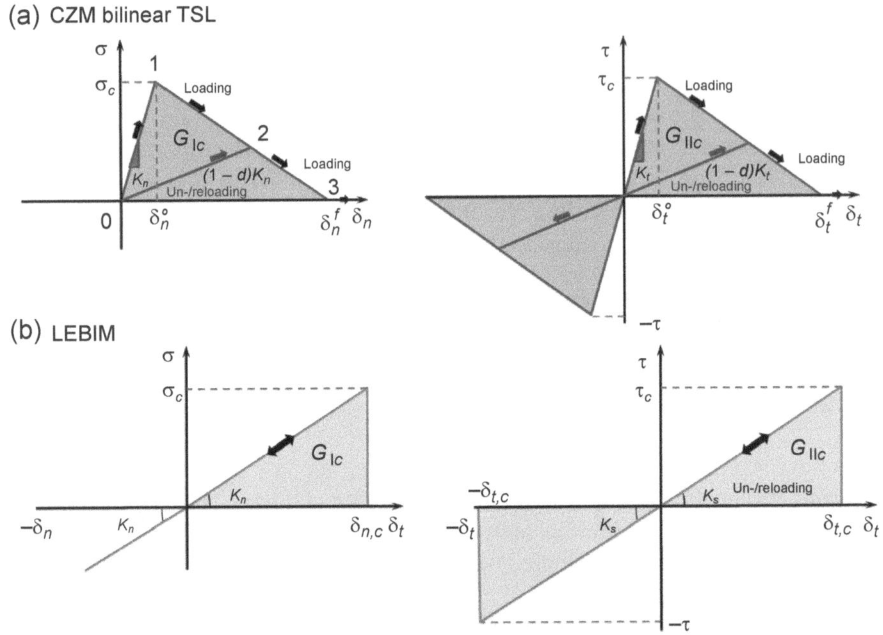

Figure 21.2 Traction-separation law (TSL) corresponding to a two-dimensional analysis: (a) bilinear cohesive zone model and (b) linear elastic brittle interface model.

Mode I (G_c close to G_{Ic}). Fig. 21.3 displays the effect of the variation of this exponent (twice, 4.5684, and one-half, 1.1421, of the reference value $\eta = 2.2842$ (Dávila et al., 2007)) on different fracture mixities of a bidimensional case ($G_{2DT} = G_{II} + G_{I}$). In this particular case, in the range of mode mixing ratio 0–0.5, the fracture toughness G_c maintains roughly the same value of the fracture toughness associated with Mode I, G_{Ic} for the highest value of η.

21.3.2 *Linear elastic brittle interface model*

The second interface type used here is the new, enhanced LEBIM (Távara et al., 2011). The key idea of this formulation can be traced back to the original model proposed by Prandtl (1933), which assumes a perfect linear elastic brittle character. Thus, the interface response can be idealized as a continuous distribution of linear springs. At any material point, the constitutive behavior takes the form:

$$
\begin{aligned}
\sigma_n(x) &= K_n \delta_n(x) \\
\tau_{t1}(x) &= K_{t1} \delta_{t1}(x) \\
\tau_{t2}(x) &= K_{t2} \delta_{t2}(x)
\end{aligned}
\tag{21.2}
$$

where σ_n, τ_{t1}, τ_{t2} represent, respectively, the normal and the tangential traction components; δ_n, δ_{t1}, δ_{t2} stand for the normal and shear relative displacements of the

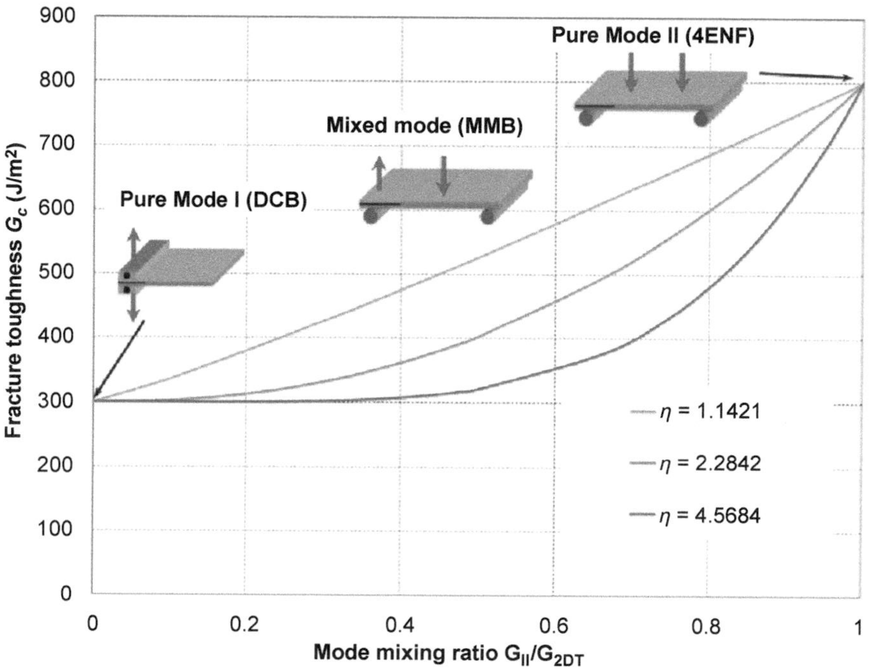

Figure 21.3 Sensitivity analysis of the effect of the η-exponent in the B−K criterion.

interface flanks; and K_n, K_{t1}, K_{t2} correspond respectively to the normal and shear stiffnesses of the springs distribution. The ERR in the LEBIM at a material point level can be obtained by means of computing the elastic strain energy (per unit area).

In this study, an enhanced 3D version of the LEBIM for modeling layers with an initial thickness h is adopted. This model (Távara et al., 2011) relies on the assumption that the adhesive is sufficiently compliant with respect to the adherents (CFRCs), so that a constant distribution of the stresses through the interface thickness can be assumed. Based on this hypothesis, the modified ERRs render:

$$G_{\mathrm{I}} = \frac{(\sigma_n)^2}{2K_n}, G_{\mathrm{II}} = \frac{(\tau_{t1})^2}{2K_{t1}}, G_{\mathrm{III}} = \frac{(\tau_{t2})^2}{2K_{t2}} \tag{21.3}$$

The interface abruptly fails when the energy-based failure criterion is fulfilled ($G(x) = G_c$), without any intermediate stage between intact and fully damaged states (Távara et al., 2011). The individual fracture toughnesses for each mode are computed as:

$$G_{\mathrm{I}c} = \frac{(\sigma_{nc})^2}{2K_n}, G_{\mathrm{II}c} = \frac{(\tau_{t1c})^2}{2K_{t1}}, G_{\mathrm{III}c} = \frac{(\tau_{t2c})^2}{2K_{t2}} \tag{21.4}$$

where σ_{nc}, τ_{t1c}, and τ_{t2c} are the interlaminar strengths of the fracture Modes I, II, and III, respectively (see Fig. 21.2 for the 2D case).

In line with the previously described CZMs, the three-dimensional version of the B−K energetic criterion, Eq. (21.1), is employed for comparison purposes. After the interface breakage, its behavior is taken as stress free. Finally, the interface model is further enhanced through the insertion of a penalty condition without friction to prevent interpenetration of the opposite flanks ($\delta_n(x) < 0$) (Távara et al., 2011). The interface stiffness for this contact condition is assumed to be equal to its original normal stiffness and it remains valid once the failure has taken place. Under these circumstances (broken interface), the constitutive law reads:

$$\begin{aligned} \sigma_n(x) &= K_n\langle\delta_n(x)\rangle_- \\ \tau_{t1}(x) &= \tau_{t2}(x) = 0 \end{aligned} \tag{21.5}$$

where the operator $\delta_n(x)$ denotes the negative Macaulay brackets. This interface model has been implemented into the commercial FE package ABAQUS by means of the user-defined material routine UMAT.

21.4 Applications

21.4.1 Stringer runout composite panels

21.4.1.1 Specimens and material description

The geometric characteristics of the different runout specimens and a geometrical sketch of the analyzed runout configurations are shown in Fig. 21.4. The structural set was primarily composed of the following common elements: (1) the skin, a flat composite panel 300 mm in width at the central region ("Central area" and "Central area (2)" in Fig. 21.4), whose width was increased up to 500 mm at the reinforcement areas ("Reinforcement" and "Reinforcement (2)" in Fig. 21.4) near the longitudinal gripping locations; (2) a composite Ω-stringer placed longitudinally along the panel whose flanges were 25 mm wide; (3) an aluminum dummy frame used with the intention of reproducing the role of the actual frame of an aircraft fuselage; and (4) a steel plate attached to the outer surface of the panel (the surface that did not face the stringer) by a series of rows of rivets at the regions "Reinforcement (2)," "Transition Zone (2)," and "Central area (2)."

The analyzed panels correspond to the following configurations (see Fig. 21.4):

- Configuration 1. The stringer flanges were extended underneath the dummy frame up the longitudinal extremes of the specimen.
- Configuration 2. Similarly to configuration 1, the stringer flanges were extended underneath the dummy frame, but it included a pair of bolts at each side of the panel near the stringer web termination (four bolts in total) that were symmetrically disposed.
- Configuration 3. In this category of runout conception, the baseline stringer flanges were terminated around the stringer web interruption, and a couple of bolts were disposed at each side of the stringer web termination.

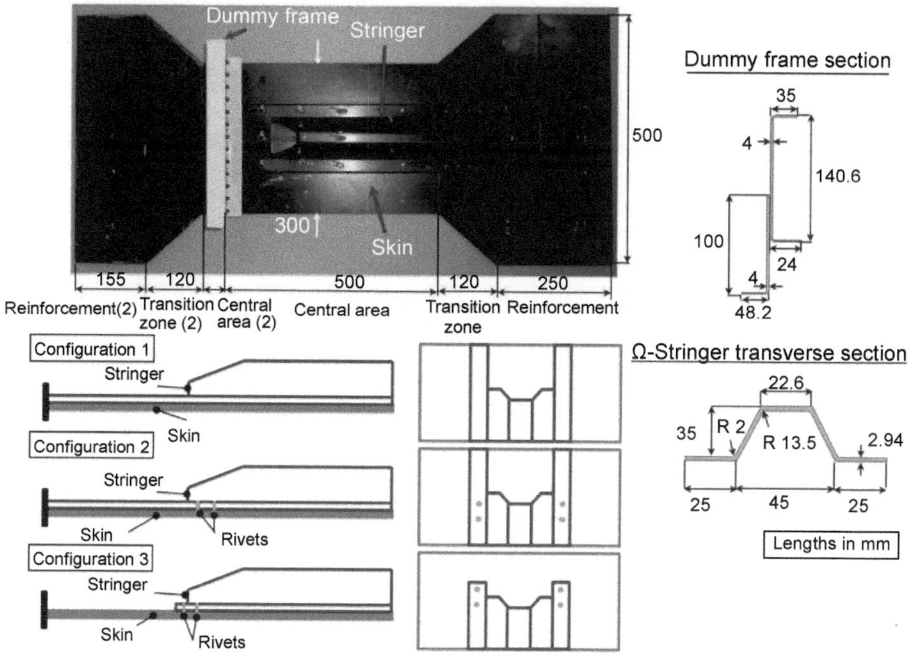

Figure 21.4 Geometric description of the analyzed runout panels.

The CFRC prepreg unidirectional material IMA/M21E was used to manufacture the skin and the stringer (see Table 21.1 for properties). The longitudinal direction identifies the zero-degree material orientation and the nominal layer thickness is equal to 0.184 mm. The stacking sequence of the stringer was $[45/-45/0/0/0/90/0/0]_S$, whereas the central skin area corresponded to $[45/-45/90/0/-45/45/90/90]_\$$, where the subscripts S and $ denote perfect symmetric laminate and symmetry, except for the last layer (one less ply), respectively. The remaining laminate dispositions coincided with those previously reported by Reinoso et al. (2012a). The stringer was cobonded to the skin using the polymeric resin EA9695.05 K, 0.2 mm in thickness (see Table 21.2 for properties). The metallic auxiliary plate and the dummy frame were manufactured using steel and the aluminum alloy 2024 T42, respectively (Table 21.2). Finally, the mechanical properties of the titanium bolts are also given in Table 21.2 (note that the set includes one steel washer for each of the bolts of the joint). Fracture properties of the adhesive layer are reported in Table 21.3.

21.4.1.2 Experimental program

Every specimen was subjected to monotonically quasistatic tensile loading conditions up to collapse. The panels were produced including an artificial defect located at one of the runout sides that was generated through the insertion of a Teflon layer whose dimensions were $25 \times 24\ mm^2$. Prior to the tests, the components were inspected in order to guarantee the absence of manufacturing defects.

Table 21.1 Stiffness properties of the unidirectional CFRC material IMA/M21E.

Stiffness properties	
Longitudinal elastic modulus E_{11}	154 GPa
Transverse elastic moduli E_{22}, E_{33}	8.5 GPa
In-plane shear modulus G_{12}	4.2 GPa
Out-of-plane shear modulus G_{13}	4.65 GPa
Out-of-plane shear modulus G_{23}	3 GPa
Poisson ratio ν_{12}	0.32

Table 21.2 Elastic properties of the adhesive, aluminum alloy, steel, and titanium materials.

Properties	Adhesive	Steel	Aluminum	Titanium
Elasticity modulus (GPa)	1.54	210	75	110
Poisson ratio	0.35	0.3	0.35	0.3

Table 21.3 Fracture properties of the material EA9695.05 K.

σ_{Ic} (MPa)	τ_{IIc}, τ_{IIIc} (MPa)	G_{Ic} (J/m^2)	G_{IIc} (J/m^2)
25.1	50	300	800

The test rig was set up to reproduce full-barrel conditions, similar to those generated around a runout region in an actual aircraft fuselage. Fig. 21.5 shows the test rig; see Reinoso et al. (2012a) for a more detailed description. The experimental supporting conditions consisted of (1) a complete displacement restriction along the longitudinal extremes of the panel (gripping edges), with the exception of the axial displacement at the mobile extreme, which corresponded to the furthest edge of the dummy frame; (2) a pair of lateral guides; and (3) an auxiliary system at the dummy frame that constrained the normal displacements at its upper flange and at its external transverse sections. A set of back-to-back strain gauges (16–24 depending upon the specimen) around the runout region monitored the response of each panel. These gauges were oriented along the axial direction (coinciding with the loading direction).

The main characteristics of the tests are given in Table 21.4, where the following aspects are reported: (1) the loading range at which the first noise and visual inspection took place, (2) the location of the identified damage features and their corresponding progression paths, (3) the collapse load, and (4) observable damage evidence after the tests.

Fig. 21.6 shows the propagation paths corresponding to the analyzed specimens, obtained through ultrasound inspection devices, of the skin–stringer debonding processes around the runout region for different panels on the pristine side of each panel. As has already been addressed, it is worth noting that the experimental damage initiation locations for each panel were strongly influenced by the runout design.

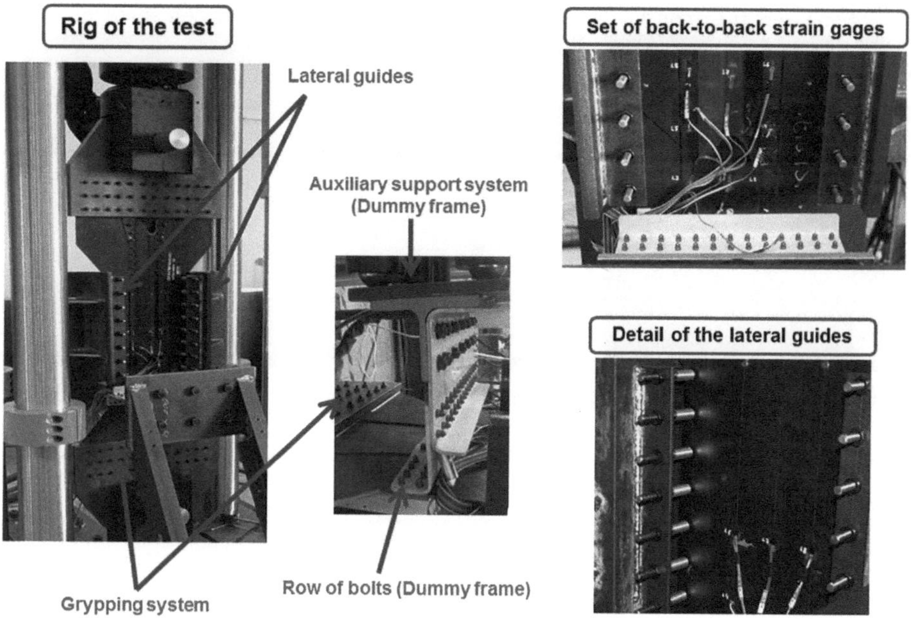

Figure 21.5 Test rig and specimen monitoring using a set of back-to-back strain gauges.

Table 21.4 Summary of the experimental program results.

Config.	First audible noise (kN)	Damage origin and path	Ultimate load (kN)	Evidence of failure mechanisms
1	80	Stringer web end + along the flanges	431	Skin–stringer debond + local fiber breakage
2	90	Lateral flange edge + along the flanges	386	Skin–stringer debond
3	50	Stringer flange end + along the flanges	310	Skin-stringer debond + local fiber breakage

21.4.1.3 Numerical results

The numerical study presented here is particularized to the configurations previously described and follows the main guidelines described by Reinoso et al. (2012a, 2013). In particular, the analysis focuses on the assessment of interface elements predictions (standard CZMs and LEBIM) regarding the initiation and development of skin–stringer debonding phenomena, especially during the initial stages of such processes for different runout concepts.

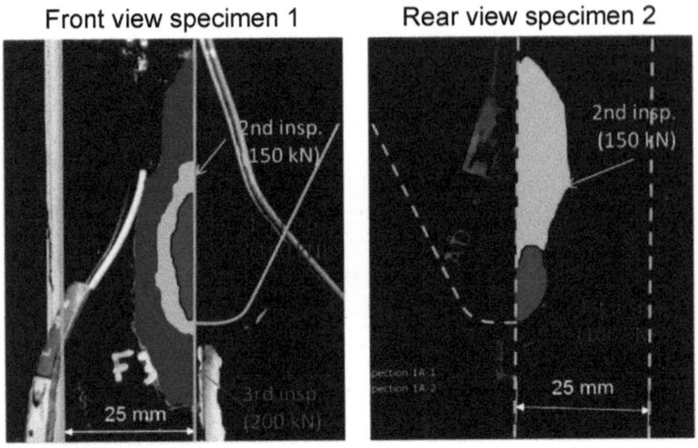

Figure 21.6 Damage measurements during the corresponding test interruptions for panels 1, 2, and 3 using ultrasound inspection devices.

Global models characteristics

The global models are composed of four different geometric entities: the skin, the stringer, the primary skin, and the dummy frame. The pairs dummy frame-skin, dummy frame-stringer (in case the flanges were prolonged underneath the first of them), and skin–stringer were numerically related using the *TIE interaction of ABAQUS, simulating ideal joints, whereas the bolts were modeled by means of the option *FASTENER. The artificial debonding of the adhesive was considered through a discrete discontinuity and imposing contact conditions between the facing surfaces of the skin–stringer pair to prevent interpenetrations.

The geometrical characteristics of the specimens suggest the development of shell-based FE models. However, standard shell models are not able to accurately estimate the full set of the stress tensor components, specifically those affecting the potential interlaminar failure. An alternative approach that can provide a plausible compromise between mechanical accuracy and computational efficiency at the global level consists of the combined use of shell (skin and stringer) and solid elements (adhesive layer) in the same model (Fig. 21.7). This methodology is denominated the shell-solid-shell approach (Blázquez et al., 2012).

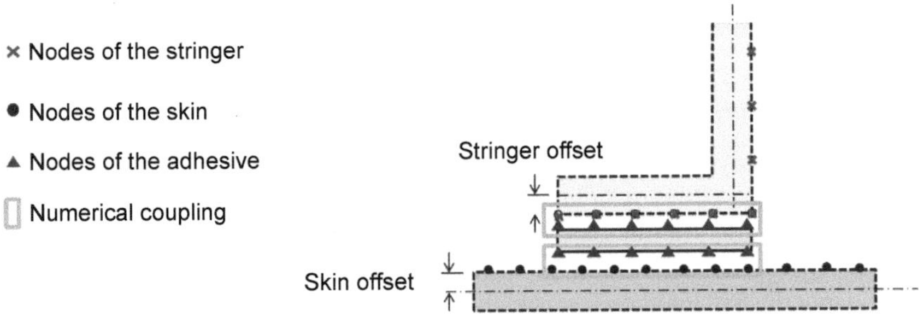

Figure 21.7 Shell-solid-shell modeling approach.

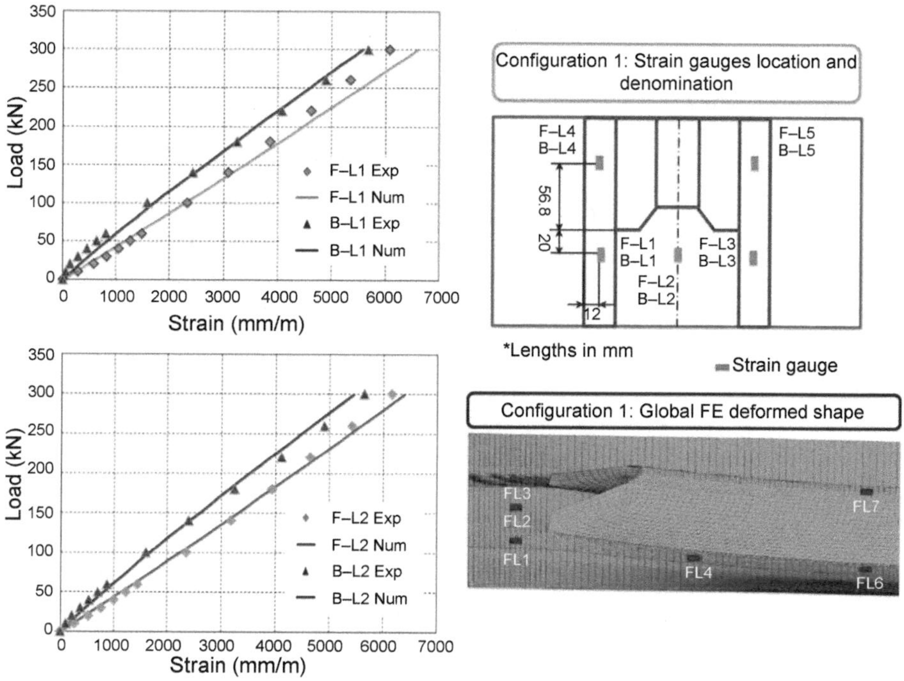

Figure 21.8 Experimental-numerical strain evolutions of some of the gauges placed near the runout region of specimen 1 where F–L and B–L denote the upper (which faces the stringer) and bottom surfaces, respectively. Details of the deformed shape-bending effects.

The applied boundary conditions replicated those thoroughly described by Reinoso et al. (2012a), aimed at reproducing the actual supporting conditions of the tests. The simulations included geometric nonlinearities using an initial and maximum step size equal to 1% of the final load for each specimen (300 kN).

The verification of the global models is carried out by comparing the numerical predictions with the recorded strain readings during the test. Figs. 21.8 and 21.9 depict the experimental–numerical correlations for specimens 1 and 2, with their corresponding test results for several gauges near the runout region. In these graphs, bending effects were registered along the entire strain evolution diagram; see the details of the deformed configuration represented in Fig. 21.8. This general response characterized the structural performance of such panels, indicating the misalignment of the load path and the stiffness center of each specimen. Referring to the experimental–numerical correlation, at initial and mid-stages (up to around 130–150 kN) of the load–strain evolution diagrams, a generally satisfactory agreement between these data can be found. However, since no damage processes were accounted for at the global level, some significant deviations can be appreciated from mid-advanced stages up to collapse. Similar correlations were obtained for the rest of the panels, but they are omitted here for the sake of brevity.

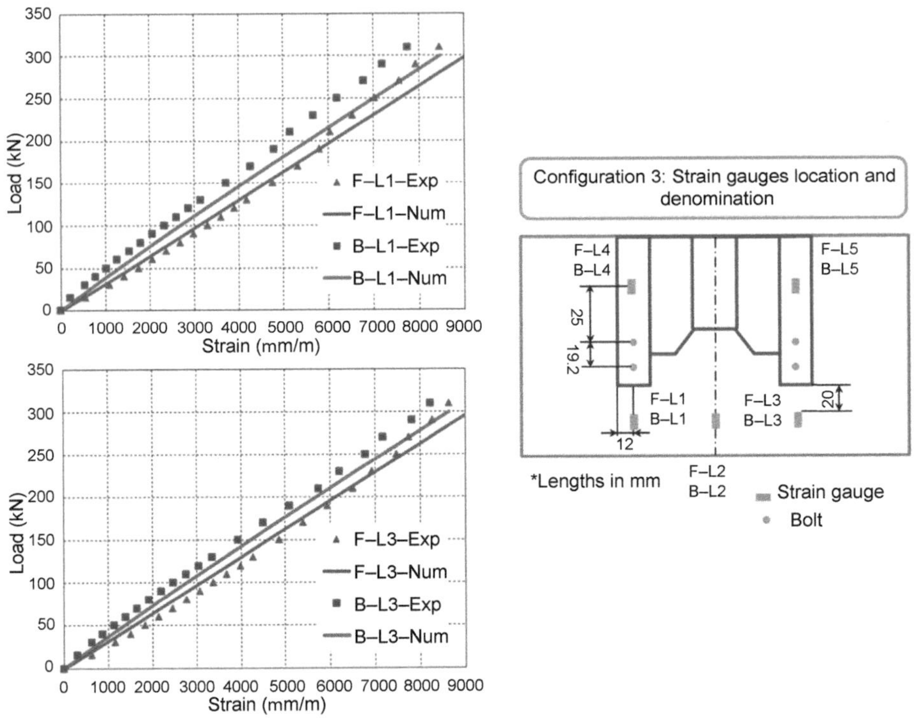

Figure 21.9 Experimental-numerical of the strain evolutions of some of the gauges placed near the runout region of specimen 3.

Local model characteristics

Motivated by the complexity of the panels, once their critical regions became identified in the global analysis through the shell-solid-shell approach, detailed 3D local models were conducted using the submodeling and shell-to-coupling techniques (ABAQUS, 2009). Local 3D models were generated using volume entities, which are basically (1) the skin portion, (2) the stringer portion, and (3) the adhesive layer. Note that some configurations additionally included the set of bolts of the hybrid bonded bolted joint.

Since damage was not modeled at a global level, as was discussed by Reinoso et al. (2012a), through the submodeling approach (one-way technique), it was necessary to evaluate the range of representativeness of the boundary conditions obtained from the global models. The reference local model must consider the same hypothesis as the global model (without damage), in such a way that the boundary conditions remain valid for the whole range of load. These boundary conditions will be representative and applicable to the local model as long as they remain not affected by the evolution of the damage. Then, it will be enough to compare the results of the local models, without damage and with damage, in the zones close to the boundaries of the local models.

Following the evaluation procedure described above (see also Reinoso et al., 2012a), it was found that the validity of the global conditions when compared to the local models ranges between 130 kN for configuration 3 and 150 kN for configuration 1. This difference is mainly attributed to the local stress conditions of each runout concept, which were substantially modified once the skin–stringer debonding was initiated. Note that this assessment was not required for the shell-to-solid coupling approach, since both the coarse and the refined (local zone) meshed regions interacted along the solution process.

With reference to the local FE discretization, although the original conception of the models included parabolic elements (C3D20R of Abaqus) for the skin and the stringer portions, the use of first-order elements (C3D8R of Abaqus, which replaced the precedent second-order elements) for contact simulations in the configuration with bolts is greatly recommended; see McCarthy et al. (2002), Egan et al. (2012), Ireman (1998). Regarding the contact options, small sliding conditions and a moderate to low value of the friction coefficient equal to 0.114 were assumed (Herrington and Sabbaghian, 1991). In addition, these latter local models (with bolts) also needed the incorporation of two extra numerical capabilities: (1) a set of a dummy springs that were attached to the head of the bolts to prevent rigid body motions, and (2) the insertion of the bolt loads through a step that preceded the application of the external loading (*PRETENSION of Abaqus) (ABAQUS, 2009). In this study, as only moderate friction was considered, a minor influence from the tightening torque was expected, which was assumed equal to 1 Nm; see Collings (1977), Olmedo and Santiuste (2012). Fig. 21.10 gives a representation of the local models analyzed in this investigation.

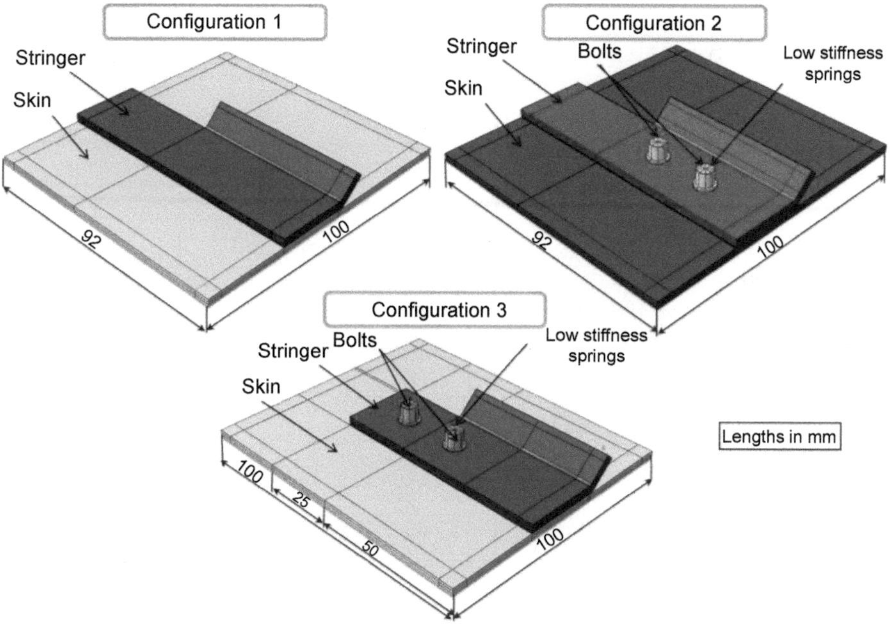

Figure 21.10 Geometric definition of the local models.

The adhesive layer was discretized using the cohesive elements COH3D8 of Abaqus, using the fracture properties given in Table 21.3. The characteristic element size for the skin and the stringer was 0.1 mm, while for the adhesive layer corresponded to 0.02 mm.

Local analysis of configurations 1 and 2

The interlaminar damage estimations obtained for panel 1 using CZM following a bilinear TSL are depicted in Fig. 21.11 for the two global/local procedures (submodeling, Fig. 21.11a, and coupling, Fig. 21.11b) at load levels 80, 100 (first ultrasound inspection), and 150 kN (second ultrasound inspection). This latter value corresponded to the load level limit at which the prescribed conditions on the submodeling approach were still representative. Both techniques predicted that the failure process was initiated at around 50 kN. As was discussed by Reinoso et al. (2012a, 2013), both methodologies estimated that the damage growth path ran mostly longitudinally around the joint as a consequence of the existing high stress concentration at the stringer web termination. These predictions accurately agreed with the ultrasound measurements, from which similar debonding patterns were obtained (Fig. 21.6). It can be observed therefore that both techniques, submodeling and coupling, lead to similar results; submodeling is used in what follows.

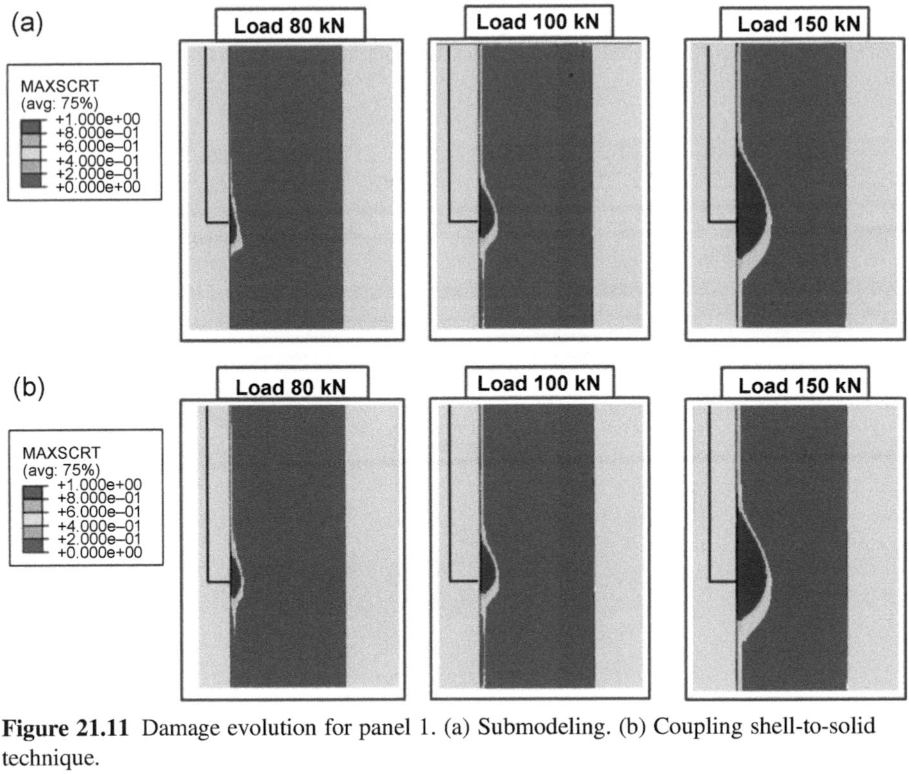

Figure 21.11 Damage evolution for panel 1. (a) Submodeling. (b) Coupling shell-to-solid technique.

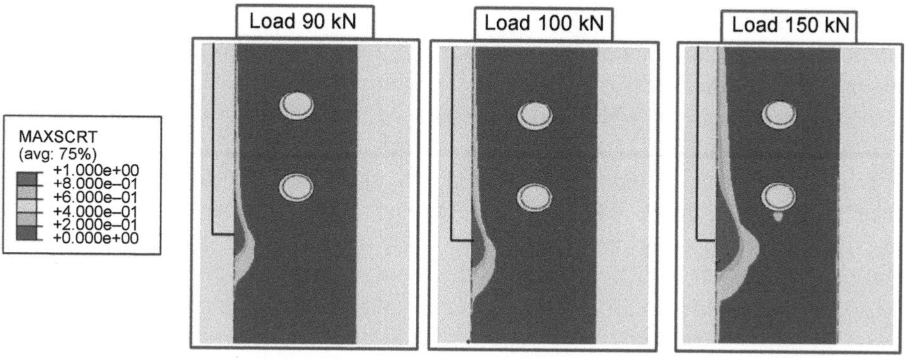

Figure 21.12 Damage evolution for panel 2 through submodeling approach and correlation with the experimental measurement.

Comparing the predictions for panels 1 and 2, it is noticeable (see Fig. 21.11a for panel 1 and Fig. 21.12 for panel 2) that for higher loads, 100 and 150 kN, damage progression was again predicted to evolve near the runout, predominantly along the longitudinal direction (Table 21.5), and satisfactorily agreeing with the experimental measurements. Note also that for these load levels, the predicted fully debonded areas for panel 2 were slightly smaller than that for panel 1 (Table 21.5). However, at 150 kN, an additional damage location was foreseen around the bolt closer to the web termination, which was also measured experimentally. This fact established a clear difference between panels 2 and 1. Analyzing the results, it can be interpreted that the presence of the bolts in the joint initially acted in favor of delaying the damage initiation; that is, strengthening the joint. However, for this configuration, once damage has substantially progressed, additional stress concentrations exist due to the bolts, which induce faster damage progression.

Table 21.5 Comparison between experimental and numerical damaged areas for panels 1 and 2 at different load levels.

Config.	Load level (kN)	Experimental damaged area	Numerical partially damaged area	Numerical fully damaged area
1	80	–	0.1 × 5 (S, C)	0.051 × 2 (S, C)
1	100	3 × 7	4 × 8 (S), 4 × 10 (C)	1.5 × 6.2 (S), 1.6 × 5.5 (C)
1	150	5 × 15	5 × 14 (S), 5 × 12 (C)	2 × 10 (S), 2 × 9.5 (C)
2	90	–	5.57 × 19.7 (S)	3.2 × 6.1 (S)
2	100	7 × 8	8 × 37.4 (S)	6.4 × 7.2 (S)
2	150	12.3 × 52	13.1 × 82.6 (S)	11.2 × 48.3 (S)

C, coupling technique; *S*, submodeling approach. Dimensions in mm (width × length).

Local analysis of configuration 3

The local damage analysis of specimen 3 using the submodeling approach is carried out in what follows. CZM and LEBIM damage models will be compared for this particular configuration, which will be also used to check the role of the parameter η of B−K criterion (Eq. 21.1), by means of a sensitivity analysis.

Initially, the skin−stringer separation at both sides was modeled by means of the linear cohesive element COH3D8 of Abaqus, replicating the FE discretization described above. A deformed shape of the configuration under study is displayed in Fig. 21.13a, whereas Fig. 21.13b depicts the damage variable of the cohesive elements (MAXCRT) computed at 50 kN for different values of the η-exponent of the B−K criterion. In line with previous analyses, this load level (50 kN) corresponded to the load at which the first audible noise was registered during the test.

Two damaged zones were predicted to occur: (1) near the ending side of the stringer flange, following a stretch band parallel to this edge (0.072 mm in width, which was undetectable in practice); and (2) around the stringer web interruption due to the stress concentration at this location. Relying on the numerical results, the deformation pattern obtained showed clear bending effects.

Continuing the computations, Fig. 21.14 depicts the comparison between the debonding regions that were experimentally measured at 100 kN (previously shown in Fig. 21.6), with respect to the numerical estimations. An overall reasonable correlation can be observed between these depictions, although the predictions for the extent of the damage were slightly smaller. Note that the ultrasound devices do not make a distinction between completely and partially damaged areas. Fig. 21.14 also

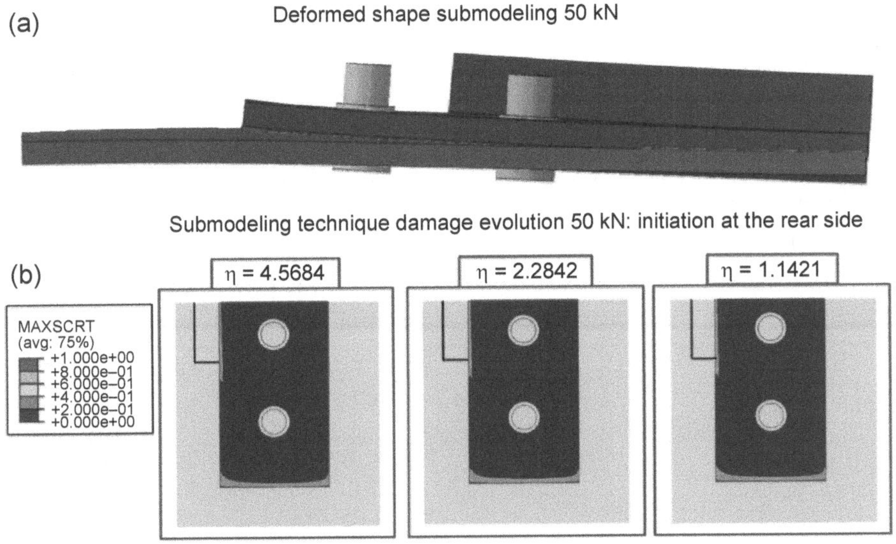

Figure 21.13 (a) Local deformation pattern at 50 kN with a scale factor of 5. (b) Numerical estimations of damaged areas at 50 kN by means of CZ elements using different B−K exponents (see Fig. 21.6 for comparison with experimental measures).

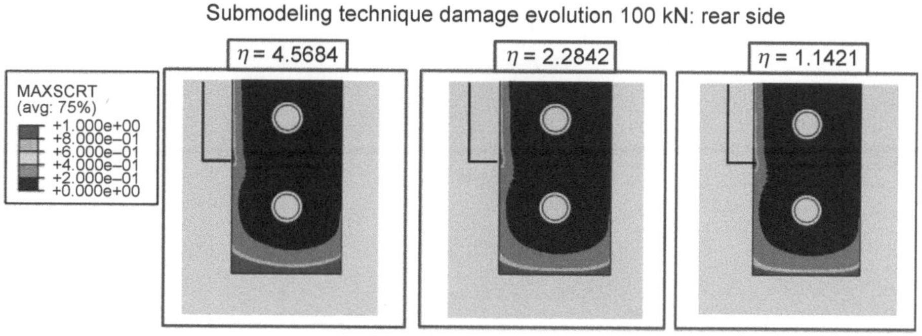

Figure 21.14 Numerical estimations of damaged areas at 100 kN by means of CZ elements using different B−K exponents (see Fig. 21.6 for comparison with experimental measures).

shows that, in good agreement with the experimental measurements, the debonding evolved in a nonsymmetric way near the flange termination, where a larger damage extent was obtained along the inner edge of the flange (the closest longitudinal edge to the web interruption). This issue might be attributed to interferences between the two aforementioned stress concentrators. However, again, at 100 kN, this debonding near the web end was not experimentally detected. In addition, the existence of remarkable differences regarding the initiation location and the geometric characteristics of the debonding growth pattern in comparison with alternative runout definitions is noticeable.

Concerning the sensitivity analysis with respect to the B−K exponent, it can be observed (Figs. 21.13 and 21.14) that the higher this parameter was, the larger the damaged areas were estimated. Regarding this, Table 21.6 summarizes (for the results shown in Fig. 21.14) the numerical predictions corresponding to each of the values of this exponent (taking into account the average value of the damaged band in the longitudinal direction of the panel) in comparison with the experimental measurements. The best agreement was achieved for the model including the largest value of the B−K exponent.

Finally, a series of subsequent analyses that replicated those previously described were also conducted by replacing the CZM approach by the LEBIM approach. Fig. 21.15 presents the comparison between the damage extent estimations corresponding to CZM and LEBIM approaches using the η-exponent equal to 2.2842 for

Table 21.6 Effect of the B−K exponent at 100 kN: comparison with experimental measurements.

B−K exponent	Partially damaged area	Completely damaged area	Experimental damaged area
1.1421	11.12 × 25	1.461 × 25	14.2 × 25
2.2842	11.58 × 25	1.615 × 25	14.2 × 25
4.5684	12.62 × 25	2.203 × 25	14.2 × 25

Dimensions in mm (length × width).

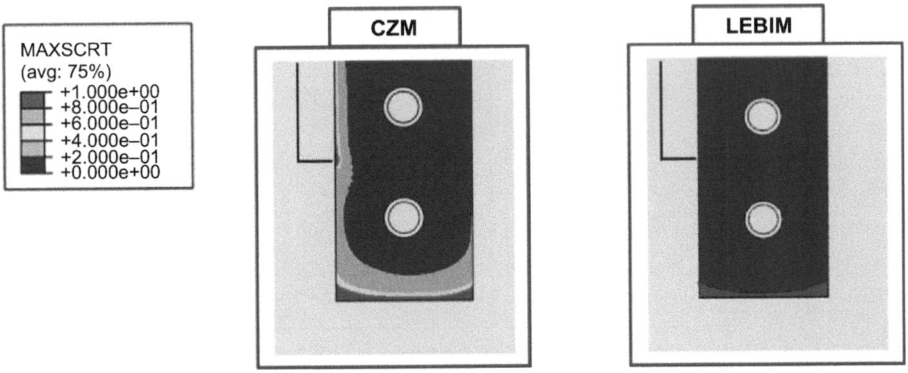

Figure 21.15 Comparison of damage estimations of CZM and LEBIM at 100 kN for $\eta = 2.2842$.

a load level of 100 kN. This graph shows that, at the stringer flange termination, a satisfactory correspondence between both models for triggering interlaminar damage was achieved. These predictions were in close agreement with the experimental measurements. The comparison of both approaches for the damaged areas was performed by considering fully deteriorated areas of the CZM approach. This is due to the fact that the LEBIM only considers totally damaged or intact interfaces without any intermediate state. Additionally, note that the LEBIM did not predict the appearance and development of interlaminar failure near the stringer web termination.

21.4.2 Postbuckling analysis of a composite Ω-stiffened panel

The analysis of a cylindrical stiffened panel loaded in compression reaching the postbuckling regime is performed in this section. The specimen was designed and manufactured by Airbus-Spain, and tested by Airbus-Germany within the framework of the ALCAS CONSORTIUM. Some specific details regarding the experimental devices are omitted here for confidentiality reasons.

21.4.2.1 Specimen and experimental test description

The cylindrical panel had the following characteristic dimensions: axial length 2708 mm, curvature radius 1900 mm, and arc length 1100 mm. The panel possesses three Ω-stringers, with a distance between their central lines equal to 220 mm, and four frames, 635 mm separated in the axial direction. The frames were riveted to the skin at the bottom flange, whereas the stringers were cobonded. Each frame had three cut-outs to secure that the stringers ran through them. A general sketch of the panel and details of the transverse sections of the frames and the stringers are shown in Fig. 21.16.

The skin and the stringers were manufactured using the unidirectional carbon fiber epoxy prepreg material IMA/M21E; see Table 21.1. Table 21.7 shows the stacking sequence for each part, zero-degrees being the axial direction of the cylinder.

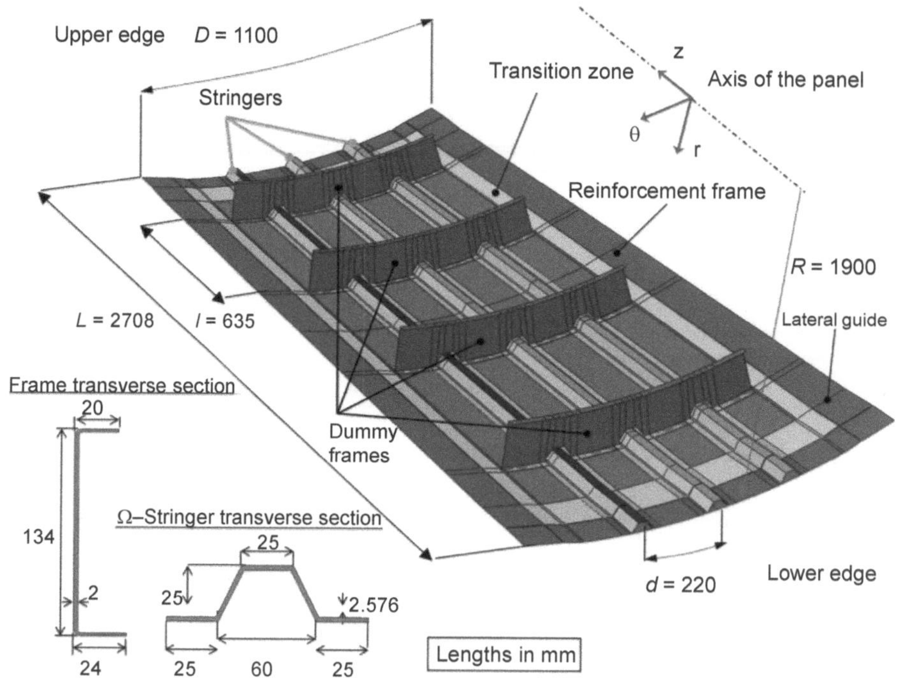

Figure 21.16 Geometric description of the Ω-stiffened panel.

Table 21.7 Panel layup disposition.

Specimen region	Laminate	Thickness (mm)
Skin central area	$[45/-45/0/90/-45/45]_s$	2.208
Skin reinforcement frame	$[45/-45/0/-45/45/90/45/-45/0/45/45/90/45/-45/0/-45/45/90/45/-45/0/90/0]_\$$	8.280
Stringer	[45/−45/0/0/0/90/0]	2.576

The subscript \$ denotes symmetry except the last layer.

A 0.3-mm layer of Loctite-Henkel Hysol 9695 (Z15435) was used to bond the stringers and the skin; its mechanical properties are given in Table 21.8.

A simplified scheme of the loading introduction system is depicted in Fig. 21.17. The external compressive loading was applied by a vertical hydraulic jack of 35 MN, mounted at the mobile base of the test unit. Four load cells controlled the magnitude of the applied solicitation. The longitudinal ends of the panels were embedded in aluminum boxes (Clamping Box in Fig. 21.17b), which were filled with a polymeric resin to favor uniform load introduction conditions.

Table 21.8 Mechanical properties of the adhesive (Loctite-Henkel Hysol 9695).

E (GPa)	ν	σ_{Ic} (MPa)	τ_{IIc}, τ_{IIIc} (MPa)	G_{Ic} (J/m^2)	G_{IIc} (J/m^2)
4.940	0.3	29.7	50	325	850

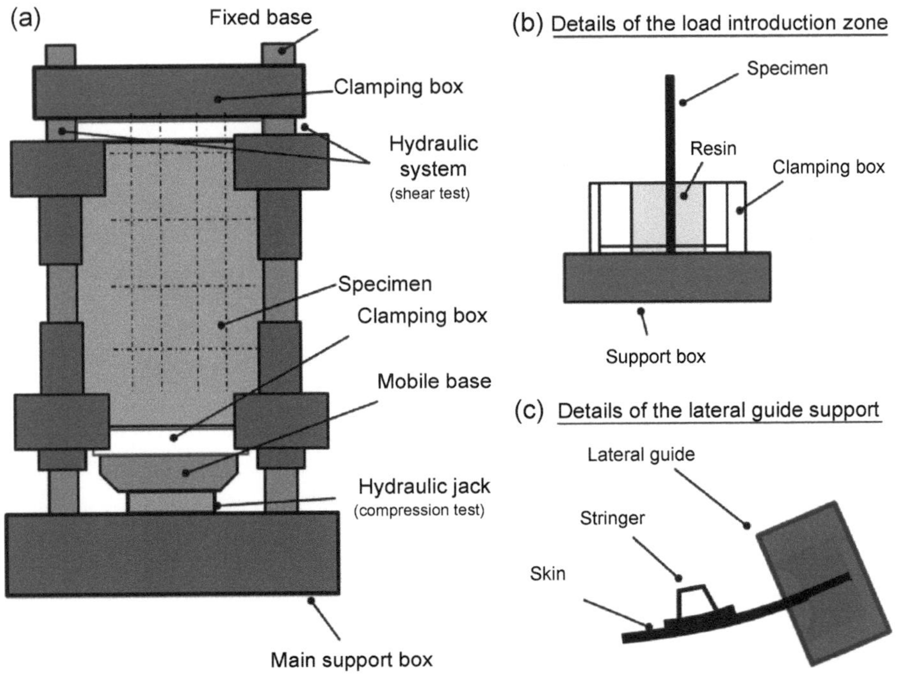

Figure 21.17 (a) sketch of the loading introduction system. (b) Details of the clamping boxes at the longitudinal ends. (c) Details of the lateral guide support.

The specimen was monitored with 20 unidirectional strain gauges (oriented along the axial direction) and four rosettes whose disposition and orientation are sketched in Fig. 21.18. Additionally, an linear variable displacement transducer was used to record the axial displacement of the specimen.

The experimental results obtained from the test will be shown at the same time as the predictions for the sake of brevity.

21.4.2.2 Global/local analysis

The numerical simulation of the present panel was carried out using global/local methodologies in the same way as the FE analysis of the runout specimens. Since no reliable information about the actual geometry of the panel, derived from the manufacturing process, was available during the simulation phase, the computations had to be based on the nominal geometry.

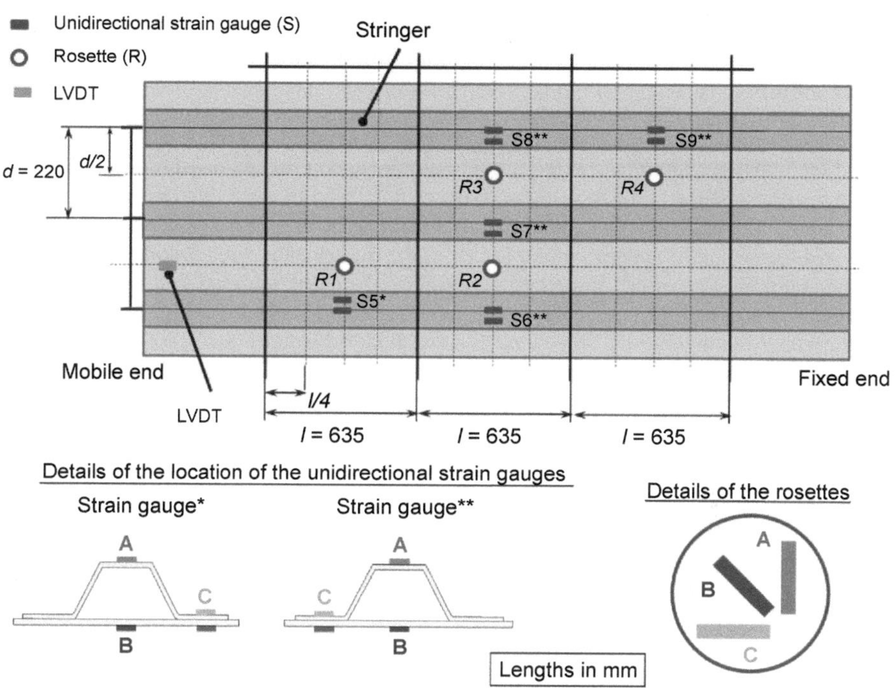

Figure 21.18 Monitoring of the test.

Global FE analysis: model details

The global model was composed of (1) the panel skin, (2) the stringers, (3) the dummy frames, and (4) the adhesive layer. The skin, stringers, and dummy frames were meshed using first-order reduced integration shell elements S4R of Abaqus, whereas solid elements C3D8R of Abaqus were employed for the adhesive layer. With respect to the mesh size, in postbuckling simulations at least the use of 10 elements at each buckling semiwave is recommended (Blázquez et al., 2012; Zimmermann et al., 2006; Degenhardt et al., 2006; Blázquez and Picón, 2010). Thus, after preliminary investigations regarding the convergence of the results for increasing mesh densities, the average element side length turned out to be equal to 1.83 mm. Table 21.9 summarizes the mesh used at the global level.

Table 21.9 Global mesh summary.

Component	No. of elements	No. of nodes	Element type	DOF
Skin	53,156	53,625	S4R	321,750
Stringer (×3)	35,360	36,309	S4R	217,854
Dummy frame (×4)	4,900	5,111	S4R	30,666
Adhesive	14,682	32,806	C3D8R	98,418

According to the previous description of the test, several numerical restrictions were introduced into the models, with the intention of reproducing the actual supporting conditions:

- A multipoint constraint was used at the lower edge of the panel ("Mobile end" in Fig. 21.18), and the nodal displacements at the master point were constrained except for the axial one, which corresponded to the movement of the mobile head of the testing machine.
- An additional multipoint constraint was used at the upper edge ("Fixed end" in Fig. 21.18). At this location, all the degrees of freedom of the master node were restrained.
- Lateral guides. The action of the lateral guides on the panel was modeled by restricting the radial displacement, perpendicular to the panel surface; see details in Fig. 21.17c.
- Transverse dummy frames. These entities were modeled independently and were connected to the rest of the model (skin and stringers) through ABAQUS numerical tie interface constraints. In addition, the boundary conditions considered for the frames restricted the radial displacement of the top flange of each frame. These conditions try to simulate in the model the effect and role that the frame plays in the whole structure.

The mechanical response of composite stiffened panels in the postbuckling regime generally exhibits a high sensitivity with respect to initial geometric imperfections (Reinoso et al., 2012b). In the absence of other types of information, combinations of the buckling modes were used to introduce geometric imperfections:

$$x_i(n) = x_i^o(n) + \sum_{k=1}^{M} a_k u_i^k(n) \tag{21.6}$$

where $x_i^o(n)$, $(i = 1{,}3)$ represent the ideal coordinates of node n; M is the number of buckling modes considered; $u_i^k(n)$ are the components of the displacement field of node n corresponding to the kth numerical buckling mode; $x_i(n)$ is the perturbed position of node n; and finally, a_k is the coefficient of the kth mode. Thus, the first step of the analysis was a linear buckling analysis in order to compute the buckling loads and shapes. Results of this step are omitted here for the sake of brevity.

The most accurate model found, among several combinations tried, with respect to the experimental data, employed the combination of buckling modes 1, 3, and 9, scaled by 10% of the skin thickness. This scaling factor is in the range of values (between 5% and 20% of the skin thickness) recommended in the literature (Lauterbach et al., 2010; Zimmermann et al., 2006; Stevens et al., 1995; Blázquez and Picón, 2010).

The main step in the global analysis was a fully geometrically nonlinear FE simulation. Global computations were conducted using the aforementioned shell-solid-shell approach in order to determine the most critical region susceptible to a subsequent local analysis, all the time bearing in mind that we are using the nominal geometry as a reference. The analysis was conducted using an initial and maximum load increment equal to 10 kN, and the implicit Newton–Raphson solution method with displacement-control was used for the simulations.

Global FE analysis: experimental–numerical correlation

Fig. 21.19 portrays the experimental (markers) and the numerical (solid line) load–displacement evolution diagram, showing also the numerical deformation patterns at several load levels. The structure collapsed abruptly at 848 kN. Then, the analysis was designed to reach 850 kN. The lowest buckling load obtained in the numerical predictions was 722 kN. A generally good agreement up to around a load level of 750 kN can be observed. It is worth mentioning the small load range existing between the specimen's global buckling (nominally 722 kN) up to collapse (found at 848 kN), a fact indicating that there is, for this particular structure, a short period of postbuckling behavior.

Fig. 21.20 shows a picture of the panel after collapse, where a combination of interlaminar and intralaminar damage events can be observed. Severe skin–stringer separation processes took place at the central frame-bay, which are identifiable at each of the stringers, although they were especially drastic for the left-most stringer. In addition, notable intralaminar failure can also be appreciated.

Figs. 21.21 and 21.22 display exemplary evolutions of the back-to-back rosettes mounted on the skin and strain gauges mounted on the stringers, respectively. In Fig. 21.21, the letter "i" indicates the location of the rosette on the inner (concave) side of the panel, whereas the letter "o" indicates the location on the outer (convex) side of the panel. A satisfactory general agreement can be appreciated between experimental measures (markers) and numerical results (solid lines) up to the buckling load of the panel. The mechanical response of the panel was fundamentally governed by a membrane dominated state up to the buckling load, where notable bending effects

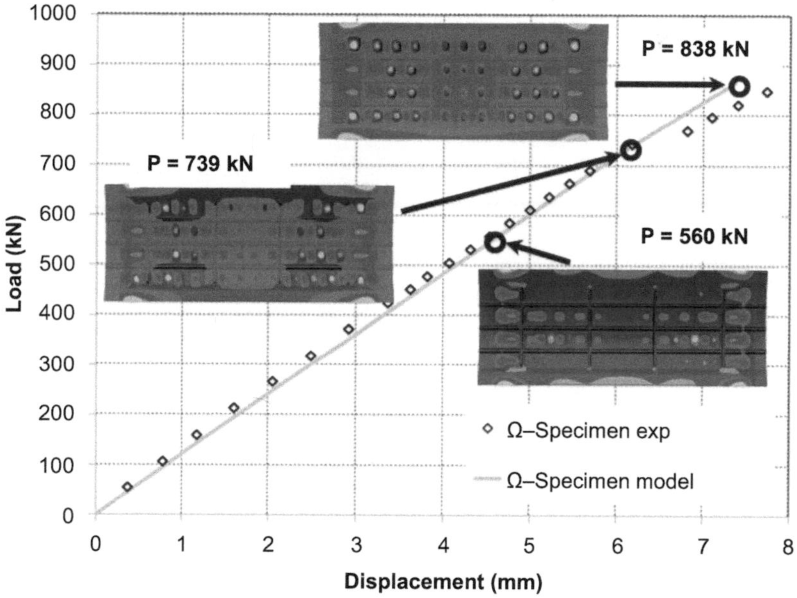

Figure 21.19 Experimental–numerical correlation of the load-shortening evolution curve.

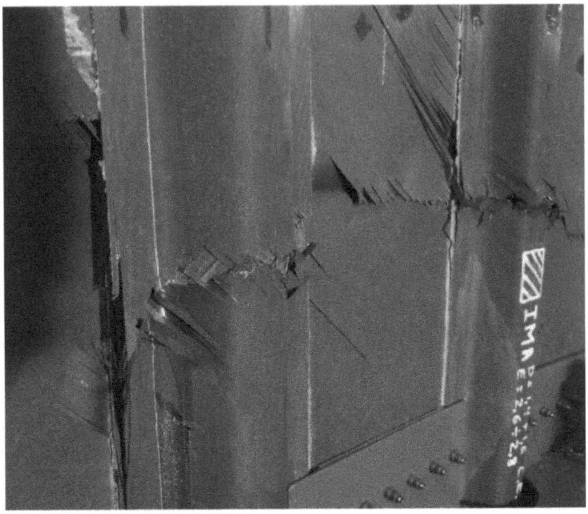

Figure 21.20 Experimental evidence of damage in the central skin bay after collapse.

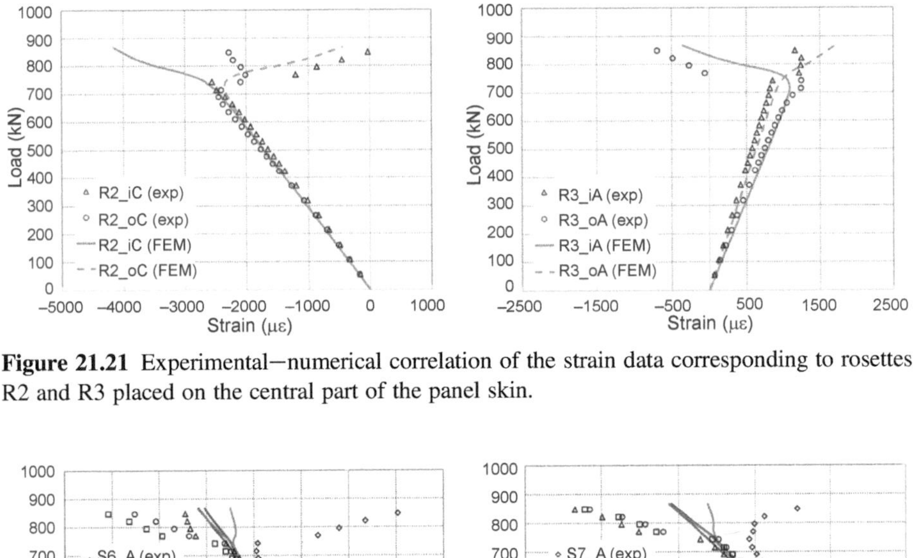

Figure 21.21 Experimental–numerical correlation of the strain data corresponding to rosettes R2 and R3 placed on the central part of the panel skin.

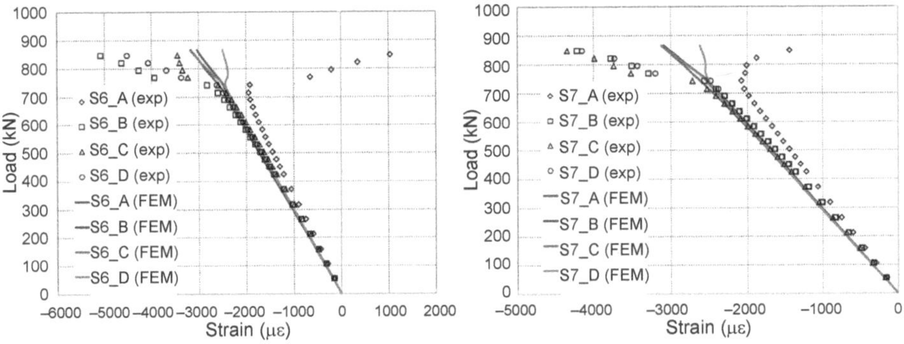

Figure 21.22 Experimental–numerical correlation of the strain data corresponding to the unidirectional gauges S6 and S7 placed on the stringers in the central part of the panel.

began to be recorded by the back-to-back gauges that were mounted on the skin bays (rosettes R2 and R3 in Fig. 21.21) and on the stringer flanges (strain gauges S6 and S7 in Fig. 21.22).

21.4.2.3 Local FE analysis

The local analysis of the panel under study in this second application followed the general guidelines reported for the runout specimens by means of the submodeling approach (ABAQUS, 2009). Attending to the panel failure and to the global numerical results, three separated and nonoverlapping high-fidelity models (L1, L2, and L3 in Fig. 21.23) from the central bay of the panel between the frames were generated. Geometries of these models were identical: a portion of 511 mm in length and 218.8 mm in width with a stringer in the middle (Fig. 21.23).

The local models were composed of three different solid-based entities: (1) the skin portion, (2) the stringer portion, and (3) the adhesive layer. Layered first-order reduced integrated continuum elements (C3D8R of ABAQUS) with a characteristic mesh size of 0.1 mm were employed for the discretization of the skin and the stringer. Cohesive elements (COH3D8 of ABAQUS) were placed at the skin−stringer interface to model interlaminar failure at the adhesive part; their material definition is given in Table 21.8. This inelastic element type was not considered at the regions close to the longitudinal edges of the local models (particularly at a distance of 5 mm), where the nominal elastic properties of the adhesive were kept, in order to preclude potential pollution

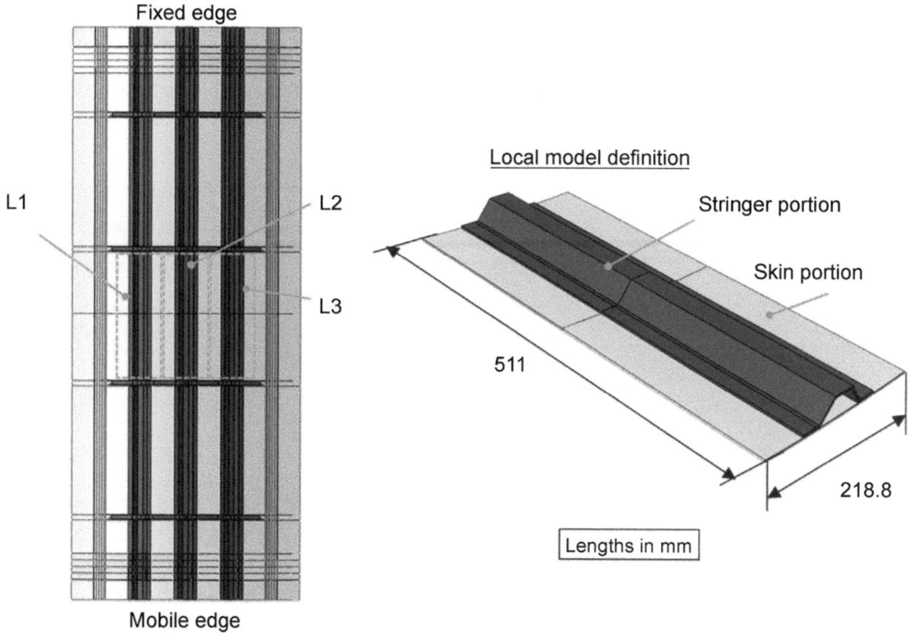

Figure 21.23 Local models: Location and geometric definition.

Table 21.10 Local mesh characteristics.

Component	No. of elements	No. of nodes	Element type	DOF
Skin portion	113,664	143,920	C3D8R	431,760
Stringer portion	120,064	139,808	C3D8R	419,424
Adhesive (cohesive)	94,200	–	COH3D8	–
Adhesive (elastic)	8,000	–	C3D8R	–
Adhesive total	102,200	208,692	–	626,076

of damage estimations due to undesirable and fictitious boundary effects. The characteristic side length of the interface elements was 0.02 mm.

Table 21.10 summarizes the main characteristics of the meshes for each of the local models generated. Meshes were nonconforming due to the different skin and stringer size versus adhesive element size used. Tie interface constraints of ABAQUS were used to joint parts: skin-adhesive and stringer-adhesive.

Local simulations were conducted using an initial and maximum load step of 1 kN, including geometrical nonlinear effects. The simulations corresponding to L1, L2, and L3 reached the specified final load of the computations (850 kN) without encountering notable difficulties in achieving converged equilibrium solutions along the process.

Fig. 21.24 shows the deformed shapes (out-of-surface displacement maps) corresponding to the multiple local models for the load levels 750 and 850 kN. Local deformation patterns were characterized by three waves in the stringer bays. As the applied load was increased, the stringer buckling was developed, which could be easily noticed by a deep valley in the final predicted deformed shape of the L1 local model. This buckling shape correlates satisfactorily with the experimental observations from

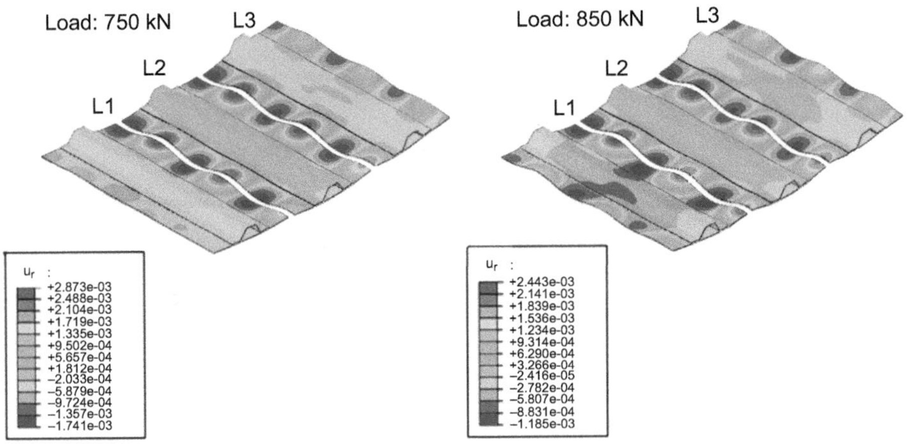

Figure 21.24 Computational estimation of the out-of-plane displacements of the panel through the multilocal model definition at 750 and 850 kN.

qualitative and quantitative points of view. It is important to notice that although we are dealing with a symmetric structure, the combinations of buckling modes 1, 3, and 9 used in this case may lead to nonsymmetric solutions, allowing different predictions for local models L1 and L3.

Numerical predictions for the interlaminar failure are shown in Figs. 21.25 and 21.26 for 700, 750, 800, and 850 kN. In these depictions, the variable MAXSCRT corresponds to the damage accumulated in the cohesive elements in accordance with the

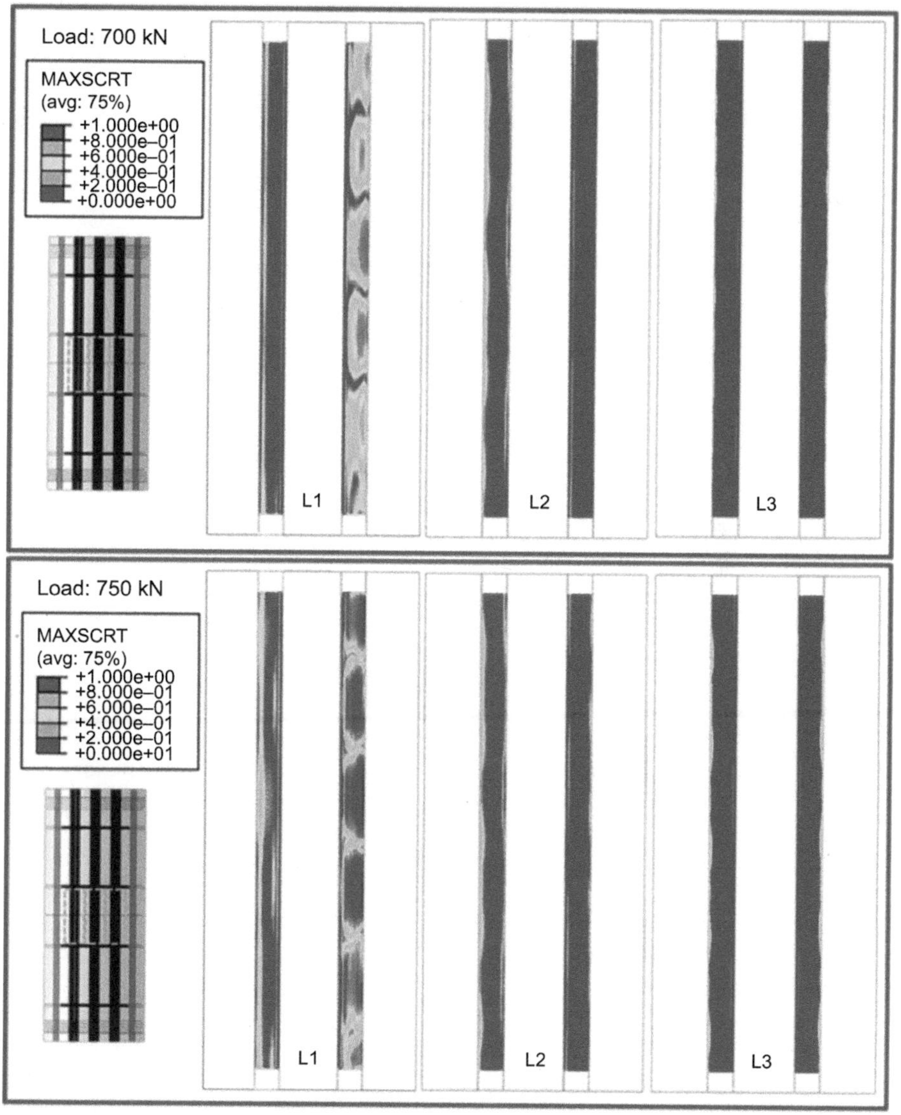

Figure 21.25 Skin–stringer damage estimations through CZMs at 700 and 750 kN.

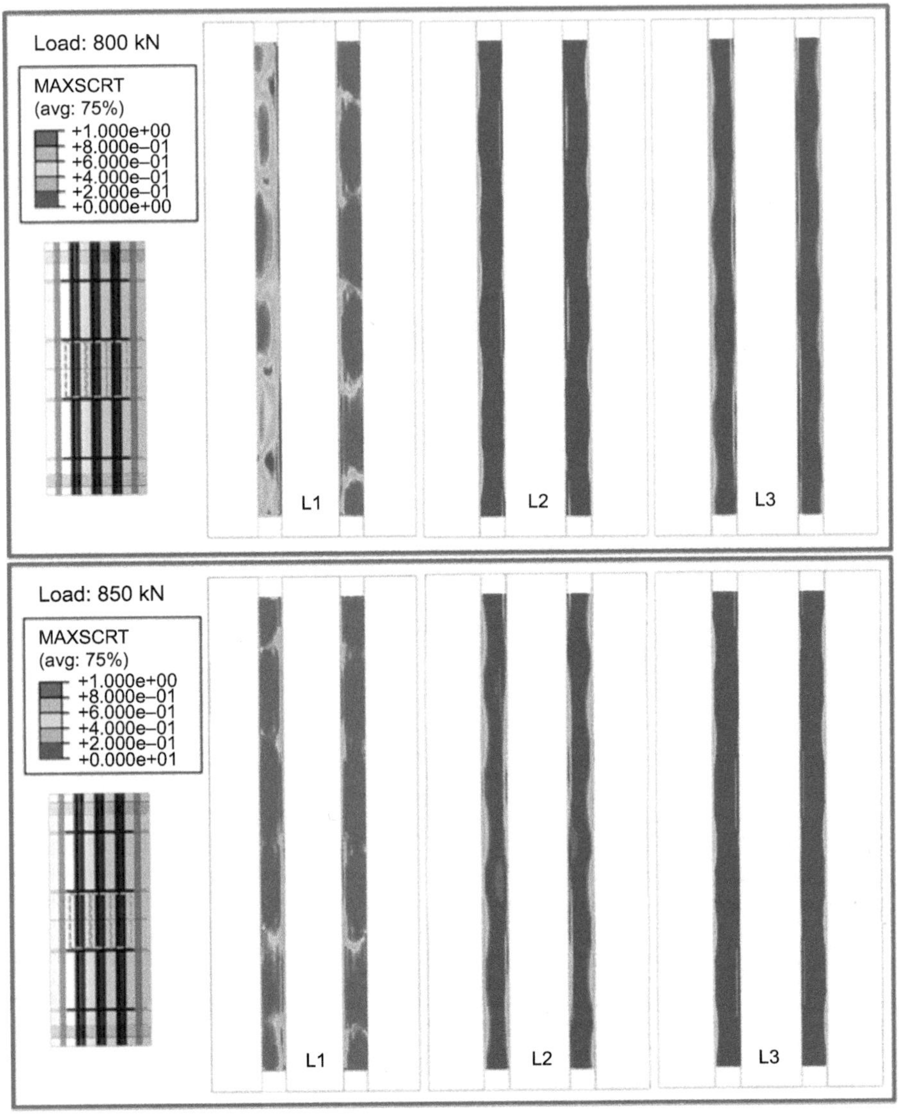

Figure 21.26 Skin−stringer damage estimations through CZMs at 800 and 850 kN.

cohesive law used. A value of MAXSCRT = 0 means no damage (between points 0 and 1 in Fig. 21.2a), whereas a value of MAXSCRT = 1 implies full damage (starting at point 3 in Fig. 21.2a).

It can be observed that the development of skin−stringer debonding was predicted to be initiated at around 700 kN for the L1 local model. Note that nearly intact skin−stringer joints were predicted for the L2 and L3 models. No damage observations were accounted for experimentally at this load level. Notice that 700 kN is lower than the numerically predicted buckling load. A prospective cause for this different behavior might be attributed to the effect of the initial geometric imperfection, which modified the nominal specimen geometry.

Progressing on the loading application, these interlaminar damage events at the skin–stringer joints were further propagated, especially for the case of the L1 model where the damage extent was predicted to cover almost the entire domain of one of the adhesive sides at 750 kN. At higher load levels (800 kN), interlaminar damage estimations were mostly present on the inner side of the adhesive layer of the L1 model, in line with the experimental stringer-based buckling pattern described above, although localized damage was also predicted for the numerical analysis at the other two local models (see Figs. 21.25 and 21.26). These failure estimations were in close agreement with the experimental observations.

21.5 Conclusions and future developments

A methodology to study damage accumulation in composite structures in the presence of stress gradients has been presented. Two interface-based models of damage have been used: a standard CZM following a bilinear TSL and an enhanced LEBIM. From a numerical simulation point of view, a global/local strategy based on the FEM has been followed, with two techniques, submodeling and coupling shell-to-solid, used for the sake of comparison.

Two composite stiffened panel configurations with stress gradients were analyzed from both numerical and experimental perspectives: (1) three different runout panel conceptions and (2) a multistringer panel in the postbuckling regime. Both panel types were numerically simulated by means of the aforementioned global/local techniques, to trigger skin–stringer debonding processes, which were identified as the predominant failure mechanisms of the panels.

From an experimental point of view, regarding the runout panels, special attention was devoted to the identification of the damage initiation locations and their corresponding growth paths during the tests. Damage predictions were only included at the local level of analysis, using the two interface damage models already mentioned. The correlation of damage estimation with the experimental measurements showed a satisfactory level of accuracy for this modeling, at the initial stages of the debonding processes. Both the CZM and LEBIM produced similar results, although in general terms, the LEBIM leads to a quicker solution process.

With reference to the second application analyzed here, a detailed description of the generated FE model at the global level was performed. Special emphasis was given to the role of auxiliary components of the structural set, as well as to the reproduction of the actual supporting conditions in the test. Satisfactory experimental–numerical correlation was achieved at the global level (load–displacement evolution curve), even during the first postbuckling stages. At postbuckling stages close to collapse, poorer correlations were encountered. These discrepancies were attributed to two main reasons: (1) the lack of reliable information regarding the actual geometric imperfections of the panel and (2) the absence of damage in the model at the global level. With regard to the FE local analysis, multi-submodeling computations including CZMs were generated. These models predicted that the panel collapse was principally motivated by the buckling of one of the stringers of the panel, carrying out the complete decohesion of this entity. This prediction was in satisfactory agreement with the experimental data. The practicability of the local model has been discussed in terms of the representativeness of the applied boundary conditions, coming from the global models to the local models.

Acknowledgments

The study was supported by the Spanish Ministry of Economy and Competitiveness/FEDER (DPI2012-37187) and the Andalusian Government (Project of Excellence No. TEP-7093). The authors are grateful to AIRBUS/Spain, who provided the analyzed panels studied, and to the staff of TEAMS for their help and strong collaboration during the experimental program.

References

ABAQUS, 2009. V6.13 Analysis User's Manual. Dassault Systèmes Simulia Corp.

Benzeggagh, M.L., Kenane, M., 1996. Measurement of mixed-mode delamination fracture toughness of unidirectional glass/epoxy composites with mixed mode bending apparatus. Compos. Sci. Technol. 56 (4), 439–449.

Blázquez, A., Picón, R., 2010. Analytical and numerical models of postbuckling of orthotropic symmetric plates. J. Eng. Mech. 136 (10), 1299–1308.

Blázquez, A., Reinoso, J., París, F., Cañas, J., 2012. Analysis in the postbuckling regime of a pressurized stiffened panel Part II: numerical analysis and effect of the geometric imperfections. Compos. Struct. 94, 1544–1554.

Boni, L., Fanteria, D., Lanciotti, A., 2012. Post-buckling behaviour of flat stiffened composite panels: experiments vs. analysis. Compos. Struct. 94 (12), 3421–3433.

Chrupalla, D., Berg, S., Kärger, L., Doreille, M., Ludwig, T., Jansen, E.L., Rolfes, R., Kling, A., 2011. A homogenization based two-way multiscale approach for composite structures. In: Rolfes, R., Jansen, E.L. (Eds.), Proceedings of the 3rd ECCOMAS Thematic Conference on the Mechanical Response of Composites. Hannover, Germany, pp. 263–270.

Collings, T.A., 1977. The strength of bolted joints in multi-directional CFRP laminates. Composites 8 (1), 43–55.

Dávila, C.G., Camanho, P.P., Turón, A., 2007. Cohesive Elements for Shells. NASA/TP-2007-214869.

Degenhardt, R., Rolfes, R., Zimmermann, R., Rohwer, K., 2006. Cocomat – improved material exploitation of composite airframe structures by accurate simulation of postbuckling and collapse. Compos. Struct. 73 (2), 175–178.

Egan, B., McCarthy, C.T., McCarthy, M.A., Frizzell, R.M., 2012. Stress analysis of single-bolt, single-lap, countersunk composite joints with variable bolt-hole clearance. Compos. Struct. 94, 1038–1051.

Ernst, G., Vogler, M., Hühne, C., Rolfes, R., 2010. Multiscale progressive failure analysis of textile composites. Compos. Sci. Technol. 70 (1), 61–72.

Falzon, B.G., Davies, G.A.O., 2003. The behaviour of compressively loaded stiffener runout specimens – Part I: experiments. J. Compos. Mater. 37 (5), 381–400.

Falzon, B.G., Hitchings, D., 2003. The behaviour of compressively loaded stiffener runout specimens – Part II: finite element analysis. J. Compos. Mater. 37 (6), 481–501.

Falzon, B., Stevens, K., Davies, G.A.O., 2000. Postbuckling behaviour of a blade-stiffened composite panel loaded in uniaxial compression. Compos. Part A Appl. Sci. Manuf. 31 (5), 459–468.

Falzon, B.G., Davies, G.A.O., Greenhalgh, E., 2001. Failure of thick-skinned stiffener runout sections loaded in uniaxial compression. Compos. Struct. 53, 223–233.

Feyel, F., 2003. A multilevel finite element method (FE^2) to describe the response of highly non-linear structures using generalized continua. Comput. Meth. Appl. Mech. Eng. 192 (28–30), 3233–3244.

Fish, J., Yu, Q., Shek, K., 1999. Computational damage mechanics for composite materials based on mathematical homogenization. Int. J. Numer. Methods Eng. 45 (11), 1657–1679.

Ghosh, S., Lee, K., Raghavan, P., 2001. A multi-level computational model for multi-scale damage analysis in composite and porous materials. Int. J. Solid Struct. 38 (14), 2335–2385.

Greenhalgh, E., García, M.H., 2004. Fracture mechanisms and failure processes at stiffener run-outs in polymer matrix composite stiffened elements. Compos Part A Appl. Sci. Manuf. 35, 1447–1458.

Herrington, P.D., Sabbaghian, M., 1991. Factors affecting the friction coefficients between metallic washers and composite surfaces. Composites 22 (6), 418–424.

Hur, S.-H., Son, H.-J., Kweon, J.-H., Choi, J.-H., 2008. Postbuckling of composite cylinders under external hydrostatic pressure. Compos. Struct. 86, 114–124.

Ireman, T., 1998. Three-dimensional stress analysis of bolted single-lap composite joints. Compos. Struct. 43, 195–216.

Krueger, R., Ratcliffe, J.G., Minguet, P.J., 2009. Panel stiffener debonding analysis using a shell/3D modeling technique. Compos. Sci. Technol. 69 (14), 2352–2362.

Labeas, G., Belesis, S., Diamantakos, I., Tserpes, K., 2012. Adaptative progressive damage modeling for large-scale composite structures. Int. J. Damage Mech. 21 (3), 441–462.

Lauterbach, S., Orifici, A.C., Wagner, W., Balzani, C., Abramovich, H., Thomson, R., 2010. Damage sensitivity of axially loaded stringer-stiffened curved CFRP panels. Compos. Sci. Technol. 70, 240–248.

McCarthy, M.A., Lawlor, V.P., Stanley, W.F., McCarthy, C.T., 2002. Bolt-hole clearance effects and strength criteria in single-bolt, single-lap, composite bolted joints. Compos. Sci. Technol. 62, 1415–1431.

Mittelstedt, C., Beerhorst, M., 2009. Closed-form buckling analysis of compressively loaded composite plates braced by omega-stringers. Compos. Struct. 88 (3), 424–435.

Olmedo, A., Santiuste, C., 2012. On the prediction of bolted single-lap composite joints. Compos. Struct. 94 (6), 2110–2117.

Orifici, A.C., Thomson, R.S., Degenhardt, R., Kling, A., Rohwer, K., Bayandor, J., 2008a. Degradation investigation in a postbuckling composite stiffened fuselage panel. Compos. Struct. 82 (2), 217–224.

Orifici, A.C., Thomson, R.S., Herszberg, I., Weller, T., Degenhardt, R., Bayandor, J., 2008b. An analysis methodology for failure in postbuckling skin-stiffener interfaces. Compos. Struct. 86 (1–3), 186–193.

Prandtl, L., 1933. Ein Gedankenmodell für den Zerreißvorgang spröder Körper. (A thought model for the fracture of brittle solids). Z. Angew. Math. Mech. 13 (2), 129–133.

Psarras, S., Pinho, S.T., Falzon, B.G., 2011. Design of composite stiffener run-outs for damage tolerance. Finite Elem. Anal. Des. 47, 949–995.

Reinoso, J., Blázquez, A., Estefani, A., París, F., Cañas, J., Arévalo, E., Cruz, F., 2012a. Experimental and three-dimensional global–local finite element analysis of a composite component including degradation process at the interface. Compos. B Eng. 43 (4), 1929–1942.

Reinoso, J., Blázquez, A., París, F., Cañas, J., 2012b. Postbuckling behaviour of a pressurized stiffened composite panel. Part I: experimental analysis. Compos. Struct. 94, 1533–1543.

Reinoso, J., Blázquez, A., Estefani, A., París, F., Cañas, J., 2013. A composite runout specimen subjected to tension–compression loading conditions: experimental and global–local finite element analysis. Compos. Struct. 101, 274–289.

Romeo, G., Frulla, G., 1997. Post-buckling behaviour of graphite/epoxy stiffened panels with initial imperfections subjected to eccentric biaxial compression loading. Int. J. Nonlinear Mech. 32 (6), 1017–1033.

Stevens, K., Ricci, R., Davies, G.A.O., 1995. Buckling and postbuckling of composite structures. Compos. B Eng. 26 (3), 189–199.

Távara, L., Mantic, V., Graciani, E., Paris, F., 2011. BEM analysis of crack onset and propagation along fiber-matrix interface under transverse tension using a linear elastic-brittle interface model. Eng. Anal. Bound. Elem. 35, 207–222.

Vescovini, R., Bisagni, C., 2012. Single-mode solution for post-buckling analysis of composite panels with elastic restraints loaded in compression. Compos. B Eng. 43 (3), 1247–1258.

Zimmermann, R., Klein, H., Kling, A., 2006. Buckling and postbuckling of stringer stiffened fibre composite curved panels – tests and computations. Compos. Struct. 73 (2), 150–161.

Failure models for composite joints: an approach based on singular stress states

22

A. Barroso, V. Mantič and F. París
School of Engineering, University of Seville, Seville, Spain

22.1 Motivation

Failure initiation prediction in adhesively bonded joints of composite materials is not a straightforward task, the number of approaches for stress characterization and failure proposals being large.

The different approaches for failure prediction of adhesive joints can be roughly divided into three groups. The first group uses nominal values of the stress state, typically a stress, a strain, or an energy value, but does not take into account the local details (Volkersen, 1938; de Bruyne, 1944; Goland and Reissner, 1944; Hart-Smith, 1974 are classical references). In the second group, the local details, namely abrupt changes in geometry and material properties, are taken into account, giving rise to singular stress states (Malyshev and Salganik, 1965; Gradin and Groth, 1984; Fernlund and Spelt, 1991; Hattori, 1991 are classical references of this group). In the third group, other approaches can be included, such as continuum damage models or degradation models or models taking into account the local details but avoiding the stress singularities (using, for instance, cohesive constitutive laws) (Laschet and Stas, 1992; Sheppard et al., 1998; Mohammed and Liechti, 2000).

In the authors' opinion, the failure proposals included in the first group are reasonably well established and several experimental results are available in the literature. Proposals included in the third group are more orientated toward failure progression and typically need more model parameters. Proposals included in the second group are less numerous for the simple reason that the stress characterization of the singular stress states at these local geometries is very complex. This study presents past and current research work on this topic carried out by the authors.

The first objective of this study is to describe, in a semianalytical way, the singular stress field in the surrounding area of these critical points that appear in adhesive joints with composite materials. The second objective is to develop new tests for the experimental determination of the generalized fracture toughness, and the third and final objective is to propose a practical design tool based on a failure envelope and plots.

In structures that include adhesive joints of composite materials where fatigue loading is responsible for failure initiation at these critical points (Fig. 22.1), and where plasticity effects in the area of the corner tip are not significant, the stress characterization is a key factor for the failure initiation prediction.

Modeling Damage, Fatigue and Failure of Composite Materials. https://doi.org/10.1016/B978-0-443-18489-5.00010-2

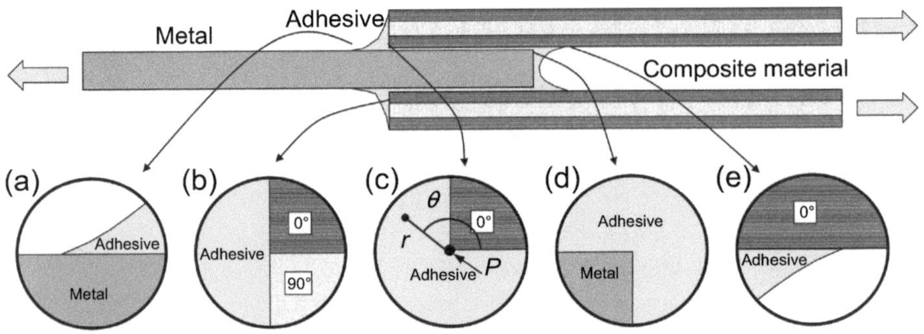

Figure 22.1 Multimaterial corners in a composite-metal double-lap joint.

In Section 22.2, a summary of the detailed description of multimaterial corners will be presented. In Section 22.3, a new test will be introduced to evaluate the generalized fracture toughness of the corner, and results from real double-lap joints will be presented. In Section 22.4, a failure envelope to be used as a failure criterion for these types of joints is achieved. Finally, in Section 22.5, a practical tool designed from the approach presented is proposed.

22.2 Stress characterization

To predict the failure initiation at multimaterial corners controlled by singular parameters of the stress state, a detailed stress characterization must be performed.

Except in particular cases (see Sinclair, 1999 for further details), the asymptotic displacement and stress states in the area of a multimaterial corner can be expressed, in a polar coordinate system (r, θ), centered at the corner tip (Fig. 22.1, corner (c)), by a series expansion with variable separation as shown:

$$u_i(r,\theta) = \sum_{n=1}^{N} K_n r^{\lambda_n} g_i^{(n)}(\theta) + \ldots \tag{22.1}$$

$$\sigma_{ij}(r,\theta) = \sum_{n=1}^{N} K_n r^{\lambda_n - 1} f_{ij}^{(n)}(\theta) + \ldots \tag{22.2}$$

where λ_n are the characteristic exponents; $g_i^{(n)}(\theta)$ and $f_{ij}^{(n)}(\theta)$ are the characteristic angular shape functions for displacements and stresses, respectively; and K_n are the generalized stress intensity factors (GSIFs). Rigid body motions are included in Eq. (22.1) for $\lambda_n = 0$ and $\lambda_n = 1$ with the appropriate definitions of $g_i^{(n)}(\theta)$ and $f_{ij}^{(n)}(\theta) = 0$. The angular functions $g_i^{(n)}(\theta)$ and $f_{ij}^{(n)}(\theta)$ are continuous and smooth functions inside each material, but may be discontinuous at material interfaces. Terms with $0 < \lambda_n < 1$ give rise to unbounded (singular) stresses with a finite value of the elastic strain energy.

In Section 22.2.1, the details for the calculation of λ_n, $g_i^{(n)}(\theta)$, and $f_{ij}^{(n)}(\theta)$ are outlined, and in Section 22.2.2, the procedure proposed for the evaluation of K_n is presented.

22.2.1 Stress singularities and characteristic angular functions

Characteristic exponents λ_n and angular shape functions $g_i^{(n)}(\theta)$ and $f_{ij}^{(n)}(\theta)$ in Eqs. (22.1) and (22.2) depend only on the local geometry, local boundary conditions, and local material properties of the multimaterial corner. Therefore, a general solution can be obtained for each specific corner configuration (Fig. 22.1). Changes in the global geometrical parameters of the adhesive joint, such as the overlap length or thickness of the adhesive or adherents, do not affect qualitatively the local configuration of the corner or the asymptotic displacement and stress fields. Typically, the most critical corner for failure initiation in a lap joint, like the one shown in Fig. 22.1, is the one at the end of the overlap region (denoted by point P in Fig. 22.1c), where the shear and peel stresses are nominally infinite and additionally peel stresses are positive in the thickness direction. In the aforementioned corner, a change in the stacking sequence of the laminate does not affect λ_n, $g_i^{(n)}(\theta)$ or $f_{ij}^{(n)}(\theta)$, provided that the last ply of the laminate remains with the same fiber orientation (0° in the example of Fig. 22.1).

The characteristic exponents λ_n can be obtained in a semianalytic way, using the Lekhnitskii–Stroh formalism of anisotropic elasticity (Lekhnitskii, 1938; Stroh, 1958, 1962; Eshelby et al., 1953; Ting, 1996). The characteristic exponents are the roots of the determinant of a characteristic equation, in a complex variable, an analytic expression that is solved numerically to evaluate λ_n. The argument principle is also implemented in the code to assess the searching of the complex roots of the equation.

The procedure developed by the authors allows multimaterial corners to be studied without limiting the number of material wedges in the corner or the nature of the material constitutive law (namely isotropic, transversely isotropic, orthotropic, or mathematically degenerate forms in the framework of the Stroh formalism). The tool also allows a wide range of boundary conditions to be defined, both between material wedges (perfect adhesion or friction) and external boundary conditions (any combination of prescribed displacement and stress, such as symmetry, antisymmetry, clamped, and free).

The mathematical aspects and the details of the computer code, developed in *Mathematica* Wolfram (1991), fall outside the scope of this chapter, but are available in Barroso et al. (2003) and Mantič et al. (2014).

Characteristic angular shape functions, $g_i^{(n)}(\theta)$ and $f_{ij}^{(n)}(\theta)$, are computed numerically for the desired angular positions. It must be stressed that λ_n, $g_i^{(n)}(\theta)$, and $f_{ij}^{(n)}(\theta)$ have to be computed only once for each corner configuration. Changes in the global geometry, in boundary conditions and/or in far-field loadings, maintaining

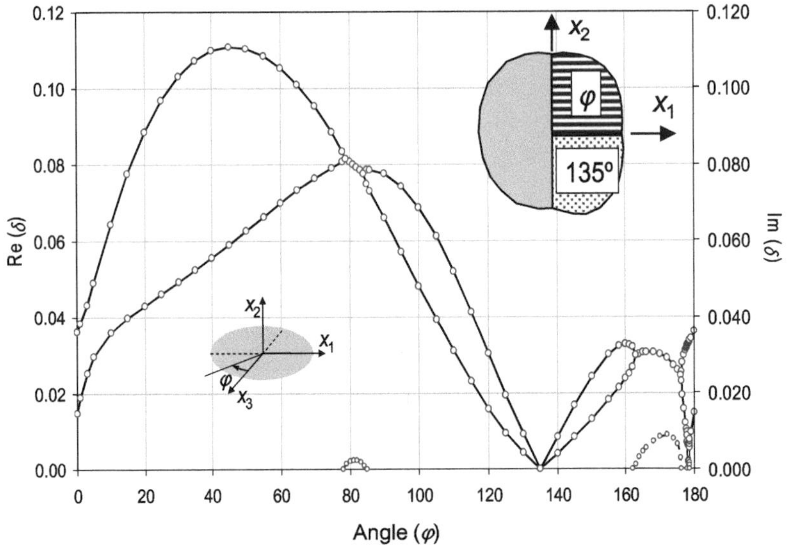

Figure 22.2 Stress singularities δ ($\delta = 1 - \lambda$) for a three-material corner of type (b) of Fig. 22.1.

the local configuration of the multimaterial corner, will alter the displacement and stress fields only by means of changes in the values of the GSIFs (K_n).

An example of the computer code's potential is shown in Fig. 22.2 for a three-material corner of type (b) of Fig. 22.1, where one of the plies has a fixed fiber orientation (135°), while the other is variable (ϕ). The values of the stress singularities $\delta = 1 - \lambda$ versus the orientation angle (ϕ) are shown in Fig. 22.2. Notice that for some specific values of ϕ, the stress singularities become a complex conjugate (for example, around $\phi = 80°$).

Fig. 22.3 shows the results of the characteristic angular shape functions $f_{ij}^{(n)}(\theta)$ corresponding to mode 1 for the three in-plane stress components $\sigma_{\theta\theta}$, σ_{rr}, and $\sigma_{r\theta}$ for the specific case of the bimaterial corner (c) in Fig. 22.1. The circumference line denotes the zero value in the polar representation of the angular shape functions, positive and negative values of the functions being, respectively, outside and inside the circumference.

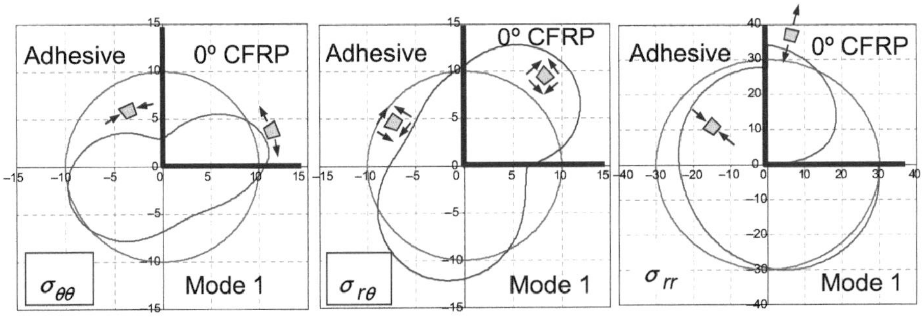

Figure 22.3 Angular shape functions for the stress components in the bimaterial corner (c) of Fig. 22.1.

Notice the lack of symmetry in the three stress components due to the lack of symmetry in the corner configuration (the composite material breaks the symmetry plane at 45°). Notice also the discontinuity in the values of the σ_{rr} component at the material interfaces.

22.2.2 Generalized stress intensity factors

GSIFs K_n depend on global geometry, boundary conditions, and far-field loading. Their calculation typically needs the numerical modeling of the problem and a postprocessing of the results. The GSIFs represent the weight of each term of the asymptotic series expansion in Eqs. (22.1) and (22.2) in the complete solution. Unlike crack problems in homogeneous isotropic materials, each K_n ($n = 1, \ldots, N$), in uncracked multimaterial corners with nonisotropic materials, has different dimensions due to the fact that each singularity order has different values.

There are several procedures to compute the GSIFs, for example, conservative integrals (Banks-Sills, 1997; Omer and Yosibash, 2005), or incorporating singularity shape functions into the problem discretization, using quarter point elements (Henshell and Shaw, 1975; Gray et al., 2003) or functions in the whole domain (Helsing and Jonsson, 2002; Marin et al., 2004). There are also procedures based on least squares adjustments (Chan et al., 1970; Liu et al., 2008). See Barroso et al. (2012b) for a review of available procedures to calculate GSIFs.

The authors of the present chapter have developed a simple and reliable procedure, based on a least squares fitting of the displacement/stress solution near the corner, for the simultaneous evaluation of various K_n. In Eqs. (22.1) and (22.2), once λ_n, $g_i^{(n)}(\theta)$, and $f_{ij}^{(n)}(\theta)$ have been calculated, the unique unknowns in the least squares fitting procedure are the K_n.

A numerical solution, denoted by superscript "NUM" and obtained in this study using the BEM (boundary element method), is used as a reference (París and Cañas, 1997). The analytical solution in Eqs. (22.1) and (22.2) is denoted by the superscript "series." The sum of squares of differences in values of the displacements, $u_\alpha^{\text{series}} - u_\alpha^{\text{NUM}}$ (J_u in Eq. 22.3a) and tractions, $t_\alpha^{\text{series}} - t_\alpha^{\text{NUM}}$ (J_σ in Eq. 22.3b), at nodes placed along the radial edges of the corner are added together (J in Eq. 22.3c):

$$J_u(K_1, \ldots, K_M) = \sum_{\alpha=r,\theta} \sum_{i=1}^{N_r} \sum_{j=1}^{N_\theta} a_\alpha \left[u_\alpha^{\text{series}}(r_i, \theta_j, K_1, \ldots, K_M)\right.$$

$$\left.-u_\alpha^{\text{NUM}}(r_i, \theta_j)\right]^2 (a_\alpha \geq 0) \tag{22.3a}$$

$$J_\sigma(K_1, \ldots, K_M) = \sum_{\alpha=r,\theta} \sum_{i=1}^{N_r} \sum_{j=1}^{N_\theta} b_\alpha \left[t_\alpha^{\text{series}}(r_i, \theta_j, K_1, \ldots, K_M)\right.$$

$$\left.-t_\alpha^{\text{NUM}}(r_i, \theta_j)\right]^2 (b_\alpha \geq 0) \tag{22.3b}$$

$$J(K_1, \ldots, K_M) = aJ_u(K_1, \ldots, K_M) + bJ_\sigma(K_1, \ldots, K_M)(a, b \geq 0) \tag{22.3c}$$

In Eqs. (22.3a) and (22.3b), N_θ denotes the number of radial edges to be taken into account, and N_r denotes the number of nodes taken at different distances along each radial edge.

In Eq. (22.3c), a and b are weighting factors, which allow us to just consider only displacements $(a, b) = (1, 0)$, only stresses $(a, b) = (0, 1)$, or both $(a, b) = (l^{-2}, \sigma^{-2})$. l and σ are characteristic length and stress values, chosen in order to force the terms given by the displacements and stresses to have values of the same order of magnitude. Independently of the option followed, the dimensionless weighting factors in Eqs. (22.3a) and (22.3b), a_α and b_α, allow the isolated components of the nodal displacements and/or tractions to be used (e.g., if only the radial component of the displacements u_r is to be used: $a_r = 1$, $a_\theta = 0$, $b_r = 0$, $b_\theta = 0$).

An approximate solution for the set of GSIFs K_n ($n = 1, \ldots,$ M) can be obtained by solving the following linear system of equations:

$$\frac{\partial J(K_1, \ldots, K_M)}{\partial K_j} = 0 \quad (j = 1, \ldots, M) \tag{22.4}$$

The details of the implementation of the procedure, its robustness, and its accuracy (in comparison with available reference benchmark problems), can be found in Barroso et al. (2012b). Owing to the nature of the numerical values involved in the calculations (elastic solution near a stress singularity, with numerical errors due to the discretization), the coefficient matrix of the linear system is typically ill-conditioned. The system is solved by means of the QR decomposition of the coefficient matrix, which is known to be more accurate than directly solving the system (see Barroso et al., 2012b, for further details). The procedure, also implemented in a *Mathematica* code, has shown to be quite robust (not dependent on extremely precise results of the numerical model close to the corner tip) and accurate (errors below 1% in most cases, using well-known benchmark problems) (see, for example Helsing and Jonsson, 2002).

When λ is a complex number, as in the case of interface cracks, $\lambda = \lambda_R + i\lambda_I$ (where λ_R and λ_I are real numbers), the associated GSIF is also a complex number $K = K_R + iK_I$ (where K_R and K_I are real numbers). In such a case, two terms can be included in Eqs. (22.1) and (22.2). In the representation of the stresses in Eq. (22.2), one term would be equal to $K_R\mathrm{Re}\left[r^{\lambda-1}f_{ij}(\theta)\right]$ and the other would be equal to $K_I\mathrm{Im}\left[r^{\lambda-1}f_{ij}(\theta)\right]$, while in the representation of the displacements in Eq. (22.1), one term would be equal to $K_R\mathrm{Re}\left[r^{\lambda}g_i(\theta)\right]$ and the other equal to $K_I\mathrm{Im}\left[r^{\lambda}g_i(\theta)\right]$.

In all calculations performed in the present study, GSIFs have been standardized as proposed by Pageau et al. (1996) in such a way that the stress component $\sigma\theta$-$\theta|$ θ=0° $= K/(2\pi r)^{1-\lambda}$.

22.3 Generalized fracture toughness determination

Once the asymptotic elastic stress state in the area of the multimaterial corner is fully defined, a proposal for the failure initiation at these corners needs the determination of the allowable values with which to compare the singularity stress field. Using a fracture Mechanics approach, the GSIFs, which control the singularity stress field, are the key parameters whose allowable values (defined as generalized fracture toughness) have to be determined.

The main point here is that, while crack problems in the stress field of isotropic homogeneous materials have certain geometrical symmetries, in a general multimaterial corner involving nonisotropic materials, this symmetry is lost. This lack of symmetry implies that the standard test configurations (e.g., K_{IC} using a symmetric loading of the compact tension specimen) are no longer valid.

To overcome this issue, the authors have conceived a new test, based on the Brazilian disk specimen (Barroso et al., 2012a; Vicentini et al., 2012). The proposal is valid for closed corners (corners with all materials bonded, without external wedge faces) and up to two singularity terms in the series expansion. These two aspects are not a significant limitation for the problem under study, because typically the most critical point from which failure initiates is the closed corner appearing at the end of the overlap (see corner (c) in Fig. 22.1). Thus, it is at the aforementioned corner, at the overlap end, where shear and peel stresses take their maximum values, the peel stresses being positive. Notice that in corner (d), peel stresses are negative, compressive stresses. Few configurations present more than two singularity terms at the corner tip (see Barroso et al., 2003; Barroso, 2007).

The proposed geometry has the multimaterial corner at the center of the disk, and the disk is diametrically loaded in compression at the external boundary, as depicted in Fig. 22.4a, with different load angles (α). A numerical model of the test configuration using a unit load value allows the GSIFs (K_1 and K_2) at the corner tip to be calculated for each angle orientation. All previously developed tools (calculation of characteristic exponents λ_n, characteristic angular shape functions $g_i^{(n)}(\theta)$ and $f_{ij}^{(n)}(\theta)$, and the least squares fitting procedure for GSIFs) have to be used for the tested configuration.

At those values of the load angle α in which a K_n ($n = 1, 2$) curve cuts the horizontal axis, one singularity term in Eq. (22.2) vanishes (as its corresponding K_n value is zero). This allows the other singularity term to be isolated and a test configuration in which the stress field is dominated by only one singularity term defined.

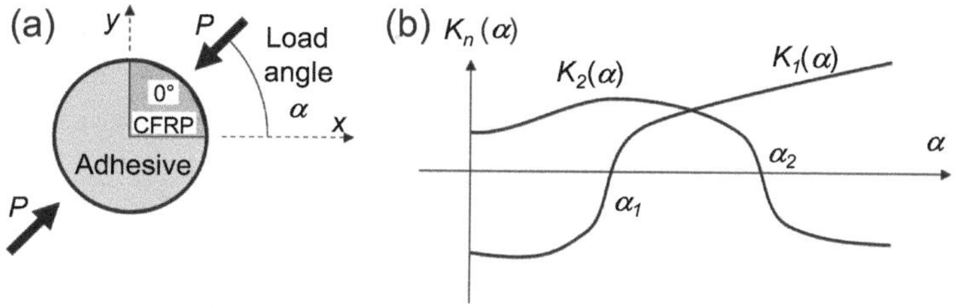

Figure 22.4 Scheme of the test to determine generalized fracture toughness values.

The size-scale effect has been taken into account when evaluating GSIF in geometrically similar samples. From a dimensional analysis, a GSIF can be expressed as (Barroso et al., 2012a):

$$K_n = \sigma_{\text{nom}} R^{1-\lambda_n} A_n \tag{22.5}$$

where $\sigma_{\text{nom}} = P/(R \cdot t)$ is a nominal stress in the sample, R is a characteristic length of the specimen (the radius), t is the thickness, and A_n is the dimensionless shape factor that takes into account the geometry and material properties of the specimen.

Let P^{exp} denote the experimental load for which failure initiation in the corner is observed, t^{exp} the real specimen thickness, and R^{exp} the real specimen radius. Then, the following equation obtained in view of Eq. (22.5) is used to determine $K_{1\text{C}}$ and $K_{2\text{C}}$:

$$K_{n\text{C}} = K_n^{\text{NUM}} \frac{t^{\text{NUM}}}{t^{\text{exp}}} \frac{R^{\text{NUM}}}{R^{\text{exp}}} \frac{P^{\text{exp}}}{P^{\text{NUM}}} \left(\frac{R^{\text{exp}}}{R^{\text{NUM}}} \right)^{1-\lambda_n} = K_n^{\text{NUM}} \frac{\sigma_{\text{nom}}^{\text{exp}}}{\sigma_{\text{nom}}^{\text{NUM}}} \left(\frac{R^{\text{exp}}}{R^{\text{NUM}}} \right)^{1-\lambda_n} \tag{22.6}$$

where superscript "NUM" denotes values from the above-defined numerical model, $\sigma_{\text{nom}}^{\text{NUM}}$ and $\sigma_{\text{nom}}^{\text{exp}}$ denote some nominal stresses in the numerical model and in the real specimen, respectively.

As an example, the bimaterial closed corner (c) in Fig. 22.1 has been analyzed, and the results are shown in Fig. 22.5. The GSIF values of the two singular terms of the corner are plotted versus the compressive load angle α. To isolate each one of the singular terms, there are two angle orientations; K_2 vanishes for $\alpha \approx 13°$ and $\alpha \approx 115°$, any of them allowing the determination of $K_{1\text{C}}$, whereas K_1 vanishes for $\alpha \approx 60°$ and $\alpha \approx 143°$, defining load orientations to determine $K_{2\text{C}}$.

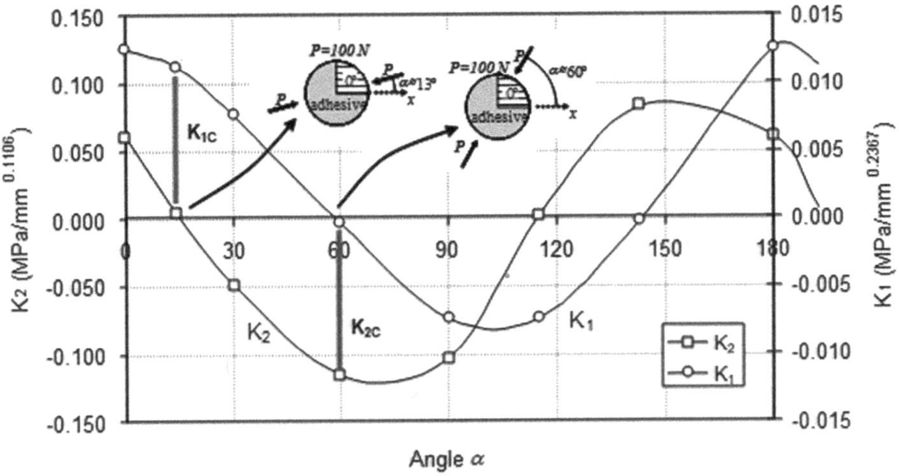

Figure 22.5 Test configurations for the determination of $K_{1\text{C}}$ and $K_{2\text{C}}$. Modified from Barroso et al. (2012a).

22.4 Failure envelope for failure initiation prediction

Once a procedure for the singular stress characterization is developed (in terms of GSIF values) and generalized fracture toughness values have been experimentally determined, a failure envelope can be generated. This failure envelope is a closed curve in the space ($K_1/K_{1C}-K_2/K_{2C}$) as depicted in Fig. 22.6.

First of all, it is important to point out that the axes of Fig. 22.6 (K_1/K_{1C} and K_2/K_{2C}) are nondimensional. The fact that the characteristic exponents for each singular term (λ_1 and λ_2) have different values, giving rise to different dimensions in the GSIF values K_1 [MPamm$^{(\lambda_1-1)}$] and K_2 [MPamm$^{(\lambda_2-1)}$]. Therefore, a coherent representation in the GSIF space needs the axis to have the same dimensions (or to be dimensionless, as in the present case).

The curve (represented by straight lines in Fig. 22.6), which divides the K_1-K_2 space into safe and nonsafe zones, cuts each axis twice as a consequence of the results obtained in Fig. 22.5. Intermediate results, in each quadrant, can be obtained loading the same Brazilian disks in intermediate compressive orientations. By means of the numerical simulation of each corresponding angle, α, K_1 and K_2 can be obtained for the experimental failure load and included in the representation in Fig. 22.6. This

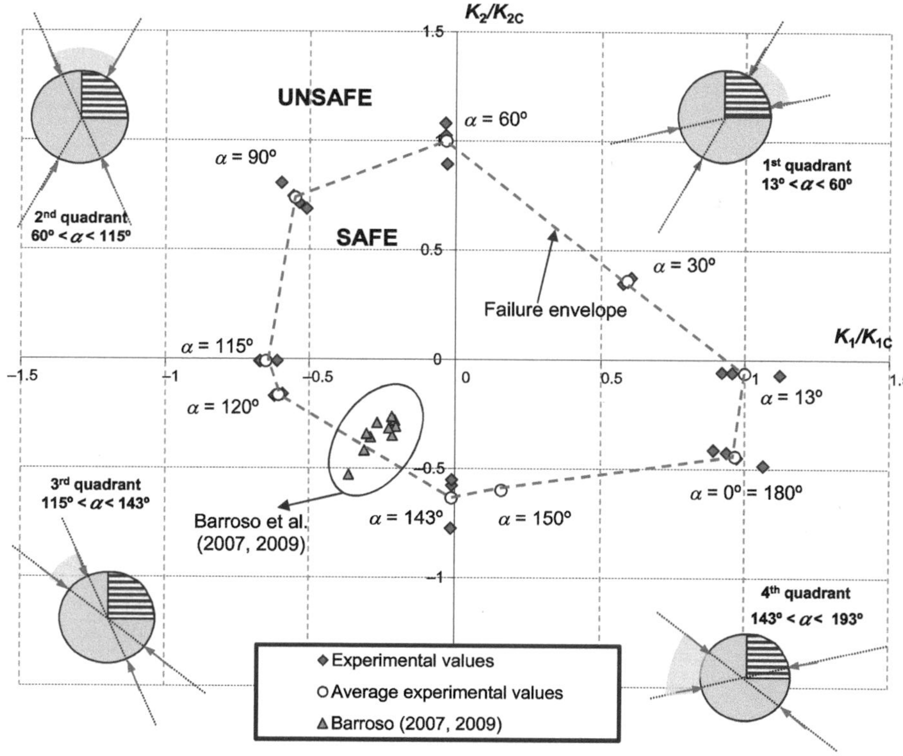

Figure 22.6 Failure envelope based on GSIF values.
From Vicentini et al. (2012).

experimental task was carried out by the authors for at least one point inside each quadrant, generating in this way a failure curve (see Vicentini et al. (2012) for detailed information of the whole manufacturing and testing procedure).

22.4.1 Generation of the failure envelope and failure criterion

Once the failure curve is defined, the failure initiation prediction in any structure that has a bimaterial closed corner under consideration (corner (c) in Fig. 22.1) can be addressed by simply calculating the GSIF values of the global structure and checking whether or not the pair (K_1, K_2) falls inside the SAFE zone. This calculation is straightforward but may be tedious as it requires a detailed numerical model of the structure and the correct postprocessing of the results. This is the reason why a practical and fast approach is proposed by the authors in Section 22.5.

Tests on real double-lap shear composite-metal samples, which are addressed in Section 22.4.2, have been shown to fail (the initiation of the failure) at the same critical point, where the last ply meets the adhesive layer and adhesive fillet. If this lamina in contact with the adhesive layer is 0°, the results for the failure envelope are those depicted in Fig. 22.6. Nevertheless, changes in the main parameters defining the joint—adherent thickness, overlap length, composite stacking sequence (keeping the last ply at 0°), etc.—do not significantly alter the load path of these types of joints, consequently giving rise to the pairs (K_1, K_2) in a narrow area of the third quadrant.

Thus, in order to generate a more precise failure envelope curve in this area, new tests were performed, using compression disks every 5°. In Fig. 22.7, the solid dots represent the samples tested under compression for each load orientation ($115° < \alpha < 145°$, every 5°) where 115° and 143° are configurations corresponding to pure

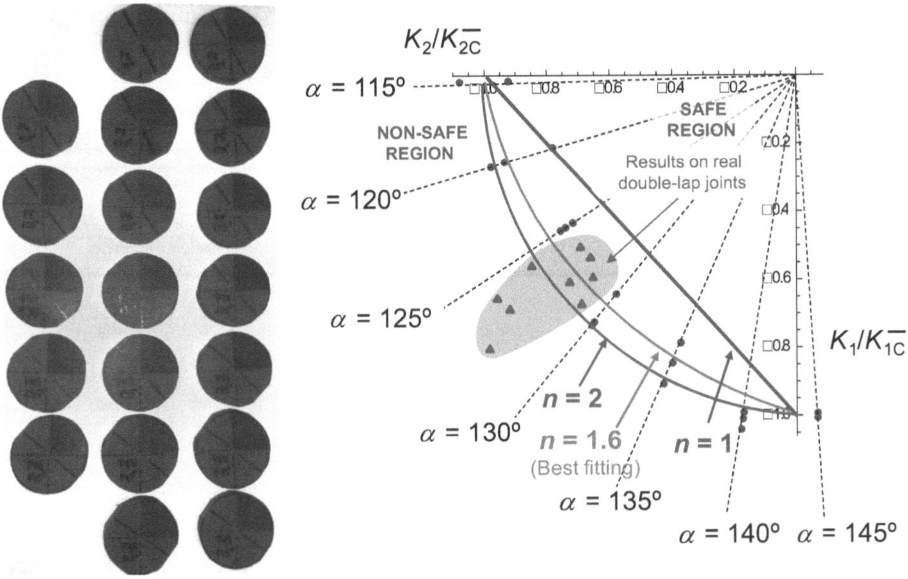

Figure 22.7 Failure envelope for the third quadrant.

modes ($K_1 \approx 0$ for 115° and $K_2 \approx 0$ for 143°). Also the samples tested are shown in Fig. 22.7. Test results on double-lap joints are also included in Fig. 22.7 (solid triangles in the shaded area). These tests are described in Section 22.4.2. The superscripts (−) at both axes denote that the reference values are associated to the negative part of the axes.

With such a density of experimental points, a more accurate failure initiation envelope can be obtained. A form for the failure criterion has been proposed as follows:

$$\left(\frac{K_1}{K_{1C}}\right)^n + \left(\frac{K_2}{K_{2C}}\right)^n \leq 1 \tag{22.7}$$

Three values for exponent n are represented in Fig. 22.7. The value $n = 1$ is simply a linear interpolation between the generalized fracture toughness values for pure terms K_1 and K_2. The value $n = 2$ gives a circumference in the K_1–K_2 space, and $n = 1.6$ is the least squares value that best fits the experimental results.

22.4.2 Tests on real-bonded joints

To generate a more precise failure envelope in the third quadrant, tests were carried out on composite-metal double-lap joints, all of them having at the critical corner point at the end of the overlap the same local bimaterial closed corner depicted in Fig. 22.1c. The aforementioned tests are static tensile tests of the samples, inducing shear and peel stresses along the overlap until the final failure of the coupons.

Different overlap lengths, adherent thicknesses, stacking sequences of the composite laminate (always keeping the last ply in contact with the adhesive layer with the same orientation, 0°, which is the longitudinal direction, the orientation of the tensile load), and surface preparations were tested (Fig. 22.8). Each joint configuration was numerically computed by means of a FEM model, and GSIFs calculated accurately at the experimentally registered failure load.

Figure 22.8 Metal-composite double-lap joints with different joint configurations (overlap length, adherent thickness, and composite stacking sequence).

The two GSIF values associated with the two singular terms, for each joint configuration, have been included in Fig. 22.7. Each point (red (gray in print version) solid triangles) represents the average value of five samples for each configuration.

Although the changes in the geometry configuration lead to changes in the load path transmission, this change is minimal and, as mentioned previously, the GSIF pairs (K_1, K_2) associated with these samples fall in a narrow area of the third quadrant of the K_1–K_2 space, as can be observed in Fig. 22.7.

It is important to stress that, although the geometric configurations, the characteristic lengths, and the loads applied to the samples used to generate the failure envelope (the Brazilian disk samples) are completely different from those corresponding to the double-lap joints, the prediction of the failure is reasonably accurate.

It is also interesting to note that, in the case of the Brazilian disk samples, the failure could be detected at a very early stage, due to the size of the samples, whereas the failure in the double-lap joint samples was only visible after the complete failure of the coupons. With all these differences in mind, the failure formula proposed can be considered to be accurate enough to predict failure initiation in the samples.

22.5 Practical application for the design of adhesive joints

Taking all previous considerations into account, it is clear that a detailed description of the singular stress field at a multimaterial corner is not a straightforward task, involving different tools and careful manipulation of the data. This is far from being a practical tool for engineers in the design process. Nevertheless, due to the fact that adhesive joints have relatively few geometrical and mechanical parameters defining the joint configuration, it is possible to generate some types of master curves from which to estimate, without complex calculations, the GSIF values of the critical corner points. This practical approach is described in what follows and needs the geometrical parameters shown in Fig. 22.9 to be defined: e_1 and e_2 are the adherent thicknesses, L is the overlap length, and E_1 and E_2 are the Young's moduli of the adherents (in the case of the composite laminate, it corresponds to the equivalent laminate stiffness in the longitudinal direction x_1).

The idea consists of using all previous tools and procedures to accurately calculate GSIF values for different values of the geometry (parametrical study), mechanical properties, and unit values of different basic loads (axial force, shear force, and bending moment). Then, a plot from which to interpolate the results for the particular geometrical, mechanical, and loading conditions is generated. The superposition principle is used here, assuming linear elastic behavior.

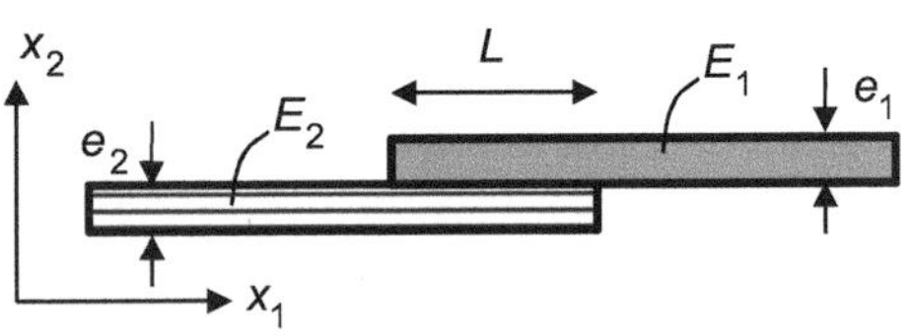

Figure 22.9 Joint data included in the parametric analysis.

22.5.1 Parametric study

Three main geometrical parameters have been included in the parametric analysis:

a. The ratio of the overlap length versus adherent reference thickness (L/e). Notice that in the presence of two different adherents, there are two adherent thicknesses (one has to be chosen as the reference thickness).
b. The ratio imbalance of the joint (d). Measured as the ratio of total axial stiffness of the adherents: $d=(E_1 \cdot t_1)/(E_2 \cdot t_2)$ for the single-lap joint and $d=(E_1 \cdot t_1)/(2 \cdot E_2 \cdot t_2)$ for the double-lap joint. Subindex 1 (or 2), in the case of double-lap joints, is linked to the inner (or outer) adherent.
c. The reference adherent thickness (e).

The parameters (L/e) and (d) are nondimensional, while (e) has the dimension of length.

The parametric study carried out in the present investigation consists of the analysis of 108 cases, corresponding to the product of:

- Two joint configurations (single- and double-lap joints)
- Three different unit loads (tensile force, shear force, and bending moment)
- Three (L/e) values, $L/e = 10$, 50, 100
- Three (d) values, $d = 0.5$, 1, 2
- Two (e) values, $e = 2$ mm, 4 mm

In the numerical models of the joint, the unit values of the load are always applied at a distance of one overlap length (Fig. 22.10), for standardization purposes. On the other side of the joint, zero displacements are imposed, giving rise to the corresponding axial force, shear force, and bending moment in equilibrium.

22.5.2 Plots of generalized stress intensity factors

GSIFs have been calculated for the bimaterial corner at the end of the overlap of the joint.

For the mechanical properties of the materials: $E = 68{,}670$ MPa (Young's modulus), $\nu = 0.33$ (Poisson's ratio) for aluminum 2024-T3; $E_{11} = 141{,}300$ MPa, $E_{22} = E_{33} = 9580$ MPa, $\nu_{12} = \nu_{13} = 0.3$, $\nu_{23} = 0.32$, $G_{23} = 3500$ MPa,

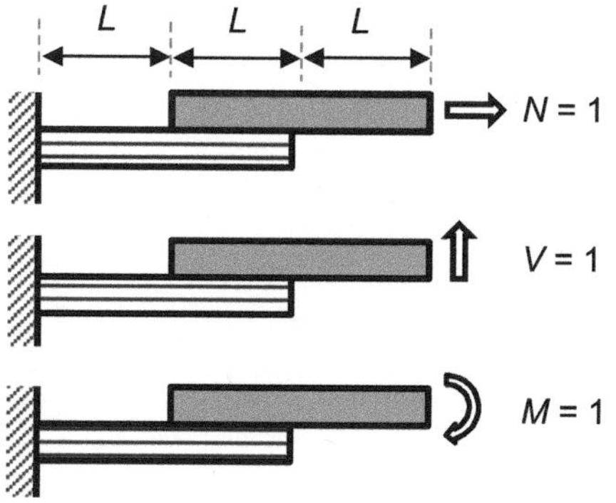

Figure 22.10 Unit load cases.

$G_{12} = G_{13} = 5000$ MPa for CFRP AS4/8552 unidirectional fiber laminate $[0]_{12}$ (1 being the fiber direction), and $E = 3000$ MPa, $\nu = 0.35$ for the adhesive FM73 OST (toughened epoxy film, from Cytec). The local singular stress field has two singular terms, whose weights are controlled by K_1 and K_2 (GSIFs):

$$\sigma_{\alpha\beta}(r,\theta) = \frac{K_1}{r^{0.236764}} f_{\alpha\beta}^{(1)}(\theta) + \frac{K_2}{r^{0.110611}} f_{\alpha\beta}^{(2)}(\theta) + \ldots \tag{22.8}$$

The calculations of the stress singularity orders and characteristic angular functions are unique for each global selection of the adherents/adhesive system, as they only depend on the local geometrical and local material configurations. The GSIF calculation has to be carried out for each parameter value *L/e*, *e*, *d* (as the GSIFs depend on the global loading and geometrical characteristics), for each unit load (*N*, *V*, *M*), and for each configuration (single, double-lap joint). The tool used for the calculations is the one developed by the authors, Barroso et al. (2012b), using the least squares procedure, this task representing the most time-consuming part of the parametric analysis.

Although not completely necessary (as already observed in a previous study conducted by Barroso et al., 2012b), a fine mesh was used at the corner (see Fig. 22.11, for a particular set of parameters in a single-lap joint configuration).

The mesh refinement used in the calculations is not practical for design purposes, and this is the reason why this part of the study is conducted. Making the calculations only once, for a set of different geometrical parameters and representing the results in a plot to be used in later calculations, removes the need to repeat tedious numerical modeling and the corresponding manipulation of results.

An example of the GSIF (K_1, K_2) results is shown in Figs. 22.12–22.14, where GSIFs result for unit values of axial load (Fig. 22.12), shear load (Fig. 22.13), and unit-bending moment (Fig. 22.14) are plotted. In these figures, the two GSIFs, K_1 and K_2, associated with the two singular terms of the asymptotic series expansion of the stress representation at the corner point (see Eq. 22.8), have been included. In Figs. 22.12–22.14, (e_1, e_4) indicate the thickness, ($d_{0.5}$, d_1, d_2) indicate the imbalance factor, and the horizontal axis represents the *L/e* parameter.

Figure 22.11 Details of the mesh refinement at the bimaterial corner tip for the GSIF calculation.

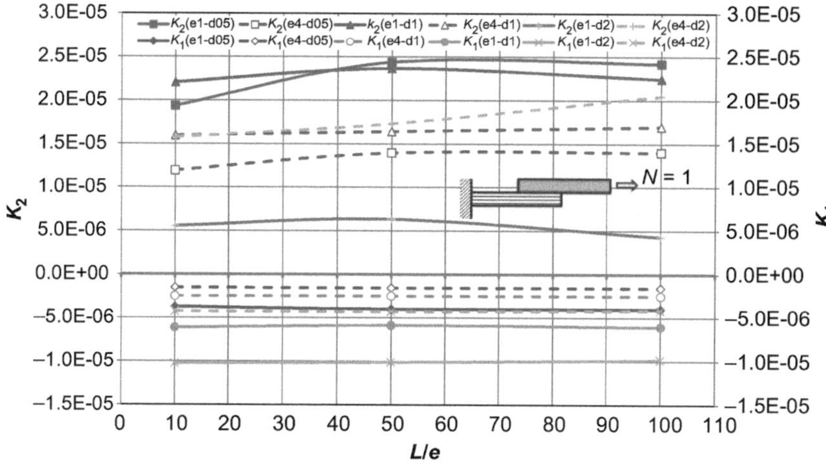

Figure 22.12 Results for GSIFs (K_1, K_2) for the unit axial load case.

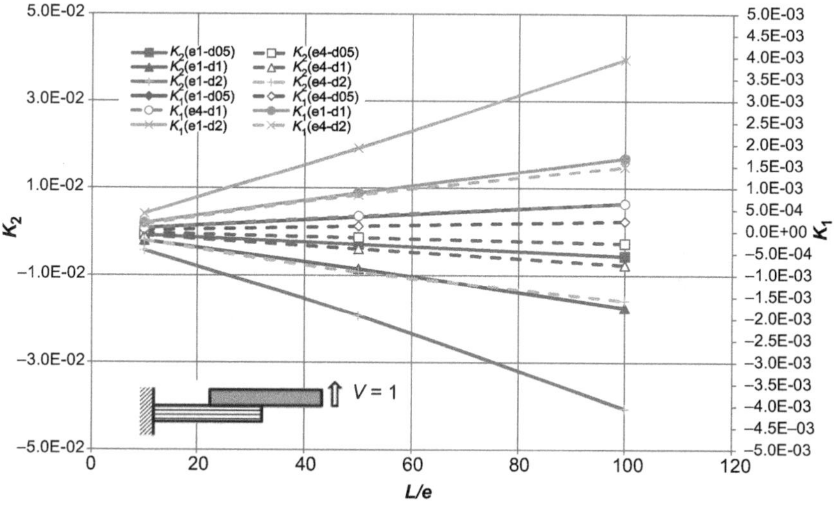

Figure 22.13 Results for GSIFs (K_1, K_2) for the unit shear load case.

22.5.3 *Example*

Once the parametric analysis is complete and plots for each unit load value are available, the way to use them is simple, fast, and reliable. Fig. 22.15 represents a structure including bonded joints. If plots as in Figs. 22.12–22.14 have been obtained for the specific bimaterial corner arising at the critical point at the end of the overlap in this joint, a simple structural analysis, obtaining the axial force (N), shear force (V), and bending moment (M) at a distance equal to the overlap length of the joint (L), is sufficient to estimate the GSIF values.

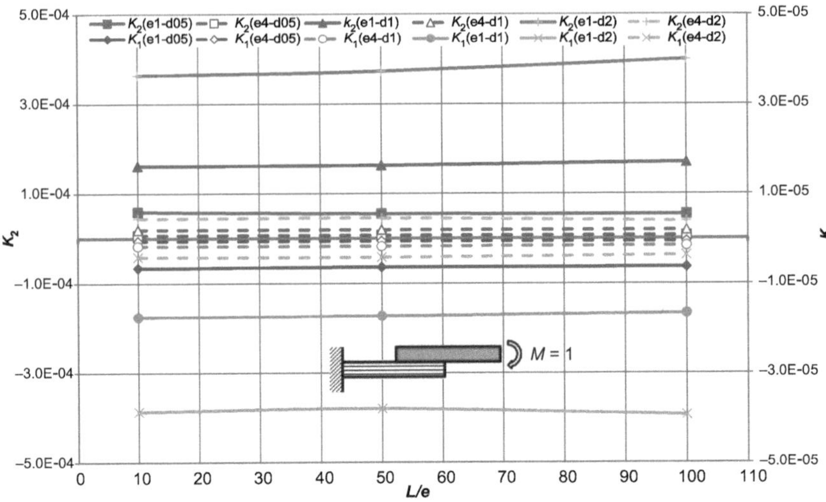

Figure 22.14 Results for GSIFs (K_1, K_2) for the unit bending moment load case.

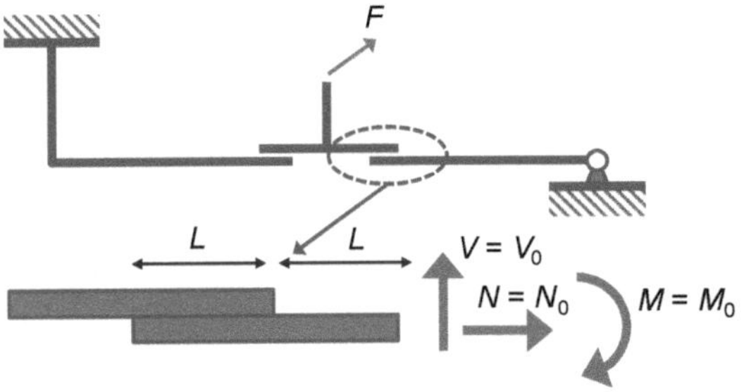

Figure 22.15 Schematic example of a structure including a bonded joint.

Each GSIF, K_k ($k = 1, 2$), can be obtained, assuming a linear elastic behavior, by means of the superposition principle. And as a result, the GSIF can easily be calculated for the specific geometrical parameters, using the results of the unit load plots:

$$K_k = N_0 \cdot K_{k(N=1)} + V_0 \cdot K_{k(V=1)} + M_0 \cdot K_{k(M=1)} (k = 1, 2) \tag{22.9}$$

where $K_{k(N=1)}$, $K_{k(V=1)}$, and $K_{k(M=1)}$ are the interpolated results, obtained from the uniaxial loads (N, V, and M) in Figs. 22.12–22.14, respectively. N_0, V_0, and M_0 are the real axial force, shear force, and bending moment at a distance of one overlap length (L) from the joint end, in the actual structure.

These GSIF values can be used in the failure envelope (Fig. 22.7) and the corresponding failure criterion (Eq. 22.7) to estimate the reserve factor of the joint.

22.6 Conclusions and future developments

A new approach for failure initiation prediction in composite joints has been presented in terms of singularity parameters of the critical points of the adhesive joint.

A procedure for the singularity analysis of linear elastic anisotropic multimaterial corners in generalized plane strain, including sliding friction contact surfaces, has been developed and implemented. A semianalytic procedure to calculate the stress singularities, which covers any kind of linear elastic material in an explicit form, avoiding numerical uncertainties and instabilities that may appear when degenerate materials are treated as limit cases of nondegenerate materials, has been developed. The general procedure developed for corner singularity analysis is well suited for use in a straightforward computational implementation.

A general postprocessing procedure for the extraction of multiple GSIFs in multimaterial corners from a numerical solution obtained by BEM (or FEM) has also been presented, implemented, and tested. This procedure uses a least squares adjustment of the finite asymptotic series expansion of the singular solution at the corner, with the numerical results in displacements and/or stresses along the boundaries and interface edges at the same corner. The procedure has no special requirements regarding the accuracy of the numerical results close to the corner tip, where higher errors in the numerical solution are expected.

A novel, experimental procedure for the determination of generalized fracture-toughness values and the associated failure criterion of closed anisotropic multimaterial corners with two singularities has been proposed and tested. The procedure is based on the use of the Brazilian disk sample with the bimaterial corner in its center and loaded at any position along the whole external perimeter. A failure envelope curve based on the GSIFs, normalized by generalized fracture-toughness values, has been determined experimentally and used in the formulation of a failure criterion for corners of this kind. A satisfactory agreement has been observed between the predictions of this failure criterion and the failure loads of a real double-lap joint of composite with metal including such a corner.

All previously developed tools have been used to accurately calculate the GISFs at the most critical corner of the adhesive joint, which appears at the end of the overlap area. These calculations have been made for unit value loading cases and for a set of three geometrical parameters, covering a wide range of geometrical configurations. All results have been presented in a graphical form, allowing future designers to perform a simple structural analysis of the joint in order to calculate the axial force, shear force, and the bending moment at the end of the overlap and then apply superposition using the unit value results from the graphs to estimate the GSIF values and the severity of the stress state using the proposed failure criterion.

In a future study, other material systems (adherents and adhesives) can be explored to create a database for design purposes.

Acknowledgments

The authors are grateful to Dr Enrique Graciani for the BEM code used in some of the calculations, to Dr Daniane Vicentini who has carried out and analyzed tests with the new Brazilian disk sample, and to Alfredo Soler and Carlos Herrera, who completed their Final Degree Projects within the framework of the present study.

References

Banks-Sills, L., 1997. A conservative integral for determining stress intensity factors of a bimaterial strip. Int. J. Fract. 86, 385–398.

Barroso, A., Vicentini, D., Mantič, V., París, F., 2012a. Determination of generalized fracture toughness in multimaterial closed corners with two singular terms. Part I: procedure and numerical results. Eng. Fract. Mech. 89, 1–14.

Barroso, A., Graciani, E., Mantič, V., París, F., 2012b. A least squares procedure for the evaluation of multiple generalized stress intensity factors at 2D multimaterial corners by BEM. Eng. Anal. Bound. Elem. 36, 458–470.

Barroso, A., Mantič, V., París, F., 2003. Singularity analysis of anisotropic multimaterial corners. Int. J. Fract. 119, 1–23.

Barroso, A., 2007. Characterization of Singular Stress Fields in Multimaterial Corners. Application to Adhesive Joints with Composite Materials (Ph.D. thesis) (in Spanish). University of Seville. http://fondosdigitales.us.es/media/thesis/485/D_T.319.pdf.

Chan, S.K., Tuba, I.S., Wilson, W.K., 1970. On the finite element method in linear fracture mechanics. Eng. Fract. Mech. 2, 1–17.

de Bruyne, N.A., 1944. The strength of glued joints. Aircraft Eng. 16, 115–118.

Eshelby, J.D., Read, W.T., Shockley, W., 1953. Anisotropic elasticity with applications to dislocation theory. Acta Metall. 1, 251–259.

Fernlund, G., Spelt, J.K., 1991. Failure load prediction of structural adhesive joints. Int. J. Adhesion Adhes. 11, 213–220.

Goland, M., Reissner, E., 1944. The stresses in cemented joints. J. Appl. Mech. 11, A17–A27.

Gradin, A., Groth, H.L., 1984. A fracture criterion for adhesive joints in terms of material induced singularities. In: Proceedings of the Third International Conference on Numerical Methods in Fracture Mechanics. Pineridge Press, Swansea.

Gray, L., Phan, A.V., Paulino, G.H., Kaplan, T., 2003. Improved quarter-point crack tip element. Eng. Fract. Mech. 70, 269–283.

Hart-Smith, L.J., 1974. Analysis and Design of Advanced Composite Bonded Joints. NASA CR-2218.

Hattori, T., 1991. A stress-singularity-parameter approach for evaluating the adhesive strength of single lap joints. JSME Int. J. Ser. I 34, 326–331.

Helsing, J., Jonsson, A., 2002. On the computation of stress fields on polygonal domains with V-notches. Int. J. Numer. Methods Eng. 53, 433–453.

Henshell, R.D., Shaw, K.G., 1975. Crack tip finite elements are unnecessary. Int. J. Numer. Methods Eng. 9, 495–507.

Laschet, G., Stas, A., 1992. Finite element failure prediction of adhesive joints using a simple damage model. Report SA-158, Structures Aerospatiales. University de Liege, Belgique.

Lekhnitskii, S.G., 1938. Some cases of the elastic equilibrium of a homogeneous cylinder with arbitrary anisotropy (in Russian). Appl. Math. Mech. 2, 345–367.

Liu, Y., Wu, Z., Liang, Y., Liu, X., 2008. Numerical methods for determination of stress intensity factors of singular stress field. Eng. Fract. Mech. 75, 4793–4803.
Mantič, V., Barroso, A., París, F., 2014. Computational procedure for singularity analysis of anisotropic elastic multimaterial corners - Applications to composites and their joints. In: Mantič, V. (Ed.), Mathematical Methods and Models in Composites. Imperial College Press, pp. 613–695, 2023.
Malyshev, B.M., Salganik, R.L., 1965. The strength of adhesive joints using the theory of cracks. Int. J. Fract. 1, 114–128.
Marin, L., Lesnic, D., Mantič, V., 2004. Treatment of singularities in Helmholtz type equations using the boundary element method. J. Sound Vib. 278, 39–62.
Mohammed, I., Liechti, K.M., 2000. Cohesive zone modeling of crack nucleation at bimaterial corners. J. Mech. Phys. Solid. 48, 735–764.
Omer, N., Yosibash, Z., 2005. On the path independency of the point-wise J integral in three-dimensions. Int. J. Fract. 136, 1–36.
Pageau, S.P., Gadi, K.S., Biggers Jr., S.B., Joseph, P.F., 1996. Standardized complex and logarithmic eigensolutions for n-material wedges and junctions. Int. J. Fract. 77, 51–76.
París, F., Cañas, J., 1997. Boundary Element Method, Fundamentals and Applications. Oxford University Press.
Sheppard, A., Kelly, D., Tong, L., 1998. A damage zone model for the failure analysis of adhesively bonded joints. Int. J. Adhesion Adhes. 18, 385–400.
Sinclair, G.B., 1999. Logarithmic stress singularities resulting from various boundary conditions in angular corners of plates in extension. J. Appl. Mech. 66, 556–559.
Stroh, A.N., 1958. Dislocations and cracks in anisotropic elasticity. Phil. Mag. 3, 625–646.
Stroh, A.N., 1962. Steady state problems in anisotropic elasticity. J. Math. Phys. 41, 77–103.
Ting, T.C.T., 1996. Anisotropic Elasticity: Theory and Applications. Oxford University Press.
Vicentini, D., Barroso, A., Justo, J., Mantič, V., París, F., 2012. Determination of generalized fracture toughness in multimaterial closed corners with two singular terms. Part II: experimental results. Eng. Fract. Mech. 89, 15–23.
Volkersen, O., 1938. Die Niektraftverteilung in Zugbeanspruchten mit Konstanten Laschenquerschritten. Luftfahrtforschung 15, 41–47.
Wolfram, S., 1991. A System for Doing Mathematics by Computer. Addison-Wesley, Redwood City.

Further reading

Barroso, A., París, F., Mantič, V., 2009. Representativity of the singular stress state in the failure of adhesively bonded joints between metals and composites. Compos. Sci. Technol. 69, 1746–1755.

Index

Note: 'Page numbers followed by "f" indicate figures and "t" indicate tables.'

G

H

I

J